HANDBOOK OF MICROSCOPY FOR NANOTECHNOLOGY

HANDBOOK OF MICROSCOPY FOR NANOTECHNOLOGY

Edited by

NAN YAO
Princeton University
Princeton, NJ, USA

ZHONG LIN WANG
Georgia Institute of Technology
Atlanta, GA, USA

KLUWER ACADEMIC PUBLISHERS
BOSTON / DORDRECHT / NEW YORK / LONDON

Library of Congress Cataloging-in-Publication Data

Handbook of microscopy for nanotechnology / edited by Nan Yao. Zhong Lin Wang.
p. cm.
Includes index.
ISBN 1-4020-8003-4 e-ISBN 1-4020-8006-9 Printed on acid-free paper.
1. Nanostructured materials—Handbooks, manuals, etc. 2. Nanotechnology—Handbooks, manuals, etc. I. Yao, Nan. II. Wang, Zhong Lin.

TA418.9.N35H35 2005
6209′.5—dc 22

2004056504

Printed in the United States of America.

9 8 7 6 5 4 3 2 1 SPIN 11129776

springeronline.com

Dedicated to Professor John M. Cowley, our graduate study advisor, in memory of his outstanding contribution to science and education

CONTENTS

PREFACE

Science and technology ever seek to build structures of progressively smaller size. This effort at miniaturization has finally reached the point where structures and materials can be built through "atom-by-atom" engineering. Typical chemical bonds separate atoms by a fraction of a nanometer (10^{-9} m), and the term nanotechnology has been coined for this emerging area of development. By manipulating the arrangements and bonding of atoms, materials can be designed with a far vaster range of physical, chemical and biological properties than has been previously conceived. But how to characterize the relationship between starting composition, which can be controlled, with the resulting structure and properties of a nanoscale-designed material that has superior and unique performance? Microscopy is essential to the development of nanotechnology, serving as its eyes and hands.

The rationale for editing this Handbook now has never been more compelling. Among many pioneers in the field of nanotechnology, Dr. Heinrich Rohrer and Dr. Gerd Binnig, inventors of the scanning tunneling microscope, along with Professor Ernst Ruska, inventor of the world's first electron microscope, were awarded the Nobel Prize in Physics in 1986, for their invaluable contribution to the field of microscopy. Today, as the growth of nanotechnology is thriving around the world, microscopy will continue to increase its importance as the most powerful engine for discovery and fundamental understanding of nanoscale phenomena and structures.

This Handbook comprehensively covers the state-of-the-art in techniques to observe, characterize, measure and manipulate materials on the nanometer scale. Topics described range from confocal optical microscopy, scanning near-field optical microscopy, various scanning probe microscopies, ion and electron microscopy, electron energy loss and X-ray spectroscopy, and electron beam lithography, etc. These are tremendously important topics for students and researchers in the field of nanotechnology. Our aim is to provide the readers a practical running start, with only enough theory to understand how best to use a particular technique and the situations in which it is best applied. The emphasis is working knowledge on the full range of modern techniques, their particular advantages, and the ways in which they fit into the big picture of nanotechnology by each furthering the development of particular nanotechnological materials.

Each topic has been authored by world-leading scientist(s), to whom we are grateful for their contribution. Our deepest appreciation goes to Professor John M. Cowley, who advised our graduate study. More than a great scientist, educator and pioneer in electron microscopy, diffraction and crystallography, he was a humble and kind man to whom we are very much indebted.

June 2004

Nan Yao
Princeton University
e-mail: nyao@Princeton.edu
http://www.princeton.edu/~nyao/

Zhong Lin Wang
Georgia Institute of Technology
e-mail: zhong.wang@mse.gatech.edu
http://www.nanoscience.gatech.edu/zlwang/

LIST OF CONTRIBUTORS

Alexandre Bouhelier
Center for Nanoscale Materials
Chemistry Division
Argonne National Laboratory,
9700 South Cass Avenue
Argonne, IL 60439 USA
E-mail: bouhelier@anl.gov

Narae Cho
Physics and NANO Systems Institute
Seoul National University,
Seoul, 151–747 Korea
E-mail: birdwings@phya.snu.ac.kr

Christian Colliex
Laboratoire de Physique des Solides,
UMR CNRS 8502
Bâtiment 510, Université Paris Sud
91405 ORSAY, France
E-mail: colliex@lps.u-psud.fr

John M. Cowley
Arizona State University, Box 871504
Dept. of Physics and Astronomy
Tempe, AZ 85287–1504 USA
E-mail: cowleyj@asu.edu

Peter A. Crozier
Center for Solid State Science
Arizona State University
Tempe, AZ 85287–1704 USA
E-mail: Peter.crozier@asu.edu

Rafal E. Dunin-Borkowski
Department of Materials Science
University of Cambridge, Pembroke
Cambridge CB2 3QZ, UK
E-mail: rafal.db@msm.cam.ac.uk

Achim Hartschuh
Universit%ot Siegen,
Physikalische Chemie I
Adolf-Reichwein-Strasse 2
D-57068 Siegen, Germany
E-mail: hartschuh@chemie.uni-siegen.de

Seunghun Hong
Physics and NANO Systems Institute
Seoul National University,
Seoul 151–747 Korea
E-mail: shong@phya.snu.ac.kr

Jiwoon Im
Physics and NANO Systems Institute
Seoul National University
Seoul, 151–747 Korea
E-mail: jwim@phya.snu.ac.kr

Jin-Feng Jia
Institute of Physics
The Chinese Academy of Sciences
Beijing, 100080 China
E-mail: jfjia@aphy.iphy.ac.cn

William A. Lamberti
ExxonMobil Research &
Engineering Company
Advanced Characterization Section
Route 22 East
Annandale, New Jersey 08801 USA
E-mail: william.a.lamberti
@exxonmobil.com

Minbaek Lee
Physics and NANO Systems Institute
Seoul National University, Seoul,
151–747 Korea
E-mail: lmb100@snu.ac.kr

Jingyue Liu
Science & Technology,
Monsanto Company
800 N. Lindbergh Blvd., U1E
St. Louis, Missouri 63167, USA
E-mail: Jingyue.liu@monsanto.com

Peter J. Lu
Harvard University,
Department of Physics
Jefferson Laboratory, 17 Oxford Street
Cambridge, MA 02138 USA
E-mail: plu@fas.harvard.edu

Sergei N. Magonov
Veeco Instruments
112 Robin Hill Rd., Santa Barbara, CA
93117 USA
E-mail: smagonov@veeso.com

Martha R. McCartney
Center for Solid State Science, Arizona
State University
Tempe, Arizona 85287, USA
Phone: 480-965-4558;
Fax: 480-965-9004
E-mail: molly.mccartney@asu.edu

Joseph. R. Michael
Sandia National Laboratories
Albuquerque, NM 87185-0886 USA
E-mail: jrmicha@sandia.gov

Paul A. Midgley
Department of Materials Science
and Metallurgy,
University of Cambridge,
Pembroke Street, Cambridge,
CB2 3QZ UK
E-mail: pam33@cus.cam.ac.uk

M. K. Miller
Metals and Ceramics Division
Oak Ridge National Laboratory
P.O. Box 2008,
Building 4500S, MS 6136
Oak Ridge, TN 37831-6136, USA
E-mail: millermk@ornl.gov

Dale E. Newbury
National Institute of
Standards and Technology
Gaithersburg, MD 20899-8371 USA
Email: dale.newbury@nist.gov

Lukas Novotny
The Institute of Optics,
University of Rochester
Wilmot Building, Rochester NY,
14627 USA
E-mail: novotny@optics.rochester.edu

John Henry J. Scott
National Institute of
Standards and Technology
Gaithersburg, MD 20899-8371 USA
E-mail: johnhenry.scott@nist.gov

Renu Sharma
Center for Solid State Science,
Arizona State University
Tempe, AZ 85287-1704 USA
E-mail: Renu.sharma@asu.edu

Li Shi
Department of Mechanical Engineering
The University of Texas at Austin
Austin, TX 78712 USA
(512) 471-3109 (phone),
(512) 471-1045 (fax)
E-mail: lishi@mail.utexas.edu

John A. Small
National Institute of Standards
and Technology
Gaithersburg, MD 20899-8371 USA
E-mail: john.small@nist.gov

David J. Smith
Center for Solid State Science and
Department of Physics and Astronomy
Arizona State University, Tempe,
Arizona 85287, USA
Phone: 480-965-4540;
Fax: 480-965-9004
E-mail: david.smith@asu.edu

Odile StÈphan
Laboratoire de Physique des Solides,
UMR CNRS 8502
Bâtiment 510, Université Paris Sud
91405 ORSAY, France
Phone : +33 (0)1 69 15 53 69
Fax : +33 (0)1 69 15 80 04
E-mail: stephan@lps.u-psud.fr

Takayoshi Tanji
Department of Electronics,
Nagoya University
Chikusa, Nagoya 464-8603, Japan
E-mail: tanji@nuee.nagoya-u.ac.jp

Zhong Lin Wang
School of Materials Science and
Engineering
Georgia Institute of Technology
Atlanta GA 30332-0245 USA
E-mail: zhong.wang@mse.gatech.edu

Scott Wight,
National Institute of
Standards and Technology
Gaithersburg, MD 20899-8371 USA
E-mail: scott.wight@nist.gov

Qi-Kun Xue
Institute of Physics,
the Chinese Academy of Sciences
Beijing, 100080 China
E-mail: qkxue@aphy.iphy.ac.cn

Wei-Sheng Yang
Institute of Physics,
the Chinese Academy of Sciences
Beijing, 100080 China
E-mail: wsyang@pku.edu.cn

Nan Yao
Princeton University
Princeton Institute for the Science and
Technology of Materials
70 Prospect Avenue, Princeton,
New Jersey 08540 USA
E-mail: nyao@princeton.edu

Natalya A. Yerina
Veeco Instruments,
112 Robin Hill Rd.,
Santa Barbara CA 93117 USA
E-mail: NErina@veeco.com

Zhiping (James) Zhou
Microelectronics Research Center
Georgia Institute of Technology

791 Atlantic Drive,
Atlanta GA 30332-0269 USA
E-mail: james.zhou@mirc.gatech.edu

Jian-Min (Jim) Zuo
Department of Materials
Science and Engineering
University of Illinois
at Urbana-Champaign,
1304 West Green Street,
Urbana, Illinois 61801 USA
E-mail: jianzuo@uiuc.edu

I. OPTICAL MICROSCOPY, SCANNING PROBE MICROSCOPY, ION MICROSCOPY AND NANOFABRICATION

1. CONFOCAL SCANNING OPTICAL MICROSCOPY AND NANOTECHNOLOGY

PETER J. LU

1. INTRODUCTION

Microscopy is the characterization of objects smaller than what can be seen with the naked human eye, and from its inception, optical microscopy has played a seminal role in the development of science. In the 1660s, Robert Hooke first resolved cork cells and thereby discovered the cellular nature of life [1]. Robert Brown's 1827 observation of the seemingly random movement of pollen grains [2] led to the understanding of the motion that still bears his name, and ultimately to the formulation of statistical mechanics. The contributions of optical microscopy continue into the present, even as the systems of interest approach nanometer size. What makes optical microscopy so useful is the relatively low energy of visible light: in general, it does not irreversibly alter the electronic or atomic structure of the matter with which it interacts, allowing observation of natural processes in situ. Moreover, light is cheap, abundant, and can be manipulated with common and relatively inexpensive laboratory hardware.

In an optical microscope, illuminating photons are sent into the sample. They interact with atoms in the sample, and are re-emitted and captured by a detection system. The detected light is then used to reconstruct a map of the sample. An ideal microscope would detect each photon from the sample, and measure with infinite precision the three-dimensional position from which it came, when it arrived, and all of its properties (energy, polarization, phase). An exact three-dimensional map of the sample could then be created with perfect fidelity. Unfortunately, these quantities can be known only to a certain finite precision, due to limitations in both engineering and fundamental physics.

One common high-school application of optical microscopy is to look at small objects, for example the underside of a geranium leaf. Micron-scale structure is easily revealed in the top layer of plant cells. But structure much smaller than a micron (such as individual macromolecules in the plant cell) cannot be seen, and looking deep into the sample (e.g. tens of cell layers) leads only to a nearly featureless blur. Clearly this is a far cry from the ideal microscope above.

Microscopes with improved resolution fall into two broad categories, near-field and far-field. Near-field techniques rely on scanning a nanoscale optical probe only nanometers above the surface of interest. Spatial resolution is then physically limited only by the lateral size of the tip of the probe, and information can only be gathered from the surface. This technique is the subject of another chapter in this text. In far-field microscopy, a macroscopic lens (typically with mm-scale lens elements) collects photons from a sample hundreds of microns away. Standard microscopes, like the one used in high-school, are of this type. The light detected often comes from deep within the sample, not just from the surface. Moreover, there are often enough photons to allow collection times sufficiently brief to watch a sample change in real time, here defined to be the video rate of about 25 full frames per second.

But all far-field techniques encounter the fundamental physical diffraction limit, a restriction on the maximum spatial resolution. In the present parlance, the precision with which the location of the volume generating a given detected photon (here termed the illumination volume) can be determined is roughly the same size as the wavelength of that photon [3]. Visible light has a wavelength of roughly a half micron, an order of magnitude greater than the feature size of interest to nanotechnology.

At first blush, then, the idea that far-field optical microscopy can contribute much to nanotechnology may appear absurd. However, a number of techniques have been developed to improve the precision with which the spatial position of an illumination volume can be determined. The most prevalent of these is confocal microscopy, the main subject of this chapter, where the use of a pinhole can dramatically improve the ability to see small objects. Other techniques have the potential for further improvements, but none so far has been applied widely to systems relevant to nanotechnology.

Several terms are commonly used to describe improvements in "seeing" small objects. Resolution, or resolving power, is the ability to characterize the distribution of sample inhomogeneities, for example distinguishing the internal structure of cells in Hooke's cork or the geranium leaf. Resolution is ultimately restricted by the diffraction limit: no optical technique, including confocal, will ever permit resolution of single atoms in a crystal lattice with angstrom-scale structure. On the hand, localization is the determination of the spatial position of an object, and this is possible even when the object is far smaller than the wavelength. Localization can be of an object itself, if there is sufficient optical contrast with the surrounding area, or of a fluorescent tag attached to the object. The former is generally more common in the investigation of nanoscale materials, where in many instances (e.g. quantum dots) the nanomaterials are themselves fluorescent. The latter is common in biology, where the confocal microscope is often used to localize single-molecule fluorescent probes attached to cellular substructures. But in many of these systems, the tags can be imaged without confocality, such as in thin cells where three-dimensional sectioning is unneeded, or when the tags are spaced out by microns or more.

Precise localization is of tremendous utility when the length scale relevant to the question at hand is greater than the wavelength being used to probe the sample, even if the sample itself has structure on a smaller length scale. For example, Brown observed micron-scale movements of pollen grains to develop his ideas on motion, while the nanoscale (i.e. molecular) structure of the pollen was entirely irrelevant to the question he was asking. The pollen served as ideal zero-dimensional markers that he could observe; their position as a function of time, not their structure, was ultimately important. In many instances, the confocal plays a similar role, where fluorescent objects serve as probes of other systems. By asking the right questions, the diffraction limit only represents a barrier to imaging resolution, not a barrier to gathering information and answering a properly formulated scientific question.

Ultimately, the confocal is not a fancy optical microscope that through special tricks allows resolution of nanoscale objects. Rather, the confocal makes the greatest contribution to nanotechnology with rapid, non-destructive three-dimensional nanoscale localization of the sample area generating a given detected photon, and the analysis (spectroscopy) of that photon. This localization property of the confocal allows real-time spectroscopy of *individual* nanoscale objects, instead of ensemble averages. As such, the confocal plays a singularly important role in the investigation of structure and dynamics of systems relevant to nanotechnology, complementing the other techniques described in this volume.

This review begins with a qualitative overview (no equations) of confocal microscopy, with a brief discussion of recent advances to improve resolution and localization. Following that is a survey of recent applications of confocal microscopy to systems of interest to nanotechnology.

2. THE CONFOCAL MICROSCOPE

2.1. Principles of Confocal Microscopy

Several texts comprehensively review the confocal microscope, how it works, and the practical issues surrounding microscope construction and resolution limitations [4–7]. This section is a brief qualitative overview to confer a conceptual understanding of what a confocal is, namely how it differs from a regular optical microscope, and why those differences are important for gaining information from structures relevant to nanotechnology. All of the applications of confocal microscopy described here rely on fluorescence. That is, the incoming beam with photons of a given wavelength hits the sample, and interactions between illumination photons and sample atoms generates new photons of a *lower* wavelength, which are then detected. The difference in the two wavelengths must be large enough to allow separation of illumination and detection beams by mirrors, called dichroics, that reflect light of one color and pass that of another. In practice, the separation is usually tens of nanometers or more.

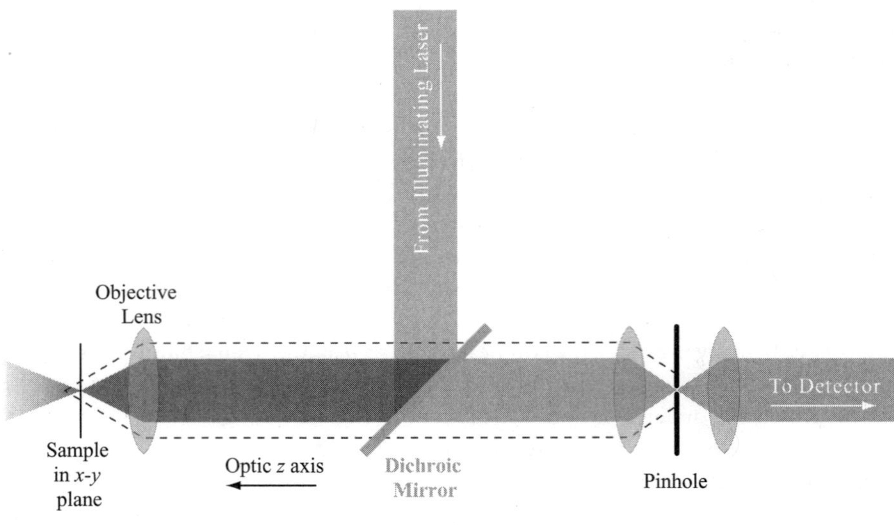

Figure 1. Confocal schematic. Laser light (blue) is reflected by the dichroic, and illuminates the sample at the focus of the objective. This excites fluorescence, and the sample then emits light at a lower wavelength (red), which goes through the objective, passes through the dichroic, and is focused down to a spot surrounded by a pinhole. Light from other locations in the sample goes through the objective and dichroic, but is rejected by the pinhole (red dotted line). (See color plate 1.)

The noun "confocal" is shorthand for *confocal scanning optical microscope*. Parsing in reverse, *optical microscope* indicates that visible radiation is used, and confocals are often based on, or built directly as an attachment to, optical microscopes with existing technology. Unlike traditional widefield optical microscopes, where the whole sample is illuminated at the same time, in confocal a beam of laser light is *scanned* relative to the sample, and the only light detected is emitted by the interaction between the illuminating beam and a small sample illumination volume at the focus of the microscope objective; due to the diffraction limit, the linear extent of this volume is approximately the wavelength of light. In a confocal, light coming back from the illumination volume is focused down to a another diffraction-limited spot, which is surrounded by a narrow pinhole. The pinhole spatially filters out light originating from parts of the sample *not* in the illumination volume. Because it is positioned at a point conjugate to the focal point in the sample, the pinhole is said to be *confocal* to it, and the pinhole allows only the light from the focused spot (that is, the illumination volume) to reach the detector.

A schematic of a typical confocal is given in figure 1. Light from a laser beam is reflected by a dichroic and focused onto a spot on the sample in the x-y plane by the microscope objective. The optic axis is along the z direction. Light from the sample, at a lower wavelength, comes back up from the illumination volume via the objective, passes through the dichroic, and is focused onto a point, surrounded by a pinhole, that is confocal with the objective's focal point on the sample. The detected light then passes to the detector. The laser beam illuminates parts of the sample covering a range of depths, which in an ordinary microscope contribute to the detected signal, and blur the image out; this is the reason that, tens of cell layers deep, the image of the geranium is blurry. In the confocal, however, the pinhole blocks all light originating from points not at the focus of the microscope objective, so that only the light from the illumination volume is detected; this effect is also known as optical sectioning. Translating the sample relative to a fixed laser beam, or moving the laser beam relative to a fixed sample, allows the point-by-point construction of the full three-dimensional map of the sample itself, with resolution limited by the size of the excitation volume, itself limited by the diffraction limit of the illuminating light.

2.2. Instrumentation

The different implementations of a confocal microscope differ primarily in how the illumination volume is moved throughout the sample. The simplest method from an optical standpoint is to keep the optics fixed, and translate the sample (figure 2a); modern piezo stages give precision and repeatability of several nanometers. Ideal from an image quality standpoint, as the optical path can be highly optimized and specific aberrations and distortions removed, sample translation is also the slowest; moving the piezo requires milliseconds, precluding the full-frame imaging at 25 frames/sec needed to achieve real-time speeds.

For higher speeds, the beam itself must be moved. Two galvanometer-driven mirrors can be used to scan the laser beam in x and y at up to a kilohertz, while maintaining beam quality (figure 2b). While not quite fast enough to achieve real-time full-frame imaging, commercial confocal microscopes based on galvanometers can reach about ten full images a second, each with about a million pixels. Beam scanning is usually accomplished much like that of a television, by first quickly scanning a line horizontally, then shifting the beam (at the end of each horizontal scan) in the vertical direction, scanning another horizontal line, and so on. Replacing the galvanometer mirror that scans horizontally with an acousto-optical device (AOD) significantly increases speed (the galvanometer is fast enough to keep up with the vertical motion). However, the AOD severely degrades the quality of the beam, and image quality correspondingly suffers. AOD-based confocals are primarily useful where gathering data at high speed is more important than achieving high resolution, as is the case in dynamical situations with relatively large (i.e. greater than micron-sized) objects.

Another major approach to increasing beam-scanning speed is to split the main laser beam into thousands of smaller laser beams, parallelizing the illumination (figure 2c). Each individual mini-beam then needs only to be moved a small amount in order for the total collection of beams to image an entire frame. This typically involves a Nipkow disk, where thousands of tiny microlenses are mounted in an otherwise opaque disk. These focus down to thousands of points, surrounded by thousands of tiny pinholes created in another disk. The laser light is thus split and focused, and then the multiple tiny beams are focused onto the sample with a single objective lens. Light from the multiple illumination volumes comes back up first via the objective and then through

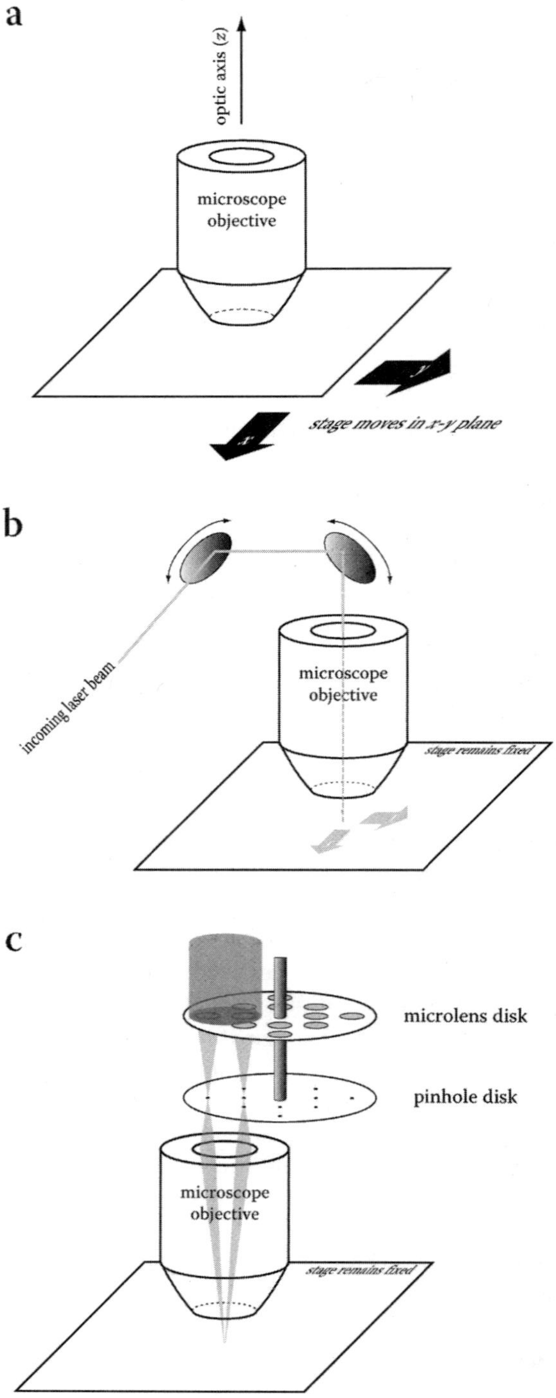

Figure 2. Confocal microscope instrumentation. (a) stage-scanning, in which the optical train remains fixed and the stage is moved. (b) beam scanning, with two moveable mirrors that move the beam itself. (c) Nipkow disk, where rotating disks of microlens and pinholes parallelize the illumination beam.

the pinholes, then goes to the camera detector, where the thousands of mini-beams are simultaneously imaged. By spinning the disk and arranging the holes in a spiral pattern, full coverage of the frame can be achieved. The main advantage of this technique is that image quality can remain high (no AOD, for instance), and speed can be increased simply by spinning the disk faster. From an engineering standpoint, Nipkow disks are durable and easy to fabricate with existing technology; their major drawback is a total lack of flexibility: Nipkow disk systems are usually optimized for only one magnification, and after fabrication, the size of the pinholes cannot be changed to accommodate different conditions.

2.3. Techniques for Improving Imaging of Nanoscale Materials

2.3.1. 4-Pi Confocal

The biggest recent development in confocal microscopy has been the use of two objectives, focused on the same point, to collect light. The name 4Pi microscopy has been applied to this general technique, and is meant to evoke the idea that all of the light is collected from a sample simultaneously (i.e. the 4 pi steradians of a complete sphere); in reality, while most of the light is collected by the two objectives, they cannot image the whole sphere [8]. A full discussion of the principles and advances in 4Pi confocal microscopy is beyond the scope of this article (see [7], [8]); only a brief qualitative discussion to convey the underlying ideas behind the superior resolution of 4Pi confocal is included here.

A regular confocal rejects light coming from parts of the sample outside of the illumination volume by means of spatial filtering through a pinhole, but even if it is made arbitrarily small, the pinhole cannot localize the light coming from the sample to better than within the typical size of this region (i.e. the wavelength) because of diffraction. In addition, there is still a small contribution to the detected signal from light outside of the focal point, though that contribution decreases with greater distance from the focal point. Limitations to resolution therefore come from a combination of the finite size of the excitation volume in the sample, and the imperfect discrimination of the pinhole itself, both fundamental physical constraints inherent to the design of a confocal microscope; they cannot be overcome simply with better implementation of the same ideas. 4Pi confocal relies on coherent illumination or detection from both objectives simultaneously, effectively doubling amount of light involved and creating an interference pattern between the two beams. This allows a dramatic increase in axial resolution, often around five-fold, though lateral resolution is unchanged.

From an instrumentation standpoint, there are three different types of 4Pi confocal microscopes, A, B and C (figure 3). In type-A 4Pi confocal, illumination beams are sent through both objectives and interfere in the sample; the light coming out of only one objective is used for detection. This is the earliest, and simplest, system, and has thus far been most widely used. In type-B, illumination occurs through just one objective, but detection of interfering light from the sample comes through both objective lenses, [9] and thus its theoretical optical properties are identical to that of type-A 4Pi [10]. In

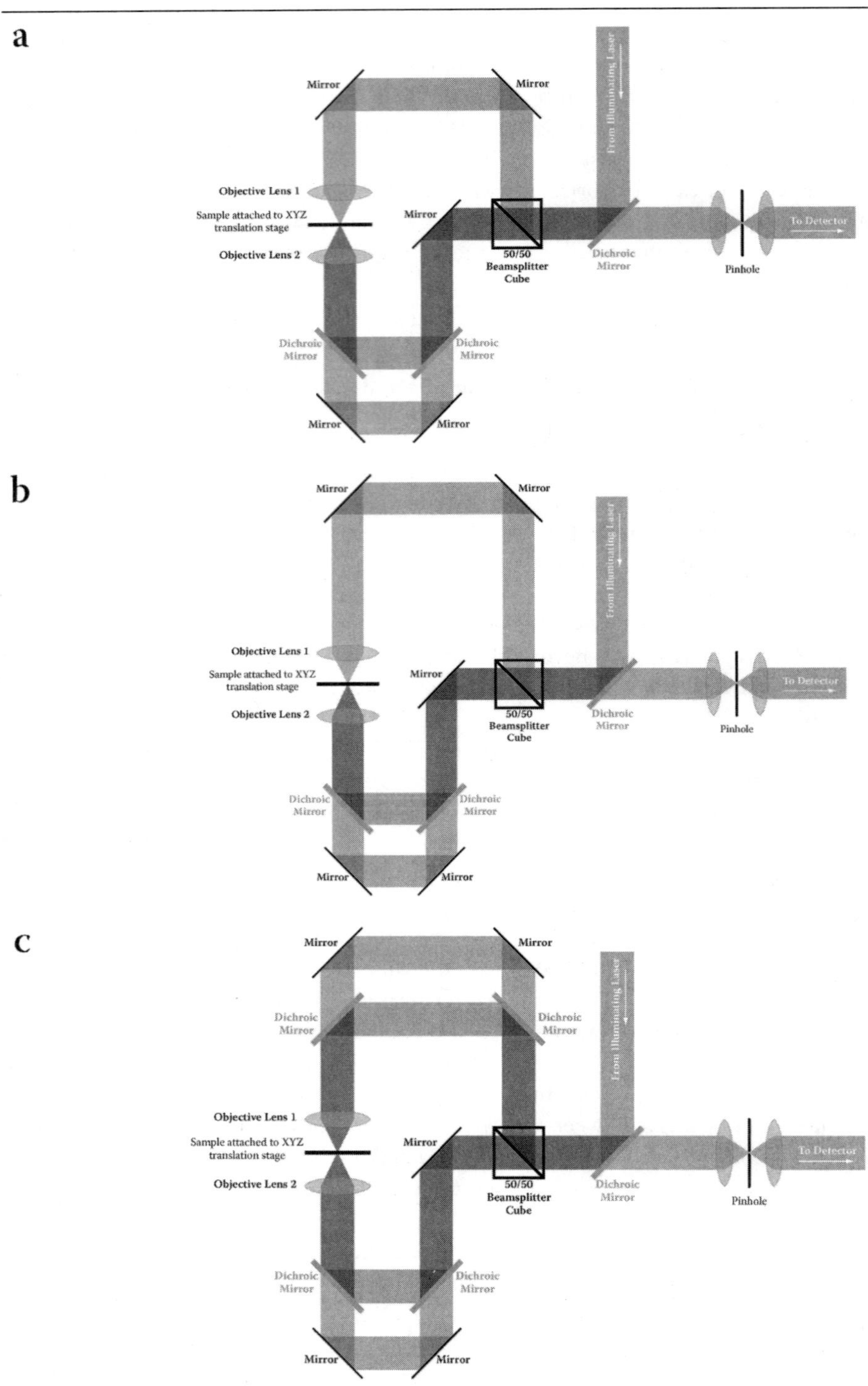

Figure 3. 4Pi Confocal configurations. (a) 4Pi-A configuration, with two illumination paths, but only one detection path. (b) 4Pi-B configuration, with only one illumination path, but two detection paths. (c) 4Pi-C configuration, with two illumination and two detection paths. (See color plate 2.)

type-C, both illumination and detection are of interfering light in the sample volume, through both objectives, [8] permitting even greater resolution [10].

Resolution is best understood in the context of the axial optical transfer function (OTF), also called the z-response function. Qualitatively, the OTF shows the contribution to the detected light from different depths in the sample (i.e. points along the optical axis). An ideal microscope would have only light from a single point in the focal plane contributing to the detected signal; in that case, the OTF would be delta function at the focus of the microscope objective (figure 4a). In a regular confocal, instead of a single delta function, the effects of finite-sized illumination volume and imperfect pinhole discrimination combine to smear out the delta function into a nearly gaussian OTF (figure 4b); with 633-nm HeNe laser illumination, the OTF of a regular confocal has a full-width at half-maximum (FWHM) of 500 nm (theory and experiment) [10]. In 4Pi confocal microscopy, the counter propagating light waves of the same frequency and intensity that illuminate the sample create an interference pattern (a standing wave). Instead of a simply gaussian shape, the OTF now has one central peak and several so-called "side-lobes" (figure 4c,d) The main advantage is that this central peak has a far narrower FWHM, theoretically calculated to be 130 nm for type-A (and thus for the optically equivalent type-B) and 95 nm for type-C, and measured at 140 nm and 95 nm, respectively [10]. The width of the central peak is independent of the relative phase between the two illuminating wavefronts (i.e. constructive or destructive interference are equivalent), [11] but nevertheless comes at the cost of having prominent side-lobes. That is, there is now a greater contribution to the light detected through the pinhole from some points farther away along the optic axis from the focal point than from some points closer, which creates artifacts. Almost all of the more recent technological developments in the 4Pi area have focused on optical "tricks" to eliminate the effects of those side bands: spatially filtering illuminating light beams with specifically-placed dark rings [12, 13] or illuminating with two photons [14, 15] to cut off the light that contributes mainly to side lobes, and computational modeling of an ideal microscope to reconstruct an "ideal" image from real data in a process known as deconvolution [15–17]. Such techniques have yielded a confocal with an effective point-spread function with a width as small as 127 nm for a type-A 4Pi confocal, with *no* significant contribution from the side lobes (figure 4e), [12] allowing sub-10 nm distances between test objects to be measured with uncertainties less than a single nanometer [18].

Such high resolution may finally allow direct imaging of nanoscale structures, and Leica Microsystems has just introduced the first commercial 4Pi system, the TCS 4PI, in April 2004 (figure 5). Nonetheless, there still remain some limitations to current 4Pi technology. The number of optical elements to be aligned and controlled in a 4Pi setup is at least twice that of a regular confocal, and since the stage is usually scanned in a 4Pi setup, scanning speeds are much lower, requiring minutes to image a full frame. While fast enough to image stationary samples like fixed cells, [19] or even slow-moving live ones, [20] this is too slow to monitor most real-time dynamics at present, though scanning speed can be improved by using multiple beam scanning techniques in setups similar to the Nipkow disk, cutting imaging time down to seconds [21].

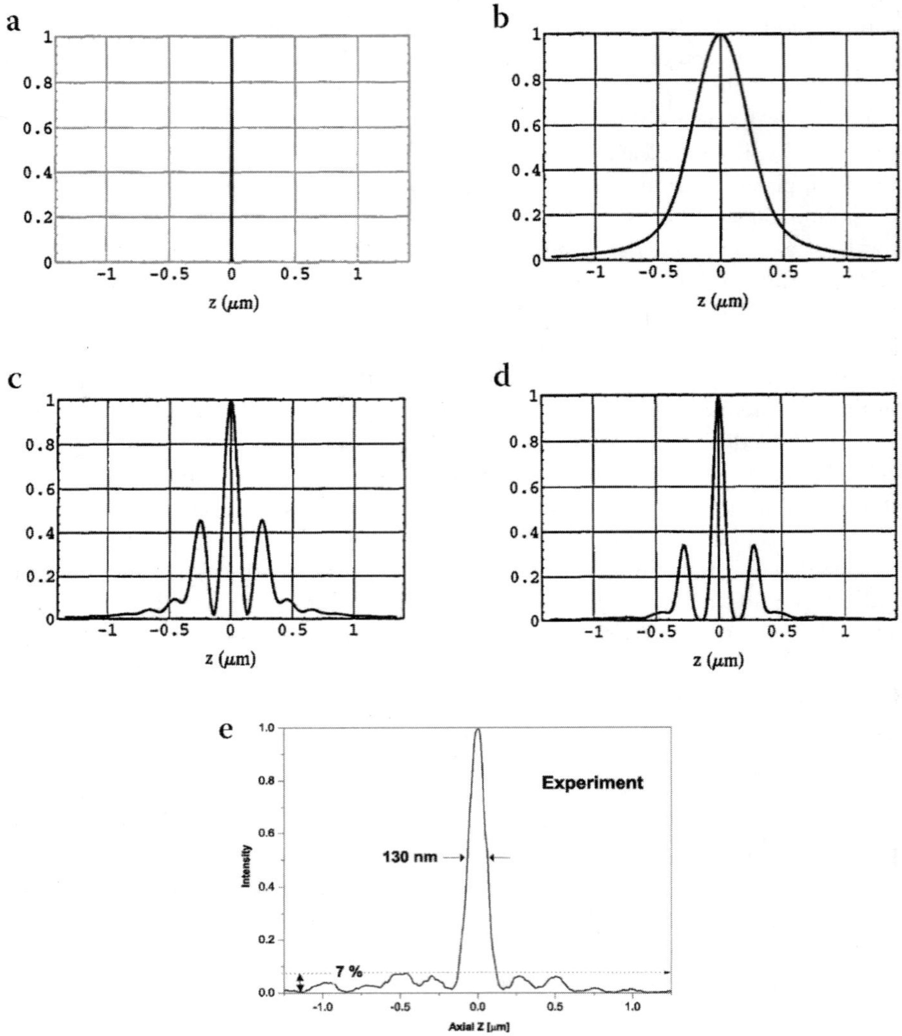

Figure 4. Z-response functions for various types of microscopes. (a) ideal imaging system, with a deltafunction at $z = 0$. (b) typical confocal microscope, with a gaussian profile. (c) 4Pi-A microscope. (d) 4Pi-C microscope. (e) 4Pi-A with Dark Ring to reduce side lobes. Reproduced from [8], [12]

2.3.2. Other Optical Techniques to Increase Resolution

Several other far-field optical techniques have achieved high resolution *without* spatial filtering by means of a pinhole. As they are neither confocal techniques, nor have been widely applied to systems relevant to nanotechnology, they will receive only brief mention.

Removing the pinholes and illuminating with an incoherent (non-laser) source in the 4Pi-A, 4Pi-B and 4-Pi-C geometries results in a setups known as I^3M, I^2M, and

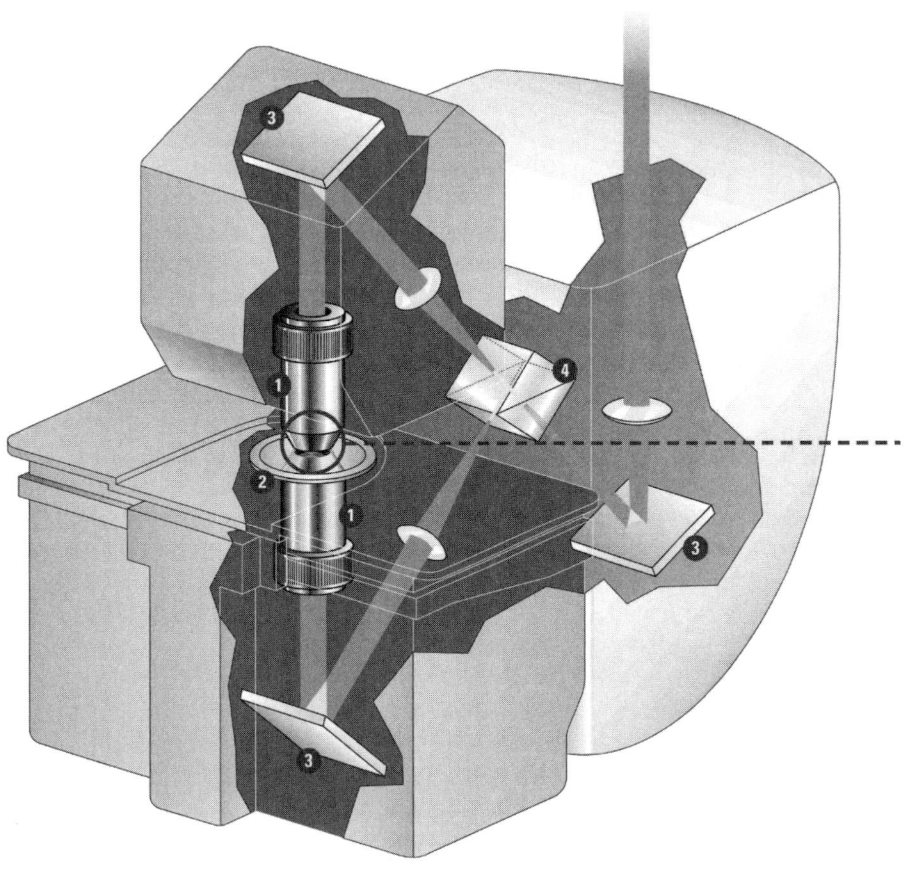

Figure 5. Leica TCS 4PI confocal microscope. (1) objective lenses, (2) sample holder, (3) mirrors, (4) beam splitter. Courtesy of Leica Microsystems, Heidelberg GmbH.

I^5M, respectively [22, 23]. Compared with 4Pi, these widefield techniques show an equivalent increase in axial resolution, though the lateral resolution is not as great. The main advantage is collection speed: light is collected from the entire imaging plane at once, as there is no beam to be scanned. The major drawback is the requirement for a large amount of computationally intense deconvolution to obtain images. Other techniques have used different geometries, objectives, mirrors, or multiple photons for illumination, but none thus far has achieved better resolution than 4Pi or I^5M, and have not been applied widely to systems of interest to nanotechnology; an excellent survey comparing the techniques is given in [24].

A couple of non-traditional optical techniques have also increased resolution in novel ways. Placing a solid hemispherical lens against the surface of the sample (figure 6a) can improve resolution to a few times better than can be achieved with only a regular objective, with light collection efficiency improved five-fold. Interestingly, these

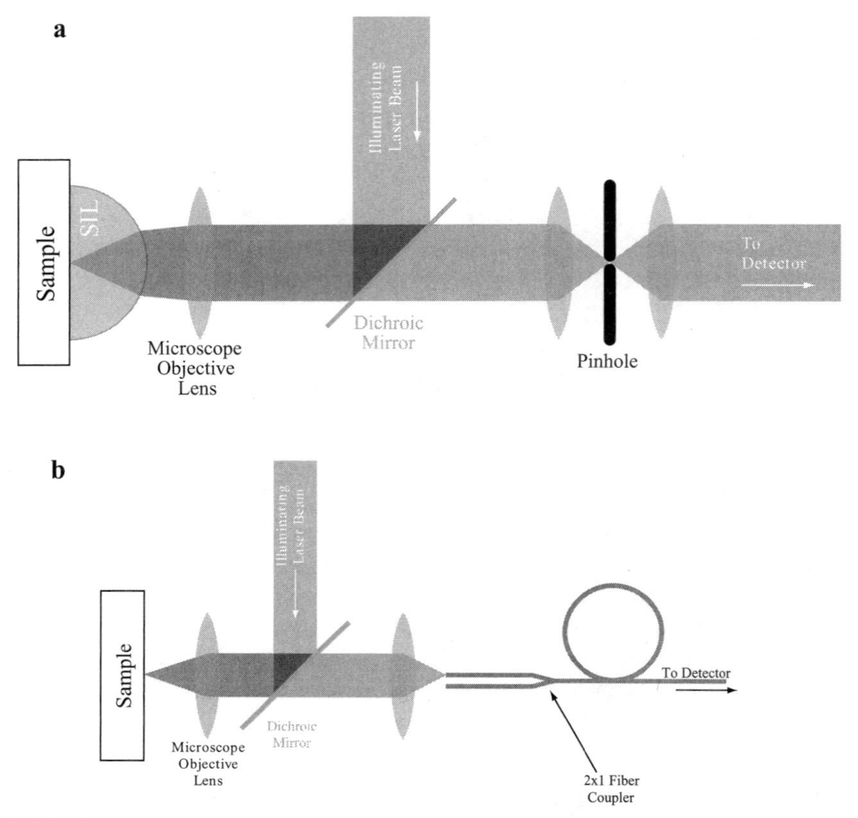

Figure 6. Other components to increase resolution. (a) Solid immersion lens, placed up against the sample. (b) 2×1 optical coupler to interfere the light from the two fibers.

improvements still persist even if the lens is slightly tilted, or there is a small air gap between the lens and the sample [25]. Also, common light detectors (PMT, APD, CCD) collect only intensity information, and can not measure phase directly. Interfering two beams, however, creates the a single output beam whose intensity is directly dependent on the phase difference of the two interfering beams. In practice, light can collected from two optical fibers (in place of the pinhole at the detector of the confocal), one along the optic axis, and one slightly displaced in the lateral direction. The signals from the two fibers are then interfered in a 2×1 optical fiber coupler (figure 6b), which creates a single output beam whose intensity is measured. This interferometric technique is sensitive to single nanometer displacements on millisecond timescales [26]. Though not strictly an optical technique, another way to increase localization precision is to use objects that emit several colors. By detecting the different colors in separate channels, then combining the position data from different colors, the final position of the objects can be determined to an accuracy of better than 10 nm; [27, 28]

the technique has also been used in 4Pi confocal setups [29] to achieve localization with single nanometer precision [30].

3. APPLICATIONS TO NANOTECHNOLOGY

As previously mentioned, the main contribution of confocal microscopy has not been to image nanoscale objects with light, but rather to use the capability to analyze the photons that have interacted with nanoscale structures, looking at their energies, temporal distribution, polarization, etc. As a result, the confocal can play a unique role in gathering information that can be obtained with no other technique. Many of the advances have come from the confocal's ability to characterize the properties of individual nanoscale objects, where previously only bulk ensemble averages could be measured. The discussion of how the confocal has been harnessed to gain information from nanoscale systems is organized by dimensionality of the system, in decreasing order.

3.1. Three-dimensional Systems

3.1.1. Nanoemulsions

An emulsion is a mixture of two immiscible liquids: small droplets of the first liquid are dispersed in the second one, called the continuous phase. Such a mixture is intrinsically unstable, and droplets will coalesce unless a surfactant is added to the continuous phase. The surfactant helps to stabilize the interface between the two liquids by reducing the surface tension between them. Except in the case of microemulsions, emulsions are thermodynamically unstable [31]. Aging leads to a change in the size distribution of the droplets, and occurs via two mechanisms: coalescence, where smaller droplets combine to form larger ones, and Ostwald ripening, where larger droplets grow at the expense of smaller ones via diffusion of molecules through the continuous phase.

Confocal observation of the changing fluorescence intensity of single nanodroplets (as small as 50 nm) flowing through a capillary tube permitted investigation into the fundamental process of emulsion coarsening in a way not accessible to bulk measurement: the rate of drop growth could be measured for single nanoscale drops in the confocal, not a statistical average for the emulsion as a whole. The work demonstrated that Ostwald ripening and coalescence were occurring at different stages of emulsion coarsening, [32] and dye diffusion was further studied by looking at fluorescence dynamics in mixtures of undyed and dyed emulsions [33]. Potential applications as isolated containers make nanoemulsions particularly interesting: encapsulation of a water-insoluble drug compound into the oil droplets of a nanoemulsion may allow controlled delivery of a substance to a designated target area in the body. Confocal has been used to monitor the uptake of dyed diblock copolymer nanoemulsions into cells, [34] and the targeted delivery to cell organelles, each dyed a different color [35].

3.1.2. Nanocapsules

Nanocontainers can also be created by coating colloidal spheres, nanocrystals or other templates, then dissolving out the core to leave a rigid hollow shell, contrasting the "soft" surface of an emulsion. The templates have designable properties and are typically micron-sized; the term nanocapsule refers to the controllable organization of wall layers: starting with a charged core, layers of alternating charged polyelectrolytes can deposited with nanometer thicknesses in a technique known as layer-by-layer (LBL) self-assembly. The chemistry of the polyelectrolytes can be varied significantly, and they can also serve as hosts for a variety of other materials, particularly ones with interesting photonic properties [36].

Confocal characterization has been used for real-time, three-dimensional imaging of the growing nanocapsules and subsequent core dissolution in a number of systems: single-walled containers templated around a cadmium carbonate core, [37] concentric sphere-in-sphere nanocapsules more mechanically robust than their single-wall counterparts, [38] nanocapsules with silver ions in the walls for designable catalysis, [39] biocompatible nanocapsules labeled with luminescent CdTe nanocrystals, [40] and nanocapsules constructed by LBL assembly of dendrimers, large branching macromolecules with fractal structure [41].

Nanocapsules can also provide a controllable local environment for investigating nanoscale chemical reactions. LBL assembly and tuning the external environment allow very fine control over the permeability of the capsule to small molecules and ions, so that large enzyme molecules can be held inside nanocapsules whose walls are permeable to a fluorescent substrate [42]. By monitoring the fluorescence changes in real-time, the confocal gives a unique, quantitative, single-molecule view on enzyme activity, [43] where previous techniques have only allowed bulk measurement of average activity; without the confocality, there is no way to isolate a single nanocapsule for study. Similarly, a pH gradient can be created between the inside and outside of a nanocapsule, and the confocal has monitored selective pH-induced precipitation of iron oxide nanocrystals inside single nanocapsules [44].

3.1.3. Other Three-dimensional Nanostructures

The confocal's ability to spatially resolve spectral information in three dimensions has been used to characterize the nanostructure of other heterogeneous materials. These systems may be constructed of different phases, such as the low-temperature Shpol'skii systems, where confocal spectroscopy was used to quantify preferential alignment of individual aromatic hydrocarbons molecules in a host of alkanes [45]. The confocal can also image the negative space in a porous material (e.g. nano-scale holes or pores) filled with dye, such as the spaces between layers of hydrotalcite-like compounds, including anionic clays and layered double hydroxides [46].

3.2. Two-dimensional Systems

3.2.1. Ferroelectric Thin Films

Even in the absence of an external electric field, ferroelectric materials exhibit an electric dipole below a certain transition temperature, and they are the electric-field analog of a ferromagnet [47]. Above this transition temperature, the net electric dipole is no longer present, but ferroelectric materials still have a nonlinear dielectric constant useful for creating elements (such as capacitors and phase shifters) in microwave integrated circuits for insertion into wireless and satellite communications devices [48]. These materials, often based on the barium/strontium titanate (BST) perovskite structure, have different properties whether in bulk or in a thin film.

Here, the confocal microscope has not contributed as a light-based imaging device in the traditional sense; rather, its capabilities to place a probing electric field precisely have been leveraged in a unique way to probe the physics of ferroelectrics [49–51]. First, pumping a BST thin film with a micron-scale capacitor aligns the film's molecular dipoles with an external electric field. After a controllable delay time, during which these dipoles begin to relax, the confocal microscope places a diffraction-limited spot of light at a particular location on the surface of a thin film. The electric field component of the illuminating laser beam serves as a sensitive probe, with emitted light having a different polarization or intensity as a result of its interaction with the ferroelectric material. Position and delay time are varied, with submicron and picosecond control. Changes in the emitted light from BST films grown under different pressures of oxygen suggest that reorientation of polarity on the nanoscale level is ultimately responsible for changes in ferroelectric behavior [52].

3.2.2. Nanopores, Nanoholes and Nanomembranes

The confocal has also been used to characterize the spatial distribution of negative space in two-dimensional systems, such as nanopores in titania thin film solar cell electrodes, [53] luminescent conjugated polymers in nanoporous alumina, [54] and pieces of fluorescently-labeled cell membrane stretched over nanoscale holes in silicon nitride [55]. This last technique is favored for AFM and other scanning-probe investigations of membranes, as isolated suspended membrane patches have improved stability and access relative to whole cells, and nanoholes are easily created with standard photolithography techniques. While SEM can characterize coverage of a cell membrane suspended over a nanohole, it is a two-dimensional technique that cannot discriminate between a suspended cell membrane and a pile of cell debris sitting on the silicon nitride surface. The three-dimensional sectioning ability of the confocal plays a critical role here: monitoring the height-dependence of fluorescence intensity yields a depth profile of fluorescent cell material, quickly distinguishing freely suspended cell membranes as small as 50 nm on a side [55].

3.3. One-dimensional Systems

3.3.1. Carbon Nanotubes

The bulk processes (e.g. carbon vapor deposition) that create carbon nanotubes typically yield a mixture of diameters, lengths, and structures, each with different physical properties. A major goal of nanoengineering is narrowing the distribution of sizes and structures to create materials with better-defined properties. Although the typical nanotube diameter of a few nanometers is well below the threshold of optical visibility, differences in nanotube structure measurably change Raman spectral profiles. This confers upon the confocal a singularly important role in nanotube characterization, as its combination of spatial and spectral resolving capacity allow it to characterize individual nanotubes at speeds far greater than those accessible to other techniques. Characterizable attributes include nanotube diameter, (n,m) chirality, [56, 57] and electrical conductivity, [58] as well as distinguishing single-walled from multiple-walled, [59] and free-standing from bundled nanotubes on insulating and conducting surfaces [60]. Confocal studies have been combined with AFM to better characterize the diameter-dependence of Raman peaks, [58] and with HRTEM to precisely correlate atomic structure with spectral properties of the same nanotubes [61]. The confocal's high-speed, single nanotube characterization ability has also been harnessed as a screen in parallel microarray applications, including carbon vapor deposition on top of a microarray of different liquid catalysts to optimize synthesis, [62] and a microarray constructed by linking DNA covalently to nanotubes to test sequence-specific binding interactions [63].

3.3.2. Nanowires

Confocal Raman spectroscopy has also characterized other one-dimensional systems. Raman spectra of silicon nanowires (5–15 nm) collected at *low* laser power match those of bulk silicon. But as power of the confocal's illuminating laser was increased, the Raman peaks shifted from those of bulk silicon in a way not consistent with quantum confinement effects, but rather suggesting that the laser is heating the wire itself [64].

3.4. Zero-dimensional Systems

3.4.1. Luminescent Nanocrystals (Quantum Dots)

Nanocrystals are crystals ranging in size from nanometers to tens of nanometers, large enough so that single atoms do not drive their dynamics, while still small enough for their electronic and optical properties to be governed by quantum mechanical effects. Under confocal microscopy, where their size precludes resolution of their features, they effectively behave as zero-dimensional points. Of the common synonyms, including 'nanoparticle' and 'quantum dot,' only nanocrystal will be used here. The confocal has been used in two broad areas: spectrally characterizing individual nanocrystals, where confocality is required to isolate individual particles, and localizing nanocrystals as point tracers in other systems, utilizing the capability of three-dimensional, real-time localization.

Recent examples of confocal characterization of the energy spectra of individual nanocrystals include cryogenic (20 K) imaging of single 7-nm ZnS nanocrystals, [65] and characterization of the optical extinction properties of nanocrystals created by conventional nanosphere lithography [66]. Quantifying the temporal dynamics of the intermittent fluorescence (blinking) typical of nanocrystals yields information on their electronic structure, and the fast on-and-off fluorescence can be captured with high-speed optical detectors, like the photomultiplier tube (PMT) or the avalanche photodiode (APD), attached to the confocal. Studies of individual CdSe nanocrystals overcoated with ZnS have shown that several energy levels may drive the optical behavior, [67, 68] with similar results found for InP nanocrystals [69]. Measuring energy spectra over time has quantified changes due to oxidation of nanocrystals in air (versus no change in pure nitrogen), [70] and the size and surface-property dependence of the behavior of silicon quantum dots, both porous [71] and crystalline stabilized with an organic monolayer [72]. By splitting the light coming back from the sample, sending half to an APD and the remainder through a prism onto a CCD, high resolution time and spectral data can be collected simultaneously. This analysis, in combination with TEM of the same individual nanocrystals to correlate optical properties with atomic structure, has shown that a single crystalline domain is not required to achieve fluorescence [73]. Finally, to characterize the heavy dependence of fluorescence behavior on nearby conductors, confocal microscopy imaged a fluorescent dye attached to both bulk gold and gold nanocrystals on the same substrate. While the dye attached to bulk gold was quenched, since the energy absorbed by the fluorophore gets transferred to the sea of electrons in the metal, dyes attached to nanocrystals remained bright, as there is no bulk into which to transfer electrons [74].

The confocal has also been used to localize nanocrystals embedded in other systems. This has aided synthesis of organic nanocrystals grown in inorganic sol-gel coatings, with confocal characterization of their size and distribution allowing systematic exploration of the phase space of the main synthesis parameters [75]. Embedding nanocrystals within polyelectrolyte layers in LBL assembly allows nanometer control of thin-film coatings that can be applied to three-dimensional objects of complex geometry, whose luminescent properties are then determined by the nanocrystals. Three-dimensional sectioning in the confocal has been crucial to characterizing these coatings, which have been applied to cylindrical optical fibers[76] and spherical latex colloidal particles [77]. In addition, the confocal has been used to characterize the three-dimensional structure of micron-sized domains of nanocrystals, including silver nanocrystals embedded between two layers in a thin film and coalesced by irradiation with a high-intensity laser beam to create planar diffractive and refractive micro-optics, [78] and silicon nanocrystals patterned onto surfaces by directing a stream of silicon atoms through a mask for nanofabrication of light sources from all silicon with pre-existing tools from the electronics industry [79].

In addition to these static applications, the real-time imaging capability of the confocal is useful for monitoring nanocrystal dynamics at higher speeds. Confocal microscopy has been combined with flow cytometry to image and count fluorescent nanoparticles in a fluid flow, yielding real-time information on their concentration [80]. Fluorescent colloidal nanospheres have been coated on one side with gold, yielding an opaque hemispherical metal coating that appears dark, and floated on the surface of a liquid. Light intensity levels of individual nanospheres can be correlated with angular orientation, allowing real-time imaging of Brownian rotational diffusion to study molecular interactions, particularly the preferential attraction of fluid molecules toward one hemisphere over the other [81]. Metal oxide nanocrystals and their halogen adducts have been shown to kill bacteria, and a combination of confocal microscopy and electrostatic measurements has demonstrated that the particles and bacteria attracted each other on account of their electric charge, shedding light on nanoscale electrostatics in solution [82]. Finally, the active transport of single nanocrystals by a dynein motor protein along microtubules in live cells has been imaged in real time with a Nipkow disk confocal; interestingly, contrasting the blinking typical of nanocrystals in free solution, these nanocrystals in live cells have nearly constant intensity [83].

3.4.2. Viruses

Far-field confocal microscopy has been combined with a near-field scanning ion microscopy in water, with a micropipette carrying an ion current tens of nanometers from an optically invisible cell membrane surface to map topography. This technique has been used to image virus-like particles, nanospheres composed of a few hundred viral proteins enclosing DNA, and their absorption into cells. The biological motivation is to understand how viruses infect living cells, but more generally, the investigation demonstrates confocal three-dimensional, real-time localization of a nanoscale object relative to an optically invisible surface [84].

By manipulating the genetic code of viruses, proteins on their surfaces can be modified to achieve desired properties in a technique called phage display, which early on was used to optimize highly-specific binding of viruses to a variety of semiconductor surfaces [85]. More recently, phage display has been used to control the morphology of calcium carbonate crystals precipitated from solution in the presence of viruses that template the crystals, in an effort to better understand biomineralization. While scanning electron microscopy can image and characterize the inorganic calcium carbonate after growth was stopped, confocal microscopy allowed three-dimensional localization of the fluorescently-labeled viruses relative to the growing crystals [86].

3.4.3. Single Molecule Studies

Confocal studies of single molecules fall into two broad groups: spectral characterization of single molecules, and localization of fluorescent molecules as tags. Static single molecule systems characterized include individual dye molecules over a range of temperatures from liquid helium to room temperature, [87] and optimized mutations of the fluorescent protein GFP in various three-dimensional substrates; [88] dynamically, time-correlation spectroscopy in the confocal has been used to measure solution concentrations down to 10^{-15} M [89]. While confocality may not be strictly necessary for these studies, increased resolution helps to isolate individual molecules. The confocal has also been used for three-dimensional localization of single fluorescent dye molecules attached as tags to another object of interest, such as single molecules inside a living cell, [90] or correlating emission spectra of molecules on a mica substrate with AFM data to develop a way measure topography optically [91].

4. SUMMARY AND FUTURE PERSPECTIVES

Confocal microscopy extends the characterization of ever smaller objects with optical microscopy to the nanometer scale. By using a pinhole to restrict detected light to only that coming from the focus of the microscope objective, the confocal allows three-dimensional sectioning and localization, often at rapid rates with proper scanning techniques. Spatially resolved spectroscopy has allowed investigation of a broad variety of systems of interest to nanotechnology, providing information accessible to no other technique.

Further developments may be anticipated in several areas. 4Pi-C confocal has already achieved resolution in the sub-100 nm range that defines "nanoscale;" it is possible that novel optical techniques will improve this even further. Other evolutionary engineering improvements will surely increase speed, perhaps allowing real-time 4Pi imaging comparable to regular confocal by using beam parallelization in the same spirit as the Nipkow disk.

The capabilities of the confocal to answer new kinds of questions are also being developed rapidly, in no small part due to the low cost and relative ease of construction of off-the-shelf laboratory optical components used to build instrumentation ancillary to the confocal. A large fraction of the applications described in this review utilized some form of home-built hardware, a trend which will surely continue. Spectroscopy will likely become faster, allowing the characterization of the changes in spectra on shorter timescales, with the ultimate goal of studying changes in single molecules, a new field with some early applications described. Better understanding of the behavior of isolated nanoscale objects will use the confocal microscope's three-dimensional, real-time localization capability to study not only the dynamics of single particles, but also the behavior of systems of thousands, or perhaps even millions, of particles with controllable interactions. Clearly, the unique capabilities of the confocal microscope will ensure its contribution to the development of nanotechnology for the foreseeable future.

ACKNOWLEDGEMENTS

The author would like to thank I. Cohen, L. Kaufman, N. Tsapis and E. Dufresne for comments on the manuscript, and B. Calloway of Leica Microsystems for providing figure 5.

REFERENCES

1. Hooke, R., *Micrographia*. 1667, London: Royal Society.
2. Brown, R., *A Brief Account of Microscopical Observations Made in the Months of June, July and August 1827 on the Particles Contained in the Pollen of Plants; and on the General Existence of Active Molecules in Organic and Inorganic Bodies*. 1828, London: Taylor.
3. Born, M. and E. Wolf, *Principles of optics: electromagnetic theory of propagation, interference and diffraction of light*. 6th corr. ed. 1997, Oxford; New York: Cambridge University Press c 1997. xxviii, 808.
4. Sheppard, C. J. R. and D. M. Shotton, *Confocal Laser Scanning Microscopy*. 1997, Oxford: BIOS Scientific Publishers.
5. Corle, T. R. and G. S. Kino, *Confocal Scanning Optical Microscope and Related Imaging Systems*. 1996, San Diego: Academic Press.
6. Diaspro, A., ed. *Confocal and Two-Photon Microscopy: Foundations, Applications, and Advances*. 2002, Wiley-Liss: New York.
7. Gu, M., *Principles of Three-Dimensional Imaging in Confocal Microscopes*. 1996, Singapore: World Scientific.
8. Hell, S. W. and E. H. H. Stelzer, *Properties of a 4Pi confocal fluorescence microscope*. J. Opt. Soc. am. A, 1992. **9**(12): p. 2159–2166.
9. Hell, S. W., *et al.*, *Confocal microscopy with an increased detection aperture: type-B 4Pi confocal microscopy*. Optics Letters, 1994. **19**(3): p. 222–224.
10. Schrader, M., *et al.*, *Optical transfer functions of 4Pi confocal microscopes: theory and experiment*. Optics Letters, 1997. **22**(7): p. 436–438.

11. Hell, S. W. and M. Nagorni, *4Pi confocal microscopy with alternate interference.* Optics Letters, 1998. **23**(20): p. 1567–1569.
12. Blanca, C. M. and S. W. Hell, *Sharp Spherical Focal Spot by Dark Ring 4Pi-Confocal Microscopy.* Single Mol., 2001. **2**(3): p. 207–210.
13. Blanca, C. M., J. Bewersdorf, and S. W. Hell, *Determination of unknown phase difference in 4Pi-confocal microscopy through the image intensity.* Optics Communications, 2002. **206**: p. 281–285.
14. Hell, S. W. and E. H. H. Stelzer, *Fundamental improvement of resolution with a 4Pi-confocal fluorescence microscope using two-photon emission.* Optics Communications, 1992. **93**(5–6): p. 277–282.
15. Soini, J. T., *et al.*, *Image formation and data acquisition in a stage scanning 4Pi confocal fluorescence microscope.* Applied Optics, 1997. **36**(34): p. 8929–8932.
16. Schrader, M., S. W. Hell, and H. T. M. van der Voort, *Potential of confocal microscopes to resolve in the 50–100 nm range.* Applied Physics Letters, 1996. **69**(24): p. 3644–3646.
17. Schrader, M., S. W. Hell, and H. T. M. van der Voort, *Three-dimensional super-resolution with a 4Pi-confocal microscope using image restoration.* Journal of Applied Physics, 1998. **84**(8): p. 4033–4042.
18. Albrecht, B., *et al.*, *Spatially modulated illumination microscopy allows axial distance resolution in the nanometer range.* Applied Optics, 2002. **41**(1): p. 80–87.
19. Schrader, M., *et al.*, *4Pi-Confocal Imaging in Fixed Biological Specimens.* Biophysical Journal, 1998. **75**: p. 1659–1668.
20. Bahlmann, K., S. Jakobs, and S. W. Hell, *4Pi-confocal microscopy of live cells.* Ultramicroscopy, 2001. **87**: p. 155–164.
21. Egner, A., S. Jakobs, and S. W. Hell, *Fast 100-nm resolution three-dimensional microscope reveals structural plasticity of mitochondria in live yeast.* Proceedings of the National Academy of Sciences, 2002. **99**(6): p. 3370–3375.
22. Gustafsson, M. G. L., *Surpassing the lateral resolution limit by a factor of two using structured illumination microscopy.* Journal of Microscopy, 2000. **198**(2): p. 82–87.
23. Gustafsson, M. G. L., D. A. Agard, and J. W. Sedat, *I5M: 3D widefield light microscopy with better than 100 nm axial resolution.* Journal of Microscopy, 1999. **195**(1): p. 10–16.
24. Gustafsson, M. G. L., *Extended resolution fluorescence microscopy.* Current Opinion in Structural Biology, 1999. **9**: p. 627–634.
25. Moehl, S., *et al.*, *Solid immersion lens-enhanced nano-photoluminescence: Principle and applications.* Journal of Applied Physics, 2003. **93**(10): p. 6265–6272.
26. Bae, J. H., *et al.*, *High resolution confocal detection of nanometric displacement by use of a 2 × 1 optical fiber coupler.* Optics Letters, 2000. **25**(23): p. 1696–1698.
27. Lacoste, T. D., *et al.*, *Ultrahigh-resolution multicolor colocalization of single fluorescent probes.* Proceedings of the National Academy of Sciences, 2000. **97**(17): p. 9461–9466.
28. Michalet, X., T. D. Lacoste, and S. Weiss, *Ultrahih-Resolution Colocalization of Spectrally Separable Point-like Fluorescent Probes.* Methods, 2001. **25**: p. 87–102.
29. Kano, H., *et al.*, *Dual-color 4-Pi confocal microscopy with 3D-resolution in the 100 nm range.* Ultramicroscopy, 2002. **90**: p. 207–213.
30. Schmidt, M., M. Nagorni, and S.W. Hell, *Subresolution axial distance measurements in far-field fluorescence microscopy with precision of 1 nanometer.* Review of Scientific Instruments, 2000. **71**(7): p. 2742–2745.
31. Safran, S., *Statistical Thermodynamics of Surfaces, Interfaces and Membranes.* 1994, Boulder: Westview.
32. Sakai, T., *et al.*, *Monitoring Growth of Surfactant-Free Nanodroplets Dispersed in Water by Single-Droplet Detection.* Journal of Physical Chemistry B, 2003. **107**: p. 2921–2926.
33. Sakai, T., *et al.*, *Dye Transfer between Surfactant = Free Nanodroplets Dispersed in Water.* Journal of Physical Chemistry B, 2002. **106**: p. 5017–5021.
34. Nam, Y. S., *et al.*, *New micelle-like polymer aggregates made from PEI-PLGA diblock copolymers: micellar characteristics and cellular uptake.* Biomaterials, 2003. **24**: p. 2053–2059.
35. Savic, R., *et al.*, *Micellar Nanocontainers Distribute to Defined Cytoplasmic Organelles.* Science, 2003. **300**: p. 615–618.
36. Decher, G., *Fuzzy Nanoassemblies: Toward Layered Polymeric Multicomposites.* Science, 1997. **277**: p. 1232–1237.
37. Silvano, D., *et al.*, *Confocal Laser Scanning Microscopy to Study Formation and Properties of Polyelectrolye Nanocapsules Derived From $CdCO_3$ Templates.* Microscopy Research and Technique, 2002. **59**: p. 536–541.
38. Dai, Z., *et al.*, *Nanoengineering of Polymeric Capsules with a Shell-in-Shell Structure.* Langmuir, 2002. **18**: p. 9533–9538.

39. Antipov, A. A., *et al.*, *Fabrication of a Novel Type of Metallized Colloids and Hollow Capsules.* Langmuir, 2002. **18**: p. 6687–6693.
40. Gaponik, N., *et al.*, *Labeling of Biocompatible Polymer Microcapsules with Near-Infrared Emitting Nanocrystals.* Nano Letters, 2003. **3**(3): p. 369–372.
41. Khopade, A. J. and F. Caruso, *Electrostatically Assembled Polyelectrolyte/Dendrimer Multilayer Films as Ultrathin Nanoreservoirs.* Nano Letters, 2002. **2**(4): p. 415–418.
42. Lvov, Y., *et al.*, *Urease Encapsulation in Nanoorganized Microshells.* Nano Letters, 2001. **1**(3): p. 125–128.
43. Chiu, D. T., *et al.*, *Manipulating the biochemical nanoenvironment around single molecules contained within vesicles.* Chemical Physics, 1999. **247**: p. 133–139.
44. Radtchenko, I. L., M. Giersig, and G. B. Sukhorukov, *Inorganic Particle Synthesis in Confined Micron-Sized Polyelectrolyte Capsules.* Langmuir, 2002. **18**: p. 8204–8208.
45. Bloess, A., *et al.*, *Microscopic Structure in a Shpol'skii System: A Single-Molecule Study of Dibenzanthanthrene in n-Tetradecane.* Journal of Physical Chemistry A, 2001. **105**: p. 3016–3021.
46. Latterini, L., *et al.*, *Space-resolved fluorescence properties of pheolphthalein-hydrotalcie nanocomposites.* Phys. Chem. Chem. Phys., 2002. **4**: p. 2792–2798.
47. Kittel, C., *Introduction to Solid State Physics.* 7th ed. 1996, New York: Wiley.
48. Van Keuls, F. W., *et al.*, *A Ku-Band Gold/BaxSr1-xTiO3/LaAlO3 Conductor/Thin Film Ferroelectric Microstrip Line Phase Shifter for Room-Temperature Communications Applications.* Microwave and Optical Technology Letters, 1999. **20**(1): p. 53–56.
49. Hubert, C. and J. Levy, *New optical probe of GHz polarization dynamics in ferroelectric thin films.* Review of Scientific Instruments, 1999. **70**(9): p. 3684–3687.
50. Hubert, C., *et al.*, *Confocal scanning optical microscopy of BaxSr1-xTiO3 thin films.* Applied Physics Letters, 1997. **71**(23): p. 3353–3355.
51. Hubert, C., *et al.*, *Mesoscopic Microwave Dispersion in Ferroelectric Thin Films.* Physical Review Letters, 2000. **85**(9): p. 1998–2001.
52. Hubert, C., *et al.*, *Nanopolar reorientation in ferroelectric thin films.* 2001.
53. Tesfamichael, T., *et al.*, *Investigations of dye-sensitised titania solar cell electrode using confocal laser scanning microscopy.* Journal of Materials Science, 2003. **38**: p. 1721–1726.
54. Qi, D., *et al.*, *Optical Emission of Conjugated Polymers Adsorbed to Nanoporous Alumina.* Nano Letters, 2003. **3**: p.?????
55. Fertig, N., *et al.*, *Stable integration of isolated cell membrane patches in a nanomachined aperture.* Applied Physics Letters, 2000. **77**(8): p. 1218–1220.
56. Jorio, A., *et al.*, *Structural (n,m) Determination of Isolated Single-Wall Carbon Nanotubes by Resonant Raman Scattering.* Physical Review Letters, 2001. **86**(6): p. 1118–1121.
57. Azoulay, J., *et al.*, *Polarised spectroscopy of individual single-wall nanotubes: Radial-breathing mode study.* Europhysics Letters, 2001. **53**(3): p. 407–413.
58. Zhao, J., *et al.*, *Diameter-Dependent Combination Modes in Individual Single-Walled Carbon Nanotubes.* Nano Letters, 2002. **2**(8): p. 823–826.
59. Ecklund, P. C., *et al.*, *Large-Scale Production of Single-Walled Carbon Nanotubes Using Ultrafast Pulses from a Free Electron Laser.* Nano Letters, 2002. **2**(6): p. 561–566.
60. Sangaletti, L., *et al.*, *Carbon nanotube bundles and thin layers probed by micro-Raman spectroscopy.* The European Physical Journal B, 2003. **31**: p. 203–208.
61. Jiang, C., *et al.*, *Combination of Confocal Raman Spectroscopy and Electron Microscopy on the Same Individual Bundles of Single-Walled Carbon Nanotubes.* Nano Letters, 2002. **2**(11): p. 1209–1213.
62. Chen, B., *et al.*, *Heterogeneous Single-Walled Carbon Nanotube Catalyst Discovery and Optimization.* Chem. Mater., 2002. **14**: p. 1891–1896.
63. Hazani, M., *et al.*, *Confocal Fluorescence Imaging of DNA-functionalized Carbon Nanotubes.* Nano Letters, 2003. **3**(2): p. 153–155.
64. Gupta, R., *et al.*, *Laser-Induced Fano Resonance Scattering in Silicon Nanowires.* Nano Letters, 2003. **3**(5): p. 627–631.
65. Tittel, J., *et al.*, *Fluorescence Spectroscopy on Single CdS Nanocrystals.* J. Phys. Chem. B., 1997. **101**(16): p. 3013–3016.
66. Haynes, C. L. and R. P. van Duyne, *Dichroic Optical Properties of Extended Nanostructures Fabricated Using Angle-Resolved Nanosphere Lithography.* Nano Letters, 2003. **3**(7): p. 939–943.
67. Kuno, M., *et al.*, *Nonexponential "blinking" kinetics of single CdSe quantum dots: A universal power law behavior.* Journal of Chemical Physics, 2000. **112**(7): p. 3117–3120.
68. Kuno, M., *et al.*, *"On"/"off" fluorescence intermittency of single semiconductor quantum dots.* Journal of Chemical Physics, 2001. **115**(2): p. 1028–1040.

69. Kuno, M., *et al.*, *Fluorescence Intermittency in Single InP Quantum Dots.* Nano Letters, 2001. **1**(10): p. 557–564.
70. van Sark, W. G. J. H. M., *et al.*, *Photooxidation and Photobleaching of Single CdSe/ZnS Quantum dots Probed by Room-Temperature Time-Resolved Spectroscopy.* J. Phys. Chem. B., 2001. **105**: p. 8281–8284.
71. Mason, M. D., *et al.*, *Luminescence of Individual Porous Si Chromophores.* Physical Review Letters, 1998. **80**(24): p. 5405–5408.
72. English, D. S., *et al.*, *Size Tunable Visible Luminescence from Individual Organic Monolayer Stabilized Silicon Nanocrystal Quantum Dots.* Nano Letters, 2002. **2**(7): p. 681–685.
73. Koberling, F., *et al.*, *Fluorescence Spectroscopy and transmission electron microscopy of the same isolated semiconductor nanocrystals.* Applied Physics Letters, 2002. **81**(6): p. 1116–1118.
74. Levi, S. A., *et al.*, *Fluorescence of Dyes Adsorbed on Hihgly Organized, Nanostructured Gold Surfaces.* Chem. Eur. J., 2002. **8**(16): p. 3808–3814.
75. Sanz, N., *et al.*, *Organic nanocrystals grown in sol-gel coatings.* Journal of Materials Chemistry, 2000. **10**: p. 2723–2726.
76. Crisp, M. T. and N. A. Kotov, *Preparation of Nanoparticle Coatings on Surfaces of Complex Geometry.* Nano Letters, 2003. **3**(2): p. 173–177.
77. Susha, A. S., *et al.*, *Formation of luminescent spherical core-shell particles by the consecutive adsorption of polyelectrolye and CdTe(S) nanocrystals on latex colloids.* Colloids and Surfaces A: Physicochemical and Engineering Aspects, 2000. **163**: p. 39–44.
78. Martin, J., *et al.*, *Laser microstructuring and scanning microscopy of plasmapolymer-silver composite layers.* Applied Optics, 2001. **40**(31): p. 5726–5730.
79. Ledoux, G., *et al.*, *Nanostructured films composed of silicon nanocrystals.* Materials Science and Engineering C, 2002. **19**: p. 215–218.
80. Ferris, M. M. and K. L. Rowlen, *Detection and enumeration of single nanometric particles: A confocal optical design for fluorescence flow cytometry.* Review of Scientific Instruments, 2002. **73**(6): p. 2404–2410.
81. Choi, J., *et al.*, *Patterned Fluorescent Particles as Nanoprobes for the Investigation of Molecular Interactions.* Nano Letters, 2003. **3**(8): p. 995–1000.
82. Stoimenov, P. K., *et al.*, *Metal Oxide Nanoparticles as Bactericidal Agents.* Langmuir, 2002. **18**: p. 6679–6686.
83. Chen, P., *et al.*, *High Imaging Sensitivity and Blinking Suppression of Single Quantum Dots in A Live Cell: Visualization of Dynein Mediated Active Transport at the Video Rate.* Nature Biotechnology, 2004.
84. Gorelik, J., *et al.*, *Scanning surface confocal microscopy for simultaneous topographical and fluorescence imaging: Application to single virus-like particle entry into a cell.* Proceedings of the National Academy of Sciences, 2002. **99**(25): p. 16018–16023.
85. Whaley, S. R., *et al.*, *Selection of peptides with semiconductor binding specificity for directed nanoparticle assembly.* Nature, 2000. **405**: p. 665–668.
86. Li, C., G. D. Botsaris, and D. L. Kaplan, *Selective in Vitro Effect of Peptides on Calcium Carbonate Crystallization.* Crystal Growth and Design, 2002. **2**(5): p. 387–393.
87. Segura, J.-M., A. Renn, and B. Hecht, *A sample-scanning confocal optical microscope for cryogenic operation.* Review of Scientific Instruments, 2000. **71**(4): p. 1706–1711.
88. Chirico, G., *et al.*, *Dynamics of green fluorescent protein mutant2 in solution, on spin-coated glasses, and encapsulated in wet silica gels.* Protein Science, 2002. **11**: p. 1152–1161.
89. Nie, S. and R. N. Zare, *Optical Detection of Single Molecules.* Annual Reviews of Biophysics and Biomolecular Structure, 1997. **26**: p. 567–596.
90. Byassee, T. A., W. C. W. Chan, and S. Nie, *Probing Single Molecules in Single Living Cells.* Analytical Chemistry, 2000. **72**(22): p. 5606–5611.
91. Kolody, L. A., *et al.*, *Spatially Correlated Fluorescence/AFM of Individual Nanosized Particles and Biomolecules.* Analytical Chemistry, 2001. **73**(9): p. 1959–1966.

2. SCANNING NEAR-FIELD OPTICAL MICROSCOPY IN NANOSCIENCES

ALEXANDRE BOUHELIER, ACHIM HARTSCHUH, AND LUKAS NOVOTNY

1. SCANNING NEAR-FIELD OPTICAL MICROSCOPY AND NANOTECHNOLOGY

The skills of Swiss watchmakers a couple of centuries ago started a trend of miniaturization aiming at controlling the material world on increasingly smaller scales. This trend never stopped ever since. The emerging field of nanoscience is leading to unprecedented understanding and control over the fundamental building blocks of matter. What was a playpen for physicists, chemists, and biologists, is rapidly evolving to mature technology and engineering. The possibility of artificially synthesized nanodevices that could become the basis for completely new technologies is the main driving force of today's investors. The way individual units organize into nanoscale patterns determines important material functionalities, including electrical conductivity, mechanical strength, chemical specificity, and optical properties. To map out the landscape of this nanoscale territory, new characterization tools have to be developed. The family of scanning probe microscopes offered scientists to zoom in on previously hidden nanoscale features. While the nanometric stylus of a scanning tunneling microscope travels over a sample surface like a blind's person stick, much could be learn if this stick was equipped with a "nano-eye", which optically look at the surface.

Extending optical perception to increasingly finer scales permitted early scientists to discover natural laws otherwise invisible. Over nearly 4 centuries, the basic design of a microscope did not really change conceptually. The legacy of Van Leeuwenhoek and Hooke is still a prime scientific tool. With the improvement of lenses in the 18th and 19th centuries, especially the effort done to correct aberrations, microscopists rapidly pushed the optical microscope to its fundamental limit: the maximum resolving power of a microscope is governed by diffraction of light. The image of a point source is not a point! This limitation, imposed by the very nature of light, was predicted by E. Abbe [1]. Lord Rayleigh in the 19th century expressed the resolution barrier in a very concise form known as the Rayleigh criterion [2]. In a visionary idea, E. H. Synge predicted in 1928 that the diffraction limit could be overcome by reducing the illuminating light source to a volume smaller than the wavelength [3]. By raster scanning this source in close proximity over a sample surface provides an image with lateral resolution better than the one imposed by the diffraction limit. The experimental realization of the forgotten idea was independently achieved in Switzerland [4] and in the U.S [5] in the early eighties, soon after the discovery of the scanning tunneling microscope. Scanning near-field optical microscopy (SNOM or NSOM) was born and adds its name to the growing family of scanning probe microscopes.

Optics provides a wealth of information not accessible to other proximal techniques. The photon, as an observable, can reveal the identity of molecules, their chemical, material, and electronic specificity. Scanning near-field optical microscopy provides the "nano-eye" in the form of a nanometric light source confined at the end of a tip. An obvious advantage of the technique over many others is its versatility. The method has been successfully used in various environmental conditions, including low temperature, high vacuum, and in liquids depending on sample requirements. There are no fundamental restrictions on the object under investigation either. The sample can be transparent or opaque, flat or corrugated, organic or semiconductor...

Although technical issues rather than fundamental limits prevented near-field optics to advance to arbitrary high resolution, recent advances in nanofabrication and a better understanding of optical interactions on the nanoscale will potentially enable SNOM to be a prime characterization tool, in the same way as optical microscopy in the past centuries.

2. BASIC CONCEPTS

The angular representation of the field in a plane $z = z_0$ near an arbitrary object can be written as:

$$E(x, y, z_0) = \iint A(k_x, k_y) \exp i\left(k_x x + k_y y + z_0\sqrt{\left(k_0^2 - k_x^2 - k_y^2\right)}\right) dk_x\, dk_y, \quad (1)$$

where $A(k_x, k_y)$ represents the complex amplitude of the field, and $k_0 = \omega/c$ is the vacuum wave vector. Equation (1) is basically the sum of plane waves and evanescent waves propagating in different spatial directions. Wave vectors k_x and k_y smaller than k_0 constitute homogenous plane waves that propagate in free space. Wave vectors satisfying this condition have low spatial frequencies. Typically, a lens of a microscope collects wave vectors that are confined to $[0..k_{max} = n\omega \sin\theta/c]$, where θ is the semi-aperture angle of the lens and n is the refractive index. In terms of resolution, the distance Δx

between two point-like objects that can just be resolved with a conventional optical microscope using coherent illumination is given by:

$$\Delta x = 0.82 \frac{\lambda_0}{n \sin\theta}, \tag{2}$$

where λ_0 is the illumination wavelength in vacuum. This equation, derived from Abbe's theory, represents the theoretical resolution given by the diffraction of light. According to Eq. (2), the separation Δx can be decreased by using shorter illumination wavelengths (UV microscopy), and/or by increasing the index of refraction (oil or water immersion objectives, solid immersion lenses) and/or by increasing the collection angle θ.

The integration in Eq. 1 runs also over k_x and k_y values that are larger than ω/c. Consequently, the field components become evanescent. The electric field of evanescenet waves propagates in the x, y plane but is exponentially attenuated in the z-direction. These fields, associated with high spatial frequencies (fine details of an object), are not detected by the objective of a classical microscope. In order to achieve superresolution, the variations of the field in the immediate vicinity of the object have to be collected. The collection of evanescent waves is the basis of scanning near-field optical microscopy.

Evanescent waves can be converted to propagating radiation by local scattering. The smaller the scatterer is and the closer it is placed near the surface of an object, the better the conversion will be. According to Babinet's principle, local scattering is analogous to local illumination. This means that the small details of an object can be accessed by either scattering the evanescent fields created by the object with a small scattering center or by illuminating the object with evanescent fields created by a local source. The field produced by the local source is converted into farfield components by the minute dimensions of the object. The combination of the reciprocity theorem and Babinet's principle is at the origin of numerous possible arrangements used in near-field optical microscopy. Although they are phenomenologically different, they are conceptually identical and lead to similar results. The choice of one arrangement over another is mainly driven by the optical characteristics of the sample. A more detailed description of common configurations will be described later.

3. INSTRUMENTATION

Scanning near-field optical microscopy belongs to the family of proximal probe microscopy such as scanning tunneling (STM) and atomic force microscopy (AFM). In these techniques, high resolution is achieved by minimizing the interaction volume between a probe and an object. The probe takes form of a tip where only the very apex is responsible for the interaction. Similarly, near-field optics uses a small probe to confine an optical interaction between probe and sample. Control over probe manufacturing is a prerequisite for routine high-resolution imaging. Unlike the successful batch microfabrication of AFM cantilevers, the fabrication of near-field probes is the

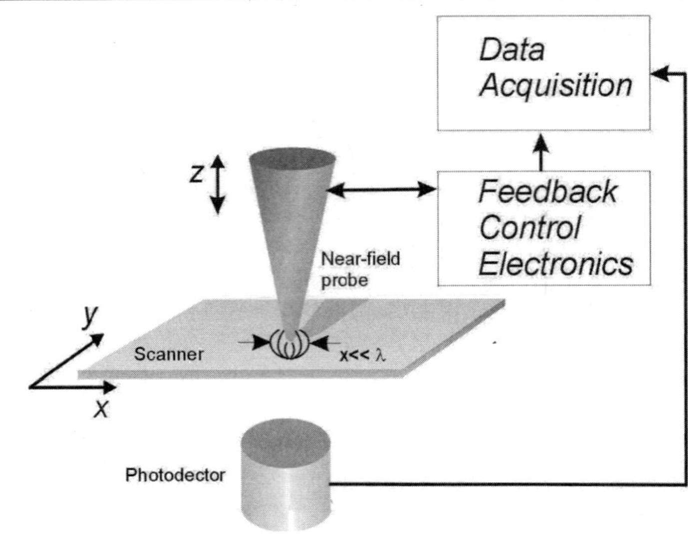

Figure 1. Schematic of a near-field optical microscope. A near-field probe confines an optical interaction to dimensions much smaller than the wavelength. A photodector collects the optical response for each probe position. The signal is read by an acquisition module, which reconstructs a two-dimensional optical image of the surface. A feedback system controls the distance between probe and sample surface to a few nanometers. Finally, a scanning stage translates the sample (or the tip) in a lateral direction.

Achille's heel of SNOM. A discussion of the role of nanotechnology in probe manufacturing will be reviewed in the next section.

The basic units forming a near-field microscope are very similar to an AFM or a STM. As depicted in Figure 1, it consists of (i) a near-field probe confining an optical interaction to dimensions smaller than the wavelength, (ii) a scanning stage permitting to move the sample or the tip laterally, (iii) a photodetector to collect the response of the optical probe-sample interactions, and finally (iv) an acquisition software to reconstruct an optical image. SNOM specifications require that the tip sample distance should be controlled with sub-nanometer precision and shear-force regulation [6] based on quartz tuning fork is extensively used [7]. The damping of a laterally oscillating tip caused by the mechanical interactions with the surface (shear forces) depends on the tip-sample separation. This damping signal is fed to a feedback mechanism (v), which controls the distance between tip and sample via a piezoelectric actuator (not shown).

3.1. Probe Fabrication

In order to retrieve the high spatial frequencies of an object, a probe is brought in close proximity of the object's surface. An image is reconstructed by scanning the probe in the plane of the sample. The lateral resolution in a near-field optical image is determined, to first approximation, by the size of the optical probe. However, the fabrication of optical probes with sub-wavelength dimensions is technically challenging, and nanotechnology plays an important role in the fabrication processes.

3.1.1. Optical Fiber Probes

Sharply pointed optical fibers are widely employed in scanning near-field optical microscopy. These probes can be used as local scatterers or as nanosources. Optical fibers have the advantage of low fabrication costs and low propagation losses. The guiding properties of these waveguides fulfill the conditions needed for near-field applications to a large extent. Polarization for instance, can be accurately controlled in the fiber. It finds important applications for the interpretation of contrast [8]. Furthermore, the operation wavelength of fibers spans from the visible to telecom wavelengths, which can be useful for recording spectroscopic information.

The central step in probe manufacturing is the formation of a taper region to form a nanometric glass tip terminating one end of the fiber. Two methods are usually employed. One approach, the so-called pulling technique, is adopted from microbiology where it is used to produce micropipettes. A pulled fiber is obtained by locally melting the glass with the help of a CO_2 laser or a hot coil. Springs attached to both ends pull the fiber apart resulting in two tapered ends [9]. In elaborate commercially available puller, parameters such as heating time, pulling force or delay time can be adjusted to control the shape of the taper. The surface of the taper is usually very smooth as a result of the local melting during the pulling process. The end of the fiber is commonly terminated by a plateau, which can vary in size depending on the pulling parameters. Alternatively, glass is a material that can be etched by chemicals such as hydrofluoric acid (HF). A bare glass fiber (without the protective polymer jacket) is dipped into a bath of HF. The surface tension of the liquid forms a meniscus at the interface between air, glass, and HF. A taper is formed due to the variation of the contact angle at the meniscus while the fiber is etched and its diameter decreases [10]. The surface tension can be modified by the addition of surface layer atop the HF. As a result, the cone angle of the taper can be slightly varied. This function is more difficult to control with the pulling technique. A striking difference, as compared to a pulled fiber is the surface roughness of the taper. The acid leads to irregular surfaces because of inhomogeneities in the glass and because of the sensitivity of the technique to outside perturbations (temperature, vibrations . . .). Alternative etching techniques have been developed to improve the surface roughness. Among them, is the tube-etching technique [11, 12] or buffered HF solutions [13] that tend to reduce the influence of external perturbations. A comparison between mechanically pulled and chemically etched fibers is provided in Refs [14, 15]. Without going into details, pulled fibers have small cone angles which are not favorable for high throughput [16] and have a flat end face that somewhat limits the size of the tip. On the other hand, their smooth surface benefits the deposition of homogeneous metal coatings as discussed later. Etched fibers have a rough surface caused by the acid attack. Consequently, the quality of a deposited metal coating is not as good. The main advantage of etching is that the cone angle of the taper can be varied and the transmission of metal-coated probes can be drastically improved [16].

3.1.2. Aperture Formation

The guiding properties of an optical fiber are well understood [17]. In the taper region, a propagating mode becomes increasingly delocalized, *i.e.,* the field reaches out of the fiber core [18, 19]. The consequence is that while the extremity of the fiber can be a few tens of a nanometer, the spatial extension of the mode exceeds this dimension by far. To confine the mode and to guide it to the very tip, a metal coating is deposited on the outside surface of the fiber. The layer acts as a reflector and therefore prevents the light to spread out of the fiber. Aluminum, silver and gold are the most commonly used materials owing to their good optical properties at visible wavelengths (small skin depth and high reflectivity). If the metal coating covers the entire taper region, the fiber will be opaque and no light will be transmitted or collected. For the light to escape, a sub-wavelength hole-or aperture-is needed at the very apex of the coated tip. The fabrication of such an aperture with nanometer sized dimensions poses a variety of technical difficulties. These are discussed in the next sections.

A) SQUEEZING TECHNIQUE In the early work of D. W. Pohl *et al.*, a completely metal-coated optical waveguide was squeezed towards a hard surface in order to press the metal away from the foremost end. Applying an offset voltage on an extendable piezoelectric tube can control the pressure on the tip. Cold deformation and abrasion take place at the end of the tip eventually forming a tiny aperture [4]. Monitoring the light throughput of the tip controls the formation of the aperture. The obtained apertures are fairly small (~80 nm) and the end faces are flat. This technique provided high-resolution images on individual fluorophores [20] and seems to regain some attention recently.

B) SHADOWING TECHNIQUE The most widespread technique to produce a small aperture uses the so-called shadowed evaporation scheme. In this approach, the apertures are produced at the time of metallization in an evaporation chamber. A metal coating is deposited such that the very apex of a fiber tip is left uncoated. This is accomplished by evaporating the metal in a direction slightly inclined to the tip axis. A homogeneous film thickness is deposited around the fiber by rotating the fiber during the evaporation process. This method is well suited for the fabrication of apertures of many tips at once. Unfortunately, the aperture shape and diameter are not reproducible between successive evaporations or even between tips evaporated at the same time. The main reason of the poor reproducibility is the number of parameters involved: tilt angle, distance to the source, rotation speed, evaporation rate, base pressure and film thickness. Furthermore, the intrinsic roughness of the metal surface also influences the quality of the apertures. An example is shown in Figure 2(a) where an electron microscope was used to image the end face of a tip. The aluminum layer is readily seen in the image. It is surrounding a dark center representing the physical aperture. An aluminum grain is obscuring part of the opening leading to an asymmetric aperture.

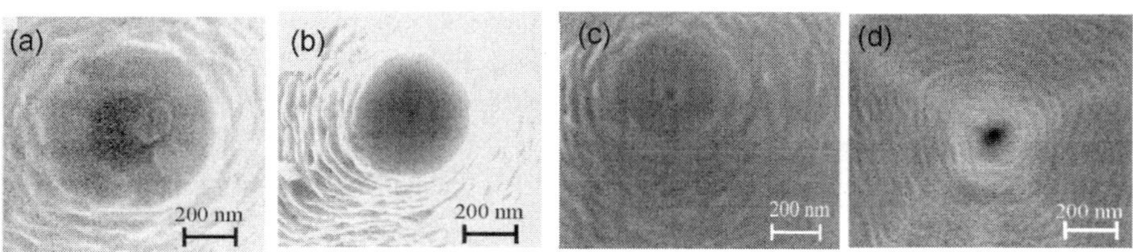

Figure 2. Gallery of apertures produced with different approaches listed in the text. (a) Shadowing technique. Courtesy of B. Hecht, University of Basel. (b) Electroerosion. (c) Focused ion beam milling. Courtesy of Th. Huser, Lawrence Livermore National Laboratory. (d) Cantilever probe. Courtesy of L. Aeschimann and G. Schürmann, IMT Neuchâtel.

C) ELECTRO-EROSION The properties of solid amorphous electrolytes such as $AgPO_3$:AgI compounds provide a unique capability of material transport through a nanoscale area. It has been successfully demonstrated that the high ionic conductivity can be employed to remove the overcoating metal layer from the very apex of fiber tips [21, 22]. In this approach, a tiny electrolytic contact is formed between a tip that is entirely overcoated with metal and the surface of the electrolyte. A bias voltage triggers the erosion of the metal layer. A feedback mechanism is used to keep the ionic current constant, thereby controlling the metal removal rate from the tip. Similar to the squezzing technique, aperture formation is monitored by collecting the light transmitted through the tip. Near-field apertures with diameters below 50 nm can be manufactured by electro-erosion (see Figure 2(b)).

D) FOCUSED ION BEAM MILLING Focused ion beam (FIB) milling is a technique that uses a beam of accelerated ions to modify materials with nanometer precision. In standard instruments, a Gallium source is ionized and the emitted ions are focused into a spot of a few nanometers on the surface of the sample. The energy of the ions is sufficient to ablate material from the sample surface. In the context of aperture formation, the ion source is directed at the apex of a pre-coated tip at an angle of 90° to the tip axis. The end of the tip is sliced away such that the slicing beam cuts through the metal coating as well as the fiber core [23, 24]. The aperture definition, the flatness of the tip end face, the polarization behavior and the imaging capabilities of such processed probes make them suitable for high resolution imaging as demonstrated by investigations of single fluorescent molecules [24]. A scanning electron micrograph of the end face of a FIB-fabricated aperture is shown in Figure 2(c). Although, this approach seems to overcome most of the problems associated with aperture fabrication, the instrument cost is considerable.

E) OTHERS METHODS A plethora of alternative techniques has been used with varying success. Among them is the so-called triangular probe or T-probe. The triangular shape originates from a tetrahedral hollow waveguide coated with an aluminum layer and squeezed against a hard surface to produce an aperture [25]. Another interesting concept described by Fischer and Zapletal was introduced to potentially overcome some of the limitations of aperture-based near-field probes [26]. The authors suggested the use of a coaxial tip as a SNOM probe. In theory, a coaxial waveguide is able to focus electromagnetic fields down to a spot much smaller than the wavelength without any cutoff of propagating modes. Despite the suitability of such optical properties, technical difficulties are still limiting the fabrication of coaxial probes.

3.1.3. Cantilever Probes

Recently, a new probe concept was introduced that is based on a micro-machined tip. It makes use of technology developed for atomic force microscopy (AFM), which is well established in various fields of science. The interaction between probe and sample is well understood, making AFM a rather easy to operate instrument even for non-specialists. AFM probes are batch fabricated which reduces the cost of fabrication. In order to be useful for optical applications, a cantilever has to be transparent for the light to pass. Silicon nitride or quartz is usually the chosen material. While the tips can be made using standard micromachining techniques, cantilever-based optical probes face the same difficulties as those associated with optical fiber based probes. Two main approaches are applied for the fabrication of AFM probe apertures: reactive ion etching [27] and direct-write electron beam lithography [28]. Figure 2(d) shows a scanning electron micrograph of a cantilever tip. Recently, a self-terminated corrosion process of the aluminum film was applied to produce successfully sub-50 nm apertures [29].

3.1.4. Metal Tips

So far, we considered probes that act as optical waveguides, i.e. they guide light to or from a nanoscale area. A second class of optical probes utilizes bare metal tips as used in STM. The technique is referred to as apertureless scanning near-field optical microscopy. A metal tip, usually tungsten, locally perturbs the electromagnetic field surrounding the specimen. The locally scattered information is discriminated from the unavoidable farfield scattered signal with lock-in and demodulation techniques. This scattering approach demonstrated material specificity with outstanding resolution both in the infrared and the visible region [30].

Alternatively, one can benefit from the strong enhancement of the electric field created close to a sharply pointed metal tip under laser illumination. This phenomenon originates from a combination of surface plasmon resonances and an electrostatic lightening rod effect [31]. The energy density close to the metal tip can be orders of magnitude larger than the energy density of the illuminating laser light. This enhancement effect is mainly used to increase the response of spectroscopic interactions such as fluorescence or Raman scattering. Examples of such applications are discussed in section 4. More recently, some experimental and theoretical research was directed at the combination of fiber-based near-field probes with field enhancing metal tips [32, 33]. Preliminary results showed lateral resolutions better than 30 nm. The main advantage of this combination is the reduction of the excitation area as well as the absence of alignment procedures between tip and laser beam.

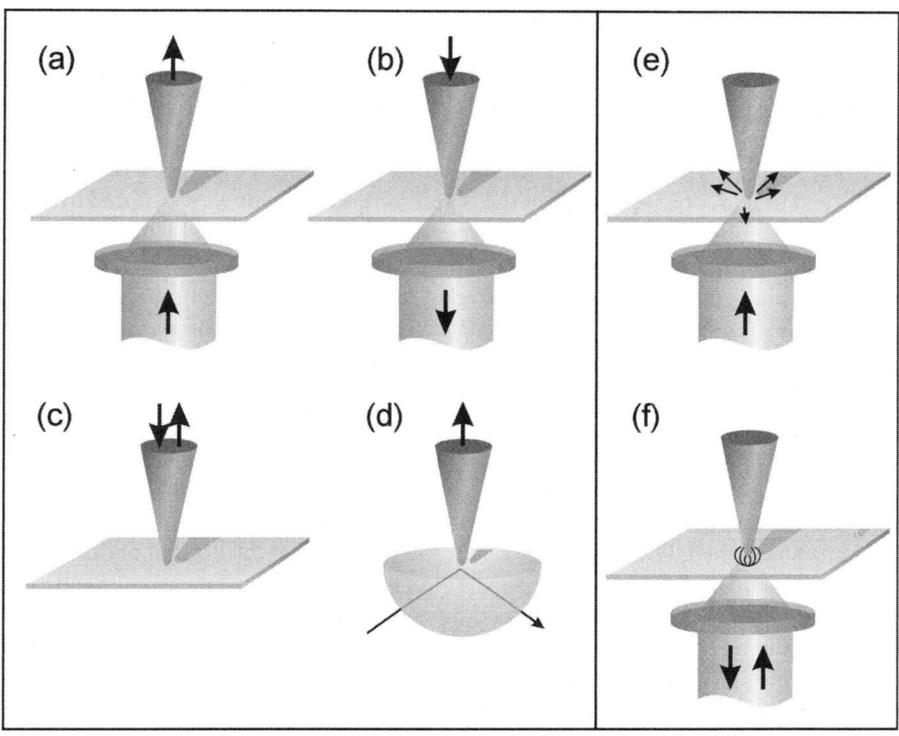

Figure 3. Schematic of the most common SNOM configurations. (a–d) aperture-based probes. (a) Farfield illumination and local probing of the near-field. (b) Near-field illumination, and farfield collection. (c) Local illumination and collection. (d) Dark field illumination. (e–f) Metal tips. (e) Farfield illumination and local scattering at the tip. (f) Local field enhancement created by farfield illumination.

3.2. Flexibility of Near-Field Measurements

As discussed in the previous section, there is no fundamental difference between locally illuminating an object and locally probing the field near it. The essence is that the confinement of the photon flux between probe and sample defines the optical resolution. In turn, many experimental variations can be employed to locally create a confined optical interaction and, for a non-specialist, the literature can be quite confusing.

Figure 3 depicts the most common configurations, emphasizing the flexibility of the technique. Near-field instruments are usually separated into two categories depending on the type of probe used: aperture-based (optical fiber, AFM cantilever) or metal-based (scattering and field enhancement). Sketches (a–d) represent configurations where a nano-aperture is used either as a nanocollector or as a local source. In Figure 3(a), an object is illuminated from the farfield using standard optics. An aperture-based tip is placed close to the surface to collect the near-field. Figure 3(b) depicts the opposite situation. The aperture now illuminates the specimen, and the response is collected with conventional farfield optics. These two configurations are commonly used for thin transparent samples due to the fact that the signals are *transmitted* through the object. However, alternative farfield illumination\collection systems can be implemented from the side to overcome some of the sample restrictions. Figure 3(c) represents a combination of the two previous configurations. Illumination and collection are both performed locally in the near-field of an object. The last configuration based on aperture probes is shown in Figure 3(e). The design can be viewed as the near-field analogue of a dark-field microscope. Illuminating the object with evanescent waves created by total internal reflection (TIR) drastically reduces the background farfield light. The technique emphasizes the fact that a near-field probe *frustrates* the evanescent field [34]. This configuration is widely used for the imaging of non-radiative electromagnetic fields. Applications range from waveguide characterization to imaging of planar plasmonic and photonic structures as discussed in the next section.

The second class of near-field microscope is depicted in Figure 3(e–f). Here, the aperture is replaced by a metal tip, which performs two functions, scattering and/or enhancing a near-field signal, depending on tip material and illumination conditions. In the first example, Figure 3(e), a tip locally scatters the near-field of an object that is illuminated by farfield means. The locally scattered signal is collected also by conventional optics. The illumination, here performed from the bottom of the object, can be implemented from the side of the tip. The choice between one illumination scheme and another is mainly governed by sample requirements. Finally, Figure 3(f) schematically represents a metal tip used as a signal enhancer. Favorable illumination of the tip creates an enhanced field at the tip end that is used as a local light source. The increased sample response is usually collected with the same optics used to illuminate the tip.

4. APPLICATIONS IN NANOSCIENCE

The trend toward nanoscience and nanotechnology is mainly motivated by the fact that the underlying physical laws change from macroscopic to microscopic. As we move to smaller and smaller length scales, new characterization techniques have to be developed to probe the properties of novel nanostructures. There is a continuing demand for new measurement methods that will be positioned to meet emerging measurement challenges.

4.1. Fluorescence Microscopy

Optical spectroscopy provides a wealth of information on structural and dynamical characteristics of materials [35]. Combining optical spectroscopy with near-field optical microscopy is especially desirable because spectral information can be spatially resolved. The need for improved spatial resolution currently limits the ability of industry to answer key questions regarding the chemical composition of surfaces and interfaces.

The detection and manipulation of a single molecule represents the ultimate level of sensitivity in the analysis and control of matter. Measurements made on an individual molecule are inherently free from the statistical averaging associated with conventional

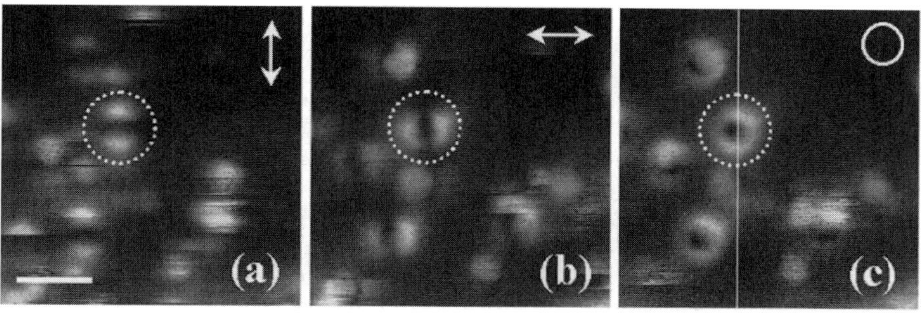

Figure 4. Series of three successive SNOM fluorescence images of the same area (1.2 by 1.2 μm) of a sample of $DiIC_{18}$ molecules embedded in a 10-nm thin film of PMMA. The excitation polarization (as measured in the farfield) was rotated from one linear polarization direction (a) to another (b) and then changed to circulat polarization (c). The fluorescence rate images of the molecules change accordingly. The molecule inside the dotted circle as a dipole axis perpendicular to the sample plane. Scale bar: 300 nm. From Ref. [24].

ensemble experiments. Single molecule techniques have a diverse range of existing applications in physics, chemistry, and biology [36]. Potential applications include single photon sources for quantum computing and massively parallel DNA sequencing. Furthermore, the local environment influences the molecular properties of species and monitoring the behavior of the fluorescence provides a sensitive probe for the molecule's local environment.

The potential of scanning near-field optical microscopy for imaging molecular systems has been recognized by two groups in 1986 [37, 38]. However, the first observation of single molecules using optical near-field technique came 7 years later [39]. In this pioneering experiment, single carbocyanine $DiIC_{12}$ dye molecules were embedded in a polymer matrix and dispersed on a glass substrate. A near-field probe made from an aluminum-coated glass fiber optically excited the molecules. The probe was raster scanned over the sample plane. The fluorescence of individual molecules was collected by a large numerical aperture objective to ensure high collection efficiency. In this remarkable paper, the authors could determine the orientation of the absorption dipole moment of each molecule by recording the spatial variation of the fluorescence as the aperture moved over the molecules. A molecule is excited only if a component of the optical electric field is polarized along its transition dipole moment. Because of the laterally and longitudinally polarized electric fields near the aperture of the probe, randomly oriented molecules can be efficiently excited. Emission patterns of single molecules turned out to be sensitive probes for the electromagnetic field distribution near a near-field probe and be suitable for the determination of molecular orientations. Veerman *et al.* have published a systematic study of this effect [24]. Figure 4 represents a series of near-field images of individual emitters for three consecutive excitation polarizations. Each pixel of the images represents the fluorescence rate for a particular tip position. By changing the polarization state of the near-field excitation, the orientation of the dipole moment can be fully determined.

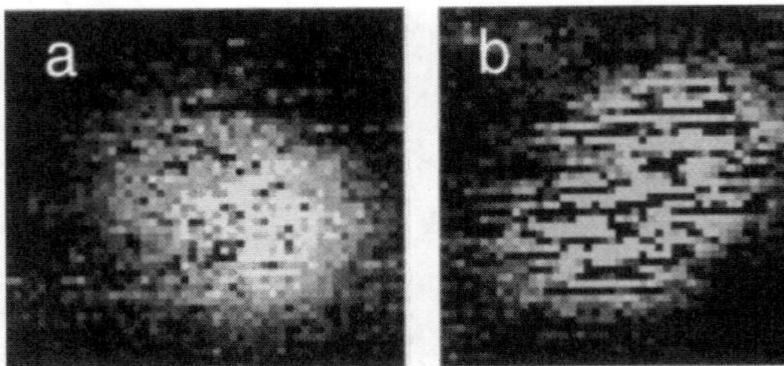

Figure 5. (a) and (b): Near-field fluorescence image of two different single DiIC18 molecules embedded in a 10 nm PMMA layer (370 by 370 nm). Dark pixels are due to temporal quantum jumps to the first excited triplet state. From Ref. [44].

Because of the dependence of a molecule's fluorescence on the local environment, the molecule's properties are influenced by the proximity of an aluminum-coated fiber tip. The fluorescence lifetime was found to be dependent on the relative position between a molecule and the near-field probe [40]. Furthermore, the radiation pattern of a dipole can be modified by the presence of a near-field tip; much alike a dipole radiation pattern is distorted by the presence of a nearby interface [41]. It was found that the radiation pattern for in-plane oriented molecules are distorted in the direction of the center of the near-field aperture while for out-of-plane oriented molecules the distortion is in the direction of the metal coating covering the glass fiber tip [42].

Measurements on individual molecules on a nanometer scale give unique insight on their complex dynamic behavior. Photobleaching of single emitters was observed for the first time, revealing on and off state of the molecule before its final and irreversible photochemical modification [43]. Time-resolved investigations attributed the dynamics of the instable molecular emission to the occupation of the lowest excited triplet state. As long as this state remains populated, the fluorescence is interrupted momentarily leading to a photon-bunching effect as seen in Figure 5 [44]. The lifetime of the triplet state is intimately linked to the environment of the molecule providing thereby a local probe for the nanoenvironment. An interesting side note is that these measurements satisfy the ergodic principle: the same triplet state lifetime distribution is measured on a single molecule at different instants in time and for a set of molecules at the same instant in time.

Near-field optical fluorescence microscopy is successfully used to investigate more complex systems than single molecules. In particular, SNOM is a promising method to study biological systems where resolution is an important parameter. An example of such biological system is nuclear pore complexes (NPCs). Single NPCs cannot be images with conventional microscopy due to the fact the nearest neighbor distance is in the range of 120 nm. The high lateral optical resolution of SNOM can provide detailed insight into the transport dynamics of a single NPC [45].

The lateral resolution obtained with optical near-field techniques is, to a good approximation, given by the diameter of the sub-wavelength aperture. Fluorescence

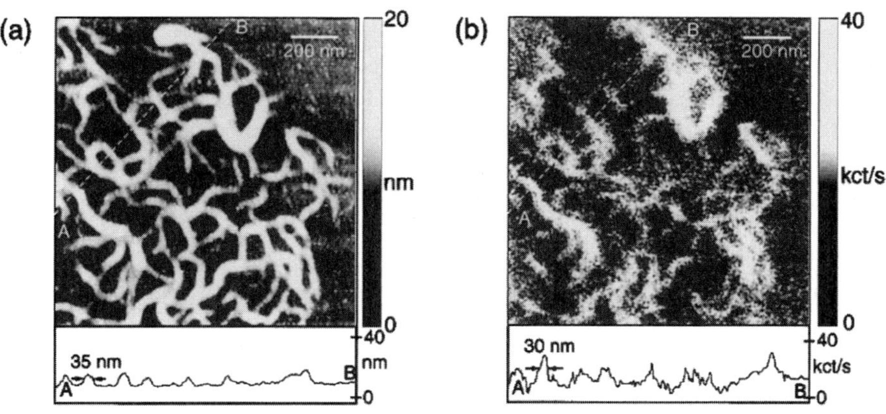

Figure 6. Simultaneous topographic image (a) and near-field two-photon excited fluorescence image (b) of *J*-aggregates of PIC dye in a PVS film on a glass substrate. The topographic cross section along the dashed line (A–B) has a particular feature of 35-nm FWHM (indicated by arrows) and a corresponding 30-nm FWHM in the fluorescence emission cross section. From Ref. [49].

images with 15 nm resolution have been reported [46]. However, there are two fundamental limits preventing higher resolutions. The first one is the penetration of light into the metal coating surrounding the aperture. The skin depth defines the ultimate excitation volume of an aperture, and aluminum is the best coating choice in the visible region. A second limitation is the throughput of a nanometer-sized aperture. The farfield transmission of a sub-wavelength aperture inversely scales with the sixth power of the radius [47]. For a small aperture, the signal-to-noise ratio becomes an important factor in the contrast formation of an image and alternative techniques might be necessary. One way to overcome the signal-to-noise limitation is to increase the fluorescence yield of individual molecules through a local field enhancement effect.

Metal nanostructures are successfully used to enhance the response of particularly small scattering and fluorescence cross-sections [48]. A metal tip can strongly enhance the electric field at its apex through a combination of an electrostatic rod effect and surface plasmon resonances [31]. If the tip is held over a fluorescent sample, the emission yield can be enhanced manifold. This technique was successfully applied to photosynthetic membranes and molecular aggregates, revealing spectroscopic information with a spatial resolution less than 30 nm [49] as shown in Figure 6.

Another promising mechanism that could potentially increase the resolution of a near-field microscope further is the so-called fluorescence resonance energy transfer (FRET) technique. FRET occurs on intermolecular distances of 1 to 8 nm. The energy from a donor molecule can be transferred nonradiatively to an acceptor molecule. The transfer efficiency depends on the inverse sixth-power of the distance between donor and acceptor and on their spectral overlap. This sensitivity can be used to further extend the fluorescence imaging capability of SNOM. Single pair FRET was demonstrated on DNA-linked donor-acceptor where the excitation of the donor was performed by a near-field probe [50]. Development towards true molecular resolution involves the combination of a functionalized tip, where donor (or acceptor) molecules have been attached, with an acceptor (or donor)-doped sample. The group of Dietler and Dunn carried out first attempts simultaneously [51, 52]. To date, the marriage of SNOM and FRET remains a technical challenge, mainly driven by the difficulties to functionalize the probing tips.

4.2. Raman Microscopy

Combined with near-field techniques, Raman spectroscopy is a promising tool for identifying and analyzing the molecular composition of complex materials. Vibrational spectra directly reflect the chemical composition and molecular structure of a sample. By raster scanning the sample and pointwise detection of the Raman spectra, chemical maps with nanoscale resolution can be obtained. A main drawback of Raman methods is the low scattering cross-section, typically 14 orders of magnitude smaller than those of fluorescence. Furthermore, the high spatial resolution achieved in near-field optics is linked to tiny detection volumes containing only a very limited number of Raman scatterers. The weakness of Raman signals in combination with the limited transmission of typical aperture probes [16] requires extended integration times even intermediate aperture sizes of about 150 nm [53, 54, 55]. An essential improvement of the spatial resolution below 50 nm with smaller apertures appears to be unfeasible.

A more promising approach makes use of the enhancement of the electric field in the proximity of nanometer sized metal structures known as surface enhanced Raman scattering (SERS). SERS is known since more than 20 years and enormous enhancement factors of up to 14 orders of magnitude have been reported allowing even for single molecule Raman measurements (see e.g. [48]). In these cases, a combination of different enhancement mechanisms is discussed. The strongest contribution is the electromagnetic enhancement caused by locally enhanced electric fields. An additional contribution results from a chemical effect that requires direct contact between scatterer and metal surface. By using a sharp metal tip, the enhancement effect can be confined to a very small volume at the end of the tip. This localized enhancement allows to selectively address different parts of the sample area and tip-enhanced spectroscopy of nanoscale sample areas can be achieved.

Tip-enhanced Raman spectroscopy has been used on a variety of different systems such as dyes and fullerene films, KTP crystals as well as single-walled carbon nanotubes (SWNT) [56, 57, 58, 59, 60] for a recent review see [61]. In this section, we review some of these studies to demonstrate the three key advantages of the method: high spatial resolution, signal enhancement (enhanced sensitivity), and chemical specificity.

High-resolution near-field Raman imaging of SWNTs on glass is demonstrated in Fig. 6. The Raman image (Fig. 6(a)) clearly shows the characteristic one-dimensional features of SWNTs, which are also seen in the simultaneously detected topographic image (Fig.6 (b)). In the topographic image however, additional circular features are present caused by water condensation. Importantly, these features are not observed in the Raman image demonstrating the chemical specificity of the method. The chemical

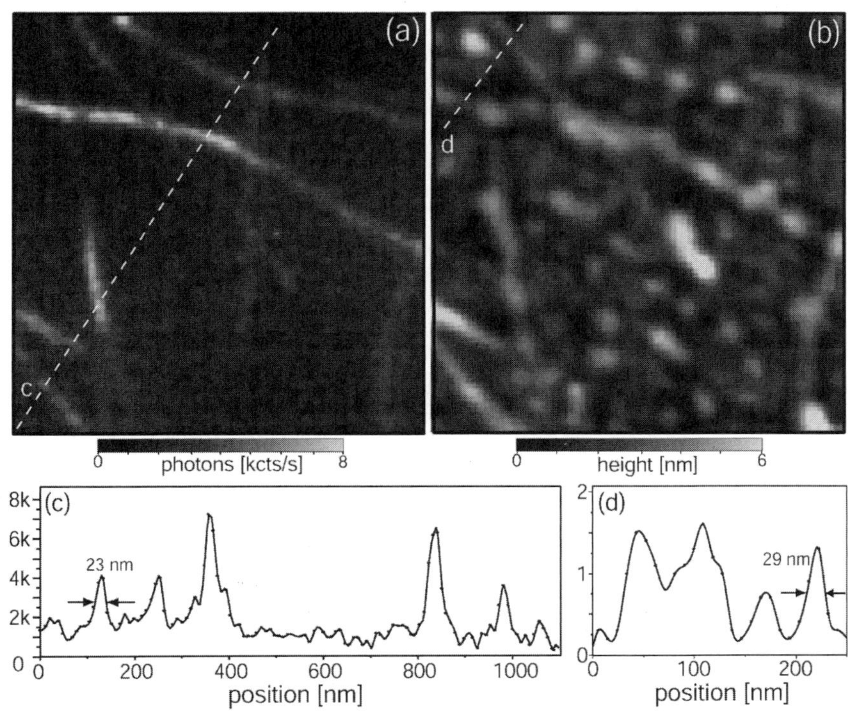

Figure 7. High spatial resolution near-field Raman image (a) and simultaneously detected topographic image (b) of single walled carbon nanotubes (SWNT) on glass. Scan area $1 \times 1\ \mu m^2$. The Raman image is acquired by detecting the intensity of the G′ band upon laser excitation at 633 nm. No Raman scattering is detected from humidity related circular features present in the topographic image indicating excellent chemical specificity (see text). (c) Cross section taken along the indicated dashed line in the Raman image. (d) Cross section taken along the indicated dashed line in the topographic image. The height of individual tubes is 1.4 nm. Vertical units are photon counts per second for (c) and nanometers for (d). From Ref [62].

specificity of the Raman scattering was also used to study local variations in the Raman spectra at the end of a SWNT [62]. In [57], spatial fluctuations within a mixture of dye molecules were investigated. In Fig. 6(c) and (d) cross-sections taken along the dotted lines in Fig. 6(a) and (b) are presented. The signal width (FWHM) observed in the Raman cross section (Fig. 6(c)) is about 23 nm, far below the diffraction limit of light used in the experiment.

The signal enhancement achieved in tip-enhanced Raman spectroscopy can be demonstrated by comparing the Raman spectra detected in presence and absence of the enhancing metal tip. In Fig. 7(a) and (b) examples for different organic molecules are shown. In order to determine the actual signal enhancement factors, the different sample volumes probed in either case have to be considered. The farfield spectrum detected without tip results from a diffraction limited area while the near-field spectrum with tip originates from the much smaller near-field area only. For a quantitative

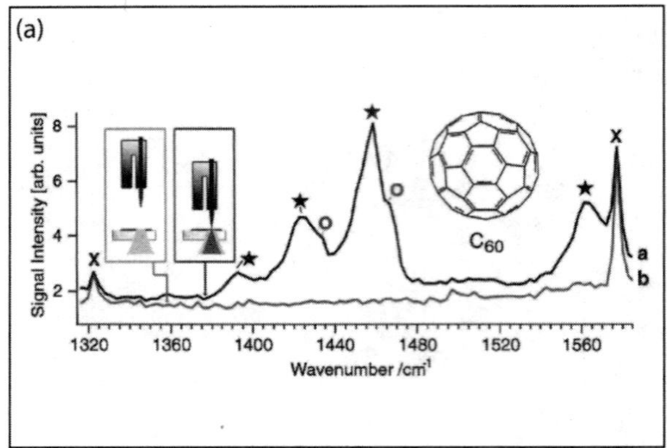

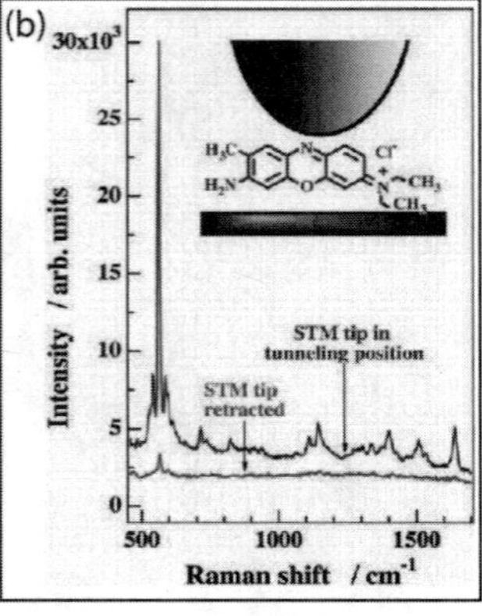

Figure 8. Signal enhancement achieved by tip enhanced Raman spectroscopy for a film of (a) C60 molecules on glass and (b) dye molecules (Brilliant Cresyl Blue) on top of a smooth gold film. From Ref. [56] [61].

discussion of the enhancement factors, these areas as well as the numbers of Raman scatterers therein have to be known. While this is difficult for dye films, the number of SWNTs can be determined based on their topography [62]. For the enhancement factors achieved, values ranging from 40 to 40000 have been reported by different groups. A listing can be found in [61].

An important question is whether the tip acts as a uniform amplifier of the complete Raman spectrum or whether the tip-enhancement varies for different Raman bands. Modifications of the relative strength of Raman lines could arise from different distributions of the filed polarization components in the near-field and farfield [59, 53]. In [63], new Raman signals are explained by significantly altered selection rules caused by the large field-gradients close to metallized aperture probes. Additional effects can also occur depending on the nature of the tip—sample distance control. In the case of AFM tapping mode and direct contact between tip and scatterer, additionally observed Raman bands have been attributed to chemical enhancement effects [56, 57]. For the larger distances of 1–2 nm (used in shear-force mode), no chemical enhancement is expected [62] while the possibility of sample material pick-up by the tip is reduced [61].

Raman methods are applicable to a large variety of systems since Raman scattering is an intrinsic property of all molecular structures. In contrast to fluorescence microscopy, there is no need for highly fluorescent samples or labeling with additional dye molecules. A further increase of the signal enhancement through optimization of the tip shape, tip material [64] and illumination mode [32] will open up fascinating insights into complex materials and might even enable for imaging of single biomolecules such as individual membrane proteins.

4.3. Plasmonic and Photonic Nanostructures

Nanoscale processes are physical interactions at the elementary level and their understanding is primordial for the design of nano-devices. The miniaturization of optoelectronic circuits, for instance, is justified if the operating time can be made sufficiently short for fast computing. The combination of ultrafast and localized phenomena is therefore of considerable interest. In this regard, surface plasmons are very attractive: they can be well localized [65] and they exhibit ultrafast dynamics [66].

4.3.1. Surface Plasmon Polaritons

At optical frequencies, the electromagnetic properties of metals are far from being ideal. The concept of the electrons gas, however, is still valid and instrumental for the understanding of plasmon phenomena, *i.e.*, collective electron density oscillations. On metal surfaces or nano-structures a surface plasmon causes a variety of optical effects, among them strong local field enhancement and confinement of charge fluctuations at the interface [67]. Surface plasmons are a class of polaritons: a coupled matter-electromagnetic mode in which electromagnetic energy is carried by electrons that behave collectively. An interesting property of surface plasmons is that their mean free path, in the visible frequency range, can extend to several tens of micrometers while being strictly confined to the surface [68]. The non-radiative nature of surface plasmons and their long propagation length open the possibility to design planar structures for a variety of applications including biological sensors and miniaturized photonic circuits [69]. On a flat metal surface, surface plasmons cannot be excited with direct radiation. The momenta of surface plasmons are larger than the momentum of a free propagating photon of the same energy. However, light traveling in a higher index medium (e.g. glass) can transfer its energy to a surface wave and can be coupled to surface plasmon [70, 71]. Another technique that can provide the missing momentum is scattering of light by subwavelength structures [72] or holes [73] as discussed later.

Until recently, surface plasmons were detected by indirect means such as reflection measurements. Because of the non-radiative nature of surface plasmons one cannot rely on conventional imaging techniques to gain insights on their intrinsic properties. Fortunately, the development of optical near-field techniques and their capability to image evanescent fields triggered a rapid development of surface plasmon understanding.

In most studies, surface plasmons are excited by a laser beam that is reflected at a glass\metal\air interface (e.g. total internal reflection inside a prism). A dip in the angular reflection measurement indicates that the photon energy is coupled to a surface wave on the metal film. A near-field probe can be used to collect the near-field intensity associated with the plasmon. A spatial intensity map can be constructed by raster scanning the tip over the surface. An example is depicted in Figure 9. The authors compared two near-field intensity distributions. Figure 9(a) represents the near-field distribution created by a Gaussian beam that is totally reflected from a prism surface. Figure 9(b) pictures a surface plasmon intensity distribution excited on a thin silver film. The main difference resides in a tail in Figure 9(b) due to the surface plasmon propagation [74]. The same method was applied to the influence of surface corrugations on plasmon propagation [75]. It was shown that scattering by film imperfection causes non-negligible damping and thus a shortening of the plasmon mean free path [76].

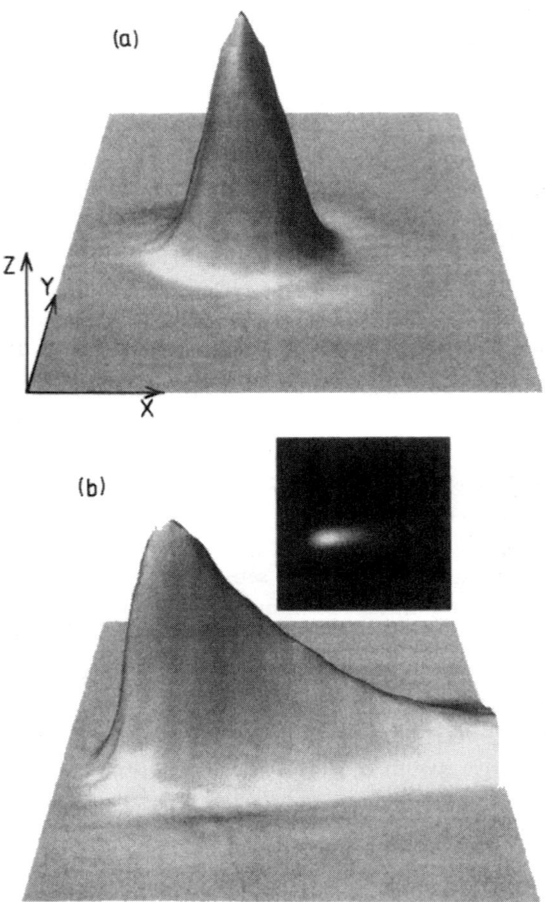

Figure 9. (a) Image of evanescent field intensity of a total internally reflected beam on prism surface. Scan range is 40 μm by 40 μm. (b) Image of the same field but with an additional 53 nm thick silver layer deposited atop the prism surface. The exponentially decaying tail is due to surface plasmon propagation. Inset: two-dimensional view of the image. From Ref. [74].

As already discussed above, the momentum mismatch between a free propagating photon and a surface plasmon can be overcome by diffracting a light beam through a subwavelength opening. The opening can take the form of holes in a thin silver film [77] or can be a near-field aperture located at the end of an optical fiber [78]. The field emitted by a sub-wavelength optical probe has a broad spatial spectrum. When such a probe is placed in close vicinity of a thin metal film, the large wave vectors ($k_{\|}$) fulfill the dispersion relation of surface plasmons and hence light can be coupled to a collective electron oscillation on the metal surface. The obvious advantage of this configuration is that surface plasmons are locally excited in a region of interest, e.g. next to nanostructures on the film surface. In many experimental configurations

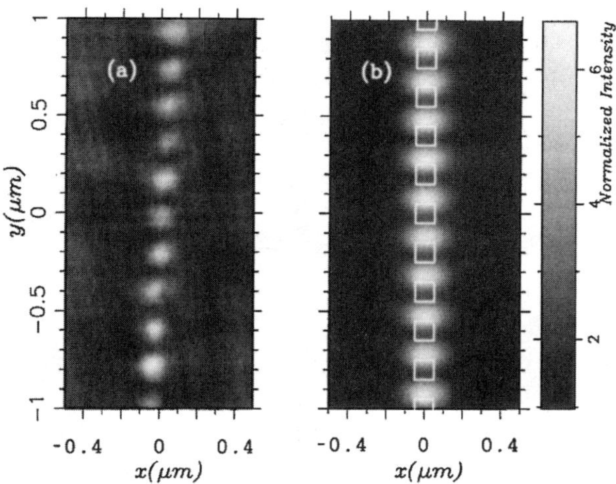

Figure 10. Image of the field distribution above a chain of Au particles (100 nm by 100 nm by 40 nm) acquired in collection mode. The field localization originated from particle-particle interactions. A comparison with a numerical simulation (b) shows that the bright spots are not on top of the Au particles (the surface projections of the particles correspond to the white squares). The intensity scale of experimental data (a) is normalized to the one of the numerical calculation (b). From Ref. [81].

the radiative decay surface plasmons can be imaged at any point in time [78, 79]. This approach permitted the quantitative evaluation of basic surface plasmon optical properties such as reflection and transmission coefficients [76, 80].

4.3.2. Plasmonic and Photonic Nanostructures

Surface plasmons are not only modes supported by thin metal films, but also by small metal particles where the cause peculiar optical phenomenon. Optical responses of individual particles are well understood and can be described accurately by Mie theory. The properties of mutually interacting particles are, however, of great current of interest, notably in the context of photonic crystals and biosensors. Photonic crystals have the unique ability to inhibit light propagation at certain wavelengths, to bend light without losses and to localize light in periodic, multidimensional arrangements. More recently, non-radiative energy transfer between nanostructures attracted some interest for the design of subwavelength optical devices. The mapping of the electromagnetic field bound to such devices is important for the assessment of device functionality. As an example, Figure 10 shows the field distribution surrounding a chain of gold nanoparticles that was deposited atop a glass prism. The entire structure was illuminated in total internal reflection. A glass fiber tip was used to locally probe the electromagnetic field and to reveal that the field is localized near individual particles [81]. Further investigations demonstrated propagation of light along heterogeneous optical nano-waveguides for several microns as well as electromagnetic coupling between separated telecommunications channels [82].

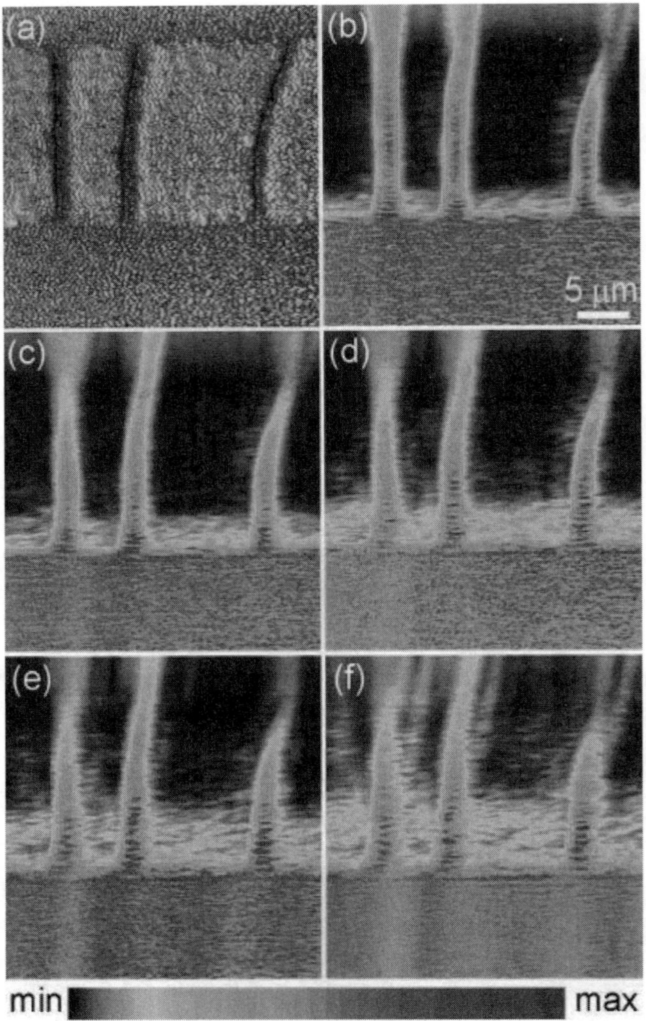

Figure 11. Surface plasmon guiding along Au nanochannels formed by surface corrugation. (a) Topographic image (30 μm by 30 μm) of the channels. (b–f) Surface plasmon intensity maps recorded with a glass fiber probe, excitation wavelengths: 713 nm, 750 nm, 785 nm, 815 nm, and 855 nm, respectively. From Ref. [83].

Near-field studies of metal structures supporting surface plasmons have shown that the plasmon field can be confined along a well-defined path such as line defect. Furthermore, the transmission of such a path can be influenced by varying the excitation wavelength. Figure 11 demonstrates this wavelength dependence for a series of gold channels [83]. In this experiment, a surface plasmon was excited by total internal reflection. By artificially corrugated the metal surface, plasmon propagation can be inhibited,

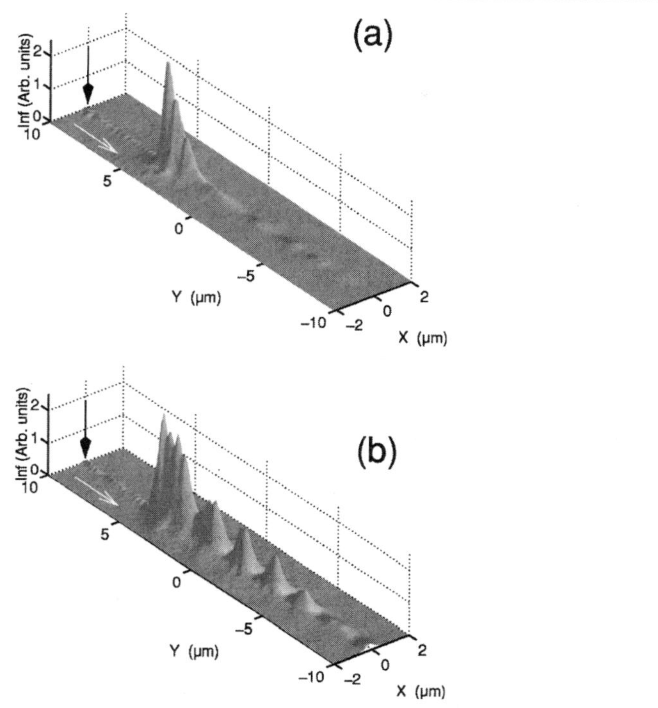

Figure 12. Field distribution along dielectric heterowires as measured with a glass tip. (a) Heterowire parameters do not allow the field to propagate. (b) The wire parameters have been changed and the field is transmitted for more than 10 μm. The black arrow shows the location of the micro-channel conducting the light to the heterowire. The white arrow represents the propagation direction. From Ref. [84].

and by leaving individual stripes uncorrugated plasmon waveguides are created. The surface plasmon wavelength was varied and its near-field intensity distribution was mapped out for each value of the wavelength. The image reveals important variations in the channel transmission as well as in the overall field distribution as the wavelength increases. Studies of this kind allow one to asses the propagation properties of surface plasmons for use as potential information lines between nanodevices.

Mesoscopic dielectric structures are another class of planar photonic crystals. An otherwise transparent reference medium is rendered opaque over some frequency range o by modulating its dielectric properties. Resonant optical tunneling can give rise to enhanced optical transfer through the medium, and near-field microscopy is an important tool for characterizing such enhanced transmission. An example is depicted in Figure 12, where two different heterowires composed of nanoscopic dielectric structures are excited in the same way. The wire in Fig. 12(a) does not support efficient propagation along the wire, whereas the wire in Fig. 12(b) exhibits enhanced transmission over 10 micrometers [84].

4.4. Nanolithography

Reducing the dimensions of structures and devices is one of the major achievements of nanotechnology. An example of the constant efforts in down-scaling is the famous Moore's law. In its 1965 prediction, G. Moore foresaw that the number of transistors on a silicon chip would increase by a factor two every second year. After nearly 40 years, Moore's vision still holds true today. However, optical projection lithography used to produce high-volume integrated circuits will soon reach its limit. It is anticipated that this technology will still be the workhorse through the 100 nm generation of devices, mainly influenced by the emergence of a new wavelength standard (193 nm). For features with smaller dimensions, a new generation of lithography will be required [85]. A variety of nanofabrication techniques have been investigated in the recent years as alternatives to classical optical projection lithography. Among them are nano-imprint lithography [86], micro-contact printing [87], electron beam [88], X-ray [89] and deep UV lithographies [90], atom manipulation [91], and finally near-field optical photolithography.

The pioneering work of Betzig and coworkers showed the feasibility of using localized fields created by a near-field probe to pattern a photoresist with sub-wavelength resolution [92]. The idea was further developed by Krausch [93] and Smolyaninov [94]. 100 nm size patterns, mostly composed of adjacent lines, were created on photoresist and subsequently transferred onto a substrate. It was pointed out that the light-sensitive polymer can also serve as a fingerprint for the field distribution near a the near-field probe. Davy *et al.* carried out a systematic study of this effect [95].

Most of near-field lithography studies use an optical fiber probe prepared as described in section 3.1. A line from an excimer or Argon laser is coupled to one end of the fiber. The fiber is held at a few nanometers from the photosensitive polymer surface to ensure that the evanescent components of the field interact with the substrate. The exposure time of the resist is usually controlled by an acousto-opto modulator. The resist is either developed to transfer the written patterns onto the substrate or analyzed directly after exposure. The analysis consists of measuring the physical photo-deformation of the surface by means of atomic force microscopy or shear-force microscopy.

The pattern size written with metal-coated fiber probes is so far limited between 60 to 100 nm. Lateral dimensions are influenced by (i) the diameter of the near-field probe plays and (ii) by the exposure time. The latter defines the aspect ratio of a given structure and careful investigations of the exposure dosage are necessary. Near-field optical lithography can be fully exploited only if the thickness of the resist is thin enough (comparable to exponential decay length of the evanescent waves) [96]. In turn, it seems that the aspect ratio of a feature is limited to 4:1. An example of a ~100 nm width photoimprinted pattern is depicted in Figure 13.

Near-field photolithography can also be accomplished by use of a metal tip that locally enhances the electromagnetic field. Similar to some of experiments described in the section on fluorescence and Raman microscopy, a suitably polarized laser beam irradiates and excites electronic resonances that give rise to an enhanced field at the extremity of the tip. Studies have shown that the confinement of the field can be

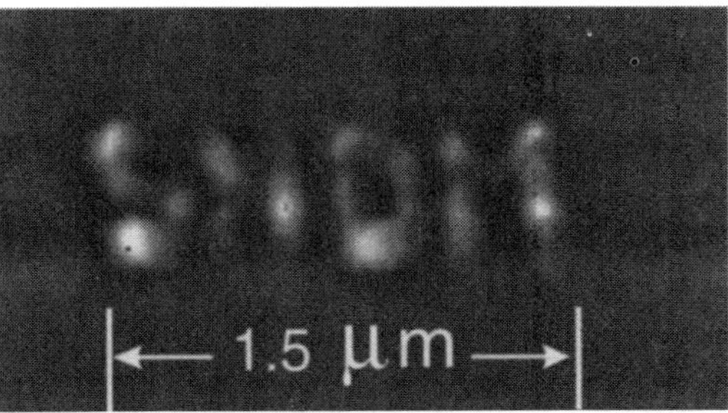

Figure 13. Photodeformation of a sensitive polymer by the proximity of near-field probe. The line widths are about 100 nm on the thinnest sections. From Ref. [95].

localized down to a few nanometers [97]. This confined field is then used to expose a photosensitive resist [98, 99]. In a related approach consists the field enhancement at a metal tip is combined with the nonlinear properties of photoresists. The probability to create a two-photon absorption process in a material scales with the field intensity squared. Therefore, by choosing appropriate pulse shape, energy, and duration, farfield sub-diffraction patterning is possible [100]. The combination of local field enhancement and nonlinear interactions is therefore very promising [101]. Preliminary results showed features with size $\lambda/10$ (see Figure 14.).

Near-field patterning has been employed in various kinds of materials and substrates. The enhanced field near a metal tip can be employed to pattern thin aluminum films or to ablate parts of it [102]. Laser-induced thermal oxidation through local heating of the metal film by the tip extremity triggers the formation of an aluminum oxide. This nano-oxidation procedure generates patterns that act as electrically insulating domains and can be used to design two-dimensional electrode patterns [103]. Another interesting class of material suitable for nanopatterning is self-assembled monolayers (SAMs). SAMs find important applications in interface science, where they are used for cellular or protein attachment. Photopatterning of SAMs opens the possibility to fabricate biomolecular structures. Well-defined chemical patterns were obtained with sizes as small as 25 nm indicating the potential use of near-field optics in nanoscale photo-chemical processes [104]. Another example where near-field optics is used to induce material modification is ferroelectric surfaces. Patterns that show 60 nm linewidths were produced using metal-coated optical fibers [105].

4.5. Semiconductors

The steady miniaturization and optimization of communication and information processing devices requires ever-smaller semiconductor structures. The more conventional top down approach involves machining of macroscopic samples down to nanometer sizes using, for example, electron beam lithography. In the bottom-up approach, nanostructures such as quantum dots are assembled to form larger assemblies. In both cases, the resulting properties of the semiconductor are dominated by quantum effects and

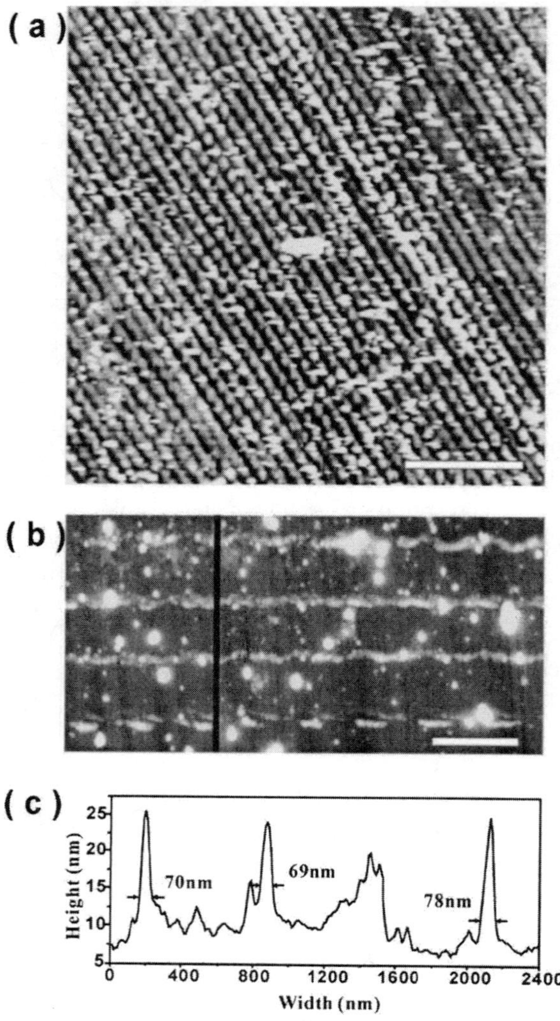

Figure 14. AFM images of two-photon produced line structures in SU-8 photoresist that was exposed by the enhanced field of metal coated silicon cantilever. The farfield peak intensities were (a) 0.9 TW/cm^2 and (b) 0.45 TW/cm^2. Panel (c) shows a cross sectional view (height profile) along the dark vertical line in (b), suggesting that two-photon apertureless near-field lithography can produce 72 nm $\pm$10 nm features using 790 nm light. The scale bars in (a) and (b) are 5 and 1 μm, respectively. From Ref. [101].

can be essentially different from those of the bulk material. For example, electronic states of quantum dots are tuned through electron confinement. The development of new devices can benefit from the high-spatial confinement of light achieved in nano-optics. The method provides new insights into sample properties and can be used to identify, distinguish and address single quantum systems and to study coupling effects of many quantum systems. The examples in this section illustrate the prospects and versatility of nano-optical methods in the field of semiconductor research.

Photoluminescence is an important tool for the study electronic of devices because it is nondestructive and nonintrusive. The optical and electronic properties of semiconductors are in fact intimately related: a quantum system exhibits quantized energy states that are identified by discrete emission wavelength. Near-field photoluminescence spectra provide a wealth of information on the spatial distribution of quantized features. In most investigations using near-field optics, charge carriers are photo-excited by the proximity of a near-field probe. Owing to the versatility of the technique, a probe is also able to locally detect sample luminescence. In the most advanced scheme, excitation and detection of photoluminescence is performed with the same near-field probe. This last scheme is usually employed for its relative simplicity of implementation in a cryogenic environment.

Near-field studies of a semiconductor system involve typically two kinds of physical measurements. In the first one, the excitation power is tuned to observe local band-filling phenomena. Spectral lines acquired at low excitation energy are characteristic of electron-hole recombinations of single exciton states whereas at higher powers biexcitons or multiexcitons are excited [106]. The second type of measurement is aimed at studying carrier diffusion by monitoring the spatial extent of the photoluminescence. This allows to assess the carrier confinement within a structure [107, 108]. An example of quasi one-dimensional excitons is depicted in Figure 15. The sample consists of a GaAs quantum wire clad between two AlGaAs barriers. Each wire segment is terminated by a quantum dot, which is characterized by a red-shifted emission from the quantum wire. The images represent a low-temperature photoluminescence map of the structure for different detection energies corresponding to peaks measured in farfield photoluminescence spectra. The spatial and spectral localization achieved by near-field microscopy permitted the assignment the emission energy to particular regions of the semiconductor structure.

The integration of devices made of different materials, *i.e.* heterostructures, is the key to modern electronics and optoelectronic technologies. Characterization of such structures is a major area of study in material science. Near-field techniques provide a means to locally address the electro-optical properties of ordered regions, domains, and dislocations [109, 110, 111, 112, 113]. Furthermore, the kinetic behavior of the carriers can be determined by locally exciting photocurrent. The photocurrent amplitude depends on the absorption length, the carrier diffusion constant, and carrier lifetime, whereas the spatial dependence is mainly governed by the diffusion length of the carriers [114, 115, 116].

Near-field techniques have also been successfully combined with time-resolved measurements to determine the dynamics of different photophysical processes. Time-resolved optical spectroscopy provides a wealth of information on the dynamic processes of free carriers and excitons. Real space transfer, trapping, dephasing, scattering and relaxation phenomena are of interest from the viewpoint of fundamental physics and device applications. The combination of femtosecond pump-probe spectroscopy and near-field optics provides direct insight into the spatio-temporal dynamic of

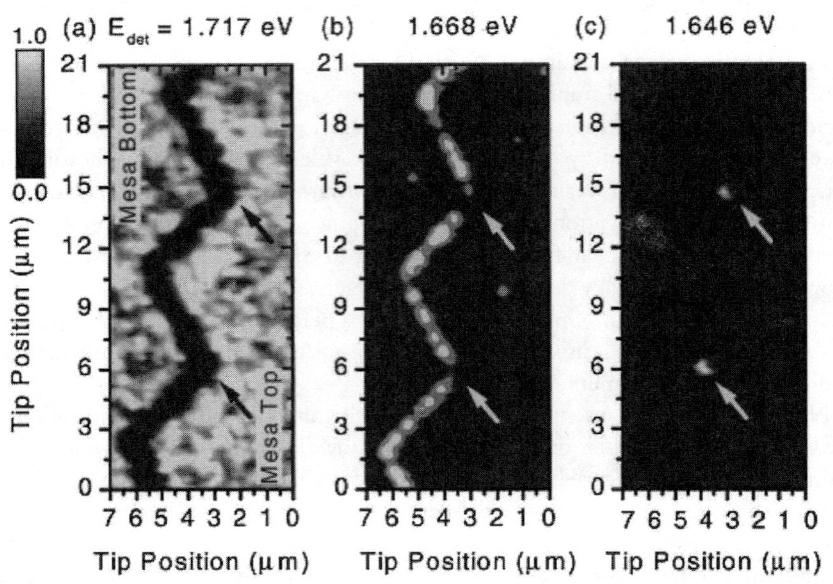

Figure 15. Low temperature (10 K) near-field images of the luminescence from a coupled quantum wire-quantum dot sample recorded at three different detection energies in an illumination/collection geometry. (a) quantum well emission, (b) quantum wire photoluminescence, and (c) quantum dot emission. From Ref. [108].

optical excitations in semiconductor structures. In these experiments, the pump and probe beam are usually provided by an ultrafast laser system. The carriers are photo-excited either in the farfield by the pump, and the change in absorption is measured locally [117, 118], or vise versa [119, 120]. Pump and probe can also be provided by the same aperture [121]. Without going into the details of these experiments, non-equilibrium carrier dynamics in low-dimension systems are investigated with good spatial (~150 nm) and temporal resolution (~200 fs).

5. PERSPECTIVES

Since the first experiment demonstrating sub-wavelength optical resolution, scanning near-field optical microscopy provided scientists with the possibility to access optical properties that were previously invisible. This unique capability triggered a rapid development of the technique and brought together scientists from various fields ranging from materials science to biology. Unfortunately, technical challenges are preventing near-field microscopy to be a routine characterization technique, as for example, atomic force microscopy. The main limiting factor remains the reliable fabrication of near-field probes. 20 years after the first realization, simple experimental factors such as the optimum tip geometry and tip material need still to be defined. Hopefully, recent advances in nanofabrication together with a better understanding of nanoscale optical phenomena will permit to overcome this technical barrier. Despite its limitations, SNOM is used in various fields of sciences owing to its unique versatility. Depending on the specific configuration, a near-field signal can be sensitive to the local electric field, magnetic field, permittivity, chemical properties, or topography. The technique is at the origin of fascinating scientific discoveries in the emerging field of nanoscience. The instrument permitted breakthrough studies of individual molecules with unprecedented details (absorption moment orientation, lifetime and dynamical properties . . .), and opened the possibility to perform vibrational spectroscopy and chemical imaging with optical resolutions down to 10 nm. It also allowed for the first time a direct observation of surface plasmons and triggered advances in plasmonic and photonic devices. There is simply no other means to map an electromagnetic field distribution with nanoscale resolution. However, interesting questions are yet to be answered. How far can we push resolution? Are we already facing fundamental limits (material skin depth . . .) preventing? How will a near-field probe look like in 5 years? These questions will be answered in the coming years and the time ahead will generate new insight into nanoscale properties. In turns, near-field microscopy will be pushed from exploratory nanoscience to a commercial nanotechnology.

REFERENCES

[1] E. Abbe, *Archiv. f. Mikroscop.* 9 (1873) 413.
[2] Lord Rayleigh, *Phil. Mag.* 42 (1896) 167.
[3] E. H. Synge, *Phil. Mag.* 6 (1928) 356.
[4] D. W. Pohl, W. Denk and M. Lanz, *Appl. Phys. Lett.* 44 (1984) 651.
[5] A. Lewis, M. Isaacson, A. Murray and A. Harootunian, *Biophys. J.* 41 (1983) 405a.
[6] E. Betzig, P. L. Finn, and J. S. Weiner, *Appl. Phys. Lett.* 60 (1992) 2484.
[7] K. Karrai and D. R. Grober, *Appl. Phys. Lett.* 66 (1995) 1842.
[8] O. J. F. Martin, *J. Microscopy.* 194 (1999) 235.
[9] E. Betzig and J. K. Trautman, *Science* 257 (1992) 189.
[10] D. R. Turner, US-Patent, 4.469.554 (1983).
[11] P. Lambelet, A. Sayah, M. Pfeffer, C. Philipona, and F. Marquis-Weible, *Appl. Opt.* 37 (1998) 7289.
[12] R. Stöckle, Ch. Fokas, V. Deckert, R. Zenobi, B. Sick, B. Hecht and U.P. Wild, *Appl. Phys. Lett.* 75 (1999) 60.
[13] T. Saiki, S. Mononobe, M. Ohtsu, N. Saito and J. Kusano, *Appl. Phys. Lett.* 68 (1996) 2612.
[14] P. Hoffmann, B. Dutoit and R.-P. Salathé, *Ultramicroscopy* 61 (1995) 165.
[15] P. Moar, F. Ladouceur and L. Cahill, *Appl. Opt.* 39 (2000) 1966.
[16] L. Novotny, D. W. Pohl and B. Hecht, *Opt. Lett.* 20 (1995) 970.
[17] A. W. Snyder and J. D. Love, *Optical waveguide theory*, Chapman and Hall, 1983.
[18] L. Novotny and C. Hafner, *Phys. Rev. E.* 50 (1995) 4095.
[19] Th. Pagnot and Ch. Pierali, *Opt. Comm.* 132 (1996) 161.
[20] T. Saiki and K. Matsuda, *Appl. Phys. Lett.* 74 (1999) 2773.
[21] D. Mulin, D. Courjon, J.-P. Malugani and B. Gauthier-Manuel, *Appl. Phys. Lett.* 71 (1997) 437.
[22] A. Bouhelier, J. Toquant, H. Tamaru, H.-J. Güntherodt, D. W. Pohl and G. Schider, *Appl.Phys. Lett.* 79 (2001) 683.
[23] Th. Lacoste, Th. Huser, R. Prioli and H. Heinzelmann, *Ultramicroscopy* 71 (1998) 333.
[24] J. A. Veerman, A. M. Otter, L. Kuipers and N. F. Hulst, *Appl. Phys. Lett.* 72 (1998) 3115.
[25] A. Naber, D. molenda, U. C. Fischer, H.-J. Maas, C. Hüppener, N. Lu and H. Fuchs, *Phys. Rev. Lett.* 89 (2002) 210801.
[26] U. Ch. Fischer and M. Zapletal, *Ultramicroscopy* 42–44 (1992) 393.
[27] G. Schürmann, P. F. Indermühle, U. Staufer and N. F. de Rooij, *Surf. Inter. Anal.* 27 (1998) 299.

[28] H. Zhou, B. K. Chong, P. Stopford, G. Mills, A. Midha, L. Donaldson and J. M. R. Weaver, *J. Vac. Sci. Technol. B* 18 (2000) 3594.
[29] D. Haefliger and A. Stemmer, *Appl. Phys. Lett.* 80 (2002) 3397.
[30] R. Hillenbrand, and F. Keilmann, *Phys. Rev. Lett.* 85 (2000) 3029.
[31] J. Wessel, *J. Opt. Soc. Am. B* 2 (1985) 1538.
[32] H. G. Frey, F. Keilmann, A. Kriele and R. Guckenberger, *Appl. Phys. Lett.* 81 (2002) 5030.
[33] A. Bouhelier, J. Renger, M. R. Beversluis, and L. Novotny, *J. Microscopy* 210 (2003) 220.
[34] D. Courjon, K. Sarayeddine and M. Spajer, *Opt. Commun.* 71 (1989) 23.
[35] M. Vacha, S. Takei, K.-I. Hashizume, Y. Sakakibara and T. Tani, *Chem. Phys. Lett* 331 (2000) 387.
[36] For a review see W. E. Moerner and D. P, Fromm, *Rev. Sci. Instr.* 74 (2003) 3597.
[37] E. Betzig, A. Lewis, A. Harootunian, M. Isaacson and E. Kratschemer, *Biophys. J.* 49 (1986) 269.
[38] U. Fischer, *J. Opt. Soc. Am. B* 3 (1986) 1239.
[39] E. Betzig and R. Chichester, *Science* 262 (1993) 1422.
[40] R. X. Brain, R. C. Dunn, X. S. Xie and P. T. Leung, *Phys. Rev. Lett.* 75 (1995) 4771.
[41] W. Lukosz and R. E. Kunz, *J. Opt. Soc. Am.* 6 (1977) 1615.
[42] H. Gersen, M. F. Garcia-parajo, L. Novotny, A. Veerman L. Kuipers and N. F. van Hulst, *Phys. Rev. Lett.* 85 (2000) 5315.
[43] W. P. Ambrose, P. M . Goodwin, J. C. Martin and R. A. Keller, *Phys. Rev. Lett.* 72 (1994) 160.
[44] J. A. Veerman, M. F. Garcia-Parajo, L. Kuipers, and N. F. van Hulst, *Phys. Rev. Lett.* 83 (1999) 2155.
[45] C. Höppener, D. Molenda, H. Fuchs and A. Naber, *J. Micrcoscopy* 210 (2003) 288.
[46] N. Hosaka and T. Saiki, *J. Microscopy* 202 (2000) 362.
[47] C. J. Bouwkamp, *Rep. Phys.* 5 (1950) 321.
[48] S. Nie and S. R. Emory, *Science* 275 (1997) 1102.
[49] E. J. Sanchez, L. Novotny and X. Xie, *Phy. Rev. Lett.* 82 (1999) 4014.
[50] T. Ha, Th. Enderle, D. F. Ogletree, D. S. Chemla, P. R. Selvin and S. Weiss, *Proc. Natl. Aca. Sci.* 93 (1996) 6264.
[51] G. T. Shubeita, S. K. Sekatsii, M. Chergui, M. Dietler and V. S. Letokhov, *Appl. Phys. Lett.* 74 (1999) 3453.
[52] S. A. Vickery and R. C. Dunn, *Biophysis. J.* 76 (1999) 1812.
[53] C. L. Jahncke, H. D. Hallen and M. A. Paesler, *J. Raman Spectrosc.* 27 (1996) 579.
[54] J. Prikulis, K. V. G. K. Murty, H. Olin and M. Käll, *J. Microscopy* 210 (2003) 269.
[55] P. G. Gucciardi, S. Trusso, C. Vasi, S. Patanè and M. Allegrini, *J. Microscopy* 209 (2003) 228.
[56] R. M. Stöckle, Y. D. Suh, V. Deckert and R. Zenobi, *Chem. Phys. Lett.* 318 (2000) 131.
[57] N. Hayazawa, Y. Inouye, Z. Sekkat and S. Kawata, *J. Chem. Phys.* 117 (2002) 1296.
[58] L. T. Nieman, G. M. Krampert and R. E. Martinez *Rev. Sci. Instrum.* 72 (2001) 1691.
[59] A. Hartschuh, E. J. Sánchez, X. S. Xie and L. Novotny, *Phys. Rev. Lett.* 90 (2003) 095503.
[60] M. S. Anderson, *Appl. Phys. Lett.* 76 (2000) 3130.
[61] B. Pettinger, G. Picardi, R. Schuster and G. Ertl, *Single Mol.* 5 (2002) 285.
[62] A. Hartschuh, N. Anderson and L. Novotny, *J. Microscopy* 210 (2003) 234.
[63] E. J. Ayars, H. D. Hallen and C. L. Jahncke, *Phys. Rev. Lett.* 85 (2000) 4180.
[64] J. T. Krug II, E. J. Sánchez and X. S. Xie, *J. Chem. Phys.* 116 (2002) 10895.
[65] S. Grésillon, L. Aigouy, A. C. Boccara, J. C. Rivoal, X. Quelin, C. Desmarest, P. Gadenne, V. A. Shubin, A. K. Sarychev and V. M. Shalaev, *Phys. Rev. Lett.* 80 (1999) 4520.
[66] M. I. Stokman, *Phys. Rev. Lett.* 84 (2000) 1011.
[67] H. Raether, *Excitation of Plasmons and Interband Transitions by Electrons*, Springer Tracts in Modern Physics Vol. 88, edited by G. Höhler, (Springer-Verlag, Berlin 1980).
[68] H. Raether, *Surface Plasmons*, Springer Tracts in Modern Physics Vol. 111, edited by G. Höhler, Springer-Verlag, Berlin (1988).
[69] For a review see W. L. Barnes, A. Dereux and T. W. Ebessen, *Nature* 424 (2003) 824.
[70] A. Otto, *Z. Angew. Phys.* 27 (1968) 207.
[71] E. Kretschmann and H. Raether, *Z. Naturforsch.* 23a (1968) 2135.
[72] H. Diltbacher, J. R. Krenn, N. Fleidj, B. Lamprecht, C. Schider, M. Salermo, A. Leitner and F. R. Aussenegg, *Appl. Phys. Lett.* 80 (2002) 404.
[73] B. Hecht, H. Bielefeldt, L. Novotny, Y. Inouye and D. W. Pohl, *Phys. Rev. Lett.* 77 (1996) 1889.
[74] P. Dawson, F. de Fornel and J.-P. Goudennet, *Phys. Rev. Lett* 72 (1994) 2927.
[75] I. I. Smolyaninov, D. L. Mazzoni, J. Mait and C. C. Davis, *Phys. Rev. B.* 56 (1997) 1601.
[76] A. Bouhelier, Th. Huser, H. Tamaru, H.-J. Güntherodt, D. W. Pohl, F. Baida and D. Van Labeke, *Phys. Rev. B.* 63 (2001) 155404.

[77] C. Sönnichsen, A. C. Duch, G. Steininger, M. Koch, G. von Plessen and J. Feldmann, *Appl. Phys. Lett.* 76 (2000) 140.
[78] B. Hecht, H. Bielefeldt, L. Novotny, Y. Inouye, and D. W. Pohl, *Phys. Rev. Lett.* 77 (1996) 1889.
[79] A. Bouhelier, T. Huser, J. M. Freyland, H. J. Güntherodt and D. W. Pohl, *J. Microscopy* 194 (1999) 571.
[80] B. Dragnea, J. M. Szarko, S. Kowarik, T. Weimann, J. Feldmann and S. R. Leone, *Nano Lett.* 3 (2003) 3.
[81] J. Krenn, A. Dereux, J. C. Weeber, E. Bourillot, Y. Lacroute, J. P. Goudonnet, G. Schider, W. Gotschy, A. Leitner, F. R. Aussenegg and C. Girard, *Phys. Rev. Lett.* 82 (1999) 2590.
[82] D. Mulin, M. Spajer, D. Courjon, F. Carcenac and Y. Chen, *J. Appl. Phys.* 87 (2000) 534.
[83] S. I. Bozhevolny, V. S. Volkov, K. Leosson and A. Boltasseva, *J. Microscopy* 209 (2003) 209.
[84] R. Quidant, J.-C. Weeber, A. Dereux, D. Peyrade, Ch. Girard and Y. Chen, *Phys. Rev. E* 62 (2002) 036616.
[85] For a review see for instance S. Wittekoek, *Microelectron. Eng.* 23 (1994) 43.
[86] S. Y. Chou, P. R. Krauss and P. J. Renstrom, *Science* 272 (1996) 85.
[87] A. Kumar and G. M. Whitesides, *Appl. Phys. Lett.* 63 (1993) 2002.
[88] P. Rai-Choudhury, *Handbook of Microlithography, and Microfabrication,* Spie Optical Engineering Press, 1994.
[89] J. R. Sheats and B. W. Smith *Microlithography Science & Technology*, (1988).
[90] M. D. Levenson, P. J. Silverman, R. George, S. Wittekoek, P. Ware, C. Sparkes, L. Thompson, P. Bischoff, A. Dickinson and J. Shamaly, *Solid State Technol.* 38 (1995) 81.
[91] D. M. Eigler and E. K. Schweizer, *Nature* 344 (1990) 524.
[92] E. Betzig, J. K. Trautman, T. D. Harris, J. S. Weiner and R. L. Kostelak, *Science* 251 (1991) 1468.
[93] G. Krausch, S. Wegscheider, A. Kirsch, H. Bielefeldt, J. C. Meiners and J. Mlynek, *Opt. Commun.* 119 (1995) 283.
[94] I. I. Smolyaninov, D. L. Mazzoni and Ch. C. Davis, *Appl. Phys. Lett.* 67 (1995) 3859.
[95] S. Davy and M. Spajer, *Appl. Phys. Lett.* 69 (1996) 3306.
[96] A. Naber, H. Kock and H. Fuchs, *Scanning* 18 (1996) 567.
[97] L. Novotny, R. X. Bian and X. Sunney Xie, *Phys. Rev. Lett.* 79 (1997) 645.
[98] F. H.'dhili, R. Bachelot, G. Lerondel, D. Barchiesi and P. Royer, *Appl. Phys. Lett.* 79 (2001) 4019.
[99] A. Tarun, M. Rosendo, H. Daza, N. Hayazawa, Y. Inouye and S. Kawata, *Appl. Phys. Lett.* 80 (2002) 3400.
[100] S. Kawata, H. B. Sun, T. Tanaka and K. Takada, *Nature* 412, 697 (2001).
[101] X. Yin, N. Fang, X. Zhang, I. B. Martini and B. Schwartz, *Appl. Phys. Lett.* 81 (2002) 3663.
[102] S. Nolte, B. N. Chichkov, H. Welling, Y. Shani, K. Lieberman and H. Terkel, *Opt. Lett.* 24 (1999) 914.
[103] D. Haefliger and A. Stemmer, *Ultramicroscopy* 209 (2003) 150.
[104] S. Sun, K. S. L. Chong and G. J. Leggett, *J. Am. Chem. Soc.* 124 (2002) 2415.
[105] J. Massanell, N. Garcìa and A. Zlatkin, *Opt. Lett.* 21 (1996) 12.
[106] M. Brun, S. Huant, J. C. Woehl, J.-F. Motte, L. Marsal and H. Mariette, *J. Microscopy* 202 (2001) 202.
[107] Y. Toda, M. Kourogi, M. Ohtsu, Y. Nagamune and Y. Arakawa, *Appl. Phys. Lett.* 69 (1996) 827.
[108] F. Intonti, V. Emiliani, C. Lienau, T. Elsaesser, R. Noetzel and K. H. Ploog, *J. Microscopy* 202 (2001) 193.
[109] J. W. P. Hsu, E. A. Fitzgerald, Y. H. Xie and P. J. Silverman, *Appl. Phys. Lett.* 65 (1994) 344.
[110] H. F. Hess, E. Betzig, T. D. Harris, L. N. Pfeiffer and K. W. West, *Science* 264, 1740 (1994).
[111] M. J. Gregor, P. G. Blome, R. G. Ulbrich, P. Grossmann, S. Grosse, J. Feldmann, W. Stolz, E. O. Göbel, D. J. Arent, M. Bode, K. A. Bertness and J. M. Olson, *Appl. Phys. Lett.* 67 (1995) 3572.
[112] T. D. Harris, D. Gershoni, R. D. Grober, L. Pfeiffer, K. West and N. Chand, *Appl. Phys. Lett.* 68 (1996) 988.
[113] A. A. McDaniel, J. W. P. Hsu and A. M. Gabor, *Appl. Phys. Lett.* 70 (1997) 3555.
[114] S. K. Buratto, J. W. P. Hsu, E. Betzig, J. K. Trautman, R. B. Bylsma, C. C. Bahr and M. J. Cardillo, *Appl. Phys. Lett.* 65 (1994) 2654.
[115] M. S. Ünlü, B. B. Goldberg, W. D. Herzog, D. Sun and E. Towe, *Appl. Phys. Lett.* 67 (1995) 1862.
[116] T. Saiki, N. Saito, J. Kusano and M. Ohtsu, *Appl. Phys. Lett.* 69 (1996) 644.
[117] J. Levy, V. Nikitin, J. M. Kikkawa, A. Cohen, N. Samarth, R. Garcia and D. D. Awschalom, *Phys. Rev. Lett.* 76 (1996) 1948.

[118] B. A. Nechay, U. Siegner, F. Morier-Genoud, A. Schertel and U. Keller, *Appl. Phys. Lett.* 74 (1999) 61.
[119] Richter, *J. Microscopy* 194 (1999) 393.
[120] M. Achermann, B. A. Nechay, U. Siegner, A. Hartmann, D. Oberli, E. Kapon and U. Keller, *Appl. Phys. Lett.* 76 (2000) 2695.
[121] V. Emiliani, T. Guenther, C. Lienau, R. Nötzel and K. H. Ploog, *J. Microscopy* 202 (2000) 229.

3. SCANNING TUNNELING MICROSCOPY

JIN-FENG JIA, WEI-SHENG YANG, AND QI-KUN XUE

1. BASIC PRINCIPLES OF SCANNING TUNNELING MICROSCOPY

In spite of its conceptually simple operation principle, scanning tunneling microscope (STM) can resolve local electronic structures on an atomic scale in real space on virtually any kind of conducting solid surface under various environments, with little damage or interference to the sample [1]. It has been invented for more than 20 years. Over the years, the STM has been proved to be an extremely versatile and powerful technique for many disciplines in condensed matter physics, chemistry, material science, and biology. In addition, STM can be used as a nano-tool for nano-scale fabrication, manipulation of individual atoms and molecules, and for building nanometer scale devices one atom/molecule at a time.

STM was originally developed to image the topography of surfaces by Binnig and Rohrer in 1982 [1]. For this great invention, they were awarded the Nobel Prize in Physics in 1986. The principle of STM is very simple, in which electron tunneling is used as the mechanism to probe a surface. In the following, in order to understand the operation principle of STM, we first give a brief introduction to the electron tunneling phenomenon.

1.1. Electronic Tunneling

Tunneling phenomena have been studied for long time and can be well understood in terms of quantum theory. As shown in Fig. 1, considering an one-dimensional vacuum barrier between two electrodes (the sample and the tip) and assuming their

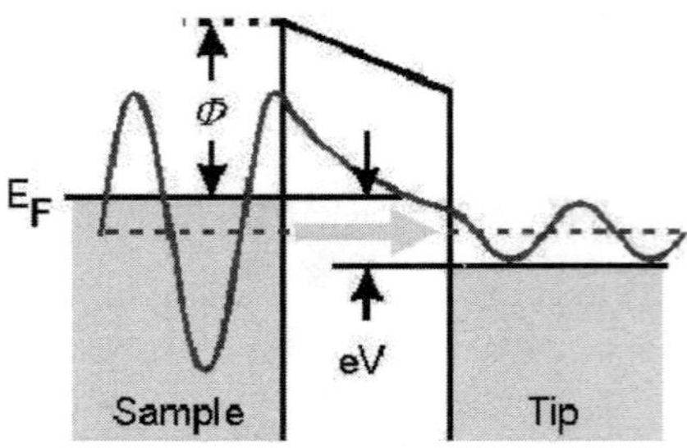

Figure 1. A one dimensional barrier between two metal electrodes. A bias voltage of V is applied between the electrodes.

work functions to be the same and thus the barrier height to be Φ, if a bias voltage of V is applied between the two electrodes with a barrier width d, according to quantum theory under first-order perturbation [2], the tunneling current is

$$I = \frac{2\pi e}{\hbar} \sum_{\mu,\nu} f(E_\mu)[1 - f(E_\nu + eV)]|M_{\mu\nu}|^2 \delta(E_\mu - E_\nu), \tag{1}$$

where $f(E)$ is the Fermi function, $M_{\mu\nu}$ is the tunneling matrix element between states ψ_μ and ψ_ν of the respective electrodes, E_μ and E_ν are the energies of ψ_μ and ψ_ν, respectively. Under assumptions of small voltage and low temperature, the above formula can be simplified to

$$I = \frac{2\pi}{\hbar} e^2 V \sum_{\mu,\nu} |M_{\mu\nu}|^2 \delta(E_\nu - E_F)\delta(E_\mu - E_F). \tag{2}$$

Bardeen [2] showed that under certain assumptions, the tunneling matrix element can be expressed as

$$M_{\mu\nu} = \frac{\hbar^2}{2m} \int d\vec{S} \cdot (\psi_\mu{}^* \vec{\nabla} \psi_\nu - \psi_\nu{}^* \vec{\nabla} \psi_\mu), \tag{3}$$

where the integral is over all the surfaces surrounding the barrier region. To estimate the magnitude of $M_{\mu\nu}$, the wave function of the sample ψ_ν can be expanded in the generalized plane-wave form

$$\psi_\nu = \Omega_s^{-1/2} \sum_G a_G \exp\left[(k^2 + |\vec{k}_G|^2)^{1/2} z\right] \exp(i\vec{k}_G \cdot \vec{x}), \tag{4}$$

where Ω_s is the volume of the sample, $k = \hbar^{-1}(2m\phi)^{1/2}$ is the decay rate, ϕ is the work function, $\vec{k}_G = \vec{k}_{||} + \vec{G}$, $\vec{k}_{||}$ is the surface component of Bloch vector, and $\vec{G}$ is the surface reciprocal vector.

To calculate the tunneling current, it is necessary to know the tip wave function. Unfortunately, the actual atomic structure of the tip is unknown and, moreover, it is

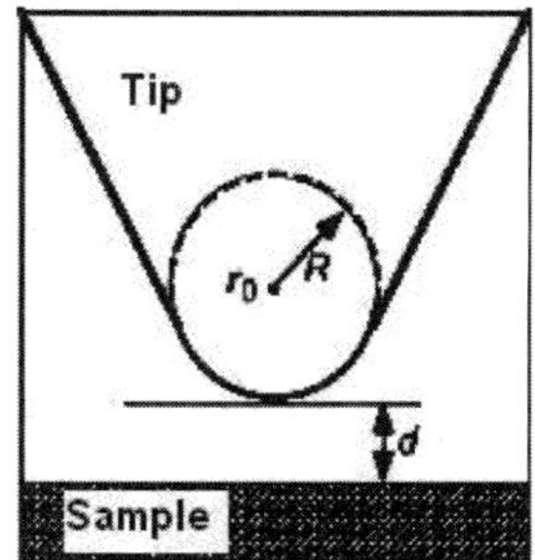

Figure 2. An ideal model for STM tip. The cusp of the tip is assumed to be a sphere with radius of R, the distance from the sample is d, the position of the center of the sphere is $\vec{r}_0$ (From Ref. 3).

very difficult to calculate the tip wave function due to its very low symmetry. However, for the tip we may adopt the reasonable model shown in Fig. 2, which was used by Tersoff *et al.* [3] to describe an ideal tip, and then the wave function of the tip is

$$\psi_\mu = \Omega_t^{-1/2} c_t k R e^{kR} (k|\vec{r} - \vec{r}_0|)^{-1} e^{-k|\vec{r} - \vec{r}_0|}, \tag{5}$$

where Ω_t is the volume of the tip, c_t is a constant determined by the sharpness of the tip and its electronic structure. For simplicity, only the s-wave function of the tip is used in the calculation. Because of

$$(k\vec{r})^{-1} e^{-k\vec{r}} = \int d^2 q b(\vec{q}) \exp\left[-(k^2 + q^2)^{1/2} |z|\right] \exp(i\vec{q} \cdot \vec{x}), \tag{6}$$

$$b(q) = (2\pi)^{-1} k^{-2} (1 + q^2/k^2)^{-1/2}, \tag{7}$$

substituting these wave functions to Eq. (3), we obtain

$$M_{\mu\nu} = \frac{\hbar^2}{2m} 4\pi k^{-1} \Omega_t^{-1/2} k R e^{kR} \psi_\nu(\vec{r}_0), \tag{8}$$

where $\vec{r}_0$ is the position of the cusp center. Substitute Eq. (8) to Eq. (2), we obtain

$$I = 32\pi^3 \hbar^{-1} e^2 V \phi^2 D_t(E_F) R^2 k^{-4} e^{2kR} \sum_\nu |\psi_\nu(r_0)|^2 \delta(E_\nu - E_F), \tag{9}$$

where $D_t(E_F)$ is the density of states at the Fermi level for the tip. Substituting the typical values for metals in Eq. (9), the tunneling current is obtained

$$I \propto V D_t(E_F) e^{2kR} \rho(r_0, E_F), \tag{10}$$

$$\rho(r_0, E_F) = \sum |\psi_\nu(r_0)|^2 \delta(E_\nu - E_F), \tag{11}$$

Thus, the STM with an s-wave tip would simply measure $\rho(r_0, E_F)$, which is the local density of states (DOS) at the Fermi level E_F and at a position r_0, the curvature

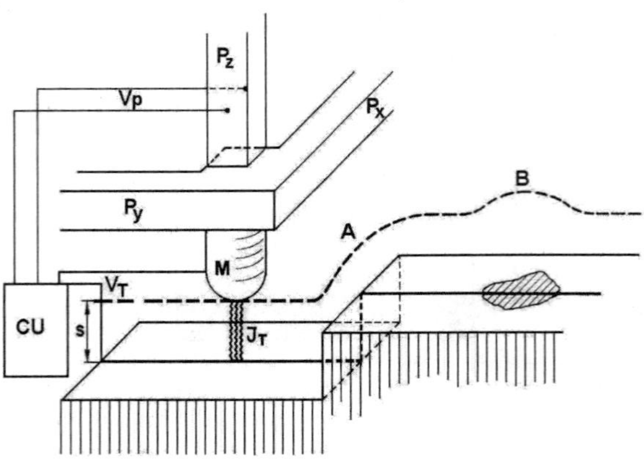

Figure 3. Operation principle of the STM (not to scale). Piezodrives P_X and P_Y scan the metal tip over the surface. The control unit (CU) applies the necessary voltage Vp to piezodrive P_Z to maintain constant tunnel current J_T at bias voltage V_T. The broken line indicates the z displacement in a scan over a surface step (A) and a chemical inhomogeneity (B). (From Ref. 1)

center of the effective tip. Tersoff *et al.* [3] also discussed the contribution of the tip wave function components of higher angular momentum, and found that these just made little difference for typical STM images. So, what the STM measures is only the property of the surface.

Because $|\psi_\nu(\vec{r}_0)|^2 \propto e^{-2k(R+d)}$, thus $I \propto e^{-2kd}$. This means that the tunneling current depends on the tunneling gap distance d very sensitively. In the typical case, the tunneling current would change one order while the gap distance changes only 1 Å. This accounts for extremely high vertical resolution of 0.1 Å of STM.

1.2. Scanning Tunneling Microscope

In March 1981, Binnig, Rohrer, Gerber and Weibel at the IBM Zürich Research Laboratory successfully combined vacuum tunneling with scanning capability, and developed the first STM in the world [1]. The basic idea behind STM is illustrated in Fig. 3. A sharp metal tip is fixed on the top of a pizeodrive (P_Z) to control the height of the tip above a surface. When the tip is brought close enough to the sample surface, electrons can tunnel through the vacuum barrier between tip and sample. Applying a bias voltage on the sample, a tunneling current can be measured through the tip, which is extremely sensitive to the distance between the tip and the surface as discussed above. Another two pizeodrives (P_X and P_Y) are used to scan the tip in two lateral dimensions. A feedback controller is employed to adjust the height of the tip to keep the tunneling current constant. During the tip scanning on the surface, the height of the tip (the voltage supplied to P_Z pizeodrive) is recorded as an STM image, which represents the topograph of the surface. This operation mode of STM is called “constant current” mode.

Constant current mode is mostly used in STM topograph imaging. It is safe to use the mode on rough surfaces since the distance between the tip and sample is adjusted by the feedback circuit.

On a smooth surface, it is also possible to keep the tip height constant above the surface, then, the variation of the tunneling current reflects the small atomic corrugation of the surface. This "constant height" mode has no fundamental difference to the "constant current" mode. However, the tip could be crashed if the surface corrugation is big. On the other hand, the STM can scan very fast in this mode for research of surface dynamic processes.

To achieve the atomic resolution, there are many requirements in STM design and instrumentation, *e.g.*, vibration isolation, scanning devices, positioning devices, electronic controller system *etc.* The details about STM design and instrumentation can be found in many review books [4–6] and will not be discussed here.

STM is so powerful that numerous researches have been done in various scientific areas since its invention. In the following sections, some representative and important applications of STM will be shown and discussed. According to the main functions of STM, the applications can be classified into three parts, *e.g.*, surface imaging, tunneling spectroscopy, and tip manipulation. In the last part, the current development in STM will also be introduced.

2. SURFACE STRUCTURE DETERMINATION BY SCANNING TUNNELING MICROSCOPY

As a microscope, STM can provide very high resolution images in real-space. These images can be used to investigate surface structures, and also surface or even subsurface atomic dynamic processes.

Before the STM was invented, surface structures were very difficult to be determined by conventional surface analysis techniques, such as low-energy electron diffraction (LEED), reflected high-energy electron diffraction (RHEED) and X-ray diffraction *etc.* Besides, these traditional techniques focus essentially only on average or collective properties. The ability to reveal the local surface atomic structure in real space make the STM very fruitful in the field of surface science, especially for structure determination.

2.1. Semiconductor Surfaces

2.1.1. Element Semiconductors

Silicon is the most important material in semiconductor industry. The 7×7 reconstruction of the Si(111) surface was first observed by Schlier and Farnsworth [7] with LEED in 1959. After then, all surface sensitive techniques have been used to determine its atomic structure, and a lot of models have been proposed to understand this complicated surface. Due to its large unit cell (49 times of the bulk unit cell), to determine its structure was a great challenge for traditional methods.

The first atomically resolved STM image of this surface was obtained by Binnig *et al.* in 1982, which marks a breakthrough in the study of Si(111)7×7 and also in the development of STM itself [1], because it was also the first atomically resolved image

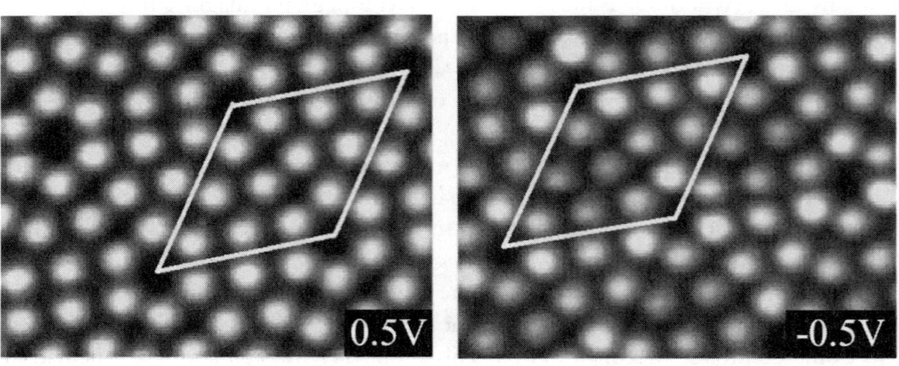

Figure 4. Atomically resolved STM images of the Si(111)7 × 7 surface. Bias voltage: +0.5 V (left), −0.5 V (right). A unit cell is outlined in the images, the size of the unit cell is 2.7 nm × 2.7 nm.

provided by STM. From then, the surface has been extensively studied with STM. As shown in Fig. 4, the STM images of Si(111)7 × 7 reveal 12 protrusions in each unit cell, and the negative biased STM image clearly shows the inequivalence between the adatoms in the two halves of a unit cell. And also, there is a corner hole in each unit cell. The information immediately helped to rule out many models proposed at that time.

The structure was finally determined by Takayanagi *et al.* in 1985 [8] on the basis of transmission electron diffraction data. The dimer-adatom-stacking-fault (DAS) model proposed by Takayanagi *et al.* is shown in Fig. 5. In the DAS model a 7 × 7 unit cell consists of 12 adatoms, 9 dimers, 6 rest atoms, and a corner hole. The atomic layers in the right triangle (or half) of the unit cell are stacked regularly and thus this half is called as unfaulted half unit cell (UFHUC), while the left half contains a stacking-fault and thus is called as fault half unit cell (FHUC), see Fig. 5(b).

For the truncated Si(111)1 × 1 surface, each surface Si atom has a dangling bond, which contribute significantly to the total surface energy. To reduce the total surface energy, the surface reconstructs to 7 × 7 and the number of dangling bond decrease from 49 to 19 per unit cell. In the DAS model, each adatom reduces 2 dangling bonds by saturating 3 dangling bonds and leading to a single dangling bond due to the fourfold coordination of Si atom. The other 7 dangling bonds are located on the 6 rest atoms and the atom at the bottom of the corner hole. The DAS model can explain the images very well. Since the dangling bonds on the adatom are partially filled, each adatom is imaged as a bright protrusion at both positive and negative biases. The inequivalence between the adatoms in two different triangles in the negatively biased STM images can be explained by the slight electronic difference caused by the stacking-fault.

In many cases, STM could not be used solely to determine surface structure since it probes only the structural information of the topmost surface layer. Moreover, it generally lacks chemical specificity. Below, we can see that the mixed topographic and

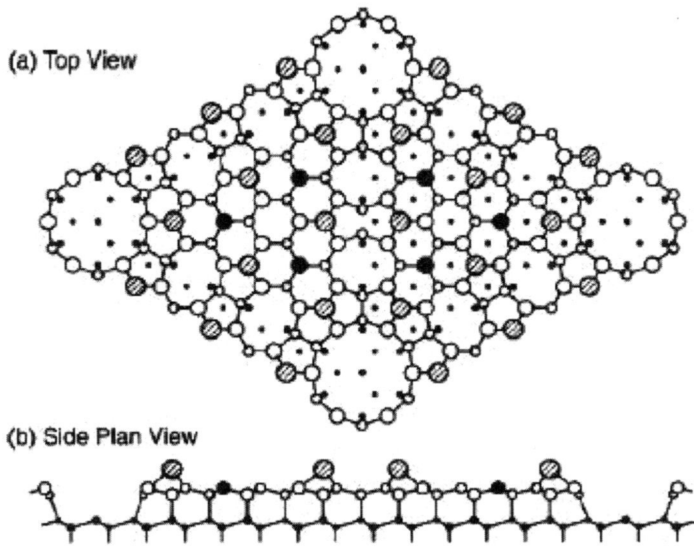

Figure 5. Top (a) and side (b) views of the dimer-adatom-stacking-fault (DAS) model of the Si(111)7 × 7 surface. The large striped circles designate the adatoms, the large solid circles designate the rest atoms, the large and small open circles the Si atoms in the 2nd and 3rd bilayers, and the small solid circles the atoms in 4th and 5th bilayer, respectively. (Proposed by Takayanagi *et al.*)

electronic features cause difficulties to determine atomic structures by STM. For this purpose, it is very important to combine STM with other relative techniques.

2.1.2. Compound Semiconductors

GaAs is a very important compound semiconductor since many electronic and optoelectronic devices are made of it. Because of its zincblend crystal structure with a tetrahedral coordination in the bulk, the polar GaAs(001) surface could be terminated with either As or Ga atoms. As a function of growth temperature, As/Ga flux ratio and preparation conditions, the (001) surface displays a number of reconstructions, starting with the most As-rich phase which has a c(4 × 4) symmetry, through the 2 × 4/c(2 × 8), 2 × 6, 4 × 6, ending with the 4 × 2/c(8 × 2) Ga-stabilized phase.

Among them, the As-rich 2 × 4 phases are the most important structures commonly used in the technological applications. It is generally accepted that the top layer of the As-rich 2 × 4 phase consists of As dimers [9]. Farrell and Palmstron analyzed their experimental results for the 2 × 4 phase and classified them into three (α, β, and γ) phases depending on the RHEED spot intensities [10]. According to different experiments, many structure models were proposed for each phase [11, 12]. Four different models are shown in Fig. 6. To solve the controversy, Hashizume *et al.* performed a comprehensive study on the surface with STM and RHEED [13, 14]. The typical STM images together with atomic resolved zoom-in images and line profiles along [110] direction are shown in Fig. 7. From the atomic resolved STM images, they

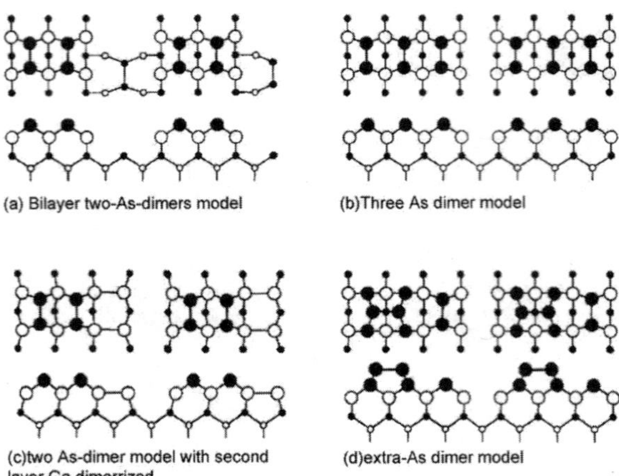

Figure 6. Four structure models proposed for the GaAs(001) 2 × 4 reconstruction. Filled (open) circles denote As (Ga) atoms. (From Ref. 13)

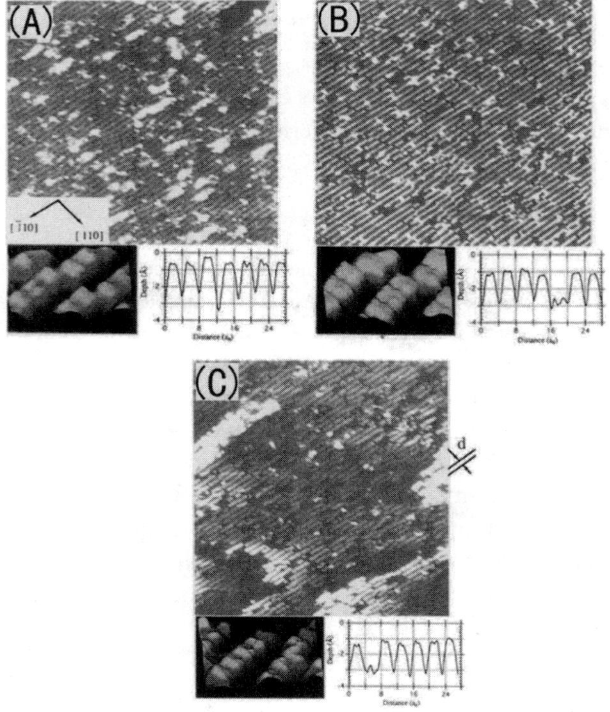

Figure 7. Typical STM images (800 Å × 800 Å) of the (A) α, (B) β and (C) γ phases together with the zoom-in images and line profiles along [110] direction of the GaAs(001) 2 × 4 reconstruction. (From Ref. 13)

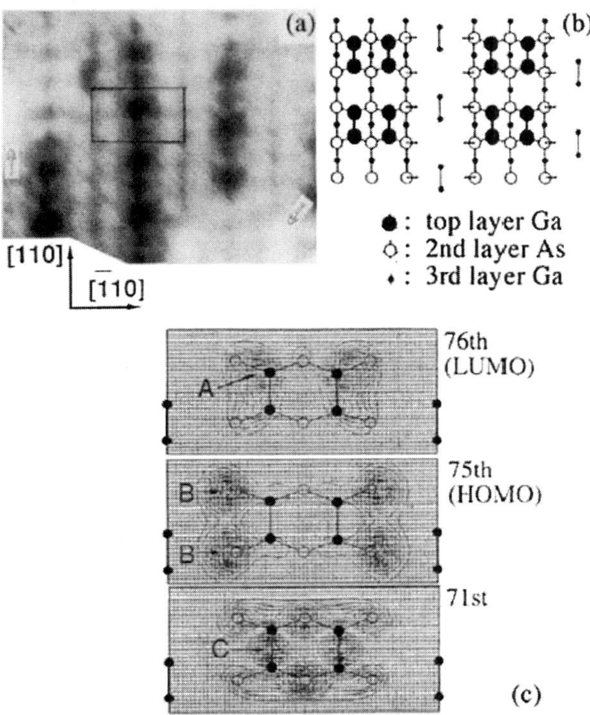

Figure 8. (a) Atomic resolution filled state STM image of the GaAs(001) 4 × 2 phase. (b) The Ga-dimer-model for 4 × 2 phase. (c) The charge distributions of the local density of the states calculated based on the Ga-model in (b) at 0.9 Å above the first layer Ga-dimer position for the 76 (LUMO), 75th (HOMO) and 71st bands. (From Ref. 15)

concluded that the outermost surface layer of the unit cell of the 2 × 4 α, β, and γ phases all consists of two As dimers and that the α and β phases are different in the atomic arrangements of the second and third layers exposed by the dimer vacancy rows. The γ phase is the less ordered β phase with "open areas" exposing the underneath disordered c(4 × 4) phase. To fully understand the structures of the α, β, and γ phases, the RHEED spot intensities for the possible 2 × 4 models were calculated using the dynamical theory. According to the calculations, they proposed a unified model: the two As-dimer model by Chadi [11] (Fig. 6a) for the most stable β phase, and the two As-dimer model incorporated with the relaxation of the second layer Ga atoms proposed by Northrup and Froyen [12] (Fig. 6c) for α phase, while the γ phases is the locally ordered β phase with the disordered c(4 × 4) unit in the open area [13].

For the GaAs(001) Ga-rich 4 × 2/c(2 × 8) and 4 × 6 phases, *Xue et al.* performed a systematical investigation with an MBE-STM system [15]. Fig. 8 shows the high-resolution filled-state STM images of the 4 × 2 surface. The 4 × 2 unit cells are highlighted in the STM images. In the filled state image, a pair of rows separated by 5.1 Å along the [−110] direction is observed, whereas the row itself is a chain of bright protrusions separated by 4 Å along the [110] direction. A new finding here is faintly imaged features which are located in the outskirt of the paired row. The weak features always couple together to form a pair-like structure in parallel to the bright rows. The separation between the neighboring pair-like features along the [110] direction was determined to be 8 Å, resulting in the 4×2 symmetry. The out-of phase arrangement of the 4×2 sub-unit gives rise to the $c(8 \times 2)$ symmetry.

Several models have been proposed for this phase, however, none of them can explain the observed STM images straightforwardly since the overlapping first layer Ga and the second layer As orbits are both accessible to the STM in the range of applied negative bias voltage to the sample and the STM is probing the local density of states near the Fermi level, not merely the surface geometry [15]. In order to resolve this discrepancy, first-principles total energy calculations of the surface charge density distribution based on the Ga-bilayer model (see Fig. 8(b)) have been performed. The calculated results are shown in Fig. 8(c). Under the filled states STM imaging condition at −1.8 V, it is found that all local densities of the states between the 71st and the 76th bands contribute to the tunneling current to form the STM image [Fig. 8(a)]. Because of the smaller potential barrier height for tunneling from the 75th band, the 75th HOMO makes the most significant contribution to the tunneling together with contributions from the overlapping 74th, 73rd and 72nd bands with the decreasing contribution, all of which are basically imaging of the second layer As atoms as individual brighter protrusions. On the other hand, the contribution for the top layer Ga dimer becomes only appreciable down at the 71st band at the middle of the Ga dimer. Thus, the top layer Ga dimer is observed as single faint hump (instead of pair-like feature) even though they are located in the top layer. Thus, the calculated results agree with the STM observation well.

Very recently, this surface was studied by theory and other techniques. A different model (called as $\zeta(4 \times 2)$) was proposed by Lee, Moritz, and Scheffler, as shown in Fig. 8(d) [16]. This model well explains the STM images, particularly the empty state image. Later, more theories and experiments support this model [17]. But, regarding to the significant rearrangement of the surface atoms, more evidences are needed to justify the model.

The Ga-rich 4×6 phase can be obtained by a higher Ga flux ratio in migration enhanced epitaxy or annealing the 2×6 phase for longer time (>15 mins) [15]. An atomic resolved STM image of 4×6 reconstruction is shown in Fig. 9, which is uniquely characterized by the array of large oval protrusions regularly located at each corner of the unit cell. The oval features are ∼0.1 Å higher than the Ga dimers. By compared the image with the Fig. 8, it was concluded that the pair of bright rows running in the [110] direction in Fig. 9 is the first layer Ga-dimers, instead of second layer As atoms, unlike in the case of 4×2 phase. The large bright oval features occupy the middle of the As rows, by overlapping with them. In Fig. 9, every individual Ga dimer is clearly resolved. Such high contrast imaging of the Ga dimers is likely due to charge transfer from the oval protrusions to the Ga dimers. After careful analysis, Xue *et al.* concluded

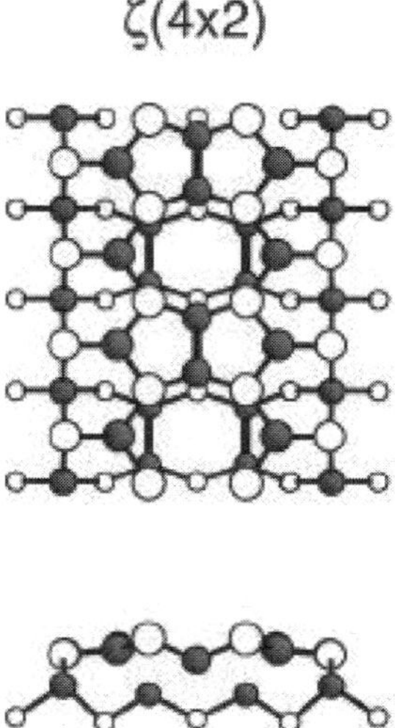

Figure 8. (d) Top views (upper row) and side views (lower row) of the $\zeta(4 \times 2)$ structure of Ga-rich GaAs(001)-4 × 2 surface. Solid spheres denote Ga atoms and open spheres As atoms. The sphere sizes re flect the distance from the surface. Dimer bonds are marked by thicker lines. (From Ref. 16)

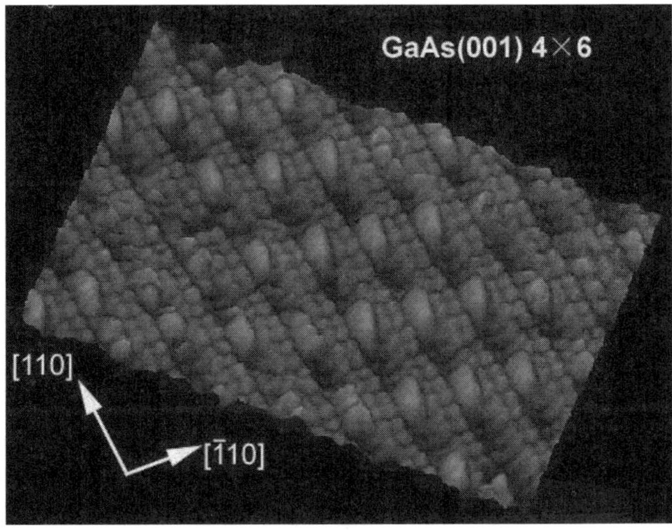

Figure 9. Atomic resolved STM image of GaAs(001) 4 × 6 surface obtained with $V_b = -1.8$ V and $I_t = 40$ pA. (From Ref. 15)

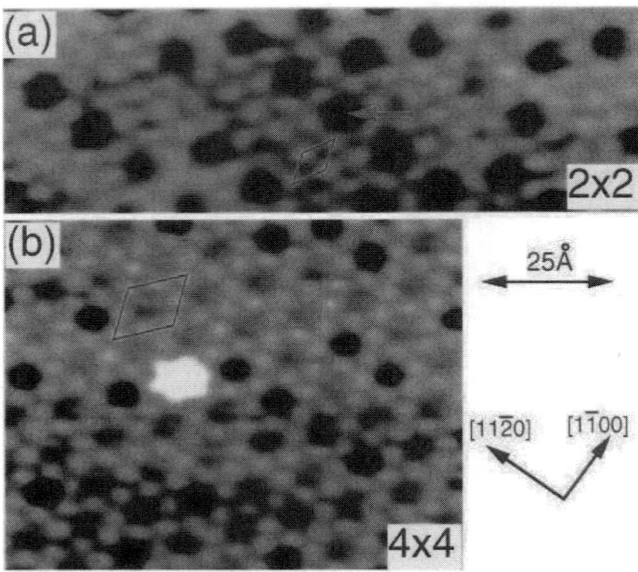

Figure 10. Filled states STM images showing (a) 2 × 2 phase (at −3.0 V), (b) 4 × 4 phase (at −2.8 V). The red arrow in (a) depicts a missing 2 × spot which transfers the 2 × 2 into the 4 × 4 structure. (From Ref. 21)

that the Ga-rich 4 × 6 phase accommodates the periodic array of Ga clusters at the 4 × 6 unit corner on top of the 4 × 2 phase. Further theoretical study does not seem to support the model, and thus the nature of the big oval protrusion keep unresolved [17].

Wide band-gap III–V nitrides (Ga/In/Al/N) have attracted much interest because of their enormous applications in short wavelength optoelectronic devices [18–21]. Absence of reversion symmetric center in hexagonal GaN crystal gives rise to a freedom in its thin film polarity; the (0001) polar surface terminated with a Ga-N bilayer known as the Ga polarity and the (000$\bar{1}$) polar surface terminated with a N-Ga bilayer known as the N polarity [20]. As the present device application depends on controlled heteroepitaxy of the GaN thin film, which is essentially a surface process, complete knowledge of the surface atomic structure is highly desirable. A study of its surface reconstructions is also of great interest since GaN, a special case of the III–V compound semiconductors, is made up of the species possessing large differences in atom radius, electronegativity, and cohesive energy, and contains both covalent and ionic bonds. GaN is also the only III–V that crystallizes in the hexagonal form [21].

The 2 × 2 and 4 × 4 reconstructions of the Ga-polar GaN(0001) surface have been studied with STM first by Xue *et al.* [21] A typical filled state STM image of the 2 × 2 phase is shown in Fig. 10(a). The 2 × 2 symmetry is evident by a regular array of bright spots separated by 6.4 Å along both the close-packing directions. The Ga-adatom model and the Ga-vacancy model are proposed for this reconstruction. However, the

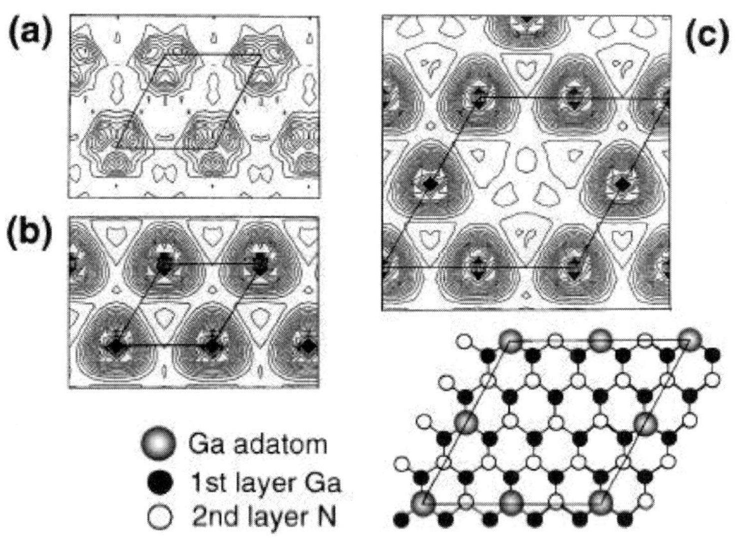

Figure 11. Surface charge density distribution calculated for (a) 2×2 Ga-vacancy, (b) 2×2 Ga-adatom, and (c) 4×4 Ga-adatom models. The local density of states is integrated from the valence bands covering about 2 eV below the highest occupied molecular orbital band, which is cut at 1.3 Å above the outermost surface layer. (From Ref. 21)

correct model cannot be established solely by the STM images. First-principles total-energy calculations again are carried out to resolve this problem [21]. In the charge density calculation, the charge is a sum of valence bands covering a range of about 2 eV below the highest occupied molecular orbital and is a reasonable approach to the STM data (~3 eV). An excellent agreement is obtained for the Ga-adatom model [Fig. 11(b)]. On the other hand, despite an expected coupling of the $2p$ orbits of three threefold coordinated N atoms in the (0001) basal plane, the charge distributions of the Ga-vacancy structure are split spatially [Fig. 11(a)], and do not agree with the experiment.

As for the 4×4 phase [Fig. 10(b)], some individual 4×4 units are observed due to missing spots from the 2×2 phase [as indicated by the arrow in Fig. 10(a)]. During annealing from 200 to 300 °C, the 2×2 and 4×4 phases always coexist. The change to the 4×4 phase with increasing temperature, which results in Ga atom/adatom loss, suggests that the 4×4 forms by the Ga desorption from the 2×2 surface. A missing adatom model is proposed for the 4×4 and investigated it theoretically [Fig. 11(c)]. The agreement between the experiment and theory is excellent [Figs. 10(b) and 11(c)]. Despite this, a model for the 4×4 reconstruction containing three As adatoms and one Ga adatom per 4×4 cell is present in [22]. Therefore, the correct model for Ga-polar GaN(0001) 4×4 structure is still under dispute.

Reconstructions of 2×2, 5×5, 6×4 and pseudo-1×1 appeared on Ga-polar GaN(0001) surface were reported by Smith *et al.* [23, 24]. It indicates that the

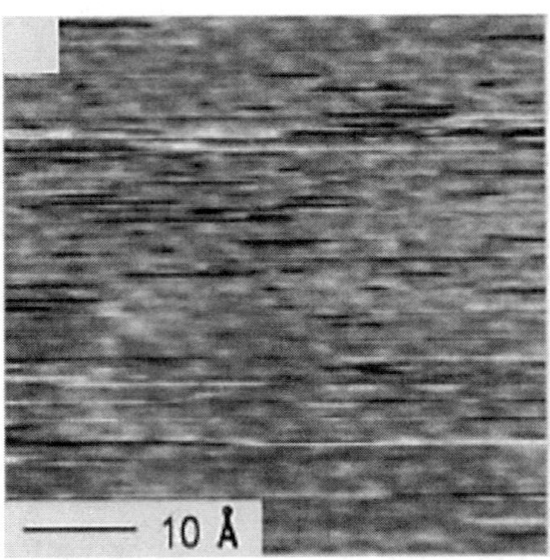

Figure 12. Atomic resolution STM image of pseudo-1 × 1 reconstruction (at −0.25 V) (From Ref. 23)

morphology of GaN(0001) surface will vary with different Ga concentration and substrate temperature. In previous study [23, 24], they showed from total energy calculations that both 2 × 2 N-adatom in H3 site model and 2 × 2 Ga-adatom in T4 site model are more stable where Ga and N adatoms are proposed to bond to three underlying Ga atoms in the Ga terminated Ga-N bilayer. Since the 2 × 2 reconstruction can be obtained by nitriding Ga-polar surface at about 600 °C, they proposed that this reconstruction may be composed of N atoms. Later, they poined out that the 2 × 2 reconstruction results from unintentional contamination of As [25].

Under Ga-rich condition the most stable phase is pseudo-1 × 1 structure, which shows sideband in RHEED pattern and satellite spots in LEED pattern. Pseudo-1 × 1 structure can be obtained either by terminating GaN growth and cooling under 350 °C, or by depositing 2 ∼ 3 ML Ga on the Ga-polar surface and annealing for a period of time. A laterally contracted Ga bilayer model is proposed by Northrup *et al.* [26]. Due to the satellite spots in LEED pattern Ga atoms in pseudo-1 × 1 Ga-bilayer are proposed to experience a rapid moving process. Therefore, the STM image of pseudo-1 × 1 reconstruction is a time-average result which probably indicates the underlying corrugation of GaN(0001) substrate (see Fig. 12).

The 5 × 5 reconstruction can be obtained by the following process: first annealing pseudo-1 × 1 phase at 750 °C, then depositing 1/2 ML Ga and reannealing at 700 °C. The 6 × 4 reconstruction is obtained by depositing 1/2 ML Ga on the 5 × 5 reconstruction and annealing at 700 °C. The 5 × 5 reconstruction and the row-like 6 × 4 reconstruction both depend on the bias voltage as seen in Fig. 13, which were suggested to be semiconducting. For the 5 × 5 reconstruction a structure model that contains Ga adatom in T4 site, N adatom in H3 site and Ga vacancies has been proposed [24].

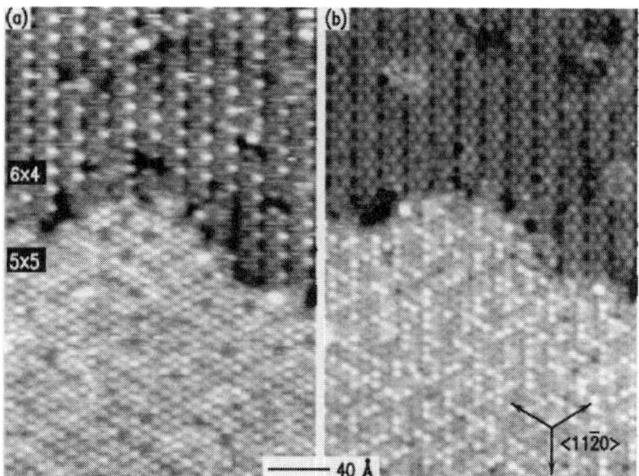

Figure 13. Dual bias images of the 5 × 5 and 6 × 4 reconstructions. The average height difference between the two reconstructions is 0.3 Å for empty states (+1.0 V sample voltage) shown in (a) and 0.4 Å for filled states (−1.0 V sample voltage) shown in (b), with the 5 × 5 being higher in each case. In both images, the total gray scale range is about 1.3 Å. (From Ref. 24)

The adsorption behavior of Ga on Ga-polar GaN(0001) was studied by specular RHEED intensity analysis. It demonstrates that the Ga coverage on GaN(0001) surface during homoepitaxial growth is a function of the Ga flux and the substrate temperature. They divided Ga absorption process into three regions according to the Ga coverage that is flux dependent. The Ga coverage is increased with Ga flux less than 0.20 ML/s. When Ga flux is between 0.20 ML/s and 0.72 Ml/s, the Ga coverage is almost unchangeable. If Ga flux is larger than 0.72 ML/s, Ga droplets form and there will be no finite equilibrium Ga coverage under higher Ga flux. Thus, the transition fluxes vary exponentially with the substrate temperature [27, 28].

Reconstructions of the N-polar GaN(0001) were investigated by STM first by Smith *et al.* [29]. They observed four reconstructions: 1 × 1, 3 × 3, 6 × 6, and c(6 × 12). The 3 × 3, 6 × 6 and c(6 × 12) reconstructions can be obtained by depositing submonolayer Ga atoms on the 1 × 1 structure. The STM images of these reconstructions are shown in Fig. 14. The 1 × 1 reconstruction appears to be hexagonal which has the same lattice to that of GaN. The 3 × 3 reconstruction also shows similar hexagonal arrangement. The 6 × 6 reconstruction displays a ring-like structure. Each ring has threefold symmetry with lobes from three neighboring rings coming close together, which results in two different height "holes" around the rings. The row-like c(6 × 12) reconstruction shows a bias-dependent characteristic, which is different from other reconstructions.

They have proposed the structure models of 1 × 1 and 3 × 3 reconstructions as shown in Fig. 15. The 1 × 1 reconstruction is suggested to contain 1ML Ga atoms bonded to the top N atoms in the N-terminated GaN bilayer. For the 3 × 3 reconstruction, the Ga adatoms are supposed to bond on top of the 1 × 1 Ga adlayer.

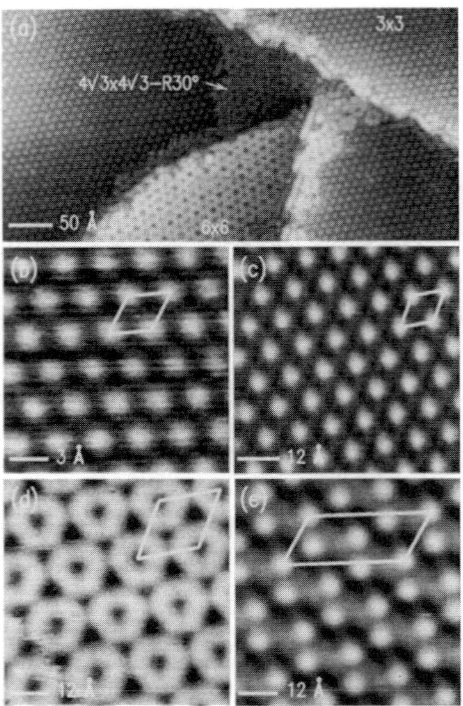

Figure 14. STM images of the N-polar GaN(0001) surface displaying (a) mixed reconstructions, with dislocation near center of image, (b) 1 × 1, (c) 3 × 3, (d) 6 × 6, and (e) *c*(6 × 12) reconstructions. Sample bias voltages are +1.0, −0.75, −0.1, +1.5, and +1.0 V, respectively. Tunnel currents are in the range 0.03—0.11 nA. Gray scale ranges are 4.2, 0.17, 0.88, 1.33, and 1.11 Å, respectively. Unit cells are indicated with edges along <1120> directions. (From Ref. 29)

2.1.3. Metal Adsorption on Semiconductors—In Nanoclusters

In the last decade, fabrication and understanding of nanoclusters have become one of the most exciting areas of research. This is driven by their great potential applications in technology and scientific importance to bridge our understanding between molecular and condensed matter physics. Recently, Xue's group explored a method of surface-mediated magic clustering and successfully fabricated the artificial cluster crystals, i.e., the periodical array of identical nanoclusters by using the ordered reconstructed semiconductor surface-the Si(111)-7 × 7 as a template [30–32].

The STM image of periodical In nanocluster array on Si(111)7 × 7 is shown in Fig. 16(a). All In nanoclusters are completely identical and also in a perfect ordering since In clusters only occupy the FHUC of Si(111)-7 × 7. The atomic resolution STM images of the In clusters at different sample biases (+0.5 V, +0.3 V and –0.3 V) are shown in Fig. 16(b)–(d), respectively. In the empty state images, the In clusters appear as hollow-centered six-spot equilateral triangles with a distance between the spots of ~5.0 ± 0.5 Å, which is much larger than the surface lattice constant 3.84 Å of the

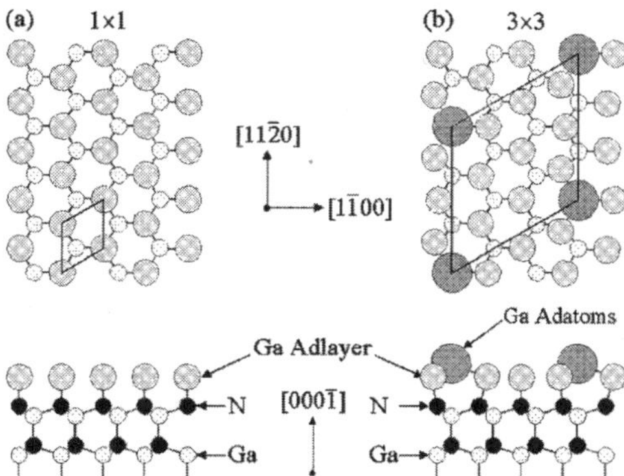

Figure 15. Schematic view of the structures for the (a) 1 × 1 Ga adlayer and (b) 3 × 3 adatom-on-adlayer reconstructions of GaN(0001). For the 3 × 3 structure, the lateral (in-plane) displacement of the adlayer atoms bonded to the Ga adatom is 0.51 Å away from the adatom. All other lateral or vertical displacements of the adlayer atoms are less than 0.1 Å. (From Ref. 29)

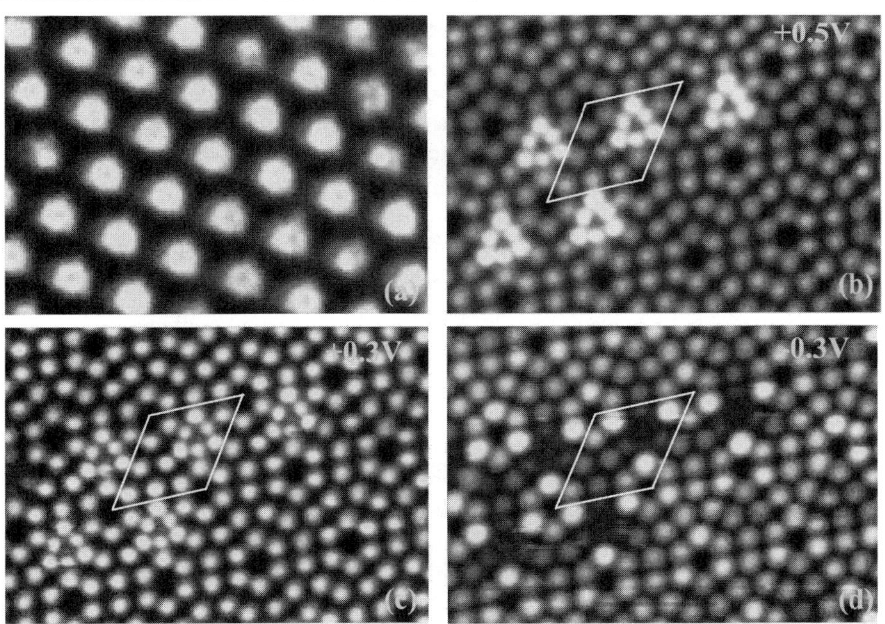

Figure 16. (a) STM image of a periodical In nanocluster array. (b–d) Atomic resolved STM images of In nanoclusters at different bias voltages, showing a pronounced bias voltage dependence of the observed images. (From Ref. 30)

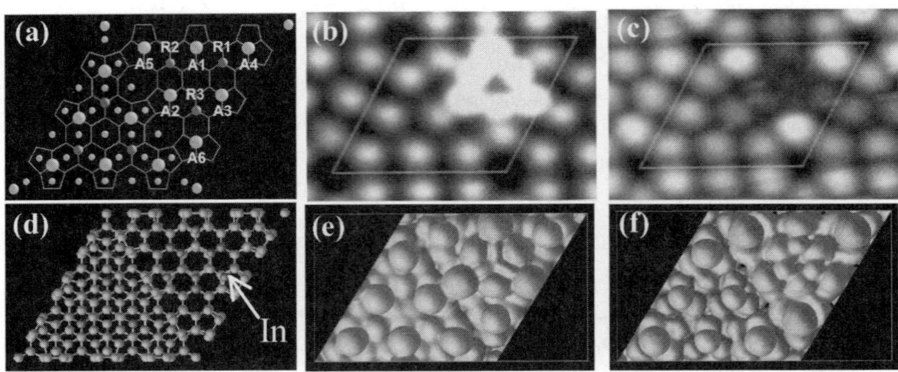

Figure 17. (a) The DAS model of Si(111)-7 × 7 surface. The FHUC is to the upper-right corner. The sites relevant to the discussion are indicated as R1–R3 for Si rest atoms and A1–A6 for Si adatoms. The yellow balls are Si atoms in the substrate, the blue balls are Si adatoms, and the red balls are Si rest atoms. (b) and (c) The STM images of the In clusters recorded at sample bias voltages of +0.6 V and –0.3 V, respectively. (d) Top view of the calculated atomic structure of the six-In cluster on Si(111)-7 × 7. The dark blue balls are In atoms. The calculated STM images are shown in (e) (for positive bias +0.6 V) and (f) (for negative bias at –0.3 V with respect to the Fermi energy) for the atomic structure in (d). The color code indicates the height of the images: dark blue being low and red being high. At typical experimental tip height of about 1 nm above the surface, only the most protruding features can be seen.

Si(111)-1 × 1 and the In-In nearest neighbor distance 3.25 Å. The triangular pattern is quite unusual in terms of normal close-packed structures observed previously. In the filled state images, however, the six-spot equilateral triangles disappear completely and the most protrusive features are the corner adatoms. The strong bias dependence of the images makes it very difficult to deduce the atomic structure of In clusters although it can be concluded that there are six In atoms in each cluster [30].

First-principles total energy calculations are employed to solve the problem. After optimization, the model in Fig. 17(d) is obtained. In this model, the six threefold-coordinated In atoms form a triangle [Fig. 17(d)]. For those In atoms at the corners of the triangle, the bond lengths are 2.57 Å, 2.64 Å, and 2.64 Å, whereas the bond angles are 113°, 113°, and 88°, respectively. For those In atoms on the edges, the bond lengths are 2.67 Å, 2.60 Å, and 2.60 Å, whereas the bond angles are 113°, 116° and 116°, respectively. Angles larger than the 109.5°-tetrahedral angle are preferential as threefold In prefers planar 120° bond angles. Both the three Si adatoms [A1–A3 in Fig. 17(a)] and the three Si rest atoms [R1–R3 in Fig. 17(a)] become fourfold coordinated. Noticeably, Si adatoms A1–A3 are displaced towards the triangle center considerably, which strengthens their bonds with the substrate atoms by resuming the 109.5°-tetrahedral angles. Each Si adatom has two 80°, one 83°, and three close-to-tetrahedral angles. Thus, by displacing Si adatoms not only can the perceived steric strain be avoided, but also the displaced Si adatoms serve as the "missing" links between the otherwise loosely packed In atoms. The calculation also shows that an In cluster on the UFHUC is 0.1 eV/cluster higher in energy than that on the FHUC, which also agrees with the experimental result that most In clusters occupy the FHUC of Si(1110-7 × 7) preferentially.

The calculated STM images in Fig. 17(e) and (f) are in remarkable qualitative agreement with experiment [Fig. 17(b) and (c)]. Interestingly, in the empty state image [Fig. 17(e)], the three brightest spots are from the lowest In atoms, which are 0.6 Å lower than Si A1–A3 with an average bond angle of 105° (thus sp^3-like). The three second-brightest spots are from the other In atoms, which are 0.3 Å lower than Si A1–A3 with an average bond angle of 115° (thus sp^2-like). Si adatoms A1–A3 are almost invisible, as they do not involve any dangling bond. Another striking feature in Fig. 17(c) is the disappearance of the six-In triangle spots under small *reverse* bias, whereas the three Si corner adatom spots (A4–A6) become significantly brighter. The calculation reveals that this change is not due to In diffusion but has an electronic origin. The calculated density of states reveals a 0.33 eV band gap 0.2 eV below the Fermi energy (E_F). States below the gap have mainly the Si/In bonding character. States above the gap but below E_F have mainly the dangling-bond character and are predominantly on Si A4–A6. The In dangling bond states are found to be above E_F thus can only be seen in the empty state image.

This application also demonstrates that STM combined with first-principles total energy calculations is a very powerful method to determine the atomic structure of surfaces with/without adsorbates. The atomic structures of Al and Ga nanoclusters have also been determined by this method [31, 32]. Some other metal (including alkali metals) clusterss were also fabricated this way [33].

2.2. Metal Surfaces

2.2.1. Metal Surfaces

Metal surfaces had been studied for more than two decades before STM was invented. In 1982, the Au(110)-2 × 1 surface was first imaged with STM by Binnig *et al.* [34]. This work confirmed the missing-row model proposed for this surface although atomic resolution was not achieved.

For non-reconstructed metal surfaces, the charge density corrugation amplitudes are typically on the order of 0.1 Å since STM usually probes the delocalized s- or p-type states, while the corrugations for semiconductor surfaces are often of several Å due to the presence of dangling bonds. Metal corrugations are usually 50–100 times smaller than those on the Si(111) 7 × 7 surface, and thus, it is much more difficult to obtain atomic resolution on metal surfaces than on semiconductor surfaces. With the development of STM instrumentation, more and more investigations on metal surfaces were reported. In 1987, the first atomic resolution STM image was observed on the Au(111) surface [35], which strongly impacts the STM investigation on metal surfaces. Au(111) has been the most widely used metal substrate in STM studies because the surface is inert and atomic resolution STM images can be obtained even in air.

Clean Au(111) surface reconstructs to $(23 \times \sqrt{3})$ reconstruction. The "herringbone structure" reconstruction can be easily observed by STM, as shown in Fig. 18(a). An atomically resolved image of a bending point is shown in Fig. 18(b) [36, 37]. This structure is explained by the stacking-fault-domain model, which involve stacking faults between fcc and hcp orderings induced by surface strain.

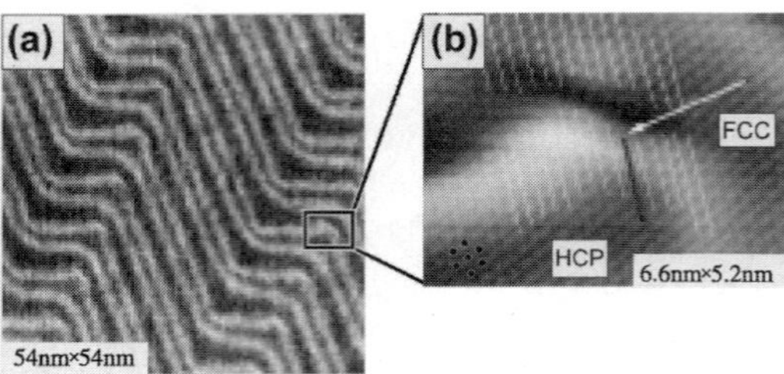

Figure 18. STM images of Au(111)-(23 × $\sqrt{3}$) surface. (a) "herringbone structure", image size: 54 nm × 54 nm. (b) atomic resolution image at the bending point (6.6 nm × 5.2 nm). (From Ref. 37)

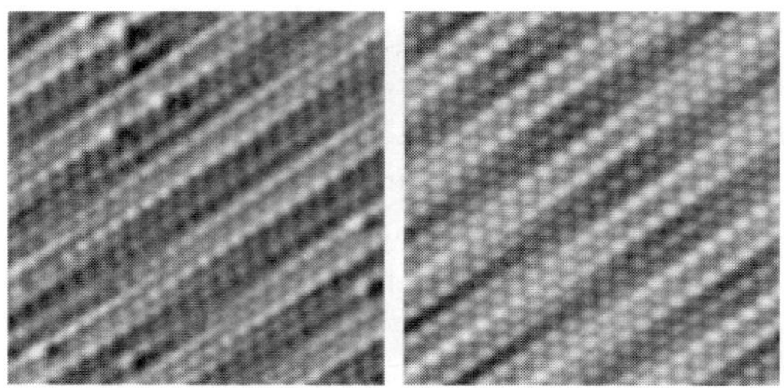

Figure 19. STM (left) and simulated (right) images of Au(001). (From Ref. 38)

For the clean reconstructed Au(001) surface, by comparing the atomic resolution STM images with simple simulations (see Fig. 19), it was found not only that the topmost atomic layer is, qualitatively, quasi-hexagonal and incommensurate, but also that it is, quantitatively, rotated by 0.1° relative to the substrate and contracted by 3.83% and 4.42% compared to a perfect (111) layer of Au, in the vertical and horizontal orientation, respectively [38].

STM has been applied to low-index surfaces of many other metals, *e.g.*, Pt, Pd, Cu, Ag, Al, *etc.* and atomic resolution has been achieved for all of them. In contrast to semiconductor surfaces, the bias voltage dependence usually is not observed on clean metal surfaces, which makes the interpretation of the STM images rather simple.

2.2.2. Adsorption on Metal Surfaces

Metal surfaces with adsorbates, especially with molecular adsorbates nowadays become increasingly important, because of their application potential in nano- and bio-science and technology [39]. Since amino acids are building blocks of proteins, adsorption of amino acids on metal surfaces, as a biological model system, has been receiving much attention [40]. Despite that in most cases each amino acid molecule can only be imaged as one protrusion, many important results about the amino acid adsorbates were obtained in a series of recent STM investigations by Zhao and coworkers, and are summarized briefly as follows.

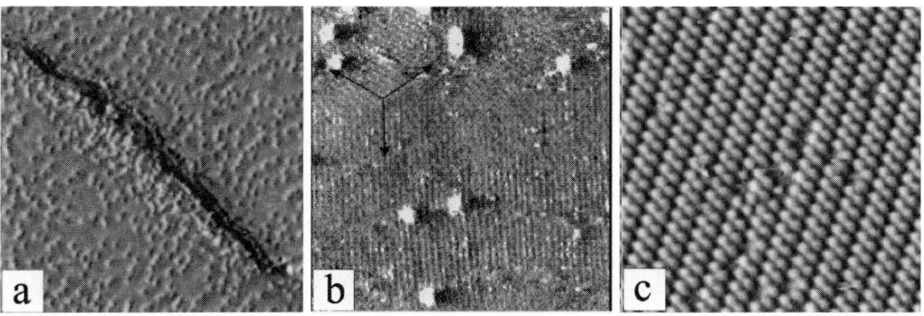

Figure 20. Three different phases of glycine adsorbed on Cu(111): (a) the 2D gas phase; (b) the chain phase; (c) the 2D solid phase. (From Ref. 41)

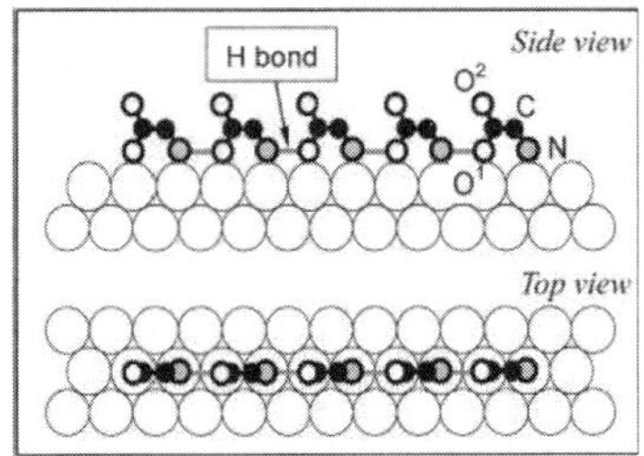

Figure 21. Schematic drawing of the chains formed by glycine molecules adsorbed on the Cu(111) surface. (From Ref. 45)

(i) Through "cook-and-look" or "anneal-and-image", it was able to determine the desorption temperature and, in turn, the binding energy of the adsorbates, and then to find if the adsorbates are chemisorbed or physisorbed on the surface. For instance, glycine was found to be chemisorbed on Cu(001) [41], while to be physisorbed on Au(110) [42].

(ii) Depending on the coverage and deposition rate, amino acid adsorbates on Cu and Au surfaces may form three different phases, i.e., the 2D gas phase, the chain phase, and the 2D solid phase (Fig. 20). Some amino acids are able to form all the three phases on Cu(001), while some others can form only one or two of the three. In the 2D gas phase the molecules are "standing" on the surface and can diffuse frequently on the surface at room temperature. The activation energy barrier was determined to be around 0.85 eV [41]. In the 2D solid phase the molecules are connected by H bonds to form different ordered structures, depending on their side chain structure [43, 44]. However, in the chain phase, different amino acids (*i.e.*, with different side chains) adsorbed on different substrates are connected by H bonds to form, surprisingly, always the same kind of 1D chains (see Fig. 21 and Ref. 45).

(iii) Moreover, amino acid adsorbates were found to be able to modify the substrate morphology significantly. For instance, adsorbates of the smallest amino acid, *i.e.*,

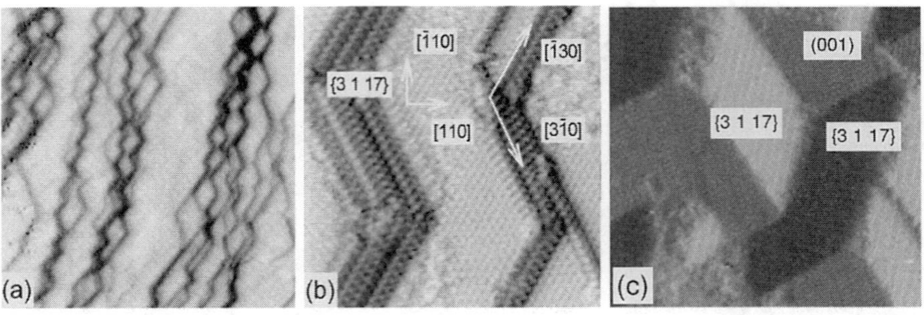

Figure 22. (a, b) Adsorption of glycine, which is not chiral, on Cu(001) makes surface steps faceting to all eight possible <310> directions and then bunching into all eight possible {3 1 17} facets [Zhao *et al.*, Surface Science 424, L347 (1999)]. (c) Adsorption of L-lysine on Cu(001) surface, in contrast, makes steps bunching only into the four {3 1 17} facets that have the same chirality. [From Zhao et al., Chinese Physics 10 (supplement), (S84 2001).]

glycine, can make all steps on the Cu(001) surface faceted (or reoriented) into eight equivalent <310> directions and then bunching into all eight equivalent {3 1 17} facets.

However, adsorption of homochiral amino acids, such as L-lysine, on the same Cu(001) surface makes the steps bunching into only the four of all eight {3 1 17} facets that have the same chirality, which is determined by the chirality of the molecules [46]. The possibility of using homochiral adsorbates to fabricate homochiral facets on a substrate is of current interest because of its potential application in chiral separations.

2.3. Insulator Surfaces

In principle, insulators cannot be studied with STM since tunneling current cannot be established between a conducting tip and an insulator. However, under some special conditions, STM observation can be performed on insulating materials. For example, BN thin film on Rh(111) surface have been investigated with STM [47]. The STM images of 2ML BN grown on Rh(111) surface by high-temperature decomposition of borazine are shown in Fig. 23. Ordered BN nanomesh is observed in the large scale image [Fig. 23(a)]. In high resolution image [Fig. 23(b)], it is clearly seen that the nanomesh consists of two layers of BN and they are offset in such a way as to expose a minimum metal surface area. NaCl(111), ZnO(0001) and TiO_2 etc. have also been reported to be studied with STM [48–50].

Recently, K. Bobrov *et al.* demonstrated that STM can be used in an unconventional resonant electron injection mode to image insulating diamond surfaces and to probe their electronic properties at the atomic scale [51]. The hydrogen-free diamond surface is insulating, no tunneling current could be obtained at any bias voltages between −6 V and +4 V. The STM tip crashed on the surface if trying to establish a tunnel current in the range 0.05–1 nA. However, at very high sample bias (+5.9 V), *i.e*, above the diamond work function (5.3 V), atomic resolution STM image can be obtained on the diamond surface as shown in Fig. 24. In Fig. 24(a), terraces rotated by 90° are clearly visible. The periodic structure of bright and dark lines is observed on every

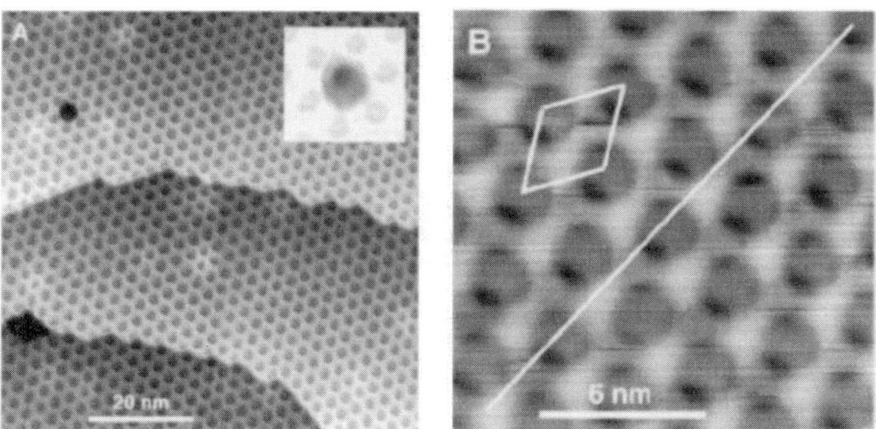

Figure 23. Constant-current STM images of the boron nitride nanomesh formed on a Rh(111) surface. (a) Large-area image taken with a bias voltage of $V_b = -1.0$ V and a tunneling current of $I_t = 2.5$ nA. The black features are defects in the mesh, one of which is shown with different contrast in the inset. (b) High-resolution image (–2.0 V and 1.0 nA) clearly showing the presence of two layers of mesh that are offset such as to cover most of the Rh(111) surface. The mesh unit cell is indicated (From Ref. 47).

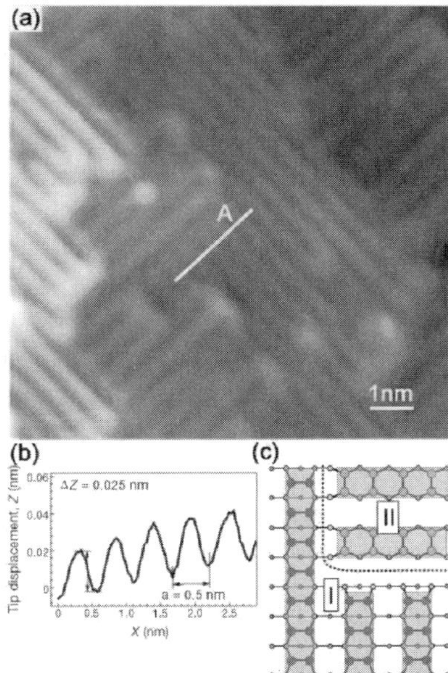

Figure 24. Clean diamond C(100)-(2 × 1) surface. (a) The STM topography (10 nm × 10 nm) of the clean diamond surface recorded in the near-field emission regime (Ub = 5.9 V, I = 1.1 nA). (b) Height variation of the STM tip along the line A. (c) Topview of a monoatomic step on the two-domain (2 × 1) reconstructed surface. The circles represent the carbon atoms belonging to the top four surface layers; the biggest circles represent the carbon-carbon dimers. The domains labelled as I and II represent the upper and lower terrace, respectively. The dimer rows are highlighted by shading. The dashed line shows schematically the boundary between the domains. (From Ref. 51)

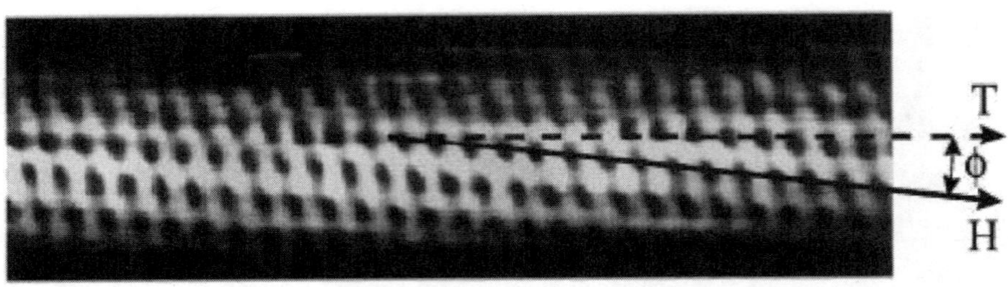

Figure 25. Atomically resolved STM image of individual single-walled carbon nanotubes. The lattice on the surface of the cylinders allows a clear identification of the tube chirality. Dashed arrows represent the tube axis T and the solid arrows indicate the direction of nearest-neighbour hexagon rows H. From the image, it can be determined that the tube has a chiral angle $\varphi = 7°$ and a diameter $d = 1.3$ nm. (From Ref. 53)

terrace. The periodicity of ~0.5 nm, as measured from the scan profile shown in Fig. 24(b), agrees well with the distance (0.504 nm) between the C–C dimer rows of the (2×1) reconstructed diamond surface. This work suggests that STM can be operated in the near-field emission regime and this method can be applied to investigate other insulating materials.

2.4. Nanotubes and Nanowires

Carbon nanotubes have attracted much attention since their discovery in 1991 due to their peculiar properties [52]. STM has been widely used to study the structure and electronic properties of carbon nanotubes [53–55]. An atomically resolved STM image of individual single-walled carbon nanotubes is shown in Fig. 25, from which, the structure (chiral angel and diameter) can be easily determined. Combined with scanning tunneling spectroscopy (STS, which will be introduced in the next section), their local electronic properties can also be related to the local structures [53].

Shown in Fig. 26 is an STM image of an oxide-removed Si nanowire, another kind of interesting nanowires [56]. The study also showed that the electronic energy gaps of Si nanowires increase with decreasing Si nanowire diameter from 1.1 eV for 7 nanometers to 3.5 eV for 1.3 nanometers, in agreement with previous theoretical predictions.

In these studies, highly ordered pyrolytic graphite (HOPG), Au(111) or Au film are often used as substrates, whereas nanotubes and nanowires are usually deposited on the surface from dilute solutions. The outmost structure of the nanotubes and nanowires can be determined from atomically resolved STM images. In addition, STS is often used to probe their electronic properties and the relationship between electronic properties and structures. Such information is very difficult to obtain with other techniques.

2.5. Surface and Subsurface Dynamic Processes

So far, it has been shown that the atomic structure of many different surfaces can be studied or even determined on the basis of high-resolution STM images. With atomic

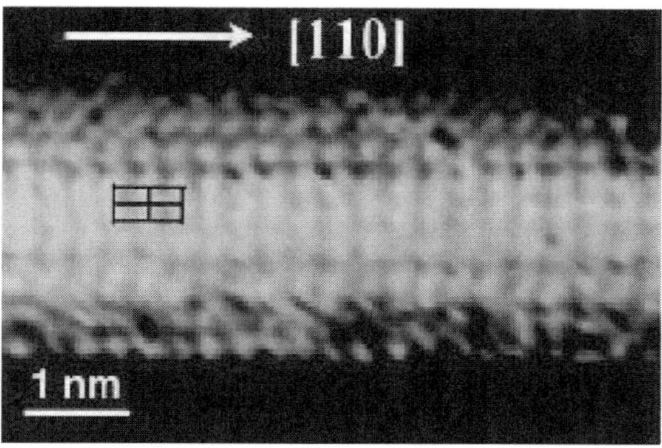

Figure 26. STM image of a Si nanowire with a Si(001) facet. The wire's axis is along the [110] direction. (From Ref. 56)

resolution STM images, it is also possible to study surface or even subsurface atomic dynamic processes.

2.5.1. Surface Diffusion

Although it is possible to study surface diffusion by checking the position of individual atoms, to find out, among hundreds or more atoms, the few that diffused from one image to another is nevertheless tedious. However, difference images obtained from a set of sequential images can make the job much easier (see Fig. 27). It was thus found that on the Ge(111) surface individual adatoms neighboring to some defects are able to diffuse even at room temperature [57]. Moreover, it was also found that, if the domains are not very large, adatoms forming a string or closed loop lying along domain walls may diffuse one after another (see Fig. 27). In addition, the mean lifetime of the diffusing adatoms can also be determined and from which the diffusion energy barrier of the adatoms was deduced to be 0.83 ± 0.02 eV, in good agreement with its theoretical value.

2.5.2. Subsurface Migration

Despite that STM is a very surface sensitive technique, from the difference images of a set of sequential images of the Ge(113) surface it was also able to find that the subsurface self-interstitial atoms are migrating frequently even at room temperature, making the local surface structure changing back and forth between (3×2) and (3×1) Fig. 28(a) and (b). Interestingly, migration of a subsurface self-interstitial atom into or out of a place results in a quite large and complicated feature around that place in the difference image, as shown in Fig. 28(c). However, considering that each subsurface atom is bound to several surface atoms this is actually quite reasonable. Furthermore, the lifetime of the self-interstitials was determined to be 400 s at room temperature, and their migration energy barrier was deduced accordingly to be 0.93 ± 0.02 eV [58].

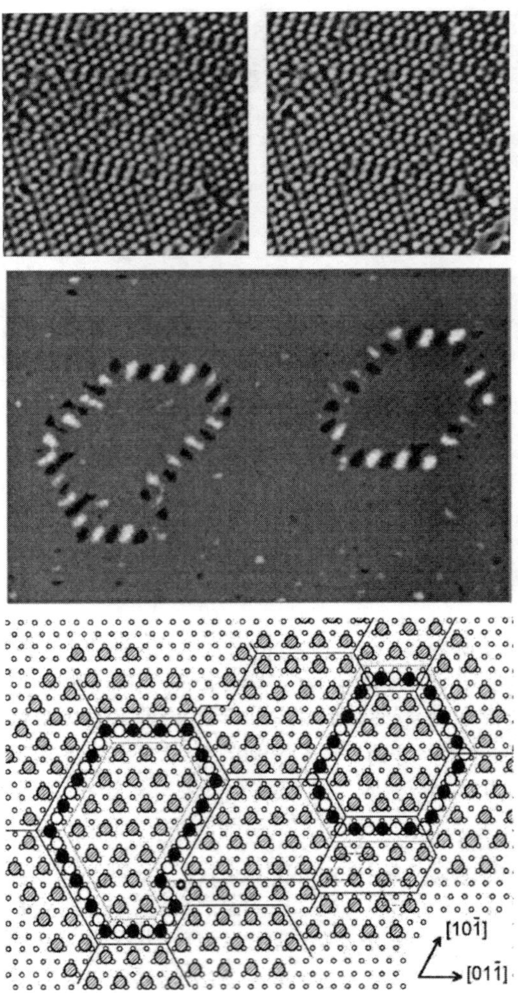

Figure 27. (a) STM image obtained from a Ge(111) surface (b) STM image obtained from the same place as in (a) but 6 minutes later. (c) Portion of the difference image obtained by subtracting (b) from (a), showing shifts of tens of the adatoms forming two closed loops. (d) Schematic drawing of (c), showing the details relevant to the adatom shifts. (From Ref. 57)

2.5.3. Movement of Subsurface Dislocations

Another type of subsurface defects, subsurface dislocations, can be found with STM as small regular bumps on many annealed metal surfaces, although argon ion bombardment or STM tip touching can induce more of them. In the case of Au(001) where, as mentioned above, the topmost atomic layer is incommensurate with the substrate and thus Moiré fringes appears on the surface. It was shown that such Moiré fringes can be used as "magnifier" to study the details of such subsurface dislocations, including

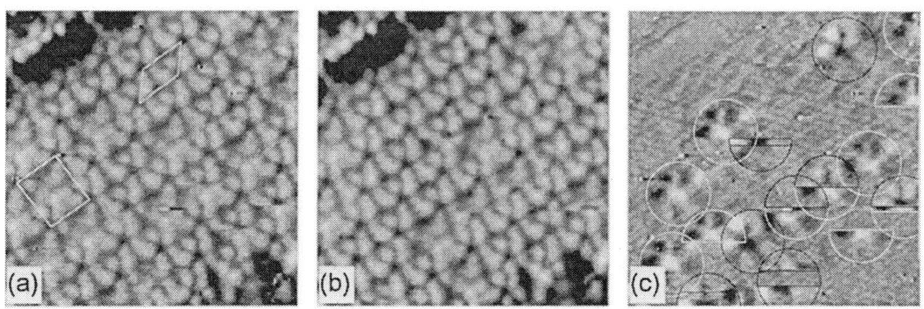

Figure 28. (a, b) Two consecutive STM images (70 × 70 Å^2) obtained from a Ge(113) facet, with a (3 × 1) and (3 × 2) unit cell outlined in (a). (c) The differential image obtained by subtracting (b) from (a). The circled areas are those where a subsurface self-interstitial atom migrated in or out between or during imagings. (From Ref. 58)

their extremely slow movement and the weak strain fields that push them to move. Specifically, on the basis of STM and simulated images, a precision of better than 0.1 Å was achieved in determination of the lateral strain fields and a dislocation speed lower than 1 Å/min was measured [38]. As Moiré fringes exist in many surfaces and adsorbate systems, the method is expected to have wide applications.

3. SCANNING TUNNELING SPECTROSCOPIES

As mentioned above, bias-dependence of STM images is often observed, particularly for semiconductor surfaces, which makes it difficult to explain the STM images. However, very useful spectroscopic information can be extracted from the bias-dependence of tunneling current. In fact, tunneling spectroscopy had been used with fixed tunneling junctions before STM was invented. More important information can be obtained by measuring tunneling spectroscopy with an STM. The scanning ability of STM makes it possible to probe local spectroscopic signals with atomic spatial resolution. By changing the tip-sample distance, the potential barrier can also be investigated with STM.

3.1. Scanning Tunneling Spectroscopy (STS)

From Eq. 10, tunneling current at a finite bias voltage V can be expressed as:

$$I \propto \int_0^{eV} \rho_S(E_f - eV + \varepsilon)\rho_T(E_f + \varepsilon)d\varepsilon, \tag{12}$$

where, ρ_S and ρ_T are the DOS of sample and tip respectively. If ρ_T is constant, then:

$$\frac{dI}{dV} \propto \rho_S(E_f - eV + \varepsilon), \tag{13}$$

i.e., the structure in dI/dV as a function of V represents the structure in the DOS of sample, which is called as scanning tunneling spectroscopy (STS).

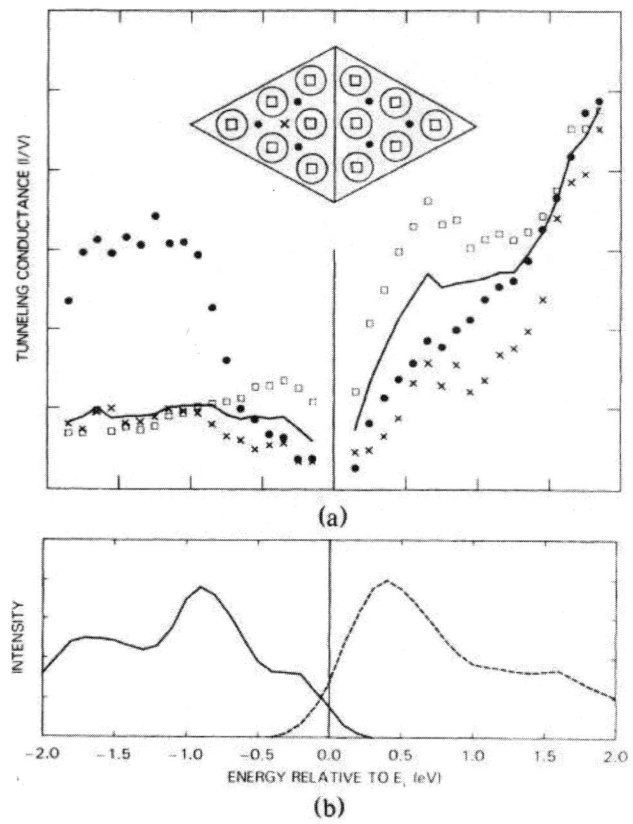

Figure 29. (a) Constant-distance $I/V \sim V$ spectra for the Si(111)-7 × 7 surface averaged over one unit cell (solid line) and at selected locations in the unit cell (other symbols). (b) Spectra obtained with UPS (solid line) and IPS (dashed line). (From Ref. 59)

The first spatial resolved tunneling spectroscopy was demonstrated by Hamers *et al.* on Si(111) 7 × 7 surface [59]. The site-selected conductance curves ($I/V \sim V$) within a Si(111)-7 × 7 unit cell are shown in Fig. 29. The physical origin and the nature of the surface states of Si(111)-7 × 7 surface, including the states due to dangling bonds on twelve adatoms, the states localized on rest atoms, the states due to Si-Si backbonds, and the states localized in the deep corner hole were directly identified. The $I/V \sim V$ spectra averaged over one unit cell is comparable with the results of ultraviolet photoemission spectroscopy (UPS) and inverse photoemission spectroscopy (IPS). Better agreement between the spectrum averaged over an area encompassing many unit cells and the data from UPS and IPS were achieved later [60]. These studies showed that the electronic structure of the tip is relatively unimportant in STS measurements.

Current imaging tunneling spectroscopy (CITS) was also proposed, which allows real-space imaging of surface electronic states. By measuring constant separation I–V

curves at each point during scanning, current images at sample voltages within a range can be obtained simultaneously with STM topographic image. The resulting real-space current images directly reflect the spatial distribution of the surface states without interference from geometric structure contributions [59]. The atomic resolved CITS images on Si(111)-7 × 7 surface are shown in Fig. 30, from which the atomic origins of the various electronic states can be easily determined. The electronic states near −0.35 eV are from the 12 adatoms, the states near −0.8 eV arise from the 6 rest atoms, whereas the states near −1.7 eV are from backbond states.

The capability of identifying surface states in real-space with atomic resolution greatly extends the utility of STM as a spectroscopic tool. STM combined with STS has been widely used to study the structure, electronic properties and their relationship of various materials.

3.2. Inelastic Tunneling Spectroscopy

By now, we only considered the elastic electron tunneling process, in which the electrons keep conservation of energy during tunneling. In fact, inelastic tunneling can also occur if the tunneling electrons couple to some excitation modes in the tunneling junction. In 1966 it was discovered that inelastic electron tunneling spectroscopy (IETS) can be obtained from molecules adsorbed at the buried metal-oxide interface of a metal-oxide-metal tunneling junction [61]. With the development of STM, it was apparent that IETS might be performed on a single molecule in the junction of a STM (STM-IETS) [62]. The metal-oxide-metal tunnel junction is replaced by the STM tunnel junction: a sharp metal tip, a vacuum gap of several angstroms, and a surface with the adsorbed molecules. The combination of atomic resolution and IETS allows the creation of atomic-scale spatial images of the inelastic tunneling channel for each excitation mode, in a manner similar to that used to map out the electronic density of states with the STM [59]. Unfortunately, the conductance changes caused by inelastic tunneling are less than 10% for the STM. Therefore, the extreme mechanical stability is necessary to obtain reasonable IETS with the STM. In addition, low temperature is required to keep thermal line-width broadening small compared with the inelastic exciting energy.

Single-molecule vibrational spectroscopy was first obtained with STM-IETS by Stipe *et al.* in 1998 [62]. To measure the IETS, a small ac modulation was added to the dc sample bias voltage, the tunneling current was fed into a lock-in amplifier to determine the first and second harmonics of the modulation frequency which are proportional to dI/dV and d^2I/dV^2, respectively. These signals were recorded as the sample bias voltage was swept from 0 to 500 mV. As shown in Fig. 31, obvious difference was found in the STM-IETS for C_2H_2 and C_2D_2 although they could not be identified in the atomic resolved STM topographic image. The C-H stretch at 358 mV for C_2H_2 was observed to shift to 266 mV for C_2D_2 (Fig. 31). These values are in close agreement with the results obtained by EELS.

By doing IETS, it is possible to identify molecules with the STM, which permits to implement chemically sensitive microscopy. Vibrational imaging of the adsorbed

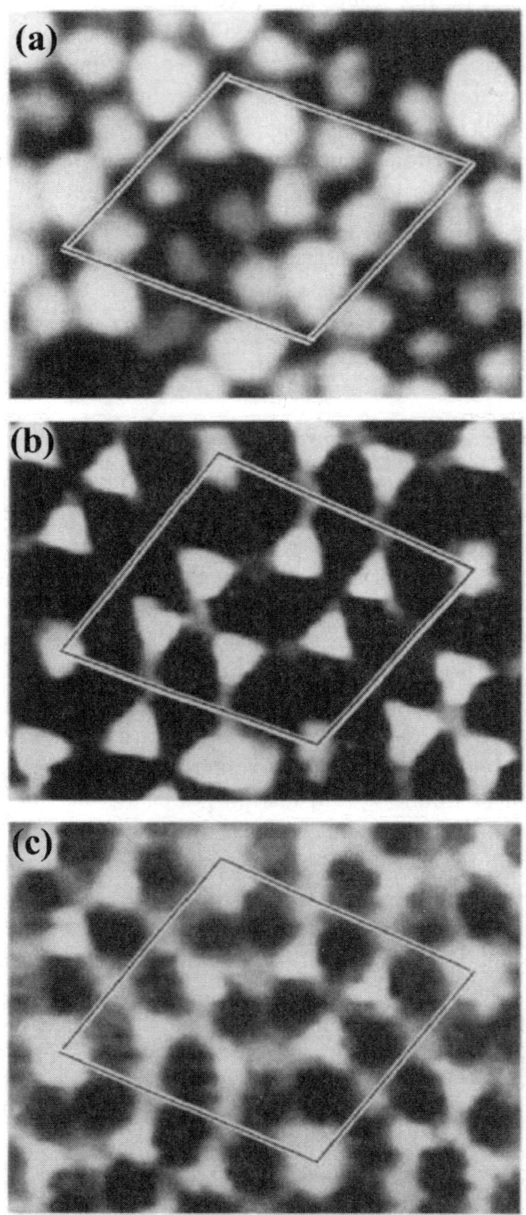

Figure 30. CITS images of occupied Si(111)-7 × 7 surface states. (a) adatom states at −0.35 V, (b) dangling-bond state from rest atoms at −0.8 V, (c) backbond state at −1.7 V. (From Ref. 59)

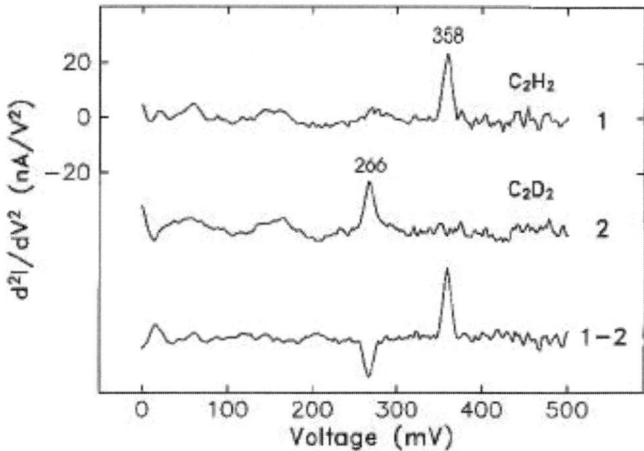

Figure 31. Background difference d^2I/dV^2 spectra for C_2H_2 (1) and C_2D_2 (2), taken with the same STM tip, show peaks at 358 mV and 266 mV, respectively. The difference spectrum (1–2) yields a more complete background subtraction. (From Ref. 62)

molecule was obtained by recording dI/dV and d^2I/dV^2 at each data point with the feedback off and the bias modulation on while scanning the tip in constant-current mode. This procedure results in three images of the same area. In a constant-current image, no contrast was observed for both acetylene isotopes [Fig. 32(A)]. When the dc bias voltage was fixed at 358 mV, only one of the two molecules was revealed in the image constructed from the d^2I/dV^2 signal [Fig. 32(B)]. By changing the dc bias voltage to 266 mV, the other molecule was imaged [Fig. 32(C)]. Two small identical depressions observed at 311 mV [Fig. 32(D)] were attributed to the change in the electronic density of states on the sites of the two molecules [62].

STM-IETS extends the vibrational spectroscopy to the single-molecule limit and provides the STM with chemical sensitivity. Combination of the high spatial resolution of STM and IETS permits to correlate variations in molecular spectra with changes in the local environment on an atomic scale [63].

3.3. Local Work Function Measurement

The general definition of the work function, i.e., the minimum energy needed to remove an electron from a metal to infinity, is clear but cannot be used to measure the local work function. Wandelt, considering that surface dipole potentials reach their saturation value already within ~2 Å from the surface, defined the local work function (LWF) as the local surface potential measured from the Fermi level E_F, which allows us to measure LWF variations induced by surface dipole patches [64, 65]. In this definition, the LWF probe can be put close to the surface compared to the dimensions of the surface patch under study. Obviously, the closer the probe is to the surface the smaller the surface patch of interest can be, provided that presence of the probe has no

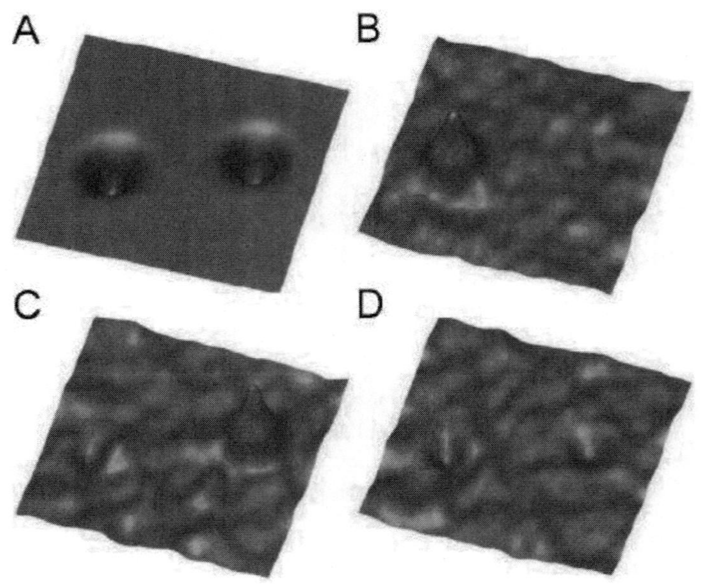

Figure 32. Spectroscopic spatial imaging of the inelastic channels for C_2H_2 and C_2D_2. (A) Regular (constant current) STM image of a C_2H_2 molecule (left) and a C_2D_2 molecule (right). The imaged area is 48 Å by 48 Å. d^2I/dV^2 images of the same area recorded at (B) 358 mV, (C) 266 mV, and (D) 311 mV. All images were scanned at 1 nA dc tunneling current. (From Ref. 62)

influence on the local surface potential, and hence this LWF definition is suitable for studying LWF variation with STM.

Work function is important in STM because it determines the height of the tunneling barrier. The tunneling current I depends exponentially on the tip-sample distance s:

$$I \propto \exp(-2\kappa s), \text{ with } \kappa = \hbar^{-1}(2m\phi)^{1/2}, \tag{14}$$

where ϕ is the effective local potential barrier height. From the above formula, we have

$$\phi_A[eV] = \frac{\hbar^2}{8m}^{-1}\left(\frac{d\ln I}{ds}\right)^2 \approx 0.95\left(\frac{d\ln I}{ds\,[\text{Å}]}\right)^2. \tag{15}$$

Binnig and Rohrer have shown that, at least in the image force range, the s dependence enters ϕ_A in second order only, or ϕ_A is nearly independent of the tip-sample separation. Moreover, it has also been pointed out that for homogeneous surfaces ϕ_A is work function [66], while for patchy surfaces, ϕ_A is equal to the LWF [65].

In an experiment, the height of the tunneling barrier or work function can be obtained by measuring the response of the tunneling current when changing the gap

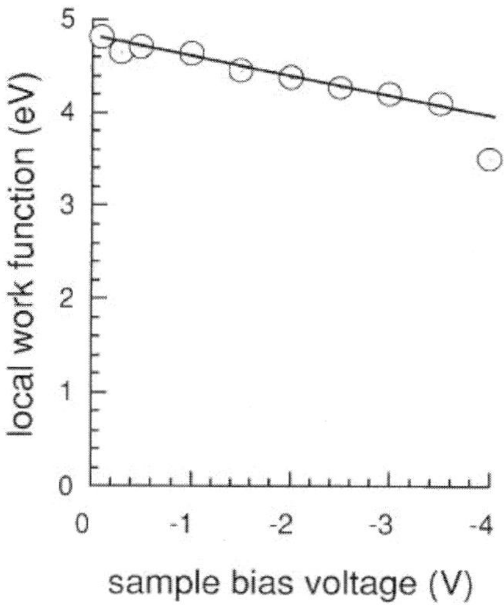

Figure 33. Measured work function dependence on sample bias voltage for Cu(111) surface. (From Ref. 69)

distance. Actually, at the beginning of the STM invention by Binnig and Rohrer, they already pointed out the possibility of measuring the work function and taking its image [67]. Jia *et al.* have measured the LWF on metal surfaces quantitatively [65, 68, 69]. In their experiments, the modulation frequency was set at 2.0 kHz, higher than a cut-off frequency of the feedback loop of the STM system they used (~1 kHz) but lower than the response frequency of the current amplifier of the STM. The frequency dependence of the work function on the Cu(111) substrate showed that the modulation frequency (2.0 kHz) is in a plateau range. The amount of modulation in the gap distance is 0.23 Å, much smaller than the gap distance, 5.5–6.0 Å.

The LWF dependence on bias voltage measured using a Cu(111) surface is shown in Fig. 33. From these measurements, it turns out that the work function drops slowly with a ratio of ~0.2 eV/V as the bias voltage increases gradually up to −3.5 V. Variation of work function with a bias voltage is quite reasonable because applying a bias voltage lowers the barrier height in the STM gap. It is qualitatively consistent with the results of previous experimental and one-dimensional numerical simulation. At a low bias voltage limit, it reaches around 4.8 eV, close to an average value of work function of Cu(111) and W(111), which is used for the probing tip. Image potential does not seem to contribute so much to the work function [69].

By measuring LWF at each point during scanning, a LWF image can be obtained simultaneously with a STM image. Figure 34(b) is the LWF image taken simultaneously with the STM image in Fig. 34(a). From the STM image alone it is rather difficult to distinguish the Au-covered areas from those uncovered, although from the former one can vaguely see the quasi-periodic triangular features. As the Au terraces have a

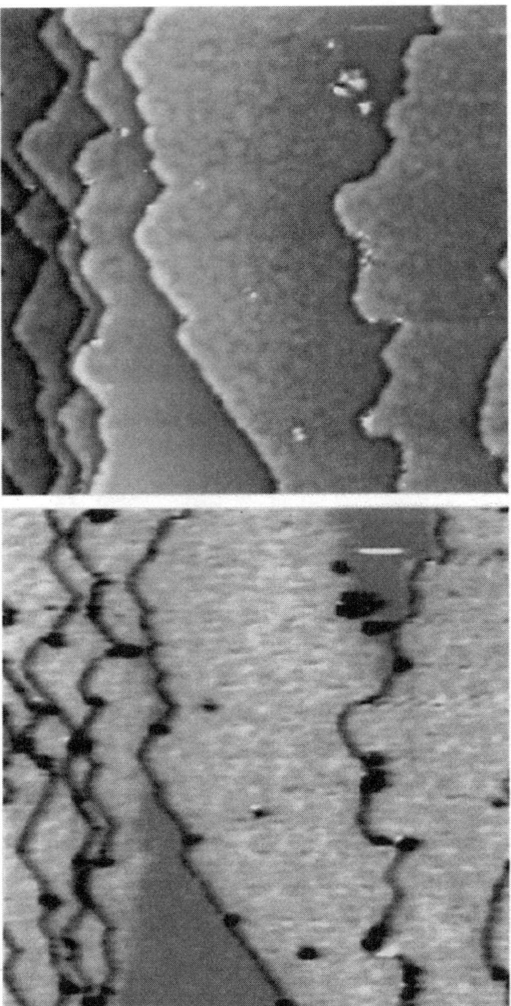

Figure 34. (a) STM images of a Au/Cu(111) surface (580 Å × 580 Å). (b) Simultaneously obtained work function image. The mean value of the work function on the Au overlayer (bright area) is 7% higher than that on the Cu(111) substrate (dark area). The dark lines correspond to a low work function zone at step edges. (From Ref. 68)

higher WF than that of the Cu terraces we identify the brighter areas in the WF image as covered by a Au layer, while the darker areas as being nude.

Similar measurements have been carried out using a Pd/Cu(111) surface [68]. An STM image obtained from the surface is given in Fig. 35(a), and the corresponding work function image [Fig. 35(b)] shows that work function measured on Pd overlayers is larger than that of the Cu substrate. The conclusion of the statistical analysis is that the first Pd layer has a larger work function than Cu(111) by 19 ± 5%. The dark contrast

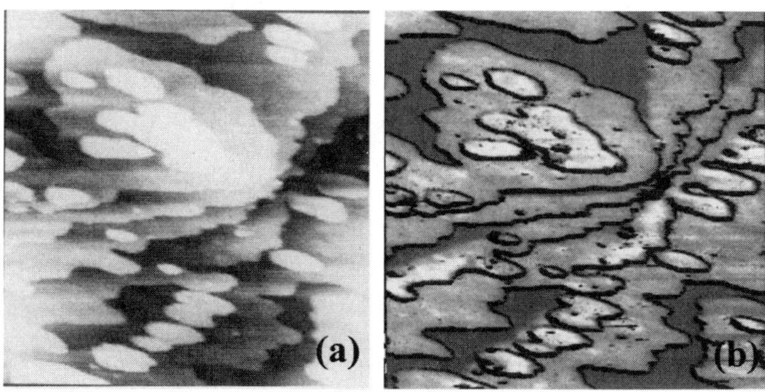

Figure 35. (a) STM and (b) work function images obtained on a Pd/Cu(111) surface. The applied sample bias voltage is −2.0 V, and the tunneling current is 0.1 nA. The size of the observed area is ~570 Å × 570 Å. The coverage of Pd is ~1.0 ML. It shows that the Pd layer has a higher work function than the Cu substrate and that the second Pd layer has a higher work function than the first Pd layer. (From Ref. 68)

along step edges is observed in the work function images taken on the Pd/Cu(111) surface as well.

Different from the results on Au/Cu(111) surface, the second layer of Pd shows a higher work function than the first layer of Pd. In this image, islands of the first Pd layer are observed on a wide terrace of the Cu substrate, and several small islands of the second Pd layer are observed on them. In the corresponding work function image [Fig. 35(b)], islands of the second layer look brighter than those of the first layer, indicating a higher work function on the second Pd layer than on the first Pd layer. According to their statistical analysis, the work function of the second layer Pd is larger than the first layer Pd by 6 ± 5%.

Quantitative analysis shows that the work function measured for the first Pd layer is already larger than that of bulk Pd(111), and it further increases with increasing thickness of Pd. This kind of overshooting of LWF measured for Pd film could be the quantum size effect on the work function since film thickness of the overlayers is smaller than the Fermi wavelength of the metals [68].

In the LWF images obtained from both the Au/Cu(111) surface [Fig. 34(b)] and Pd/Cu(111) surface [Fig. 35(b)] dark valleys along steps can be observed, indicating that the LWF at steps is much lower than that on terraces. This agrees with the fact that the work function decreases with increasing step density [70]. To show more details, a line scan crossing a step that separates two Au terraces is shown with a solid line in Fig. 36(a). According to the statistic based on more than 100 WF images like Fig. 34(b), the mean full width at half maximum and depth of the LWF valley for Au-Au monatomic steps are 6.5 ± 1 Å and 0.9 ± 0.3 eV, respectively, while for Cu-Cu monatomic steps are 10 ± 1 Å and 1.9 ± 0.3 eV, respectively.

Dipoles induced by Smoluchowski smoothing effect [71] at steps are very likely responsible for the formation of such LWF valleys. A simple simulation has thus been

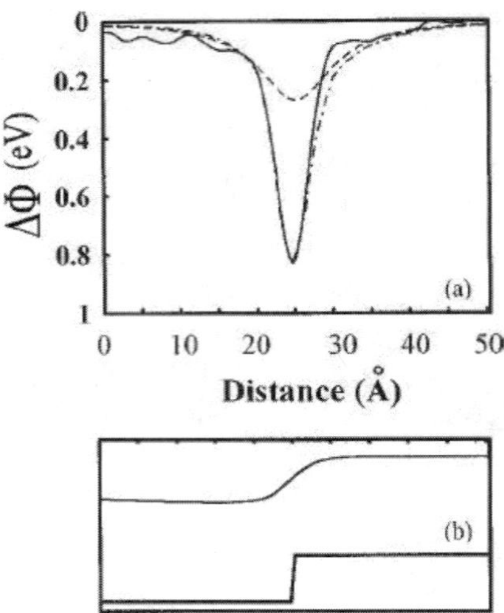

Figure 36. (a) Comparison of the experimental local work-function profile crossing an Au-Au monatomic step (solid line) with its simulated counterpart (dashed-dotted line). The reduction induced by the step dipoles alone is also shown (dashed line). (b) STM line scan (top) obtained simultaneously with the solid line in (a), and the schematic step profile showing the location of the step. (From Ref. 65)

made accordingly, where an infinite row of equal dipoles is used to simulate a step. If a right-hand coordinate system is set such that the axis of the dipole row lays long the y-coordinate axis with the positive end of the dipoles pointing to the outside of the surface, i.e., the $+z$ direction, then the local surface potential at a point (x, z) induced by the dipole row is given by:

$$\phi_D(x, z) = \left(\frac{Q}{4\pi\varepsilon_0}\right) \ln \frac{(z + l/2)^2 + x^2}{(z - l/2)^2 + x^2}, \tag{16}$$

where Q is the linear density of charge, l is the distance between the positive and negative charges. Let d be the spacing of the step atoms; then the induced dipole moment can be calculated as $\mu = Qdl$ per step atom. Since it is the constant-current rather than the constant height mode that was used in the experiment, to simulate the LWF line scan shown in Fig. 36(a) with Eq. 16, what they have to calculate is not $\phi_D(x, z_0)$ but $\phi_D[x, z(x)]$, where $z(x)$ is the real STM line scan [the top curve in Fig. 36(b)], along which the LWF is probed. Note that the line scan is quite different from the schematic step profile, which is expected to be more like the profile of the real step. The reason for this is twofold: Smoluchowski smoothing [71] as mentioned above, and obviously, convolution with the tip. However, as pointed out by Binnig and Rohrer [72], the step topography has one more effect on the measured values of LWF because what is measured, as mentioned above, is the response $d(\ln I)$ to the modulation of the gap distance ds, which ought to be in the normal direction of the surface. If at a point the normal of the surface is not in the z direction but tilted away by an angle θ

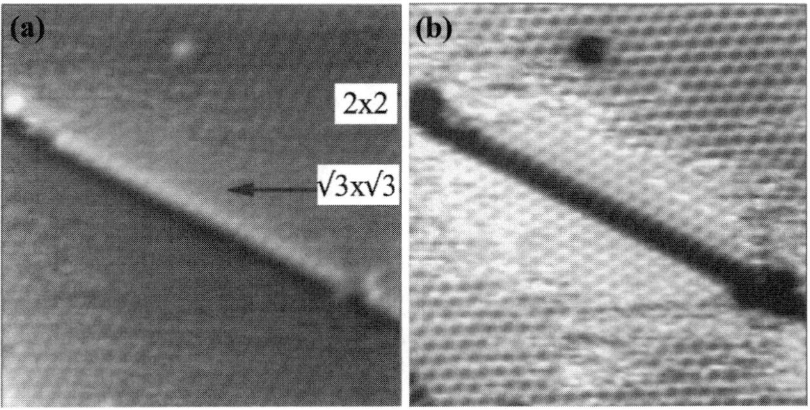

Figure 37. Atomic resolved STM (a) and work function (b) images on S/Pt (111) surface. The scanning area is 120 Å × 120 Å, containing both 2 × 2 and $\sqrt{3} \times \sqrt{3}$ reconstruction. It is demonstrated that the $\sqrt{3} \times \sqrt{3}$ structure has larger work function than 2 × 2 structure. (From Ref. 73)

then the real *ds* is reduced by a factor of $\cos\theta$ even if the modulation of the tip height *dz* is constant. As a result, the measured local work function of that point is reduced by a factor of $\cos^2\theta$. So, after taking this into account in the simulation, the LWF variation around a step is then calculated as

$$\Delta\phi(x) = \phi_{AT} - [\phi_{AT} - \phi_D(x, z)]\cos^2\theta, \tag{17}$$

where ϕ_{AT} is the measured LWF of the terraces that are separated by the step, and $\phi_D(x, z)$ is the potential of the dipole row along the step and hence is given by Eq. 16. In the calculation the value of θ at each point was determined from the real STM line scan and the distance between the positive and negative charges l was set to be the step height (l has almost no effect on the final results). By optimizing the gap distance and the linear density of dipole moment, a good agreement between the calculated and experimental curves has been achieved. The calculated $\Delta\phi(x)$ and $\phi_D(x, z)$ are shown in Fig. 36(a) as the dotted-dashed and dashed lines, respectively, along with the experimental curve (solid line) for comparison. The tip height is 4.9 Å, and the dipole moment is $\mu = 0.16 \pm 0.05$ D/step atom, in agreement with the value of 0.2–0.27 D/step atom derived by Besoke *et al.* from a stepped Au(111) surface [70]. A similar simulation has also been carried out for Cu-Cu monatomic steps, and the result is $\mu = 0.5 \pm 0.15$ D/step atom, which is about twice as large as that of Au-Au steps [65].

Atomic resolution can also be achieved with LWF measurement. As shown in Fig. 37, atomic resolved STM and LWF images were obtained on S/Pt(111) surface. At this coverage, 2 × 2 and $\sqrt{3} \times \sqrt{3}$ reconstructions coexist on the surface [Fig. 37(a)], the LWF image [Fig. 37(b)] shows that the $\sqrt{3} \times \sqrt{3}$ structure has a larger work function than 2 × 2 structure [73].

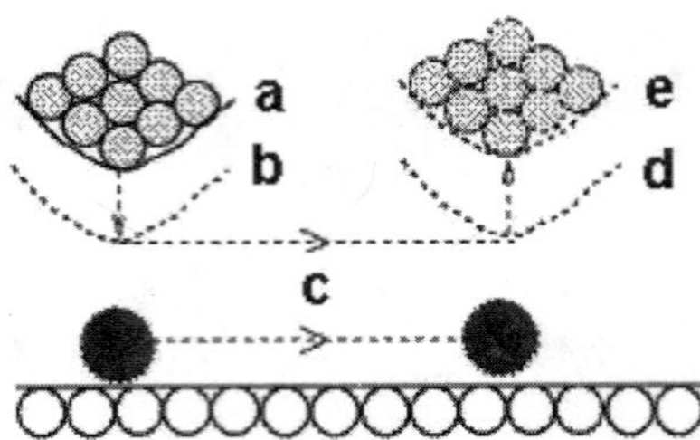

Figure 38. A schematic illustration of the process for sliding an atom across a surface. The atom is located and the tip is placed directly over it (a). The tip is lowered to position (b), where the tip-atom attractive force is sufficient to keep the atom located beneath the tip when the tip is subsequently moved across the surface (c) to the desired destination (d). Finally, the tip is withdrawn to a position (e) where the tip-atom interaction is negligible, leaving the atom bound to the surface at a new location. (From Ref. 74)

It has been shown that STM is indeed a powerful technique for measurement of LWF, and that measuring LWF with STM is very useful for elemental identification on metal surfaces. This technique provides unique information on how the atomic structure of a surfaces is related to the work function and thus is very useful for elucidating processes on solid surfaces.

4. STM-BASED ATOMIC MANIPULATION

As discussed above, STM is very powerful in studying atomic structure and electronic properties of various surfaces. In these studies, the tip-sample interaction is usually kept as small as possible so that the investigations are non-destructive. However, if one adjusts the parameters to increase the tip-sample interaction in a controlled way, STM can also be used to fabricate nano-structures down to the atomic level. Various nano-structures can be constructed by different methods, including manipulation of single atoms [74], scratching [75], oxidation [76], tip-induced chemical reactions [77–78], heating [79] and *etc.* [4]. Below, we will introduce some of them.

4.1. Manipulation of Single Atoms

Eigler and colleagues at IBM succeeded in writing "IBM" with xenon atoms in 1990 and pioneered the new field of manipulation of single atoms [74]. Toggling a single atom and pulling/pushing it on a surface were first demonstrated on the Xe adsorbed Ni(110) surface using a low-temperature UHV STM. The process to move an adsorbed Xe atom is shown in Fig. 38. The STM scanning is first stopped and the tip is placed directly above the atom (a). Then lower the tip toward the atom to increase the tip-atom interaction (b); this is achieved by changing tunneling current to a higher value (typically ~30 nA). This step is critical, the tip-atom interaction has to be strong enough to allow the atom to overcome the energy barrier to slide to neighboring place on the substrate. On the other hand, the tip-atom interaction has to be smaller than the interaction between atom and substrate so that the atom cannot be transfer from

Figure 39. "Quantum corral" built with 48 Fe atoms on Cu(111) surface. (From Ref. 80)

substrate to tip. The tip (dragging the atom together) is then moved under closed-loop conditions to the desired destination slowly (c) and stops there (d). Finally, the tip is withdrawn by reducing the tunneling current to the value used for imaging (~1 nA) and leaving the atom at the destination.

By repeating this procedure to position other adsorbed atoms, structures of ones own design can be fabricated atom by atom. Using this method, "quantum corrals" was built with 48 Fe atoms on Cu(111) surface [80]. As shown in Fig. 39, the interference effects of electron waves can be clearly observed in the corral. From the dependence of periodicity of the wave on bias voltage, they could determine the effective mass of electrons in Cu(111) surface states to be about 0.37 m_e(m_e, the mass of a free electron), which is in good agreement with the value obtained by other techniques [80].

Recent, the "quantum mirage" effect was demonstrated using an elliptical corral built with Co atoms on Cu(111) surface [81]. Conventional image projection relies on classical wave mechanics and the use of natural or engineered structures such as lenses or resonant cavities. This work demonstrates that the electronic structure surrounding a magnetic Co atom can be projected to a remote location on the Cu(111) surface; electron partial waves scattered from the real Co atoms are coherently refocused to form a spectral image or "quantum mirage". The focusing device is an elliptical quantum corral, assembled on the Cu surface. The corral acts as a quantum mechanical resonator, while the two-dimensional Cu surface state electrons form the projection medium. When placed on the surface, Co atoms display a distinctive spectroscopic signature, known as the many-particle Kondo resonance, which arises from their magnetic moment. Fig. 40 shows that when a magnetic cobalt atom is placed at a focus point of elliptical corrals (a, b), some of its properties also appear at the other focus (c, d), where no atoms exists. When the interior Co atom is moved off focus, the mirage vanishes. Over 20 elliptical resonators of varying size and eccentricity were made to search for the formation of a quantum mirage. It was found that as *a* (the semimajor axis length) is increased monotonically while *e* (eccentricity) is fixed, the mirage

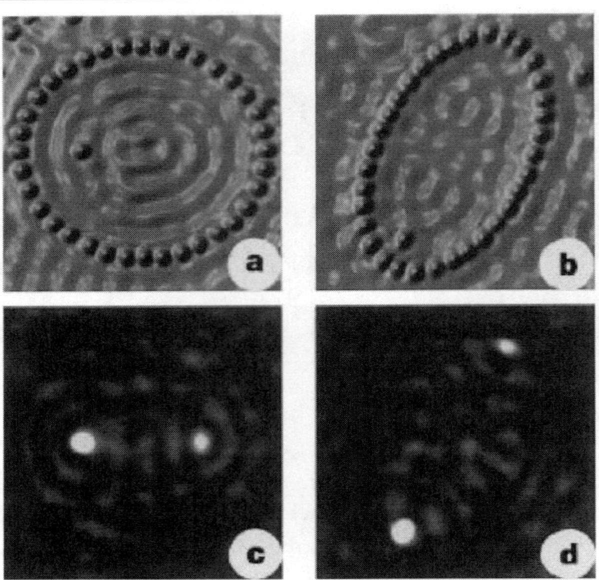

Figure 40. Visualization of the quantum mirage. a, b, Topographs showing the $e = 1/2$ (a) and $e = 0.786$ (b) ellipse each with a Co atom at the left focus. c, d, Associated dI/dV difference maps showing the Kondo effect projected to the empty right focus, resulting in a Co atom mirage. (From Ref. 81)

is switched on and off. In each period of this switching, the classical path length $2a$ changes by a half Fermi wavelength [81].

Because the quantum mirage effect projects information using the wave nature of electrons rather than a wire, it has the potential to enable data transfer within future nanometer scale electronic circuits so small that conventional wires do not work.

4.2. STM Induced Chemical Reaction at Tip

The finely focused electron beam from STM tip can also be used to induce local chemical reaction, which provides another method to fabricate various pre-designed nano-structures on the surface.

In 1992, Dujardin *et al.* demonstrated that individual $B_{10}H_{14}$ molecule adsorbed on Si(111)7 × 7 surface could be dissociated by electrons emitted from STM tip at a bias voltage of 8 V [77]. In 1997, Stipe *et al.* dissociated single O_2 molecules on the Pt(111) surface in the temperature range of 40 to 150 K using tunneling current from an STM tip [82]. Fig. 41 shows that two O_2 molecules are dissociated by voltage pulses of 0.3 V. The dissociation rate as a function of current was found to vary as $I^{0.8\pm0.2}$, $I^{1.8\pm0.2}$, and $I^{2.9\pm0.3}$ for sample biases of 0.4, 0.3, and 0.2 V, respectively. These rates are explained using a general model for dissociation induced by intramolecular vibrational excitations via resonant inelastic electron tunneling [82].

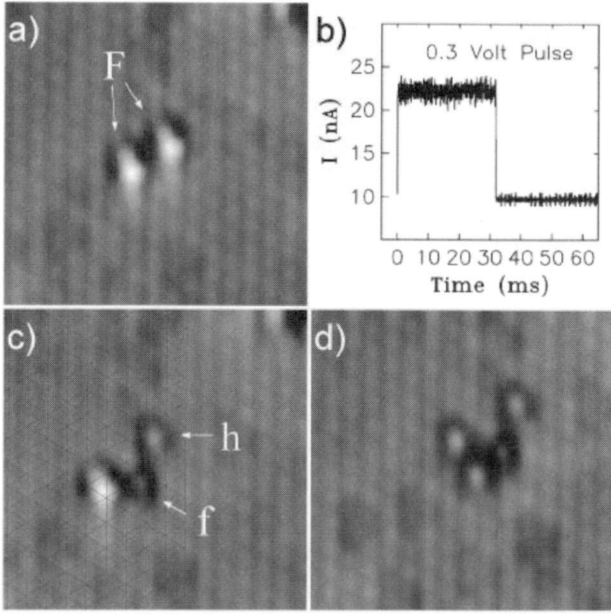

Figure 41. (a) STM image of two adjacent pear shaped O2 molecules on fcc sites. (b) Current during a 0.3 V pulse over the molecule on the right showing the moment of dissociation (step at $t \sim 30$ ms). (c) After pulse image with a grid fit to the platinum lattice showing one oxygen atom on an fcc and one on an hcp site along with the unperturbed neighboring molecule on an fcc site. (d) STM image taken after a second pulse with the tip centered over the molecule showing two additional oxygen atoms on hcp sites. Raw data images scanned at 25 mV sample bias and 5 nA tunneling current. (From Ref. 82)

Recently, it was demonstrated that with an STM in a controlled step-by-step manner utilizing a variety of manipulation techniques, all elementary steps of a complex chemical reaction can be induced on individual molecules and new individual molecules can be synthesized [83]. The reaction steps involve the separation of iodine from iodobenzene by using tunneling electrons, bringing together two resultant phenyls mechanically by lateral manipulation and, finally, their chemical association to form a biphenyl molecule mediated by excitation with tunneling electrons. The reaction process is schematically illustrated in Fig. 42.

The first reaction step, iodine abstraction from iodobenzene [Figs. 42(a) and 1(b)], was performed by positioning the STM tip right above the molecule at fixed height and switching the sample bias to 1.5 V for several seconds. From the linear dependence of the dissociation rate on the tunneling current, they concluded that the energy transfer from a single electron causes the breaking of the C-I bond [83]. As shown in Fig. 43, after dissociation (a–b), the iodine and phenyl are spaced closely (c). So, the iodine atoms were pulled by the tip to further separate them from the phenyls (d). To clear the manipulation path (e), the iodine atom located between the two

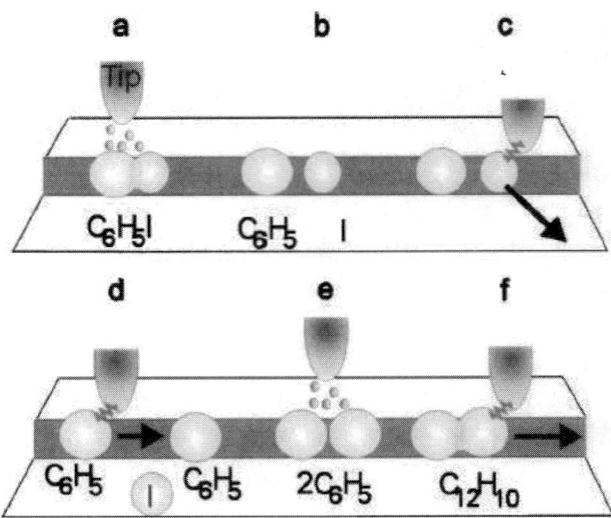

Figure 42. Schematic illustration of the STM tip-induced synthesis steps of a biphenyl molecule. (a), (b) Electron-induced selective abstraction of iodine from iodobenzene. (c) Removal of the iodine atom to a terrace site by lateral manipulation. (d) Bringing together two phenyls by lateral manipulation. (e) Electroninduced chemical association of the phenyl couple to biphenyl. (f) Pulling the synthesized molecule by its front end with the STM tip to confirm the association. (From Ref. 83)

phenyls is removed onto the lower terrace. Lateral manipulation was continued until two phenyls were located close to each other (f). The shortest achievable distance between the centers of two phenyls is 3.9 ± 0.1 Å, as determined from the STM images. Even though the two phenyls are brought together spatially they do not join at the temperature of 20 K unless further measures are taken. To induce the last reaction step, association, molecular excitation by inelastic tunneling was used. The STM tip was stopped right above the center of the phenyl couple and the bias was raised to 500 mV for 10 s. Then the voltage was reduced to its original value of 100 mV and the STM tip continued scanning. The distance between the phenyl centers changes upon association with 4.4 ± 0.05 Å, which is consistent with the distance of 4.3 Å between the two centers of the p rings in gas-phase biphenyl [83].

This work opens up new fascinating routes to the individual assembly of novel man-designed molecules or construction of nanoscale molecular-electronic and molecular-mechanical devices from a variety of building blocks which might also be prepared *in situ*.

More recently, Moresco *et al.* showed that STM tip could be used to rotate single legs of a single Cu-tetra-3,5 *di-terbutyl-phenyl porphyrin* (Cu-TBPP) molecule in and out of the porphyrin plane in a reversible way on a stepped Cu(211) surface [84] and they found the internal configuration modification drastically changed the tunneling current passing through the molecule. This work demonstrated that the controlled

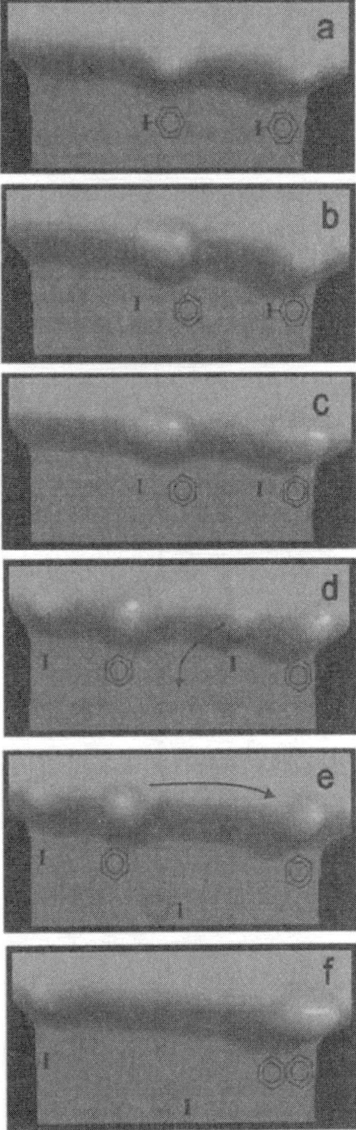

Figure 43. STM images showing the initial steps of the tip-induced Ullmann synthesis. (a) Two iodobenzene molecules are adsorbed at a Cu(111) step edge. (b),(c) Iodine is abstracted from both molecules using a voltage pulse. (d) Iodine atoms (small protrusions) and phenyl molecules (large) are further separated by lateral manipulation. (e) The iodine atom located between the two phenyls is removed onto the lower terrace to clear the path between the two phenyls. (f) The phenyl molecule at the left side is moved by the STM tip close to the right phenyl to prepare for their association. (Image parameters: +100 mV, 0.53 nA; 70×30 Å^2.) (From Ref. 83)

rotation of the legs induced by the STM tip realizes the principle of a conformational molecular switch [84].

5. RECENT DEVELOPMENTS

In traditional surface analysis techniques, the sample is probed by means of electrons, photons, ions, and other particles with a spatial resolution determined by the spatial extent of the probe beams. Therefore, atomic resolution is very difficult to achieve with the conventional techniques. In contrast, with atomic-resolution, STM is based on a totally different principle, in which a local probe (very sharp tip), precise scanning, and an electronic feedback are combined subtly. To achieve the atomic resolution, the tip is brought very close to the sample, in near-field regime, and is controlled precisely by monitoring the tunneling current. Following the basic idea of STM, many novel scanning probe microscopes have emerged based on the piezoelectric scanning, feedback control and various interactions between probe tip and sample. Some important techniques have been summarized in Table I by Wickramasinghe [5]. Below, some current developments will be reviewed.

5.1. Spin-Polarized STM (SPSTM)

In the STM/STS discussed above, the spin of the tunneling electrons has not been considered. If a magnetic tip is used, the tunneling current will be spin-dependent. This means that the STM tip is sensitive to the spin of the tunneling electrons and can be used to investigate the magnetism structure of a sample with high spatial resolution. This idea was first proposed by Pierce in 1988 [85] and it eventually led to the invention of the spin-polarized STM (SPSTM).

For SPSTM, a magnetic tip is required to provide a highly efficient source or detector for spin-polarized (SP) electrons. The ideal tip for SPSTM must meet several conditions: First of all, the apex atom must exhibit a high spin polarization in order to achieve a good signal-to-noise ratio. Second, dipolar interaction between tip and sample due to the stray fields should be as low as possible because it may modify or destroy the intrinsic domain structure of the sample. Third, in order to separate magnetic from topographic and electronic contributions to the tunnel current it should be possible to reverse the quantization axis periodically. Finally, in order to be able to image the domain structure of any sample-no matter whether its easy axis is in-plane or out-of-plane, one should be able to control the orientation of the quantization axis of the tip parallel or perpendicular to the sample surface [86]. Several possible tip materials have been discussed in Ref. 4. The details on how to prepare an SPSTM tip can be found in Ref. 86. In the following, some applications of SPSTM are reviewed.

Using CrO_2 tip and a Cr(001) sample, Weisendanger *et al.* observed the vacuum tunneling of SP electrons in SPSTM for the first time in 1990 [87]. The topological antiferromagnetism of the Cr(001) surface with terraces alternately magnetized in opposite directions and separated by monatomic steps provides an ideal test structure for SPSTM experiments. With a normal nonmagnetic W tip, the monatomic step

Table I. SXM Techniques and Capabilities (From Ref. 5)

1. Scanning Tunneling Microscope (1981)
   – G. Binnig, H. Rohrer
   – Atomic resolution images of conducting surfaces
2. Scanning Near-Field Optical Microscope (1982)
   – D. W. Pohl
   – 50 nm (lateral resolution) optical images
3. Scanning Capacitance Microscope (1984)
   – J. R. Matey, J. Blanc
   – 500 nm (lat. res.) images of capacitance variation
4. Scanning Thermal Microscope (1985)
   – C. C. Williams, H. K. Wickramasinghe
   – 50 nm (lat. res.) thermal images
5. Atomic Force Microscope (1986)
   – G. Binning, C. F. Quate, Ch. Gerber
   – Atomic resolution on conducting/nonconducting surfaces
6. Scanning Attractive Force Microscope (1987)
   – Y. Martin, C. C. Williams, H. K. Wickramasinghe
   – 5 nm (lat. res.) non-contact images of surfaces
7. Magnetic Force Microscope (1987)
   – Y. Martin, H. K. Wickramasinghe
   – 100 nm (lat. res.) images of magnetic bits/heads
8. "Frictional" Force Microscope (1987)
   – C. M. Mate, G. M. McClelland, S. Chiang
   – Atomic-scale images of lateral ("frictional") forces
9. Electrostatic Force Microscope (1987)
   – Y. Martin, D. W. Abraham, H. K. Wickramasinghe
   – Detection of charge as small as single electron
10. Inelastic Tunneling Spectroscopy STM (1987)
    – D. P. E. Smith, D. Kirk, C. F. Quate
    – Phonon spectra of molecules in STM
11. Laser Driven STM (1987)
    – L. Arnold, W. Krieger, H. Walther
    – Imaging by non linear mixing of optical waves in STM
12. Ballistic Electron Emission Microscope (1988)
    – W. J. Kaiser (1988)
    – Probing of Schottky barriers on nm scale
13. Inverse Photoemission Force Microscope (1988)
    – J. H. Coombs, J. K. Gimzewski, b. Reihl, J. K. Sass, R. R. Schlittler
    – Luminescence spectra on nm scale
14. Near Field Acoustic Microscope (1989)
    – K. Takata, T. Hasegawa, S. Hosaka, S. Hosoki, T. Komoda
    – Low frequency acoustic measurements on 10 nm scale
15. Scanning Noise Microscope (1989)
    – R. Moiler, A. Esslinger, B. Koslowski
    – Tunneling microscopy with zero tip-sample bias
16. Scanning Spin-precession Microscope (1989)
    – Y. Manassen, R. Hamers, J. Demuth, A. Castellano
    – 1 nm (lat. res.) images of paramagnetic spins
17. Scanning Ion-Conductance Microscope (1989)
    – P. Hansma, B. Drake, O. Marti, S. Gould, C. Prater
    – 500 nm (lat. res.) images in electrolyte

(continued)

Table I. (continued)

18. Scanning Electrochemical Microscope (1989)
 –O. E. Husser, D. H. Craston, A. J. Bare
19. Absorption Microscope/Spectroscope (1989)
 –J. Weaver, H. K. Wickramasinghe
 –1 nm (lat. res.) absorption images/spectroscopy
20. Phonon Absorption Microscope (1989)
 –H. K. Wickramasinghe, J. M. R. Weaver, C. C. Williams
 –Phonon absorption images with nm resolution
21. Scanning Chemical Potential Microscope (1990)
 –C. C. Williams, H. K. Wickramasinghe
 –Atomic scale images of chemical potential variation
22. Photovoltage STM (1990)
 –R. J. Hamers, K. Markert
 –Photovoltage images on nm scale
23. Kelvin Probe Force Microscope (1991)
 –M. Nonnenmacher, M. P. O'Boyle, H. K. Wickramasinghe
 –Contact potential measurements on 10 nm scale

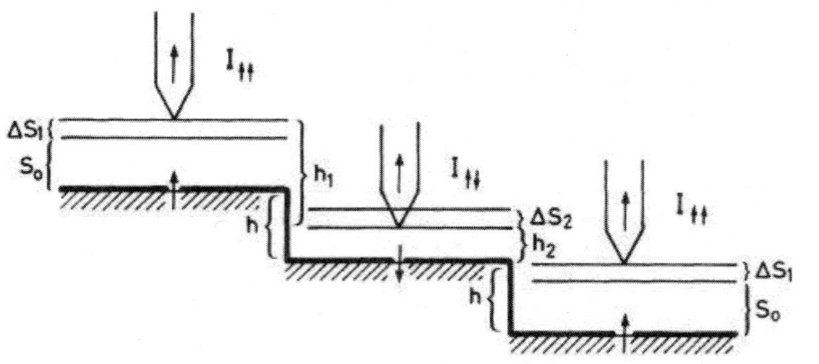

Figure 44. Schematic drawing of a ferromagnetic tip scanning over alternately magnetized terraces separated by monatiomic steps of height h. An additional contribution from SP tunneling leads to alternating step heights $h_1 = h + \Delta s_1 + \Delta s_2$ and $h_2 = h - \Delta s_1 - \Delta s_2$. (From Ref. 87)

height of Cr(001) was determined to be 1.49 ± 0.08 Å from STM topographic images, which is in good agreement with 1.44 Å for bcc Cr(001). The CrO_2 tip was prepared in such a way that the preferred magnetization direction of the tip is perpendicular to the sample surface [87]. After replacing the W tip by a CrO_2 tip, a periodic alternation of the measured monatomic step heights between larger and smaller values compared to the mean single step height value of 1.44 Å is observed. The deviation from the single step height value determined with a CrO_2 tip can be as large as ±15% which is much larger than the experimental error with a nonmagnetic W tip. An additional contribution from SP tunneling was employed to explain the periodic alternation of the monatomic step height values. As sketched in Fig. 44, assuming that the CrO_2 tip is first scanning over a terrace with the same direction of magnetization as the CrO_2

tip, the tunneling current $I_{\uparrow\uparrow}$ will then be increased due to a contribution from SP tunneling: $I_{\uparrow\uparrow} = I_0(1 + P)$, where I_0 is the tunneling current without this contribution and P is the effective spin polarization of the tunneling junction. Since the STM is operated at constant current, an additional contribution to the tunneling current leads to a corresponding increase Δs_1 of the mean distance s_0 between the tip and the sample surface. If the CrO_2 tip is scanning over a terrace with the opposite direction of magnetization, the tunneling current $I_{\uparrow\downarrow}$ will be decreased: $I_{\uparrow\downarrow} = I_0(1-P)$, leading to a corresponding decrease Δs_2 of the tip-sample distance. The measured single step height values therefore alternate between $h_1 = h + \Delta s_1 + \Delta s_2$ and $h_2 = h - \Delta s_1 - \Delta s_2$, where h is the topographic monatomic step height.

The effective polarization of the tunneling junction is given by [Ref. 87]:

$$P = \frac{I_{\uparrow\uparrow} - I_{\uparrow\downarrow}}{I_{\uparrow\uparrow} + I_{\uparrow\downarrow}} = \frac{\exp(A\sqrt{\phi}\Delta s_1) - \exp(-A\sqrt{\phi}\Delta s_2)}{\exp(A\sqrt{\phi}\Delta s_1) + \exp(-A\sqrt{\phi}\Delta s_2)} = \frac{\exp(A\sqrt{\phi}\Delta s) - 1}{\exp(A\sqrt{\phi}\Delta s) + 1}, \tag{18}$$

where $A \approx 1.025$ eV$^{-1/2}$ Å^{-1}, ϕ is the average local tunneling barrier height, $\Delta s = \Delta s_1 + \Delta s_2$. According to the experimental results, $\Delta s = 0.2 \pm 0.1$ Å, taking $\phi = 4.0 \pm 0.5$ eV, the effective polarization of the tunneling junction P was derived to be $(20 \pm 10)\%$.

The first SPSTM studies of the Cr(001) surface were performed on a nanometer scale. The CrO_2 tips were too blunt to achieve atomic resolution. The first atomic resolution SPSTM experiment was done on $Fe_3O_4(001)$ surface. The different spin configurations of $3d^{5\uparrow}3d^{\downarrow}$ for Fe^{2+} and $3d^{5\uparrow}$ for Fe^{3+} were identified by using a sharp Fe tip prepared *in situ* [88, 89].

Real-space imaging of two-dimensional antiferromagnetism with the atomic resolution was achieved also by Wiesendanger's group in 2000 [90]. The experiment was done on Mn/W(110) surface with an SPSTM at 16 K. A monolayer Mn grows pseudomorphically on W(110) surface. An STM image with W tip is shown in Fig. 45(A). First-principles calculations shows that the so-called c(2 × 2) antiferromagnetic state is energetically favourable and that the magnetocrystalline anisotropy energy favours an in-plane spin orientation. In SPSTM experiments, Fe-coated probe tips was used to fulfill the condition that the experiment required a magnetic tip with a magnetization axis in the plane of the surface. Figure 45(B) shows an STM image taken with such a tip. Periodic parallel stripes along the [001] direction of the surface can be recognized. The periodicity along the [110] direction amounts to 4.5 ± 0.1 Å, which corresponds well to the size of the magnetic c(2 × 2) unit cell. The inset in Fig. 45(B) shows the calculated STM image for the magnetic ground state, i.e., the c(2 × 2)-AFM configuration. The theory and experiment are in a very good agreement [86, 90].

These stuties demonstrate that SPSTM is a powerful technique for understanding of complicated magnetic configurations of nanomagnets and thin films engineered from ferromagnetic and antiferromagnetic materials used for magnetoelectronics.

Recently, it was demonstrated that SPSTM images can be observed with an antiferromagnetic probe tip. The advantage of its vanishing dipole field is most apparent

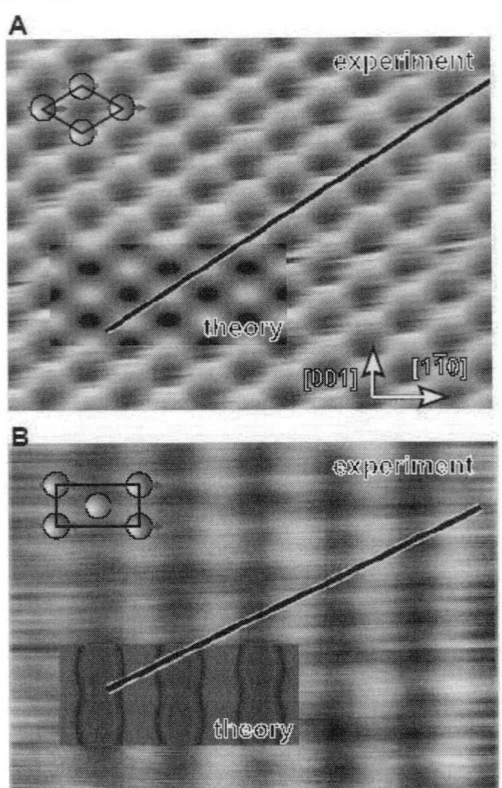

Figure 45. Comparison of experimental and theoretical STM images of a Mn ML on W(110) with (A) a nonmagnetic W tip and (B) a magnetic Fe tip. The unit cell of the calculated magnetic ground-state configuration is shown in (A) and (B) for comparison. Tunneling parameters for both images are $I_t = 40$ nA and $U = 23$ mV. The image size is 2.7 nm by 2.2 nm. (From Ref. 90)

in external magnetic fields. This new approach resolves the problem of the disturbing influence of a ferromagnetic tip in the investigation of soft magnetic materials and superparamagnetic particles [91].

In order to overcome the difficulties of separating topographic, electronic, and magnetic information one may measure the local differential conductivity dI/dV with a magnetic tip [86]. In Fig. 46, the dI/dV spectra measured on Gd(0001) at a sample temperature $T = 170$ K is compared with (inverse) photoemission spectroscopy (IPES) data on a similar sample at the same temperature. It is known from previous experiments that the Gd(0001) surface state is exchange-split into a filled majority and an empty minority spin contribution. While the occupied majority spin part (↑) appears as a peak in the PES data the unoccupied minority spin part (↓) is observed in IPES measurements. Indeed, tunneling dI/dV spectra exhibit a peak at a sample bias value of $V = +430$ mV and a shoulder at $V = -200$ mV (Fig. 46, top panel), being in

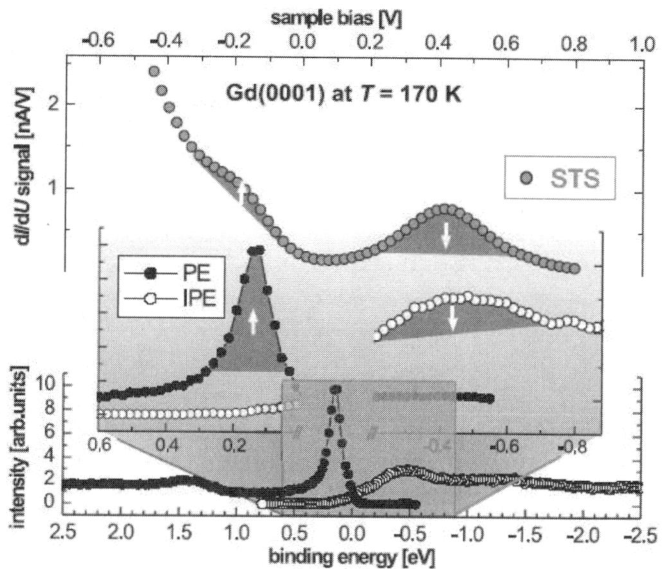

Figure 46. PE (left) and IPE (right) spectra of the spin-split Gd(0001) surface state measured at 170 K (bottom panel). The occupied part of the surface state appears in the PES while the empty part is weakly visible in the IPES (bottom and inset). In contrast, tunnelling spectroscopy allows the measurement of occupied and empty electronic states within a single experiment (top). The peak position derived with both experimental techniques correspond well. (From Ref. 86)

good agreement with the binding energies for the empty and the occupied parts of the surface state as determined by PES and IPES (see Fig. 46, bottom panel).

dI/dV mapping with nanometer resolution can be also performed with a magnetic tip. Fig. 47 shows the STM topographic image and spatially resolved dI/dV images measured at $T = 70$ K with a W tip coated with 5–10 ML Fe on a sample prepared by depositing 10 ML of Gd on the W(110) substrate held at 530 K. This preparation procedure leads to partially coalesced Gd islands ($\theta_{loc} \approx 20$ ML) with a single Gd wetting layer on the W(110) substrate as shown in the STM image of Fig. 47(a). Fig. 47(b) and (c) show dI/dV images at $V = -0.2$ V and $+0.45$ V, i.e. sample biases which correspond to filled and empty parts of the surface state, respectively, measured within the box of Fig. 47(a). Both images show a domain wall crossing the island from top to bottom [86, 92]. This work demonstrates that the domain structure of Gd(0001) islands with a resolution below 20 nm can be imaged by dI/dV mapping with a magnetic tip.

Recently, Wulfhekel and Kirschner showed that magnetic contrast could be obtained in a similar way of measuring the tunnel magnetoresistance (MR) of planar junctions [93]. By applying an alternating current of frequency f through a small coil wound around the magnetic tip, the longitudinal magnetization of the tip was switched periodically. The tip material, a metallic glass, was chosen to have a low coercivity, vanishing

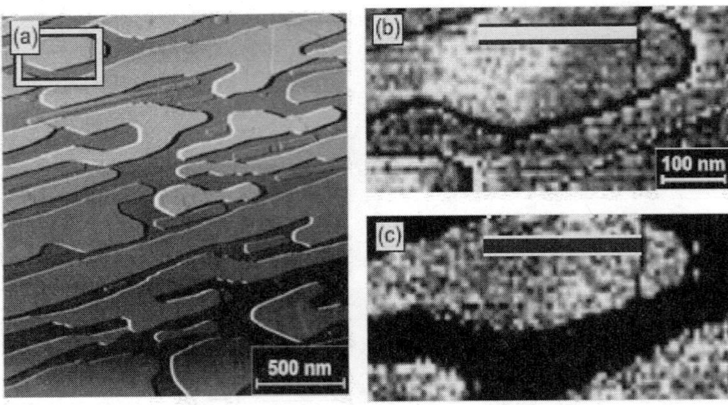

Figure 47. (a) Topographic image of 10 ML Gd(0001)/W(110). dI/dU maps measured on the island indicated by the box in (a) at (b) $U = -0.2$ V and (c) $U = +0.45$ V, i.e. the peak position of the majority and minority parts of the surface state, respectively. In (b) the left part of the island appears brighter (high conductivity) than the right part (low conductivity). In (c) the contrast is reversed. (From Ref. 86)

magnetostriction, low saturation magnetization, and low magnetization losses. These parameters allow a rapid switching of the magnetization of the tip without mechanical vibrations of the tip due to magnetostriction or magnetization losses. Furthermore, they minimize the influence of the field of the coil on the sample magnetization. The frequency f was chosen far away from any mechanical resonances of the STM and well above the cutoff frequency of the feedback loop. Variations of the tunnel probability due to the magnetotunnel effect, i.e., maximal probability for parallel and minimal for antiparallel orientation between tip and sample magnetization, result in variations of the tunnel current with the frequency f. These variations were detected with a lock-in amplifier. Since the tip is magnetized along its axis and perpendicular to the sample surface, sensitivity for the perpendicular magnetic component of the sample was obtained [93]. In contrast to the previously described spectroscopy of the differential conductivity dI/dV, which requires that different domains are simultaneously visible in a single image this dI/dm_T method allows the identification of a magnetic contrast even if the sample is in a single-domain state. Since the local MR method allows a rather direct detection of the sample's domain structure a detailed knowledge of the spin-averaged electronic structure of the surface under investigation is no longer required [86, 93].

The high spatial resolution and surface magnetic sensitivity of SPSTM allow it be a powerful tool to study the unsolved basic magnetic problems. There is no doubt that SPSTM will play a major role in the field of magnetic microscopy in the following years.

5.2. Ultra-Low Temperature (ULT)-STM

Ultra-low temperature STM (ULTSTM) is an important direction in the development of scanning tunneling microscopes. STM working in the millikelvin temperature range allows the study of physical phenomena that only occur at very low temperatures, for example, superconducting phase transitions in heavy fermion materials. Very low temperatures can also dramatically improve the energy resolution in the STM measurements. In, 1999, Pan *et al.* succeeded in building an ULTSTM for operation in a magnetic field with very high spatial and spectroscopic resolution [94].

Although low temperatures bring the benefits of low thermal drift and low thermal noise, which are required for high-resolution measurements, ultra-low temperature refrigeration techniques often hamper the efforts to achieve high-resolution measurements due to the introduction of mechanical vibrations, e.g. vibrations due to evaporation of the liquid, and from pumps. Furthermore, the physical space within the cryostat, especially when a high magnetic field is required, is often too limited to allow an effective cryogenic vibration-isolation stage. Therefore, to design an ULTSTM which can achieve atomic resolution is challenging. These challenges were overcome by the efforts in the following three elements: (1) a very rigid STM head that is less susceptible to vibration, (2) a refrigeration scheme that has very low intrinsic vibrational noise, and (3) a good external vibration-isolation system to reduce transmission of vibrations from the external environment to the STM cryostat. Finally, they demonstrated that the ^{3}He refrigerator based very low temperature STM they constructed can reliably operate at temperatures down to 250 mK and in magnetic fields of up to 7 T with high spatial and spectroscopic resolution.

Recently, a dilution-refrigerator-based STM with sample temperatures of 20 mK was achieved [95]. The unconventional superconductor, $Sr_2Ti_xRu_{1-x}O_4$ with $x = 0.00125$, was studied with this ULTSTM. Fig. 48(a) shows an atomic resolution topographic image of the SrO plane of Ti-doped Sr_2RuO_4 with the square lattice of Sr atoms clearly visible. The four dark spots correspond to the Ti atoms replacing the Ru atoms one layer below the surface. A complicated gap-like structure in local density of states was measured at all locations on the surface, with some modifications caused by the Ti atoms. The superconducting gap was not clearly visible, possibly due to surface termination effects, but other gap-like structures were found at $\sim$5 and $\sim$50 meV [95].

5.3. Dual-tip STM

Dual-tip STM (DTSTM) was first suggested by Niu *et al.* [96] in 1995. Normal single-tip STM can only probe static properties of electronic system, the transport properties are out of its capability. A DTSTM can solve this problem easily. Niu *et al.* also proposed that a DTSTM can be applied to: (1) deduce useful information about the band structure of surface states; (2) measure scattering phase shifts of surface defects; (3) observe transition from ballistic to diffusion transport to localization; and (4) measure inelastic mean free paths [96].

Actually, Tsukamoto *et al.* had built a twin-tip STM in 1991. However, the two tips were fixed together and could not scan independently [97]. In 2001, Watanabe *et al.* reported that they had constructed a DTSTM using multiwall carbon nanotubes (NT) as STM probes. They also developed an active damper system for DTSTM to reduce

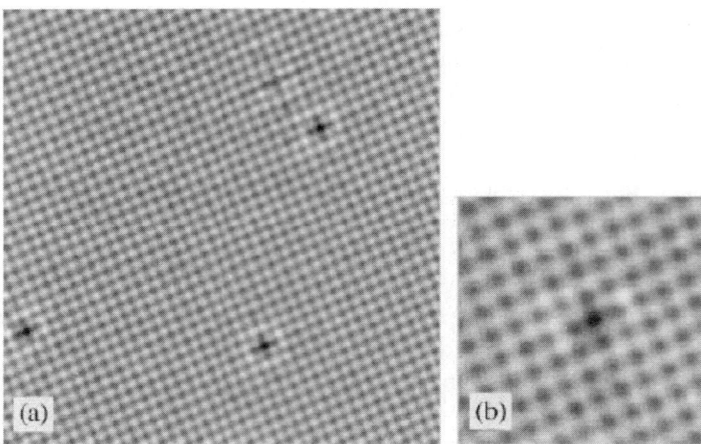

Figure 48. Topographic image of $Sr_2Ti_xRu_{1-x}O_4$ where $x = 0.00125$. The white lattice is the SrO plane, and the black cross-like objects are the Ti atoms located one layer below the surface. (a) 120 Å x 120 Å image, (b) enlarged (35 Å square) view of single impurity. (Images taken at 0.1 nA and −100 mV). (From Ref. 95)

mechanical vibration, which is dominated by the characteristic vibration of the two-probe system. The DTSTM allows to elucidate the electric property of a sample with a spatial resolution of ~1 nm. Using this system, the current–voltage curves of a single NT ring have been measured. The electrode configuration is shown in Fig. 49(a). Fig. 49(b) shows a DTSTM constant current image of the NT ring and the first probe. The I–V curves for various V_G (0, 1, 2, 3, 4, and 5 V, respectively) measured by DTSTM at room temperature in dry-N_2 atmosphere are shown in Fig. 49(c). These results show that the small NT ring on the Si substrate is a field-effect transistor having sharp switching behavior and the possibility of nanometer-scale electronic circuits composed of NT devices [98].

Also in 2001, an ultra-high vacuum DTSTM with two mechanically and electrically independent probes was built by Boland's group [99].

Meanwhile, an ultrahigh vacuum compatible cryogenic DTSTM was constructed by Chen's group [100]. The microscope is attached at the bottom of a low-loss liquid helium Dewar and can be operated down to 4.2 K. The two tips can be manipulated independently and positioned as close as one desires limited only by their radius of curvature. The coarse positioning system consists of five linear steppers driven by piezo-tubes. The displacement of each stepper can be monitored by its own embedded capacitive position sensor with a submicron resolution, thus allowing accurate control of the tip navigation process. An alignment procedure, using a specimen made of three mutually nonparallel planes, is introduced to bring the two tips into overlapped scan ranges without the help of an additional guiding device such as an electron microscope. The overall system exhibits good mechanical rigidity and atomic resolution

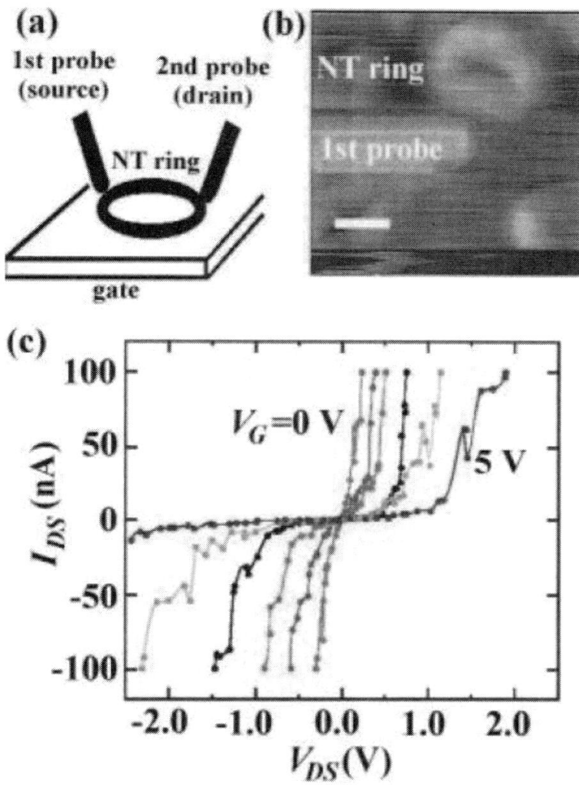

Figure 49. (a) Electrode configuration: two NT probes were connected with a NT ring on the poly-Si(~3.5 nm, *n*-type)/SiO_2 (~2 nm)/Si(100)/Au substrate, where the Au layer acted as a gate electrode. (b) The DTSTM constant current image of the NT ring and the first probe (scale bar, 10 nm). The image was recorded by scanning the second probe. (c) *I–V* curves for various V_G measured by DTSTM at room temperature in dry-N_2 atmosphere. V_G from left to right are 0, 1, 2, 3, 4, and 5 V, respectively. (From Ref. 98)

has been achieved with either tip. This instrument is well suited for investigating low temperature quantum properties of atomically clean nanostructures in a three-terminal configuration [100].

We have reason to believe that, with the fast development of DTSTMs, they will play more important roles in the research of surface science, nano-materials, nano-devices and *etc.*

5.4. Variable Temperature Fast-Scanning STM

STM has been proved to be very successful in studies of the static structural or electronic properties of surfaces. However, it is often desirable to investigate dynamic processes, i.e., surface diffusion, chemical reactions, nucleation and growth, or phase

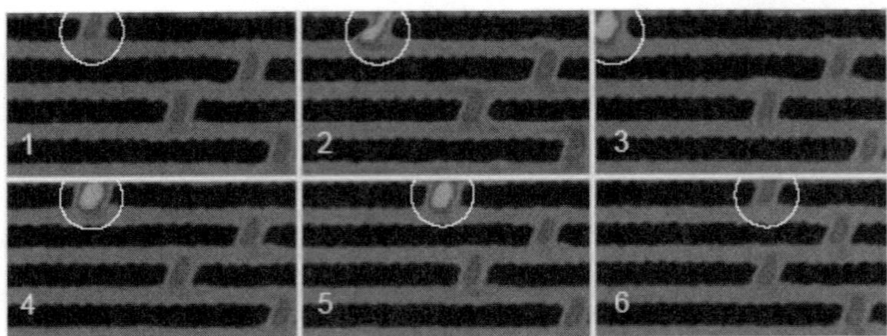

Figure 50. Six STM images (56 × 31 Å^2) extracted from a typical STM movie. Shown is the development of a normal Pt adatom into a bright Pt-H intermediate complex, and then back to a normal Pt adatom again. All STM images were obtained in the constant-current mode with tunnel resistances above 100 MΩ in which case the influence of the tip was found to be negligible. The Pt adatom of interest is marked by a white circle. (From Ref. 110)

transformations. Therefore, high-speed STM is needed for these applications. Since temperature influences the rate of such kinetic processes strongly, control of the sample temperature may allow one to adjust the rate for the processes to the accessible time scale of the STM. Thus, a variable temperature fast-scanning STM is a unique tool to observe dynamic processes. Many variable temperature fast-scanning STMs have been developed in various groups [101–108] with the scanning speed up to 20 frames/s [103], and temperature range from 25 K to 900 K (each instrument can only change temperature in a different smaller range). To achieve the high scanning speed, constant-height scanning mode is often used [103].

A lot of successful applications have been performed with variable temperature fast-scanning STMs.

The one-dimensional diffusion of Pt adatoms in the missing row troughs of the reconstructed Pt(110)(1 × 2) surface is monitored directly from atomically resolved fast-scanning STM images by Besenbacher's group [109]. It is found that not only jumps between nearest neighbor sites but also long jumps, i.e., jumps between next nearest neighbor sites, take place. The hopping rate for these long jumps is found to follow an Arrhenius dependence on temperature. The activation barriers for single and double jumps are determined to be $E_{d1} = -0.81$ eV and $E_{d2} = -0.89$ eV, respectively. This energy difference may be interpreted as a measure of the energy dissipation of the Pt adatoms on the Pt(110)-(1 × 2) surface [109].

Later, they also found that surface self-diffusion of Pt adatoms can be enhanced by adsorbed hydrogen and observed the formation of Pt-H complex which has a diffusivity enhanced by a factor of 500 at room temperature, relative to the other Pt adatoms [110]. As shown in Fig. 50, the formation of intermediate Pt-H complexes has been directly imaged by a fast-scanning STM at 303 K and a hydrogen pressure of 7×10^{-7} mbar. These Pt-H complexes show up in the STM images as brighter Pt adatoms (increase in apparent height ~0.4 Å). After some time, the brighter adatom reverts to a normal brightness. The density functional calculations indicated that the

Pt-H complex consists of a hydrogen atom trapped on top of a platinum atom, and that the bound hydrogen atom decreases the diffusion barrier.

The catalytic oxidation of carbon monoxide (CO) on a platinum (111) surface was studied by Ertl's group with a variable temperature fast-scanning STM [111]. The adsorbed oxygen atoms and CO molecules were imaged with atomic resolution, and their reactions to carbon dioxide (CO_2) were monitored as functions of time. From temperature dependent measurements, they obtained the kinetic parameters, which agree well with the data from macroscopic measurements. In this way, kinetic description of a chemical reaction was achieved that is based solely on the statistics of the underlying atomic processes observed by STM.

Traveling reaction fronts in the oxidation of hydrogen on a Pt(111) surface were also investigated by them [112–113]. These fronts were observed during dosing of the oxygen-covered surface with hydrogen at temperatures below 170 K. The fronts represented 10 to 100 nm wide OH-covered regions, separating unreacted O atoms from the reaction product H_2O. O atoms were transformed into H_2O by the motion of the OH zone. Their investigations revealed the velocity and the width of the fronts as a function of temperature. A simple reaction–diffusion model has been constructed, which contains two reaction steps and the surface diffusion of water molecules, and qualitatively reproduces the experimental observations.

REFERENCES

1. G. Binnig, H. Rohrer, Ch. Gerber, and E. Weibel, Phys. Rev. Lett. **49**, 57 (1982).
2. J. Bardeen, Phys. Rev. Lett. **6**, 57 (1961).
3. J. Tersoff and D. R. Hamann, Phys. Rev. Lett. **50**, 1998 (1983) and Phys. Rev. **B31,** 805 (1985).
4. R. Wiesendanger, "Scanning probe microscopy and spectroscopy-methods and applications" Cambridge University Press, 1994.
5. J. A. Stroscio, W. J. Kaiser, "Scanning tunneling microscopy", Academic Press, 1993.
6. C. J. Chen, "Introduction to scanning tunneling microscopy", Oxford University Press, 1993.
7. R. E. Schlier and H. E. Farnsworth, J. Chem. Phys. **30**, 917 (1959).
8. K. Takayanagi, Y. Tanishiro, M. Takahashi and S. Takahashi, J. Vac. Sci. Technol. **A3**, 1502 (1985).
9. A. T. Cho, J. Appl. Phys., **47**, 2841 (1976).
10. H. H. Farrel and C. J. Palmstrom, J. Vac. Sci. Technol. **B8**, 903 (1990).
11. D. J. Chadi, J. Vac. Sci. Technol. **A5**, 834 (1987).
12. J. E. Northrup, and S. Froyen, Phys. Rev. Lett. **71**, 2276 (1993).
13. T. Hashizume, Q. K. Xue, J. M. Zhou, A. Ichimiya, and T. Sakurai, Phys. Rev. Lett. **73**, 2208 (1994).
14. T. Hashizume, Q. K. Xue, A. Ichimiya, and T. Sakurai, Phys. Rev. **B51**, 4200 (1995).
15. Q. K. Xue, T. Hashizume, J. M. Zhou, T. Sakata, T. Ohno, and T. Sakurai, Phys. Rev. Lett. **74**, 3177 (1995).
16. S. H. Lee, W. Moritz, and M. Scheffler, Phys. Rev. Lett. **85**, 3890 (2000).
17. D. Paget, Y. Garreau, M. Sauvage, P. Chiaradia, R. Pinchaux, and W. G. Schmidt, Phys. Rev. B 64, R161305 (2001); C. Kumpf, L. D. Marks, D. Ellis, D. Smilgies, E. Landemark, M. Nielsen, R. Feidenhans'l, J. Zegenhagen,O. Bunk, J. H. Zeysing, Y. Su, and R. L. Johnson, Phys. Rev. Lett. **86**, 3586 (2001); W. G. Schmidt, F. Bechstedt, and J. Bernholc, Applied Surface Science **190**, 264 (2002); W. G. Schmidt, Appl. Phys. **A 75**, 89 (2002); M. Pristovsek, S. Tsukamoto, A. Ohtake, N. Koguchi, B. G. Orr, W. G. Schmidt, and J. Bernholc, Phys. Stat. Sol. **240**, 91 (2003).
18. S. Nakamura, T. Mukai, M. Senoh, Appl. Phys. Lett. **64**, 1687 (1994).
19. R. Smith, R. M. Feenstra, D. W. Greve, J. Neugebauer, and J. Northrup, Phys. Rev. Lett. **79**, 3934 (1997).
20. A. R. Smith, R. M. Feenstra, D. W. Greve, M.-S. Shin, M. Skowronski, J. Neugebauer, and J. Northrup, Appl. Phys. Lett. **72**, 2114 (1999).

21. Q. K. Xue, Q. Z. Xue, R. Z. Bakhtizin, Y. Hasegawa, I. S. T. Tsong, T. Sakurai, and T. Ohno, Phys. Rev. Lett. **82**, 3074 (1999).
22. V. Ramachandran *et al.,* preceding Comment, Phys. Rev. Lett. **84**, 4014 (2000).
23. A. R. Smith, R. M. Feenstra, D. W. Greve, M. S. Shin, M. Skowronski, J. Neugebauer, and J. E. Northrup, J. Vac. Sci. Technol. B **16**, 2242 (1998).
24. A. R. Smith, R. M. Feenstra, D. W. Greve, M. S. Shin, M. Skowronski, J. Neugebauer, J. E. Northrup, *Surf. Sci.* **423**, 70 (1999).
25. A. R. Smith, V. Ramachandran, R. M. Feenstra, D. W. Greve, A. Ptak, T. Myers, W. Sarney, L. Salamanca-Riba, M. Shin, M. Skowronski, *MRS Internet J. Nitride Semicond Res.* **3**, 12 (1998).
26. J. E. Northrup, J. Neugebauer, R. M. Feenstra, A. R. Smith, Phys. Rev. B **61**, 9932 (2000).
27. C. Adelmann, J. Brault, D. Jalabert, P. Gentile, H. Mariette, Guido Mula, B. Daudin, J. Appl. Phys. **91**, 9638 (2002).
28. Christoph Adelmann, Julien Brault, Guido Mula, Bruno Daudin, Phys. Rev. B **67**, 165419 (2003).
29. A. R. Smith, R. M. Feenstra, D. W. Greve, J. Neugebauer, and J. E. Northrup, Phys. Rev. Lett. **79**, 3934 (1997).
30. J. L. Li, J. F. Jia, X. J. Liang, X. Liu, J. Z. Wang, Q. K. Xue, Z. Q. Li, J. S. Tse, Z. Y. Zhang, and S. B. Zhang, Phys. Rev. Lett. **88**, 066101 (2002).
31. J. F. Jia, X. Liu, J. Z. Wang, J. L. Li, X. S. Wang, Q. K. Xue, Z. Q. Li, Z. Y. Zhang and S. B. Zhang, Phys. Rev. **B66**, 165412 (2002).
32. J. F. Jia, J. Z. Wang, X. Liu, Q. K. Xue, Z. Q. Li, Y. Kawazoe and S. B. Zhang, Appl. Phys. Lett. **80**, 3186 (2002).
33. M. Y. Lai and Y. L. Wang, Phys. Rev. B **64**, 241404 (2001); V. G. Kotlyar, *et al.*, Phys. Rev. B **66**, 165401 (2002); H. H. Chang, *et al.*, Phys. Rev. Lett. **92**, 066103 (2003); K. Wu, *et al.*, Phys. Rev. Lett. **91**, 126101 (2003).
34. G. Binnig, H. Rohrer, Ch. Gerber, and E. Weibel, Surf. Sci. **131**, L379 (1983).
35. V. M. Hallmark, S. Chiang, J. F. Rabolt, J. D. Swallen, and R. J. Wilson, Phys. Rev. Lett. **59**, 2879 (1987).
36. Ch. Wöll, S. Chiang, R. J. Wilson, and P. H. Lippel, Phys. Rev. **B39**, 7988 (1989).
37. J. A. Stroscio, D. T. Pierce, R. A. Dragoset, and P. N. First, J. Vac. Sci. Technol. **A10**, 1981 (1991).
38. Z. Gai, Y. He, X. Li, J. F. Jia, and W. S. Yang, Surf. Sci. **365**, 96 (1996).
39. E. Umbach, K. Glöckler, and M. Sokolowski, Surf. Sci. **402–404**, 20 (1998).
40. B. Kesemo, Surf. Sci. **500**, 656 (2002).
41. X. Zhao, H. Wang, R. G. Zhao, and W. S. Yang, Mater. Sci. Eng. **C16**, 41 (2001).
42. X. Zhao, H. Yan, R. G. Zhao, and W. S. Yang, Langmuir **18**, 3910 (2002).
43. X. Zhao, R. G. Zhao, and W. S. Yang, Langmuir **16**, 9812 (2000).
44. X. Zhao, R. G. Zhao, and W. S. Yang, Langmuir **18**, 433 (2002).
45. X. Zhao, H. Yan, R. G. Zhao, and W. S. Yang, Langmuir **19**, 809 (2003).
46. X. Zhao, J. Am. Chem. Soc. **122**, 12584 (2000).
47. M. Corso, W. Auwärter, M. Muntwiler, A. Tamai, T. Greber, J. Osterwalder, Science **303**, 219 (2004).
48. W. Hebenstreit, M. Schmid, J. Redinger, R. Podloucky, P. Varga, Phys. Rev. Lett. **85**, 5376 (2000).
49. O. Dulub, U. Diebold, and G. Kresse, Phys. Rev. Lett. **90**, 016102 (2003).
50. D. A. Bonnell, Prog. in Surf. Sci. **57**, 187 (1998).
51. K. Bobrov, A. J. Mayne, and G. Dujardin, Natrue **413**, 616 (2001).
52. S. Iijima, Nature **354**, 56 (1991).
53. J. W. G. Wildöer, L. C. Venema, A. G. Rinzler, R. E. Smalley and C. Dekker, Nature **391**, 59 (1998).
54. T. W. Odom, J. L. Huang, P. Kim and C. M. Lieber, Nature **391**, 62 (1998).
55. M. Ouyang, J. L. Huang, and C. M. Lieber, Phys. Rev. Lett. **88**, 066804 (2002).
56. D. D. D. Ma, C. S. Lee, F. C. K. Au, S. Y. Tong, S. T. Lee, Science **299**, 1874 (2003).
57. Z. Gai, H. Yu, and W. S. Yang, Phys. Rev. **B53**, 13547 (1996).
58. Z. Gai, R. G. Zhao, and W. S. Yang, Phys. Rev. **B56**, 12303 (1997).
59. R. J. Hamers, R. M. Tromp, and J. E. Demuth, Phys. Rev. Lett. **56**, 1972 (1986).
60. R. J. Hamers, Ann. Rev. Phys. Chem. **40**, 531(1989).
61. R. C. Jaklevic and J. Lambe, Phys. Rev. Lett. **17**, 1139 (1966).
62. B. C. Stipe, M. A. Rezaei, W. Ho, Science **280**, 1732 (1998).
63. L. J. Lauhon and W. Ho, Phys. Rev. **B60**, R8525 (1999).
64. K. Wandelt, in "Thin Metal Film and Gas Chemisorption", edited by P. Wissmann (Elsevier, Amsterdam, 1987).

65. J. F. Jia, K. Inoue and Y. Hasegawa, W. S. Yang, T. Sakurai, Phys. Rev **B58**, 1193 (1998).
66. G. Binnig, N. Garcia, H. Rohrer, J. M. Soler, F. Flores, Phys. Rev. **B30**, 4816 (1984); J. Tersoff, and D. R. Hamann, Phys. Rev. Lett. **50**, 1998 (1983).
67. G. Binnig, H. Rohrer, Surf. Sci. **126**, 236 (1983).
68. Y. Hasegawa, J. F. Jia, K. Inoue, A. Sakai, T. Sakurai, Surf. Sci. **386**, 328 (1997).
69. J. F. Jia, K. Inoue and Y. Hasegawa, W. S. Yang, T. Sakurai, J. Vac. Sci. Technol. **B15**, 1861 (1997).
70. K. Besocke, B. Krahl-Urban, H. Wagner, Surf. Sci. **68**, 39 (1977); B. Krahl-Urban, E. A. Niekisch, H. Wagner, Surf. Sci. **64**, 52 (1977).
71. R. Smoluchowski, Phys. Rev. **60**, 661 (1941).
72. G. Binnig and H. Rohrer, IBM J. Res. Dev. **30**, 355 (1986).
73. Y. Hasegawa, J. F. Jia, T. Sakurai, Z. Q. Li, K. Ohno, and Y. Kawazoe, in "Advances in scanning probe microscopy", edited by T. Sakurai and Y. Watanabe, Springer, 2000, p. 167.
74. D. M. Eigler, and E. K. Schweizer, Nature **344**, 524 (1990).
75. M. A. McCord, and R. F. W. Pease, Appl. Phys. Lett. **50**, 569 (1987).
76. J. A. Dagata, J. Schneir, H. H. Harary, J. Bennett and W. Tseng, J. Vac. Sci. Technol. **B9**, 1384 (1991).
77. G. Dujardin, R. E. Walkup, and Ph. Avouris, Science **255**, 1232 (1992).
78. S. T. Yau, D. Saltz, M. H. Nayfeh, Appl. Phys. Lett. **57**, 2913 (1990).
79. U. Staufer, R. Wiesendanger, L. Eng, L. Rosenthaler, H. R. Hidber, H.-J. Güntherodt, and N. García, Appl. Phys. Lett. **51**, 244 (1987).
80. M. F. Crommie, C. P. Lutz and D. M Eigler, Science **262**, 218 (1993).
81. H. C. Manoharan, C. P. Lutz, D. Eigler, Nature **403**, 512 (2000).
82. B. C. Stipe, M. A. Rezaei, W. Ho, S. Gao, M. Persson, and B. I. Lundqvist, Phys. Rev. Lett. **78**, 4410 (1997).
83. S.-W. Hla, L. Bartels, G. Meyer, and K.-H. Rieder, Phys. Rev. Lett. **85**, 2777 (2000).
84. F. Moresco, G. Meyer, K.-H. Rieder, H. Tang, A. Gourdon, and C. Joachim, Phys. Rev. Lett. **86**, 672 (2001).
85. D. T. Pierce, Phys. Scr. **38**, 291 (1988).
86. M. Bode, Rep. Prog. Phys. **66**, 523 (2003).
87. R. Wiesendanger, H.-J. Güntherodt, G. Güntherodt, R. J. Gambino, and R. Ruf, Phys. Rev. Lett. **65**, 247 (1990).
88. R. Wiesendanger, I. V. Shvets, D. Bürgler, G. Tarrach, H.-J. Güntherodt, and J. M. D. Coey, Z. Phys. **B86**, 1(1992).
89. R. Wiesendanger, I. V. Shvets, D. Bürgler, G. Tarrach, H.-J. Güntherodt, J. M. D. Coey and S. Gräser, Science **255**, 583 (1992).
90. S. Heinze, M. Bode, A. Kubetzka, O. Pietzsch, X. Nie, S. Blügel, R. Wiesendanger, Science **288**, 1805 (2000).
91. Kubetzka, M. Bode, O. Pietzsch, and R. Wiesendanger, Phys. Rev. Lett. **88**, 057201 (2002).
92. M. Bode, M. Getzlaff, and R. Wiesendanger, Phys. Rev. Lett. **81**, 4256 (1998).
93. W. Wulfhekel and J. Kirschner, Appl. Phys. Lett. **75** 1944 (1999).
94. S. H. Pan, E. W. Hudson, and J. C. Davis, Rev. Sci. Instrum. **70**, 1459 (1999).
95. B. I. Barker, S. K. Dutta, C. Lupien, P. L. McEuen, N. Kikugawa, Y. Maeno, J. C. Davis, Physica **B329–333**, 1334 (2003).
96. Q. Niu, C. Chang, C. K. Shih, Phys. Rev. **B51**, 5502 (1995).
97. S. Tsukamoto, B. Siu, and N. Nakagiri, Rev. Sci. Instrum. **62**, 1767 (1991).
98. H. Watanabe, C. Manabe, T. Shigematsu, and M. Shimizu, Appl. Phys. Lett. **78**, 2928 (2001).
99. H. Grube, B. C. Harrison, J. F. Jia, and J. J. Boland, Rev. Sci. Instrum. **72**, 4388 (2001).
100. H. Okamoto and D. Chen, Rev. Sci. Instrum. **72**, 4398 (2001).
101. S. Hosaka, T. Hasegawa, S. Hosoki, and K. Takata, Rev. Sci. Instrum. **61**, 1342 (1990).
102. L. Kuipers, R. W. M. Loos, H. Neerings, J. ter Horst, G. J. Ruwiel, A. P. de Jongh, and J. W. M. Frenken, Rev. Sci. Instrum. **66**, 4557 (1995).
103. J. Wintterlin, J. Trost, S, Renisch, R. Schuster, T. Zambelli, G. Ertl, Surf. Sci. **394**, 159 (1997).
104. R. Curtis, M. Krueger, and Eric Ganz, Rev. Sci. Instrum. **65**, 3220 (1994).
105. R. Curtis, T. Mitsui, and E. Ganz, Rev. Sci. Instrum. **68**, 2790 (1997).
106. C. Y. Nakakura, V. M. Phanse, G. Zheng, G. Bannon, E. I. Altman, and K. P. Lee, Rev. Sci. Instrum. **69**, 3251 (1998).
107. D. Croft and S. Devasia, Rev. Sci. Instrum. **70**, 4600 (1999).
108. L. Petersen, M. Schunack, B. Schaefer, T. R. Linderoth, P. B. Rasmussen, P. T. Sprunger, E. Laegsgaard, I. Stensgaard, and F. Besenbacher, Rev. Sci. Instrum. **72**, 1438 (2001).

109. T. R. Linderoth, S. Horch, E. Lægsgaard, I. Stensgaard, and F. Besenbacher, Phys. Rev. Lett. **78**, 4978 (1997).
110. S. Horch, H. T. Lorensen, S. Helveg, E. Lægsgaard, I. Stensgaard, K. W. Jacobsen, J. K. Nørskov and F. Besenbacher, Nature **398**, 134 (1999).
111. J. Wintterlin, S. Völkening, T. V. W. Janssens, T. Zambelli, G. Ertl, Science **278**, 1931 (1997).
112. C. Sachs, M. Hildebrand, S. Völkening, J. Wintterlin, G. Ertl, Science **293**, 1635 (2001).
113. C. Sachs, M. Hildebrand, S. Völkening, J. Wintterlin, and G. Ertl, J. Chem. Phys. **116**, 5760 (2002).

4. VISUALIZATION OF NANOSTRUCTURES WITH ATOMIC FORCE MICROSCOPY

SERGEI N. MAGONOV AND NATALYA A. YERINA

INTRODUCTORY REMARKS

Scanning tunneling microscopy (STM) and Atomic Force Microscopy (AFM) were introduced about 20 years ago [1, 2]. Since this time these techniques have revolutionized surface analysis by providing high-resolution visualization of structures at the atomic- and nanometer-scales. The remarkable feature of STM and AFM instruments is their ability to examine samples not only in an ultrahigh vacuum but also at ambient conditions and even in liquids. In both methods, the localized interaction between a sharp probe and a sample is employed for surface imaging. STM is based on detection of tunneling current between a sharp metallic tip and a conducting surface. This circumstance limits STM applications, and it is applied mostly to studies of atomic structures and atomic-scale processes on different conducting and semiconducting samples, primarily in UHV conditions. Therefore, the use of STM is confined to research laboratories at Universities and Government Institutions dealing with fundamental problems of surfaces, whereas industrial laboratories are using AFM exclusively which can be applied for characterization of materials of any kind. This functionality is inherent to AFM, which is based on detection of more universal tip-sample mechanical forces.

The scope of AFM applications includes high-resolution examination of surface topography, compositional mapping of heterogeneous samples and studies of local mechanical, electric, magnetic and thermal properties. These measurements can be performed on scales from hundreds of microns down to nanometers, and the importance of AFM, as characterization technique, is further increasing with recent developments in nanoscience and nanotechnology. In studies of surface roughness, AFM complements optical and stylus profilometers by extending a measurement range towards the sub-100 nm scale and to forces below nanoNewton. These measurements are valuable in several industries such as semiconductors, data storage, coatings, etc. AFM together with scanning electron microscopy of critical dimensions is applied for examination of deep trenches and under-cut profiles with tens and hundreds of nanometers dimensions, which are important technological profiles of semiconductor manufacturing. AFM capability of compositional imaging of heterogeneous polymer systems (blends, block copolymers, composites, filled rubbers) attracts the attention of researchers working in industries, which are dealing with synthesis, design and formulation of plastic materials as well as their applications. In this function, AFM assists other microscopic and diffraction techniques (light, X-ray, and neutron scattering). Nanoscale objects such as mineral and organic filler particles, carbon nanotubes or individual macromolecules of biological and synthetic origin are distinguished in AFM images. Studies of these objects and their self-assemblies on different substrates are addressing important problems of intermolecular interactions in confined geometries. Better understanding of these interactions and the ways they might be controlled are needed for a preparation of functional surfaces, nano-scale patterning and manipulation of nanoscale objects.

Local probing of mechanical properties is another important function of AFM that offers unique capabilities for studies of structure-property relationships at the nanometer scale. A recording of force curves and performing nanoindentation at surface locations of tens of nanometers in size are routinely employed for such measurements. At present, this is only a comparative analysis of mechanical responses of different samples or different sample components. In addition to mechanical properties, examination of local electric properties at the sub-micron scales will be welcomed by many applications. Electric force microscopy, which is most known AFM technique for mapping of conducting regions of various samples, is based on measurements of electric field gradients acting between a metal-coated probe and conducting sample regions. Detection of local electric properties such as current-voltage characteristics of the nanoscale objects is a more challenging task and requires substantial instrumental improvements to became a routine procedure.

At present AFM became a mature characterization technique that is in permanent development. Intensive efforts are underway in AFM instrumentation and its applications. The design of novel probes with various geometries and unique dynamic properties has already enhanced the technique's dynamic capabilities, mechanical measurements and image resolution. The use of piezoceramic actuators as scanners in AFM instruments has such drawbacks as non-linearity and creep, which are related to polycrystalline nature of these materials. An introduction of high-precision scanners based on closed-loop positioning systems is addressing this problem. The recognized AFM limitation is its low efficiency due to slow scanning. The development of new approaches to fast scanning will enable high throughput capabilities of imaging and screening for combinatorial approaches in material science and technology. Nanomechanical measurements become crucial for characterization of nanomaterials, which offer the promise of breakthrough longstanding limits of material performance. Most importantly, the proper characterization of these materials and their performance is impossible without quantitative studies of nanomechanical properties. Therefore, there are strong incentives for development of reliable approaches toward quantitative nanomechanical analysis. The AFM-based techniques (nanoindentation, scratching, etc.) have intrinsic advantages for overcoming fundamental difficulties of indenters, which are routinely used for micro-mechanical testing, and which are not suitable at scales below a half of micron and for operation at low forces. Various attempts are on the way to make nanomechanical measurements with AFM more quantitative, with unique spatial resolution and also to provide such measurements in broad frequency range.

This chapter presents a short review of contemporary AFM and main issues related to its instrumentation and practical imaging at the nanometer scale. AFM applications will be illustrated by examples taken from studies of single macromolecules and their self-assemblies on different surfaces and compositional mapping of semicrystalline polymers, block copolymers, polymer blends and composites. The choice of practical examples reflects the fact that AFM studies of polymers are the field most familiar to the authors.

BASICS OF ATOMIC FORCE MICROSCOPY

Main Principle and Components of Atomic Force Microscope

In AFM, mechanical force interactions acting between a sharp probe and a sample are used for surface imaging. The probe, which represents a micromachined cantilever with a sharp tip at one end, is brought into interaction with the sample surface. The interaction level between the tip apex and the sample is determined through precise measurements of the cantilever displacements. Initial attempts to apply STM for gauging the cantilever deflection had little success. An optical level detection, which had been originally suggested for gravimeters [3], appeared invaluable for precise measurements of the cantilever deflection in most commercial atomic force microscopes [4]. In this procedure, a laser beam, which is deflected from the backside of the cantilever, is directed to a 4-segment positional photodetector, which is divided into segments for measurements of normal and lateral deflections of the cantilever. At present, the optical level detection is the most reliable way to measure the tip-sample force interactions, Figures 1a–b. This approach does not completely free of problems related with the use of light, such as parasitic interference at the cantilever-sample confinement, heating of a cantilever and a sample by the laser beam. Therefore, the microscope designers are looking for alternative approaches. Among them is the AFM based on a microfabricated piezocantilever, in which the cantilever itself provides not only the deflection sensing but also the actuation [5].

The surface imaging is realized by detecting the tip-sample force in different locations while the probe is rastering the sample surface with the help of a piezoelectric actuator. A feedback control applied during imaging ensures that the tip-sample force is preserved at a constant level. The error signal, which is used for feedback control,

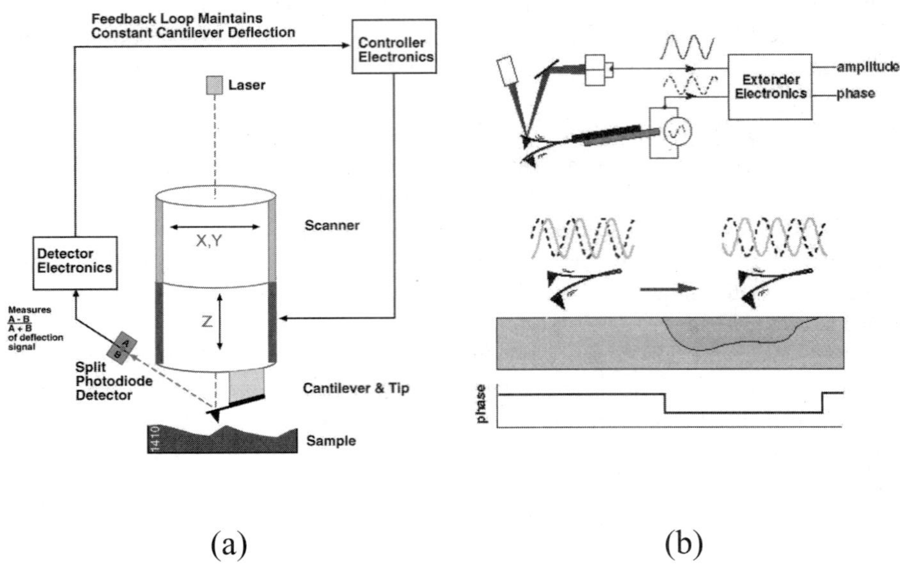

Figure 1. (a) Sketch demonstrating main components of atomic force microscope working in the contact mode in which the tip is permanently engaged into the sample. A cantilever deflection responding to tip-sample forces is measured with the optical lever scheme. (b) Sketch illustrating phase detection and phase imaging in tapping mode. Phase of the probe oscillation changes when an AFM probe comes into interaction with the sample. Phase can be different when the probe interacts with different components of a heterogeneous sample.

is amplified to generate height images, which reflect surface corrugations. The height image, in which brighter contrast is assigned to elevated surface locations, represent the vertical translations of the piezo-scanner needed to eliminate the error signal when the probe is moved from one sample location to the other. The error signal images, which, might be considered as maps of derivatives of height corrugations, emphasize fine surface features that are poor resolved in the height images.

From a brief description of the method it becomes clear that the main components of atomic force microscope are probes, optical detection system, piezo-scanners and electronics for a management of scanning procedures and data acquisition, Figures 1a–b. In the microscope, these components are assembled into a microscope stage, which must satisfy the requirements of minimum vibrational, acoustic and electronic noise as well as small thermal drift. Basic information about these components could be useful for better understanding the performance of AFM instruments, their unique features and limitations.

Scanners, which are applied for 3D movement of the sample or probe in AFM, are made of piezoelectric materials, which provide the precise positioning and ability to transport the objects in the micron range with sub-angstrom precision. Yet due to polycrystalline nature of these materials, the motion of real scanners deviates from linear dependence on applied voltage, especially at voltages generating large translations. In addition, the motion along the different axes is not completely independent. Therefore, a careful design, precise construction and calibration are important objectives that should be addressed during manufacturing of the scanners and their use. These efforts will allow the real scanners to approach a desirable performance, yet an additional electronic control is still needed. In an open-loop scanner, the controller drives the scanner using a non-linear voltage profile that produces a linear motion. This profile is taught during the calibration procedure on surface gratings with the known pitch in lateral dimensions and height steps in vertical direction, Figures 2a–b. Such calibration has limited precision because the scanner response to a particular voltage depends on the material history. Minimal distortions are expected when scanning is performed at the small range near the scanner rest point and the distortions will increase substantially when high voltages are applied for large-scale scanning. The situation is worse when the scanner is applied for small scans far away from the rest point immediately after its use for large scans.

To address the open-loop control problems, the scanner can be fitted with independent position sensors. In this case of a closed-loop system, the controller reads the sensor outputs and adjusts the drive voltage in order to achieve the desired motion. The microscopes with the closed-loop control of the scanners became popular recently to address problems of object manipulation, surface lithography and patterning in the micron and sub-micron scales. It is worth noting that despite the improved precision of the probe or sample translation the performance of the closed-loop systems is subjected to influence of thermal drift and additional noise that hurts quality of high-resolution imaging. Scan accuracy of both systems (open-loop and close-loop) depends on their calibration using appropriate standards. Man-made standards are available for the lateral scales of hundreds of nanometers and larger, Figures 2a–b. For calibration at the nanometer and atomic scale, one can apply periodical patterns of natural materials such as alkanes and the lattice spacing of crystalline surfaces of mica and highly-ordered pyrolitic graphite, Figures 2c–d.

AFM has been introduced for visualization of structures at the atomic-scale. However, with development of its applications the technique became useful for many other purposes and the size of the samples and structures to be examined has varied tremendously. This need led to the development of AFM instruments that can be used for studies of large objects (e.g. 12-inch Si wafers) with the instrument operation fully automated. In the automated microscopes, in addition to piezoscanners, different translation XY stages are applied. In addition to motorized stages, flexure stages are also used in AFM instruments. The flexure stages are closer to the performance of piezo-scanners and offer some additional capabilities. The close-loop flexure stages are also developed for commercial microscopes.

An introduction of microfabricated Si_3N_4 and Si probes, which consists of the cantilevers with a sharp tip at one end that can be prepared in batch processes, was one of the key events that led to the broad use of AFM instruments. Major parameters of the AFM probes are the cantilever shape and stiffness, oscillatory parameters (resonance frequency, Q-factor), tip geometry (a shape and size of its apex), and specific functionality.

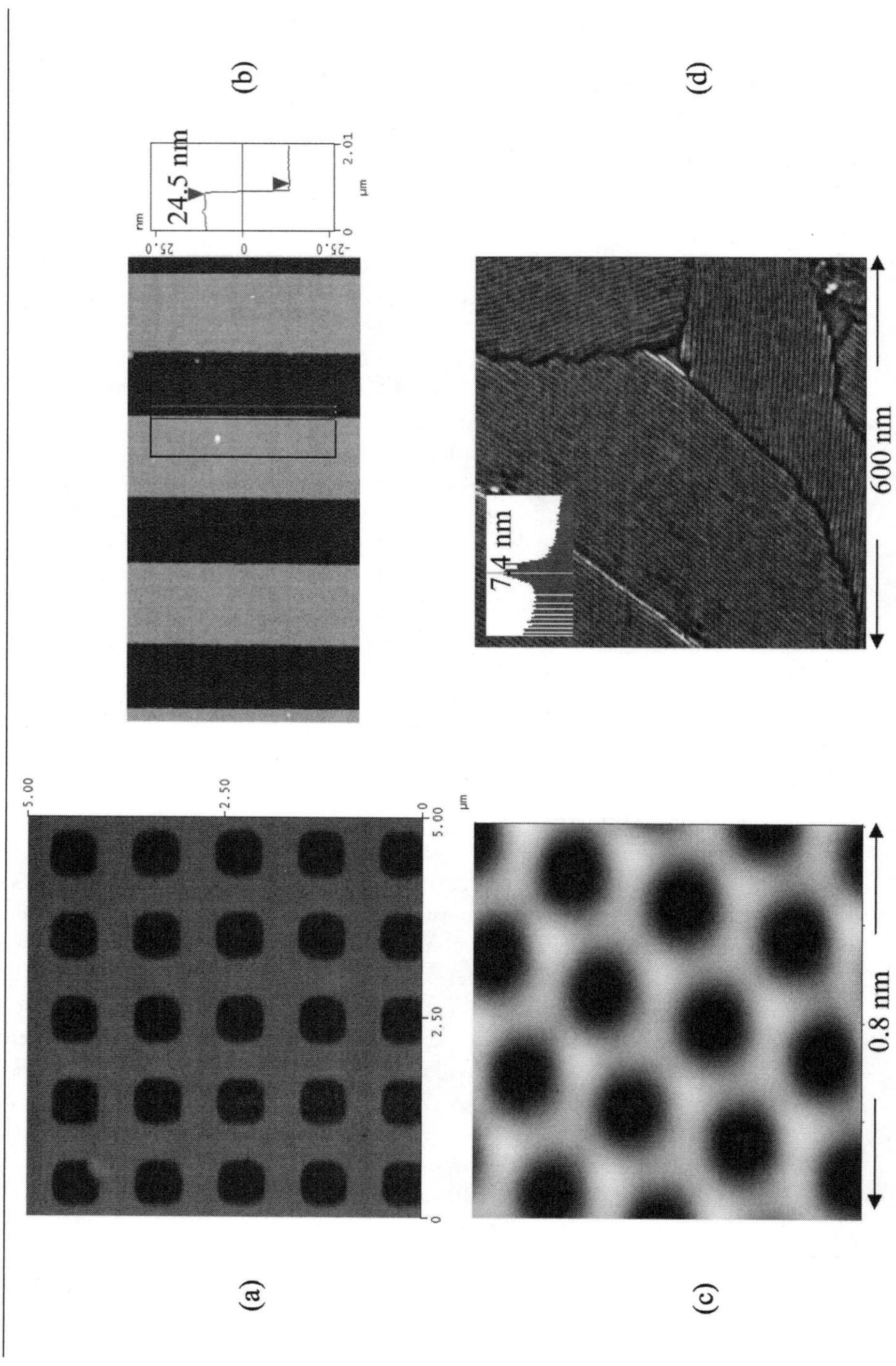
(a)
(b)
(c)
(d)
24.5 nm
7.4 nm
0.8 nm
600 nm
µm
nm

Most of the probes have rectangular or triangular cantilevers and a sharp pyramidal tip at the end. Practically, it is more feasible to make thinner and softer cantilevers out of S_3N_4. Therefore these probes, which are traditionally made with triangular cantilevers, are applied for studies of soft biological samples and are used primarily for contact mode measurements in air and under water. Stiffness of Si_3N_4 probes depends on dimensions of the triangular cantilevers and varies in the 0.01 N/m–0.6 N/m range. Si probes usually have rectangular cantilevers and the range of stiffness is much broader: from 0.1 N/m to 400 N/m. The softest probes can be used for the contact mode measurements whereas tapping mode [6] imaging requires stiffer probes because one should be able to retract the probe from a sample in every cycle of its oscillation. This can be achieved only with probes whose stiffness overcomes adhesive interactions with the sample.

Before an experiment it might be quite difficult to determine what minimal stiffness of the probe is needed for successful measurements of a particular sample. For studies of soft materials (polymers and biological objects), a broad range of probes can be used. On one hand, softer probes will facilitate gentle imaging of these materials. On another hand, visualization of the composition of the heterogeneous samples with high contrast requires the probe with optimal stiffness. Therefore, the probes, whose stiffness varies in the range from 0.1 N/m to 400 N/m, can be employed for compositional mapping. This is related to the fact that stiffness of polymeric materials differs in a broad range and matching the probe stiffness to that of a polymer sample or its different components helps to visualize individual components of multicomponent materials. The probe choice also depends on the operation mode and environmental conditions. For imaging in air, Si probes with stiffness of 3–5 N/m are most useful, especially when high-resolution and low-force imaging is required. For compositional imaging, which is commonly conducted at elevated tip-forces, Si probes with stiffness 30–40 N/m will be a good choice. Imaging under liquids can be done with soft Si probes (0.3–1 N/m). The resonant frequency and Q-factor of the probes are essential dynamic parameters that influence the scanning rate of imaging in oscillatory modes and soft probes with high-resonance frequency are ideal for fast scanning of biological objects.

Tip geometry is the crucial parameter for many AFM applications such as measurements of narrow trenches and rectangular surface steps, profiling of single lying objects, as well as visualization of atomic-scale features on crystalline surfaces. The overall shape of Si_3N_4 tip is a square pyramid with the half-angles of its faces $\sim 35^\circ$. The nominal radius of curvature at the tip is <20 nm. Si probes are etched in the shape of an irregular pyramid with the nominal apex radius <10 nm. Near the apex the shank is triangular with the half angles of 17° (sides), 25° (front), and 10° (rear),

Figure 2. (a)–(b) Height images of the calibration standards for lateral (X, Y) and vertical (Z) directions, respectively. (c) STM image of highly ordered pyrolitic graphite. (d) Phase image of normal alkane $C_{60}H_{122}$ layer on graphite. The insert in the top left corner shows the power spectral density plot and a value of the most pronounced peak, which corresponds to the length of the alkane molecules in the extended all-trans conformation.

where the "front" is closest to the end of the cantilever. For critical measurements of surface features one should consider the absolute orientation of the sample surface and tip to avoid an incorrect judgment. The quality of commercial AFM probes might vary, therefore, for reliable imaging one can preliminary check the probes by imaging test samples such as Au colloid spheres on a smooth substrate, ridged structures of the $SrTiO_3$ (305) surface, edges of the TiO_2 surface, and sharp pyramids. Special care should be exercised during such measurements to avoid undesirable tip damage.

The described probes are most common in routine AFM applications, and the tip apex size of Si probe is one of the factors determining the imaging resolution in tapping mode. Therefore, there are ongoing efforts of design and manufacturing of novel probes with sharp extremities. Two kinds of new probes [7, 8], which might be useful for high-resolution imaging, are shown in Figures 3a–b. The first one was prepared by plasma-assisted deposition of carbon materials on the apex of Si tip. The probe of the second type has a diamond tip with a mechanically sharpened apex. The radius of the curvature at the end of these probes is approaching 1 nm. Recent results demonstrated that tapping mode imaging with true molecular resolution could be achieved with the spiky probes [7]. Unfortunately, multiple spikes, which grow at the Si apex, limit the ease-of-use of these probes. The diamond tip does not have this drawback, however the cost of these probes is significantly higher than the probes with spikes that are produced in batch process. Carbon nanotubes with nanometer-scale diameter were also suggested for use as AFM probes, and their fabrication has advanced from a manual assembling to the catalytic growth of nanotubes at the apex of AFM probes [9, 10]. Yet the images obtained with CNT probes so far do not show the resolution improvement. They also show mechanical instabilities that limit the use of the CNT probes.

Probes with specific functionality can be prepared by coating the cantilever or tip with different materials. Metallic coatings are deposited on the cantilevers in order to increase their optical reflectivity and electric conductivity. The probes with ferromagnetic coatings are applied for magnetic force microscopy. Unfortunately, the coating can make the tip apex less sharp. It is worth noting that the AFM probes with piezoelectric coatings might offer exceptional capabilities for this technique in the near future. A possible application of such cantilevers for self-actuation and detection of the tip-sample interactions might eliminate the optical detection and its related restrictions for some AFM applications. The dynamic characteristics of the piezoelectric cantilevers are superior to those of the regular cantilevers, which are driven externally. This circumstance has been utilized in the development of fast scanning mode, which is essential for high efficiency of AFM and high throughput measurements. So far, due to some instrumental and practical hurdles, this approach has not been broadly accepted. Instead it is possible to make use of these cantilevers for dynamic mechanical measurements of polymer samples. The preliminary results show that this approach allows extending the mechanical studies to high frequencies (up to 100 kHz), which are not accessible to conventional dynamic mechanical analysis [11].

For many years, the probes with chemically modified tips were employed for selective detection of surface locations with different chemical properties. Typical chemical

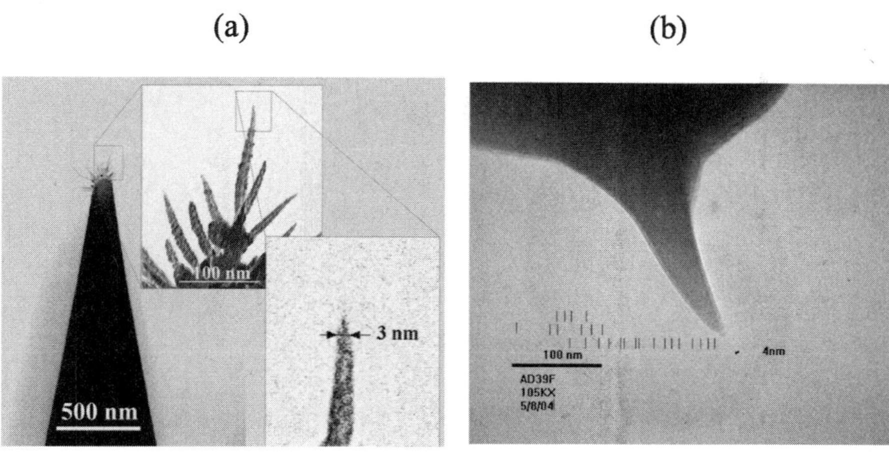

Figure 3. Electron microscopy micrograph of the tips of novel AFM probes. (a) The probe with the carbon spikes (the picture – courtesy of D. Klinov, Institute of Biorganic Chemistry, Moscow, Russia). (b) The diamond probe with a mechanically sharpened apex (the picture – courtesy of B. Mesa, MicroStar Technology, Inc., Huntsville, TX, USA).

modification includes a coating of the tip with a gold layer followed by adsorption of alkylthiols with various functional groups. The use of these modified probes is quite a challenging task. An adequate control of the coating integrity, which might be damaged by tip-sample forces, as well as statistical approaches for data collection and analysis are required for getting reliable results with these probes. Alternatively, AFM probes with different chemical nature can be microfabricated of polymeric materials. Such probes, which are prepared by either etching of photoresist [12] or by extrusion, offer a desirable diversity of the tip material. They also are much softer than the Si and Si_3N_4 probes that facilitate measurements with lower forces.

Concluding the short description of AFM instrumentation it is worthwhile to note the importance of the overall mechanical design of the microscopes, which substantially influence the quality of images. Visualization of the nanoscale structures, particularly in the sub-100 nm scale, essentially depends on the thermal drift of the microscope. Thermal drift harms imaging because the rate of AFM scanning is limited by the physics of tip-sample force interactions and dynamics of the probes, especially, when oscillatory modes are applied. Fast scanning approaches, so far, require an increase of tip-sample interactions that might not be useful for imaging of soft samples. Therefore, special care should be taken to minimize thermal drift by a proper microscope design and rational choice of construction materials for instrument components and enclosures, which are required for acoustic noise isolations. Note that the close-loop scanners are also suffering from thermal drift that limits the value of this approach. An example of imaging with low thermal drift microscope Dimension 5000, in which room temperature drift is as small as 0.5 nm per minute, given in Figures 4a–d. These four images of an array of single macromolecules (polyphenylacetylene with mini-dendritic groups) on graphite were collected in sequential scans of the same area,

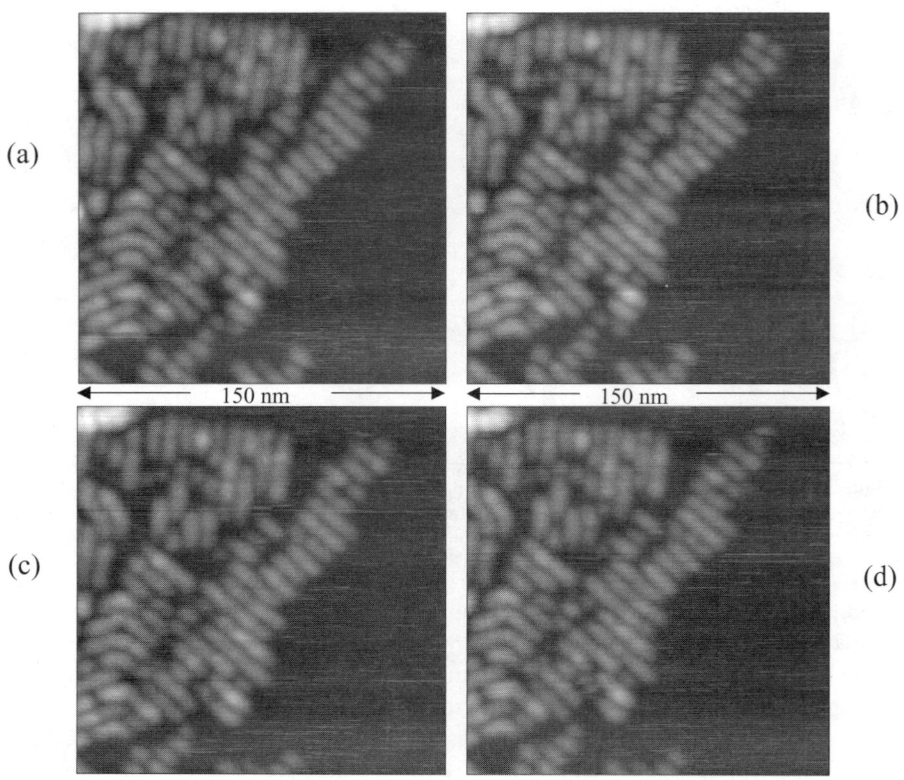

Figure 4a–d. Sequential height images of a stack of single chains of polyphenylacetylene with mini-dendritic groups on graphite.

when the probe was rastering the surface in alternative directions (up to down, down to up, etc.).

Operational Modes, Optimization of the Experiment and Image Resolution

There are two main operation modes in AFM: contact mode and tapping or intermittent contact mode. In the contact mode, which was introduced in practice first, the probe comes into a permanent contact with a sample surface. A product of the cantilever stiffness on its deflection determines the tip-sample force. For many samples, this mode should be applied with caution and the cantilevers with low spring constants are needed for gentle profiling of soft surfaces. Imaging with high-resolution was demonstrated with the contact mode AFM on many crystalline surfaces [13]. Besides surface imaging, AFM in its force modulation mode [14] has been effectively used for evaluation of sample mechanical properties by modulating the tip-force with an additional actuator. Lateral tip-sample forces accompany scanning of surfaces with the tip being in contact, and these forces can be recorded for evaluation of surface friction.

Unfortunately, lateral forces applied to soft samples might induce a strong shearing deformation and sample damage. This limits the contact mode applicability to studies of polymers and biological objects.

In AFM, local tip-sample forces can be measured using deflection-versus-distance curves [15]. These measurements are also helpful for choosing appropriate set-point deflections for surface imaging with different forces. By immersing the sample and the probe in liquid, one can eliminate capillary forces applied to the tip by a liquid contamination layer, which presents on surfaces in air. Therefore, imaging in liquids can be performed at small forces below 1 nN. For biological samples, aqueous media is essential and most of AFM studies of these objects are done underwater. For other materials, imaging in liquids is only an optional, not a routine operation.

This situation changed drastically with the introduction of the AFM oscillatory mode known as the tapping mode [6]. Tapping is performed by the probe, which is driven into oscillatory motion at its resonant frequency by an additional piezoactuator, Figure 1b. A drop of the cantilever amplitude when the tip comes into interaction with a sample is used as a measure of these interactions, and the amplitude drop is kept at a pre-set level during scanning. In tapping mode, permanent shearing forces are almost eliminated and the intermittent tip contact with the sample surface occurs at a high frequency (tens and hundreds of kHz) that also restrict material damage. Such operation is gentler than the contact mode, despite the fact that stiffer probes are used in tapping mode. This mode has revolutionized AFM applications because a broad range of samples and materials of industrial importance has become accessible for studies at ambient conditions. For example, the contact mode measurements of Si wafers caused a surface damage that can be avoided by using tapping mode.

The advantages offered by tapping mode for imaging of soft materials are balanced by the complexity of dynamic tip-sample force interactions that in some cases makes the analysis and interpretation of tapping mode images quite challenging. The main problem is that a change of amplitude of an oscillating cantilever while it interacts with a sample does not solely determine tip sample forces. Alterations of the resonant frequency and phase of the cantilever are more sensitive for this purpose. Corrugated surface features as well as contamination traces make the sensitive phase and frequency measurements less applicable for feedback than the amplitude changes. Therefore the amplitude is used for feedback and the vertical adjustments of the piezoscanner needed for keeping the amplitude drop constant are reflected in the height image. The phase changes of the interacting probe during imaging are presented in the phase image, Figure 1b. Usually in the tapping mode, the height and phase images are recorded simultaneously. Phase imaging is most valuable for compositional analysis of heterogeneous samples [16]. Differences in adhesive and mechanical properties of different components are responsible for various phase contrast observed on these samples during imaging. The tapping mode with phase or frequency detection also made possible broad applications of electric force microscopy and magnetic force microscopy.

Qualitative differences of mechanical, adhesive, electric, magnetic and other properties are sufficient for compositional mapping, yet the demand of local quantitative measurements of these properties is increasing with developments of nanoscience and nanotechnology. The local measurements have been a challenge since the introduction of AFM, and there are a number of relevant problems. They include preparation of specialized probes, optimization of instrumentation, and a lack of appropriate theoretical analysis of the tip-sample forces at the nanometer scale.

At present, tapping mode is the most common AFM mode. In its applications, researchers are always facing a problem in getting the most valuable information, which can be rationally understood. The crucial issue is a correct choice of instrumental parameters for high-resolution surface imaging and compositional mapping of heterogeneous samples, which are usually quite different to achieve these goals. The principal issue of AFM is tip-sample forces and their control. Minimization of the tip-force allows avoiding surface damage and reducing the tip-sample contact area that facilitates high-resolution imaging. Compositional mapping will benefit from an increase of tip-forces because differences between mechanical properties will be better manifested in this case.

High-resolution imaging is the unique feature of STM and AFM. STM observations of atomic defects on semiconductor and metallic surfaces have proved true atomic-scale resolution with this technique. Image resolution in AFM is a complicated issue. In the contact AFM mode, images of crystal surfaces show atomic-scale patterns identical to the crystallographic lattices [13]. The absence of atom-size defects in such images suggests that the tip contact area in this mode is larger than the atomic size. This does not exclude a possibility of lattice imaging based on periodic variations of the normal and lateral forces, which are exercised by a tip when it is moving along a periodical surface [17]. The imaging resolution of tapping mode and of another oscillatory mode—frequency modulation, used for AFM in UHV has different issues. It is clear that contact area of the tip should be comparable with a size of features to be resolved. In general, the contact area is determined by mechanical characteristics of the tip and sample as well as by the apex radius and the tip-force. In UHV conditions of clean sample surface and high quality factor of the oscillating probe, a fine tip-force control is realized by detection of small frequency shifts. Such measurements in frequency modulation mode can be performed with true atomic resolution as confirmed by images of atomic-scale defects [18]. In air, the precise force control is limited, and the best image resolution achieved in tapping mode applications with etched Si probes (apex radius $\sim$10 nm) is around 1 nm. Recent experiments with the sharper probes – carbon spikes grown at the end of a commercial Si probes, have been successful, and true molecular scale resolution was obtained in imaging of surface of polydiacetylene crystal [7]. The tapping mode image of 20 nm on side (Figure 5a) revealed a well-defined pattern with almost vertical rows and a few molecular-size defects of 3–5 angstroms in size. This observation confirms true molecular resolution of the image. Bright spots along the rows are assigned to individual side groups of the polymer chain forming the crystal. In general, the regular molecular order is consistent with the dimensions of the crystallographic surface lattice ($b = 0.491$ nm, $c = 1.410$ nm, $\alpha = 89.5°$, Figure 5b) and with the pattern observed in the contact mode images of the same surface, Figure 5c. The use of the probes with carbon spikes that allow imaging with true molecular resolution requires definite precautions such as a gentle engagement and the low-force operation to avoid a fracture of sharp spikes. Imaging

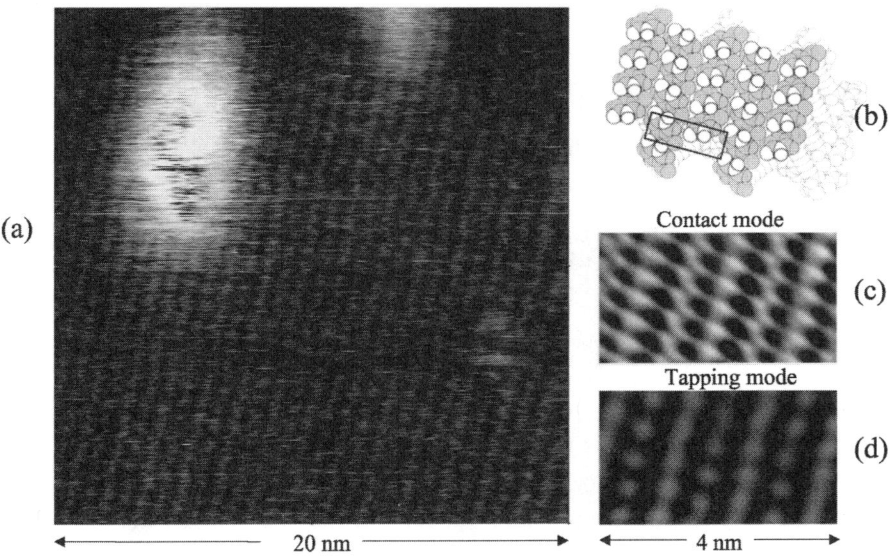

Figure 5. (a) Height AFM images, which were recorded in tapping mode with the novel carbon probe on the *bc*-plane of the polydiacetylene crystal. (b) Molecular arrangement at the *bc*-plane of the crystals of polydiacetylene [2,4-hexadienylenebis(p-fluorobenzenesulfonate)]. Most elevated F, H and C atoms at the *bc*-surface are shown as the unfilled circles and other atoms of the side groups, which are slightly lower, are shown as the gray circles. (c) Contact mode image of the *bc*-plane of the polydiacetylene crystal. (d) Tapping mode image of the *bc*-plane of the polydiacetylene. The 2D FFT filtering (only the most pronounced reflexes of the 2D power spectra were applied for a reconstruction of the surface lattice) led to the periodical patterns in (c) and (d), which were obtained from the images in Ref. S. N. Magonov, G. Bar, H.-J. Cantow, H.-D. Bauer, I. Müller and M. Schwoerer, *Polym. Bull.* 26 (**1991**) 223 and in (a).

of corrugated surfaces with these tips is rather challenging due to possible interference of several spikes, which grow at the same apex.

Two other successful examples of high-resolution imaging with the spikes are shown in Figures 6a–f. In these figures, the images of Si wafer surface and low-K polymer coating [19], which were obtained with these probes, are compared with the images of the same samples recorded with regular etched Si probes. It is obvious that nanoscale roughness of Si surface is much higher in the image obtained with the sharper probe, Figures 6a–b. This is even better seen in the cross-sectional height profiles seen in Figures 6c–d. There are also evident differences of the size of dimples and grains in the images of the nanoporous low K coating. Again the application of the shaper probe provides the height image with much finer surface structure than the use of the regular probe.

An example, which illustrates compositional mapping, is given in Figures 7a–d, where one sees height and phase images of a layer of ultra long alkane $C_{390}H_{782}$, which were obtained at low-force (light tapping) and high-force (hard tapping) imaging. In light tapping, the height image, Figure 7a, shows a number of linear elevations, which

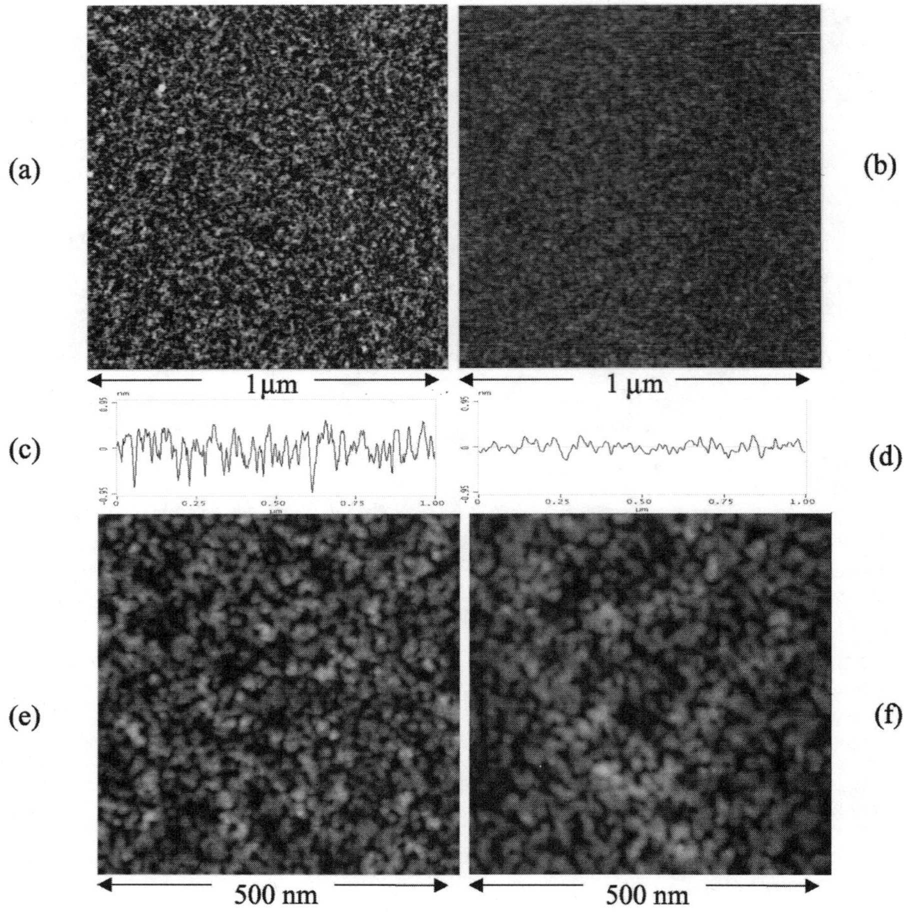

Figure 6. (a)–(b) Height images of Si wafer obtained with the probe with carbon spikes and with regular etched Si probe, respectively. (c)–(d) Cross-section height profiles taken across the images in (a) and (b), respectively. (e)–(f) Height images of a nanoporous low K material obtained with the probe with carbon spikes and with regular etched Si probe, respectively.

are running from left to right. The corresponding phase image (Figure 7b) exhibits only minor contrast at the elevations similar to one expected from the error image. In ideal case of light tapping, the probe should track a samples surface lightly enough that the phase response is not different from the phase of the non-interacting probe. This condition can be achieved when the set-point amplitude, A_{sp}, is close to the amplitude of a non-interacting probe, A_0, in immediate vicinity of the sample surface. The tip-sample force increases with decrease of A_{sp} and hard tapping conditions is usually correspond to $A_{sp} = 0.4$–$0.5\ A_0$. Height and phase images of the alkane layer in Figures 7c–d, which were recorded at hard tapping, are different from those in

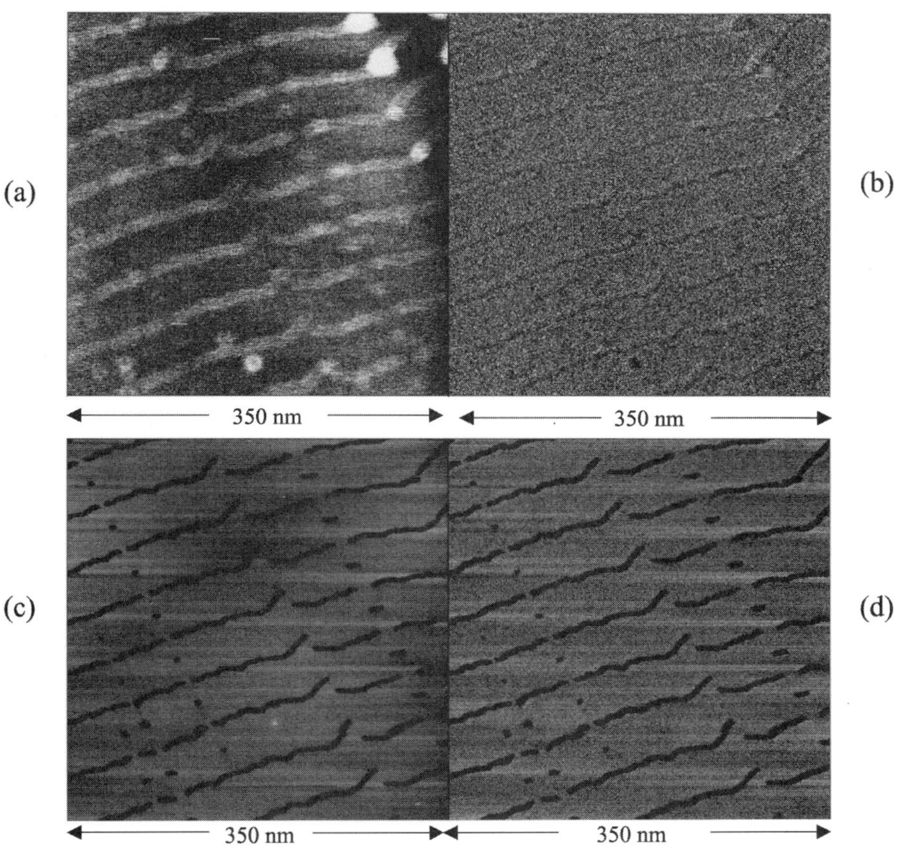

Figure 7. (a)–(b) Height and phase images of ultra long alkane $C_{390}H_{782}$ layer, which were obtained at light tapping. (c)–(d) Height and phase images of the same area as in (a)–(b), which were obtained at hard tapping.

Figures 7a–b, and these changes were reversible. The most pronounced variations are in the contrast of linear features, which are either brighter or darker compared with the rest of the area. The fact that the repeat distance across the stripped pattern (~48 nm) is close to the length of the extended chain of $C_{390}H_{782}$ alkane, suggests that the linear features present the chain ends. On one hand, the end -CH_3 groups of the alkane are more bulky than –CH_2- of the alkane chain and they also are mobile being the chain ends. Therefore, these locations are seen as elevated in height image obtained in light tapping. On other hand, the mobile chain ends do not resist the probe penetration through the alkane layer close to the substrate when hard tapping conditions are applied. This is consistent with the contrast change at these locations from bright to dark in the height image and also with the appeared dark contrast in the phase image. Note that the width of the linear features assigned to the chain ends is larger than the size of two individual $-CH_3$ groups which come to this interface from the neighboring lamellae. We could not exclude that, in addition to $-CH_3$ groups, a few neighboring $-CH_2-$ groups may contribute to the pattern changing in the AFM images. The finding that the width of the interlamellar interfaces seen in AFM images is increasing as temperature raises supports this suggestion. Molecular dynamic simulations of alkanes on substrates predicted similar effect [20]. The assignment of the dark strips seen in the height and phase images Figures 7c–d to the interlamellar interfaces led to another important result. The dark strips in these images are not always continuous, and one can see one linear dislocation and few locations where short dark strips are shifted a half lamellar size with respect to the main strip. The latter defects are most likely associated with bridging molecules or their blocks in which chains are shifted along their main direction with respect to the other molecules in the same lamellae. These and other defects, which are resolved in AFM of various alkanes at different temperatures will be discussed elsewhere [21]. Here we state that imaging at various tip-forces is useful procedure that provided valuable data about structural organization of alkane layers on graphite.

Our experience shows that on transition from light to hard tapping, the well-defined changes in height and phase images, similar to those seen in Figures 7a–d, are quite rare. Usually, they occur when differences in mechanical properties of surface locations or sample components are large. In many other cases, imaging in hard tapping leads to pronounced contrast variations in phase images whereas height images might change insubstantially. The high sensitivity of the phase images to sample heterogeneities made phase imaging an invaluable constituent of AFM.

A simple protocol can be applied for variable force imaging in the tapping mode. In the beginning of the experiment, the resonant frequency of the probe is determined in immediate vicinity of a sample surface, and the driving frequency is usually chosen at the resonant frequency. By changing the voltage applied to the piezoactuator, an operator chooses A_0, which typically varies in the 1–150 nm range. An engagement of the probe to a sample surface is important procedure especially when very sharp probes such as those with the spikes are applied. A sewing engagement procedure common for MultiMode and Dimension types of microscopes is best suited for the gentle engagement. After this, scanning of the probe over the sample surface can be performed at different A_{sp}, which is typically is in the 0.9–0.2 A_0 range. The magnitude of A_0 also influences the level of the tip-sample force interactions and larger amplitudes provide high-force operation. In some cases, alterations of A_{sp} and A_0 are not sufficient for compositional mapping of heterogeneous samples and one should apply probes with different stiffness to reach this goal. The stiffer probes are needed when a layer of rubbery material covers a sample surface and a tip penetration through this layer is required to reach underlying structures.

Imaging in Various Environments and at Different Temperatures

For over 10 years, AFM measurements have been typically performed at ambient conditions. An exception is the investigation of biological systems, where samples should be examined in their natural environment, i.e. in water. Obviously by extending AFM studies to different environments and temperatures the value of this technique will increase enormously and we should expect the progress and interesting discoveries in this field in near future.

Contact AFM mode imaging in water has an advantage due to the removal of a surface capillary layer that increases tip-sample forces. Tapping mode operation in liquid faces a substantial lowering of the cantilever resonant frequency and its quality factor. This is evident from a comparison of the amplitude-versus-frequency curves shown in Figures 8a–b. After immersion in water, the resonant frequency of the cantilever with a spring constant of $\sim$0.4 N/m, which we often applied for imaging of polymers under liquid, has dropped more than a factor of two and Q-factor changed from 130 to 4. These effects should be taken into account while imaging in liquids, and related adjustments of imaging parameters e.g. an increase of feedback gains will help avoid an unnecessary tip-force increase.

One can make use of imaging in liquid in different ways. For example, a selective swelling or etching of individual components of heterogeneous materials assists in revealing the microphase separation in such systems. This is especially valuable if the material components have similar mechanical properties that making their recognition in phase images obtained by hard tapping difficult. Such capability is illustrated by the images shown in Figures 8c–d. A microphase pattern in thin film of polystyrene-block-polyvinylpyridine, PS-*b*-PVP, block copolymer, which has been subjected to a long-term annealing at high temperature, is barely seen in the height and phase images recorded at ambient conditions. Immersion of this sample into acidic water (pH = 2.2) is followed by a slow development of bright spheres, which reflect the swelling of PVP blocks. With time, the number of spheres has increased and their hexagonal order becomes evident. These observations confirm the build up of this morphology during temperature-induced microphase separation. Selective swelling combined with *in situ* AFM observations can be also applied visualization of morphology development during crosslinking of rubber materials.

The possibilities of AFM imaging in liquid for industrial research are enormous because they provide access to materials behavior in real life cases. Studies of biomaterials such as implants, contact lenses, drug release systems, etc. will benefit from such applications. Placing a sample in liquid might lead to its excessive softening; therefore, imaging in vapors of different solvents might be a practical and useful alternative. Both studies in vapor and in liquid can be also combined with temperature variations that further expand the range of AFM applications.

AFM measurements at different temperatures have advanced in last 5 years but they still far from routine. The main reason is instrumental difficulties in performing experiments at high and low temperatures. The hurdles associated with temperature control of a sample and of the microscope components surrounding have been only partially solved. The extra functions are usually needed for variable temperature AFM imaging include cooling of the piezoscanner, dual heating of the sample location using sample and probe heaters, monitoring of the surface temperature with 1°C accuracy by using the temperature dependence of the Si probe resonant frequency,

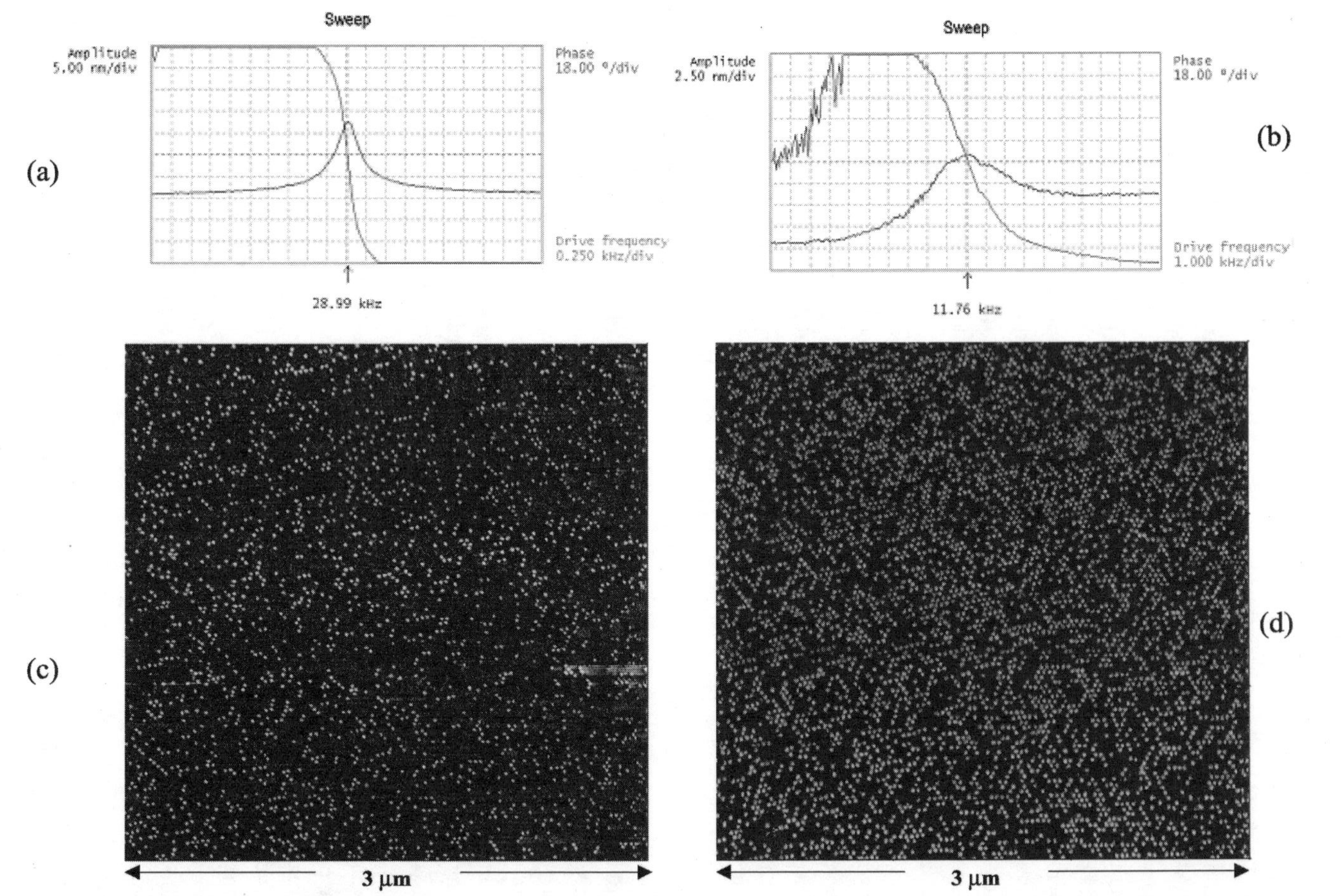

Figure 8. (a)–(b) Amplitude-versus-frequency curves of tapping mode in air and in liquid, respectively. (c)–(d) Height images of polystyrene-block-polyvinylpyridine block copolymer, obtained in acidic water 10 and 40 minutes after immersion.

and environmental control of the sample by purging with inert gases, such N_2, Ar, He [22]. The inert atmosphere is required for high-temperature imaging of some polymer samples to avoid their oxidation. This achieved by purging with He gas, whose presence is easy monitored by increase of the cantilever resonant frequency and quality factor.

At present, researchers are performing AFM experiments at elevated temperatures up to 250°C–300°C and capability of imaging at sub-zero (°C) temperatures is under development. Temperature measurements can be also conducted in vacuum although the AFM studies in vacuum are much less common than those in air. Imaging at elevated temperatures is usually aimed at monitoring structural changes in materials related to different thermal transitions such as melting, crystallization, recrystallization, glass transition, etc. Measurements at elevated temperatures are not much different from those at room temperature yet when approaching melting temperature one might use larger oscillation amplitudes of the probe in order to overcome increasing adhesive tip-sample interactions.

Imaging of polymer samples at elevated temperatures provides some new and unexpected data helping to address important questions concerning polymer structures and their behavior. For example, in the AFM study of structural organization of liquid crystalline carbosilane dendrimers, heating and cooling of thick dendrimer films led to the counterintuitive discoveries of shrinkage and expansion of some ordered domains [22]. This behavior has been explained by anisotropic thermal expansion of these liquid crystalline systems. Visualization of structural changes, which accompany heating of an ultra thin film of low-density polyethylene (PE), has revealed a recrystallization process [23] that was not obvious from differential scanning calorimetry (DSC) data. Spherulitic morphology of the LDPE film had disappeared when the temperature of the sample reaches 80°C and immediately after that large lamellar platelets start to grow. A continued raise of the sample temperature leads to complete melting at 115°C, expected from DSC data. *In situ* AFM monitoring of melt crystallization of poly(ethylene terephthalate), PET, showed that stacking of polymer lamellae governs this process [24]. Visualization of the lamellar stacking is crucial for understanding of the X-ray diffraction studies of this polymer, which require reliable models for a rational interpretation of the reciprocal space data. AFM observations of morphology of semicrystalline polymers and, particularly, their nanometer-scale granular structure, which was found on polymer surfaces and in bulk, were among the factors that revitalized strong interest in the studies of polymer crystallization. Two examples of visualization of the grain structure of polymer surfaces are presented in Figures 9a–e. The images of syndiotactic polystyrene at 170°C and 240°C are shown in Figures 9a–b. The lamellar morphology, which is distinctively seen in the first image, characterizes the β-phase of this polymer. At temperatures close to the melting point, grainy structures appeared as is evident from the second image. Similar changes from lamellar to grainy morphology were also observed in samples of melt-crystallized PET, shown in Figures 9c–e. This transformation had occurred when the sample temperature was lowered below Tg of this polymer. These observations suggest that granular morphology could arise either due to a disintegration of polymer lamellae into smaller blocks near its melting or from a solidification of the amorphous material surrounding

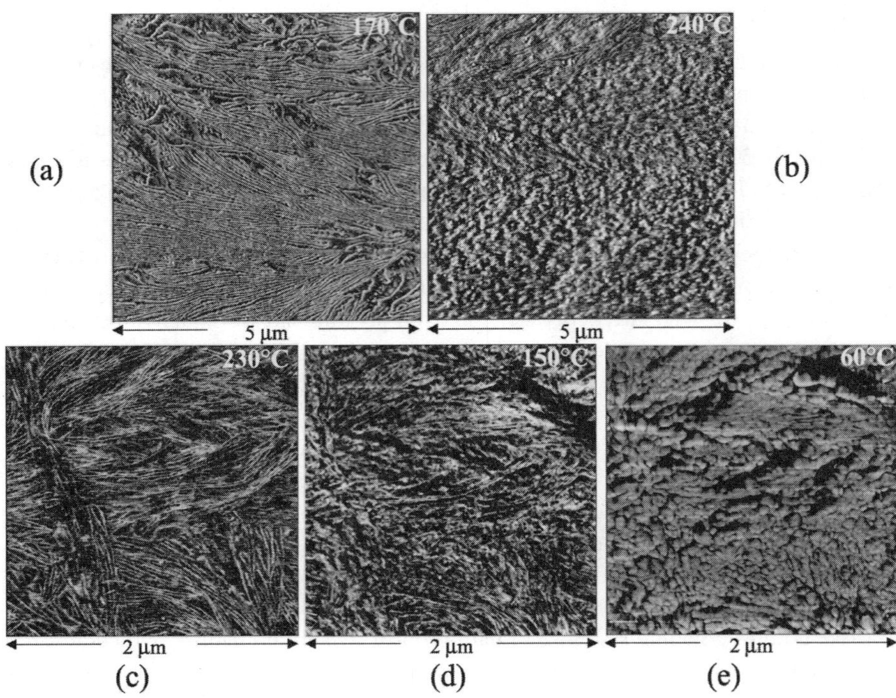

Figure 9. (a)–(b) Phase images of β-syndiotactic polystyrene obtained at 170°C and 240°C, respectively. (c)–(e) Phase images of melt-crystallized PET at 230°C, 150°C and 60°C, respectively.

the lamellar core. In the second case, an AFM probe penetrates through the top-most amorphous material when it is in rubbery state and reaches the ordered lamellar core.

To date, AFM studies at low temperatures are less common than those at elevated temperatures. Crystallization of polydiethylsiloxane from mesomorphic phase proceeds at temperatures around 0°C and it is accompanied by morphological changes shown in Figures 10a–c [25]. An array of mesomorphic lamellar aggregates, which is seen in Figure 10a, underwent gradual crystallization as the temperature of the samples was dropped to −5°C and kept at this level. The crystallized material is characterized by lamellae with sharper edges, which also exhibit a brighter contrast in the phase images. Phase images (Figures 10b–c) demonstrate that the crystallization front has moved from the top to the bottom of the scanning area.

There is another expectation, which is related with AFM studies at low temperature. Small objects such as single macromolecules deposited on different substrates can be seen in AFM images only if they adhere to the surface strongly enough to resist the tip-force they experience during imaging. This statement has been confirmed by the finding that single polymer chains, which are observed at room temperature, are not seen in AFM images at elevated temperatures. They are most likely detached from

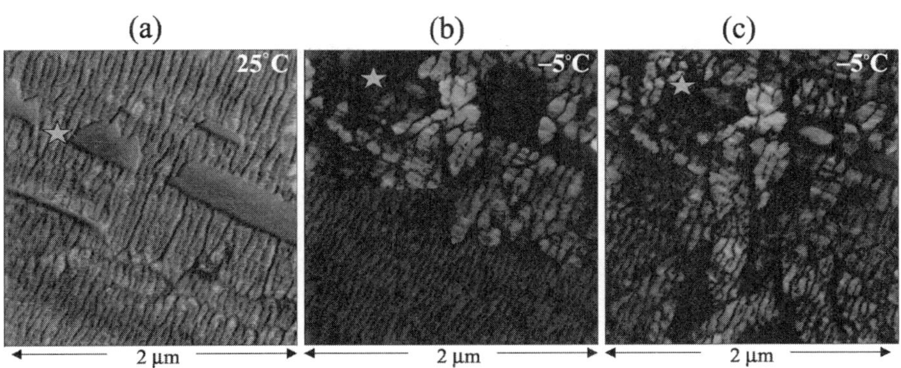

Figure 10a–c. Phase images of polydiethylsiloxane, which were taken at room temperature (a) and after cooling to −5°C (b)–(c).

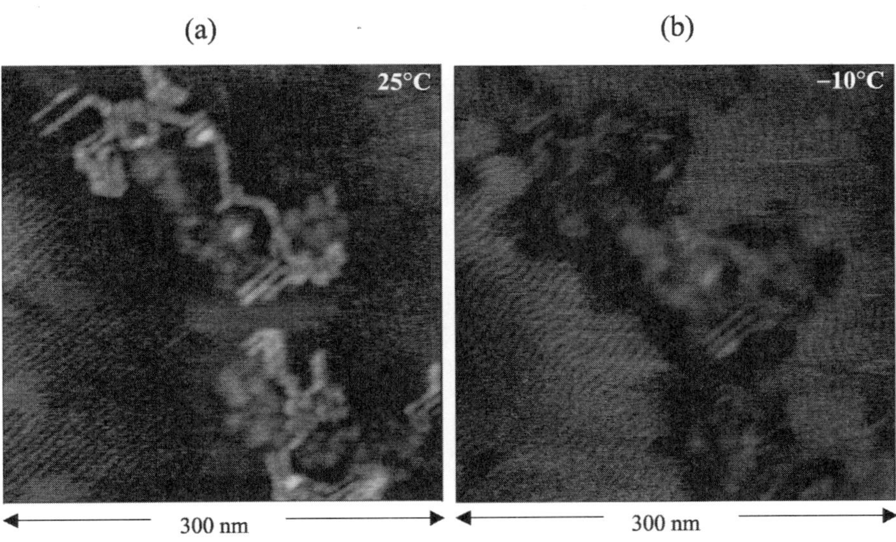

Figure 11. (a)–(b) High-resolution height images of single chains of polyphenyleneacetylene with mini-dendritic groups at 25°C and at −10°C, respectively.

the substrate due to their increased mobility. The molecules return to the substrate when the sample temperature is lowered back to room temperature. Consequently, at room temperature we might also see only part of the surface material deposited on a substrate. Therefore, observations at low temperatures can assist us in visualizing more surface species not seen at room temperature. With this in mind, we performed imaging of polymer chain molecules with mini-dendritic groups, and some of the results are shown in Figures 11a–b. Figure 11a shows an aggregate of several macromolecules, some of which exhibit straight segments, grains, which are surrounded by periodical structures. The latter structures are not rare but not well-understood guests in images of single polymer macromolecules deposited on different substrates from very dilute solutions. The second image (Figure 11b) obtained at the sub-zero temperature, shows differences in the shape of the macromolecules most likely related to conformational rearrangements induced by the temperature decline. Although in this case, we noticed only alterations of the molecules seen at room temperature, forthcoming studies will help us observe additional surface species.

Researchers, who are regularly using AFM, are aware of different factors that affect image reproducibility. Not only is calibration of piezoscanners and checking tip sharpness necessary, additional steps are needed to improve image quality. Specific topographic features and surface roughness can cause undesirable image distortion. Sharpness and symmetry of the tip apex are the main source of image distortion when surfaces have features comparable in size and shape with the probe extremities. Therefore, use of several different probes might help avoid misinterpretation. Probe shape changes might happen during scanning either as the result of a mechanical damage of a tip shape during imaging of a hard surface or due to tip contamination by surface material if a sample is sticky. Operation at low forces is useful to get away from the first problem and might also help avoid the second.

Studies of homogeneous surfaces are less subject to image variations compared to studies of heterogeneous surfaces with locations of different stiffness and adhesion. In addition to the image variations caused by geometric factors, alterations of images of heterogeneous materials are also related to different responses of individual components to the tip-force. Therefore, some images changes could be related to force variations. Images with a pronounced contrast of individual components are usually obtained in operation imaging with elevated forces. For rational imaging of heterogeneous materials, a researcher should find a set of experimental parameters (A_{sp}, A_0, the probe stiffness, etc) most suitable for compositional mapping of a particular family of materials. In this respect it useful to examine model samples which are prepared by varying a composition of a particular blend or multicomponent material.

IMAGING OF MACROMOLECULES AND THEIR SELF-ASSEMBLIES

Visualization of Single Polymer Chains

AFM provides the unique capability to visualize single macromolecules and their size, conformation, surface interactions and other phenomena can be studied by analysis of the image. Preparation of samples for such observations is straightforward; the only challenge being fixation of individual molecules to the surface. AFM images of DNA are the best known and three are presented in Figures 12a–c. The large-scale image shows numerous DNA strands spread on a mica substrate. A close look at a few of them in the height images, which were obtained in light and hard tapping, revealed variations of the shape and width of the curved structures representing the macromolecules, Figures 12b–c. These changes are reversible and can be assigned tentatively

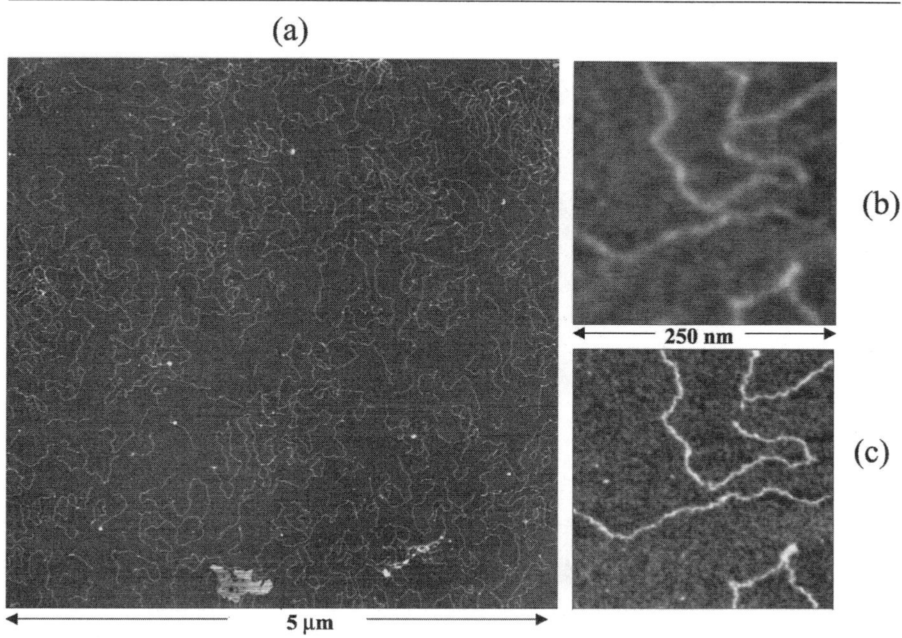

Figure 12. Height images of DNA molecules on mica, which were recorded in air. The images in (a) and (b) were obtained in light tapping, the image in (c) – in hard tapping.

to a soft "jacket" existing around the DNA core on mica, which can be depressed by the tip.

Since the first AFM studies of nano-scale objects, which also include single polymer macromolecules lying on substrates [26, 27], one of basic questions is how relevant is their height and width measured in the images compared to their real dimensions. One can foresee at least two reasons why AFM images of single macromolecules on surfaces might give the dimensions different from those in bulk or in solution, which are measured by other methods. The first reason is related to a possible relaxation of macromolecules on substrates that lead, for example, to their flattened shape. The second reason is due to peculiarities of AFM. Convolution of the tip-shape with real macromolecule dimensions is responsible for their widening in the AFM images. Also, one should not exclude tip-force induced deformation of the macromolecules that causes a height reduction. In contrast to the macromolecules' width, a contour length of extended macromolecules can be measured more precisely. For a number of polymers, whose chain molecules adopt an extended conformation on a substrate, the contour length can be considered as a measure of its molecular weight. The histograms, which reflect a distribution of contour length for large number of single macromolecules as measured from AFM images, have been compared with molecular weight distribution estimated from GPC and light scattering data [28].

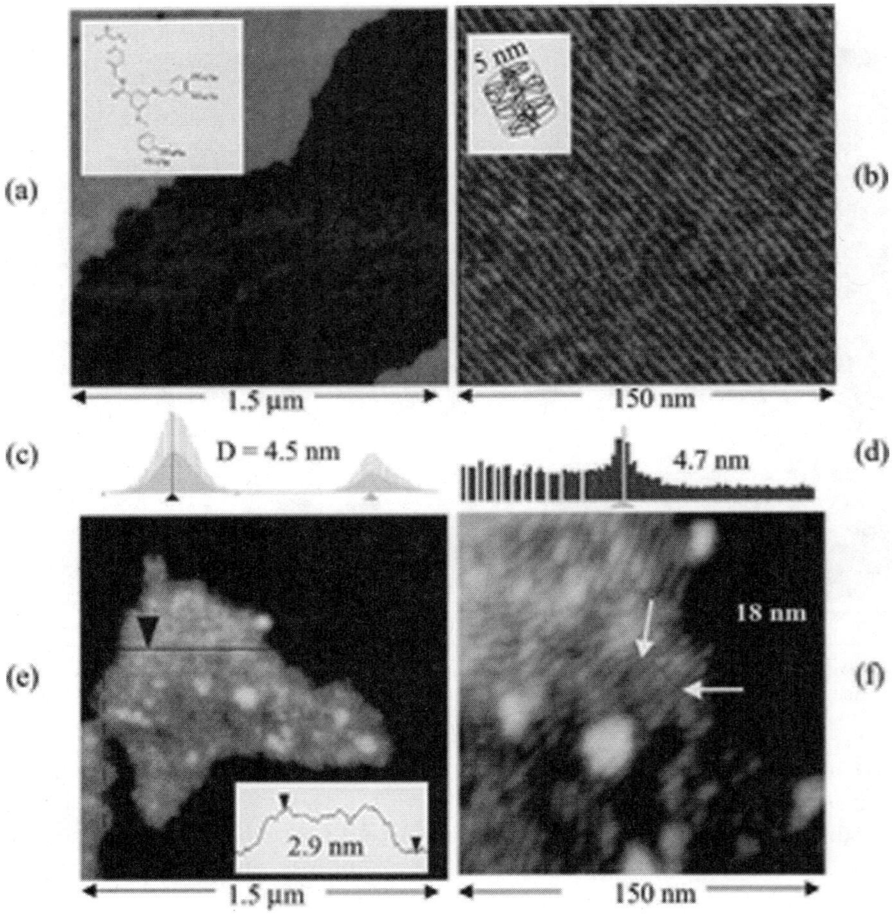

Figure 13. Height image (a) and phase image (b) of a sample of polyphenylacetylene with mini-dendritic groups representing the molecular order in bulk. The inserts in top left corner of these images show a chemical structure of the polymer and a sketch of the polymer conformation in bulk estimated from the X-ray data, respectively. The height histogram in (c) shows an average thickness of the top layer in (a). The power spectral density plot in (d) reveals the spacing of the molecular order in (b). Height and phase Images of the same polymer deposited on mica from dilute solution are shown in (e) – height image and in (f) – phase image. The insert at the bottom of the image in (e) shows a height profile across the polymer domain in the direction indicated with the arrows. The arrows in the image in (f) define the combined width of 4 polymer chains, which equals to 18 nm.

Issues related to size of polymer macromolecules in bulk and on surfaces as well as possible tip-force induced deformation have been explored in study of polyphenylacetylene with mini-dendritic groups. To prepare the sample surface, a polymer layer in melt was squeezed between two flat substrates and cooled. This sandwich was then split through the ordered material and examined with AFM. The image in Figure 13a shows a layered structure of this material in bulk. A regular packing of

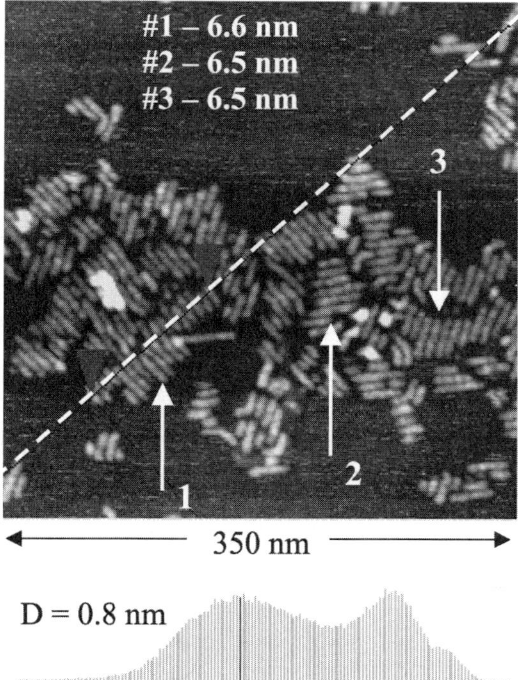

Figure 14. Height image of an aggregate of single chains of polyphenylacetylene with mini-dendritic groups, which were deposited on graphite from dilute solution. The number at the top of the image show the average width of the chains in the stacks indicated with the arrows.

extended linear structures is seen in the high-resolution image in Figure 13b. The power spectral density plot, which is presented below this image, revealed 4.7 nm spacing between these structures. This dimension is close to the 5-nm size of individual polymer chains in the hexagonal columnar arrangement of these molecules in bulk as revealed by X-ray analysis.

The chain molecules of the same polymer were observed on mica and graphite in samples prepared by spin casting of the dilute polymer solution. A domain of the macromolecules on mica is shown in Figure 13c. Its height, 2.9 nm, is slightly less than one expected based on the size of the individual polymer chains. A high-magnification image, Figure 13d, shows that the domain is formed of linear features whose estimated width is ~4.5 nm. Therefore, the size of the polymer chains on mica is close to their size in bulk. A partially reduced height of the chains on mica might be due to slight height depression by the tip. Estimates of the polymer chain dimensions on graphite from the AFM image in Figure 14 indicate that the macromolecules' height is much smaller and their width is larger than those of the chains on mica. This finding suggests substantial spreading of the single polymers on graphite.

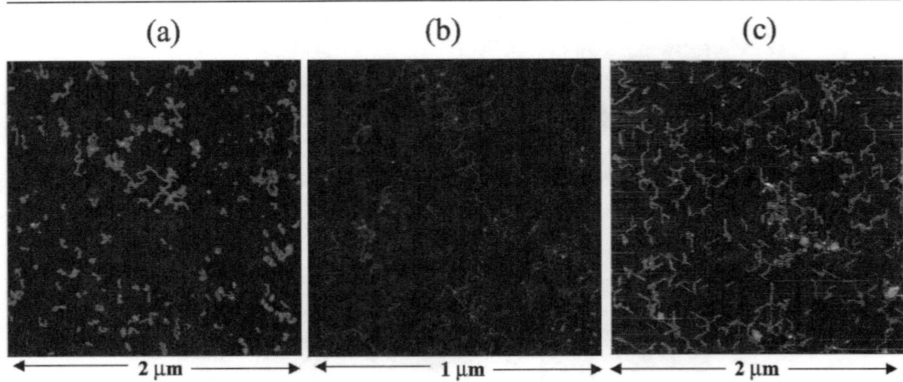

Figure 15. (a)–(c) Height images of single macromolecules of polymethylmethacrylate with mini-dendritic groups, which were deposited on mica, graphite and WSe_2, respectively.

Remarkably, the macromolecules on graphite are seen in the shape of individual straight or bent rods whose orientation reflects their epitaxy on the substrate. As a result of this epitaxy, the molecules are unraveled along surface lattice directions of the graphite. This phenomenon is common for alkanes and polyethylene, and the fact that the 0.25 nm spacing of the graphite lattice. The repeat distance along the alkyl chain in all-trans conformation is close to 0.25 nm, and this is believed to be the reason for the epitaxy. The multiple examples show that a presence of alkyls at the terminals of mini-dendritic side group of polymers also leads to the same arrangement [29]. Single macromolecules of polymethylmethacrylate (PMMA) with mini-dendritic groups, which were deposited on mica, graphite and WSe_2, are shown in Figures 15a–c. WSe_2 is the inorganic layered material with an atomically flat surface made of Se atoms in the closed hexagonal packing with the inter-atomic distance of 0.32 nm. Mica is also an atomically flat substrate, the surface atoms of which are arranged in a hexagonal order with an inter-atomic spacing of 0.52 nm. The arrangement of the chains on WSe_2 is similar to graphite, whereas on mica, the individual macromolecules are less extended and do not epitaxially orient on the substrate. This finding suggests that the epitaxy on graphite and WSe_2 is governed more by the overall symmetry than by the matching of the atomic-scale spacings of the substrate. The hydrophilic nature of mica has imposes restrictions on the unfolding of these macromolecules.

Height-temperature AFM measurements of PMMA chains with mini-dendritic groups on the above mentioned substrates as well as the experiments on the tip-force assisted transport of these objects on the substrates help understanding the individual macromolecules' adhesion [30]. Thermal motion of the unfolded chains was observed on graphite and WSe_2 as soon as the sample temperature was raised 10–20 degrees above room temperature. With further temperature increase, the macromolecules were not seen in the images because their adhesion to the substrate was too poor to resist a tip-force. Mostly likely, the chains were floating in the liquid contamination layer, which present on surfaces in air. The situation was quite different on mica, where the chains and their aggregates were observed without any positional changes even at temperatures above 200°C when partial thermal degradation has started. Strong adhesion of these polymer chains to mica is confirmed by the unsuccessful attempts to move the macromolecules across the substrate. These attempts led to cutting of the chains into pieces [30]. The tip-assisted transfer of the same macromolecules on graphite and WSe_2 was successful, and the macromolecules have been moved along the tip trajectories.

Alkanes, Polyethylene and Fluoroalkanes

To demonstrate the capabilities of AFM for studies of self-assembly and order in ultra thin layers we will consider several results obtained on alkanes of different length and single crystals of polyethylene (PE), one of the most important industrial polymers. When alkanes are deposited on graphite by spin casting of their diluted solutions, they form layers with the well-defined epitaxial order as shown in the images of $C_{36}H_{74}$, Figures 16a–b. The domains with differently oriented striped structures are seen in the larger-scale image. The other image shows the alternating bright and dark strips whose order is characterized by a spacing of 4.7 nm corresponding to the length of the extended alkane molecule. By analogy to the layers of $C_{390}H_{782}$, the darker regions can be assigned to the interlamellar interfaces where mobile $-CH_3$ groups and few their $-CH_2-$ neighbors are located. With the increase of a concentration of alkane solution, nanocrystals were formed on the top of the layers as it was observed in study of the alkanes with longer chain, $C_{60}H_{122}$ [31]. In the $C_{60}H_{122}$ samples, the nanocrystals melted around 95°C, while the epitaxial layers were observed at temperatures up to 140°C. In the measurements above the bulk melting point T_m of alkane crystals (T_m = 95°C for $C_{60}H_{122}$), an AFM probe penetrates through a melt polymer and reaches an ordered layer lying immediately on graphite. The lamellar order of this layer, which exhibits an additional thermal stability due to specific interactions with the substrate, has been observed at temperatures up to 50 degrees above T_m. This has been found for $C_{60}H_{122}$, $C_{122}H_{246}$, $C_{242}H_{486}$ and $C_{390}H_{782}$ [21, 32]. The high-temperature images of $C_{390}H_{782}$ layer in Figures 16c–d demonstrate its lamellar order at two different locations. The image at 100°C, which is the temperature below T_m (T_m = 128°C for $C_{390}H_{782}$), shows multiple defects reflecting a relative shift of the chain molecules along its main direction. The lateral extent of these defects is different at various locations, and they heal as the sample temperature approaches 130°C. At the left part of the image in Figure 16c, the remnants of the second layer are seen as single bright strips of the lamellar width. They also disappeared after heating to 130°C. The morphology of the $C_{390}H_{782}$ layer at 140°C, Figure 16d, has distinctive differences compared to the previous one. First of all, the T-type defects, in which two lamellae are merged into one, are the only defects seen at 140°C. Second, a width of brighter strips became smaller than that of the darker strips. This suggests that larger parts of the macromolecules are mobile as temperature was raised from 100°C to 140°C.

PE molecules are essentially extra long alkanes but the main difference is that polymerization leads to variations of the chain length of individual molecules that influence

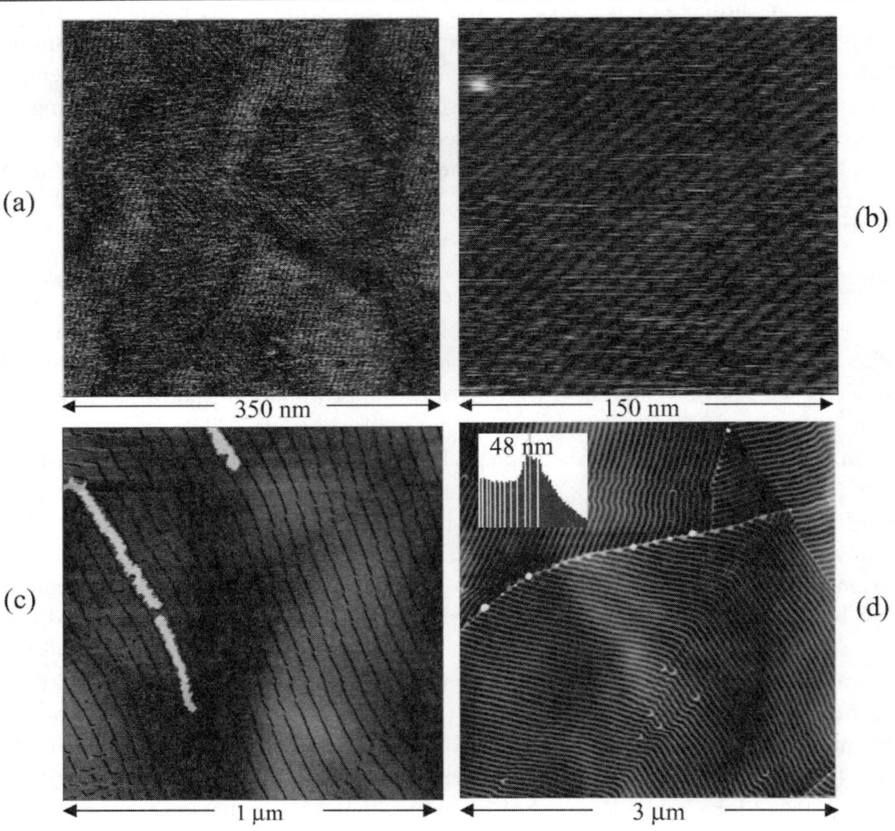

Figure 16. Height images of the layers of $C_{36}H_{74}$ alkane (a)–(b) and ultra long alkane $C_{390}H_{782}$ (c)–(d). Images in (a)–(b) were obtained at room temperature and images in (c) and (d) at 100°C and 140°C, respectively. The insert at the left top corner of the image in (d) shows the power spectral density plot with a value of most pronounced peak.

many properties of these materials. Ultra long alkanes due to their monodisperse nature are the best know models of PE, and these alkanes have been intensively examined during last 20 years [33]. The common feature of ultra long alkanes (starting with alkane with carbon atoms above 150) and PE is that their crystallization in dilute solutions proceeds by multiple folding of individual chains into thin lamellae folded chain structures which have the lozenge shape. The chain folding is the kinetically driven process, and thermal annealing of these crystals promotes a partial unfolding of polymer chains toward the more energetically favorable extended chain conformation. Morphology changes accompanying annealing of single crystals of PE have been of interest for many years [34]. Recent AFM studies of the annealing-induced structural transformations have revealed new details concerning these pathways, which depend on a sample preparation history and nature of the substrate.

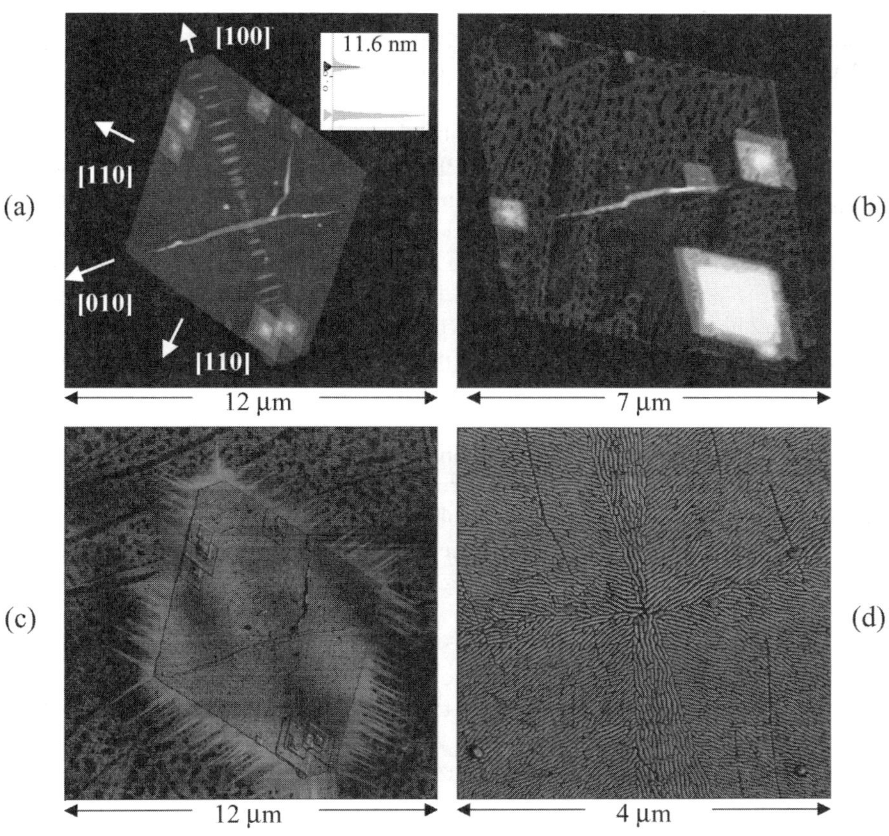

Figure 17. (a)–(b) Height images of dried single crystal of PE on Si substrate before and after annealing at 115°C, respectively. The height histogram in the top right corner in (a) define the crystal thickness. In (a), white arrows indicate different crystallographic directions. (c)–(d) Phase images of wet single crystal of PE on graphite after annealing at 75°C and 140°C, respectively.

Main structural changes, which are observed during annealing of single crystals of PE, are shown in Figures 17a–d. The crystals, which are grown in dilute solution, have a tent-like shape. After deposition on a flat substrate, such crystals collapse and adopt a shape of flat lozenge. A slightly truncated lozenge of PE crystal with a number of spiral overgrowths and a central pleat aligned in the [010] direction is shown in Figure 17a. The polymer chains are oriented in the [001] direction, i.e. almost perpendicular to a substrate. The crystal thickness (11.6 nm) corresponds to the chain stem length, and the top and bottom surfaces are formed of the chain folds. Several tiny wrinkles oriented along the [010] direction define two small (100) sectors and a slightly truncated shape of the crystal. It was found that during annealing of the PE crystals on Si and mica substrates, which were dried under vacuum before the annealing, the polymer chains unfold without a chain reorientation. This leads to a local thickening and development of the holes, Figure 17b.

Annealing of wet PE crystals (which were stored in air but not dried under vacuum) on the same substrates initiated a formation of lamellar ribbons (edge-on lamellae) in which the polymer chains reoriented parallel to the lamellar surface. The same transformation was found for single crystals of PE on graphite independent from the sample preparation history, Figures 17c–d [35]. In addition, the annealing of the crystals on graphite was accompanied by a substantial outflow of the polymer material from the crystal to the substrate. The image of the crystal after annealing at 75°C shows the outflow of the polymer in the [110] and [100] directions, Figure 17c. The outflow material most likely consists of mobile polymer chains, which were not properly incorporated into the crystal. Annealing at temperatures below 100°C was not accompanied by noticeable changes of the lozenge. The major evolution has initiated above 100°C, and it led to the formation of lamellar ribbons. The ribbons oriented along the [110] and [100] directions completely filled out the crystal sectors of the lozenge, Figure 17d. *In situ* monitoring of the initial steps of this transformation allowed us to suggest that the PE recrystallization and chain reorientation are facilitated by traces of solvent trapped underneath these objects during their collapse on a flat substrate. Analogous chain unfolding pathways were found for single crystals of $C_{390}H_{782}$ [32].

The structural transformations of the single crystals of PE, which are discussed above, were observed during their step-like annealing at different temperatures. When heating of the crystal deposited on mica to 140°C proceeded fast, it caused a complete melting of the crystal as seen by a formation of a large droplet of melt, Figure 18a–b. Quenching of this material to room temperature has been accompanied by recrystallization. The patch of recrystallized material, which remained within the frame of the initial lozenge, is illustrated by the height and phase images in Figures 18c–d, respectively. High-magnification images revealed its fibrilar and granular structures, Figures 18e–f. The fibrilar structures represent the edge-on lamellae whereas the isolated grainy structures can be assigned to the predominantly amorphous material, Figure 18e. The surface of the recrystallized material exhibits grainy morphology similar to one, which was observed on the surface of melt-crystallized low-density PE [36]. AFM investigations of the crystallization and melting processes in a droplet of semicrystalline polymer with known amount of the polymer might be a good model system to study these processes in the confined geometry.

Fluorinated materials have a number of important properties such as low surface energy that ensure their technological value. Semifluorinated alkanes with the structure $F(CF_2)_m-(CH_2)_nH$, F_mH_n, are forming micelles in solution and monolayers at the air-water interfaces [37]. Structural organization of these materials has been examined for a while yet many issues remain controversial. We have applied AFM to study self-assembly of perfluoroalkyl alkane, F14H20, and several results, which illustrate specifics of structural organization of this material, are presented below. In the extended conformation, the fluorinated part of the F14H20 due to large size of F atoms and related steric hindrance adopts a helical conformation with 1.64 nm in length and the

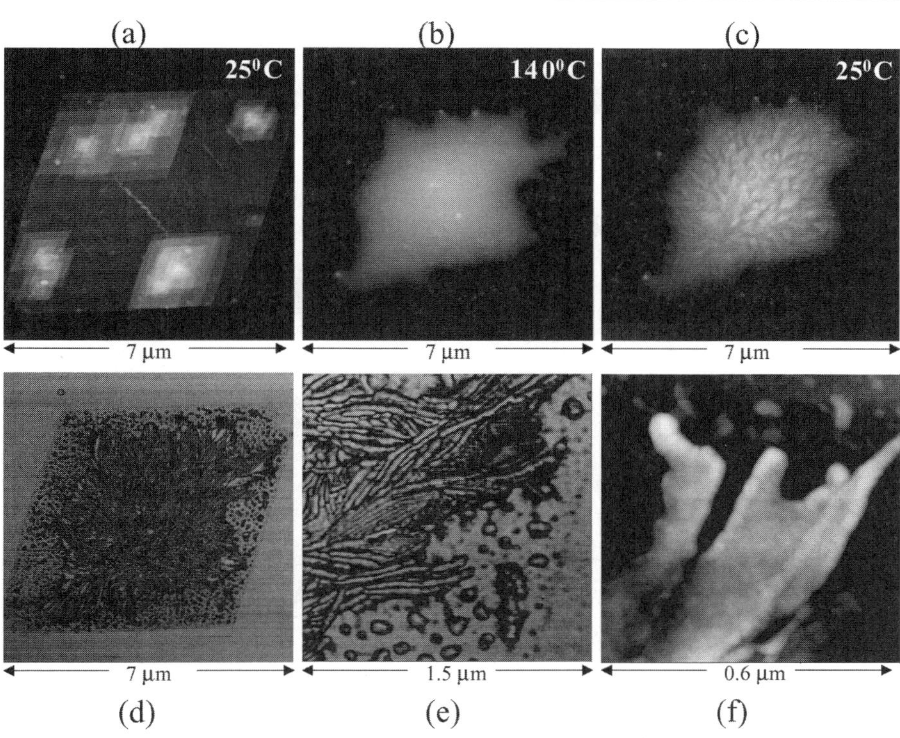

Figure 18. Height images of single crystal of PE on Si substrate at room temperature—(a), after fast heating to 140°C – (b) and after cooling to room temperature – (c). Phase images in (d)–(e) and height image in (f) show structural details of the semicrystalline PE patch formed after cooling from 140°C to room temperature.

hydrogenated part is characterized by the all-trans zigzag conformation with 2.55 nm in length. A width of these parts is also different: 0.60 nm for $-CF_2-$ sequences and 0.48 nm for $-CH_2-$ sequences. In earlier work, monodisperse surface micelles of F8H16, which were transferred from the water subface of Langmuir trough on Si substrate, were observed in AFM images [38]. The micelles have a round shape with a diameter of 30 nm and height of 2 nm in height. A hole in the center makes them look like donuts. These nanoscale 2D objects decorated the substrate being packed almost hexagonally.

When the F14H20 films were prepared by spin casting from dilute solution on different substrates, including a water subface in Langmuir trough, we have observed several types of structures, which depend on solvent and substrate nature. When the material was deposited on graphite from hydrogenated solvents such as decalin and octane, the extended ribbons of 30 nm in width and ~2 nm in height have been seen in AFM images, Figures 19a–b. The ribbons are aligned along three main directions of the substrate that assumes their epitaxial arrangement. Besides these ribbons, several

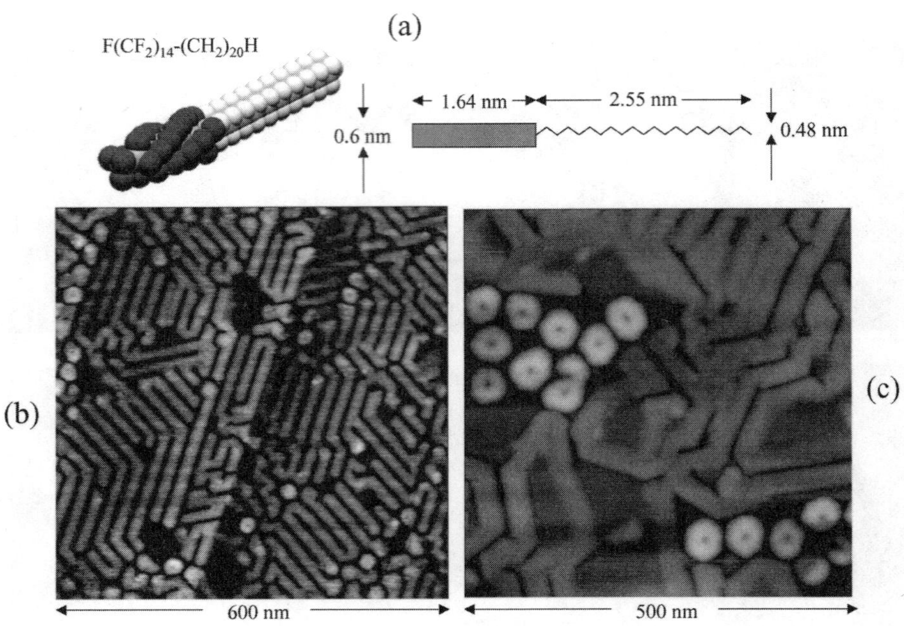

Figure 19. (a) Sketch of semifluorinated alkane F14H20 and its molecular dimensions in fully extended conformation. (b)–(c) Height images of F14H20 nanostructures deposited on graphite from decalin and octane, respectively.

donut-like clusters are seen. In general, the donuts are twice higher than the ribbons. Some of the donuts have a defective structure with a missing quarter of the donut, as seen in the left bottom of Figure 19b. This observation, as well as the fact that the external contour of many donuts exhibits a hexagonal contour, hints that the donuts and ribbons are formed of similar molecular assemblies. Clusters of different shape were formed in fluorinated solvents and deposited on water, mica and graphite as seen from AFM images in Figures 20a–c. Most likely, the circular twisting of the ribbons formed these monodisperse structures with an average diameter of 80 nm, Figure 20a. On mica surface, these clusters were packed in a hexagonal order. The power spectra of 2D FFT, which is shown in the insert in the right top corner of the image in Figure 20a, exhibit two sets of the hexagonal patterns corresponding to the cluster order and to the cluster shape. The clusters on graphite are quite alike yet they exhibit the more pronounced hexagonal shape in Figure 20b as compared to the clusters on mica. There are distinctive differences of the nanoscale morphology of the adsorbates, which were deposited from the hot and cold solutions on graphite. The clusters cast from the cold solution exhibit a fine substructure consisting of sets of linear strips with a 7 nm spacing. Akin linear strips are seen at the substrate locations in between the clusters, which were put on graphite from the hot solution. In both cases, linear structures are oriented in correspondence to the substrate symmetry.

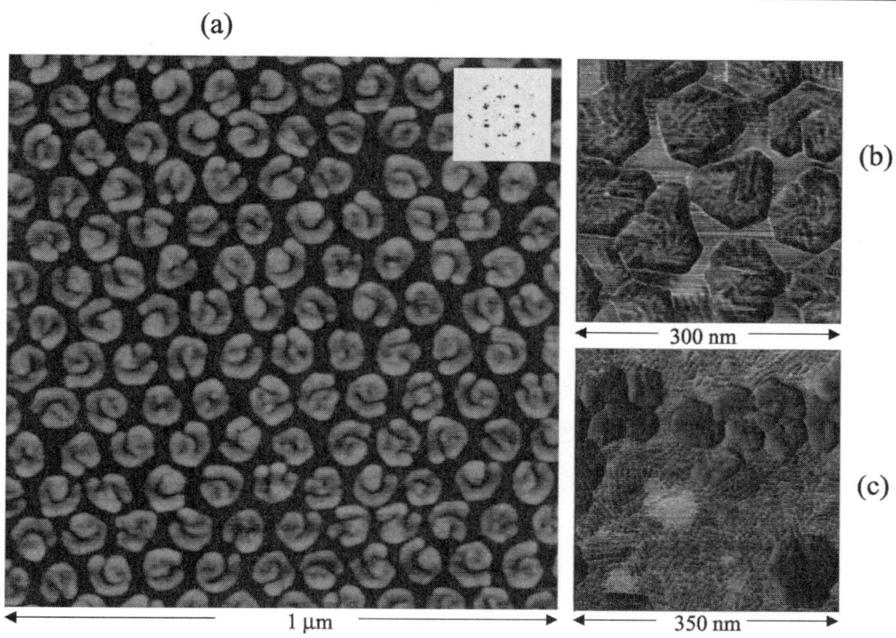

Figure 20. (a) Height image of F14H20 nanostructures deposited on mica from hexafluoroxylene. The insert in the right top corner shows the 2D FFT power spectral density pattern. (b)–(c) Phase images of F14H20 nanostructures deposited on graphite from cold and hot hexafluoroxylene, respectively.

The F14C20 adsorbates, which were deposited on graphite from octane, were examined at temperatures close and above their isotropization temperature of 95°C, Figures 21a–b. In the image recorded at 80°C, the domains with the linear strips similar to those seen in Figure 20c occupy most of the surface area. There are few straight ribbons, which did not melt yet. At higher temperature, the ribbons disappeared, and the domains have transformed into a continuous layer. As in the case of alkane layers on graphite, the F14H20 layer retains its order at temperatures above T_m. The molecular order of this layer is characterized by two spacings: 7.2 nm and 9 nm. The first one describes the linear strips, whereas the second one – the micellar arrangement, which is distinctively seen at the right edge of the image in Figure 21b.

AFM studies of the F14H20 adsorbates provide new information concerning the molecular order in these systems. The fact that the molecular arrangement on graphite shows a hexagonal symmetry implies that alkyl groups are interacting with the substrate. The spacing of the molecular order (~7 nm) is smaller that the double length of the F14H20 molecule. This suggests that either the molecules are tilted within these structures or there is a partial interdigitation of the molecules forming double layer.

AFM provides several important findings concerning self-assemblies of the semifluorinated alkane although these data alone are not enough for complete understanding

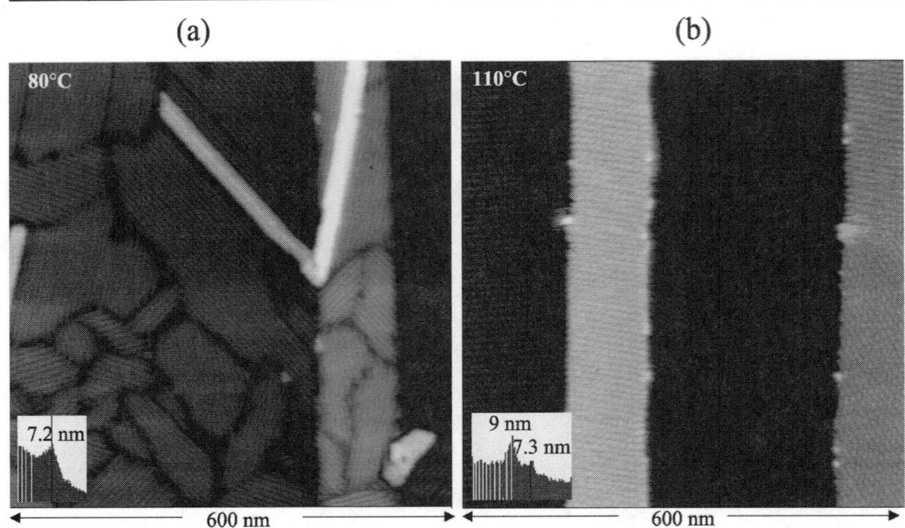

Figure 21. (a)–(b) Height images of F14H20 nanostructures deposited on graphite from octane solution, which were obtained at 80°C and 110°C, respectively. The inserts in the left bottom corner show 2D power spectral density plot and values of the most pronounced peaks.

of their molecular organization. Further interplay of AFM, X-ray diffraction and spectroscopic measurements is needed for accomplish this goal.

STUDIES OF HETEROGENEOUS SYSTEMS

Semicrystalline Polymers

In the following, we will consider AFM studies of heterogeneous polymer systems, which are the most common objects of industrial research. Multicomponent polymer systems present an important class of commercial materials that addresses variety of practical applications. The way these materials are prepared substantially influences their technological properties. Therefore, examination of morphology and composition of multicomponent polymer systems provides key evidence for optimization of their formulation and properties. Strictly saying, heterogeneous materials are those, which consist of components with different chemical components: copolymers, polymer blends, composites, etc., semicrystalline materials can be also assigned to such materials.

In electron microscopy, visualization of lamellar structures of semicrystalline polymers is usually performed on samples, which were etched or stained. These procedures are not necessary for AFM studies because the crystalline and amorphous components are differentiated due to differences of their mechanical properties that are reflected in the probe response. Practically, compositional mapping is achieved with hard tapping when the images demonstrate the distribution of individual components on the sample surface. For many polymer systems, especially those with rubbery-like components, imaging at elevated forces might result in the AFM probe penetration through top rubbery layer. In this way, more rigid subsurface structures can be seen in the images at the depths up to hundreds of nanometers.

Bulk morphology of polymer samples can be of primary interest and the application of an ultramicrotome is needed to access the bulk polymer structure with AFM. This preparation tool became the important accessory for AFM analysis of heterogeneous polymer samples. It is worth noting that polymer blocks with a smoothly cut surface, which is best prepared with a diamond knife, are usually employed for AFM imaging. Thus an elaborate and time-consuming preparation of ultrathin slices of polymer material, which is required for TEM, is eliminated. It is worth noting that quantitative estimates of the sample composition, in the analysis of AFM images, are more accurately reflect this important characteristics than the data obtained with TEM. The reason is that TEM micrographs present 2D view of a thin but still 3D material section.

First two examples are dealing with morphology of PE films. AFM images of industrial linear low-density PE film, whose surface and near-surface morphology can be examined without any sample preparation, are shown in Figures 22a–b. The height image presents the large-scale surface morphology with corrugated structures formed by micron-size bumps. At higher magnification, lamellar structures are resolved in the phase image. The lamellar edges are seen as individual bright strips or as their clusters at locations where lamellar stacks are present. In addition to the image contrast, individual components of heterogeneous materials can be also identified due to their specific shape. The corresponding example is presented in AFM images of commercial low-density PE, Figures 22c–d. The large-scale surface morphology of this material is different from that in Figure 22a. An orientation pattern is recognized by surface features, which extend from the top left corner to the right bottom corner in Figure 22c.This is a consequence of mechanical stresses during manufacturing of the film. The platelets of a mineral filler are clear seen in the phase image in Figure 22d. Such fillers are added to polymers in order to modify their properties, e.g. adhesion, mechanical strength, flammability, etc.

Block Copolymers

Block copolymers are important components of engineering plastics. In recent years, the interest to these materials is growing because of unique capabilities of tuning their nanoscale architecture through microphase separation. Among new applications areas are nanolithography and the templating of inorganic structures such as nanowires and magnetic dots. Extended ordered structures, which are needed for these applications, are developed during annealing process above glass transition temperatures (T_g) of the blocks. Visualization of microphase separation patterns of block copolymers with AFM is trivial when hard tapping imaging is performed at temperature where one component is in glassy state, another – in rubbery-like. These both issues: annealing and visualization are illustrated by AFM images in Figures 23a–b. The phase images were obtained on a film triblock copolymer of polystyrene-block-polybutadiene-block-polystyrene (SBS) immediately after spin-casting and after intensive annealing. The differences of morphology of the samples are evident. The alternative lamellar structures

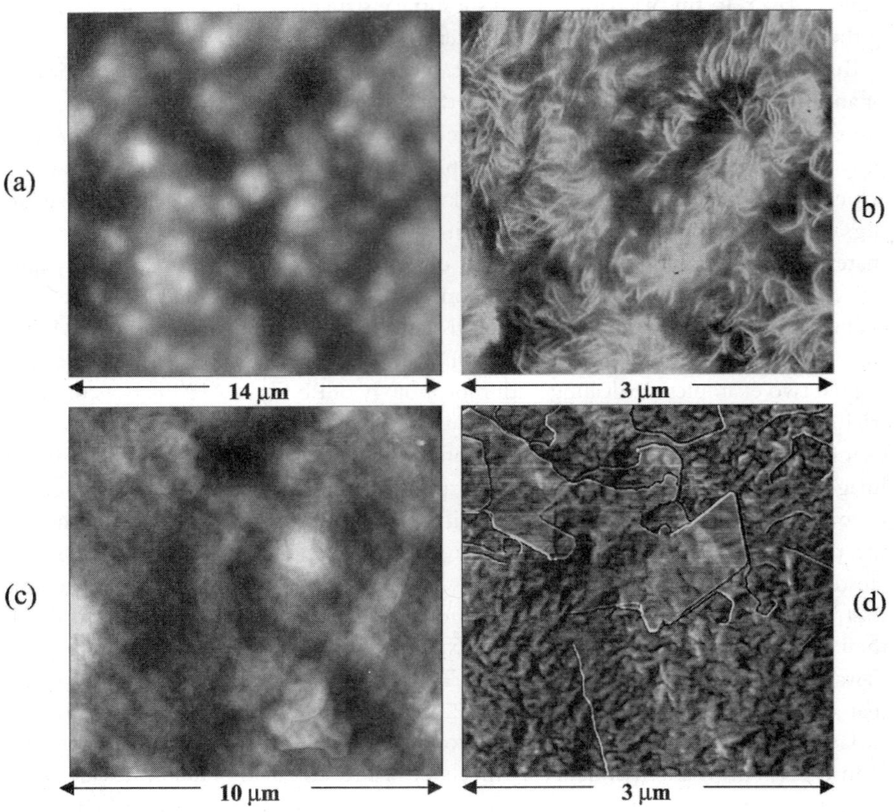

Figure 22. (a)–(b) Height and phase images of surface of linear low-density PE film. (c)–(d) Height and phase images of surface of low-density PE film.

belong to glassy polystyrene (PS) blocks, which have a bright contrast, and to rubbery polybutadiene (PB) blocks, which are recognized by darker contrast. Actually, the contrast between the components is strong at temperatures of around 100°C where softening of PS block occurs. Therefore, for some block copolymers, high-contrast imaging can be achieved at temperatures above the glass transition of one component but below the glass transition of another one [39].

An example of AFM characterization of bulk morphology of block copolymers is given in Figures 23c–d. These images were obtained on two cross-sections, which were made perpendicular and parallel to the main direction of a SBS rod, respectively. The rod was initially prepared with a viscosimeter and subjected to an extremely long term annealing. A well-ordered hexagonal pattern was detected in the AFM image of the perpendicular cross-section (Figure 23c) and the extended linear structures are seen in the image of the parallel cross-section (Figure 23d). These results unambiguously indicate that the perfect packing of polystyrene and polybutadiene cylinders was

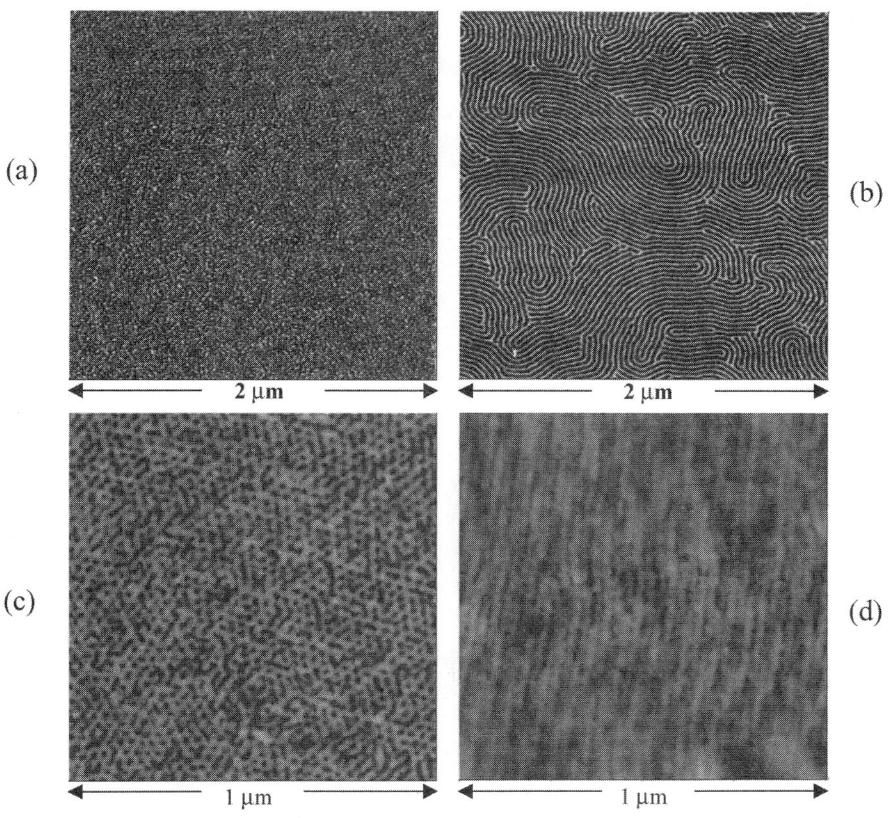

Figure 23. (a)–(b) Phase images of polystyrene-block-polybutadiene-block-polystyrene (SBS) film just after spin casting and after high-temperature annealing, respectively. (c)–(d) Phase image of cross-sections of an annealed SBS rod, which were made perpendicular and parallel to the rod main direction.

achieved in this material. As in the case of the SBS film, the dark spots in the image in Figure 23c correspond to PB cylinders and the bright spots – to PS [40].

Polymer Blends and Nanocomposites

Multicomponent rubber-like materials are often explored with AFM to tackle various problems of morphology of elastomer blends, partially cross-linked materials and visualization of filler [carbon black (CB), silica, clay particles, oil, etc] distribution, [41, 42]. Oil is usually incorporated in high viscous elastomers matrix to improve the material rheology and decreases the cost as well. Here we present the results of the ongoing study of morphology alterations of ethylene-propylene-diene rubber (EPDM) loaded with oil, which is caused by cross-linking process. The phase image of uncured EPDM filled with oil exhibits the complex morphology of this material, Figure 24a. Dark locations are related to oil component and they are distributed through the sample either

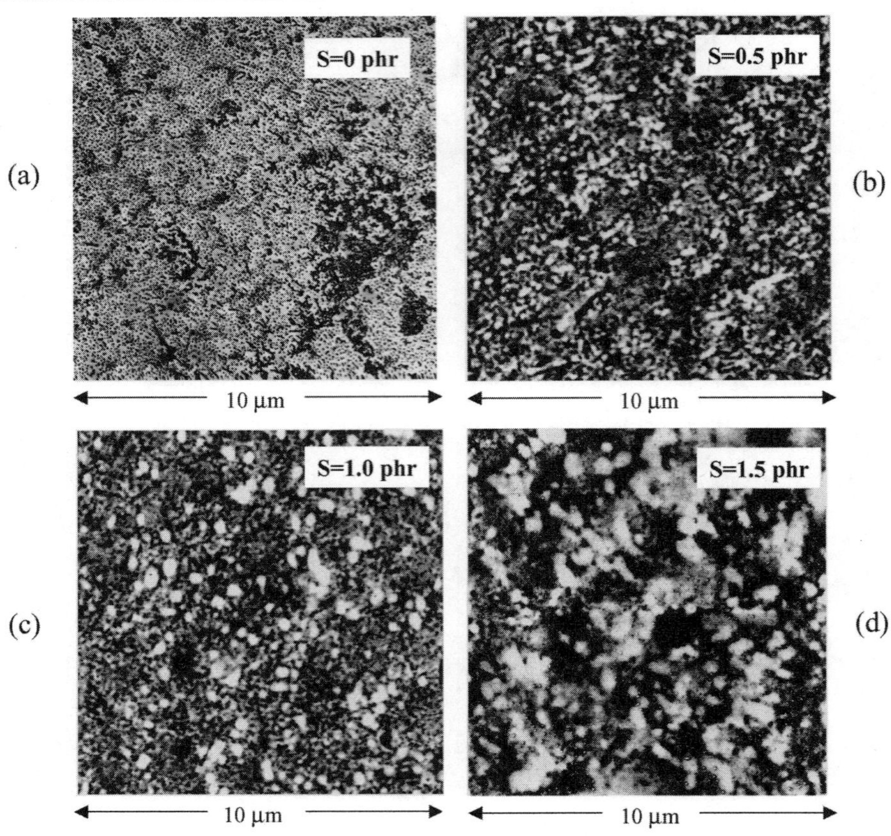

Figure 24. Phase images of EPDM rubber loaded with oil in an uncured state (a) and after curing with different amount of sulfur: S = 0.5 phr – (b), S = 1.0 phr – (c) and S = 1.5 phr – (d).

as homogeneously spread nanoscale inclusions or as larger clusters of various size and shape. Bright domains, which appear in the images of the samples after curing, represent the cross-linked rubber material. With the increase of concentration of curing agent (sulfur) from 0.5 phr to 1.0 phr the size of these domains enlarges from 52 to 82 nm, Figures 24b–c. As sulfur content increased further (S = 1.5 phr), the cross-linked material is seen as aggregates of 100–200 nm in size, and they occupied 28% of the image area, Figure 24d. This is twice more than the 13% area coverage by the cured rubber at S = 0.5 phr. that is twice large than in Figure 24b.

New engineering materials—thermoplastic vulcanizates (TPV) are more easily processed than traditional cross-linked rubbers but exhibit similar performance. They are essentially blends of rubber components with plastics, which are loaded with different fillers. Therefore, analysis of morphology of such TPV is helpful for optimization of their performance. Electrically conducting TPV, which are filled by CB particles, are can be used as sensors, switches, and electromagnetic shields. A percolation threshold

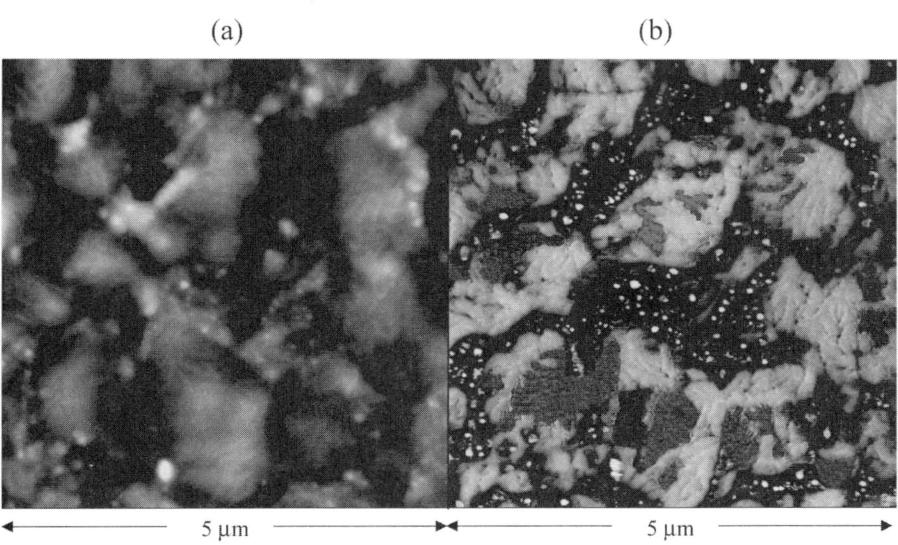

Figure 25. (a)–(b) Height and phase images of thermoplastic vulcanizate made by mixing EPDM, isotactic polypropylene and carbon black.

in these materials could be reached at small CB loading because of selective localization of these conducting particles in one of the components or at the interfaces. Electric force microscopy (EFM) in also attracted for imaging of CB particles [43]. Height and phase images in Figures 25a–b show morphology of the material made of isotactic polypropylene (iPP), EPDM and CB. The distribution of individual components is well distinguished in the phase image. The bright patches are allocated to EPDM domains in which the contrast variations point to locations with different cross-linking density. There are no indications of the presence of CB particles in the rubber domains. The darker regions with 40–50 nm bright particles represent iPP domains filled by CB. The described morphology has been the result of the optimization of mixing conditions and allows achieving low specific resistance (2.5 Ohm × cm) with minimal CB loading.

Composite materials with filler particles of tens and hundreds of nanometers in size are known and utilized for a long time. Recent attention to such composite materials has been motivated by research efforts to design nanocomposites with improved mechanical, adhesion, thermal and other properties. Polymers filled with mineral layered materials, such as different clays, graphite, etc. are of special interest. One of the possible developments in these systems is based on exfoliation of clay clusters into individual sheets of molecular thickness. Success of these efforts will give considerable increase of the filler surface that will impact properties, which depend on polymer-filler interfacial interactions. The intercalation and exfoliation processes can be monitored with high-resolution TEM and AFM that offer visualization of the clay particles and their individual sheets. Two AFM images of the polymer nanocomposite are presented

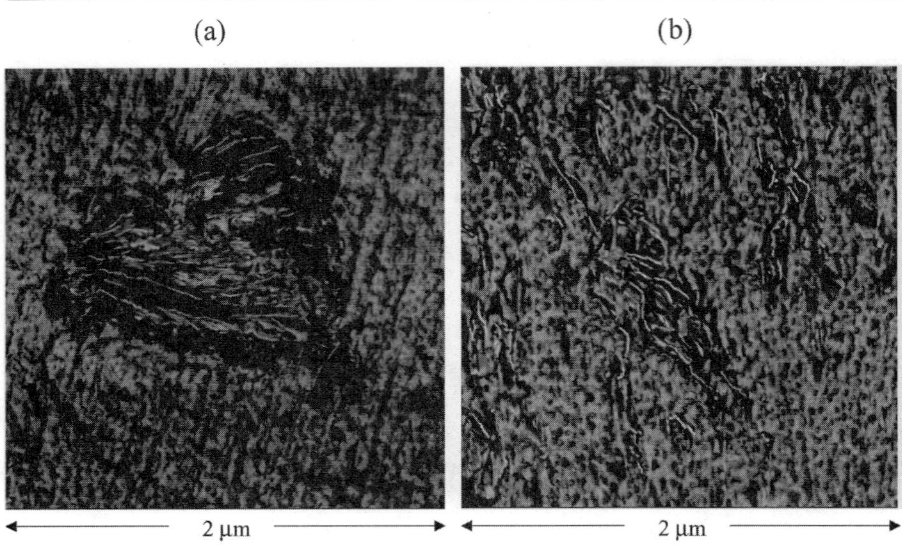

Figure 26. (a)–(b) Phase images of nanocomposite prepared of polypropylene and clay particles.

in Figures 26a–b. The images were obtained on a microtomed surface and the first of them shows a clay particle with multilayer morphology. The sample location with individual clay sheets, which are seen as edge-on structures of 40–50 nm in width, is presented in the second image. This morphology hints on successful exfoliation of the clay particles. There are several issues of sample preparation and optimization of imaging of extremely thin filler layers that should be considered in interplay between AFM and TEM.

Biomaterials are one of the fast growing areas where AFM role for material characterization will increase substantially. Obviously, dealing with materials, which performance includes interactions with living tissue and circulating blood in physiological conditions, will be more challenging for characterization. Therefore, there are a lot of hurdles to overcome, and we are just at the beginning. First steps are already made and AFM imaging of biological objects is quite established field, and there is also knowledge in studies of polymer materials, which are often used as biomaterials. Example of AFM studies of biomaterial is give by images in Figures 27a–b. These images were obtained on surface of felt matrix, which is prepared by mixing collagen and hyaluronic acid (HyA). The felt is employed for connective tissues repair and cosmetic purposes [44]. The large-scale morphology in Figure 27a shows an extended collagen aggregate incorporated into HyA. The later is recognized by dendritic morphology, which is formed of fibrilar strands. The phase image in Figure 27b demonstrates the details of the collagen aggregates, which is built of individual collagen. Collagen is a triple helix object with 300 × 1.5 nm dimensions, which consists of three extended protein chains that wrap around one another [45, 46]. The collagen molecules are also exhibit 60 nm pitch. The fact that similar pitch is observed in the extended aggregate

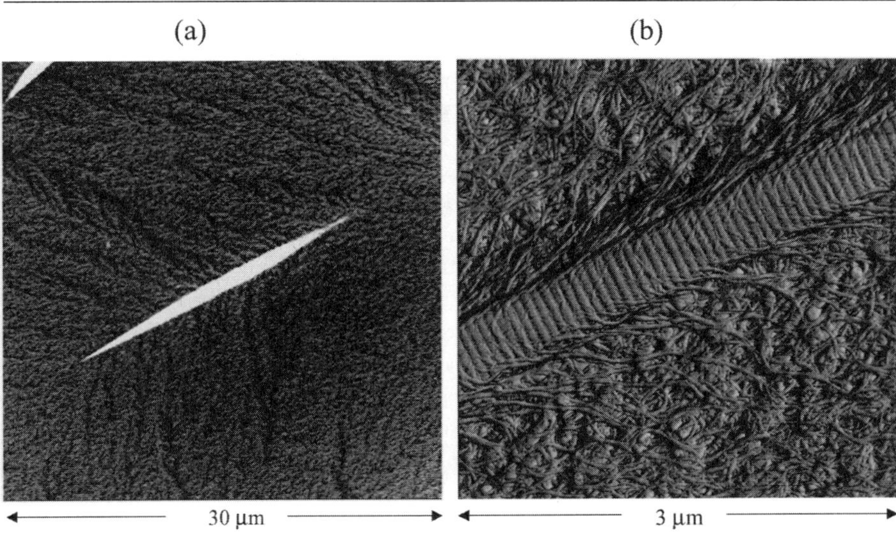

Figure 27. (a)–(b) Height and phase images of a surface of felt matrix prepared by mixing collagen and hyaluronic acid.

(Figure 26b) implies the ordered stacking of individual collagen molecules forming this structure. In addition, one sees a large number of thin strands unwinding from the aggregate into the nearby surface regions. This architecture helps of the collagen incorporation into the HyA matrix.

CONCLUDING REMARKS

Atomic force microscopy is the leading scanning probe technique and has become important in characterization of materials at the sub-micron scale. A principle AFM application is making high-resolution 3-dimensional images of surface topography. The examination of roughness of manufactured surfaces and accurate control of microscopic patterns are two important areas of industrial AFM applications. The unbeatable force sensitivity and nm-scale resolution of AFM is likely to make it practically the exclusive method for evaluation of sub-100-nm structures that will be the essence of nanotechnology. Ongoing development of softer and shaper probes will empower this technique further. This progress will enhance image resolution and will improve control over tip-sample forces at lower force level thus allow imaging of extremely soft materials impossible to date. In addition to high-resolution imaging, compositional mapping is recognized as the key feature for the majority of applications in studies of soft materials such as polymers and biological objects. The examples collected in this chapter support this conclusion. Further expansion of AFM experimental data is expected with a broader use of temperature and environmental accessories.

Progress is also expected in local measurements of mechanical properties. There are a number of different approaches (force modulation, oscillatory shearing, nanoindentation, etc.), which are applied for this purpose. These techniques provide semi-quantitative data in a relatively narrow frequency range, mostly around resonant frequencies of the piezoactuators or probes. Although such measurements are extremely important they should be extended to a broader frequency range over several decades of frequencies where molecular motion and relaxation of macromolecules take place. Such studies of polymer viscoelastic behavior have become important not only for macroscopic samples but also for functional plastic and rubbery-like structures and elements with dimensions in the millimeter and micron range. Measurements at those scales should be performed with AFM–related techniques. The demand for such studies will further increase with development of nanotechnology.

The probing of materials with AFM is basically realized through mechanical interactions. Therefore, only indirect correlation between these measurements and chemical nature of a sample or its constituents can be derived from such studies. This does not fulfill the need of compositional analysis with chemical identification. The latest achievements in this field are primarily related with soft X-ray microscopy, which provide chemically sensitive mapping with resolution of tens nanometers [65]. Yet this technique requires the use of synchrotron radiation sources. Conventional methods like FTIR-microscopy provides chemical mapping only with resolution of several microns. Recently, a combination of AFM and infrared spectrometry has been successfully applied to collecting IR spectra of polymers from the tip-sample junction [66]. In perspective, such a combination might become an extremely powerful mapping technique with lateral resolution in the tens of nanometers range.

REFERENCES

1. G. Binnig, H. Rohrer, Ch. Gerber and E. Weibel, *Phys. Rev. Lett.* 49 (1982), 57.
2. G. Binnig, C. Quate and Ch. Gerber, *Phys. Rev. Lett.* 56 (1986), 930.
3. S. Alexander, US Patent 4,369,657, filed July 25, 1981.
4. S. Alexander, L. Hellemans, O. Marti, J. Schneir, V. Elings, P. K. Hansma, M. Longmire and J. Gurley, *J. Appl. Phys.* 65 (1989), 164.
5. Y. Miyahara, M. Deschler, T. Fujii, S. Watanabe and H. Bleuler, *Appl. Surf. Sci.* 18 (2002), 450.
6. Q. Zhong, D. Innis, K. Kjoller and V. B. Elings Surf. Sci. Let. 290 (1993) L688.
7. D. Klinov and S. Magonov, *Appl. Phys. Lett.* 84 (2004), 2697.
8. The diamond probes are made by B. Mesa, MicroStar Technology, Huntsville, TX, USA, www.microstartech.com.
9. H. J. Dai, J. H. Hafner, A. G. Rinzler, D. T. Colbert and R. E. Smalley, *Nature* 384 (1996) 147.
10. J. H. Hafner, C. L. Cheung and C. M. Lieber, *Nature* 398 (1999), 761.
11. C. Su and S. Magonov, Patent Application, USA (2003).
12. R. Pechmann, J. M. Koehler, W. Fritzsche, A. Schaper and T. V. Jovin, *Rev. Sci. Instr.* 65 (1994) 3702.
13. S. N. Magonov and M.-H. Whangbo, *Surface Analysis with STM and AFM*, VCH, Weinheim (1996).
14. P. Mailvald, H.-J. Butt, S. A. C. Gould, C. B. Prater, B. Drake, V. Elings and P. K. Hansma, *Nanotechnology* 2 (1991), 103.
15. A. L. Weisenhorn, P. Maivald, H.-J. Butt and P. K. Hansma, Phys. Rev. B. 45 (1992) 11226.
16. D. Chernoff and S. Magonov in: *Comprehensive Desk Reference of Polymer Characterization and Analysis*, R. F. Brady, Jr. (Ed.) pp. 490–531ACS, Oxford Press, 2003.
17. U. Landman, W. D. Luedke and A. Nitzan, *Surf. Sci. Lett.* 10 (1989), L177.
18. F. J. Giessibl, *Science* 267 (1995), 68.

19. Mohler, C. E., Landes, B. G., Meyers G. F., Kern, B. J., Oullette K. B., Magonov S. N. AIP Conference Proceedings 683 Characterization and Metrology for ULSI: 2003 International Conference, D. G. Seiler et. al. (Eds.) pp. 562–566 American Institute of Physics, New York (2003).
20. F. Y. Hansen, L. W. Bruch and H. Taub, *Phys. Rev. B* 54 (1996), 14077.
21. S. N. Magonov, N. Yerina, G. Ungar and D. Ivanov, in preparation (2004).
22. D. A. Ivanov, R. Daniels, and S. N. Magonov, *Exploring the High-Temperature AFM and Its Use for Studies Polymers*, Application Note published by Digital Instruments/Veeco Metrology Group (2001). URL: http://www.veeco.com/appnotes/AN45_HeatingStage.pdf
22. S. A. Ponomarenko, N. I. Boiko, V. P. Shibaev and S. N. Magonov, *Langmuir* 16 (2000), 5487.
23. Yu. K. Godovsky and S. N. Magonov, *Langmuir* 16 (2000), 3549.
24. D. A. Ivanov, Z. Amalou and S. N. Magonov, *Macromolecules* 34 (2001) 8944.
25. Yu. K. Godovsky, V. S. Papkov and S. N. Magonov, *Macromolecules* 34 (2001), 976.
26. J. Kumaki, Y. Nishikawa and T. Hashimoto, *JACS* 118 (1996), 3321.
27. S. S. Sheiko and M. Möller, *Chem. Revs* 101 (2001), 4099.
28. M. Gerle, K. Fischer, M. Schmidt, S. Roos, A. H. E. Mueller, S. S. Sheiko, S. A. Prokhorova and M. Möller, *Macromolecules* 32 (1999), 2629.
29. V. Percec, M. N. Holerca, S. N. Magonov, D. J. P. Yeardley, G. Ungar, H. Duan and S. D. Hudson, *Biomacromolecules* 2 (2001), 706.
30. S. Magonov in: *Applied Scanning Probe Techniques*, H. Fuchs, M. Hosaka and B. Bhushan (Eds.) pp. 207–251, Springer, Berlin, 2004.
31. S. Magonov and N. Yerina, *Langmuir* 19 (2003), 504.
32. S. N. Magonov, N. A. Yerina, G. Ungar, D. H. Reneker and D. A. Ivanov, *Macromolecules* 36 (2003), 5637.
33. G. Ungar and X. Zeng, *Chem. Revs*. 101 (2001), 4157.
34. P. H. Geil, *Polymer Single Crystals*, John Willey & Sons, New York, (1963).
35. S. N. Magonov, N. A. Yerina and D. H. Reneker, *Macromolecules* (2004), submitted.
36. S. N. Magonov and Yu. Godovsky, *Amer. Lab.* (31) 1999 , 52.
37. G. L. Gaines, *Langmuir* 7 (1991), 3054.
38. M. Maaloum, P. Muller and M. P. Kraft, *Angew. Chem* 114 (2002), 4507.
39. M. J. Fasolka, A. M. Mayes and S. Magonov, *Ultramicroscopy* 90 (2001), 2.
40. S. N. Magonov, V. Elings, J. Cleveland, D. Denley and M.-H. Whangbo, *Surface Science* 389 (1997), 201.
41. A. A. Galuska, R. R. Poulter and K. O. McElrath, *Surface and Interface Analysis* 25 (1997) 418.
42. N. A. Yerina and S. N. Magonov S. N., *Rubber Industry and Technology* 76 (2003) 846.
43. R. Viswanathan and M. B. Heaney, *Phys. Rev. Lett.* 75 (1995), 4433.
44. C. J. Doillon, C. F. Whyne, S. Brandwein and F. H. Silver, *J. Biomed. Mater. Res.* 20 (1986), 1219.
45. K. A. Piez, in: *Encyclopedia of Polymer Science and Engineering*, Vol. 3, p. 699. Willey, New York (1985).
46. H. Yokota, F. Johnson, H. Lu, R. M. Robinson, A. M. Belu, M. D. Garrison, B. D. Ratner, J. Trask, D. L. Miller, *Nucleic Acids Research* 25 (1997), 1064.

5. SCANNING PROBE MICROSCOPY FOR NANOSCALE MANIPULATION AND PATTERNING

SEUNGHUN HONG, JIWOON IM, MINBAEK LEE AND NARAE CHO

1. INTRODUCTION

1.1. Nanoscale Toolbox for Nanotechnologists

In everyday life, we utilize various *tools* such as pens, knives, needles etc. From the beginning of human history, "tools" have been one of the key aspects that differentiate ourselves from wild animals and define our civilizations. For this reason, we are called "Homo Faber," and human history is often divided into several ages based on the tools used at that time period: stone age, bronze age, and iron age etc.

The age of modern nanotechnology began with the inventions of nanoscale precision imaging tools such as scanning probe microscopes (SPM). SPM utilizes *a nanometer-size needle-shape probe* (or simply called, "tip") to image surfaces in a close proximity somewhat like blind people use a stick to find out the road conditions [1]. In scanning probe lithography (SPL) processes, the same probes for SPM are utilized to modify substrates in various manners such as scratching, writing, burning etc. Various SPL methods and their functionalities are summarized in Table 1. As a matter of fact, SPL provides quite versatile functionalities similar, but with nanometer scale precision, to many macroscale tools. And it became a key component of modern "nanoscale toolbox." This review provides a brief overview about states-of-the-art SPL processes and their applications.

The major components of SPM systems are 1) nanoscale proximal probes that directly interact with samples and 2) piezoelectric actuator that drives the probe with nanoscale precision. Three typical probes used in SPM systems are shown in Figure 1: 1)

Table 1. List of Scanning Probe Lithography Methods and Their Capabilities.

SPL Method		Instruments	Environment	Key Mechanism	Typical Resolution	Patterning Materials	Possible Applications
Nanoscale Pen Writing	Dip-Pen Nano-lithography	AFM	Ambient	Thermal Diffusion of Soft Solid	~10 nm	SAM, Biomolecules, Sol-Gel, Metal etc.	Biochip, Nanodevice, Mask Repair etc.
	Nanoscale Printing of Liquid Ink	NSOM	Ambient	Liquid Flow	~100 nm	Etching Solution, Liquid	Mask Repair etc.
Nanoscale Scratching	Nanoscale Indentation	AFM	Ambient	Mechanical Force	~10 nm	Solid	Mask Repair etc.
	Nanografting	AFM	Liquid Cell	Mechanical Force	~10 nm	SAM	Biochip etc.
	Nanoscale Melting	AFM	Ambient	Mechanical Force and Heat	~10 nm	Low Melting Point Materials	Memory etc.
Nanoscale Manipulation	Atomic and Molecular Manipulation	STM	Ultrahigh Vacuum (Often Low Temperature)	Van der Waals or Electrostatic Forces	~0.1 nm	Metals, Organic Molecules etc.	Molecular Electronics etc.
	Manipulation of Nanostructures	AFM	Ambient	Van der Waals or Electrostatic Forces	~10 nm	Nanostructure, Biomolecules	Mask Repair, Nanodevices etc.
	Nanoscale Tweezers	Possibly AFM	Ambient	Van der Waals or Mechanical Force	~100 nm	Nanostructures	Electrical Measure. etc.
Nanoscale Chemistry	Nanoscale Oxidation	STM or AFM	Humid Air	Electrochemical Reaction in a Water Meniscus	~10 nm	Si, Ti etc.	Nanodevice etc.
	Nanoscale Desoprtion of SAM	STM or AFM	Humid Air	Electrochemical Reaction in a Water Meniscus	~10 nm	SAM	Nanodevices etc.
	Nanoscale Chemical Vapor Deposition	STM	Ultrahigh Vacuum with Precursor Gas	Nanoscale Chemical Vapor Deposition	~10 nm	Fe, W, etc.	Magnetic Array etc.
Nanoscale Light Exposure	Nanoscale Light Exposure	NSOM	Ambient	Photoreaction	~100 nm	Photosensitive Materials	Nanodevices etc.

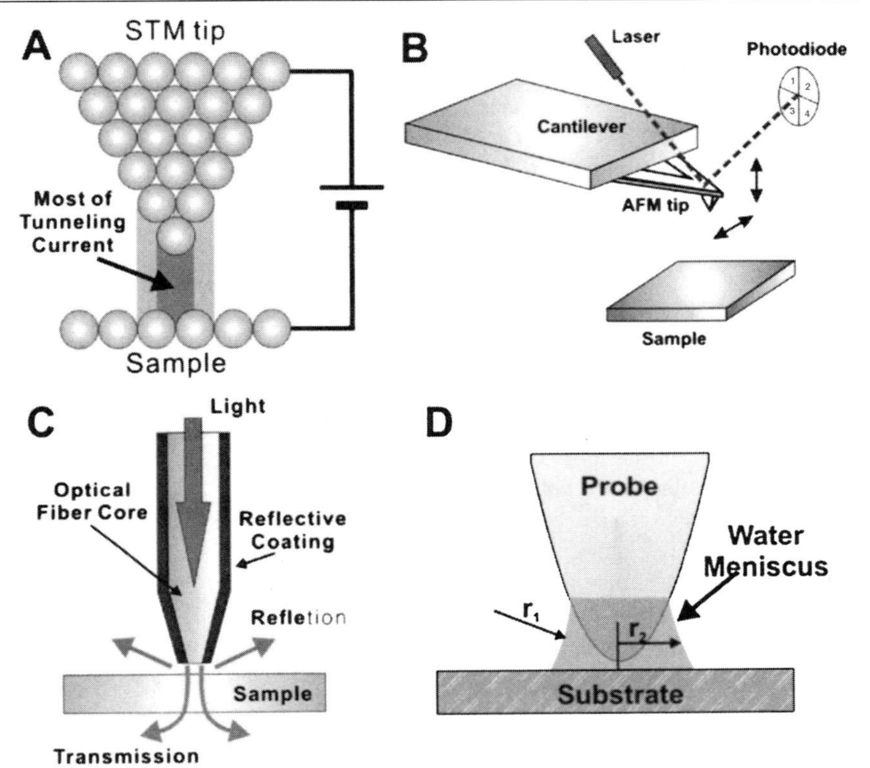

Figure 1. (A)–(C) Schematic diagram depicting three different scanning probes. (A) Scanning tunneling microscopy. (B) Atomic force microscope. (C) Near-field scanning optical microscope. (D) Schematic diagram depicting a water meniscus formed at a probe under ambient conditions.

scanning tunneling microscopy (STM) probe, 2) atomic force microscopy (AFM) probe, and 3) near field scanning optical microscope (NSOM) probe.

A STM probe is a nanoscale, often atomic scale, sharp *metallic* needle (Figure 1A). When the tip is near the surface, usually in a few nanometer, with a voltage bias, tunneling currents flow between tip and surface. At a small bias voltage V, the tunneling current I can be approximated by

$$I \propto Ve^{-\frac{\sqrt{8m\phi}}{\hbar}d}$$

where d is the distance between STM tip and sample and ϕ is the work function of the tip [2]. For a typical work function of 4 eV, the tunneling current decreases about one order of magnitude when the gap increases only by 0.1 nm. For this reason, the tunneling current is confined usually at the very end, often at one atom, of the tip (Figure 1A), which allows one to inject the current only in a nanometer scale region. Also, the current change indicates the gap distance changes. In a constant-height STM imaging mode, the tunneling current signal is monitored by the feedback circuit to maintain a constant gap distance between tip and surface.

An AFM probe is a sharp conducting or non-conducting needle with a flexible cantilever (Figure 1B). In common AFM operations, an AFM probe is either in a direct contact with sample (contact mode), or it oscillate near the surface (non-contact mode). In a contact mode, the *repulsive force* between tip and surface causes the bending of the cantilever. The bending can be measured by measuring the reflected laser beam path utilizing a quadrapole photodetector. This force detection system combined with piezoelectric actuator allows one to precisely control the contact force between tip and surface.

In a non-contact mode operation, the AFM probe oscillates near the sample surface and the oscillation amplitude of the cantilever is measured by measuring that of the reflected laser beam path via a photodetector. When the tip is near the surface, the *attractive force* between tip and surface cause the resonance frequency change and result in the change of the oscillation frequency.

A NSOM probe is optical fiber or a hollow micropipette that can transmit light in it (Figure 1C). Unlike other regular optical fibers, a NSOM probe is tapered at one end so that transmitted light comes out only from a few tens of nanometer area at the end. A NSOM probe allows us to shine light to a specific region of the sample surface.

In SPL processes, these probes are located at a desired location on the sample via piezoelectric actuator and the nanoscale region is modified in various means (e.g. tunneling current, force, light etc.). The results of SPL often strongly depend on environmental conditions. One important example is a water meniscus between tip and surface under ambient conditions (Figure 1D). When a SPM probe is in a close proximity with the surface, water from the air condenses between tip and surface. Based on Kevin Equation, the shape of water meniscus can be approximated by

$$\left(\frac{1}{r_1}+\frac{1}{r_2}\right)^{-1} = \frac{\gamma V}{RT\,\log(p/p_s)}$$

where r_2, γ, V, and p/p_s are the radius of the water meniscus, a surface tension, molar volume, and relative vapor pressure, respectively [3]. As matter of fact, it is a fair statement that, under ambient conditions, scanning probes actually interact with sample through the water meniscus. A water meniscus causes many parasitic effects such as capillary forces and affects the result of SPL.

1.2. Motivations

Two major motivations for the development of SPL techniques can be: 1) the resolution limit of microfabrication techniques and 2) needs for versatile lithographic method for new materials including biomaterials.

The major lithography technique for modern microelectronics is the microscale-resolution photographic patterning method called, "*photolithography*." In this process, solid surfaces are first coated with a photosensitive polymer resist layer, and only specific regions on the resist layer are exposed to the UV light through a photographic mask. The exposure of light breaks (or strengthens) the molecular bonding in the polymer resist layer. The exposed (or unexposed) resist layer is specifically removed by immersing the substrate into certain solvents, and the polymer resist patterns were left on the specific area of the substrate. The polymer resist pattern can be used as a resist of subsequent etching or other processing steps. By repeating these processes, many complicated microelectronic components can be created on a single substrate, which enables integrated computer chips and, eventually, modern information technology.

During the last four decades, semiconductor industry has been extensively working on improving the resolution of photolithography method so that more components can be integrated in a single chip. This size reduction and higher degree of integration are crucial to improve the performance and power efficiency of the integrated circuits, which can effectively reduce the manufacturing cost. Currently, microelectronic chips with a feature size as small as 90 nm are commercially available.

Due to the diffraction of light, one of the major parameters determining the ultimate resolution of photolithography is the wavelength of light. According to the Rayleigh Equation [4], the resolution of the photolithography method is,

$$R = k_1 \lambda / NA$$

where λ is the wavelength of the light used, NA is the numerical aperture of the lens system, and k_1 is a constant that depends on the photoresist. It is usually believed that resolution of the photolithography is approximately $\lambda/2$. Extensive efforts have been given to use shorter wavelength light to improve the resolution. Ultra violet light with a wavelength of ~160 nm has been employed to fabricate ~90 nm commercial chips. However, it is becoming increasingly difficult to find transparent optical components for shorter waverlength light. Various techniques have been studied as a next-generation lithography method. These include scanning probe-based lithography, extreme ultraviolet lithography based on reflection optics, X-ray lithography, e-beam projection system and stamping methods [5]. Many new methods successfully demonstrated the lithographic capabilities down to 10 nm size features. However, it is not yet clear which method will eventually replace the photolithography technique. SPL process also has been extensively studied as a possible candidate to overcome the resolution limit of microfabrication.

On the other hand, another crucial applications for SPL is patterning soft materials including biomaterials. Recent dramatic development of nanotechnology and biological science provides us *new nanometer scale building blocks* for functional devices such as nanoparticles, nanowires, biomolecules, organic molecules etc. By combining these new materials with conventional solid-state devices, one should be able to build a generation of new functional devices. These include biological sensors [6–10] to detect harmful virus and protein motor-based nanomechanical systems [11, 12]. These new generation devices are often termed as *hybrid devices* because they are comprised of organic materials as well as solid-state nanostructures. However, these new materials are not compatible with conventional photolithography process. For example, photolithography processes often heat samples up to 100°C, while many biomolecules lose their functionalities above the body temperature. Since SPL processes can be done under ambient conditions or even in liquid environments, they are suitable for patterning soft materials. For these reasons, SPL have been extensively utilized to pattern chemical and biological molecules on solid substrates in a nanometer-scale resolution.

In the following, we will discuss the basic mechanisms and applications of individual SPL methods.

2. NANOSCALE PEN WRITING

2.1. Dip-Pen Nanolithography

The dip-pen nanolithography (DPN) process is a direct deposition technique [13–16]. It utilizes an AFM tip as a *nanoscale quill pen*, molecular substance as ink, and solid substrates as paper. Figure 2A shows the basic mechanism of DPN as well as approximate dimensions of commonly used pen and ink molecules. When the molecule-coated tip is in contact with the substrate, ink molecules *thermally diffuse* out onto the substrate and form molecular patterns. Under ambient conditions, water condenses at the AFM tip-substrate junction and affects the molecular diffusion. Like macroscale quill pens, the molecular ink coating on the tip surface works as an ink reservoir. A number of variables including relative humidity, temperature, and the tip speed, can be adjusted to control the ink transport rate, feature size, and linewidth. The relative humidity changes the size of water meniscus between tip and surface, which affects the diffusion rate of hydrophilic molecular species. Precise control of the relative humidity is required to achieve nanoscale precision. Even with its short history, the DPN method demonstrated unprecedented $\sim$10 nm patterning resolution [14]. Considering the resolution of current commercial ink-jet printers $\sim$20 μm, the resolution of DPN is revolutionary.

Various types of organic molecules have been used for DPN experiments. The most common molecule ink is self-assembled monolayer (SAM) molecules (Figure 2B) [5]. These molecules are usually comprised of three different parts: 1) chemisorbing group, 2) end group, and 3) chemically inert spacer. When these molecules are deposited onto a proper substrate, they chemically anchor to the substrate and form a stable crystalline monolayer film. Specific binding groups can be chosen depending on the substrate. By depositing these molecules, one can completely change the chemical properties of the surface to that of end groups. Various end groups can be used for specific applications. For example, one can even use specific sequence DNA or proteins as end groups.

The best characterized systems of SAMs are alkanethiols $R(CH_2)_nSH$ adsorbed on gold surfaces. Here, R can be any end groups and thiol (-SH) is chemisorbing groups. Alkanethiols are chemisorbed spontaneously on a gold surface and form alkanethiolates. This process is assumed to occur with the loss of dihydrogen by the cleavage of the S-H bond. Sulfur atoms bonded to the gold surface bring the alkyl chains into close contact: these contacts freeze out configurational entropy and lead to an ordered structure. For alkyl chains of up to approximately 20 carbon atoms, the degree of interaction in a SAM increases with the density of molecules on the surface and the length of the alkyl backbones. This assembly is a very quick process that may occur in a couple of

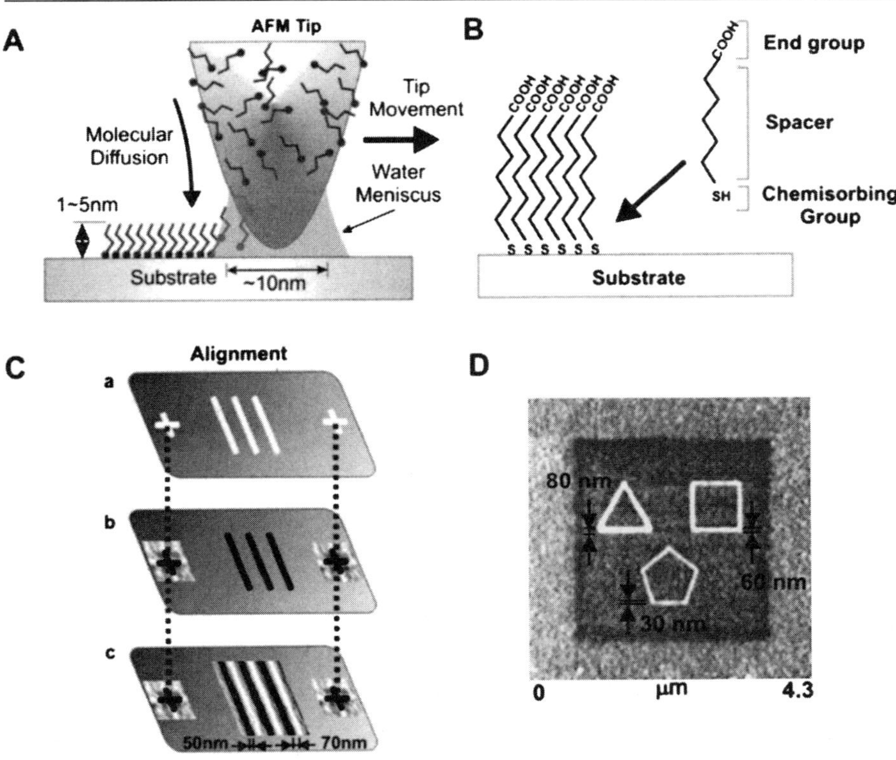

Figure 2. (A) Schematic diagram depicting dip-pen nanolithography. Adapted with permission from [13] *Science* 283 (1999) 661. (B) Schematic diagram showing the basic structure of example self-assembled monolayer (SAM) molecules. SAM molecules chemically anchor to the substrate utilizing the chemisorbing groups and expose the end groups. As a result, the surface properties are converted to the end group. (C) Schematic diagram depicting the method for generating aligned soft nanostructures. (**a**) The first pattern is generated with 16-mercaptohexadecanoic acid (MHA) (denoted by white lines), along with microscopic alignment marks (cross-hairs), by DPN. The actual lines are not imaged to preserve the pristine nature of the nanostructure. (**b**) The second set of parallel lines is generated with 1-octadecanethiol (ODT) molecules, on the basis of coordinates calculated from the positions of the alignment marks in (a). (**c**) Lateral force microscope image of the interdigitated 50-nm-wide lines separated by 70 nm. (D) SAMs in the shapes of polygons drawn by DPN with MHA on an amorphous Au surface. An ODT SAM has been overwritten around the polygons Adapted with permission from [14] *Science* **286**, 523 (1999).

seconds. This ability to form ordered structures rapidly might be one of the factors that ultimately determine the success of the DPN technique.

Even though the DPN technique shares remarkable similarities with macroscale quill pens or ink-jet printers, they also have a significant difference.

Unlike conventional pens, bulk ink coating on the pen in the DPN process remains *soft solid* because DPN writing is usually performed below the melting temperature of molecular ink (However, the readers should be cautioned. At this stage, the exact phase *at the tip/substrate junction* is not yet clearly understood. Because of the water meniscus and nanoscale dimension of the junction, the phase of molecules at the junction might be quite different from that on the tip surface). Thus, the behavior of molecular ink in the DPN process can be described by individual *molecular diffusion* rather than *capillary flow* of liquid ink. In case of macroscale pen that is based on the capillary flow of liquid ink, the ultimate resolution of pen writing is determined by the properties of ink solvent. It is often very difficult to control the liquid flow in nanoscale resolution due to unpredictable behaviors of liquid such as viscosity change, ink meniscus etc. However, in most DPN processes, ink molecules behave like non-interacting 2-dimensional gas whose behavior can be described by Fickian diffusion model [17]. Molecular diffusion based on *random walk* is usually easier to predict and control than the capillary flow of liquid. The nanometer scale precision of DPN can be partly attributed to this different phase of molecular ink and writing mechanism.

It is also worth mentioning several practically important characteristics of DPN. One aspect is a *stable ink deposition rate*. When the DPN process is stabilized, the deposition rate of molecules from the tip to the substrate remains almost constant. The stable writing speed can be attributed to small contact point between AFM tip and substrate. As a result, the amount of ink molecules deposited through the contact point is relatively small compared with the amount of molecular ink (ink reservoir) on the AFM tip surface. From a practical point of view, this stability is a very important in designing practical lithography machine.

Secondly, DPN technique, like other SPL methods, utilizes the same probe for both patterning and imaging, which allows one to image the surface structures with the probe and deposit molecular substances onto specific regions of the substrate. DPN can be used to generate and align nanostructures with pre-existing patterns on the substrate with ultrahigh registration (Figure 2C, D) [14].

Finally, it is relatively easy to develop multiple-probe patterning systems based on the DPN process (Figure 3A). Since DPN is relying on *molecular diffusion* for writing, its patterning speed does not depend on the contact force between probes and surfaces. It implies that one can easily generate multiple patterns simply by attaching multiple probes to a single AFM system and writing patterns simultaneously without worrying about the contact forces between individual probes and surfaces. Hong and Mirkin demonstrated patterning with 8 probes based on this concept [16]. Liu *et al.* demonstrated fabrication of multiple solid nanostructures utilizing a 32-probe system [18].

Patterned SAM can be utilized as an etching resist for wet chemical etching process. Weinberger *et al.* demonstrated etching of Au utilizing 1-octadecanethiol as an etching resist layer [19]. Then, the etched Au patterns are used as an etching resist for the silicon etching process. Significantly, since the thickness of SAM is ~2 nm, it can be an ideal etching resist layer. Also, the resolution of DPN is not limited by optical wavelength.

Specific assembly steps inspired by DPN-generated molecular patterns can be utilized to place nanoscale functional elements at the specific position. Here, surface regions are coated by SAMs with specific functional groups which can attract desired nanostructures (e.g. nanoparticles, carbon nanotubes etc). When the functionalized substrate is place in the solution of nanostructures, the functional groups attract the

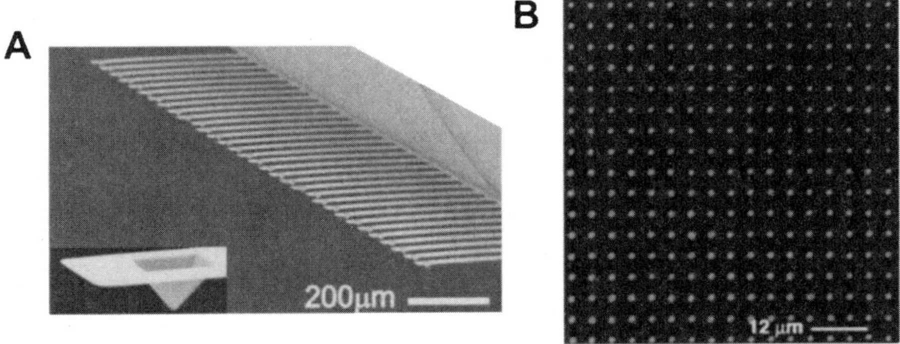

Figure 3. (A) Scanning electron micrograph of a 32-DPN-probe array. The insert shows an enlarged view of a single tip at the end of a beam. Adapted with permission from [18] *Nanotechnology* 13 (2002) 212. (B) Direct DPN transfer of DNA onto SiO_2 substrates. Epifluorescence micrograph of fluorophore-labeled DNA (Oregon Green 488-X) hybridized to a DPN-generated pattern of complementary oligonucleotides on an SiO_2 surface. The scale bar represents 12 μm. Adapted with permission from [20] *Science* 296, (2002) 1836.

nanostructures and induce the assembly onto specific locations. Various interactions can be utilized to specifically assembly nanostructures onto desired surface area. These include covalent bonding, electrostatic interactions, and biomolecular interactions. Especially, biomolecular interaction is quite intriguing because of its variety and specificity. Mirkin *et al.* demonstrated surface functionalization with single strand DNA (ssDNA) molecules via DPN (Figure 3B) [20]. When the functionalized substrate is immersed in the solution of DNA-functionalized nanoparticles, only nanoparticles with specific *complementary* sequence of ssDNAs can assemble to the desired surface area. Hong *et al.* demonstrated directed assembly of carbon nanotubes onto specific location and built carbon nanotube-based electronic devices [21].

Bio-molecular patterning via DPN is of huge practical importance. DPN allows one to directly deposit general biological molecules onto solid surfaces with a nanometer scale resolution. This new capability allows one to envision new biomolecular functional devices such as ultrahigh density gene chips or nanoscale biosensors. Wilson *et al.* demonstrated direct patterning of ~100 nm size thiol-modified *collagen* molecules on Au surface [22]. In this experiment, thiol groups bind to Au surface and form stable structures. However, they also observed that the contact mode writing is not suitable for patterning large bio-molecules because direct contact of AFM tip to the molecules can destroy the bio-molecular structures on the surface. Tapping mode patterning is first applied to generate the organic molecular patterns on the surface. In the tapping mode, the AFM tip gently taps the surface without a direct contact, which can minimize the interaction between AFM tip and surface nanostructures. It allows one to create stable bio-molecular structures with less damage. By this method, Wilson *et al.* created a collagen patterns with 30 nm line width. Mirkin *et al.* demonstrated direct deposition of thiol-modified DNA molecules onto Au surfaces [20]. To

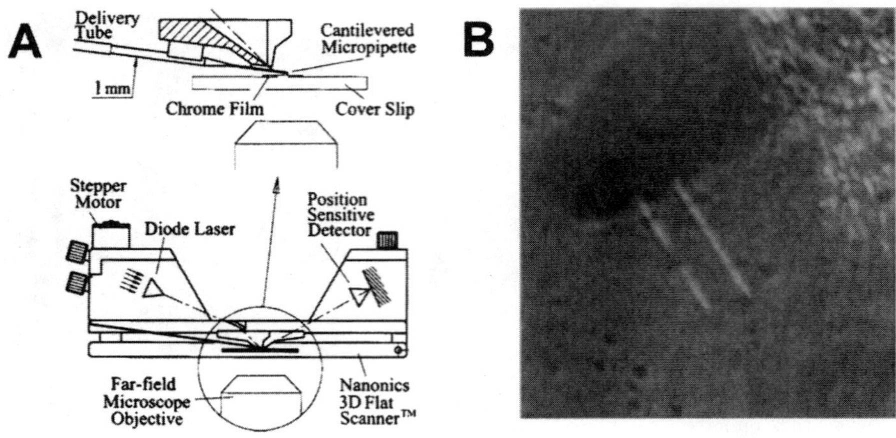

Figure 4. (A) Schematic diagram depicting nanoscale fountain pen system. A cantilevered micropipette is attached to the AFM and delivered etching solution to the substrate. (B) Frames from a video that show the progress of the controlled etching of a chrome film. The linewidth of the etched lines in these far-field images are ~1.45 μm. Adapted with permission from [24] *Appl. Phys. Lett.* 75, 2689 (1999).

enhance the ink loading on the AFM tip, the surface of the AFM tip is first modified by 3'-aminopropyltrimethoxysilane, which promotes reliable adhesion of the DNA ink to the tip surface. They also demonstrated the direct deposition of DNA onto the SiO_2 surface. Before the DPN writing, the surface of thermally oxidized Si wafer was activated with 3'-mercaptopropyltrimethoxysilane (MPTMS). Then, DNA molecules with acrylamide groups are deposited via DPN, and they form stable nanostructures on the surface (Figure 3B).

2.2. Nanoscale Printing of Liquid Ink

A common macroscale direct deposition method is ink-jet printing where liquid ink is driven to the substrate via a pressure and capillary flow. The most common ink-jet systems are office ink-jet printers that deposit dyes onto paper. However, the same strategy has been applied to deposit various materials such as metal, photoresist, DNA etc [23]. Parameters affecting the resolutions of ink-jet type printing include 1) the diameter of dispensing holes, 2) the precision of positioning system, and 3) the capillary effect of liquid inks. Common office ink-jet printers have the resolution of ~20 μm.

Lewis *et al.* utilized a micropipette attached to the AFM system to deposit Cr etching solution onto Cr substrate (Figure 4) [24]. By this way, they could etch the Cr substrate in a submicrometer scale resolution. Micropipettes are hollow glass tubes with the opening diameter often smaller than 100 nm at one end. They are usually used for NSOM systems to transmit light from or to the substrate. Lewis *et al.* bent the micropipettes to form a cantilever so that they can be used to measure forces in AFM system, which allow one to avoid unwanted scratch on the substrate during the deposition due to the excessive contact force. Due to the surface tension of the liquid solution, a relatively high pressure (~3 ATM) has been applied to dispense the ink onto the substrate.

This is a general strategy that can be used for many different species of liquid ink. However several considerations should be given to achieve a high resolution.

First, the diameter (<100 nm) of the opening of the micropipettes is usually much smaller than the sizes (~500 nm) of common dust particles or defects in ink. Extensive purity control over the substrate and solution is required to avoid any clogging.

Secondly, the dispensed liquid ink can form a meniscus at the end of the micropipette. The size of the ink meniscus, often combined with water meniscus from the air, is hard to predict, and it can be as large as the outer diameter (~1000 nm) of the micropipettes. In this case, the resolution can be limited by the outer diameter of the tube even though the hole size can be smaller than 100 nm (Figure 4B). It is crucial to understand the capillary behaviors of ink to achieve a high resolution. In fact, controlling extremely small amount of liquid is an important, though still very difficult, issue for many important applications such as bio-MEMS. Until now, the resolution of the nanopen writing with liquid ink is limited to a sub-micrometer range though one can routinely make micropipettes with sub-100 nm-diameter openings. However, its resolution is expected to improve as we have better understandings about the capillary behavior of nanoscale liquid droplets.

3. NANOSCALE SCRATCHING

3.1. Nanoscale Indentation(1)

A simplest and most common SPL technique is "nanoindentation" which utilizes a mechanically strong AFM tip to physically *scratch off* relatively soft substrates. In this process, an AFM tip is in a direct contact with the substrate and applies a strong contact force, often larger than 100 nN, to the substrate. For comparison, a common force for AFM contact mode imaging is about 1nN. Strong AFM tips such as diamond-coated tips are used for nanoindentation.

Kim and Lieber utilized silicon nitride AFM tip with a contact force >50 nN to scratch 10 nm-wide line-shape grooves on MoO_3 thin film on the MoS_2 substrate (Figure 5A–F) [25]. The experiment has been done in the nitrogen-filled glove box to minimize the effect from a water meniscus. In Figure 5B and C, specific portion of two nanocrystals are cut by nanoindentation method. Figure 5D, E, and F show manipulation of nanocrystals using the AFM tip.

The nanoindentation is also utilized to directly machine other materials by plastic deformation [26–28]. In this way a single resist layer or the top layer of a semiconductor heterostructure [29–37] has been patterned and subsequently used as etch mask. Patterning of the top layer of a bilayer [38] or trilayer resist system [39] has been demonstrated as compatible to the lift-off technique.

Important issues for nanoindentation processes are: 1) capillary forces via water meniscus and 2) wearing of the tip.

When the AFM tip is in contact with the substrate under ambient conditions, water from the air condenses, and water meniscus is formed between AFM tip and surface

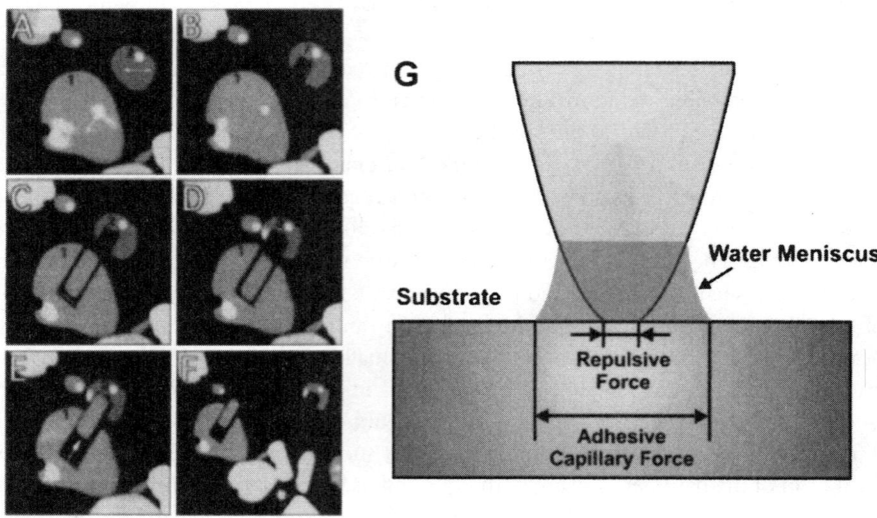

Figure 5. (A)–(F) AFM images of selected steps in the fabrication of a nanostructures (containing three interlocking pieces) by AFM lithography and manipulation. (A) Initial positions of two MoO_3 nanocrystals, crystal 1 and crystal 2 (preferred sliding directions are indicating by two-headed arrows). (B) A 52-nm notch was defined in crystal 2 by nanomachining. (C) A 58-nm free rectangle (latch) was machined in crystal 1, and crystal 2 was translated toward crystal 1. (D) Crystal 2 was translated to align the notch of crystal 2. (F) The latch was broken after a force of 41nN was applied to the latch axis. Adapted with permission from [25] Kim and Lieber, *Science* 272 (1996) 1158. (H) Schematic diagram depicting two different forces between SPM probe and surface. In addition to repulsive contact force, adhesive capillary due to water meniscus is applied to the surface.

(Figure 5G). The size of water meniscus often goes over submicrometer. Due to the surface tension of the water, attractive capillary forces are applied to both AFM tip and substrate. As shown in the Figure 5G, inside the direct contact region, the strong repulsive contact force presses down the surface, while the capillary force pulls up the substrate inside the meniscus region. When the AFM tip sweeps along the substrate under ambient conditions, both forces are applied onto the substrate. Under ambient conditions with common relative humidity (~30%), the magnitude of capillary force often goes over 100 nN, which is strong enough to destroy various materials. As a result, the indented pattern size under ambient conditions is often determined by the size of the water meniscus rather than that of the direct contact area. A low-humidity condition is preferred for a smaller feature size.

Considering common tip-sample contact area ~100 nm^2 and indentation forces over 100 nN, the pressure on the tip during nanoindentation can be over 1 GPa. For example, AFM tip is often made with silicon nitride whose tensile strength is about 0.3 GPa. An AFM tip coated with polycrystalline diamond (tensile strength of ~ 1 GPa) is often used to minimize the wearing of the AFM tip during the nanoindentation process. Since the depth of indentation depends on the pressure, the wearing of the

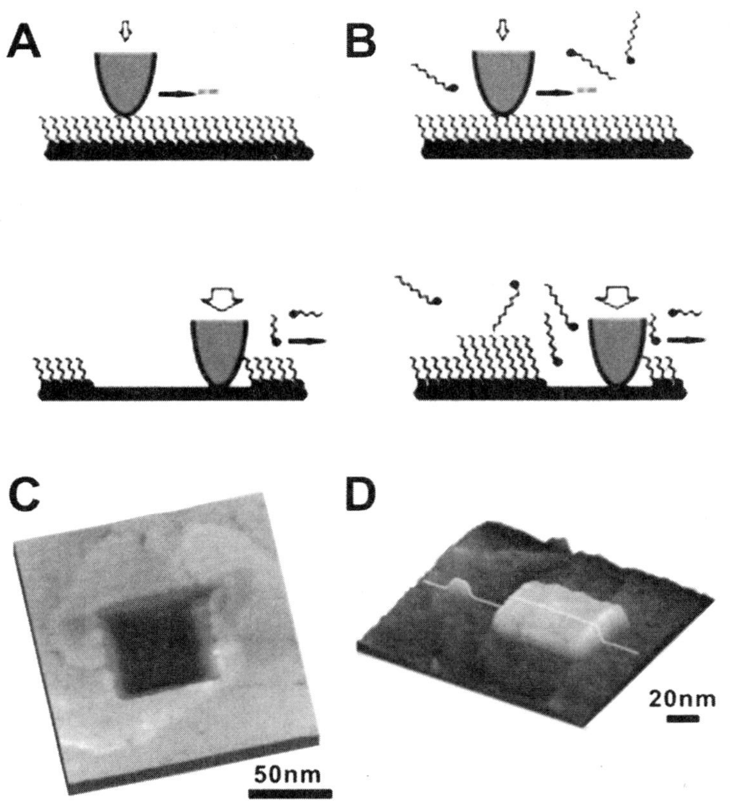

Figure 6. (A) Schematic diagrams depicting the nanoscale shaving method. SAM layer is removed using the AFM tip in solution. (B) Schematic diagrams depicting the nanografting method. SAM layer is removed in the solution of other SAM molecules so that the molecules in the solution can backfill the region where the existing SAM. (C) The result of nanoshaving. 160 × 160 nm^2 topographic images of $C_{18}S/Au(111)$ with the thiols shaved away from the central 50 × 50 nm^2 square. (D) Fabrication of two $C_{18}S$ nanoislands (3 × 5 and 50 × 50 nm^2) in the matrix of a $C_{10}S$ monolayer using nanografting. Adapted with permission from [40] *Acc. Chem. Res.* **33** (2000) 457.

AFM tip may change the contact area and eventually result in unpredictable indentation results. The tip wearing has been a major problem especially in nanoindentation processes on relatively hard metal and semiconductor substrates. Recent development of extremely strong nanostructures such as carbon nanotubes (tensile strength ~100 GPa) is expected to partially solve this problem.

3.2. Nanografting

Nanografting process utilizes an AFM tip to mechanically scratch relatively soft self-assembly monolayer (SAM) under solution environments (Figure 6A) [40]. If the solution contains other SAM molecules, they fill up the scratched regions (Figure 6B). Since

SAM is a soft material, moderate force ($5 \sim 10$ nN) is usually enough for scratching. One important advantage is that since the experiment is done under solution environments without any water meniscus, one does not have to worry about capillary forces, and final scratching result is mainly determined by the diameter of the AFM tip. It is relatively easy to achieve a high resolution. Also, wearing of AFM tip is usually much less than other nanoindentation processes due to the relatively weak indentation force.

Figure 6C shows topographic images of $C_{18}S$/Au(111) with the thiols shaved away from the central 50×50 nm^2 square. When the solution contains other SAM molecules, the final structures depend on the two SAM molecules used in the process. Figure 6D shows $C_{18}S$ nano-island in the matrix of a $C_{10}S$ monolayer using nanografting. Since $C_{18}S$ is longer than $C_{10}S$ molecules, it appears as a higher structure. The $C_{18}S$ nanoislands not only have an order and closed packed $(3^{1/2} \times 3^{1/2})\ R30°$ lattice but also have fewer defects such as pinholes or uncovered areas [41–43]. Using the same procedure, thiols with various chain lengths from 2 to 37 carbons (C_2SH, C_6SH, $C_{16}SH$, $C_{22}SH$, and $C_{18}OC_{19}SH$) have been successfully patterned [42, 43]. The observed heights and high-resolution images of these nanopatterns indicate that the thiols are close-packed within the patterns. In addition, nanopatterns terminated with various functional groups such as –OH, –COOH, $-NH_2$, and –CHO have also been produced [42–44].

One important issue in the nanographting process is replacement reaction between two SAM molecules. Since one SAM layer is kept in the solution of other SAM molecules during the process, the molecules in the solution may replace those on the surface. A special consideration should be given to avoid unwanted replace reactions.

3.3. Nanoscale Melting

In addition to mechanical forces, thermal heating can be used to speed up the indentation process. Binning *et al.* built an array of AFM tips that can be heated individually (Figure 7) [45, 46]. When the tip locally heated the PMMA resist layer up to 400°C over its glass transition temperature, an indentation mark is formed with a very weak force. The indentation mark can be used as a mean to record information. Here, each indentation mark corresponds to 1 bit information. The same tip can be used to quickly read out the indented marks. For reading, the tip is heated up to 300°C that is below the glass transition temperature and its is swept on the surface that is kept at room temperature. Since the tip-sample effect contact area is different when the tip is in the indentation mark or flat surface, the heat transfer rates are different and eventually result in different tip temperature. The tip temperature is measured by measuring the resistance of the heater line on each tip. Currently, read/write rate of a few kilobit/sec has been achieved. Theoretically, it can be as high as $1 \sim 2$ Mbits/sec. Here, the resolution of the indentation process is determined by the size of the AFM tip and the thickness of the polymer layer. Utilizing a thin PMMA layer, the density as high as 400 Gbit/inch^2 can be expected (Figure 7B).

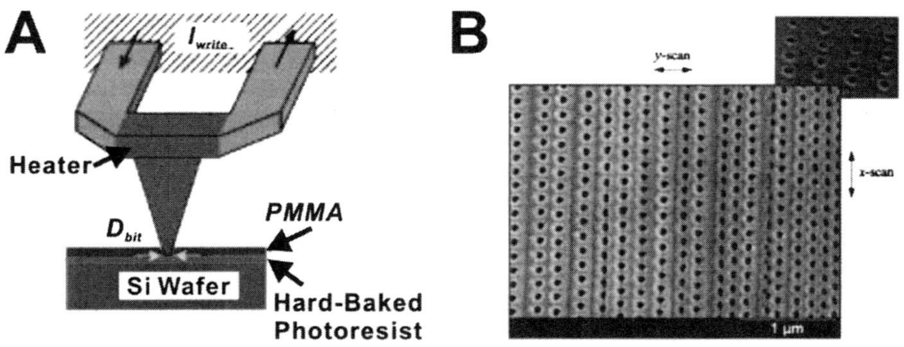

Figure 7. (A) Schematic diagram depicting the basic mechanism of nanomelting. The indentations on PMMA layer are created by the pressure from the heated AFM tip. Adapted with permission from [45] *Appl. Phys. Lett.* 74 (1999) 1329. (B) Nanoscale marks on PMMA created by the nanomelting method. Adapted with permission from [46] *IBM J. Res. Develop.* 44 (2000) 323.

4. NANOSCALE MANIPULATION

4.1. Atomic and Molecular Manipulation (1)

In 1990, the atomic letters written with individual Zenon atoms make the world understands the power of SPM lithography. In this process, an STM tip was utilized to pick up and release individual atoms on the solid substrate under a low temperature (~4K) environment [47]. Two basic manipulation mechanisms are depicted in Figure 8A [48]. Here, the tip can be approached to the atom and slide so that the atom are attracted to the tip and follow its lateral motion. On the other hand, when the tip is very close to the atom, it can pick up individual atoms via Van der Waals or electrostatic forces.

This technique can be utilized for fundamental research such as imaging electron density under the confined environment (Figure 8B) [49–51]. When atoms are arranged in a circular form, electrons are confined inside the circle and it shows a ripple of electron density (Figure 8B). In this experiment, Fe atoms are picked up and moved utilizing STM tip on the Mo surface. This experiment has been done under ultrahigh vacuum at low temperature to avoid thermal motion and oxidation.

On chemical stable surfaces, atomic manipulation can be demonstrated under ambient conditions. In 1992, Garcia *et al.* utilized a voltage pulse on the STM tip and successfully removed three Se atoms from the WSe_2 (001) surface [52]. The STM based atomic manipulation technique has been applied to manipulate various materials such as silicon, HOPG, Au, Pt etc [53, 54].

Recently, Ho *et al.* utilized a STM tip to manipulate and even change the chemical composition of individual molecules under low temperature environments. The chemical entities of the newly created molecules are characterized by inelastic tunneling spectroscopy method [55].

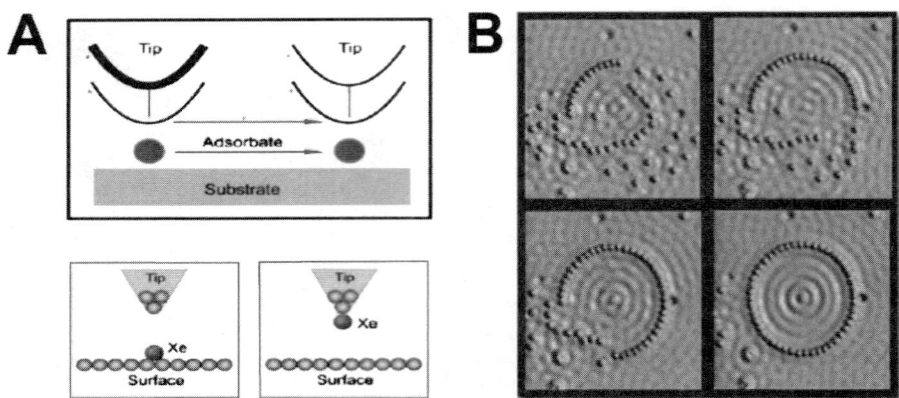

Figure 8. (A) Schematic diagram depicting two basic mechanisms of atomic manipulation via STM. The STM tip either slide (top) or pick up (bottom) individual atoms. Adapted with permission from [48] *Single Mol.* 1 (2000) 79. (B) Formation of "atomic corral." Fe atoms are placed on a circular formation so that electrons are confined in the "corral." Adapted with permission from [49] *Science* 262, 218 (1993).

Lyding *et al.* took advantage of the fact that Si surfaces become unreactive when the Si dangling bonds are saturated by hydrogen to generate chemical inert Si(100) surface [56]. STM tip is utilized to desorb the hydrogen from selected areas of the surface and thus restore locally the chemical reactivity of the Si atoms. Exposure of such a selectively depassivated surface to various gases led to reaction in only the depassivated areas.

4.2. Manipulation of Nanostructures

Atomic force microscope also can be utilized to manipulate nanostructures. The advantage is it can be used on non-conducting surfaces, while the disadvantage is it is very difficult to obtain atomic resolution. Juno *et al.* [57] and Schaefer *et al.* [58] utilized an AFM to push nanometer-sized particles over surfaces. In these experiments, the sample is imaged with non-contact mode to minimize the lateral force acting on the sample, and nanoparticles are pushed with AFM probes with the force feedback off. The problem is that normal force cannot be controlled during the manipulation as the feedback is switched off. This might result either in damaged tips due to too much normal force or insufficient force to keep the tip on surface.

Hansen *et al.* applied a scheme where they imaged nanoparticles in tapping mode and pushed the particles over a surface in contact mode, both while the feedback was switched on [59]. Lieber *et al.* used AFM to test nanobeam mechanics by bending them during contact mode imaging [60]. Biological molecules such as DNA [61–63] and chromosomes [64] have also been manipulated with an AFM.

Superfine *et al.* combined AFM with virtual reality vision and haptic controller to build a convenient nano-manipulator (Figure 9) [65–66]. The virtual reality vision provides 3D images of the surface to help users. The haptic and force feedback system

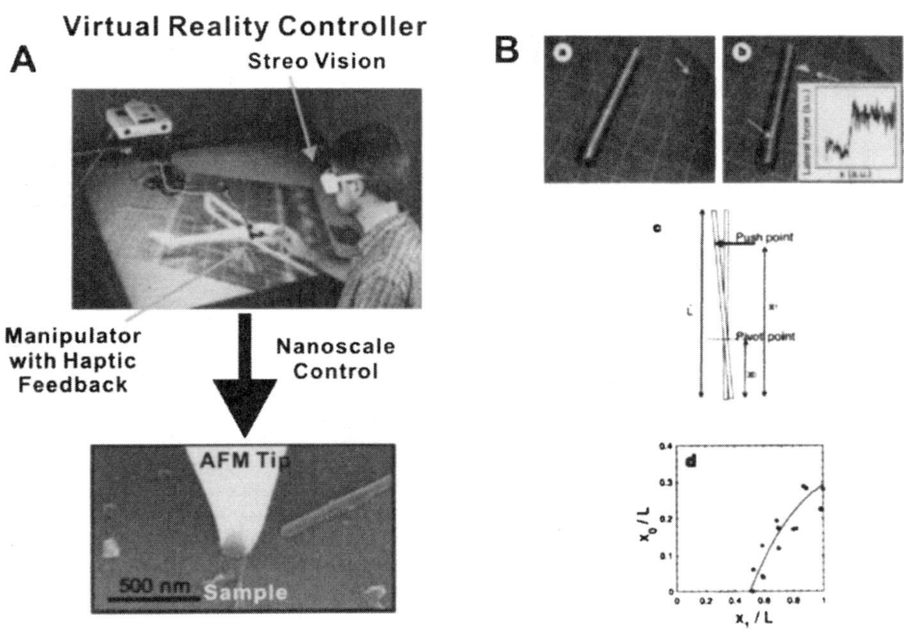

Figure 9. (A) Nanomanipulator utilizing AFM. Virtual reality controller is combined with AFM for user-friendly process. (B) Sliding carbon nanotube. **a**, The tube lies in its original position. Grid lines are overlain so that one of the grid axes corresponds to the original orientation of the tube axis. The absolute position of the grid relative the fiducial feature indicated by the arrow was adjusted to be consistent with that in **b** so that the point of rotation could be determined. **b**, The tube's orientation after AFM manipulation. The pivot point and push point are indicated by the bottom and top arrows, respectively. The white dashes to the right of the push point are markers indicating the trajectory of the AFM tip during manipulation. Inset shows the lateral force trace during a sliding manipulation. **c**, Diagram of the pushing process. Measuring from the bottom of the tube, x_1 is the push point, and x_0 is the point of rotation (or pivot point). **d**, The relation between the push point and the pivot point for several pushing trials is compared with theory (plotted as a solid curve). Adapted with permission from [65] *Nature* 397 (1999).

allows one to feel the contact and lateral forces on the AFM tip with their hands. The system has been utilized to manipulate various nanostructures such as carbon nanotubes. It is also possible to manipulate biomolecules such as Fibrin, Adnovirus, TMV, and DNA [67].

4.3. Nanoscale Tweezers

The development of new tools for manipulating and probing matter at nanometer length scales is critical to advances in nanoscale science and technology. However, the single probe tips used in SPMs limit these tools' ability to manipulate objects and measure physical properties; for example one tip cannot *grab* an object, and electrical measurements cannot be made without a second contact to structures. Two probes in the form of tweezers could overcome these limitations of single-probe manipulation

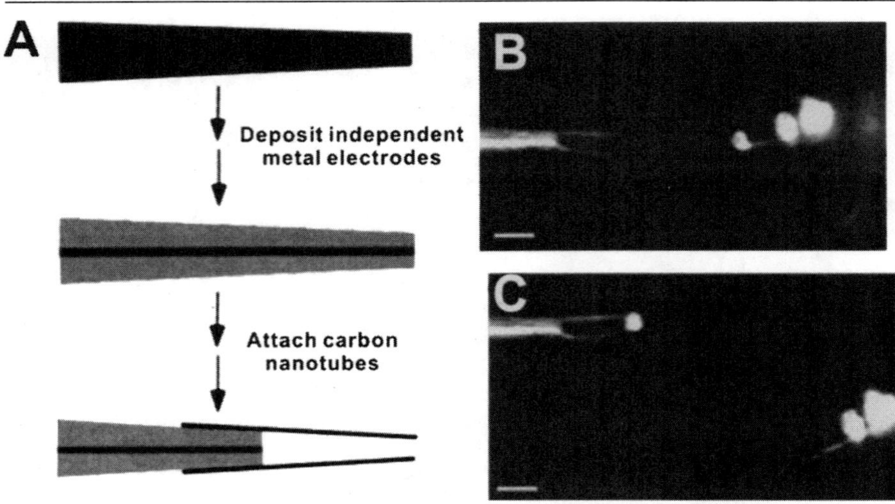

Figure 10. (A) Schematic illustrating the deposition of two independent metal electrodes and the subsequent attachment of carbon nanotubes to these electrodes. (B)–(C) Darkfield optical micrographs showing the sequential process of nanotweezer manipulation of polystyrene nanoclusters containing fluorescent dye molecules. (**B**) Approach of the nanotweezers to nanoclusters. (**C**) Alignment of the tweezer arms on a small cluster. A voltage was applied to nanotweezer arms on the nanocluster, and then the nanotweezers and cluster were moved away from the sample support. The fluorescent polystyrene nanoclusters and nanotube arms are both readily observed in the dark-field image. The SiC sample, which was deposited from a water-based suspension, consisted primarily of clusters of individual 310-nm polystyrene beads. We focused on grabbing the smallest clusters, ~500 nm in diameter, in our experiments. SEM analysis of these clusters showed that they consisted of several 310-nm nanoclusters. Scale bars, 2 μm. Adapted with permission from [68] *Science* 286 (1999) 2148.

and thus might enable new types of fabrication and easy electrical measurements on nanostructures.

Kim and Lieber utilized nanometer diameter multi-walled carbon nanotubes to create nanotweezers that can be utilized for nanoscale manipulation and measurement (Figure 10) [68]. For fabrication of nanotube nanotweezers, free-standing electrically independent electrodes were deposited onto tapered glass micropipettes, which can be routinely made with end diameter of 100 nm. Then carbon nanotubes were attached to the independent Au electrodes using an approach similar to that used for the fabrication of single-nanotube SPM tips. They utilized this nanotweezer to grab and measure the electrical properties of GaAs and SiC nanoparticles. Even though nanotweezers have not yet been attached to scanning probe systems, it provides a mean to overcome the limitation of single-probe manipulation systems.

5. NANOSCALE CHEMISTRY

5.1. Nanoscale Oxidation

Figure 11A shows an example process of nano-oxidation. When the tip is in contact with surfaces under ambient conditions, they are connected through a water meniscus.

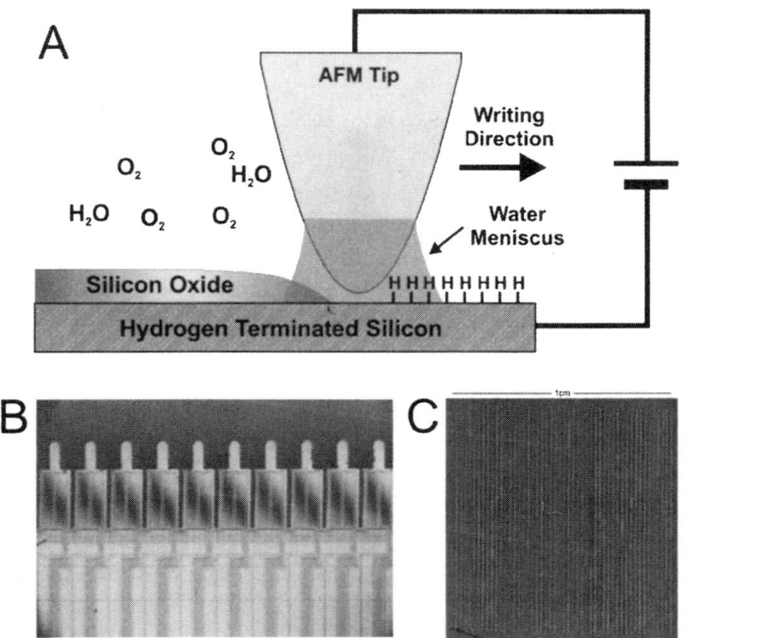

Figure 11. (A) Schematic diagram depicting nano-oxidation process on hydrogen-terminated silicon surfaces. The water meniscus works as a nanoscale reaction chamber. (B) An array of cantilevers with integrated actuators and sensors. The cantilevers are spaced by 200 μm. Adapted with permission from [85] *Appl. Phys. Lett.* 72 (1998) 2340. (C) 50 × 1 parallel AFM lithography over $1\,cm^2$. The lithography was accomplished by electric field enhanced oxidation of silicon at 15 V, and at a scan speed of 1 μm/s. The lithographed oxide pattern was transferred into the silicon using KOH. The picture was formed from 24 optical photographs taken with 5X Nomarski microscope. Adapted with permission from [86] *Appl. Phys. Lett.* 73 (1998) 1742.

When a positive bias is applied to surface (anode) relative to the tip (cathode), the water meniscus works as a small anodization chamber. Anodization and/or oxygen evolution proceed on the sample surface at the point beneath the tip. Surface (anode) reactions are:

$$M + xH_2O \rightarrow MO_2 + 2xH^+ + 2xe^-$$
$$2H_2O \rightarrow O_2 + 4H^+ + 4e^-$$

where M is the surface material (e.g. Si, Ti etc) and x is an oxidation number. For silicon surfaces, M is Si and x is 2.

Tip (cathode) reaction:

$$2H_2O + 2e^- \rightarrow H_2 + 2OH^-$$

A faradaic current flows through the tip-sample junction as a result of these electrochemical reactions. Since the nano-oxidation process utilizes a water meniscus, the process strongly depends on the amount of adsorbed water and, therefore, the atmospheric humidity.

Nano-oxidation process utilizing STM was realized shortly after its original development as a tool for atomic-scale microscopy. A STM operated was used to perform tip-induced oxidation of local regions of a H-passivated Si (111) surface for use as an etch mask [69] (Figure 11A). In this process, Si surface is first passivated by hydrogen. When local current is applied through the conducting probe in the water meniscus, hydrogen is removed, and water reacts with silicon to form silicon oxide. Recently, a STM in ultrahigh vacuum (UHV) was used to form Si oxide lines as narrow as 1 nm wide [70]. On the other hand, the AFM has become an attractive option for performing similar work because among, other reasons, it allows independent control over the oxidation mechanism that is done by a contact force. In STMs, a voltage bias is required for both the oxidation and feedback control of the imaging so the imaging of an oxidized pattern must be done carefully in order to not oxidize the surface any further. Most recently, AFM tip-induced oxide lines have been used as etch masks to demonstrate a Si metal-oxide-semiconductor field-effect transistor (MOSFET) [71] and side-gated FET [72]. On a thin Ti film, Matsumoto *et al.* have successfully fabricated a room temperature operable single electron transistor (SET) with 15 nm features [73]. One important aspect that has made their work particularly unique was that AFM-generated oxides were used as integral parts of the SET device, and not just as a step in the fabrication process.

In order to optimally use the SPM tip-induced oxidation process for the fabrication of nanoscale electronic devices, it is necessary to understand the mechanisms and kinetics of the process so that we can reliably control the featured characteristics. The mechanisms of the AFM tip-induced oxidation process have been studied by several authors in the case of Si [74–79], GaAs [80, 81] and other related semiconductors. Diebold *et al.* demonstrated local oxidation of TiO_2 surfaces [82]. Lemeshko *et al.* studied the oxidation mechanism of Ti film [83]. Blum *et al.* demonstrated the tip-induced oxidation of PbS surfaces [84]. Quate *et al.* built individually-addressable multiple AFM probes and generate multiple silicon oxide lines simultaneously in a parallel manner (Figure 11B, C). By this way, they can improve the throughput of the process [85, 86].

5.2. Nanoscale Desorption of Self-Assembled Monolayers

The resolution of the patterning method is usually determined by two major factors: the size of the proximal probe and the thickness of patterning materials. Thus, the highest resolution can be achieved by patterning monolayer materials such as self-assembled monolayer. Figure 12A shows the basic mechanism of SAM patterning process under ambient conditions via conducting SPM probes. At high humidity (Figure 12 A (b)), SAM layer is stripped off by the electrochemical reaction in water meniscus, while it remains without a water meniscus at low humidity (Figure 12 A (a)). Here, the water meniscus also plays a role as an electrochemical reaction chamber like nano-oxidation processes while in a little bit different manner. Crooks *et al.* proposed that

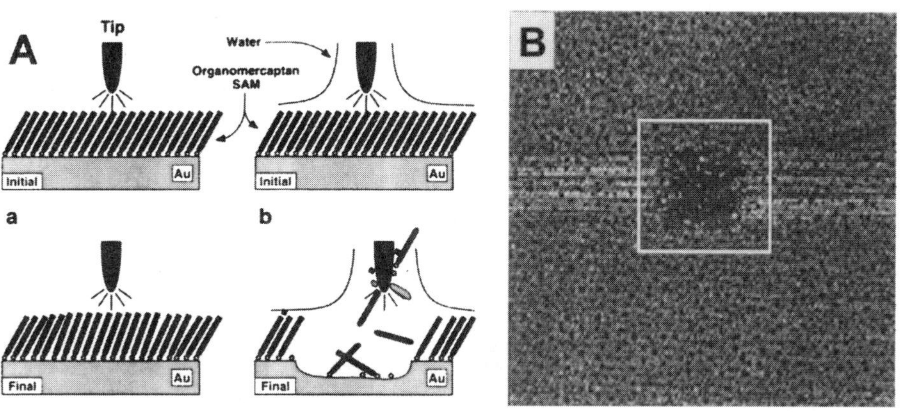

Figure 12. (A) Schematic diagram depicting nanoscale desorption process of SAM (a) with and (b) without water meniscus. With a water meniscus, the electrochemical reaction causes the removal of the SAM layer. However, the reaction does not occur without a water meniscus at low humidity. (B) The SAM in the square region is removed utilizing a STM probe. The experiment is performed in air with 75% relative humidity and 3 V tip bias. Adapted with permission from [87] *J. Phys. Chem.* 100 (1996) 11086.

the current flow in the water meniscus induce the oxidation process and remove the SAM molecules from the surface [87, 88]. Example reaction for alkanethiol SAMs on Au is:

$$CH_3(CH_2)_nS - Au + 2H_2O \rightarrow Au + CH_3(CH_2)_nSO_2H + 3e^- + 3H^+$$

Since a water meniscus works as a reaction chamber, the patterning speed strongly depends on the relative humidity. Figure 12B shows the result after stripping off the SAM in a nanoscale square region. The stripped part can be later chemically etched to transfer the pattern to the substrate.

Different self-assembled monolayers have been patterned using a conducting SPM probes on various surfaces such as Au, GaAs, Si, SiO_2 etc [89–95].

5.3. Nanoscale Chemical Vapor Deposition

In STM, a-few-volt bias is applied in a nanometer-size gap between tip and surface, which results in a very large electric field in the gap region. When the organometallic gas is supplied in the gap, the high field and field-emitted electrons in the gap can decompose the gas molecules resulting in metal deposition on the surface. Common organicmetallic compounds are metal carbonyl ($M(CO)_n$, M=Cd, W, Fe etc) compounds. Using this method, one can create various metallic nanoscale dots and line patterns on conducting substrates.

Ehrichs *et al.* placed a STM system in an ultrahigh vacuum chamber (base pressure $\sim 10^{-8}$ Torr) and provided dimethylcadmium (DMCd) gas in the chamber. DMCd has relatively low dissociation energy of 3.14 eV [96]. The pressure of DMCd in the chamber is held constant at $\sim$200 mTorr. Nanoscale deposition of Cd on the surface

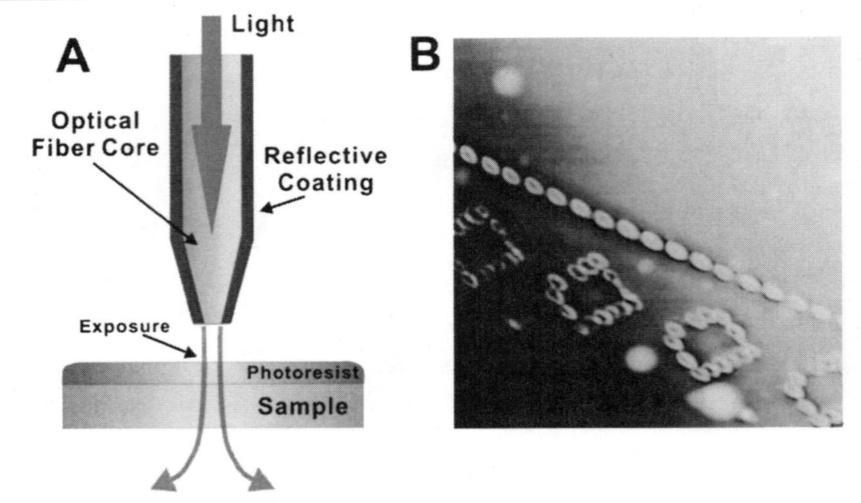

Figure 13. (A) Schematic diagram depicting nanoscale light exposure to photoresist films via NSOM. (B) Developed negative photoresist patterns created by NSOM. Adapted with permission from [102] *Appl. Phys. Lett.* 67 (1995) 3859.

near the STM tip occurred with $V_t = 11.8$ V and $I = 500$ nA. By this method, they can deposit sub-100-nm size Cd lines and dots on the substrate.

McCord *et al.* deposited tungsten and gold onto gold substrates utilizing tungsten-hexacarbonyl and dimethyl-gold-trifluoro-acetylacetonate, respectively ($Vt > 15$ V, $I \sim 100$ nA) [97]. McCord and Awschalom deposited magnetic iron nanostructures onto SQUID device pick coils utilizing iron pentacarbonyl ($Vt > 27$ V, $I = 100 \sim 500$ pA) [98]. The magnetic properties of the deposited nanoparticle array were studied via the SQUID devices.

Later, Kent *et al.* deposited array of iron nanoparticles on a gold-coated Hall magentometer and studied the magnetic properties [99]. In this process, Iron particles are formed by using a STM under UHV conditions (base pressure $\sim 2 \times 10^{-10}$) to decompose iron pentacarbonyl gas. To initiate the growth the substrate-tip bias is raised to 15 V and a current of 50 pA maintained in the presence of 30 μTorr of iron pentacarbonyl. The STM feedback loop is active and maintains a constant current and thus constant height between tip and growing deposit. When the deposit has reached the desired height above the surface the tip is retracted to stop the growth.

6. NANOSCALE LIGHT EXPOSURE

Since NSOM probes can transmit light to the nanometer scale regions on the surface, they can be used to pattern various photosensitive materials (Figure 13).

NSOM technique has been utilized as a tool for photolithography [100–102]. A major limitation of photolithography is the diffraction of light. One way to overcome this limitation is shining a light through a small aperture or holes like NSOM probes.

Here, the light exposure induces the local polymerization (or breakage of polymer links in negative photoresist), and it allows one to selectively dissolve the resist layer with a developing solution. Using this technique, a feature size typically as small as 1/6 of the wavelength can be achieved.

Madsen *et al.* utilized NSOM to oxidize the hydrogen passivated silicon surface [103]. The 457.9 nm line from an argon ion laser was used as the illumination wavelength. This wavelength matches approximately the Si-H binding energy of ~3eV. In the light-exposed area, Si-H binding is broken and silicon is oxidized to form silicon oxide. In the potassium hydroxide solution, silicon oxide works as an etching resist and only Si-H regions are etched, which allows one to transfer optically written patterns. Directly written optically induced patterns down to 110 nm in width have been demonstrated.

NSOM also have been utilized to locally modify other photosensitive materials such as photosensitive polymers [104] and ferroelectric films [105].

7. FUTURE PERSPECTIVES

Various scanning probe lithography methods have been utilized to manipulate and modify materials in a nanometer scale resolution. Since SPL does not use light for lithography, it is not limited by the diffraction limit, and its resolution can be as small as individual atoms. Another important advantage is that it can be used under quite versatile environments including ambient, liquid and vacuum (Table 1). For this reason, SPL method has been applied for patterning soft and biological materials that are very sensitive to environmental conditions. In addition, since many SPL processes can be done under ambient conditions, SPL does not require extensive instruments for environmental control and its instrumentation cost is much lower than those for other lithography techniques (e.g. e-beam lithography, focused-ion beam etc).

One major bottleneck applying SPL methods for industrial application is its throughput. Since SPL is a serial patterning process in nature, it is usually slower than parallel patterning processes such as photolithography. One obvious solution is having multiple probes patterning simultaneously. Parallel patterning systems have been realized for some SPL processes such as dip-pen nanolithography, nanomelting, and nanooxidation. As the parallel SPM technology advances, the throughput is expected to improve.

SPL is expected to have immediate applications for high-resolution rapid prototyping and custom manufacturing area (Figure 14). These include fabrication of prototype nano-devices, photomask repair, mask-less lithography, etc. In SPL processes, the patterns designed in the computer can be directed printed to the substrates without extensive preparation steps. It makes SPL strategy an ideal method for custom-design high precision manufacturing tools like e-beam lithography does for microelectronics industry. However, SPL strategy has several advantages over other custom manufacturing tools: 1) high resolution down to single atomic level, 2) versatile lithography environments friendly to soft and biological materials, and 3) low instrumentation cost.

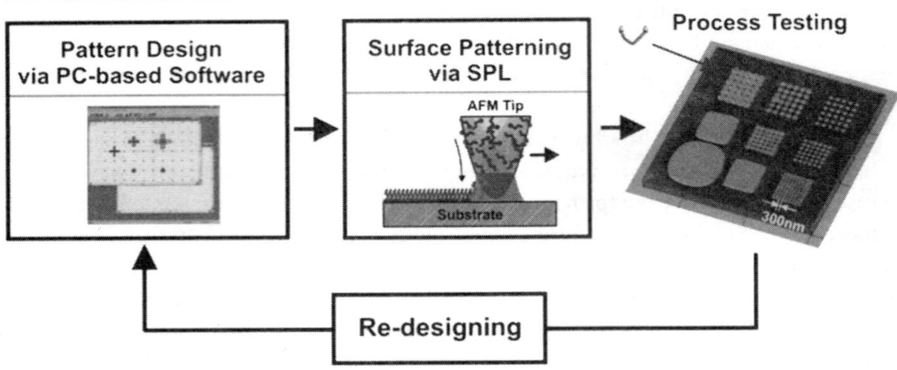

Figure 14. Schematic diagram depicting rapid-prototyping of nanodevices via SPL strategy. Designed patterns are directly printed onto the substrate without any intermediate steps. The fabricated devices are tested and the test result can be used to redesign device patterns.

Now, SPL strategy provides quite a versatile tool set for nanotechnologists, and it has become an essential tool kit for research and development in the age of nanotechnology.

REFERENCES

1. G. Binnig and H. Rohrer, *Rev. Mod. Phys.* 71 (1999) S324.
2. P. K. Hansma and J. Tersoff, *J. Appl. Phys.* 61 (1987) R1.
3. Israelachvili, *Intermolecular & Surface Forces*, Academic Press, San Diego, USA (1991).
4. S. Okazaki, *J. Vac. Sci. Technol. B* 9 (1991) 2829.
5. Y. Xia and G. M. Whitesides, *Angew. Chem. Int. Ed.* 37 (1998) 550.
6. T. A. Taton, C. A. Mirkin and R. L. Letsinger, *Science* 289 (2000) 1757.
7. Y. Cui, Q. Wei, H. Park and C. M. Lieber, *Science* 293 (2001) 1289.
8. G. S. Wilson and Y. Hu, *Chem. Rev.* 100 (2000) 2693.
9. M. M. Miller, P. E. Sheehan, R. L. Edelstein, C. R. Tamanaha, L. Zhong, S. Bounnak, L. J. Whitman and R. J. Colton, *J. Magn. Magn. Matr.* 225 (2001) 138.
10. J. Fritz, M. K. Baller, H. P. Lang, H. Rothuizen, P. Vettiger, E. Meyer, H. J. Güntherodt, Ch. Gerber and J.K. Gimzewski, *Science* 288 (2000) 316.
11. J. R. Sellers and B. Kachar, *Science* 249 (1990) 406.
12. K. Soong, G. D. Bachand, H. P. Neves, A. G. Olkhovets, H. G. Craighead and C. D. Montemagno, *Science* 290 (2000) 1555.
13. R. Piner, J. Zhu, F. Xu, S. Hong and C. A. Mirkin, *Science* 283 (1999) 661.
14. S. Hong, J. Zhu and C. A. Mirkin, *Science* 286 (1999) 523.
15. S. Hong, J. Zhu and C. A. Mirkin, *Langmuir* 10 (1999) 7897.
16. S. Hong and C. A. Mirkin, *Science* 288 (2000) 1808.
17. J. Jang, S. Hong, G. C. Schatz and M. A. Ratner, *J. Chem. Phys.* 115 (2001) 2721.
18. M. Zhang, D. Bullen, S.-W. Chung, S. Hong, K. S. Ryu, Z. Fan, C. A. Mirkin and C. Liu, *Nanotechnology* 13 (2002) 212.
19. D. A. Weinberger, S. Hong, C. A. Mirkin, B. W. Wessels and T. B. Higgins, *Adv. Mater.* 12 (2000) 1600.
20. L. M. Demers, D. S. Ginger, S.-J. Park, Z. Li, S.-W Chung and C. A. Mirkin, *Science* 296 (2002) 1836.
21. S. Rao, L. Huang, W. Setyawan and S. Hong, *Nature* 425 (2003) 36.
22. D. L. Wilson, R. Martin, S. Hong, M. Cronin-Glolmb, C. A. Mirkin and D. L. Kaplan, *Proc. Natl. Acad. Sci. USA* 98 (2001) 13660.
23. D. B. Pique (Eds.), Direct-Write Technologies for Rapid Prototyping Applications, Academic Press, San Diego, U.S.A. (2002).
24. Lewis, Y. Kheifetz, E. Shambrodt, A. Radko, E. Khatchatryan and C. Sukenik, *Appl. Phys. Lett.* 75 (1999) 2689.

25. Y. Kim and C. M. Lieber, *Science* 257 (1992) 375.
26. T. A. Jung, A. Moser, H. J. Hub, D. Brodbeck, R. Hofer, H. R. Hidber and U. D. Schwarz, *Ultramicroscoy* 42 (1992) 1446.
27. X. Jin and W. N. Unertl, *Appl. Phys. Lett.* 61 (1992) 657.
28. E. Boschung, M. Heuberger and G. Dietler, *Appl. Phys. Lett.* 64 (1994) 3566.
29. M. Wendel, S. Kuhn, H. Lorenz and J. P. Kotthaus, *Appl. Phys. Lett.* 65 (1994) 1775.
30. M. Fujihira and H. Takano, *J. Vac. Sci. Technol. B* 12 (1994) 1860.
31. S. Yamamoto, H. Yamada and H. Tokumoto, *Jap. J. Appl. Phys.* 34 (1995) 3396.
32. P. Pingue, M. Lazzarino, F. Beltram, C. Cecconi, P. Baschieri, C. Frediani and C. Ascoli, *J. Vac. Sci. Technol. B* 15 (1997) 1398.
33. Avramescu, K. Uesugi and I. Suemune *Jap. J. Appl. Phys.* 36 (1997) 4057.
34. S. F. Y. Li, H. T. Ng, P. C. Zhang, P. K. H. Ho, L. Zhou, G. W. Bao and S. L. H. Chan, *Nanotechnology* 8 (1997) 76.
35. B. Klein and U. Kunze, *Superlattices and Microstructures* 23 (1998) 441.
36. B. Klehn and U. Kunze, *J. Appl. Phys.* 85 (1998) 3897.
37. R. Magno and B. R. Bennett, *Appl. Phys. Lett.* 70 (1997) 1855.
38. L. L. Sohn and R. L. Willett, *Appl. Phys. Lett.* 70 (1997) 1855.
39. V. Bouchiat and D. Esteve, *Appl. Phys. Lett.* 69 (1996) 3098.
40. G.-Y. Liu, S. Xu and Y. Qian, *Acc. Chem. Res.* 33 (2000) 457.
41. S. Xu, P. Laibinis and G.-Y. Liu, *J. Am. Chem. Soc.* 120 (1998) 9356.
42. S. Xu, G.-Y. Liu, *Langmuir* 13 (1997) 127.
43. S. Xu, S. Miller, P. E. Laibinis and G.-Y. Liu, *Langmuir* 15 (1999) 7244.
44. K. Wadu-Mesthrige, N. Armo, S. Xu and G.-Y. Liu, *Langmuir* 15 (1999) 8580.
45. G. Binnig, M. Despont, U. Drechsler, W. H≅berle, M. Lutwyche, V. Vettiger, H. J. Mamin, B. W. Chui and T. W. Kenny, *Appl. Phys. Lett.* 74 (1999) 1329.
46. P. Vettiger, M. Despont, U. Drechsler, U. Dηrig, W. H≅berle, M. I. Lutwyche, H. E. Rothuizen, R. Stutz, R. Widmer and G. K. Binnig, *IBM J. Res. Develop.* 44 (2000) 323.
47. D. M. Eigler and E. K. Schweizer, *Nature* 344 (1990) 524.
48. G. Meyer, J. Repp, S. Zoephel, K.-F. Braun, S. W. Hla, S. Foelsch, L. Bartels, F. Moresco and K. H. Rieder, *Single. Mol.* 1 (2000) 79.
49. M. F. Crommie, C. P. Lutz and D. M. Eigler, *Science* 262 (1993) 218.
50. M. F. Crommie, C. P. Lutz and D. M. Eigler, *Nature* 363 (1993) 524.
51. M. F. Crommie, C. P. Lutz, D. M. Eigler and E. J. Heller, *Surface Review and Letters* 2 (1995) 127.
52. R. G. Garcia, *Appl. Phys. Lett.* 60 (1992) 1960.
53. S. E. McBride and G. C. Wetsel, Jr., *Appl. Phys. Lett.* 59 (1991) 3056.
54. R. M. Ostrom, D. M. Tanenbaum and A. Gallagher, *Appl. Phys. Lett.* 61 (1992) 925.
55. H. J. Lee and W. Ho, *Science* 286 (1999) 1719.
56. J. W. Lyding, T.-C. Shen, J. S. Hubacek, J. R. Tucker and G. C. Abelm, *Appl. Phys. Lett.* 64 (1994) 2010.
57. T. Juno, K. Deppert, L. Montelius and L. Samuelson, *Appl. Phys. Lett.* 66 (3627) 3627.
58. D. M. Schaefer, R. Reifenberger, A. Patil and R. P. Andres, *Appl. Phys. Lett.* 66 (1995) 1012.
59. L. T. Hansen, A. Kuhle, A. H. Sorensen, J. Bohr and P. E. Lindelof, *Nanotechnology* 9 (1998) 337.
60. E. W. Wong, P. E. Sheehan and C. M. Lieber, *Science* 277 (1997) 1971.
61. H. G. Hansma, J. Vesenka, C. Siegerist, G. Kelderman, H. Morrett, R. L. Sinsheimer, V. Elings, C. J. Bustamante and P. K. Hansma, *Science* 256 (1992) 1180.
62. E. Henderson, *Nucleic Acid Res.* 20 (1992) 445.
63. J. Vesenka, M. Guthold, C. L. Tang, D. Keller, E. Delaine and C. Bustamante, *Ultramicroscopy* 42 (1992) 1243.
64. R. W. Start, S. Thalhammer, J. Wienberg and W. M. Heckl, *Appl. Phys. A.* 66, (1998) S579.
65. M. R. Falvo, R. M. Taylor II, A. Helser, V. Chi, F. P. Brooks, Jr., S. Washburn and R. Superfine, *Nature* 397 (1999) 236.
66. M. R. Falvo, G. J. Clary, R. M. Taylor II, V. Chi, F. P. Brooks Jr., S. Washburn and S. Superfine, *Nature* 389 (1997) 582.
67. M. Guthold, G. Matthews, A. Negishi, R. M. Taylor, D. Erie, F. P. Brooks and R. Superfine, *Surf. Interf. Analys.* Vol. 27 (1999).
68. P. Kim and C. M. Lieber, *Science* 286 (1999) 2148.
69. J. A. Dagata, J. Schneir, H. H. Harary, C. J. Evans, M. T. Postek and J. Bennett, *Appl. Phys. Lett.* 56 (1990) 2001.

70. J. W. Lyding, T.-C. Shen, J. S. Hubacek, J. R. Tucker and G. C. Abeln, *Appl. Phys. Lett.* 64 (1994) 2010.
71. S. C. Minne, H. T. Soh, P. Flueckinger and P. J. McMarr, *Appl. Phys. Lett.* 66 (1995) 1388.
72. P. M. Campbell, E. S. Snow and P. J. McMarr, *Appl. Phys. Lett.* 66 (1995) 1388.
73. K. Matsumoto, M. Ishii, K. Segawa, Y. Oka, B. J. Vartanian and J. S. Harris, Jr., *Appl. Phys. Lett.* 68 (1996) 34.
74. H. Sugimura, T. Uchida, N. Kitamura and H. Masuhara, *J. Phys. Chem.* 98 (1994) 4352.
75. M. Yasutake, Y. Ejiri, and T. Hattori, *Jpn. J. Appl. Phys., Part 2* 32 (1993) L1021.
76. E. Gordon, R. T. Fayfield, D. D. Litfin and T. K. Higran, *J. Vac. Sci. Technol. B* 13 (1995) 2805.
77. T. Teuschler, K. Mahr, S. Miyazaki, M. Hundhausen and L. Ley, *Appl. Phys. Lett.* 67 (1995) 3144.
78. D. Stievenard, P. A. Fontaine and E. Dubois, *Appl. Phys. Lett.* 70 (1997) 3272.
79. Ph. Avouris, T. Hertel and R. Martel, Appl. Phys. Lett. 71 (1997) 285.
80. Y. Okada, S. Arnano, M. Kawabe, B. N. Shirnbo and J. S. Harris, *Jr., J. Appl. Phys.* 83 (1998) 1844.
81. M. Ishii and K. Matsumoto, *Jpn. J. Appl. Phys., Part 1* 34 (1995) 1329.
82. U. Diebold, J. Lehman, T. Mahmoud, M. Kuhn, G. Leonardelli, W. Hebenstreit, M. Schmid and P. Varga, *Surf. Sci.* 411 (1998) 137.
83. S. Lemeshko, S. Gavrilov, V. Shevyakov, V. Roschin and R. Solomatenko, *Nanotechnology* 12 (2001) 273.
84. S. Blum, A. J. D. Schafer and T. Engel, *J. Phys. Chem. B* 106 (2002) 8197.
85. S. C. Minne, G. Yaralioglu, S. R. Manalis, J. D. Adams, J. Zesch, A. Atalar and C. F. Quate, *Appl. Phys. Lett.* 72 (1998) 2340.
86. S. C. Minne, J. D. Adams, G. Yaralioglu, S. R. Manalis, A. Atalar and C. F. Quate, *Appl. Phys. Lett.* 73 (1998) 1742.
87. J. K. Schoer, F. P. Zamborini and R. M. Crooks, *J. Phys. Chem.* 100 (1996) 11086.
88. J. K. Schoer and R. M. Crooks, *Langmuir* 13 (1997) 2323.
89. H. U. Muller, C. David, B. Volkel and M. Grunze, *J. Vac. Sci. Tech. B* 13 (1995) 2846.
90. J. Chen, M. A. Reed, C. L. Asplund, A. M. Cassell, M. L. Myrick, A. M. Rawlett, J. M. Tour and P. G. Van Patten, *Appl. Phys. Lett.* 75 (1999) 624.
91. M. J. Lercel, G. F. Redinbo, H. G. Craighead, C. W. Sheen and D. L. Allara, *Appl. Phys. Lett.* 65 (1994) 974.
92. H. Sugimura and N. Nakagiri, *J. Vac. Sci. Technol. A* 14 (1996) 1223.
93. F. Keith Perkins, E. A. Bobisz, S. L. Brandow, J. M. Calvert, J. E. Kosaowski and C. R. K. Marrian, *Appl. Phys. Lett.* 68 (1996) 550.
94. A. Inoue, T. Ishida, N. Choi, W. Mizutani and H. Tokumoto, *Appl. Phys. Lett.* 73 (1998) 1976.
95. R. Maoz, S. R. Cohen and J. Sagiv, *Adv. Mater.* 11 (1999) 55.
96. E. E. Ehrichs, R. M. Silver and A. L. de Lozanne, *J. Vac. Sci. Tech. A* 6 (1988) 540.
97. M. A. McCord, D. P. Kern and T. H. P. Chang, *J. Vac. Sci. B* 6 (1988) 1877.
98. M. A. McCord and D. D. Awschalom, *Appl. Phys. Lett.* 57 (1990) 2153.
99. A. D. Kent, T. M. Shaw, S. von Molnar and D. D. Awschalom, *Science* 262 (1993) 1249.
100. S. I. Bozhevolnyi, I. I. Smolyaninov and O. Keller, *Appl. Opts.* 34 (1995) 3793.
101. G. Krausch, S. Wegscheider, A. Kirsch, H. Bielefeldt, J. C. Meiners, and J. Mlynek, *Opt. Commun.* 119 (1995) 283.
102. Smolyaninov, D. L. Mazzoni and C. C. Davis, *Appl. Phys. Lett.* 67 (1995) 3859.
103. S. Madsen, M. Muellenborn, K. Birkelund and F. Grey, *Appl. Phys. Lett.* 69 (1996) 544.
104. S. Davy and M. Spajer, *Appl. Phys. Lett.* 69 (1996) 3306.
105. Massanell, N. Garcia and A. Zlatkin, *Opt. Lett.* 21 (1996) 12.

6. SCANNING THERMAL AND THERMOELECTRIC MICROSCOPY

LI SHI

1. INTRODUCTION

Thermal and thermoelectric transport in nanometer scale devices and structures has become one of the foci of current efforts in nanotechnology, due to the increasing importance of nanoscale thermal and thermoelectric transport phenomena in many applications ranging from computing to energy conversion. One important example is heat dissipation in the metal-oxide field effect transistors (MOSFETs), which have been the driving force of the semiconductor industry for the past two decades. The gate length of the MOSFET has been continuously reduced in order to achieve higher switching speed and lower manufacturing cost. This critical length has been shrunk to 85–90 nm by year 2002 and will approach 20–22 nm in year 2013 [1]. The length scale is comparable to the scattering mean free paths of electrons and phonons. As a result, nanotransistors exhibit unique electron and phonon transport phenomena that have not been observed in micron-scale devices. For example, the effective thermal conductivity of the structure is drastically reduced due to increased phonon-boundary scattering. Additionally, as nanotransistors are miniaturized, the power density is increased. These two effects combine to cause localized self heating and elevated operating temperatures that can reduce device speed and time to failure.

Besides nanotransistors, a variety of other nanoscale devices and structures are being actively developed for electronic, optoelectronic, thermoelectric, electromechanical, and sensing applications. Examples include nanoelectronic devices based on carbon nanotubes [2], nanolasers based on ZnO_2 nanowire arrays [3], thermoelectric coolers

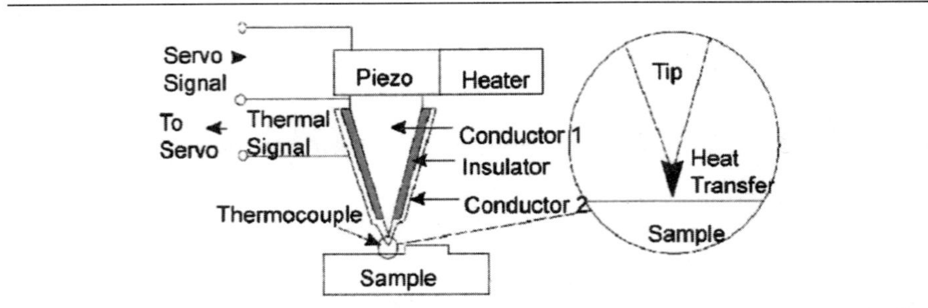

Figure 1. Schematic diagram of the scanning thermal profiler.

based on superlattices [4] and nanowires [5], nanosensors based on carbon nanotubes [6], Si nanowires [7], and metal oxide nanobelts [8]. The characteristic size of these low dimensional structures is in the range of 1–100 nm. The short length scale gives rise to unique thermal transport properties that cannot be observed using conventional experimental methods developed for bulk materials.

Scanning probe microscopy (SPM) techniques have enabled the direct observation of physical phenomena with high spatial resolution. A variety of novel SPM-based measurement techniques have been developed to investigate electronic, optical, thermal, mechanical, chemical, and acoustic properties in the nanoscale. Two of such techniques are the Scanning Thermal Microscopy (SThM) and Scanning Thermoelectric Microscopy (SThEM). The SThM and SThEM can measure temperature, thermal properties, thermopower, electronic band structure, carrier/dopant concentration with nanoscale spatial resolution. The ability of thermally probing nanoscale phenomena has made the SThM and SThEM a powerful tool for studying fundamental thermophysics and for characterizing micro-nano scale devices and materials. These two techniques have demonstrated applications in materials, microelectronic, energy, and pharmaceutical research and development.

A variety of SThM and SThEM methods have been explored since the invention of the scanning tunneling microscope (STM) and the atomic force microscope (AFM). This chapter introduces various SThM and SThEM instruments, and the theory and applications of these two techniques.

2. INSTRUMENTATION OF SCANNING THERMAL AND THERMOELECTRIC MICROSCOPY

2.1. Instrumentation of Scanning Thermal Microscopy

In 1986, Williams and Wickramasinghe pioneered a so-called scanning thermal profiler technique employing a STM probe with a thermocouple fabricated at the probe tip [9], as illustrated in Fig. 1. A thermocouple sensor was produced at the end of a STM probe tip by the junction of two dissimilar inner and outer conductors. An insulator separated the two conductors in all areas remote from the tip. The size of the thermocouple junction could be made to have dimensions on the order of

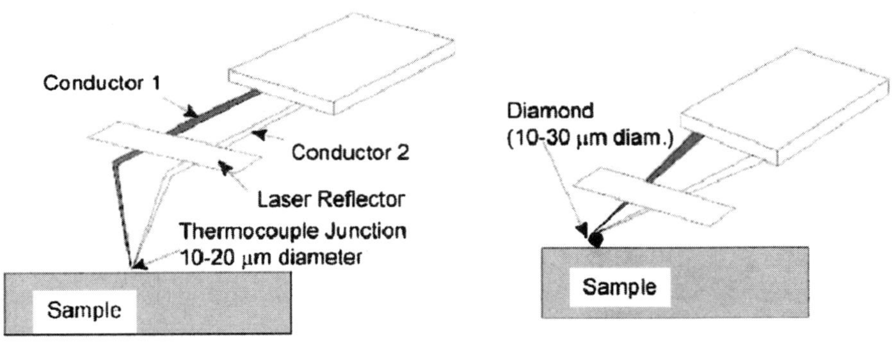

Figure 2. Schematic diagram of the wire thermocouple AFM probe.

100 nm. When a temperature difference exists between the thermocouple junction and the end of the lead wires, a voltage could be measured at the end of the lead wires with a minimum detectable tip temperature change less than 1 mK.

Although the main purpose of the scanning thermal profiler was not for mapping temperature distribution of a surface but for regulating the tip-sample distance, it stimulated intense efforts to develop a SPM technique for thermal microscopy. Because the STM requires the sample surface to be conducting, a SThM method developed in the configuration of a STM cannot be applied readily to a dielectric sample. In many applications such as mapping the temperature distribution of a microelectronic device, the sample surface is often covered by a dielectric film. Therefore, it is desirable to develop a general SThM method that can be applied to both conducting and non-conducting samples. In 1993, Majumdar *et al.* [10] introduced a wire thermocouple AFM probe, as depicted in Fig. 2. In an AFM, the tip-sample spacing is regulated by the force acting on the probe tip. As such, the AFM can map the topography of both conducting and non-conducting samples. While a thermocouple is fabricated on the probe, heat transfer between the tip and the sample changes the temperature of the probe tip during tip scanning. This allows for simultaneous mapping of topography and temperature.

The heart of the SThM is the thermal probe. A variety of thermal probes have been developed following the pioneering work of Williams and Wickramasinghe and that of Majumdar *et al.* A comment feature of these thermal probes is a thermal sensor fabricated at the end of an AFM or STM tip. The thermal sensor can be a thermocouple, a resistor thermometer, or a Schottky diode. A thermocouple measures the temperature difference between the junction of its two constituent metals and the other (remote) ends of the two metal wires. As such, there is no fundamental limitation to miniaturize the junction in order to achieve a better spatial resolution. On the other hand, a resistance thermometer can not be miniaturized readily, because the resistance of a nanoscale resistor can be too small for sensitive electronic detection. Similar problems exist for a temperature sensor based on a metal-semiconductor Schottky diode, as the I–V characteristics of a Schottky diode is affected not only by temperature change at the metal-semiconductor junction, but also by that of the body of the semiconductor. As a consequence, the Schottky diode sensor measures an average temperature of the junction and the body, instead of the localized temperature at the junction.

Significant efforts have been devoted to fabricate thermal probes with a thermocouple junction at the end of an AFM probe. In some early works, high thermal conductivity materials were used to make the probe. It was realized later that the thermal probe needs to be made of low thermal conductivity materials in order to thermally isolate the thermocouple junction at the end of the AFM tip. Otherwise, the temperature rise at the junction can be very different from that of the sample because of significant heat loss to the AFM cantilever. Based on a one-dimensional heat transfer model, Shi *et al.* addressed this issue by using low-thermal conductivity silicon dioxide (SiO_2) and silicon nitride (SiN_x) as the tip and cantilever materials [11]. It was shown by their modeling results that compared to silicon or metal probes with a similar shape, this design can improve the thermal isolation of the sensor from ambient. In addition, the thermocouple was made of Pt-Cr, Cr-Ni, or Cr-Ir to achieve low thermal conductivity and large thermopower. Furthermore, they attempted to minimize the cantilever width, metal line width and thickness, and junction size, and increased the tip height, for increasing thermal resistance of the cantilever and thermally isolating the junction.

The fabrication process of these thermal probes consisted of only wafer-stage processing steps as shown in Fig. 3, with more than 300 probes fabricated on one single wafer. First, a 0.5–1 μm thick low pressure chemical vapor deposited (LPCVD) SiN_x film was grown on both sides of 100 mm diameter double-side polished silicon wafers with (100) orientation, followed by the growth of a 8-μm-thick LPCVD silicon dioxide or low temperature oxide (LTO) film. The LTO film on the backside was stripped in 5:1 buffered hydrofluoric acid (5:1 BHF) (Fig. 3(a)). The LTO on the front side of the wafers was annealed at 1000°C for one hour. The SiN_x film on the backside of the wafer was then patterned for use as a mask in a subsequent bulk micro-machining step.

The probe tips were fabricated out of the 8-μm-thick LTO film by reactive-ion-etching (RIE) and wet etching. A chrome film was sputtered on the LTO, and patterned by photoresist into squares (Fig. 3(b)). Masked by the photoresist and chrome squares, the top 5 μm-thick LTO was etched in CF_4 and CHF_3 plasma. The photoresist plus chrome masks were undercut and the precursor of a sharp-angle tip shape was defined during this etching step. After the RIE etching, the remaining LTO film was etched in BHF 5:1 until the masks were just etched free (Fig. 3(c)). Before that, the 8 μm thick LTO film at the unprotected region was etched away. The RIE plus wet etching process could reproducibly yield oxide tips with a tip radus of about 20 nm and a half angle of 10°–20°.

After the oxide tips were fabricated, a 50–100 nm-thick Pt film (or Cr film) was sputter deposited and patterned on the front sides of the wafers (Fig. 3(d)). A 250–300 nm thick LTO was then deposited to cover the tip. Photoresist with appropriate viscosity was spun on the wafer. At a certain spinning speed, surface tension prevents a sharp tip from being covered by a spin-coated photoresist, leaving the very

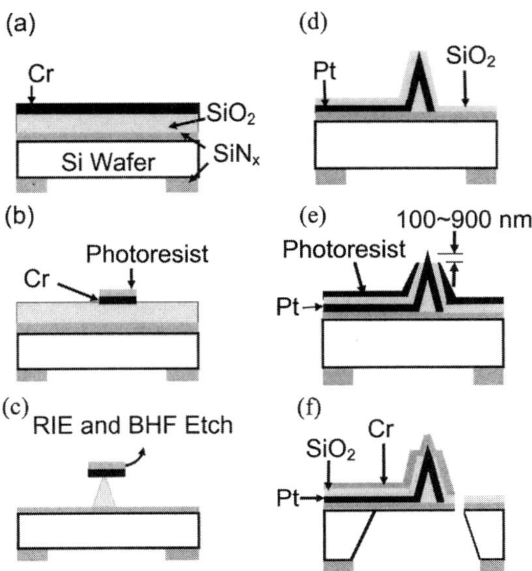

Figure 3. Fabrication process of thermally-designed thermocouple AFM probes. (a) Deposit SiN_x and then SiO_2 by LPCVD, strip SiO_2 and pattern SiN_x at the backside of the wafer, sputter Cr; (b) pattern Cr into 10 μm by 10 μm squares; (c) etch SiO_2 by RIE and BHF until the Cr and photoresist mask was etched free, resulting in a SiO_2 tip; (d) sputter and pattern Pt on the wafer, grow SiO_2 on Pt by LPCVD; (e) spin photoresist on the tip, leaving the tip end uncovered, etch SiO_2 from the tip end; and (f) sputter and pattern Cr on the wafer, pattern SiN_x on the front side of the cantilever, release the cantilever in TMAH.

ends of the tips uncovered (Fig. 3(e)). The exposed LTO film at the tip end was then etched in BHF 5:1. The height of the etched region depends on the photoresist viscosity and coating speed as well as etching time, and could be controlled in the range of 100–500 nm.

After the Pt (or Cr) was exposed at the tip end, the photoresist was stripped and 50–100 nm thick Cr (or Ni) was sputter deposited and patterned to form Pt-Cr (or Cr-Ni) junctions (Fig. 3(f)). After the thermocouples were made, the silicon nitride film was patterned into cantilever shapes. Then 5% tetramethylammonium hydroxide (TMAH) in water was used to etch grooves from the backside of the wafer until the wafer was etched through and the cantilevers were released (Fig. 3(f)). Figure 4 shows the cross section and scanning electron micrographs of a finished thermal probe.

The thermal time constant of the probe is an important parameter that determines limits the scanning speed of the thermal probe. This parameter was measured using a sample containing a 350 nm wide gold line. A sinusoidal current of a frequency f was passed through the gold line, resulting in a modulated temperature of the gold line at a frequency of $2f$. The temperature oscillation led to a $2f$ oscillation in the four-probe electrical resistance of the gold line, which was measured and used to calculate the $2f$ component in the temperature. The amplitude and phase of the $2f$ component of

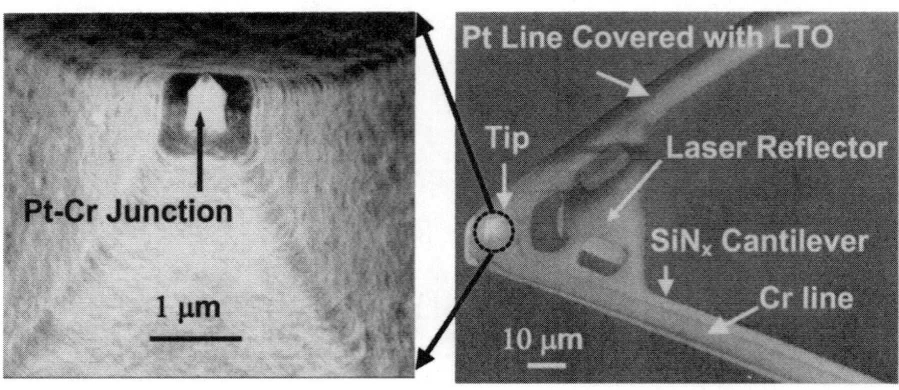

Figure 4. Scanning electron micrographs of a batch-fabricated probe. Left: a close up of the Pt-Cr junction at the tip end; Right: an overview of the probe.

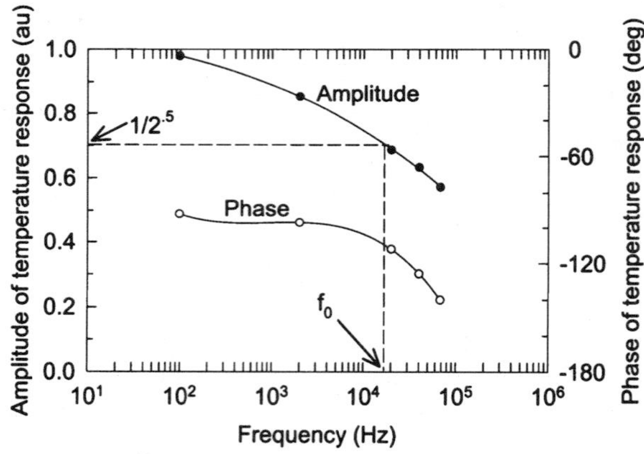

Figure 5. Dynamic temperature response of the junction as a function of heating frequency.

the temperature at the probe tip was measured by a lock-in amplifier while the probe contacted the gold line. The result is shown in Fig. 5. The AC temeperature showed a typical first order dynamic response of $T_{ac} = T_{dc}/\sqrt{1 + (2\pi f \tau)^2}$ and decreased to about $1/\sqrt{2}$ at a heating frequency of $f_0 = 18$ kHz. The time constant τ, defined as $1/2\pi f_0$, was calculated to be 8.8 μs.

2.2. Instrumentation of Scanning Thermoelectric Microscopy

After the invention of the scanning thermal profiler, in early 1990s Williams and Wickramasinghe reported a novel method called Scanning Chemical Potential Microscopy (SCMP) for mapping the local chemical potential, essentially thermoelectric power, of a MoS_2 and a graphite sample [12]. In their experiment conducted in air, an atomically sharp metal STM tip was scanned on the surface of the sample at a constant height using a unique feedback control loop. In this feedback loop, the voltage is applied to the tip through a large resistor (100 MΩ) connected in series with the tip-sample tunneling junction. In the STM mode, a constant voltage drop across the tip-sample junction was maintained by adjusting the tip height using the feedback loop. Because the sample was heated to about 30 K higher than the tip temperature held at ambient, a temperature gradient localized near the tip-sample junction was generated in the sample. This temperature difference creates a thermoelectric voltage corresponding to the local chemical potential of the sample. This thermoelectric voltage was measured intermittently between two successive STM line scans: after each STM line scan, the STM tip scanned the same line at a constant height with the feedback loop interrupted and the applied voltage reduced to zero. The obtained thermoelectric voltage map shows an atomic modulation.

In the following years, few efforts were reported on the development of thermoelectric microscopy. One notable experiment was done by Poler *et al.* using a setup similar to the SCMP for measuring the thermoelectric property of Guanine monolayers [13]. As they pointed out, the measured property by this technique is the thermopower (or Seebeck coefficient) of the sample. Therefore, they renamed this method as Scanning Thermopower Microscopy (STPM). Later on, Ghoshal *et al.* proposed to use the aforementioned thermocouple AFM probe to map the Seebeck coefficient of a sample, and named this AFM-based method as Scanning Thermoelectric Microcopy (SThEM) [14]. In this chapter, all these techniques aiming at probing surface thermoelectric properties are collectively referred as SThEM.

Very recently, Shi and Shih have further investigated the use of a SThEM method to profile Seebeck coefficient of semiconductor nanostructures including shallow p-n junctions in MOSFET devices and individual quantum well and barrier layers of superlattice thermoelectric devices [15]. Their measurement scheme is rather different from the SCPM in two aspects. First, their measurement is conducted in ultra high vacuum (UHV) because in air, heat conduction through the air gap between the STM tip and the sample dominates over that caused by electron and phonon coupling between the tip and the sample. Consequently, the temperature gradient in the sample, which is created by the colder tip, spreads over to a distance comparable to the mean free path of air molecules. This can severely limits the spatial resolution of the thermoelectric microscopy method. Second, they allowed a nanoscale elastic contact between the tip and the sample as this was found to be necessary for measuring a stable thermoelectric voltage on a semiconductor sample.

In the UHV SThEM setup, a heater wire attached to the sample holder was used to raise the sample temperature 10–30 K above ambient. The tip-sample contact was realized in the following procedure. At each point, after a tip-sample gap of about 1 nm was stabilized by feedback controlling a constant tunneling current, the control loop was disconnected by a manually-triggered relay and the STM bias was reduced to zero, as illustrated in Fig. 6. The tip and the sample were then connected to an electrometer with input impedance larger than 10^{13} Ω. Subsequently, the piezo-tube of the STM was used to drive the tip step by step at 0.5 Å per step toward the sample. Large fluctuation

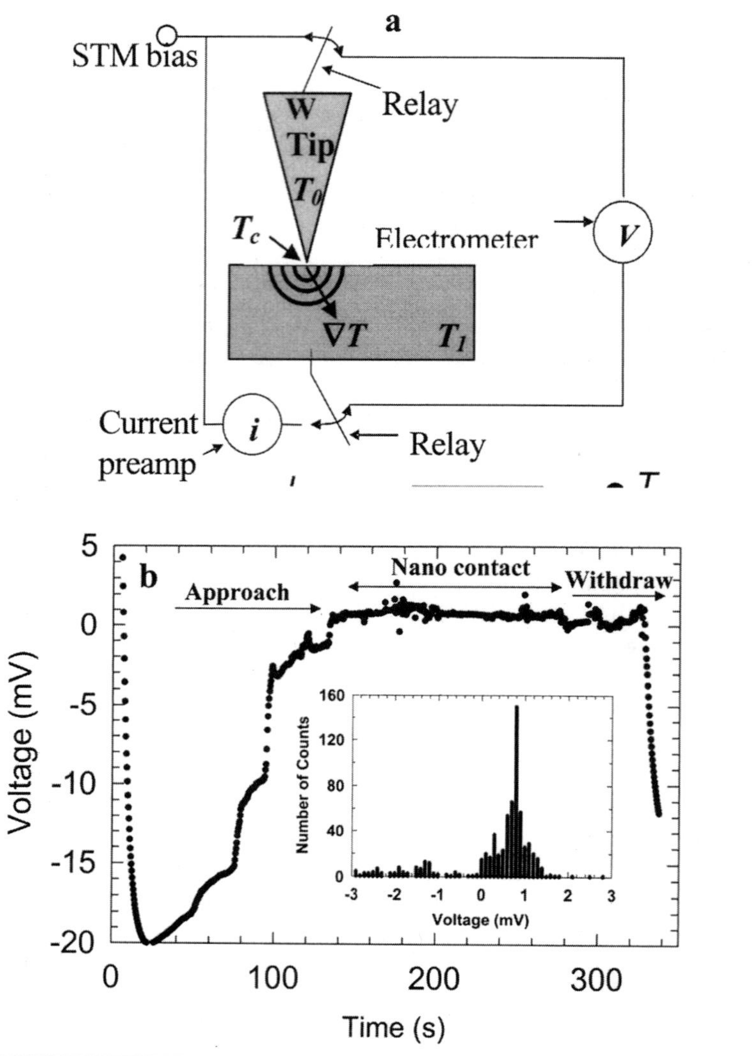

Figure 6. (a) Schematic diagram of the SThEM setup. (b) Time profile of the measured voltage as the tip approached the sample. Inset: histogram of the measured voltage.

in the measured tip-sample voltage was observed before the tip contacted the sample. The fluctuation is thought to be caused by a large and fluctuating tunneling resistance between the metal tip and the semiconductor sample. As the tip made a nano-contact with the sample, the voltage signal was found to be very stable. Further progression of the tip often caused crash. Therefore, the tip was slightly moved back and forth by up to 1 Å around the contact position to verify the stability of the measured voltage as well as to maintain an elastic contact. After the voltage measurement was made at each point, the sample was scanned to obtain a STM height image, which showed little change of image resolution. This measurement scheme is made feasible by the low drift (~10 Å/hour) and high vertical resolution (~0.1 Å) of the UHV STM.

The voltage measured between the tip and the sample during the nano-contact is caused by thermoelectric effects. As the tip contacts the sample, band bending occurs due to the formation of a Schottky junction. This leads to a built-in contact potential. Under an isothermal condition, however, the built-in potential does not contribute to the voltage measured by the electrometer. On the other hand, the contact of a colder tip and a hotter sample creates a temperature gradient in the sample concentrated near the contact (see Fig. 6). Carriers were made to drift under the temperature graident. This leads to a thermoelectric voltage proportional to the local S, a spatial average in a hemispheric sample volume of a large temperature gradient, i.e.

$$V(x, y) = S(x, y)(T_c - T_1) \tag{1}$$

where T_c and T_1 are the temperatures of the contact and the back side of the sample, respectively. The thermoelectric power of the metal tip and lead wires is much smaller than that of a semiconductor sample, and thus is ignored. This allows the SThEM to obtain a map of thermopower variation on the sample surface.

3. THEORY OF SCANNING THERMAL AND THERMOELECTRIC MICROSCOPY

3.1. Theory of Scanning Thermal Microscopy

The sensitivity and spatial resolution of the SThM largely depend on the mechanisms of heat transfer between the thermal probe and the sample. As shown in Fig. 7, heat transfers between the probe and the sample via solid-solid conduction, conduction through the air gap between the probe and the sample, radiation, and conduction through a liquid meniscus at the tip-sample contact. Air conduction and radiation are not localized at the tip-sample contact, and thus will cause poor spatial resolution. Among these two heat transfer paths, the radiation contribution is usually negligible unless the sample temperature is very high (>600 K). On the other hand, the influence by air conduction has been found to be significant, as discussed in the following paragraph.

The relative contribution of different heat transfer mechanisms between the thermal probe and the sample was examined and quantified by Shi and Majumdar [16]. They used a thermally-designed and batch-fabricated thermal probe and several calibration samples containing thin film metal lines with different width and length. The metal lines were joule heated during the experiment and its temperature was determined by measuring its temperature-dependant electrical resistance.

The temperature rise of the thermocouple junction was measured when the tip was in contact with the joule-heated metal line. It was found that the junction temperature rise was about 53%, 46%, and 5% of that of a 50 μm, 3 μm, and 0.3 μm wide line, respectively, showing a trend of decreasing sensitivity with decreasing sample line width. This suggests that the probe was heated more by air conduction between the tip and

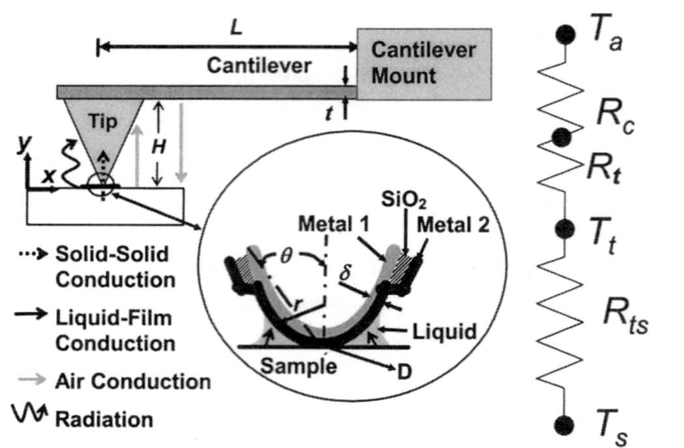

Figure 7. Schematic diagram of a thermal probe in contact with a sample. Also shown are the tip-sample heat transfer mechanisms and a thermal circuit diagram of the system.

the larger hot area of the wider lines. This trend indicates that air conduction plays an important role in tip-sample heat transfer.

A further experiment was used to determine the relative contribution of various tip-sample heat transfer mechanisms. In the experiment, the 350 nm wide line was joule heated to 5.3 K above room temperature. The cantilever deflection and temperature rise of the sensor were recorded simultaneously when the sample was raised toward and then retracted from the batch-fabricated thermal probe. When the sample approached the tip, the cantilever deflection signal remained approximately constant before the sample contacted the tip, as shown in the deflection curve in Fig. 8. In this region, the sensor temperature rise was mostly due to air conduction between the probe and the sample, and it can be estimated that radiation contribution was negligible when both the sample and the tip were close to ambient temperature. As the tip-sample distance was reduced, the sensor temperature rise due to air conduction increased slowly. Before the sample made solid-solid contact to the tip, the adsorbed liquid layers on the tip and the sample bridged each other. Initially, this liquid bridge pulled the tip down by a van der Waals force, as being seen in the dip labeled as "jump to contact" in the deflection curve. Coincidentally, there was a small jump in the sensor temperature due to heat conduction through the liquid bridge. As the sample was raised further, both the solid-soid contact force and the sensor temperature increased gradually, until the cantilever was deflected for more than 100 nm. After this point, the sensor temperature remained almost constant as the contact force increased.

As the sample was retracted from the tip, the sensor temperature again remained almost constant until at a cantilever deflection of 100 nm, the sensor temperature rise reduced roughly linearly but at a smaller slope than that found in the approaching cycle. As the sample was lowered further, the tip was pulled down together with the

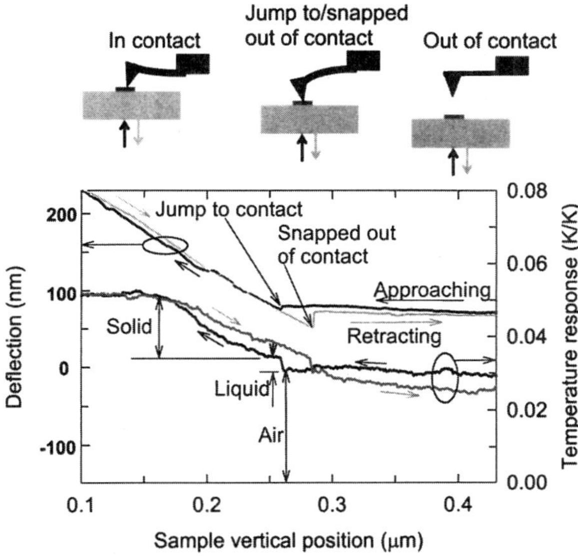

Figure 8. Cantilever deflection and temperature response of the probe as a function of sample vertical position when a 350 nm wide line was raised toward and then retracted from the tip.

sample by surface tension of the liquid bridge until after a certain point, the restoring spring force of the cantilever exceeded the surface tension and the tip "snapped out of contact" with the sample. Associated with the breaking of the liquid bridge, there was a small drop in the sensor temperature.

The above experiment shows several mechanisms. First, before the tip contacts the sample, air conduction contributed to a sensor temperature rise up to 0.03 K per K sample temperature rise, which was about 60 percent of the maximum sensor temperature rise at large contact forces. Second, conduction through a liquid meniscus was responsible for the sudden jump and drop in sensor temperature when the tip "jumped to contact" to and "snapped out of contact" from the sample, respectively. Third, solid-solid conduction resulted in the almost linear relationship between the sensor temperature rise and the contact force. This is a well understood feature for macroscopic solid-solid contacts [17]. Since the sensor temperature decreased at a slower slope during unloading (decreasing contact force), there must have been plastic deformation during loading (increasing contact force). For plastic deformation, the contact area increased with load [18], resulting in the linear increase of solid-solid contact conductance with contact force. However, since the conductance was still a function of the load during unloading, elastic recovery must have been significant.

One question remains as to why the sensor temperature saturated for cantilever deflection larger than 100 nm. To clarify this, the contact force corresponding to the 100 nm deflection need to be calculated. The spring constant of the composite

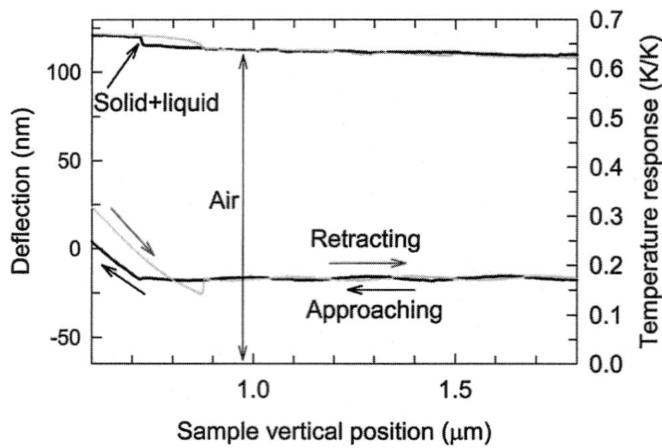

Figure 9. Cantilever deflection and temperature response of the probe as a function of sample vertical position when the 5.8 μm wide line approached and then retracted from the tip.

cantilever beam was calculated using Roark's formulas [19] to be 0.38 ± 0.11 N/m Therefore, the 100 nm deflection corresponds to a contact force $F = 38 \pm 11$ nN.

Assuming plastic deformation, this contact force resulted in a contact spot with a diameter

$$d_c = \sqrt{\frac{4F}{\pi H}}, \tag{2}$$

where H is the hardness of the tip or sample, whichever is softer. Among the tip and sample materials, the hardness of Au is the lowest and is of the order of 1 GPa. This leads to $d_c \approx 8$ nm. The tip radius was about 50 nm. However, the details on the tip end are not clearly shown in the SEM image due to the lack of resolution. The linear increase in contact area was probably due to the roughness on the surfaces or on the sample, since it is well known that in the junction of two random rough surfaces, the contact area increases linearly with contact force [18]. As the contact force increased to about 38 nN, the contact size approached the diameter of the asperity. At this point, the contact area could not increase further with the contact force, until the asperity was completely pressed into the sample by a contact force much larger than those used in the experiment. As a result, the sensor temperature in Fig. 8 remained almost constant for a deflection larger than 100 nm.

The point contact experiment was repeated for a 5.8 μm wide and 2000 μm long line and the result is shown in Fig. 9. While the increase of normalized sensor temperature due to solid and liquid conduction was similar in magnitude to those in Fig. 8, the normalized sensor temperature rise due to air conduction was about 0.6 K per K temperature rise on the sample, one order of magnitude higher than the corresponding one (0.03 K per K temperature rise on the sample) in Fig. 8. Therefore, it is clear that for the 5.8-μm wide line, air conduction dominated tip-sample conduction and was responsible for the large temperature rise.

These experiments reveal that the contribution of air conduction in tip-sample heat transfer depends on the size of the heat source on the sample. For large heat sources, air conduction dominates tip-sample heat transfer. As the size of the heat source reduces, the contribution of air conduction decreases and solid-solid conduction and liquid conduction become important. For micro/nano- devices with localized submicron heated features, such as carbon nanotube circuits, air conduction contribution can be reduced to a level smaller than that from solid-solid and liquid film conduction.

To estimate the magnitude of solid-solid and liquid film conductance, Shi and Majumdar [16] developed a one-dimensional head conduction model considering various microscale heat transfer mechanisms. Using the model, they inferred from the experimental data in Fig. 8 that the thermal conductance of the liquid meniscus was $G_{lf} = 6.7 \pm 1.5$ nW/K, and solid-solid conduction varies linearly with the contact force with a contact conductance of 0.76 ± 0.38 W/K-N. For contact forces larger than 38 ± 11 nN, the solid-solid conductance saturated at $G_{ss} \approx 30$ nW/K, because the contact size between the sample and an asperity on the tip end approached the $\sim$10 nm diameter of the asperity. For this solid-solid contact with an effective contact radius $b \approx 5$ nm, it can be shown that

$$G_{ss} = \pi k_{ef} b; \frac{2}{k_{eff}} = \frac{1}{k_s} + \frac{1}{k_{tip}(\tan\theta)^2} \tag{3}$$

where k_s and k_{tip} are the thermal conductivity of the sample and the tip, respectively, and θ is the half angle of the tip. With $\theta \approx 18^\circ$, $k_s \approx 200$ W/m-K for the thin film gold metal line, $k_{tip} \approx 15$ W/m-K for the Pt-Cr film at the tip end, and $b \approx 5$ nm, it can be estimated that $G_{ss} \approx 49$ nW/K. This is approximately the saturation value of 30 nW/K estimated from the thermal model.

Due to the small contact thermal conductance at the nanoscale tip-sample contact, the temperature rise of the thin film thermocouple probe was below 0.1 K per K sample temperature rise when micro/nano- devices with submicron localized heat source were imaged. In addition, due to the influence of air conduction, the measured temperature profile by the SThM was found to deviate from the true one. This makes it necessary to correct the SThM measurement results by careful thermal modeling. Alternatively, the influence of air conduction can be eliminated by performing the SThM in vacuum.

3.2. Theory of Scanning Thermoelectric Microscopy

The critical feature of the SThEM is to generate a localized temperature gradient at the tip-sample contact. The spatial extent of this temperature gradient, which limits the spatial resolution, not only depends on the contact size or tip radius, but is also determined by the scattering length (l), within which local thermal equilibrium cannot be obtained. Ignoring the latter microscale heat transfer effect, one can estimate the temperature in the sample at a distance r away from the contact point to be

$$T(r) \approx T_1 - \frac{a}{r}(T_1 - T_c) \tag{4}$$

As r is a few times larger than a, $T(r)$ approaches T_1 and the temperature gradient becomes negligible. This analysis holds only if $r \gg l$. For $r \leq l$, local thermodynamic equilibrium cannot be obtained within r, and the temperature gradient will have a spatial extent on the order of l. For example, if $l = 10$ nm, and $a = 1$ nm, according to eqn. 7, at $r = 5a = 5$ nm, $T(r)$ should have been very close to T_1. However, because local thermodynamic equilibrium cannot be obtained within $r = 5$ nm $< l$, it is impossible to achieve a large temperature difference between T_c and $T(r = 5$ nm), and the temperature approaches T_1 only when $r > l$. Hence, eqn. (7) derived by ignoring the microscale heat transfer effect breaks down for a length scale of the order of l. At room temperature, l is on the order of 1 nm for amorphous materials, and can be up to hundreds of nanometers for semiconductors. In degenerately doped semiconductors, the scattering length is found to scale with average inter-dopant distance to be a few nanometers [15]. In degenerately doped superlattice structures, l is expected to be further reduced by scattering at the interfaces.

Another question in SThEM is regarding the actual temperature drop across the sample. As shown in Fig. 6, the temperature T_c at the tip-sample contact is between T_0, the temperature of the tip holder, and T_1, the temperature at the backside of the sample. T_c is determined by the ratio of the thermal resistance (R_t) of the tip to that of the sample (R_s), i.e.,

$$T_c = T_0 + \frac{T_1 - T_0}{1 + \dfrac{R_s}{R_t}} \tag{5}$$

Assuming spherical symmetry in the sample and the tip and ignoring microscale heat transfer effects, one can estimate R_t and R_s as

$$R_s \approx (2\pi k_s a)^{-1} \text{ and } R_t \approx (2\pi k_t a(1 - \cos\theta))^{-1} \tag{6}$$

where k_s and k_t are the thermal conductivity of the tip and the sample, respectively, and a is the contact radius. Typically the tungsten STM tip has much higher thermal conductivity than a semiconductor sample. Hence, R_s/R_t is large and T_c is close to T_0.

The third important issue for the SThEM is the Schottky barrier formed due to the contact of a metal tip and a semiconductor sample. As a result, carriers are depleted from the contact point. Assuming spherical symmetry in the sample and $w \gg a$, where w is the depletion width, one can estimate that

$$w \approx \left(\frac{3a\varepsilon\varepsilon_0|\phi|}{Ne}\right)^{1/3} \tag{7}$$

With $a = 1$ nm, $\phi = 0.5$ V, $N = 10^{19}$ cm^{-3}, $\varepsilon = 16$, one finds $w = 5$ nm. This number is just slightly larger than the inter-dopant distance. In the derivation of

Eqn. 7, a continuous distribution of dopants has been assumed. This continuum assumption shall break down when w is shorter than the dopant-dopant distance. Nevertheless, it can be expected the depletion width is in the range of 5–10 nm. This depletion zone can have a different S from the bulk value, and affect the SThEM measurement. However, it was found that the measured S values by SThEM were very close to the bulk values [15]. This observation was attributed to a smaller radius (w) of the depletion zone than the spatial extent (l) of the hemispheric region with a large temperature gradient. As such, the bulk-like region where $w < r < l$ contributes to most of the measured thermoelectric voltage of the SThEM.

4. APPLICATIONS OF SCANNING THERMAL AND THERMOELECTRIC MICROSCOPY IN NANOTECHNOLOGY

4.1. Thermal Imaging of Carbon Nanotube Electronics

Carbon nanotubes (CNs) are made of graphitic cylinders. One distinguishes between multi-wall carbon (MW) CNs consisting of a series of coaxial graphite cylinders, and single-wall (SW) CNs that are one atom thick, usually with a small number (20–40) of carbon atoms along the circumference and several micron long along the cylinder axis.

At room temperatures, electrons near the Fermi level in a metallic SWCN can have a long mean free path on the order of microns [20]. For a metallic SWCN a few microns long, electron flows from one end of the tube to the other end without scattering with phonons, electrons, defects, and boundary. This situation is referred as ballistic electron transport. Semiconducting SWCNs have recently been observed to behave like ballistic conductors [21], while both ballistic [22] and diffusive [23] transport behaviors have been reported by MWCNs.

Experiments also found that metallic CNs can sustain high electric fields on the order of 100 kV/cm and high current densities up to 10^9 A/cm^2 [23, 24], two orders of magnitude higher than a normal metal such as copper. This excellent current carrying capacity of CNs may have potential applications for micro/nano electronics.

The potential electronic applications of CNs require a detail understanding of their electrical as well as thermal transport properties. Despite of various studies on electron transport, there remain several unanswered questions regarding heat dissipation in current carrying CNs. First, it is unclear whether heat dissipation occurs in the bulk of the CNs or at the contacts. For a SWCN at low electric fields, for example, heat dissipation is not expected to occur in the bulk. On the other hand, compared to the low field transport property, the electron-phonon scattering mechanism can be different at high fields for metallic SWCN. As shown by Yao *et al.* [24], for metallic SWCNs current saturates at about 25 μA at applied voltage exceeding a few volts. They proposed that the current saturation is due to electron scattering with optical phonons at high electric field. For MWCNs and semiconducting SWCNs, heat dissipation in the bulk is likely to occur if they are diffusive. However, there lack of direct experimental evidences to justify these predictions.

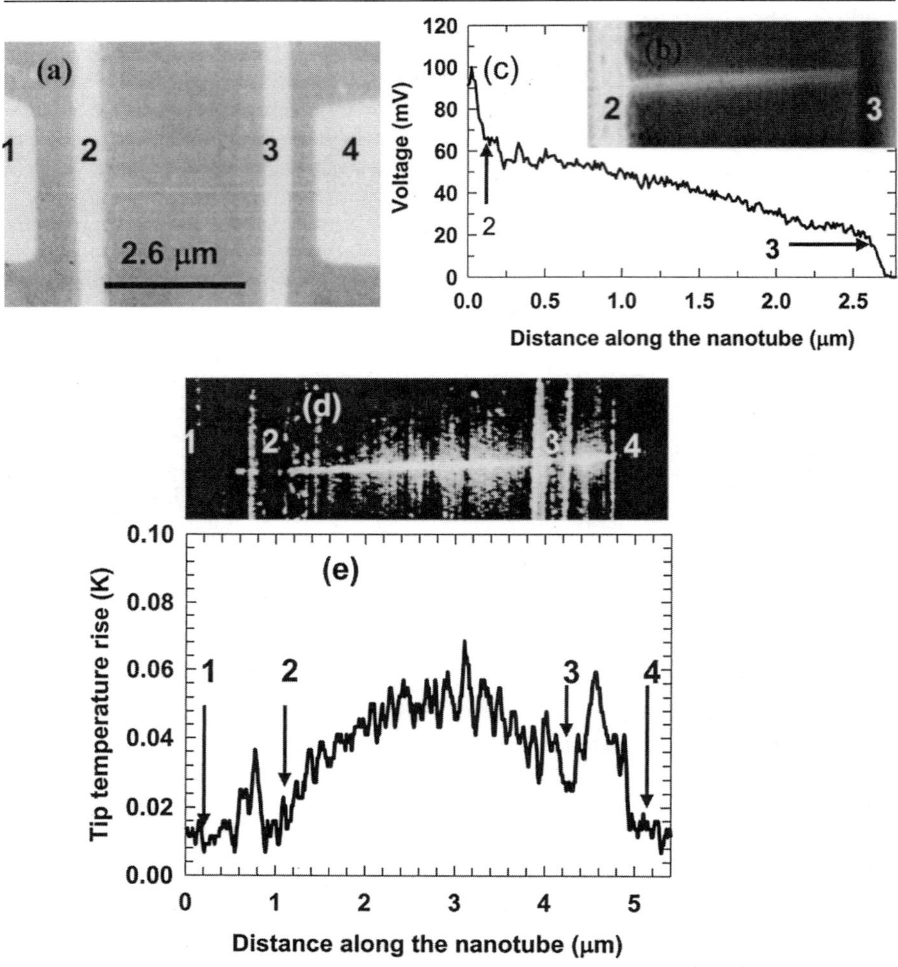

Figure 10. (a) Topographic image, (b) EFM image, (c) voltage profile, (d) thermal image, and (e) temperature profile along a 7 nm diameter MWCN.

Shi *et al.* have used the SThM technique to measure temperature distribution in current-carrying CNs and to investigate heat dissipation in the nanostructures [25, 26]. They have observed temperature profiles supporting ballistic (diffusive) transport in SWCNs and MWCNs. The measurement result on a MWCN is discussed below.

Figure 10 shows the AFM topographic image of a sample containing a 7-nm diameter MWCN on 1-μm-thick SiO_2 of a Si wafer. Four 30-nm-thick Au/Cr lines were patterned on top of the MWCN by electron beam lithography and lift off technique. Electrostatic force microscopy (EFM) was used to determine various electrical resistances in the device. A 100 mV AC voltage was applied at contact 2 with contact 3 grounded and contacts 1 and 4 floating. The frequency of the voltage was set to the resonant frequency of the EFM cantilever probe, which is a conducting AFM probe. Because of the electrostatic force between the sample and the probe, the cantilever was made to oscillate at its resonant frequency and the oscillation amplitude is proportional to the voltage of the sample in close proximity to the probe tip [22]. The sample voltage can be measured from the oscillation amplitude. The obtained EFM image was shown in Fig. 10(b). The voltage profile shown in Fig. 10(c) indicates that voltage decayed almost linearly along the MWCN and there were large voltage drops near the two contacts. The two terminal resistance measured with a bias of 100 mV between contacts 2 and 3 was 52 KΩ. From the EFM data, it was estimated that the electrical resistance was about 26 KΩ in the tube, and 16 and 10 KΩ at contacts 2 and 3, respectively. The segment of the tube between contacts 2 and 3 was 2.6 μm long. Therefore, the resistance in the tube was about 10 KΩ/μm. The linear voltage drop along the tube and the resistance value confirm the results from previous EFM study [22] that MWCNs behave as a diffusive conductor with a well-defined resistance about 10 KΩ/μm. In addition, for this device, the tube-contact interface were not clean, resulting in the much larger contact resistance than that found in the MWCN sample used in the previous EFM experiment.

SThM was further used to verify that the MWCN sample was diffusive and dissipative. A voltage of 0.7 V was applied at contact 4 with contact 1 grounded and contacts 2 and 3 floating. This resulted in a current of 9.3 μA between contacts 1 and 4. The corresponding thermal image and temperature profile were shown in Figs. 10(d) and 10(e). It is clear that the temperature at the middle was higher than those close to contacts 2 and 3 and the overall curvature of the profile is negative. This indicates that heat indeed was dissipated in the bulk of the tube.

It should be noted that on top of the metal covering the tube, the probe measured a much lower temperature than that when the probe was directly on top of the tube. It is possible that the temperature on top of the 30 nm thick metal had decreased significantly although the temperature could be high in the tube underneath the metal.

4.2. Thermal Imaging of ULSI Devices and Interconnects

SThM has been employed to map temperature distribution on top and cross sectional surfaces of MOSFETs and ULSI interconnects. Kwon *et al.* employed the batch-fabricated SThM probes to map temperature distribution on the cross sectional surface of an operating MOSFET with a gate length of 5 μm [27]. The thermal images clearly show a hot spot near the drain side. This is believed to be caused by the pinch-off effect near the drain, which results in a discontinuous channel and significant heat dissipation there. Efforts are currently underway to measure temperature distribution on the cross sectional surface of nanotransistors with a gate length shorter than 100 nm. Shi *et al.* have used the SThM to study localized heating in different VLSI submicrometer W-plug via structures provided by Texas Instruments, Inc. The via samples consisted of two levels of 0.6 μm thick Al-Cu metallization that were separated by a layer of 0.9 μm thick SiO_2 and connected through one or more W-plug vias. There was a 0.1 μm TiN layer on the top and bottom of each metal line. The samples were coated with a standard passivation layer of 1 μm thick SiO_2 followed by a capping layer of

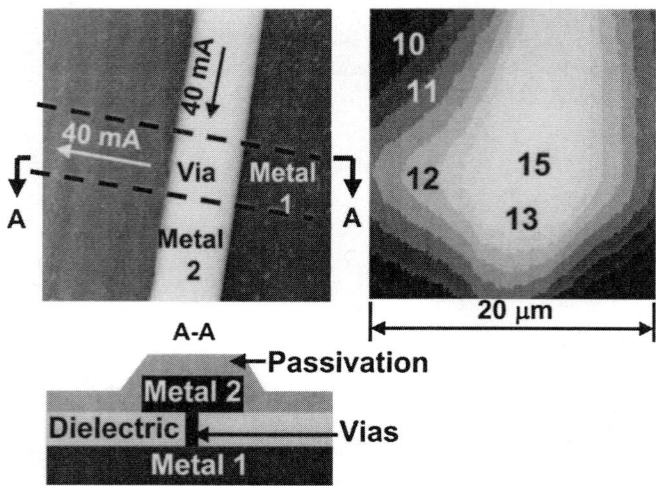

Figure 11. Topographic image (top left), thermal image (top right), and cross section of a via structure. The values shown in the topograph image are the temperature rise in degree Kelvin at the Pt-Cr junction during scanning.

0.3 μm thick Si_3N_4, as shown in Fig. 11. The topography, cross section, and SThM-measured temperature distribution of one via structure are also shown in Fig. 11 for a current of 40.5 mA. For this sample, the two metal lines were 3 μm wide and were connected by three 0.4 μm diameter vias. The temperature rise in the vias was determined from the change in electrical resistance using a temperature coefficient of resistance (TCR) of 1.01×10^{-3} K^{-1} obtained from an earlier work [28]. The via temperature was measured by the resistive thermometry as a function of current. For a current of 40.5 mA, the temperature rise in the via was 30 K. During scanning, the maximum temperature rise in the junction was found to be 15 K when the probe was directly on top of the vias. The high temperature region spread to the current flow direction.

4.3. Shallow Junction Profiling

Homo- and hetero-junctions constitute the fundamental building blocks of electronic, optoelectronic, and thermoelectric devices. Characterization of the dopant/carrier distribution and the electronic structures of semiconductor junctions is thus an important task. For example, profiling shallow p-n junctions in MOSFETs has been a prominent characterization issue that has challenged the semiconductor industry for many years. Simulation and characterization of dopant implantation, diffusion and activation is currently one of the highest-priority tasks among the research needs for extending CMOS, to its ultimate limit at or beyond the 22-nm node. In order to understand how different processing parameters and device structures will affect the dopant/carrier profiles and the electronic structures, accurate characterization data must be obtained

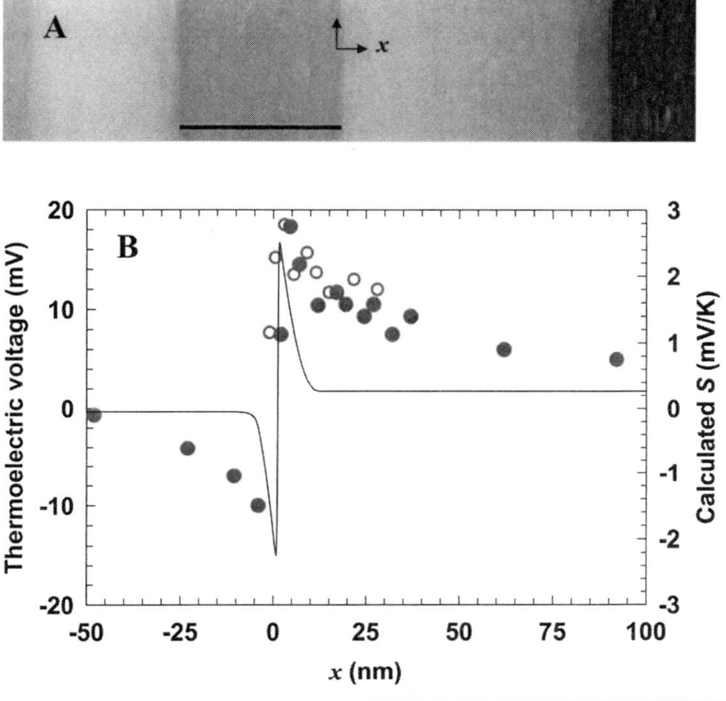

Figure 12. (A) STM image of a GaAs p-n junction array (filled state with sample bias of -2 V). The bright strips are p-doped with Be to 1.1×10^{19} cm^{-3}, and the dark strips are n-doped with Si to 9×10^{18} cm^{-3}. Scale bar: 100 nm. (B) Measured thermoelectric voltage (circles) and calculated S profile (dash line) across the p-n junction as a function of distance (x) from the metallurgical junction. The two measurement profiles (open and solid circles) were obtained at two different y locations with $T_1 - T_0 = 20$ K. The x position is obtained from STM images of a scan size of 30×30 nm.

using 2D dopant/carrier profiling tools with a 2 nm spatial resolution and 4% accuracy, as specified by the 2001 International Technology Roadmap for Semiconductors. A number of 2D dopant/carrier profiling methods have been investigated. However, the roadmap requirements have not been satisfied and 2D dopant/carrier profiling has remained essentially unsolved [29].

Lyeo *et al.* have employed the SThEM method for measuring the local S at a number of locations across GaAs p-n junctions. One set of measurement data are shown in Fig. 12. One can observe an abrupt inversion of the polarity of the measured thermoelectric voltage within a distance of 2 nm from the junction. The measured thermoelectric voltage at each point of the p-n junction is due to the diffusion of local carriers under a temperature gradient, and is coupled to the local carrier concentration. For an n-type nondegenerate semiconductor, the local thermopower is [30]

$$S = \frac{1}{-eT}\left(E_c - E_F + \frac{3}{2}k_B T\right) = -\frac{k_B}{e}\left(\ln\frac{N_c}{n} + \frac{3}{2}\right) < 0 \tag{8}$$

For a p-type nondegenerate semiconductor,

$$S = \frac{1}{eT}\left(E_F - E_V + \frac{3}{2}k_B T\right) = \frac{k_B}{e}\left(\ln\frac{N_v}{p} + \frac{3}{2}\right) > 0 \tag{9}$$

where $E_C(E_V)$ is the conduction (valence) band edge, E_F is the Fermi level, $n(p)$ is the electron (hole) concentration, $N_c(N_v)$ is the effective density of states of electrons (holes), k_B is the Boltzman constant, e is the electron charge, and T is the temperature. These equations illustrate that the magnitude and sign of S depend on carrier concentration and type, respectively. The dependence of S on the carrier type has been routinely used for determining the dopant type using the hot probe method in characterization laboratories. In fact, connecting a hot probe to a semiconductor and measuring the sign of the generated thermoelectric voltage is a textbook example for determining dopant types.

Using Eqns. 8 and 9 and those for degenerate semiconductors, Lyeo *et al.* have calculated the thermopower profile from the dopant profile obtained by SIMS. As shown in Fig. 12, the calculated thermopower profile exhibits the same trend as the measured thermoelectric voltage profile. They have further calculated the band structure and carrier profile from the STPM measurement results, as shown in Fig. 13. The calculation procedure is as following. First, they calibrated the STPM measurement using a sample with known dopant concentration and thermopower. The calibration allows them to convert the measured thermoelectric voltage to thermopower. The band structure and carrier concentration were then obtained using Eqns. 8 and 9 as well as those for degenerate semiconductors.

As a comparison, they also calculated the band structure and carrier concentration from the SIMS dopant profile. The band structure was obtained using the depletion approximation, and was used to calculate the carrier profile according to the Fermi-Dirac statistics. As shown in Figs. 13, the results from the SIMS and STPM exhibit good agreements with each other, except at the edges of the depletion region. Lyeo *et al.* have conducted a detailed study and concluded that the discrepancy shown in Fig. 13 was due to the break down of a continuum assumption at a lenthg scale comparable to the dopant-dopant distance (~5 nm) and the inaccuracy of the depletion approximation, both invoked for the calculation based on the SIMS profiles.

Nevertheless, their work demonstrates that by nanoscale profiling of S, the SThEM can be used to map out band structure and carrier concentration with nanometer spatial resolution. Furthermore, the carrier concentration is a reasonable good measure of the activated dopant density, when the dopant density varies over a scale comparable to the Debye length (13 nm at a concentration of 10^{17} cm^{-3} and 0.4 nm at 10^{20} cm^{-3}) [31].

As a 2D dopant/carrier profiling tool, the SThEM technique has two distinct features. First, for other dopant profiling techniques such as SCM and STM, the voltage applied to the tip creates an electric field penetrating into the sample. This electric field further causes band bending in the sample and shift the apparent junction location. It has been observed that the electronic junction locations obtained by SCM and STM

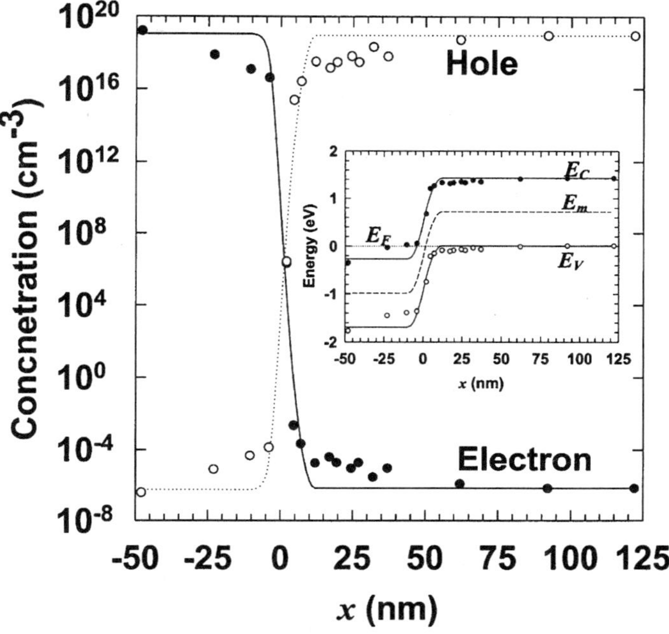

Figure 13. Carrier profile across the p-n junction shown in Fig. 12. Solid and dotted lines are electron and hole concentrations, respectively, calculated based on the energy band diagram obtained from the SIMS dopant profile using the depletion approximation. Solid and open circles are electron and hole concentrations, respectively, calculated from the measured *S* profile. To convent the measured thermoelectric voltage to the *S* profile, we calibrated the SThEM measurement using a sample with known *S*. Inset: Energy diagram of the p-n junction (Lines: calculated from the SIMS dopant profile using the depletion approximation. Circles: calculated from the measured *S* profile)

depend strongly on the tip voltage [32], and can shift 2–20 nm from that obtained from the SThEM. In SThEM, no voltage is applied at the tip during thermopower measurement. Thus, the unwanted effects of tip voltage-induced band bending and junction shifting are absent in SThEM, allowing SThEM to locate the actual electronic junction. Furthermore, the abrupt inversion of thermoelectric power across the electronic junction allows one to use SThEM to locate the electronic junction with a precision and resolution better than 2 nm.

5. SUMMARY AND FUTURE ASPECTS

The Scanning Thermal and Thermoelectric Microscopy methods discussed in the preceding sections have allowed the direct observations of thermal and thermoelectric transport phenomena in the nanometer scale. Several applications including mapping temperature distribution on nano and molecular electronics, and shallow junction profiling have been demonstrated. A variety of other applications have also been demonstrated using these techniques. For example, the SThM has the capability of mapping thermal conductivity and diffusivity of a polymer sample with sub micron spatial resolution. This unique feature has been used to image polymer blends and found applications in drug research and development [33–36]. The ability of SThEM in mapping Seebeck coefficient with nanometer scale spatial resolution can be applied to probe the superior thermoelectric properties of low dimensional thermoelectric materials, including quantum well, wire, and dot superlattices that will potentially revolutionize the energy conversion process.

Currently, new techniques are being invented for improving the spatial resolution or accuracy of these two techniques. For example, recent developments in near filed optical techniques for nano-thermometry have stimulated much interest in this direction. In addition, the possibility of using tunneling [39] or nano-contact thermometry to improve [37–38] the spatial resolution of the SThM is being investigated. The success of these new methods will allow the SThM and SThEM to become highly-powerful microscopic tools for nanotechnology research and development.

ACKNOWLEDGEMENTS

A significant portion of the meterials in this chapter are based on the results of the collaborations between the author and A. Majumdar, C. K. Shih, P. Kim, P. L. McEuen, and their students. For this research, the author was supported by the National Science Fundation (NSF) via two awards (CTS-0221180 and CTS-0239179). The opinions in this chapter are solely those of the author and do not necessarily reflect those of NSF.

REFERENCES

1. 2001 International Technology Roadmap for Semiconductors, http://public.itrs.net.
2. C. Dekker, *Physics Today* 52 (1999) 22.
3. M. H. Huang, S. Mao, H. Feick, H. Yan, Y. Wu, H. Kind, E. Weber, R. Russo, P. Yang *Science* 292 (2001) 1897.
4. L. D. Hicks and M. S. Dresselhaus, *Phys Rev B* 47 (1993) 12727.
5. J. P. Heremans, C. M. Thrush, D. T. Morelli and M. C. Wu, *Phys. Rev. Lett.* 88 (2002) 216801.
6. M. Shim, N. Wong, S. Kam, R. J. Chen, Y. Li, H. Dai, *Nano Lett.* 2 (2002) 285.
7. Y. Cui, Q. Wei, H. Park and C. M. Lieber, *Science* 293 (2001) 1289.
8. E. Comini, G. Faglia, G. Sberveglieri, Z. W. Pan and Z. L. Wang, *Appl. Phys. Lett.* 81 (2002) 1869.
9. C. C. Williams and H. K. Wickramashinghe, *Appl. Phys. Lett.* 49 (1986) 1587.
10. A. Majumdar, J. P. Carrejo and J. Lai, *Appl. Phys. Lett.* 62 (1993) 2501.
11. L. Shi, O. Kwon, A. Miner and A. Majumdar, J. *Microelectromechanical Sys.* 10 (2001) 370.
12. C. C. Williams and H. K. Wickramasinghe, *Nature* 344 (1990) 317.
13. J. C. Poler, R. M. Zimmermann, E. C. Cox, *Langmuir* 11 (1995) 2689.
14. U. Ghoshal, L. Shi, A. Miner and A. Majumdar, Proc. 19th Int. Thermoelectrics Conf., p. 221, London (2000).
15. H.-K. Lyeo, A. A. Khajetoorians, L. Shi, K. P. Pipe, R. J. Ram, A. Shakouri and C. K. Shih, *Science* 303 (2004).
16. L. Shi and A. Majumdar, *J. Heat Transfer* 124 (2002) 329.
17. M. Williamson and A. Majumdar, *J. Heat Transfer* 114 (1992) 802.
18. K. L. Johnson, *Contact Mechanics*, Cambridge University Press, New York (1989).
19. R. J. Roark and W. C. Young, *Roark's Formulas for Stress and Strain,* McGraw-Hill, New York (1989).
20. S. J. Tans, M. H. Devoret, R. J. A. Groeneveld and C. Dekker, *Nature* 394 (1998) 761.
21. A. Javey, J. Guo, Q. Wang, M. Lundstrom and H. Dai, *Nature* 424 (2003) 654.
22. A. Bachtold, M. S. Fuhrer, S. Plyasunov, M. Forero, E. H. Anderson, A. Zettl and P. L. McEuen, *Phys. Rev. Lett.* 84 (2000) 6082.
23. S. Frank, P. Poncharal, Z. L. Wang and W. A. de Heer, *Science* 280 (1998) 1744.
24. Z. Yao, C. L. Kane, and C. Dekker, *Phys. Rev. Lett.* 84 (2000) 2941.

25. L. Shi, S. Plyasunov, A. Bachtold, P. L. McEuen and A. Majumdar, *Appl. Phys. Lett.* 77 (2000) 4295.
26. P. Kim, L. Shi, A. Majumdar and P. L. McEuen, *Physica B* 323 (2002) 67.
27. O. Kwon, *Thermal Design, Fabrication, and Imaging of MEMS and Microelectronic Structures,* PhD dissertation, University of California at Berkeley (2000).
28. K. Banerjee, A. Amerasekera, N. Cheung and C. Hu, *Electron Device Lett.* 18 (1997) 405.
29. P. De Woff, R. Stephanson, T. Trenkely, T. Clarysse, T. Hantschel, and W. Vanderorst, *J. Vac. Sci. Tech. B.* 18 (2000) 361.
30. H. J. Goldsmid, *Electronic Refrigeration,* Plenum, New Yorks (1964).
31. Huang, C. C. William and J. Slinkman, *Appl. Phys. Lett.* 66 (1994) 344.
32. S. Gwo, R. Smith, C. K. Shih, K. Sadra and B. G. Streetman, *Appl. Phys. Lett.* 61 (1992) 1104.
33. R. J. Pylkki, P. J. Moyer, P. E. West, *Jpn. J. Appl. Phys.* 33 (1994) 3785.
34. A. Hammiche, D. J. Hourston, H. M. Pollock, M. Reading, M. Song, *J. Vac. Sci. Instrum.* 14 (1996) 1486.
35. A. Hammiche, M. Reading, H. M. Pollock, M. Song, D. J. Hourston, *Rev. Sci. Instrum.* 67 (1996) 4268.
36. M. Maywald, R. J. Pylkki, F. J. Reineke, L. J. Balk, *Prog. Nat. Sci.* 6 (1996) S103.
37. K. E. Goodston, M. Ashegi, *Microscale Thermophys. Eng.* 1 (1997) 225.
38. D. A. Fletcher, K. B. Crozier, C. F. Quate, G. S. Kino, K. E. Goodston, D. Simanovskii, D. V. Palanker, *Appl. Phys. Lett.* 77 (2000) 2109.
39. J. M. R. Weaver, L. M. Walpita, H. K. Wickramasinghe, *Nature* 342 (1989) 783.

7. IMAGING SECONDARY ION MASS SPECTROMETRY

WILLIAM A. LAMBERTI

1. SECONDARY ION MASS SPECTROMETRY AND NANOTECHNOLOGY

Imaging secondary ion mass spectrometry ("SIMS") is a technique that holds great potential for use in the field of nanotechnology. In much the same way that focused ion beam tools evolved to become a critical capability in the fabrication of many nano-mechanical devices, SIMS is becoming a central characterization tool within the nanotechnology community. For example within the semiconductor industry, SIMS employed at low incident ion energy is now used routinely to obtain trace information from extremely thin (1 nm) layered structures in depth, albeit in the non-imaging mode of analysis [1, 2]. Imaging SIMS lateral resolution, when employing high sensitivity primary ion species, has historically been limited to 1000 nm. This disparity in lateral resolution versus depth resolution can lead to measurement difficulties when analyzing fine structures in three dimensions. However, recent instrumental developments in ion sources and SIMS ion optics have dramatically improved lateral resolution to the regime below 100 nm while maintaining high sensitivity [3]. Due to the higher primary ion energies typically employed in these high lateral resolution instruments, depth resolution is somewhat degraded, typically on the order of 10 nm. As a result, SIMS image and depth resolutions have now converged to the point where the analyst can now routinely obtain information from nanovolumes.

The application of Imaging SIMS to mainstream nanotechnology areas such as nanosensors and micro-electromechanical systems ("MEMS") has been scarce to date. This can largely be attributed to a number of factors, including instrument availability, cost, and performance. There is also an inherent need for a highly skilled SIMS analyst. However, many examples of SIMS applications exist that do indeed qualify as examples of nano-analyses within fields as diverse as semiconductor technology, biology, metallurgy, catalysis, and cosmochemistry [4].

Many novel, *indirect*, SIMS analyses have been possible that provide information from regions that are smaller than the classical image resolution limit of SIMS (i.e. 1000 nm). The fundamental SIMS sputtering process, where a single primary ion impacts a surface and produces secondary ions, is inherently highly-localized (<5 nm radius) [5, 6]. Additionally, the probability that molecular ions are produced from a single impact is directly related to the nearest-neighbor distribution of species at the point of impact. Phenomena such as these can be used to obtain nano-scale information from materials without actually determining the precise event position with a resolution better than 1000 nm. Examples of methods utilizing this approach include coincidence spectroscopy [7, 8] and dynamic SIMS molecular spectroscopy [9].

The advantage of modern imaging SIMS instruments is that one can now *directly* probe sub 100 nanometer-scale features for compositional information, while maintaining trace sensitivity (several ppm) in these nano-volumes. The inherent ability of SIMS to detect *any* elemental species is a powerful capability for understanding the overall chemistry of a given material. Isotopic speciation is also possible with the high-resolution mass spectrometers incorporated into many SIMS instruments. Stable isotope tagging of reactant molecules permits the determination of reaction locales with high spatial resolution via SIMS, and is finding application in fields such as biology, medicine, and catalysis [10].

2. INTRODUCTION TO SECONDARY ION MASS SPECTROMETRY

2.1. Overview

The SIMS process is shown schematically below (figure 1). A beam of primary ions is generated in a suitable ion source, and accelerated towards a sample surface under vacuum. The interaction between the primary ion and the sample is complicated in nature, with many secondary particles being ejected for each incident ion. This process is referred to as "sputtering" [11, 12]. Electrons, neutral atoms, molecular fragments, and ions are generated in the sputtering process. Ionized mass fragments (monatomic and polyatomic) are subsequently accelerated through a mass spectrometer, where each fragment is separated according to its mass to charge ratio ("m/z"). Detection of the mass fragments then occurs via electron multiplier, faraday cup, or other suitable charged-particle detector. Mass spectra and ion images are among the types of data then produced.

Ion images can be produced via raster of the primary beam and synchronous detection of the mass filtered secondary ion signal, analogous to the process employed in scanning electron microscopy. This mode is referred to as the *microprobe mode* of ion imaging. It is important to note that in microprobe mode, image resolution is largely dependent upon primary ion probe diameter, with submicron lateral resolution now routinely attainable.

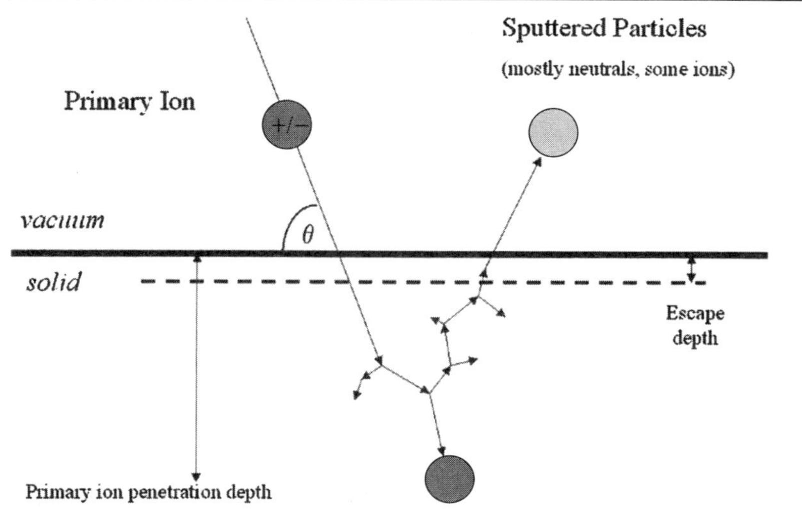

Figure 1. Schematic of the SIMS sputter process.

An alternative mode of image formation is available in some instruments known as the *microscope or direct mode* of ion imaging. The microscope mode is non-raster based, and utilizes electrostatic ion lenses in the secondary ion column. These lenses enable the formation of image planes and crossover points in the ion optical path. For final mass filtered image formation, the image planes are projected onto a position-sensitive detector. Several types of position sensitive detectors are in use today. Most commonly, a micro-channel plate coupled to a phosphor screen, with the resulting image captured by use of a suitable high-sensitivity CCD. Alternatively, ion images can be captured by use of a direct ion-imaging detector such as the resistive anode encoder ("RAE"). In microscope mode imaging, the image resolution is essentially dependent upon the ion optical lenses and their corresponding electric field strengths. In practice, image resolution is limited to roughly 1000 nm.

The choice of primary ion species will determine the efficiency of generation of specific secondary ion species, and subsequent experimental sensitivity. Primary ion energy will determine the subsequent spatial resolution in a given analysis, with lower energies favoring reduced surface mixing due to sputtering, thus improving depth resolution. However, higher primary ion beam energy can result in improved lateral resolution when in the microprobe mode.

Typical primary ion species include cesium (Cs^+), oxygen (O_2^+ or O^-), and gallium (Ga^+) with incident energies in the range of a few hundred eV to several keV. Cesium and oxygen primary beams provide significantly enhanced production of secondary ions due to the chemical interaction of each primary species with the sample surface. When used as a primary ion, oxygen enhances the production of positive secondary ions. This results from the increased oxygen concentration in the near-surface region as sputtering proceeds. This implanted oxygen tends to form metal-oxygen bonds, which when subsequently broken due to further sputtering, tend to leave the oxygen (high electron affinity) with a net negative charge and the counter ion with a net positive charge. Cesium favors the production of negative secondary ions, largely due to a reduction in the workfunction of the surface species as cesium is implanted in the sputtering process. This reduction in workfunction permits more electrons to exist in excited states above the surface potential energy barrier. This increased availability of excited electrons within the surface then favors the production of negative ions [13, 14, 15].

The SIMS process itself remains only partly understood at the fundamental level. Molecular simulations, coupled with experimental observations continue to provide new insights into the mechanisms underlying sputtering, ionization, and subsequent detection of secondary ions [16, 17]. As with many techniques, there are potential pitfalls, artifacts and quantification difficulties with SIMS. However, it remains a very powerful tool for the determination of nano-phase compositions in materials. A complete, detailed discussion of the entire SIMS process is beyond the scope of this text. However, the reader is referred to several excellent references in the field [18, 19, 20].

2.2. General SIMS Instruments

From the perspective of instrumentation, SIMS can be categorized into one of two possible regimes: dynamic SIMS or static SIMS. The simplest distinction between dynamic and static SIMS is based upon relative sputter rates. Dynamic SIMS involves the use of high fluxes of incident primary ions, with relatively rapid removal of the atomic layers in the sample being analyzed. Much molecular information is lost in the dynamic SIMS process due to the high primary ion flux. The high ion flux can also increase surface roughening and loss of interface resolution as the analysis proceeds in depth. Nonetheless, many structures can be probed in-depth readily, albeit with only limited molecular information preserved.

Static SIMS, on the other hand, refers to the sampling of only a fraction (less than 0.1%) of the topmost monolayer of a sample's surface. Very few incident primary ions impact the sample, so very few molecular bonds are broken. All information stems from the near-surface region of the sample, with extensive molecular information possible [21]. It is worth noting that recent developments in primary ion sources utilizing multiply-charged or cluster ions have blurred some of the distinctions between these regimes in specialized cases [22, 23, 24]. In practice, for the analysis of organics with a need for molecular information, static SIMS is preferred, while dynamic SIMS usually provides the ultimate in trace elemental detection sensitivity (ppb) and spatial resolution.

Each of these regimes also tends to determine the choice of preferred mass spectrometer type. The lower secondary ion flux generated in Static SIMS molecular studies require the use of a full-spectrum mass analysis (simultaneous detection and no lost ions), and high transmission as provided by a time-of-flight ("TOF") mass spectrometer. In fact, the use of TOF spectrometers has become almost synonymous with static SIMS, which is often referred to as TOF-SIMS. High-performance, commercial versions of these instruments include the Ion-TOF 5 [25] and Phi-Trift III [26]. The mass resolution (M / delta M) of a TOF spectrometer is dependent upon the temporal resolution of the secondary ion burst coming from the sample. This burst originates from a pulsed primary ion burst, so the temporal quality of primary ion pulse (in a process known as "bunching") defines the achievable mass resolution in an analysis. Pulsing of the primary beam is done with a duty factor of roughly 10^{-5}, and material removal is intentionally extremely slow. Unfortunately, a compromise often must be reached between mass resolution, spatial resolution, and compositional sensitivity. In some cases, TOF-SIMS lateral resolution can be below 1000 nm and may be quite appropriate for the study of certain nanostructures. This is especially true if organic molecular information is desired from such structures in the topmost surface layers. The choice of SIMS technique obviously depends upon the scientific question at hand. Issues of spatial resolution, surface versus bulk, detection sensitivity, and molecular versus elemental information must be carefully considered.

Dynamic SIMS analysis can, in principle, also be carried out using a TOF spectrometer, but is more often performed using a magnetic sector mass spectrometer with 100% duty cycle (Mattauch-Herzog configuration being common [27]), or a quadrupole mass spectrometer. Quadrupole mass spectrometers suffer from relatively poor mass resolution and transmission, but are generally less expensive and more compact. Magnetic sector instruments can simultaneously possess very high mass resolution and high transmission. The additional advantage of a non-TOF instrument, in the case of microprobe imaging, is that mass resolution and transmission can be maintained independently of image resolution. This result stems from the fact that the primary beam need not be pulsed to provide mass resolution, thus avoiding defocusing effects on the primary beam. The benefit of de-coupling these analytical parameters is significant in practice, as one can now independently optimize each portion of the instrument to serve its respective function. For example ion source design, primary ion species, and primary ion probe size can all be optimized to produce the best possible combination of sensitivity and image resolution for the analysis without impacting mass spectrometer performance.

2.3. High Resolution Imaging SIMS Instruments

It may be inferred from the previous discussion that imaging SIMS instruments with the highest spatial resolution and sensitivity tend to be magnetic sector based, dynamic SIMS instruments. Early attempts at optimizing spatial resolution on such instruments utilized highly focussed gallium (40 keV Ga+) primary ion beams [28]. Early commercial focussed ion beam ("FIB") instruments utilized gallium ion beams and added a quadrupole mass spectrometer to the FIB in the hopes of producing a useful high resolution imaging SIMS [29]. In fact, the use of gallium as a primary ion species in these instruments resulted in superb image resolution on the order of 10–20 nm. However, gallium produces extremely poor secondary ion yields for most species, resulting in poor SIMS sensitivity.

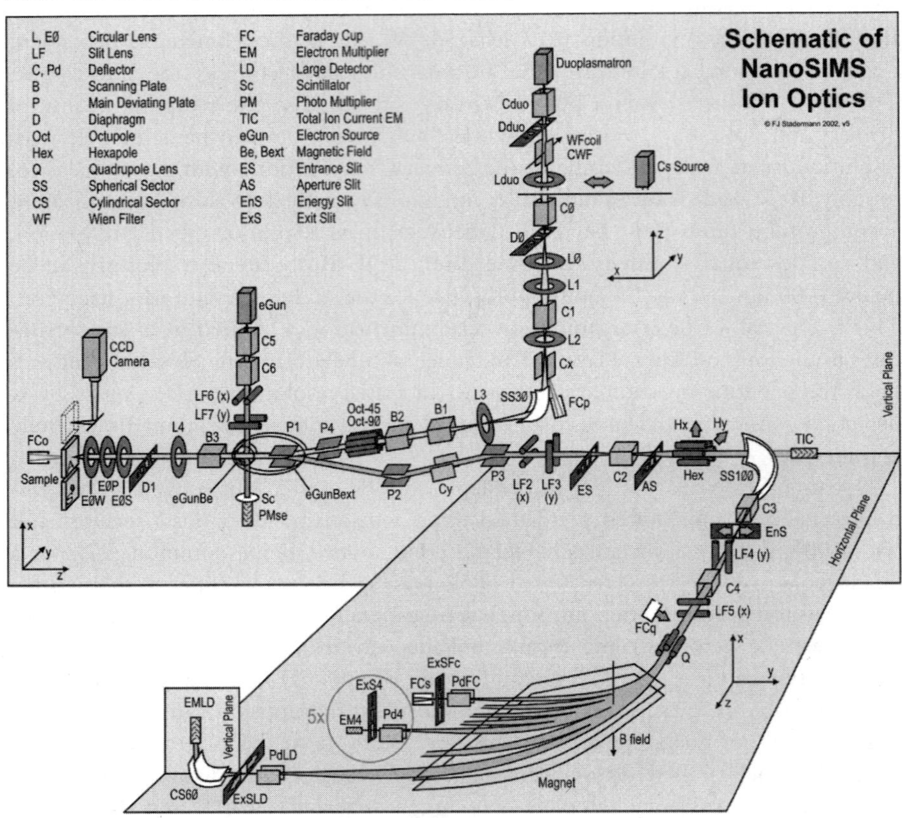

Figure 2. Schematic of ion optics for Cameca NanoSims 50. Based upon first series of commercial instruments. [reprinted with permission from F. J. Stadermann, Washington University at St. Louis. © F. J. Stadermann] (See color plate 3.)

The next major development in high resolution imaging SIMS emerged as the NanoSIMS 50 ("N50") a magnetic sector based, microprobe instrument utilizing nano-focussed cesium and oxygen ion sources [30]. Conceived in the early 1990's, the instrument design evolved over several years, with the first commercial instrument being installed in 2001. The N50 is uniquely designed for high transmission, high mass resolution, high sensitivity and high spatial resolution (50 nm), providing arguably the best overall imaging SIMS performance to date. A schematic of the N50 ion optics is provided in figure 2.

Several design features contribute to the high performance of this instrument. For example, the primary ion beam strikes the sample at normal incidence, permitting a very short distance between the sample and the probeforming lens. This geometry results in greatly reduced probe aberrations, and improved lateral resolution. Additionally, the sample is immersed in a very high electric field, facilitating efficient capture of secondary ions (i.e., a higher useful yield). The spectrometer design provides high mass resolution (M / delta (M) = 3000 minimum) without the use of beam stops. With beam stops used, mass resolution can readily exceed 10,000 while still maintaining high transmission.

One of the drawbacks of a magnetic sector instrument compared to a TOF instrument is serial mass species detection. This limitation is addressed in the N50 design by the use of multiple, parallel ion detectors that can be independently positioned at the exit of the mass spectrometer. The latest design employs seven ion detectors, along with ion-induced secondary electron imaging (when in the negative ion collection mode) [31, 32].

The ultimate spatial resolution of the N50 is typically better than 50 nm when using the cesium primary beam, and 150 nm when using the oxygen primary beam. Each primary ion species is chosen to maximize the sensitivity for detection of a particular secondary ion. As a primary ion, cesium enhances the yield of electronegative species, while oxygen tends to enhance the yield of electropositive species [33]. Thus these two primary ion sources together effectively span the entire periodic table with ppm to ppb detection limits.

3. EXPERIMENTAL ISSUES IN IMAGING SIMS

Many issues exist in SIMS analyses that can limit the ability to quantify a result in absolute terms. In some cases, even qualitative analysis is not possible. Examples of issues one may encounter include:

- Sputter rates can vary significantly for multi-phase or polycrystalline samples, from one phase to the next, even under conditions of constant and uniform primary ion beam flux.
- Matrix effects, due to differences in local chemical environment, can cause secondary ion yields to vary dramatically (by several orders of magnitude).
- Surface roughening can occur as sputtering proceeds, resulting in loss of interface resolution.
- For polycrystalline samples, grain orientation effects can lead to variable secondary ion yields due to different amounts of primary ion channeling in each grain.
- Sample charging can make an analysis completely impossible in the worst cases, and unpredictable in some others.
- Re-deposition of sputtered material onto the analysis area, during the sputter process, can cause high background levels in some situations.
- Sample preparation and handling can have significant effects on a given SIMS measurement.

Given these issues, many analyses are still, in fact, approachable. Many quantification schemes have been developed for SIMS analysis. A summary of the major SIMS quantification methods is given in Table 1. While not exhaustive, a few of the more typical

Table 1. Summary of Major SIMS Quantification Schemes

Method	Reported Error Range (+/−)	Benefits	Drawbacks
Relative Sensitivity Factor ("RSF") [31]	10–50%	Simple to employ on well-defined systems.	- Sample matrix must be constant and well-known. - Initial standards limit accuracy of subsequent analyses.
Ion-Implanted Internal Standard [32, 33, 34]	5–50%	Can be used for any species. Sample and standard are always analyzed together.	- Standards must be generated via ion implantation for each analysis. - Sample damage can occur during implantation.
Matrix Isotope Species Ratio ("MISR") [38]	3–50%	Convenient to use on homogenous samples.	- Sample heterogeneity limits accuracy. - Calibration curves need to be generated for each sample type. - Only accounts for oxide matrix effect.
CsM^+ Cluster ("CsM^+") [46, 47, 48]	10% (w/stds) 50% (w/o stds)	Relatively insensitive to most matrix effects.	- Poor sensitivity due to low signals. - Difficult on insulators.
Infinite Velocity ("IVM") [49, 50, 51]	10–300%	No standards required.	- Not useful on insulators. - High error range

methods are briefly introduced below. A more complete discussion of each method can be found within the referenced sources.

As an example, often the analysis can be facilitated through the use of external standards developed for the system of interest. Relative sensitivity factors ("RSF's") are then determined for individual species in a given matrix. This is the approach used within much of the semiconductor industry for depth profile interpretation in well-known matrices [34].

Quantitative results can also be obtained with the use of an internal standard for each analyzed area. This introduction of an internal reference can be accomplished by a variety of methods, including ion implantation [35, 36, 37] or by physically mounting a reference standard alongside the sample in each analysis area. The use of a physically separate reference standard has been used with success, for example, in the study of trace alloying elements in steels. Utilizing a reference alloy sample to be used as a relative comparison to the sample being analyzed within each field of view, imaging SIMS analysis was performed on both the reference and unknown simultaneously. This approach resulted in a very precise (<5%) relative determination of trace boron (1–10 ppm) since each analysis of reference versus unknown was done under identical conditions of instrumental alignment and primary beam dose. Useful relative compositional information can often be obtained using this method, with final measurement accuracy limited largely by the accuracy of the internal standard [38].

The method of standard additions can also be employed, via methods such as ion implantation over a range of concentrations, to establish the analytical response for a particular species within a given matrix or sample type. Implantation of minor isotopes can be used to minimize interference from naturally occurring levels in the standards [39].

The approach known as the matrix ion species ratio method ("MISR") [40] also utilizes external standards as a means to calibrate an instrument's sensitivity for a given species, in a given matrix. The MISR method attempts to account for local matrix effects (due to variable amounts of oxygen in the sample matrix).

This is accomplished by measuring the secondary ion signal from a series of standards, as a function of partial pressure of oxygen near the sample surface. Oxygen partial pressure is controlled via an imposed oxygen gas bleed. In the MISR approach, one monitors the production of two matrix species signals as well as the species of interest. In the case of a ferrous NBS-662 steel alloy [41] the authors monitored the ratio $^{112}Fe_2{}^+/^{54}Fe^+$ as well as the species to be calibrated ($^{52}Cr^+$). The ratio of the *molecular* iron fragment to the *monatomic* fragment is very sensitive to oxygen pressure, and thus provides a means to correlate the Cr^+ sensitivity factor to the local oxygen concentration at the sample surface. The $^{112}Fe_2{}^+/^{54}Fe^+$ ratio in an "unknown" sample thus is a direct means to determine the appropriate Cr^+ sensitivity factor. Here the "unknown" still must be nominally of the same composition as the calibrant.

Another approach that can be useful when absolute standards are not possible to obtain involves simply comparing samples on a relative basis. The analyst may make very reproducible measurements within a series of samples for trends that correlate with performance. For example, quantitative SIMS image analysis of heterogeneous catalysts is possible using a statistical, image-based approach when analyzing large numbers of catalyst particles [42, 43]. Here, many catalyst particles are analyzed via imaging SIMS with subsequent multi-particle image analysis. Sufficient numbers of catalyst particles are analyzed so that the mean of the SIMS ion intensity distribution approaches a statistically valid estimate of the bulk concentration for each species. Bulk concentration values can then be used to calculate an individual catalyst particle's concentration for a particular species.

An effective approach for obtaining relative quantitative information in biological structures via SIMS involves the use of isotopically tagged molecules. While not directly used for determining absolute concentrations, the distribution and quantity of isotopically tagged molecules can be determined by mapping the appropriate isotope ratio (15 N/14 N, for example). This method works well even for samples with extreme topography or sample charging due to the fact that all isotopes of a given species will experience essentially the same effects. In these experiments, the enhancement of a minor isotope over its naturally occurring abundance has been used directly to identify, for example, the uptake of a tagged molecule towards a particular biological structure in living organisms. Any observed increase is then used to quantify the relative growth or repair rate of that particular biological structure versus the surrounding tissue [44, 45].

In summary, many issues need to be considered in a given SIMS analysis to ensure that meaningful results are obtained. However, extensive experience exists within the

SIMS literature, so very often an answer exists to most of the problems encountered. Creativity and careful experimental design are often the key to a successful analysis.

4. APPLICATIONS IN NANOTECHNOLOGY

As stated previously, the use of SIMS for the determination of nanophase information is well established in classical SIMS application areas such as semiconductor technology, catalysis, cosmochemistry, biology and materials science. These studies serve as excellent examples of what may be possible for the study of modern nanotechnology. In each of the examples that follow, it is clear that the direct determination of compositional information on a scale below 1000 nm (and often below 100 nm) is readily attainable using high resolution imaging SIMS. This capability sets the stage for exciting possibilities in many areas of nanotechnology such as MEMS, nanosensors, nanomedicine, and more.

4.1. Example—Precipitate Distributions in Metallurgy

The performance of metallurgical systems is often dependent on the microstructure and microchemistry present in the material at hand. The relevant length scales of these properties can range from a few angstroms to many millimeters. Very often, light elements such as boron, nitrogen, oxygen, and carbon are added as critical alloying elements to enhance strength, toughness, or corrosion resistance of steels. For example, the unwanted segregation of certain species into precipitates or grain boundaries can degrade the performance of an alloy. Imaging SIMS is uniquely qualified as a tool to map these compositional distributions, for any element, including hydrogen. In instruments such as the N50, these distributions can now be determined with resolutions that permit meaningful comparisons with data obtained from transmission electron microscopy thus permitting improved correlation of composition with microstructure. In some high resolution imaging SIMS instruments, it is also possible to generate ion-induced secondary electron images simultaneously with secondary ion images. The sputter process also provides grain or relief contrast that can be used to advantage for the metallurgist, even in alloys that are difficult to relief etch using traditional wet chemical metallurgical etches. In these cases, the secondary electron image can often provide perfect positional correlation of microstructure with composition.

An example of SIMS compositional imaging with simultaneous secondary electron imaging is given in figure 3(a–c). The sample is a welded steel alloy forging that was examined using a Cameca N50 using a 16keV Cs^+ primary beam, with negative secondary ion detection. The sample was sectioned, mounted in a tin-bismuth, low melting point eutectic alloy, and mechanically polished to provide a cross-sectional view of the material [52].

The formation of numerous fine precipitates is seen. In this particular data set, given the pixel density (125 nm/pixel) and probe size used (100 nm), precipitates as small as 250 nm are readily revealed. Here, conservative sampling is employed, where 2 pixels/feature (in any one direction) are assumed to be the minimum sampling level

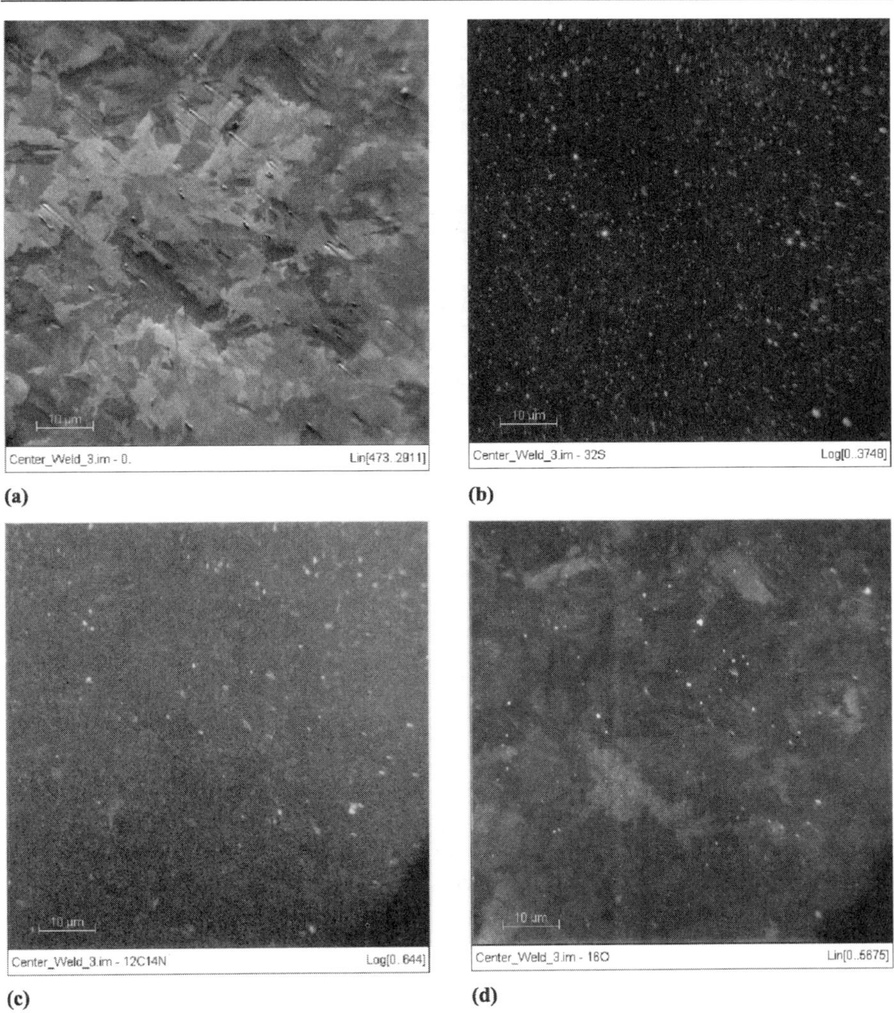

Figure 3. (a) Ion-induced secondary electron image of steel forging in welded zone, showing useful microstructural features. (Cameca N50, 16 keV Cs^+ primary ions). (b) SIMS ion image of sulfur ($32S^-$). Extensive precipitate formation detected, with 250 nm phases readily revealed. Composition is directly correlated with microstructure at left. (c) SIMS ion image of nitrogen ($26CN^-$). Evidence suggests that nitrogen is distributed both in solid solution and as precipitates. (d) SIMS ion image of oxygen ($16O^-$). Oxygen also seems to be distributed both in solid solution and as precipitates. Grain orientation effects are evident.

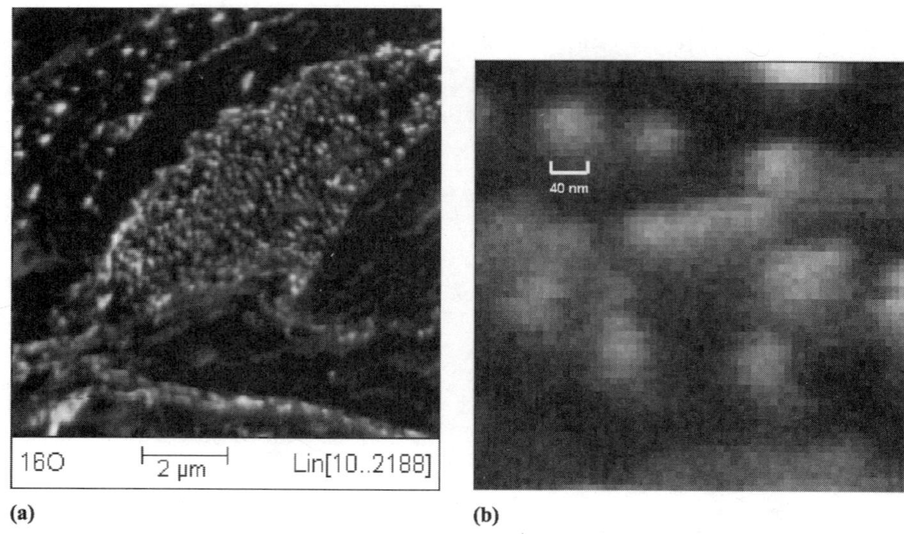

Figure 4. (a) SIMS ion image of copper grains. $16O^-$ image showing trace (bulk, ppm) distribution of copper in mechanically deformed grains. (Cameca N50, 16 keV Cs^+ primary ions). (b) Zoomed view from $16O^-$ ion image at left. Precipitate spacings well below 50 nm are clearly resolved, with precipitates on the order of 40 nm detected.

required to define a given feature [53]. As in any analysis, the analyst must strike a compromise between analyzed area, pixel density, dwell time, spatial resolution, and analysis time.

An example of higher resolution data is shown in figure 4(a–b). In this example, an 'oxygen free' copper alloy was being examined for trace oxide distributions. The sample at hand was mechanically deformed copper powder. Physical deformation occurred due to high-speed (supersonic) impact of the copper grains onto a substrate. The extreme sensitivity of dynamic SIMS in the N50 proved critical here, as both electron microprobe and TEM were unable to perform a successful analysis. Individual grains are revealed, with original grain boundaries and extent of plastic deformation apparent. Features below 50 nm are clearly resolved [54].

4.2. Example—Heterogeneous Catalyst Studies

Heterogeneous catalysis is a commercially important application area for the petrochemical industry where imaging SIMS has been applied. For example, extensive use of imaging SIMS has been made in the area of fluid catalytic cracking ("FCC") [55]. Modern FCC catalysts contain a complex mixture of submicron phases such as zeolites, clay, bulk alumina and a silica or alumina binder. From the perspective of nanotechnology, FCC catalysts can be considered as highly engineered nanoscale ceramics. From the perspective of SIMS, these systems are challenging due to their inherent compositional complexity and electrical insulating properties.

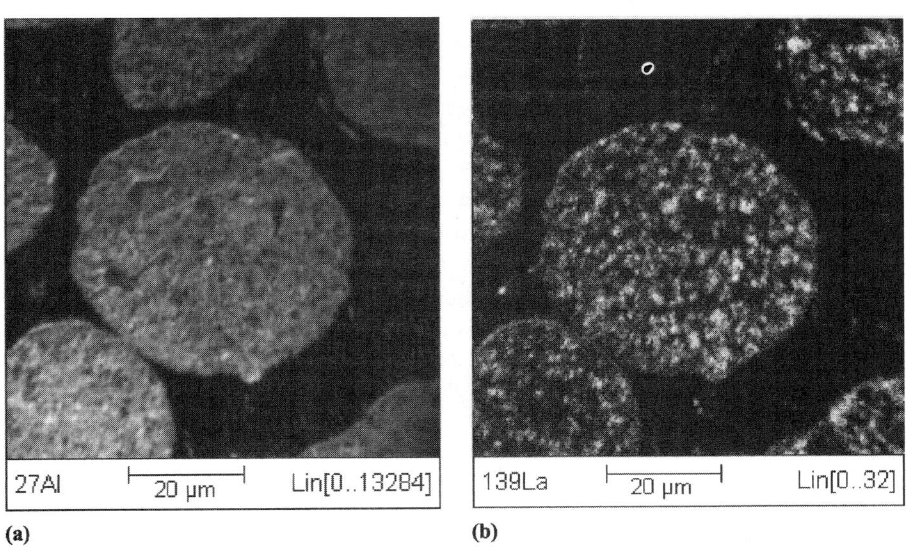

Figure 5. (a) Lower magnification aluminum SIMS ion image ($27AI^{+}$) of FCC catalyst particles as viewed in cross section. Submicron resolution is preserved. (Cameca N50, 16 keV O^{-} primary ions).
(b) Corresponding rare earth map ($139La^{+}$) showing position of zeolite phases throughout each FCC particle. Zeolites are intentionally doped with rare earth species to enhance hydrothermal stability.

As a first order approximation, the most important phase in FCC catalysis is the zeolite, which can be rendered inactive in use due to the accumulation of trace poisons such as nickel or vanadium, even at levels as low as a few ppm. The ability of imaging SIMS to identify and quantify the levels of poisons and other species on zeolites is obviously critical. Extending this capability to the complex mixture of nanoscale phases is truly a powerful tool in studying these systems.

Imaging SIMS data obtained on a typical FCC catalyst is shown in figure 5(a–b) [56]. Silicon and aluminum are contained in all phases, although at different concentrations. Silicon to aluminum difference images (or ratio images) can thus be used to effectively discriminate between the major phases. In order of decreasing silicon to aluminum ratio, we have the following typical progression: silica binder > zeolite > clay > bulk alumina or alumina binder. Note that the binder is usually either a pure silica or alumina. In addition to silicon to aluminum ratio, other trace species can often aid the analyst in major phase identification. For example, zeolites often intentionally contain rare earth species as stabilizers (i.e. lanthanum, cerium, etc). Similarly, clays are often uniquely identified by species such as titanium, iron, or calcium.

In figure 5 above, the catalyst particles have been embedded into epoxy and polished to reveal random cross sections. The particles are generally spherical, with diameters ranging between 40–100 um. Aluminum (viewed here as $27AI^{+}$) identifies the overall particle location. The lanthanum image ($139La^{+}$) gives a clear view of the zeolite distribution throughout the particles.

Using the oxygen primary beam, the N50 resolution is limited to approximately 150 nm. For higher resolution imaging, the cesium source is preferred (50 nm resolution attainable routinely) but can have some difficulty with charge compensation on difficult insulators. As can be seen in figure 6(a–f) below, high resolution data can be obtained on these catalyst samples, despite their electrical insulating properties. Features as small as 50 nm are clearly resolved. Simple image arithmetic provides a convenient tool for phase discrimination based upon appropriate compositional criteria such as silicon to aluminum ratio [57]. Color overlays as shown in figure 6(f) enable the rapid, qualitative determination of phase relationships.

Once the phases of interest have been discriminated, the use of binary masks enables the subsequent image analysis of compositional and morphological trends for these phases. For example, percent loading of zeolite within a catalyst can be readily calculated, as can the degree of poison pickup directly on to the zeolite phase versus other phases. The reader is referred to the work of Leta, *et al.* for details on the quantitative approach employed in this system [58].

4.3. Example—Nanoscale Biological Structures

One exciting area of research utilizing high resolution imaging SIMS involves the use of stable isotopic tracers to track biological activity within organisms [59]. Many biological structures possess both mechanical and chemical properties that serve as excellent analogues to the world of nanotechnology and are nano-scale in size [60].

One recent study of melanin granules within mammalian hair shafts demonstrates the utility of high resolution elemental SIMS imaging in these systems [61]. Compositional information from sub-100 nm structures within the hair shaft was readily obtained. Terrestrial isotope ratios for sulfur species were measured in the cuticle and cortex, demonstrating the additional possibility for isotopic tagging studies with a spatial resolution below 100 nm.

In a separate study, the distribution of iodo-benzamide melanoma markers within the cell ultrastructure has been studied successfully via high resolution imaging SIMS [62]. Here, the iodo-benzamide marker is used as a diagnostic tool for the detection of cancerous cells. The iodine ion ($127I^-$) was detected only in the melanosomes, and not in other structures within the cells (figure 7). This confirms the suspected preferential uptake of this molecule as a marker for cancerous cells. This also demonstrates the potential utility of imaging SIMS in the development of therapeutic agents to attack specific cancerous structures.

5. SUMMARY AND FUTURE PERSPECTIVES

Recent improvements in instrumental performance have brought imaging SIMS well into the nano-regime. The ability to map trace compositional information for structures below 100 nm is now routinely possible. In order to take full advantage of these new technologies, it is important for the analyst to learn from the extensive traditional SIMS knowledge base that already exists. This is especially important as one attempts to determine optimal analytical protocols or quantify observations.

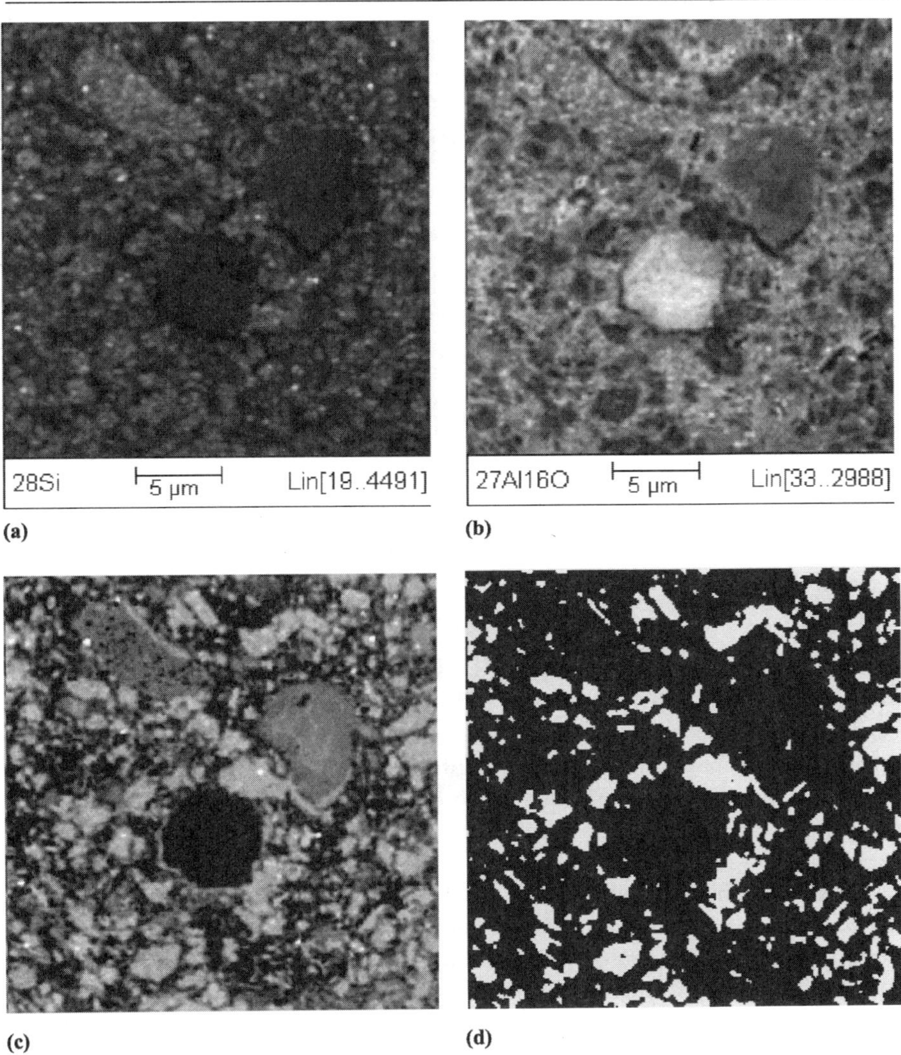

Figure 6. (a) High resolution silicon SIMS ion image ($28Si^-$) of FCC catalyst interior with 50 nm phases revealed. (Cameca N50, 16 keV Cs^+ primary ions). (b) High resolution aluminum SIMS ion image ($27AI16O^-$) of FCC catalyst interior. (c) Image processing methods can be used effectively to discriminate phases. Here, a silicon minus aluminum image (from 6a & 6b) reveals a complex phase map. (d) Simple binary map begins to define zeolite phases uniquely. Threshold function based upon silicon-aluminum intensities from image at left. Subsequent analysis can now reveal vanadium pickup on zeolite versus other phases. (e) Higher magnification image of $28Si^-$ demonstrates 50 nm phase identification. (f) Color overlay of $28Si^-$ (red) and $27AI16O^-$ (green) reveals fine scale phase relationships. (See color plate 4.)

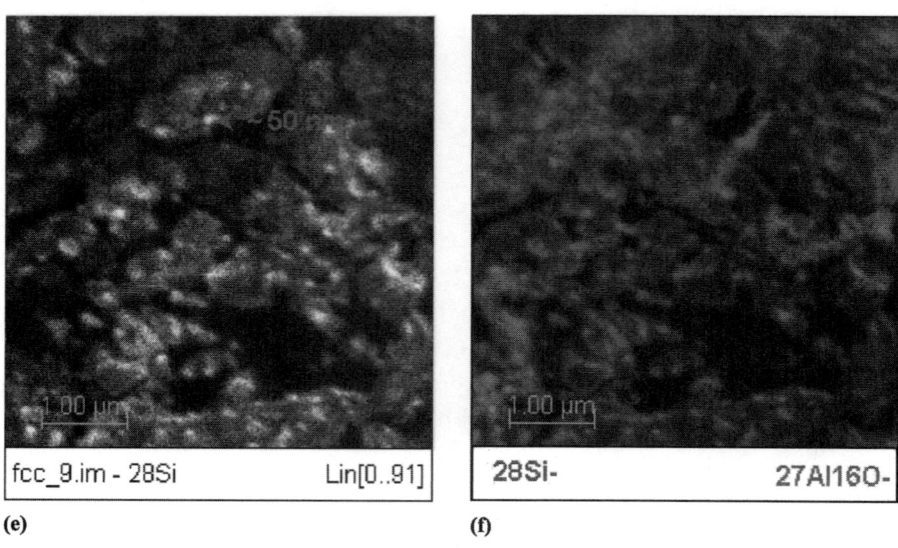

(e) (f)

Figure 6. (continued)

Both dynamic SIMS and TOF SIMS instruments are highly developed from the point of view of mass spectrometry. Still, improvements continue to emerge. For example, the use of array detectors at the exit plane of a magnetic sector has the potential for enabling full-spectrum imaging within a limited mass range [64, 65]. Ion source development is proving to be critical in both TOF and magnetic sector SIMS, pushing the limits of resolution and sensitivity, as new source designs are explored [66].

The future of high resolution imaging SIMS is equally exciting as one looks towards the wide array of applications in many fields of science. The field of nanotechnology is starting to realize the utility of SIMS, as more instruments with high spatial resolution become available. It is widely recognized that ion-beam based instruments such as FIB tools are critical in the development of nanodevices. In a similar way, high resolution imaging SIMS is poised for applications in nanotechnology where one needs to probe the composition of fine structures, at trace levels, with isotopic specificity, and a spatial resolution on the order of 50–1000 nm.

REFERENCES

1. M. G. Dowsett, Proceedings of SIMSXI, p. 259–264. John Wiley & Sons. Orlando, FL (1997).
2. P. A. Ronsheim and K. L. Lee, Proceedings of SIMSXI, p. 301–304. John Wiley & Sons. Orlando, FL (1997).
3. F. Hillion, B. Daigne, F. Girard, G. Slodzian, Proceedings of SIMS IX, p. 254, John Wiley & Sons. Yokohama (1993).
4. S. Messenger, L. P. Keller, F. J. Stadermann, R. M. Walker, and E. Zinner, *Science*, 300 (2003) 105.

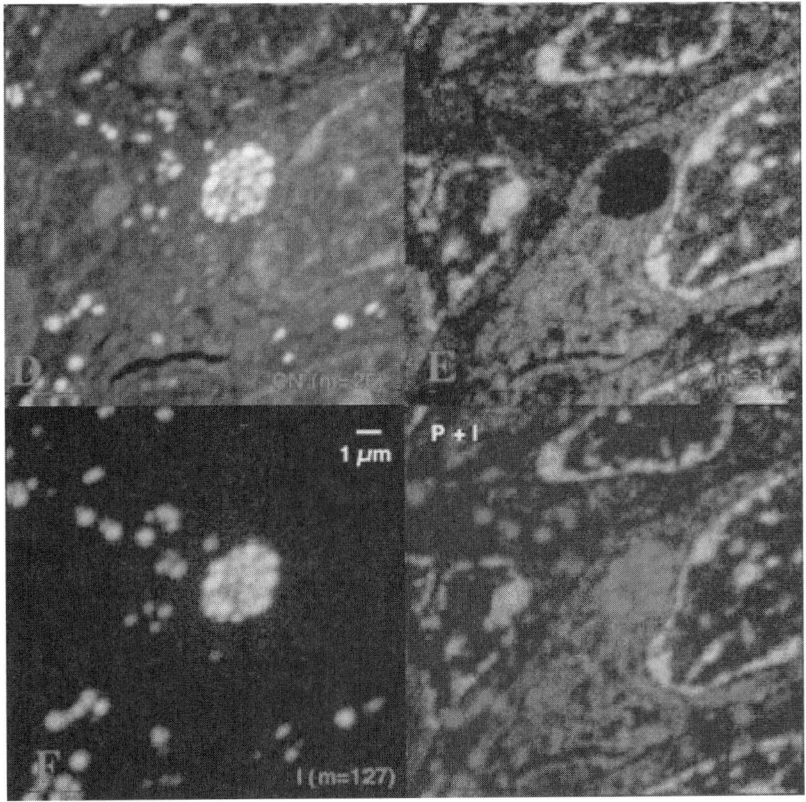

Figure 7. High resolution SIMS images of B16 melanoma cells from a mouse lung colony. Nitrogen (viewed as 12C14N- in frame "D") indicates overall cell structure. Phosphorus (31P- in frame "E") is present through the cell, being concentrated in the cell nucleus but absent from melanin. Iodine, present only in the marker, clearly targets the melanin. Reprinted with permission [63]. (See color plate 5.)

5. Z. Postawa, B. Czerwinski, M. Szewczyk, E. Smiley, N. Winograd and B. Garrison, *J. Phys. Chem.* B 2004.108.7831–7838.
6. Z. Postawa, B. Czerwinski, M. Szewczyk, E. Smiley, N. Winograd and B. Garrison, *Anal Chem.* (2003). 4402–4407.
7. S. V. Verkhoturov, E. A. Schweikert, and N. M. Rizkalla, *Langmuir*, 18 (2002) 8836–8840.
8. A. V. Hamza, T. Schenkel, A. V. Barnes, and D. H. Schneider, *J. Vac. Sci. Technol.* A 17(1) (1999) 303.
9. C. F. Klein and D. P. Leta, Proceedings of SIMSVII, p. 45–48. John Wiley & Sons. Monterey, CA (1989).
10. E. Hindié, Proceedings of SIMS XI, p. 113, John Wiley & Sons. Orlando, FL (1997).
11. M. W. Thompson, *Philos. Mag.* 18 (1968) 377.
12. P. Sigmund, *Phys. Rev.* 184 (1969) 383.
13. H. A. Storms, K. F. Brown, and J. D. Stein, *Anal. Chem.*, 49 (1977) 2023.
14. C. A. Andersen, *Int. J. Mass Spectrom. Ion Phys.*, 3 (1970) 413.
15. M. L. Yu, *Physical Review Letters*, 40, No. 9, (1978) 574.
16. R. Smith, D. E. Harrison, Jr., B. J. Garrison. *Phys. Rev. B*, 40 (1) (1989) 93.
17. I. A. Wojciechowski, S. Sun, C. Szakal, N. Winograd, and B. J. Garrison, *J. Phys. Chem. A*, 108 (2004) 2993–2998.

18. R. G. Wilson, F. A. Stevie, and C. W. Magee, *Secondary Ion Mass Spectrometry—A Practical Handbook for Depth Profiling and Bulk Impurity Analysis, Wiley-Interscience, New York* (1989).
19. J. C. Vickerman, A. Brown, and N. M. Reed (Eds.), *Secondary Ion Mass Spectrometry, Principles and Applications. International Series of Monographs on Chemistry 17.* Clarendon Press, Oxford (1989).
20. M. L. Pacholski and N. Winograd, *Chem. Rev.* p. 2977–3005 (1999).
21. T. Stephan, *Planetary and Space Science*, 49 (2001) 859–906.
22. A. Wucher, S. Sun, N. Winograd, *Appl. Surf. Sci.* (2004) in press.
23. F. Kotter, E. Niehuis, and A. Benninghoven, Proceedings of SIMS XI, p. 459, John Wiley & Sons. Orlando, FL (1997).
24. A. Tempez *et al.*, *Rapid Commun. Mass Spectrom.*, 18 (2004) 371–376.
25. ION-TOF GmbH, Gievenbecker Weg 15, 48149 Muenster, Germany.
26. PHI, Physical Electronics, Inc. 18725 Lake Drive East, Chanhassen, MN 55317.
27. J. Mattauch and R. Herzog, *Z. Physik*, 89 (1934) 786.
28. R. Levi-Setti, G. Crow, Y. L. Wang, N. W. Parker*, R. Mittleman, and D. M. Hwang, *Phys. Rev. Lett.* 54 (1985) 2615–2618.
29. L. Christman, presented at *8th Annual Workshop on Secondary Ion Mass Spectrometry*, Lake Harmony, PA (1995).
30. G. Slodzian, B. Daigne, F. Girard, F. Hillion, Proceedings of SIMS VIII, p. 169, John Wiley & Sons. Amsterdam (1991).
31. F. Horreard, presented at *17th Annual Workshop on Secondary Ion Mass Spectrometry*, Westminster, CO (2004).
32. F. Hillion, F. Horreard, F. J. Stadermann, Proceedings of SIMS XII, John Wiley & Sons. Brussels (1999).
33. H. A. Storms, K. F. Brown, and J. D. Stein, *Anal. Chem.*, 49 (1977) 2023.
34. R. G. Wilson, F. A. Stevie, and C. W. Magee, *Secondary Ion Mass Spectrometry—A Practical Handbook for Depth Profiling and Bulk Impurity Analysis, Wiley-Interscience, New York* (1989).
35. D. P. Leta and G. H. Morrison, *Anal. Chem.*, 52 (1980) 277.
36. G. L. Liu *et al.*, J. Vac. Sci. Technol. B14(1) (1996).
37. A. G. Borzenko, *et al.*, *Microscopy and Microanalysis*, 29–31 (Nov 1999).
38. W. C. Horn and W. A. Lamberti, *12th Annual Workshop on Secondary Ion Mass Spectrometry*, Gaithersburg, MD (1999).
39. D. P. Leta and G. H. Morrison, *Anal. Chem.*, 52 (1980) 277.
40. J. D. Ganjei, D. P. Leta, and G. H. Morrison, *Anal. Chem.*, 50 (1978) 285.
41. Ibid.
42. D. P. Leta and E. L. Kugler in *Characterization and Catalyst Development—ACS Symposium Series 411*, Chapter 32, p. 354 (1989).
43. D. P. Leta, W. A. Lamberti, M. M. Disko, E. L. Kugler, W. A. Varady, in *ACS Symposium Series 452—Advances in FCC Technology II*, Chapter 17, p. 269–292, (1991).
44. G. H. Morrison and S. Chandra. Proceedings of SIMS XI, p. 109, John Wiley & Sons. Orlando, FL (1997).
45. E. Hindie', Proceedings of SIMS XI, p. 113, John Wiley & Sons. Orlando, FL (1997).
46. C. W. Magee, W. L. Harrington, and E. M. Botnick, *International Journal of Mass Spectroscopy and Ion Processes*, 103 (1990) p. 45–46.
47. P. Willich and R. Bethke, Proceedings of SIMS XI, John Wiley & Sons. Orlando, FL (1998) p. 991–994.
48. T. Mootz, A. Adriens, and F. Adams, *International Journal of Mass Spectroscopy and Ion Processes*, 156 (1996) p. 1–10.
49. P. A. W. van der Heide, *et al.*, *Surface and Interface Analysis*, 21, (1994) pp. 747–757.
50. P. A. W. van der Heide, *Surface Science Letters*, 302 (1994) L312–L318.
51. R. Losing, N. Reger, F. J. Stadermann, and H. M. Ortner. Proceedings of SIMS XI, p. 1019, John Wiley & Sons. Orlando, FL (1997).
52. W. C. Horn and W. A. Lamberti, *17th Annual Workshop on Secondary Ion Mass Spectrometry*, Westminster, CO (2004).
53. C. E. Shannon, *The Bell System Technical Journal*, 27 (July, October 1948) p. 379–423, 623–656.
54. W. C. Horn and W. A. Lamberti, *17th Annual Workshop on Secondary Ion Mass Spectrometry*, Westminster, CO (2004).
55. E. L. Kugler, D. P. Leta, *J. of Catalysis*, 109, (1988) 387.
56. W. A. Lamberti and W. C. Horn, *17th Annual Workshop on Secondary Ion Mass Spectrometry*, Westminster, CO (2004).
57. Ibid.

58. D. P. Leta, W. A. Lamberti, T. E. Clark, W. A. Varady. Proceedings of SIMS VIII, John Wiley & Sons. (1992) p. 541.
59. Peteranderl, R. and C. Lechene. *Journal of The American Society for Mass Spectrometry*. In press.
60. Pickles, J., *V. Mechanisms of transduction and excitation in the cochlea. An Introduction to the Physiology of Hearing*. Academic Press. 107–153 (1982).
61. Hallegot, P., R. Peteranderl and C. Lechene, *Journal of Investigative Dermatology*. 122 (2004) 381–386.
62. J. Guerquin-Kern, F. Hillion, J. Madelmont, P. Labarre, J. Papon, and A. Croisy, *Biomedical Engineering Online* 2004, 3:10. (http://www.biomedical-engineering-online. com/content/3/1/10).
63. Copyright © 2004, Guerquin-Kern *et al.*; licensee BioMed Central Ltd. This is an Open Access article: verbatim copying and redistribution of this article are permitted in all media for any purpose, provided this notice is preserved along with the article's original URL: (http://www.biomedical-engineering-online. com/content/3/1/10). Biomed Eng Online. 2004; 3 (1): 10.
64. Barnes IV, J. H., Schilling, G. D., Sperline, R. P., Denton, M. B., Barinaga, C. J., Koppenaal, D. W., and Hieftje, G. M., *J. Am. Soc. for Mass. Spectrom.*, (2004). (In Press).
65. J. Barnes, R. Sperline, M. B. Denton, C. Barinaga, D. Koppenaal, E. Young, and G. Hieftje, *Anal. Chem.*, (2002) 74 (20), 5327–5332.
66. Guillermier, C., C. P. Lechene, J. Hill and F. Hillion. *Review of Scientific Instruments*. 74, (2003) 3312–3316.

8. ATOM PROBE TOMOGRAPHY

M. K. MILLER

1. ATOM PROBE TOMOGRAPHY AND NANOTECHNOLOGY

A general view of nanotechnology encompasses the design and fabrication of materials whose properties are controlled or influenced by features on the nanometer scale. Although nanotechnology often refers to semiconductors and other novel devices, the properties of many conventional materials and alloys are also controlled by features on the nanometer scale. In order to develop the full potential of these types of materials and understand their properties, their nanostructures must be characterized. A variety of state-of-the-art techniques are used to characterize different components of the nanostructure. Atom probe tomography (APT) and atom probe field ion microscopy (APFIM) have played important roles in the characterization of materials on the nanometer scale for more than five decades [1, 2]. Historically, the atom probe has been mainly applied to metals, alloys, and other high-conductivity specimens, but advances in both specimen preparation techniques and atom probe design have made the technique increasingly applicable to lower-conductivity specimens, with particular implications for semiconductor-based structures.

In this chapter, a description of the technique of atom probe tomography is presented. The main application of atom probe tomography is the characterization of the size, concentration and morphology of nanometer-scale solute fluctuations, such as precipitates, clusters and the levels of solute segregation to interfaces. The technique is generally referred to as atom probe tomography, since three-dimensional images of the internal structures of the specimen are generated from many slices, each containing a few atoms. An overview of the various types of atom probes and the field ion microscope is given. The types of analyses that may be commonly performed with these instruments are also described. Some examples of the capabilities of the atom probe tomography technique are also presented.

2. INSTRUMENTATION OF ATOM PROBE TOMOGRAPHY

Atom probe tomography is a unique form of microstructural characterization that has evolved from field ion microscopy and its precursor field electron emission microscopy. The many different variants of atom probes that have been produced and have culminated in the state-of-the-art local electrode atom probe are described in this section.

2.1. Field Ion Microscope

The field ion microscope was introduced by Prof. E. W. Müller in 1951 as a major conceptual change from the field electron emission microscope [3]. In contrast to the negative voltage used in the field emission microscope [4], the field ion microscope used a positive voltage applied to the cryogenically cooled specimen in the presence of an image gas. The field ion microscope required significantly sharper needle-shaped specimens to enhance the field sufficiently for the electron tunneling process used to field ionize the image gas atoms. This new instrument produced the first images of individual atoms in a tungsten specimen.

The basic components of a field ion microscope are an ultrahigh vacuum system in which a cryogenically-cooled needle-shaped specimen is positioned approximately 5 cm from a phosphor screen. A small quantity of image gas, typically $\sim 1 \times 10^{-5}$ mbar of helium for the refractory elements and neon for most other materials, is introduced into the vacuum system and then a high positive voltage is applied to the specimen. Other image gases, such as hydrogen, argon and nitrogen, have also been used for a few materials with low evaporation fields [2]. The image gas atoms close to the apex of the specimen become polarized due to the high electric field. These gas atoms are then attracted to the specimen where they become thermally accommodated to the cryogenic temperature. If the field strength at the apex of the specimen is a few volts per nanometer, these image gas atoms are ionized and the resulting positive ions are radially repelled from the positively-charged specimen towards the phosphor screen. This field ionization process occurs over the entire hemispherical apex region of the specimen and produces the field ion image on the phosphor screen. An example of a field ion image of an iridium specimen is shown in Fig. 1a. The concentric rings evident in this specimen are due to the different sets of atomic terraces of the crystal lattice and the hemispherical nature of the apex of the specimen. The arrangement of the centers of the atomic terraces, i.e. the crystallographic poles, is close to a stereographic projection. The bright and dark regions of contrast evident in the field ion micrograph of iridium are known as zone decoration and their appearance is specific to the material. In contrast, field ion images of amorphous materials exhibit a random distribution of spots, as shown in Fig. 1b for a $Pd_{40}Ni_{40}P_{20}$ bulk metallic glass. Crystallographic features such as grain and twin boundaries are evident in the field ion micrograph as abrupt disruptions in the pattern of the atomic terraces, as shown in Fig. 1c for a grain boundary in NiAl. Dislocations that satisfy a visibility criterion and emerge near crystallographic poles are evident as spirals in the atomic terraces, as shown in Fig. 1d. Solute segregation to boundaries or dislocations can also be observed by the presence of bright spots, as shown for boron segregation to a grain boundary in molybdenum in Fig. 1e. However, the identity of the segregating species cannot be unequivocally established directly from the field ion image and atom probe analysis is required.

The magnification of the image is a function of the specimen to phosphor screen distance, d, and the end radius of the specimen, r_t, and is given by $\eta = d/\xi r_t$, where ξ is a projection parameter known as the image compression factor. Since the typical end radius of the specimen is in the 10 to 50 nm range, magnifications are in the millions. The radius of the specimen increases and consequently the magnification decreases as atoms are destructively removed during the experiment. This process requires the application of an increasing standing voltage to maintain the same field strength for the field ionization process. The typical range of operation is voltages of up to ~20 kV.

In modern field ion microscopes, a microchannel plate image intensifier is positioned immediately in front of the phosphor screen to increase the gain by a factor of approximately 10^3 and also to use the more efficient output electrons to stimulate the phosphor rather than the helium or neon image gas ions. This higher gain enables the field ion images to be recorded on film or digital cameras.

Since the ionization rate is influenced by the ionization potentials, work function and sublimation energy of the specimen, microstructural phases generally exhibit different contrast in the field ion image. An example of nanometer scale precipitates in a nickel-based superalloy Alloy 718 specimen is shown in Fig. 1f.

If the field on the specimen is raised further, the surface atoms of the specimen may also be removed by field ionization. This process is termed field evaporation to distinguish it from thermal evaporation as it is generally performed at cryogenic temperatures. Field evaporation is used to smooth out any irregularities produced during specimen preparation and to remove surface contamination. Field evaporation is used to analyze the interior of the specimen by atomic scale serial sectioning thereby revealing the morphology of features present in the microstructure. Field evaporation is also used to remove atoms in the atom probe.

2.2. Types of Atom Probe

In 1967, Müller, Panitz and McLane developed the atom probe in order to identify the atoms that were imaged in the field ion microscope [5]. The original variant of the atom probe used a small hole in the center of the microchannel plate and phosphor screen assembly which served as the entrance aperture to a time-of-flight mass spectrometer, as shown in Fig. 2. The atoms were removed from the specimen at a well-defined time by the application of a short duration positive voltage pulse to the specimen. For many years, mechanical mercury-wetted reed relay based systems were used to produce these high voltage pulses at pulse repetition rates of between 50 and 200 Hz [1]. However, modern instruments now use solid-state devices that are capable of significantly higher repetition rates. The positive polarity high voltage pulse can be capacitively

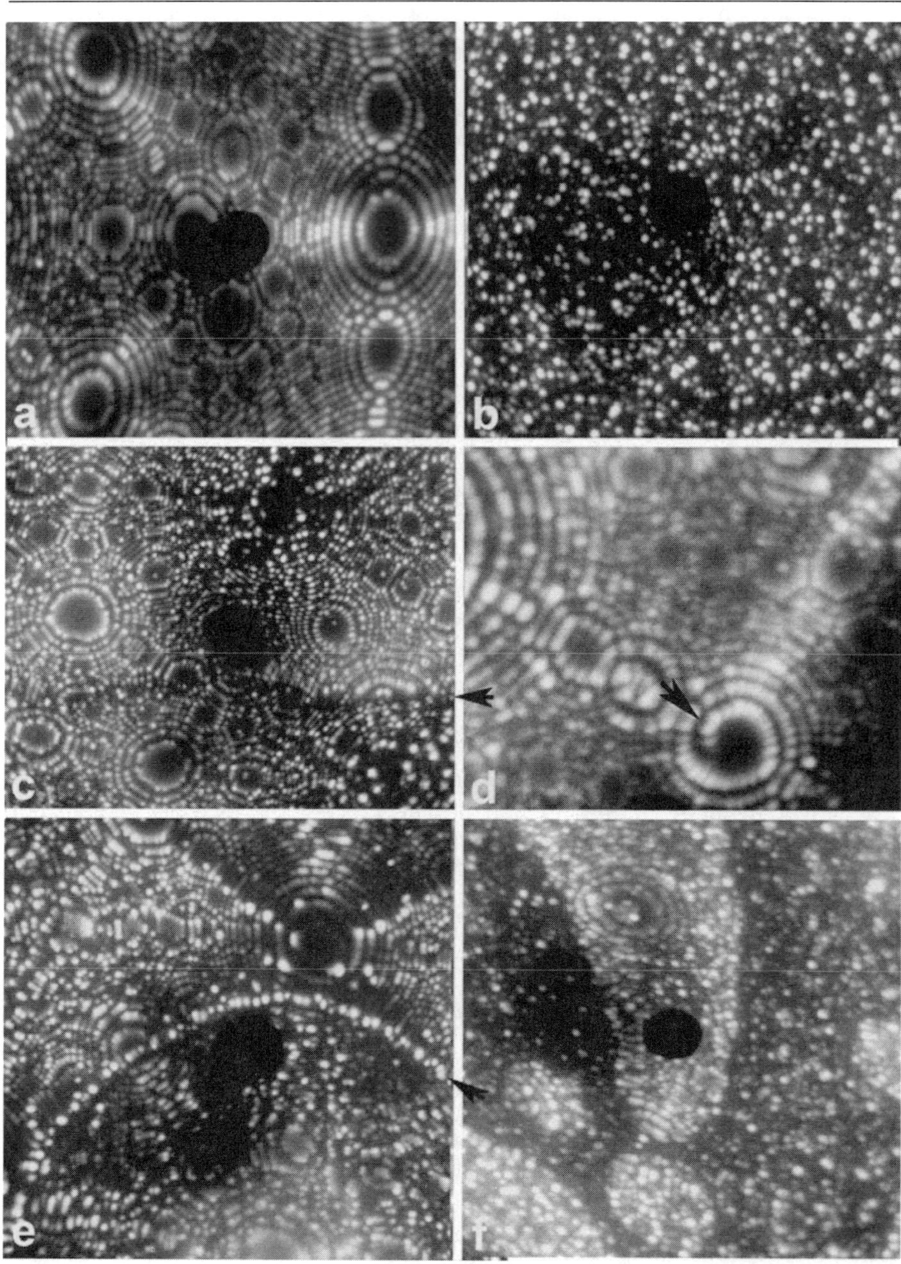

Figure 1. Field ion micrographs of a) iridium, b) a $Pd_{40}Ni_{40}P_{20}$ bulk metallic glass, c) a grain boundary in NiAl, d) a dislocation in B2-ordered NiAl, e) a boron-decorated grain boundary in the heat affected zone of a molybdenum weldment and f) brightly-imaging γ'' and γ' precipitates in Alloy 718. The dark circular region in the center of some of these images is dead region surrounding the entrance aperture of the mass spectrometer in the classical atom probe.

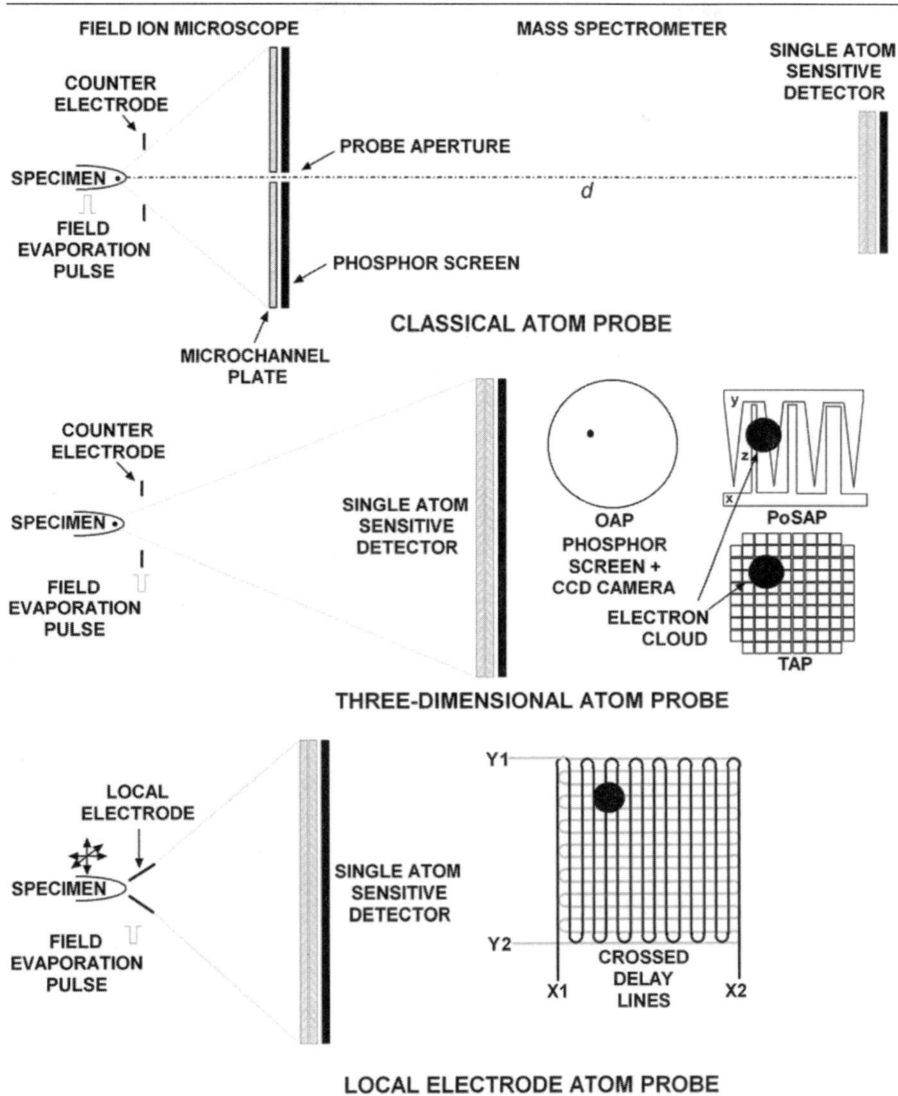

Figure 2. Schematic diagrams of the classical atom probe, the three-dimensional atom probe and the local electrode atom probe. Some of the different types of detectors used on the three-dimensional atom probe are illustrated.

coupled to the standing voltage on the specimen. Alternatively and equivalently, a negative polarity pulse can be applied to a circular counter electrode a short distance in front of the specimen. The magnitude of the pulse voltage and the specimen temperature are selected to ensure that all the atoms present in the specimen are equally likely to field evaporate on the application of the high voltage pulse. Typical conditions are a specimen temperature 50–60 K for a wide variety of materials and ~20 K for aluminum alloys and a pulse voltage of 15 to 20% of the standing voltage. The correct parameters ensure that the more strongly bound elements are not retained on the surface and the less strongly bound elements are not preferentially removed by the standing voltage between the high voltage pulses thereby influencing the accuracy of the compositional determination. Calibration experiments are used to establish the correct conditions on new materials. A pulsed laser has also been used to momentarily heat the apex of the specimen to induce field evaporation on the standing voltage [6]. This pulsed laser method has been primarily used for the analysis of semiconducting specimens where the high voltage pulse would be attenuated before reaching the apex region of the specimen.

The field evaporated ions are detected on a single atom sensitive detector at the end of the mass spectrometer, as shown in Fig. 2. In the field evaporation process, the potential energy of the atom on the surface, neE, just prior to field evaporation is converted into kinetic energy, $1/2mv^2$ when the atom leaves the surface. Therefore, the mass-to-charge ratio, m/n, of each atom removed can be determined from

$$\frac{m}{n} = \frac{c}{d^2}(V_{dc} + V_{pulse})t^2, \tag{1}$$

where d and t are the flight distance and flight time from the specimen to the detector, V_{dc} and V_{pulse} are the standing and pulse voltages on the specimen, and c is a constant used to convert the mass into atomic mass units. Although contrary to Müller's original concept of the operation of the atom probe, this process is generally conducted in the absence of the image gas to reduce the background noise. In the original atom probe, the flight time was manually measured from an oscilloscope trace. This method has been superseded with computer controlled high-speed digital timing systems. A typical sample in this type of atom probe is cylindrical volume with a diameter of up to ~2 nm that is defined by the entrance aperture and that contains approximately 10,000 to 50,000 atoms. Nowadays, this type of atom probe is referred to as a classical, conventional or one-dimensional atom probe.

The classical atom probe is very inefficient in the number of atoms collected from the specimen due to the small effective size of the aperture. Therefore, several variants have been developed to improve the collection efficiency. The 10 cm atom probe or imaging atom probe (IAP), designed by Panitz, had a significantly larger field of view [7, 8]. This variant used a pair of spherically curved channel plates and phosphor screen so that the flight distance from the specimen to all points on the detector were identical. The detector could be momentarily energized to record only the ions of a given mass-to-charge ratio. Therefore, a two-dimensional map of the solute distribution could be obtained. However, these data were difficult to quantify and alternative approaches were developed.

These newer variants all feature an analysis chamber housing the field ion microscope and mass spectrometer, a specimen storage or preparation chamber and a fast entry air lock to load interchangeable specimens. The base pressure in the analysis chamber is achieved with either a turbomolecular, diffusion or ion pump supported with a titanium sublimation pump (TSP) or non-evaporable getter (NEG) pump and is typically less than 1×10^{-10} mbar. These designs differ primarily in the type of position-sensitive single atom detector that is used, as shown in Fig. 2. All the detectors are based on a stack of two (in a chevron configuration) or three (in a Z configuration) microchannel plates providing a gain of more than 10^6 times. However, different types of position-sensing elements are used. In these instruments, the active area of the detector together with the flight distance defines the area of analysis on the specimen's surface. Typical flight distances are smaller than the 1 to 2.5 m used in the one-dimensional atom probe and are typically between 0.45 and 0.62 m. These variants include the optical atom probe (OAP) that uses a phosphor screen and a CCD camera [9–11], the position-sensitive atom probe (PoSAP) that uses a three anode wedge-and-strip detector [12, 13], the tomographic atom probe (TAP) that uses a 10×10 square array of anodes [14], the optical position-sensitive atom probe that features a primary detector with a phosphor screen and a pair of secondary image-intensified detectors consisting of an 8×10 array of anodes and a phosphor screen and a CCD camera [15], and the optical tomographic atom probe (OTAP) that uses a combination of a linearly segmented anode and a phosphor screen and a CCD camera [16]. These types of instruments are collectively referred to as three-dimensional atom probes (3DAP). A typical sampled volume in a three-dimensional atom probe has an area of $\sim 10 \times \sim 10$ nm to $\sim 20 \times 20$ nm and contains approximately 10^5 to 10^6 atoms. The typical time required to collect this number of atoms is of the order of 20 h. In both the classical atom probe and the three-dimensional atom probe, field ion microscopy is almost always performed at the start of the experiment. A region of the specimen is then selected for analysis by rotating the specimen about its apex until the field ion image of the selected region covers either the probe aperture or the single atom detector. The image gas is then removed from the system and the specimen is field evaporated until a sufficient number of atoms are collected or the specimen fails.

A new variant of three-dimensional atom probe known as a scanning atom probe (SAP) or a local electrode atom probe (LEAP®) has been introduced recently [17–20]. This instrument uses a small aperture or local electrode positioned extremely close to the apex of the specimen, as shown in Fig. 2c. A photograph of the local electrode atom probe is shown in Fig. 3. Nishikawa's original concept of the scanning atom probe was that the local electrode could be scanned across the surface of a rough specimen and a natural protrusion could be selected for analysis thereby extending the type of specimen that could be studied. However, the analysis of natural protrusions is limited to simple compositional analysis since the non-uniform end forms of these types of protrusions do not permit reliable reconstructions of the positions of the atoms. In addition, surface contamination and surface diffusion can have a serious influence on the reliability of the analysis. It is also possible to analyze traditional needle-shaped field ion specimens with this variant. In addition, in-situ analyses may be performed on many individual microtips in two-dimensional arrays of microtips that are fabricated from the surface of flat specimens by focused ion beam based techniques. An integral part of this instrument is a nano-positioning piezoelectric stage to accurately align

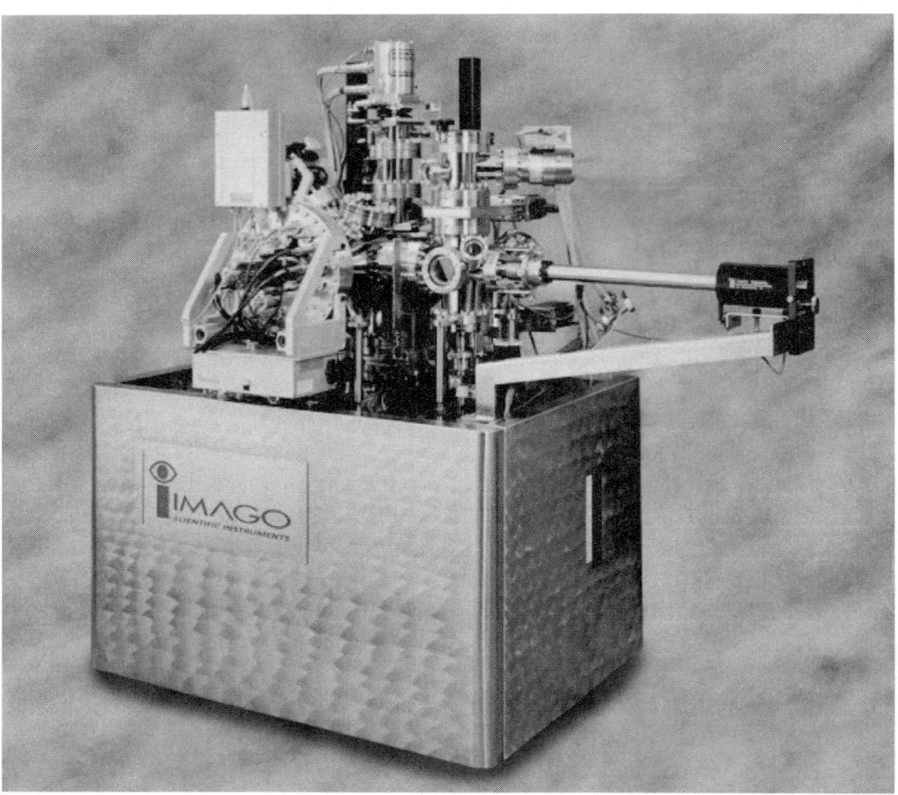

Figure 3. Photograph of the local electrode atom probe. Courtesy Imago Scientific Corporation.

the aperture in the local electrode with either the apex of a traditional needle-shaped specimen or a protrusion. The alignment of the specimen with the local electrode is performed while viewing the process with a pair of orthogonal high resolution video cameras attached to long working distance microscopes.

The primary advantage of this local electrode configuration is that the voltage that has to be applied to the specimen is significantly reduced from traditional atom probe configurations. For example, only ~50% of the voltage is required for a 30 μm aperture positioned ~10 μm in front of the apex of the specimen compared to the 5–10 mm diameter apertures placed 4–10 mm in front of the specimen in the conventional three-dimensional atom probe. This lower voltage not only extends the range of the instrument to blunter specimens but, more importantly, allows the instrument to be operated at significantly faster repetition rates that are possible due to the correspondingly lower pulse voltage. The repetition rate currently available commercially for generating these high voltage pulses is 200 kHz and the limit is likely to be in excess of 1 MHz. In order to take advantage of these high repetition rates, a different type of position-sensitive detector is used. Therefore, the local electrode atom probe uses a chevron microchannel plate with a crossed delay line coupled to an ultrafast digital timing system. Unlike other types of atom probe, the close proximity of the local electrode with the specimen dictates that the design uses a vibration isolated cryo-generator to minimize any change in the field due to small changes in the alignment of specimen to the local electrode. This local electrode configuration also enables the single atom detector to be positioned significantly closer to the specimen without the normally associated degradation in mass resolution. As this configuration has a wide field of view, the design and operation of the local electrode atom probe can also be simplified by eliminating the goniometer that is normally used to rotate the specimen to select the region of analysis. A typical sample volume in the local electrode atom probe contains 10^6 to 3×10^7 atoms and datasets containing over 10^8 atoms have been collected. Due to the faster high voltage pulser, detector and electronics, the analysis time has been reduced by several orders of magnitude (~300X) compared to the traditional three-dimensional atom probe. Typical collection rates in the local electrode atom probe are 5 to 10×10^6 ions per hour. This high collection rate enables many specimens to be characterized in a day.

Since a time-of-flight mass spectrometer is used in all of these instruments, the atom probe is able to detect and has equal sensitivity for all elements. Unlike some other techniques, no pre-selection of elemental species is required to perform an analysis. The mass resolution of the atom probe is limited by small deficits in the energy of the ions. These energy deficits arise due to the full magnitude of the pulse not being transferred to the ion as it leaves the specimen. The initial method to improve the mass resolution in the one-dimensional atom probe was the addition of a Poschenreider lens to the mass spectrometer [2, 21]. This lens is a section of a toroid with matched length field free regions. Ions with different energies take slightly different length paths in the lens and are isochronally focused on the single atom detector. One additional advantage of the Poschenreider lens is that it can filter out any residual image gas and other impurity ions that are field evaporated on the standing voltage before they strike the detector thereby improving the background noise level in the analyses. Unfortunately, the Poschenreider lens configuration cannot be used in the three-dimensional atom probes due to their large acceptance angles. Therefore, reflectron lenses or energy mirrors have been used in three-dimensional atom probes [2, 22]. In this type of lens, the ions enter a region of increasing field and are reflected back to the detector. Ions with lower energy do not travel as far into the lens as ions with the full energy and are isochronally focused on the detector. However, the reflectron also reduces the detection efficiency by ~10% due to a field defining mesh at the entrance/exit of the lens. These energy-compensation devices improve the mass resolution so that the individual isotopes of the elements can be resolved. Due to the close proximity of the counter electrode, energy compensation is not required in the local electrode atom probe to resolve the individual isotopes. However, further improvement in the mass resolution can be achieved with a method based on post acceleration [23]. This post acceleration method can also increase the field of view.

2.3. Specimen Preparation

As with many techniques, the quality of the specimen is one of the most important parameters in the success of an experiment. The aim of specimen preparation is to produce a needle-shaped specimen with an end radius that is less than approximately 50 nm. In addition, the taper angle of the shank of the needle should be no larger than 10° as this parameter defines the maximum number of atoms that may be collected from the specimen. Most atom probe specimens are produced by either electropolishing or ion milling techniques depending on the material, original shape and the type of specimen [1, 2].

In most metallic specimens, the bulk sample is reduced into square or cylindrical shaped bars that are ~0.25 mm across by ~10 mm long usually with the use of a diamond wire saw or diamond impregnated cutting wheel. Other starting forms of material such as wires or whiskers are also suitable. The bar is generally crimped into an annealed copper tube to facilitate handling in subsequent stages. Electropolishing is generally performed in two stages. In the first stage, the specimen is suspended in a 5 to 6-mm-thick layer of electrolyte floated on top of a dense inert liquid such as a polyfluorinated polyether. The electropolishing cell is configured with the specimen as the anode and a wire or circular cathode made of gold or platinum. Since only the middle portion of the specimen is in the electrolyte, a necked region is formed in the central region of the bar. For some materials, the electropolishing process is continued until the necked region is too narrow to support the weight of the bottom half of the bar and two atom probe specimens are made. However for many materials, the process is normally terminated before this point is reached so that a second stage with more controlled electropolishing conditions may be used. Since the necked region has been formed in the first stage, the specimen may be transferred to a simpler cell without the inert liquid and electropolishing continued until separation occurs. A list of suitable electrolytes for a wide variety of materials may be found elsewhere [2]. The resulting needles are then carefully cleaned in a suitable solvent to remove any traces of the electrolyte. Specimens are generally examined in an optical microscope at magnifications of 100 to 200X to check the quality of the surface finish, taper angle and the end radius before insertion into the atom probe.

Specimens that are either too blunt for further analysis or had fractured during analysis may be resharpened. This micropolishing method may also be used on freshly made or corroded specimens to remove any damaged regions or surface films. In this method, the specimen (anode) is carefully positioned in a drop of electrolyte suspended in a platinum loop (cathode). The electrolyte is held in place in the ~3-mm-diameter loop by surface tension. A voltage is then applied to the cell to electropolish the region of the specimen in the electrolyte. The position of the specimen in the electrolyte may be adjusted with the use of a micromanipulator and is generally monitored under a low power (~30X) microscope. A desired taper angle may be sculpted by adjusting position and the time spent polishing in the electrolyte.

Thin film specimens such as surface, bi- or multi-layer films are generally fabricated into the needle-shaped specimens with the use of a focused ion beam (FIB) miller.

The aim of the fabrication process is to position the area of interest in the apex of the needle. One common method is to use a silicon substrate that has been Bosch etched into an array of ~5 × ~5 μm square or rod posts [1, 24]. The resistivity of the silicon should be less than 0.05 Ω-cm to ensure that the pulse voltage is not attenuated in the specimen and thereby rendered ineffective. Alternatively, posts may be cut with the use of a dicing saw [25]. The multilayer film is then deposited on the surfaces of these posts. A 2- to 3-μm-thick platinum cap is deposited on top of the region of interest, i.e. the center of the post, to protect the underlying region from gallium implantation during milling. In specimens without this cap, concentrations of up to 30% gallium has been measured. The post is then ion milled with an annular 30 keV gallium ion beam where the outer diameter is slightly larger than the extent of the post and the inner diameter is ~2 μm. The inner diameter is progressively reduced to ~0.06 μm during milling to produce the desired taper angle and end radius. Milling is terminated when the end radius of the specimen is less than ~50 nm and the last remnant of the platinum cap is removed. Examination or monitoring of the specimen should be performed with the electron beam rather than the gallium ion beam to minimize gallium implantation, and therefore instruments featuring ion and electron columns within a single system are preferred.

In the case of the local electrode atom probe, this milling method may also be used to fabricate specimens from a specific location on the surface of a specimen by placing a platinum cap or marker at the desired location and then milling a wide moat around that cap before preparing the microtip as described above. This milling method may also be used to fabricate specimens of fine powders. For this starting form of material, an individual powder particle is attached to the end of a relatively blunt needle with either a conducting epoxy or a platinum bridge deposited in the FIB.

3. BASIC INFORMATION

The raw data from the instrument, i.e., the pixel coordinates in the case of the CCD camera based detectors, the charges from the anodes, or the times measured at the ends of the crossed delay lines, are first converted into true x and y distances on the detector. These distances are then scaled by the magnification to obtain the atom coordinates in the specimen. The accuracy of these positions is limited by small trajectory aberrations due to interactions with the neighboring atoms as the atom leaves the surface of the specimen. Consequently, the spatial resolution in this x-y plane is approximately 0.2 nm. The third or z coordinate is estimated from the position in the evaporation sequence taking into account the detection efficiency of the mass spectrometer, typically ~60%, and the atomic density of the material under analysis. The spatial resolution in this direction is typically 0.05 nm. Additional corrections are required to take into account the hemispherical nature of the specimen and the orientation of the volume of analysis relative to the main axis of the needle-shaped specimen. These resolutions only permit the reconstruction of atomic planes if their orientation is close to a suitable direction and their lattice spacing is sufficiently large, and are not sufficient to place the atoms on their atomic sites in the crystal structure.

The elemental identity of each ion is interpreted from the mass-to-charge ratio. Fortunately, the time-of-flight mass spectrometer has sufficient mass resolution to distinguish the individual isotopes of all the elements, generally produces only one or two different charge states for each element, and produces a limited number of complex molecular ions. Databases of these charge states and molecular ions have been compiled for almost all elements [1]. Therefore, most peaks in the mass spectra can be readily assigned to the correct element. In cases involving more than one possibility, tables of natural isotope abundances may be used to distinguish the elements and deconvolute overlapping peaks for concentration estimates.

The concentration, c, of a solute in a small volume is determined in atomic fraction from the number of atoms of each type, n_i, in the volume, i.e., $c = n_i/n$, where n is the total number of atoms in the sample. The accuracy of the concentration and the minimum detection level are therefore functions of the number of atoms in the volume sampled. The standard error of the concentration measurement, σ, is given from counting statistics by $\sigma = \sqrt{c(1-c)/n}$.

4. DATA INTERPRETATION AND VISUALIZATION

The initial data produced in the three-dimensional atom probe by the reconstruction process described in the previous section are the x, y, and z coordinates and the mass-to-charge ratio of each ion. These data may be investigated in a variety of methods that involve either the examination of pairs of atoms, shells of atoms surrounding an atom of interest, one-dimensional strings of atoms, and one-, two- and three-dimensional arrays of blocks of atoms. The methods may be considered as being naturally divided into two main categories: the visualization of solute inhomogenities, such as precipitates and solute segregation, and the quantification of the magnitude of the solute inhomogenities. These methods are generally applied after the experiment is completed although some simplified versions of these methods can be performed as the data is being acquired.

Due to the three-dimensional nature of the data, most visualizations are performed interactively so that the data can be viewed rapidly from different perspectives. When the first generations of three-dimensional atom probes were developed, expensive graphically-oriented workstations were required to visualize the data. Nowadays, modern personal computers generally have sufficient processing and rendering power to perform these visualizations in real time on datasets containing many million ions.

4.1. Visualization Methods

The basic and most intuitive method used to visualize the solute distribution is the atom map. In this representation, a sphere or dot is used to indicate the position of each type of solute atom, as shown in Fig. 4a for a distribution of ~3-nm-diameter copper precipitates in a model pressure vessel steel. Different colors and sizes of spheres may be used to visualize multiple types of solutes simultaneously. The atom map is normally used to view the distribution of the solute atoms rather than the solvent

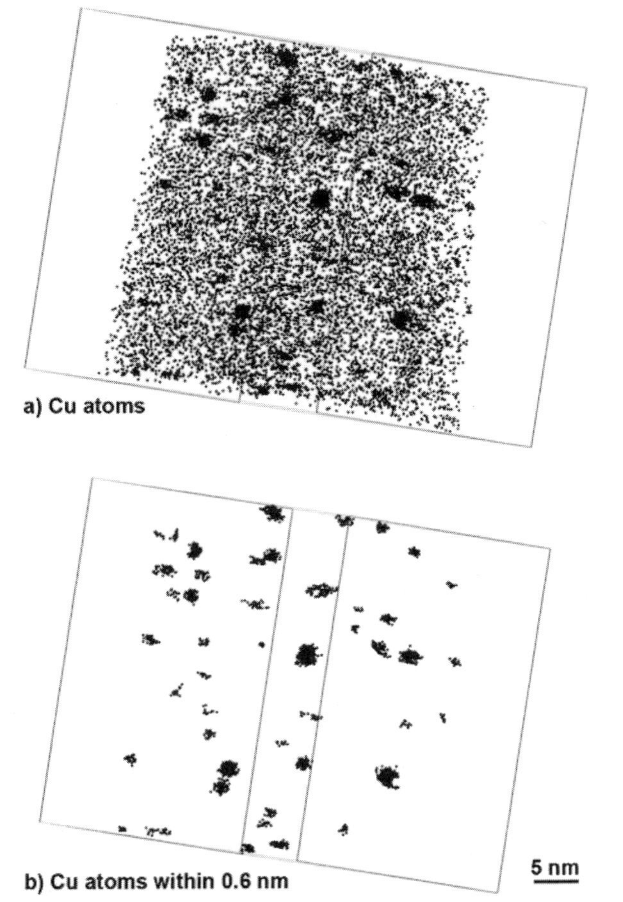

Figure 4. Atom maps of the copper atoms in a model iron-0.8% copper alloy that was annealed for 7,200 h at 290 °C to produce ultrafine copper precipitates. All the copper atoms are shown in a), whereas only the copper atoms within 0.6 nm of another copper atom are shown in b). 60 ~3-nm-diameter copper precipitates are detected in this volume. These ultrafine copper precipitates have been shown to be one of the main features responsible for the embrittlement of the pressure vessel of nuclear reactors during service.

atoms. This method is most effective in cases where there is a large difference in solute concentration between the feature of interest and the surrounding matrix and the solute level in the matrix is relatively low. Therefore, atom maps are frequently used to detect small precipitates and segregation to interfaces, grain boundaries, dislocations, etc. The significantly larger area of analysis and orders-of-magnitude larger number of atoms in the sample in data from the local electrode atom probe imposes additional visibility criterion and consideration of the longer rendering times. For higher concentration solutes, displaying only a fraction of the solute atoms or subsampling the volume can

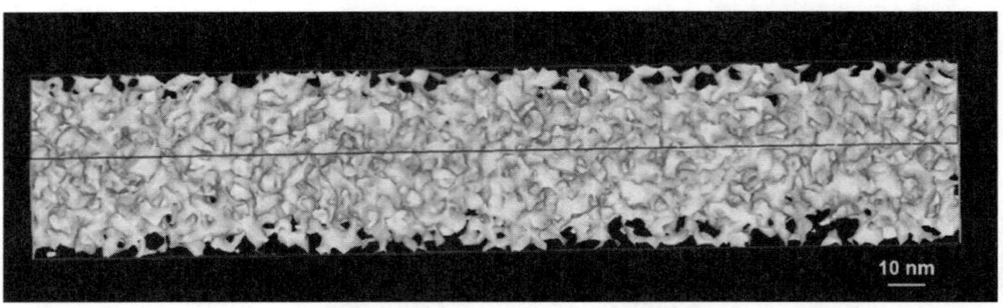

Figure 5. Isoconcentration surfaces in a $Pd_{40}Ni_{40}P_{20}$ bulk metallic glass. a) A fine-scale isotropic interconnected network structure of two amorphous phases is evident after annealing for 80 min at 340 °C.

improve the visibility of features. Alternately, the background or matrix atoms can be eliminated from the display through the use of the maximum separation method, as shown in Fig. 4b. This method is based on the principle that the solute concentration in a precipitate is significantly higher than in the surrounding matrix and therefore the solute atoms are closer together. Therefore, a maximum separation distance between solute atoms can be defined that will encompass all the solute atoms in the particle and eliminate those in the surrounding matrix.

The morphology and size of features present may be visualized with the isoconcentration surface. This method is the three-dimensional equivalent of a contour map. In this representation, the compositions of a finely spaced (~0.5 to 1 nm), regular three-dimensional array of volume elements are estimated. The composition array may be smoothed with a moving average to minimize the influence of the statistical fluctuations due to the relatively small sample size in each composition determination. Then, a surface is constructed at a selected solute concentration generally with the use of an applied marching cube algorithm [26]. The concentration level used for the isoconcentration surface influences the volume fraction of the phases and may change the degree of interconnectivity in complex structures. An example of a highly interconnected, percolated, network structure of amorphous phases in a $Pd_{40}Ni_{40}P_{20}$ bulk metallic glass that was isothermally aged at 340 °C are shown in Fig. 5.

Fine-scale concentration variations may be visualized with orthogonal or arbitrary slices through the sample. As in the previous method, a three-dimensional composition array is constructed from the data. Then, the concentrations of a two-dimensional plane through these composition data are exhibited as different levels on a grey or color scale, as shown in Fig. 6. This plane may be swept interactively through the entire volume to highlight different composition regions and to examine their morphologies.

The presence of small precipitates can also be visualized with the x-ray tracer representation. As before, a composition array is constructed from the data. Then, one of the primary axes—usually the x or y axis—is chosen. The maximum or minimum concentrations of the columns of cells perpendicular to this selected direction are exhibited on a two-dimensional intensity map, as shown in Fig. 7 for the copper

Figure 6. Arbitrary slice through a $Pd_{40}Ni_{40}P_{20}$ bulk metallic glass showing regions of high and low phosphorus corresponding to the crystalline phase and the matrix, respectively.

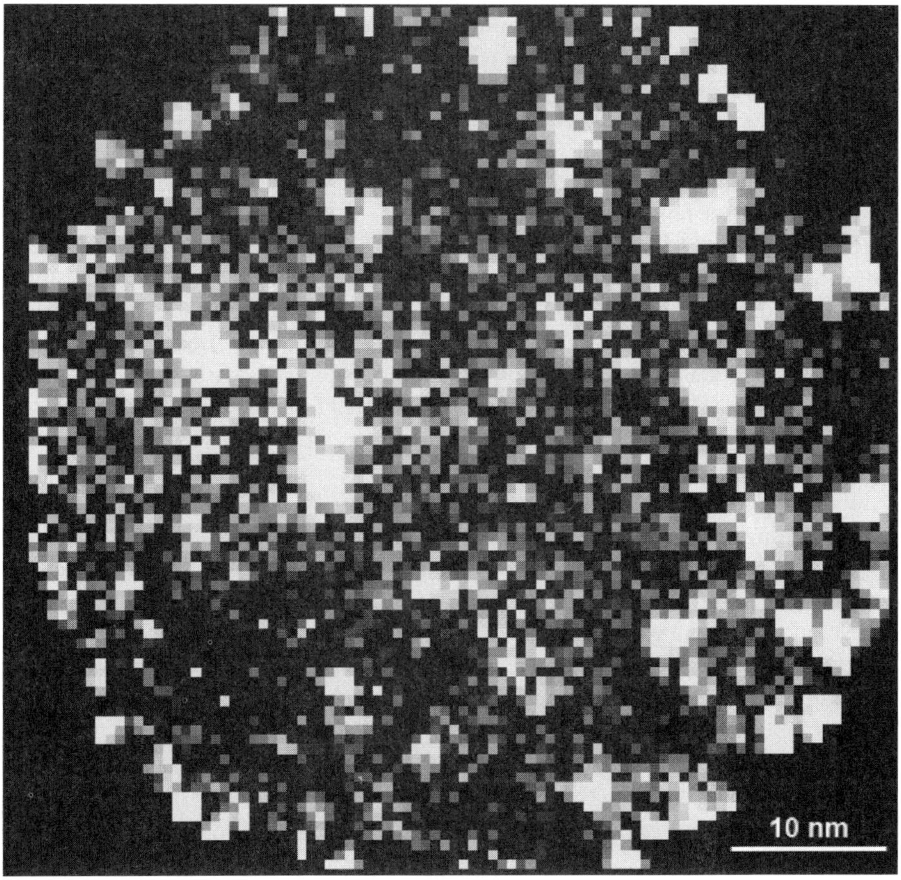

Figure 7. An x-ray tracer representation of the maximum copper concentration in a model iron-0.8% copper alloy that was annealed for 7,200 h at 290 °C. The ~3-nm-diameter copper precipitates are evident by the high contrast regions.

precipitates in the model pressure vessel steel. This method is most effective in cases in which the volume fraction of the features of interest is insufficient to produce image overlap of those features along the chosen direction.

4.2. Quantification Methods

Atom probe tomography data may be analyzed to quantify the size, composition and some topological parameters of the microstructure [1, 2]. In this section, some of the more common methods are outlined.

Several methods have been developed to determine the size of nanometer-scale features. In the previous section, the maximum separation method was applied to visualize the atoms that belong to individual features. This method yields the position of all the solute atoms in the features. Since the atom positions are known, the center of mass and the radius of gyration and the related Guinier radius can be calculated directly from standard formulae. In particles where there are multiple elements present, these parameters may be estimated for each element individually. An alternative statistical method to evaluate the size of feature is the autocorrelation function either in one or three-dimensional forms [1, 2]. The size of small particles is estimated from the position of the first minimum in the lags.

For spherical particles, radial concentration profiles may be constructed from the centers of mass of each particle into the matrix. These radius concentration profiles are constructed by calculating the compositions of thin spherical slices of increasing radius centered at the center of mass. This information may be used to estimate the extent of any interfacial segregation or solute inhomogeneities within the particles.

The number density of the features may be estimated from the number of features in the analyzed volume. The volume of the analyzed region is generally estimated from the number of atoms, detection efficiency of the mass spectrometer and the atomic density of the sample.

The original method to determine the composition of features in the data was the selected volume method. In this method, a small spherical or cuboidal volume is defined to match the size of the feature of interest and the numbers of atoms of each solute within the volume are determined. The composition of the feature is then estimated, as described above from the number of atoms of each solute. In practice, this method is extremely time consuming to perform for a high number density of small particles and is usually only applied to coarse particles. Alternative automatic methods are generally used for other cases.

One of the fastest methods to estimate the individual compositions of a high number density of small particles is the envelope method [1]. This method uses the atom positions defined by the maximum separation method to define the extent of each particle by marking the cells in a fine three-dimensional grid. The typical grid spacing is 0.1 to 0.2 nm. The composition of the particle is then determined from all the atoms in these marked cells.

In materials where there is a high volume fraction of more than one phase, as in the example of the bulk metallic glass shown in Fig. 5, the compositions of the coexisting phases can be estimated from the statistical analysis of concentration frequency distributions. The frequency distribution is constructed from the concentrations of equiaxed blocks of atoms typically containing 25 to 250 atoms. The frequency distribution is the number of blocks with each solute concentration. Phase separation is detected if the measured frequency distribution deviates statistically from a binomial distribution with the average solute content. The extent of the phase separation and the compositions of the coexisting phases can be estimated from the sinusoidal based Pa method or the Gaussian based Langer-Bar-on-Miller (LBM) method [27–30].

The co- and anti-segregation behavior of the solutes in a volume containing different phases may be examined with the use of contingency tables. The contingency table for the two solutes of interest is created from the compositions and may be regarded as a two-dimensional frequency distribution [31, 32]. The correlation between the elements is determined by comparing the numbers of composition blocks containing high or low concentrations of the two elements with those expected from a random distribution.

The early stages of phase separation, clustering and chemical short range ordering may be investigated with methods based on the analysis of Markov chains. For this type of analysis, the three-dimensional data is converted into a one-dimensional string or chain of atoms. The chain is usually taken along the z direction as it has the highest spatial resolution and is typically the longest dimension in the data. In this type of analysis, B refers to the solute of interest and A is any other type of atom. Two main methods of examining the chains have been devised. The Johnson and Klotz method is based on the examination of pairs of atoms, n_{AA}, n_{AB} and n_{BB}, and comparing the results to the numbers expected for a random solid solution [33]. If n_{AB} is higher than expected in a random solid solution, then the data exhibit short range order. Conversely, if n_{BB} is higher than expected in a random solid solution, then the data exhibit solute clustering. The Tsong method compares the number of sequences of $\mathrm{AB}^{n}\mathrm{A}$ atoms, where n is the number of consecutive B atoms [34]. If the number of ABA ($n = 1$) chains experimentally observed is larger than expected, the data exhibit solute clustering. If the number of ABBA, ABBBA, (i.e., $n \geq 2$) chains experimentally observed is larger than expected, then the data exhibit solute clustering.

The amount of segregation at interfaces and grain boundaries may be quantified directly from an analysis that contains an interface with the use of a method based on the Gibbsian interfacial excess [35]. The Gibbsian interfacial excess of element i, $\Gamma.i$, is defined as $\Gamma_i = \frac{n_{ix}}{A}$, where n_{ix} is the excess number of solute atoms associated with the interface and A is the interfacial area over which the interfacial excess is determined [35–37].

The three-dimensional data obtained in atom probe tomography can also be analyzed for topological parameters. Topological analysis has been used to determine the degree of interconnectivity of the microstructure. In order to quantify the interconnectivity, the number of loops per unit volume or "handles" is determined. The handle density may be regarded as a similar parameter to the number density of isolated precipitates. One such measure is the Euler characteristic, $E = n_{\mathrm{n}} - n_{\mathrm{e}} + n_{\mathrm{f}}$, where, n_{n} is the number of nodes, n_{e} is the number of edges and n_{f} is the number of faces in a

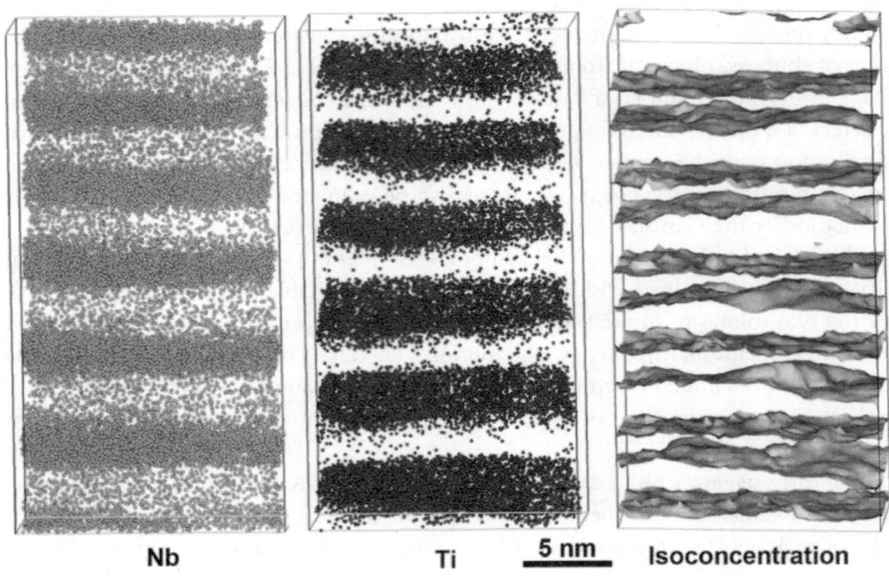

Figure 8. Niobium and titanium atom maps and isoconcentration surface from a sputter deposited niobium (3.3 nm) -titanium (2.2 nm) multilayer film.

structure [38]. The data may also be examined to determine whether a continuous path can be traced within the second phase from one surface of the volume to another i.e., the microstructure is percolated. Other topological parameters such as the fractal and fracton dimensions can also be determined [39].

5. SAMPLE ANALYSIS OF NANOMATERIALS: MULTILAYER FILMS

The characterization of the atomic chemical composition of nanostructured materials is important in applications such as magnetic data storage and semiconductor thin film devices. The performance-defining properties of these multilayer films and devices are sensitive to the nature and quality of the inter-layer interfaces. Some atom probe tomography results are presented from a multilayer film comprising of 40 layer pairs of 2.2 nm Ti/3.3 nm Nb that were sputter deposited under UHV conditions at ambient temperature onto a n-doped Si [100] wafer (resistivity = 5 Ω-cm) [40, 41]. The atom probe specimens were prepared by the focused ion beam process described previously. The niobium and titanium atom maps and the isoconcentration surface depicting the interface between the niobium and titanium layers are shown in Fig. 8. The interface between the niobium and titanium levels was relatively smooth. Niobium and titanium concentration profiles oriented perpendicular to the plane of the interface, shown in Fig. 9, indicate that significant solute interdiffusion of niobium had occurred in the titanium layers. The niobium interdiffusion into the titanium layers has facilitated the titanium phase to grow in a pseudomorphically body centered cubic form rather than in its bulk equilibrium hexagonal closed packed form [40, 41].

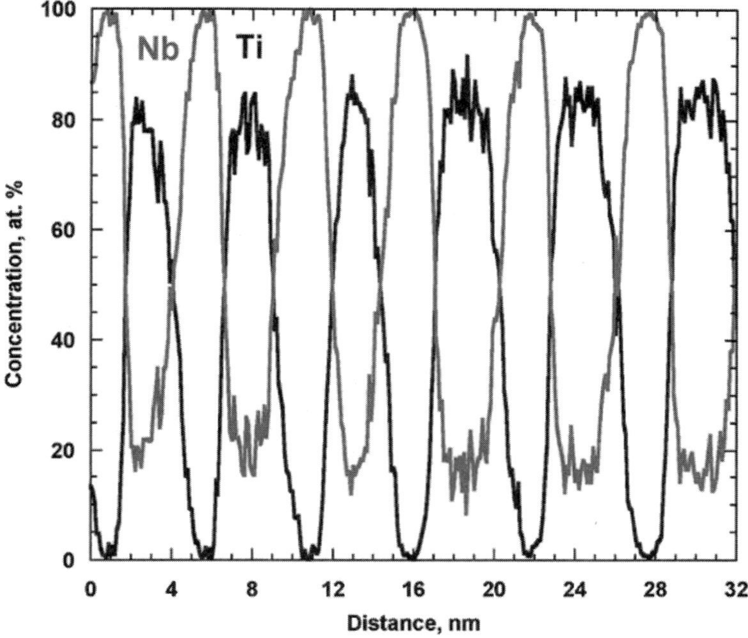

Figure 9. Niobium and titanium concentration profiles oriented perpendicular to the interfaces of the multilayer film shown in Fig. 8. Some intermixing of the solutes is evident.

6. SUMMARY AND FUTURE PERSPECTIVES

Atom probe tomography is a powerful tool for the characterization of the size, morphology and composition of ultrafine features in a variety of materials. With the development of new forms of specimen preparation especially with focused ion beam milling systems, atom probe tomography should be extended to a wider variety of applications in nanotechnology.

The local electrode atom probe is a major advance in atom probe design and has greatly simplified the operation of the instrument. There is considerable scope for its continued improvement through increases in the area of analysis, mass resolution and overall data acquisition speed. New, more efficient and automated methods to analyze the three-dimensional data should be also forthcoming as the user base for atom probe tomography increases.

7. ACKNOWLEDGEMENTS

The author would like to thank K. F. Russell for her technical assistance, and Profs. G. B. Thompson and H. L. Fraser of Ohio State University, Drs. R. B. Schwarz and T. D. Shen of Los Alamos National Laboratory, Prof. G. R. Odette of the University of California—Santa Barbara, and Prof. T. F. Kelly and his colleagues at Imago Scientific Corporation, Madison, WI for the results of the collaborations featured in this chapter. Research at the Oak Ridge National Laboratory SHaRE Collaborative

Research Center was sponsored by the Division of Materials Sciences and Engineering, U.S. Department of Energy, under contract DE-AC05-00OR22725 with UT-Battelle, LLC.

8. REFERENCES

1. M. K. Miller, *Atom Probe Tomography: Analysis at the Atomic Level*, Kluwer Academic/Plenum Press, New York, NY (2000).
2. M. K. Miller, A. Cerezo, M. G. Hetherington and G. D. W. Smith, *Atom Probe Field Ion Microscopy*, Oxford University Press, Oxford (1996).
3. E. W. Müller, *Z. Phys.*, **31** (1951) 136.
4. E. W. Müller, *Zh. Tehk. Fiz.*, **17** (1936) 412.
5. E. W. Müller, J. A. Panitz and S. B. McLane, *Rev. Sci. Instrum.*, **39** (1968) 83.
6. T. T. Tsong and G. L. Kellogg, *J. Appl. Phys.*, **51** (1980) 1184.
7. J. A. Panitz, *Rev. Sci. Instrum.*, **44** (1973) 1034.
8. J. A. Panitz, Field desorption spectrometer, US Patent No. 3,868,507 (1975).
9. M. K. Miller, presented at the 21st Microbeam Analysis Society meeting, Albuquerque, NM, 1986.
10. M. K. Miller, *Surf. Sci.,* **246** (1991) 428.
11. M. K. Miller, *Surf. Sci.,* **266** (1992) 494.
12. A. Cerezo, T. J. Godfrey and G. D. W. Smith, *Rev. Sci. Instrum.*, **59** (1988) 862.
13. C. Martin, P. Jelinsky, M. Lampton, R. F. Malina and H. O. Auger, *Rev. Sci. Instrum.*, **52** (1981) 1067.
14. B. Deconihout, A. Bostel, A. Menand, J. M. Sarrau, M. Bouet, S. Chambreland and D. Blavette, *Appl. Surf. Sci.*, **67** (1993) 444.
15. A. Cerezo, T. J. Godfrey, J. M. Hyde, S. J. Sijbrandij and G. D. W. Smith, *Appl. Surf. Sci.*, **76/77** (1994) 374.
16. B. Deconihout, L. Renaud, G. Da Costa, M. Bouet, A. Bostel and D. Blavette, *Ultramicroscopy*, **73** (1998) 253.
17. O. Nishikawa and M. Kimoto, *Appl. Surf. Sci.*, **74/75** (1994) 424.
18. T. F. Kelly, P. P. Camus, D. J. Larson, L. M. Holzmann and S. S. Bajikar, *Ultramicroscopy*, **62** (1996) 29.
19. T. F. Kelly and D. J. Larson, *Mater. Characterization*, **44** (2000) 59.
20. A. Cerezo, T. J. Godfrey, M. Huang and G. D. W. Smith, *Rev. Sci. Instrum.*, **71** (2000) 3016.
21. W. P. Poschenrieder, *Int. J. Mass Spectrom. Ion Phys.*, **9** (1972) 83.
22. B. A. Mamyrin, V. I. Karataev, D. V. Shmikk and V. A. Zagulin, *Sov. Phys. JETP*, **3** (1973) 45.
23. B. Deconihout, R. Saint-Martin, C. Jamot and A. Bostel, *Ultramicroscopy*, **95** (2003) 239.
24. D. J. Larson, D. T. Foord, A. K. Petford-Long, T. C. Antony, I. M. Rozdilsky, A. Cerezo and G. D. W. Smith, *Ultramicroscopy*, **75** (1998) 147.
25. K. R. Kuhlman and J. Wishard, *NASA Tech Brief NPO-30667*, June 2003, p. 62.
26. W. E. Lorensen and H. E. Cline, *Computer Graphics*, **21(4)**, (1987) 163.
27. J. M. Sassen, M. G. Hetherington, T. J. Godfrey, G. D. W. Smith, P. H. Pumphrey and K. N. Akhurst, Properties of Stainless Steel in Elevated Temperature Service, M. Prager, ed., American Society of Mechanical Engineers, New York, NY, 1987, p. 65.
28. T. J. Godfrey, M. G. Hetherington, J. M. Sassen and G. D. W. Smith, *J. de Phys.*, **49-C6** (1988) 421.
29. M. G. Hetherington, J. M. Hyde, M. K. Miller and G. D. W. Smith, *Surf. Sci.*, **246** (1991) 304.
30. J. S. Langer, M. Bar-on and H. D. Miller, *Phys. Rev. A*, **11** (1975) 1417.
31. B. S. Everitt, The Analysis of Contingency Tables, Chapman and Hall, London, UK, 1977.
32. E. Camus and C. Abromeit, *J. Appl. Phys.*, **75** (1994) 2373.
33. C. A. Johnson and J. H. Klotz, *Technometrics*, **16** (1974) 483.
34. T. T. Tsong, S. B. McLane, M. Ahmad and C. S. Wu, *J. Appl. Phys.*, **53** (1982) 4180.
35. J. W. Gibbs, The Collected Works of J. Willard Gibbs, Yale University Press, New Haven, CT, 1948 Volume 1.
36. B. W. Krakauer and D. N. Seidman, *Phys. Rev. B*, **48** (1993) 6724.
37. M. K. Miller and G. D. W. Smith, *Appl. Surf. Sci.*, **87/88** (1995) 243.
38. D. Hilbert and S. Cohn-Vossen, *Anschauliche Geometrie*, Springer-Verlag, Berlin, 1932, p. 254.
39. M. G. Hetherington and M. K. Miller, *J. de Phys.*, **50-C8** (1989) 535.
40. G. B. Thompson, R. Banjeree, S. A. Dregia, M. K. Miller and H. L. Fraser, *J. Mater. Res.*, **19** (2004) 1582.
41. G. B. Thompson, H. L. Fraser and M. K. Miller, *Ultramicroscopy*, **100** (2004) 25.

9. FOCUSED ION BEAM SYSTEM—A MULTIFUNCTIONAL TOOL FOR NANOTECHNOLOGY

NAN YAO

I. INTRODUCTION

With nanotechnology being at the forefront of modern day research and development, new knowledge and successful application of increasingly small technologies is in high demand. Having this in mind, it is clear that the advancement and production of new high grade tools is a necessity if we are to see industrial progress. In particular materials science, with its never ending push for smaller scale analysis, is constantly looking for new tools that will help characterize materials and phenomena in addition to being able to machine at micro and nano-scales.

One tool which is expected to continue having success in answering this plea is the focused ion beam (FIB). The machine is relatively new, with great potential in nanotechnology and material science, and so naturally, there is much interest in exploring its applications. It allows slim or even no disruption of a procedure in addition to having tremendous micro and nano-machining capabilities. This interest has resulted in the development of the two-beam FIB system, which has advanced hand in hand with the complexity of new materials. Materials now being developed have more geometric intricacies and smaller feature size than ever before. Such complexity is inherent in most biomaterials and their synthetic analogues, as well as in nanotechnology. Beyond the structural complexity, phenomena occurring on even smaller length scales often adversely affect the performance and reliability of these materials. Foremost among these is inter-diffusion among the layers, resulting in chemical and structural gradients (figure 1). Understanding and improving on the performance and durability of these

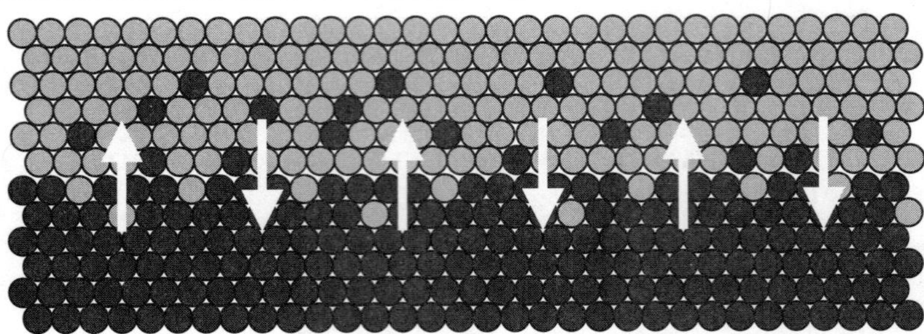

Figure 1. Schematic diagram of interlayer diffusion.

systems requires high-resolution structural, chemical and geometric analysis of cross sections through the layers, a function which the two-beam FIB system excels at.

This chapter is intended as a review of the major literature on the two-beam system. Since the FIB is an essential component of the two-beam, a survey of the FIB is presented first, followed by a discussion of the new or improved features of the two-beam system achieved by importing the scanning electron microscope (SEM). The goal is to discuss the wide spectrum of possibilities resulting from FIB technology, so as to serve as a reference guide for future work.

We begin by describing the fundamental differences between ions and electrons, and then continue to explain how these properties affect the structure and functionality of the FIB and SEM. After a brief explanation of the properties of interest and theory behind the FIB, we discuss various techniques for which it is most useful. Milling, of course, is a natural result of using relatively heavy ions in the beam. It allows precise modification of sample surfaces, as well as creating well defined cross-sections. The milling machine can easily be converted to a deposition system by using a gas delivery device to apply the desired chemical compound near the surface impact point of the ion beam. In conjunction with its milling potential, the FIB's deposition capabilities allow the creation of almost any micro-structure (figure 2). In addition, ion implantation, another technique available in the FIB system, is very useful in modifying certain materials. Of course, we also study the imaging possibilities of the FIB, where the large size of the ions can also provide advantages found in no other imaging tool. Once we've covered these aspects of the FIB, we discuss how these and other processes are improved by incorporating both the FIB and SEM into a single machine, justly named the two-beam FIB system. We will observe that in combining the two, we obtain capabilities that exceed anything found in either the FIB or SEM.

After discussing the advantages available from using the two-beam FIB system, we focus on several major areas of FIB applicability, beginning with its vast milling capabilities for surface modification of materials. Since the FIB machine is first and foremost a micro scale milling machine, its applications for material alteration are countless. We look into the ever advancing field of integrated circuits and discuss how the FIB can be used for defect analysis as well as structure modification. We also consider

Figure 2. Example of the FIB's milling and deposition capabilities. (*Courtesy of Fibics Incorporated*)

the lithographic advantages of the FIB in addition to its brute physical capabilities of cross sectioning and sharpening, and proceed to study the characterization of the interaction between certain materials and focused ions.

In addition to milling, TEM sample preparation has been a widely studied and familiar use for the FIB machine, and thus is worth noting. The FIB can perform this function significantly faster than old methods, saving valuable time. For electron transparent samples, the FIB machine is able to thin out a region of a bulk material by digging trenches and subsequent shavings to make the samples as thin as possible. We discuss the basic techniques of lift-out and micro-pillar sampling, issues such as warping, and an environmental application of the FIB pertaining to TEM sample preparation.

We also take a deep look into the many imaging applications of the FIB in the two-beam system. Images can be obtained from a secondary electron detector using either the SEM or the FIB by simply toggling between the two. Each has its advantages depending on the specific needs. For example, using electrons to image, as in the SEM, entails much less damage to the material, while providing a much better resolution. On the other hand, ion beam imaging, found in the FIB, provides better sensitivity to details such as crystal orientations and grain structure because of the depth of penetration from channeling. The two-beam system is then an exceptionally practical machine, as it combines the crisp non destructive images of the scanning electron microscope with the milling capabilities of the focused ion beam. Additionally, we explore a very useful method of obtaining three-dimensional data by shaving off thin layers, and as a result, obtaining a series of two dimensional. With this data, available only from the combining SEM and FIB applications, graphs and images can easily be interpolated with the two dimensional data. By then performing a secondary ion mass spectrometry (SIMS) analysis to yield elemental composition, we can have a complete set of data.

Finally, we comment on the sample damage that using the FIB can induce. Although this can sometimes be a problem, we will discuss ways of using this as an advantage. Even with this limiting quality, the FIB has tremendous breadth in its applicability, and deserves all the attention it has received as an important new technology.

II. PRINCIPLES AND PRACTICE OF THE FOCUSED ION BEAM SYSTEM

A focused ion beam system is an effective combination of a scanning ion microscope and a precision machining tool. The FIB was developed as a result of research by Krohn in 1961 on liquid-metal ion sources (LMIS) for use in space [1]. Interested in developing thrusters that used charged metal droplets, he first documented ion emission from a liquid metal source [2]. The discovery of LMIS found novel applications in the areas of semiconductors and materials science. Commercialization of the FIB was occurred in the 1980's, mainly geared towards the growing semiconductor industry, but also finding a niche in the materials sciences industry.

The modern focused ion beam system utilizes a LMIS at the top of its column to produce ions, usually Ga^+. From here the ions are pulled out and focused into a beam by an electric field and subsequently passed through apertures, then scanned over the sample surface (figure 3). Upon impact, the ion-atom collision is either elastic or inelastic in nature. Elastic collisions result in the excavation of surface atoms—a term called sputtering—and are the primary cause of the actual modification of the material surface (figure 4). Inelastic collisions occur when the ions transfer some of their energy to either the surface atoms or electrons. This process produces secondary electrons (electrons which become excited and are able to escape from their shell), along with X rays (the energy released when an electron drops down into a lower shell). Secondary ions are also produced through the collision, seemingly after secondary electrons have been emitted.

Valuable information can be gathered from all emission, depending on the capabilities of the machine. The signals from the ejected ions, once collected, can be amplified and displayed to show the detailed structure of the sample surface. Beyond its imaging capabilities, the FIB can be used as a deposition tool by injecting an organometallic gas in the path of the ion beam, just above the sample surface. This technique proves to be practical for certain material repair applications. In addition, by containing the ion beam over a specific region of the sample for an extended length of time, the continuous sputtering process will lead to noticeable removal of material, which is useful for probing and milling applications.

The underlying factor in the physics behind the FIB is the fact that ions are significantly more massive than electrons (table 1), a remarkably enhancing feature. The collisions between the large primary ions and the substrate atoms cause surface alterations at various levels depending on dosage, overlap, dwell time, and many other variables, in a way that electrons cannot achieve [3].

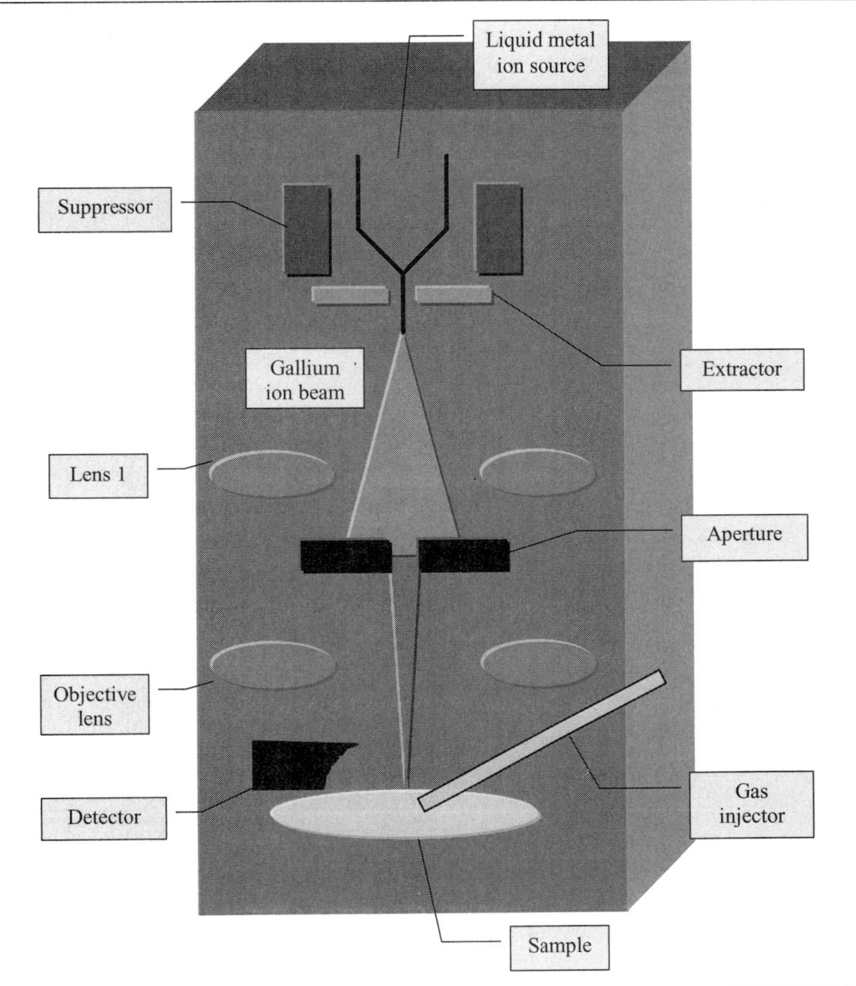

Figure 3. Schematic diagram of the FIB system.

The properties of the FIB system give it a unique ability to isolate a specific region of the sample, so as to only make necessary modifications without undermining the integrity of the entire sample. This has a wide range of applicability, from simple techniques like making probe holes to more complicated techniques such as cutting a precise three-dimensional cross-section of a sample. In addition, the FIB, having both drilling and deposition capabilities, is ideal for failure analysis and repair.

2.1. Ion Beam Versus Electron Beam

Because of the analogous nature of ion beam and electron beam microscopy, a comparison of the two can help provide a better understanding of the subject. The fundamental

Table 1. Quantitative Comparison of FIB Ions and SEM Electrons.

Particle:	FIB	SEM	Ratio
Type	Ga^+ ion	Electron	
Elementary charge	+1	−1	
Particle size	0.2 nm	0.00001 nm	20,000
Mass	1.2×10^{-25} kg	9.1×10^{-31} kg	130,000
Velocity at 30 kV	2.8×10^5 m/s	1.0×10^8 m/s	0.0028
Velocity at 2 kV	7.3×10^4 m/s	2.6×10^7 m/s	0.0028
Velocity at 1 kV	5.2×10^4 m/s	1.8×10^7 m/s	0.0028
Momentum at 30 kV	3.4×10^{-20} kgm/s	9.1×10^{-23} kgms	370
Momentum at 2 kV	8.8×10^{-21} kgm/s	2.4×10^{-23} kgm/s	370
Momentum at 1 kV	6.2×10^{-21} kgm/s	1.6×10^{-23} kgm/s	370
Beam:			
Size	nm range	nm range	
Energy	up to 30 kV	up to 30 kV	~
Current	pA to nA range	pA to μA range	~
Penetration depth:			
In polymer at 30 kV	60 nm	12000 nm	0.005
In polymer at 2kV	12 nm	100 nm	0.12
In iron at 30 kV	20 nm	1800 nm	0.11
In iron at 2 kV	4 nm	25 nm	0.16
Average signal per 100 particles at 20 kV:			
Secondary electrons	100–200	50–75	
Back-scattered electron	0	30–50	0
Substrate atom	500	0	infinite
Secondary ion	30	0	infinite
X-ray	0	0.7	0

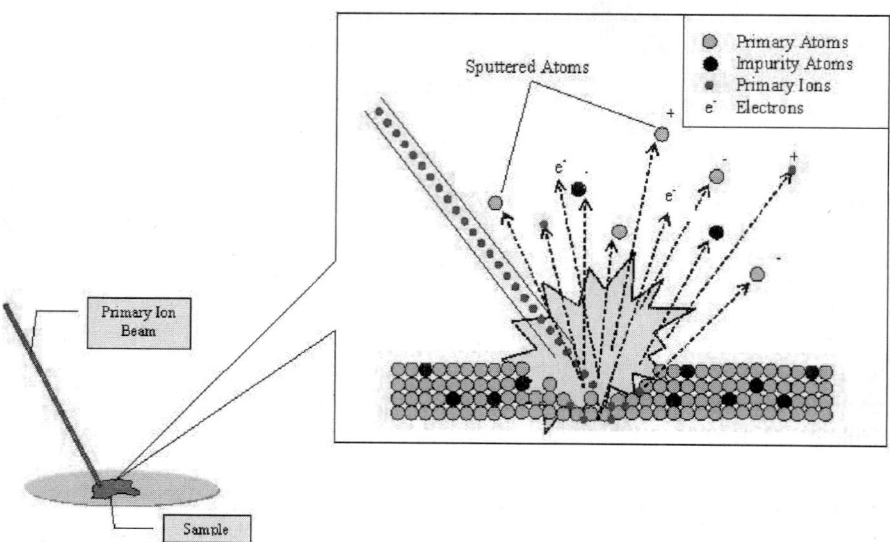

Figure 4. Schematic diagram of the ion beam sputtering process.

difference between these two techniques lies in the use of different charged particles for the primary incident beam; as their names suggest, the former utilizes ions while the latter relies on electrons. Ions and electrons have three major differences. To begin with, electrons are negatively charged while ions can be positively charged. This alone does not have any significant consequence, but the other two differences—size and mass—can appreciably alter the interactions between the beam and the sample. These differences are shown quantitatively in table 1.

When a beam of energetic particles, be it ions or electrons, strikes a solid surface, several processes get underway in the area of interaction. A fraction of the particles are backscattered from the surface layers, while the others are slowed down within the solid. Unlike electrons though, the relatively large ions have a difficult time penetrating the sample by passing between individual atoms. Their size increases the probability of interaction with atoms, causing a rapid loss of energy. As a result, ion-atom interaction mainly involves the outer shell, resulting in atomic ionization and breaking of the chemical bonds of the surface atoms. This is the source of secondary electrons and the cause of changes in the chemical state of the material. Similarly, since the inner electrons of the sample cannot be reached by the incoming ion beam, inner shell excitation does not occur, and as a result, usable X-rays are not likely to be generated as is the case when using an electron beam.

The total length the ion travels is known as its "penetration depth," a term which also applies to electrons, although they often pass much deeper into the sample (table 1). The atomic collision has a statistical nature, and therefore, the penetration depth adheres to a symmetric Gaussian distribution around the mean value. The moving ion recoils one of the atoms in the sample, which then causes another constituent atom to recoil. As a result, the ions create many atomic defects along their path. This generation of defects plays an important role when using an ion beam for material modifications.

As for the difference in mass, ions—being many times heavier—can carry about 370 times the momentum of electrons. Upon contact, this momentum can be transferred to the atoms in the sample, causing enough motion to remove the atoms from their aligned positions. This sputtering effect, not present when an electron beam is used, has important milling applications which are discussed later. The key idea to remember is that as a result of the ions' large size and mass, an ion beam surpasses the functional range of an electron beam by removing atoms from the sample surface in a precise, controlled manner.

In particular, gallium (Ga^+) ions are used in the FIB for several reasons. Firstly, its low melting point makes gallium convenient in that it requires only limited heating, and therefore is in the liquid phase during operation. Also, its mass is just about ideal, as it is heavy enough to allow milling of the heavier elements, but not so heavy that it will destroy a sample immediately. Finally, since some of the ions used in the beam will be implanted into the sample, it is important to use an element that will not interfere with the analysis of the sample. Gallium can easily be distinguished from other elements, and therefore does not cause trouble in this regard. Although gallium is not the only possible choice, it has become the popular ion for use in the FIB system.

Unlike with electrons, the collision using a gallium ion beam induces secondary processes such as recoil and sputtering of constituent atoms, defect formation, electron excitation and emission, and photon emission. Thermal and radiation-induced diffusion contribute to various phenomena of interdiffusion of constituent elements, phase transformation, amorphourization, crystallization, track formation, permanent damage, and so on. Ion implantation and sputtering change the surface morphology; craters, facets, grooves, ridges, pyramids, blistering, exfoliation, and a spongy surface may develop. All of these are interrelated in a complicated way, so that a single phenomenon cannot be understood without discussing several of these processes. Therefore, it is necessary to quantitatively understand the experimental observations and to have stringent design abilities for sophisticated applications of these versatile processes in the field of nanotechnology. With that, we can aim at material modification, deposition, implantation, erosion, nanofabrication, surface analysis, and a seemingly endless list of other applications.

2.2. Focused Ion Beam Microscope Versus Scanning Electron Microscope

The differences between ions and electrons can be extended to their respective machines, the FIB system and the scanning electron microscope (SEM). The two are designed and work very similarly, but the FIB's use of gallium ions from a field emission liquid metal ion (FE-LMI) source rather than electrons provides functionality and applicability different from that of the SEM. As discussed in the previous section, this focused primary beam of gallium ions is rastered on the surface of the material to be analyzed. As it hits the surface, a small amount of material is sputtered, or dislodged, from the surface. The dislodged material may be in the form of secondary ions, atoms, and secondary electrons. These are then collected and analyzed as signals to form an image on a screen as the primary beam scans the surface. This image forming capability allows high magnification microscopy in the FIB system.

In both the FIB and SEM machines, a source emits charged particles which are then focused into a beam and scanned across small areas of the sample using deflection plates or scan coils. In the SEM, the focusing is accomplished using magnetic lenses. Ions though, are much heavier and therefore slower than electrons, so the corresponding Lorenz force is lower. As a result, magnetic lenses are less affective on ions, so the FIB system is equipped with electrostatic lenses which prove much more effective in focusing the ions into a beam (figure 5).

Both instruments are used for high resolution imaging by collecting the secondary electrons produced by the beam's interaction with the sample surface. Contrast is formed by the fact that raised areas of the sample (hills) produce more collectable secondary electrons than depressed areas (valleys). A viewing monitor is synched to the scan coils controlling the beam so that as the beam scans across the sample surface, its image is reproduced on the screen. For the FIB, imaging resolution below 10 nm is possible. This image will show both topographic information and materials contrast and often provides information complementary to that obtained from an SEM image. Since the ion-solid interaction also depends on the crystal grain orientation,

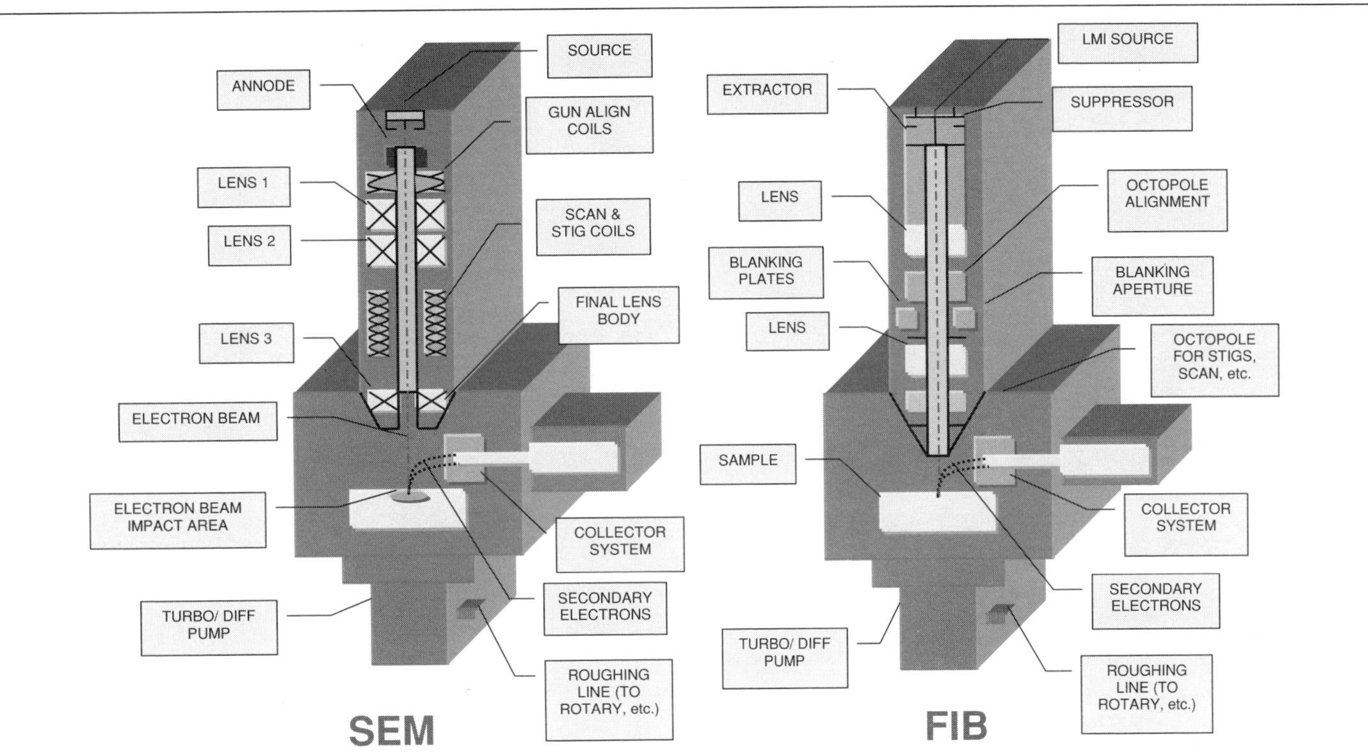

Figure 5. Schematic comparison of the SEM and FIB machines.

information concerning grain size and orientation can be obtained simply by collecting an image of the surface using the FIB. Note that although ions move much slower than electrons of the same energy, they are still fast compared to the image collection mode; in practice this has only negligible consequences, taking into account an image shift of a few pixels.

As for the sample itself, the FIB can accommodate wafer samples up to 8 inches (200 mm) in diameter; there is no minimum sample size. The sample must be vacuum compatible and conductive samples are preferred. Both individual and packaged parts can be used. Unlike in the SEM though, the sample will be altered during examination under the FIB, as the large, heavy ions will cause sputtering on the sample surface. Also, when using ions the sample often develops a positive charge which can distort the image. To combat this effect, an electron flood gun is used in the FIB to keep the sample more neutral.

2.3. Milling

By far the most powerful and versatile capability of the FIB system is milling. It adds a third dimension to microscopy, allowing us to modify the material surface, create cross sections, and carve materials into any shape we desire. For example, this method can provide electron-transparent, site-specific cross-sections of heterogeneous catalysts (figure 6). Again, when the relatively heavy and high-momentum ions collide with atoms on the sample surface, significant amounts of energy are transferred to the atoms, causing them to break bonds and leave their place in the atom matrix. Unlike electrons, which are hardly heavy enough to cause any atom movement in the sample (like trying to move a soccer ball by throwing ping pong balls at it), the ions, acting like soccer balls themselves, easily disrupt the placement and alignment of the atoms and create a sputtering effect. So when using an ion beam to bombard a sample, milling is a natural and continuous consequence. By controlling aspects the sputtering such as rate, location, and depth, we can take advantage of the FIB system's remarkable milling abilities.

In general, the higher the primary beam current, the faster material is sputtered from the surface. Therefore, if only high-magnification imaging is desired, a low-current beam must be used. For milling applications though, high-current beam operation is used to sputter or remove material from the surface initially, then a lower beam current is used for fine polishing. The sputtering rate can be easily and accurately controlled by altering the beam current or using smaller spot sizes. In addition, material sputter rate selectivity can be achieved through a process known as gas assisted etching (GAE). In this technique, one of several halogen gases is introduced to the work surface in the immediate vicinity of the desired hole or cut. The material-specific absorbency rates enhance the formation of volatile products under ion bombardment. This allows oxides to be cleared without damage to conductors, and conductors to be cut without damage to underlying insulators. Typical resolution for cutting is around 0.1 μm. The holes and cuts can be very accurately placed (usually within 20 nm) [4] and can reach buried layers provided that the depth to width aspect ratio is kept below 4:1. GAE,

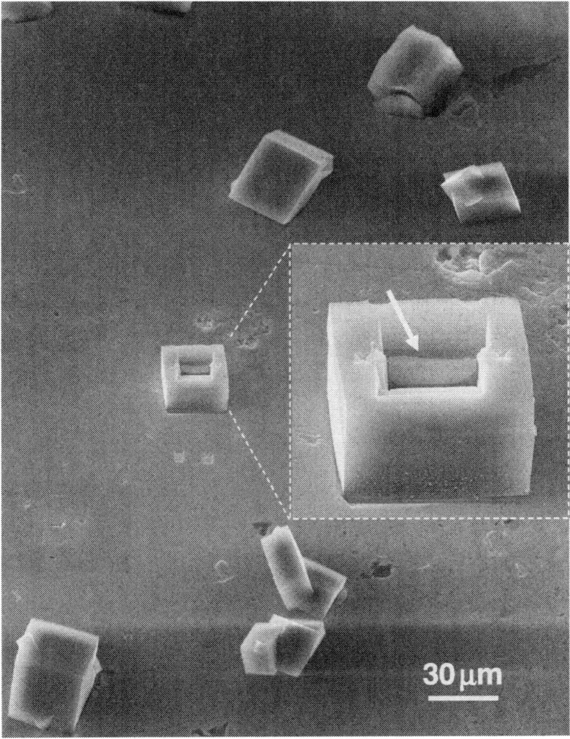

Figure 6. FIB sections from a heterogeneous catalyst. Arrow shows a thin sample slice that can be extracted from an individual small crystal.

often using iodine or xenon difluoride, can increase the aspect ratio to 8:1 or better. As a result, the milling efficiency is typically a few $\mu m^3/nC$ but varies in different materials. The actual rate depends on the mass of the target atom, its binding energy to the atom matrix, and its matrix orientation with respect to the incident direction of the ion beam.

The GAE technique is more efficient in removing large volumes of material that would be prohibitively time-consuming otherwise. It is important to keep in mind that GAE is quite chemistry-specific, which gives us the ability to selectively etch one material without affecting the others. For example, materials like Si, Al and GaAs can have their removal rate increase 20 to 30 times through GAE, while oxides like SiO_2 and Al_2O_3 are not affected [5]. Similarly, in the presence of water vapor, carbon based materials show a highly enhanced removal rate whereas materials that oxidize, such as Al metallization lines and silicon, will actually show dramatically reduced rates [6].

In addition to changing the rate of material removal, GAE can be "neater" than ordinary ion milling, as redeposition of etched materials does not occur when using the former. However, the GAE technique has its problems as well. It contaminates the sample deeper into the surface than simple ion milling [7]. Also, GAE does not have the pinpoint precision that ion milling can achieve. Oftentimes, a combination of initial GAE followed by ion milling is used when an operation requires bulk volumes of material removal as well as high precision.

With the FIB system, we have both lateral and depth control over the beam. Of course, if the beam is held over one area for a length of time, more material will be removed from that area. This process can be used to expose buried lines for eventual cutting or attaching to other circuit elements, depending on the desired modification of the structure. Unfortunately, an ion beam will not etch through unlimited thicknesses of material. By injecting a reactive gas into the mill process, the aspect ratio of the ion beam's cutting depth can be dramatically altered, depending on such variables as sample composition, mill area, beam parameters, and whether an enhanced etch process is used; this allows us to reach lower levels in the sample disturbing the upper layer. Having this precision and control allows applications unique to the FIB system.

2.4. Deposition

The FIB demonstrates its versatility by how easily it can be converted from a micromilling system to a deposition system. A simple adjustment allows the addition of material instead of the usual sputtering and removal of material. To create this change in function, a gas delivery system is added in order to supply a chemical compound just above the sample and directly in the path of the ion beam. The gas, usually an organometallic compound, is adsorbed onto the sample surface. This precursor is then struck by the gallium ion beam (and additionally by some secondary emission products), causing it to decompose as its bonds are broken. The volatile organic impurities are released and removed by the FIB's vacuum system. The desired metal, say platinum, is heavy enough to remain deposited on the surface, creating a thin film that can act as an electrical conductor (figure 7).

Although the chemical deposition just described is the predominant method of deposition with the FIB, there are two others which in certain situations may be advantageous. The first method, direct deposition, uses large diameter beams with low energy to deposit thin films, usually of gold. Low energy microbeams need very small currents and are very slow. The other alternate deposition method is nucleation with chemical vapor deposition (CVD), usually used to deposit carbon, aluminum, and iron. As an extension of chemical deposition, this technique uses the same gas injection system to create nucleation on the substrate. Once that is done, CVD is used to grow a thin film in the nucleation regions. The main advantage of this method is that it produces a film of uniform height [8].

Once again, when an organometallic gas such as $W(CO)_6$ is introduced in the area of the ion beam, the ion beam can be controlled to decompose the molecules which adsorb onto the surface and form a metal layer. This metal conduction layer can function as a probe pad or be used to make a new connection for nano-circuitry. Unfortunately, the film is often impure, causing a high resistivity. Long wiring runs requiring lower

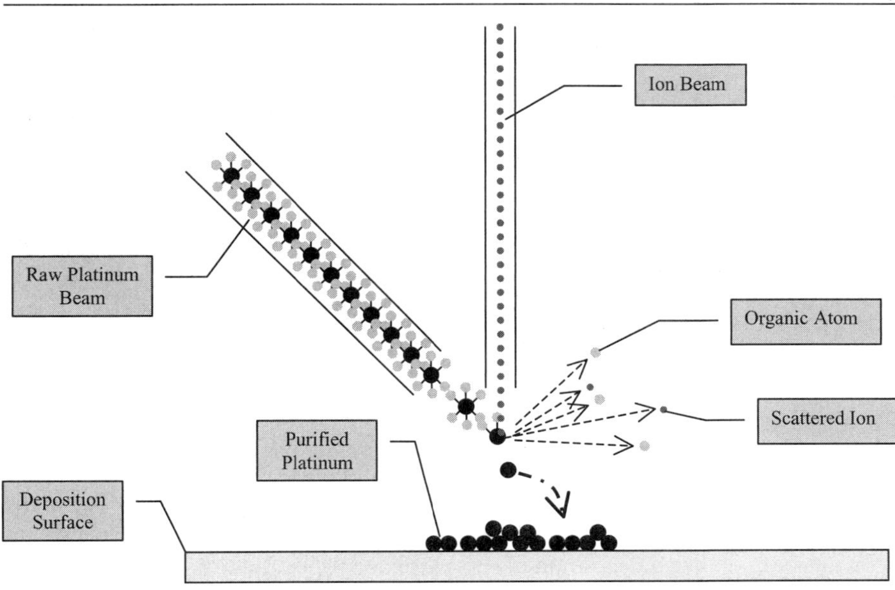

Figure 7. Schematic diagram of the platinum deposition process.

resistivity can be achieved by combining FIB local connections with long laser CVD deposited gold conductors. In this technique, a gold carrier gas is decomposed by a scanning laser beam, leaving a 5–10 μm wide gold line on the work surface with a relatively low sheet resistance. In addition, just as it can deposit a conducting thin film, the FIB can be used to deposit a high quality oxide insulator. A siloxane type gas and oxygen are introduced into the path of the ion beam at the impact area. Decomposition of the complex silicon bearing molecule in the presence of oxygen leads to absorption into the surface and formation of a silicon dioxide insulator layer. With the aid of oxide deposition, the re-insulation of cut IC wiring and FIB deposited conductors is now possible, making complex multilevel wiring repair practical.

The FIB deposition system, although very useful, is not perfect. In general, the cracking of the organometallic molecule is not complete, leaving some organic impurities deposited in the thin film. In addition, some lingering gallium atoms can hurt the insulating ability of a deposited layer. Still, although CVD deposits are often more pure, the advantage of using the FIB system it its ability for precise, localized deposition, and its capacity for controlling the different heights of the depositions. Even so, the use of ion beams may not always be ideal; research has shown that in certain cases, electron nanofabrication has been able to match the work of the ion beam without the impurities that come along with using the FIB [9].

Nevertheless, the FIB is unparalleled in its patterning capability. By adjusting parameters of time, gas flux, and ion current, it allows deposition of three-dimensional structures with complete control over size, position, and height. For example, Khizroev, *et al.*

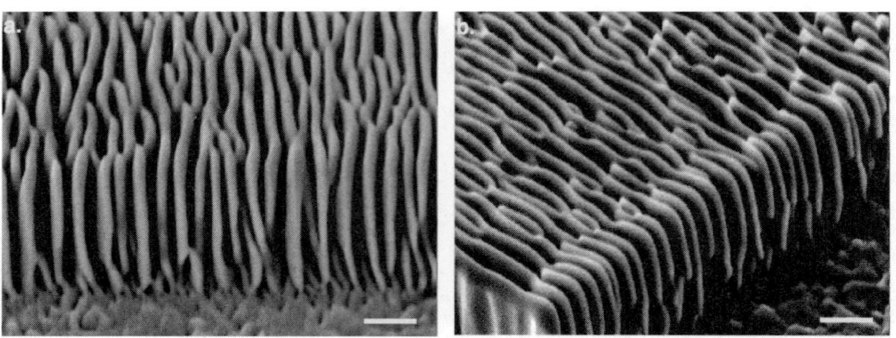

Figure 8. SE images of a two-beam FIB fabricated Pt nano-fin structure on Si surface, where (a) and (b) are viewed from different directions (scale bar: 0.3 μm) [11].

have successfully deposited non-magnetic tungsten at the nanoscale range to aid in the fabrication of nanomagnetic probes [10]. In addition, we have performed experiments with ion-beam induced deposition of self-assembled nanostructures using a technique that formed nanoscale hollow bulk structures. These nanofins were deposited perpendicular to the substrate (figure 8) [11]. Using deposition in combination with its milling capabilities, the FIB can accurately build almost any nanostructure, a feature which is invaluable in today's research.

2.5. Implantation

Another critical feature of the FIB system is its ability to produce maskless ion implantation, a technique fundamental in the fabrication of semiconductor devices. In traditional semiconductor fabrication, ions are allowed to bombard the entire wafer, and regions which are not to be implanted are covered by a patterned film, or mask [3]. With the FIB, whose level of localized control and specificity has already been discussed, we now can discard the mask and simply aim the beam at the areas where implantation is desired. This provides new abilities to control gradients in doping and the control the depth of the implantation [3]. Like many of the other capabilities of the FIB system, its use for ion implantation will minimize costs, both in time and money.

Ion implantation has been used for materials modification as well. In general, defect formation following ion implantation plays unpleasant roles when using the FIB system, but occasionally it can have a positive effect. For example, recent experiments have shown that implanted atoms can be induced to self-assemble into nanoclusters through thermal annealing or radiation. To understand these properties, a Monte Carlo model has been developed with the ability to simulate diffusion, precipitation and interaction kinetics of implanted ions [12].

With all its advantages in accuracy and precision, an almost directly related disadvantage in using the FIB for ion implantation is its slow processing rate. Even so, there are situations where this rate is not an issue, and in these cases, the FIB becomes an crucial tool.

2.6. Imaging

Although we have already discussed some aspects of FIB imaging, they are worth repeating and expanding upon. Again, the method used in FIB imaging is similar to that in the SEM, except that the incident particles are ions instead of electrons. When the ion beam strikes the sample surface, one of the resulting processes is the emission of secondary particles, including low energy secondary electrons, neutral atoms, secondary ions, and photons. In general, the most prevalent among these emissions is secondary electrons, whose signal is usually used to obtain the image. In some cases though, images can only be obtained from secondary ion signals [13].

As mentioned earlier, imaging resolution below 10 nm is possible with the FIB, but what makes FIB images particularly useful is their ability to display detailed surface topography. The key mechanism for displaying topographic contrast is the fact that the emission rate of secondary particles (ions or electrons) will vary with the angle between the incident beam and the normal to the surface of the sample. Since the FIB has a shallow penetration depth relative to an electron beam, particles are generated from only a limited portion of the sample, and therefore the FIB allows much higher surface sensitivity, and in some cases is able to display a more detailed topographical map [4, 13].

In addition to topographical contrast, another merit of FIB imaging is its ability to contrast between grounded conductors and insulating areas or isolated conductors. The incident Ga^+ ions will place a positive charge on the insulating or isolated conducting surface areas, suppressing the yield of secondary electrons collected for imaging from these regions. Insulators and isolated conductors will therefore appear darker in the images, while grounded conductors will be bright. Although moderate surface charging can enhance voltage contrast in ion beam images, too much charge can severely depress secondary electron emission, and so with insulator surfaces, secondary ions are often used to obtain the image. In general, the voltage contrast imaging technique is useful for locating failure sites in devices as well as detecting endpoints when etching (figure 9), making the FIB an attractive tool in many integrated circuit applications [14].

The FIB system is also proficient in providing materials contrast. This results from the fact that different materials produce different yields of secondary particles because of the complex way the ion beam energy is lost within the sample. Materials contrast is frequently the dominating effect in FIB images. It is worth mentioning that the yield of secondary ions from metallic surfaces in greatly increased by the presence of oxides of carbides, making these regions much brighter. This property gives the FIB applications in studies involving corrosion or grain boundary oxidation of metals [4].

The last type of contrasting we will mention is channeling, or crystallographic orientation contrasting. Some of the incident ions are "channeled" down between the lattice planes of the specimen, with the penetration depth depending on the beam

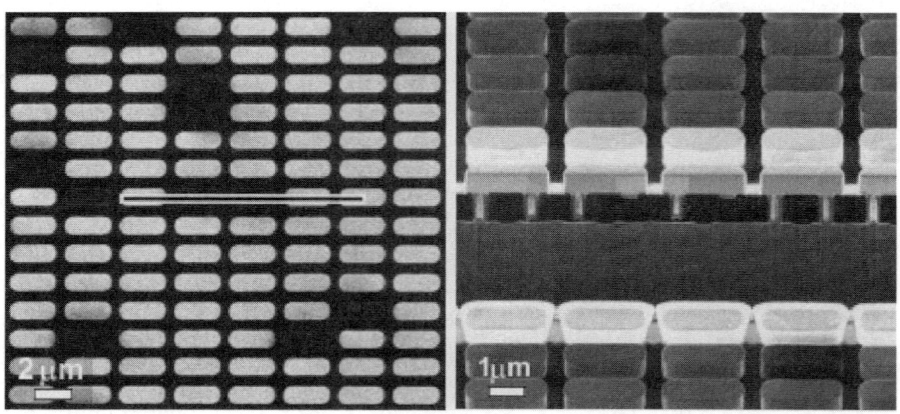

Figure 9. Voltage contrast reveals open via positions in a contact chain structure. The cross-section image at the position of the line (right) shows deep etching of the vias. (*Courtesy of IMEC*)

current, the relative angle between the ion beam and the lattice plane, and the interplanar spacing of the lattice. Channeled ions yield fewer secondary particles, causing the channeling grains to appear darker in FIB images. This technique is often used in monitoring the grain size distribution within Al-Cu films as well as in the study of electromigration effects in metallization [13, 15].

In conjunction with the high resolution picture, the FIB can be equipped for secondary ion mass spectrometry (SIMS), which allows detailed elemental analysis (figure 10) [16]. With this and the various contrasting techniques discussed, the FIB can produce an elemental map with a nearly complete set of vital data for almost any microscopy application.

2.7. The Two-Beam System

So far, we have focused on the properties and capabilities of the FIB system, and certainly a plethora of applications can be derived from these capabilities alone. Still, by combining the FIB's imaging and sample interaction abilities with the SEM's high resolution, non-destructive imaging, new heights in microscopy research can be achieved. The two-beam system accomplishes just that; by putting FIB and SEM technology together in a single machine, the two work symbiotically, achieving tasks beyond the limitations of either individual system.

In a two-beam system, the ion beam and electron beam are placed in fixed positions, with the former coming in at an angle (figure 11). The two beams are co-focused at what is called the "coincidence point," typically with a 5 mm working distance; this point is an optimized position for the majority of operations taking place within the machine. For best performance, the ion beam is tilted 45°–52° from the electron beam, allowing SEM imaging and FIB sample modification without having to move the sample. In addition, the stage can be controlled to tilt, allowing changes in the

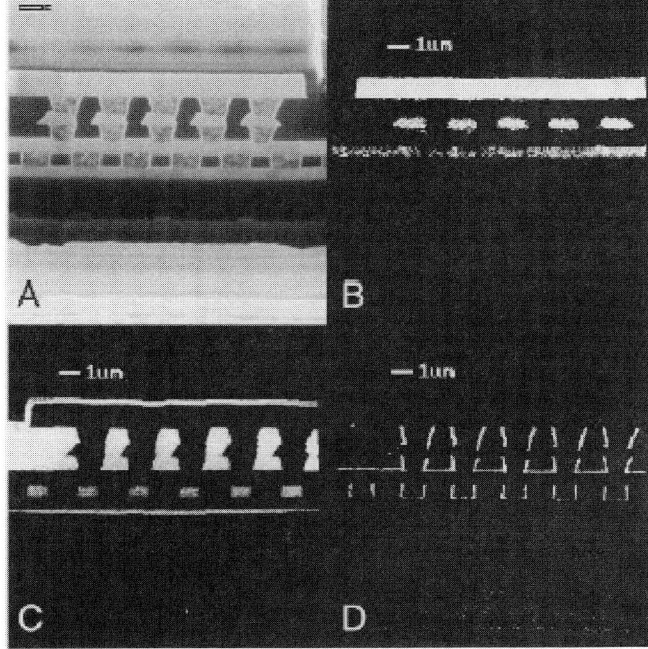

Figure 10. Chemical map of a cross-sectioned face of an IC. (A) FIB secondary electron image. (B) Mass 27 Al. (C) Mass 28 (Si). (D) Mass 48 (Ti). The maps are 256 × 256 pixels. Primary beam current = 40 pA [16].

sample-beam orientation. As with the FIB system, the two-beam can be controlled using integrated software with a single user interface.

Within the two-beam system, the FIB and SEM complement each other perfectly, providing new advantages which can simplify and improve microscopy research. For example, FIB imaging is very useful because of its high contrasting abilities, but it can cause damage to the sample. The SEM, on the other hand, has relatively lower contrast, but its images have higher resolution, and its use will not damage the sample. Depending on the specific needs of the user, the FIB, SEM, or a combination of the two can be used to obtain whatever imaging data is required. For example, in a study done at Portland State University, both the FIB and SEM were used to image and characterize carbon nanotubes, each providing useful information [17]. Similarly, reconstruction of 3D structure and chemistry of a sample will be simplified by the use of the two-beam system, as this technique is achieved by interpolating the 2D SEM/FIB images and ion-assisted SIMS chemical maps of successively exposed layers milled by the ion beam [18].

In addition to the combined imaging benefits, the two-beam system allows precise monitoring of FIB operation through the SEM. By using the slice-and-view technique for observing the progress of an ion-beam cross section, the operator can precisely stop

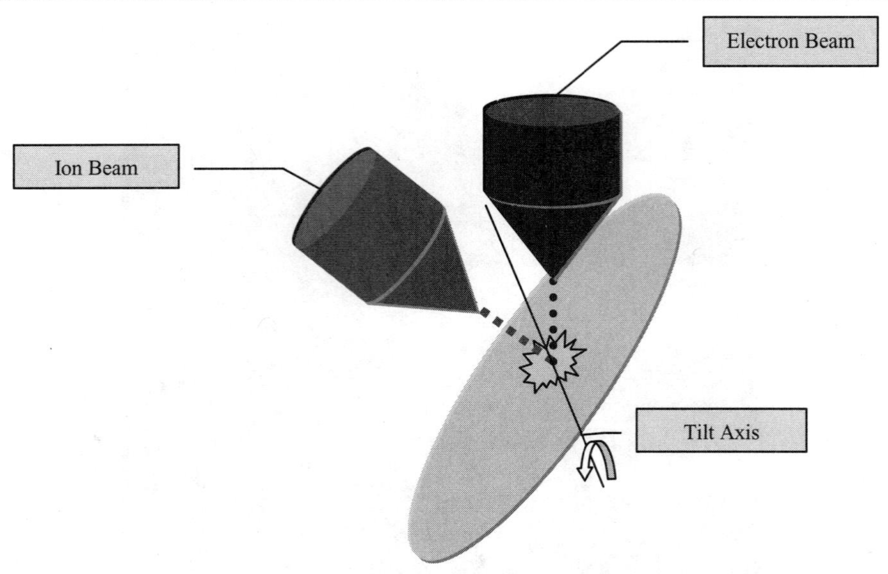

Figure 11. Schematic diagram of two-beam system.

the milling process at any point to obtain local information. Furthermore, state of the art technology now allows us to use the ion beam and electron beam simultaneously without interference, eliminating the trouble of switching back and forth. In the "CrossBeam" system, the sample can be imaged in real time with the SEM while the FIB is being used, giving the operator complete control over the milling process. This provides even higher levels of accuracy when creating cross-sections [19]. The SEM's damage-free monitoring is especially useful in the final phase of TEM sample preparations. It solves the dilemma found when using the FIB system alone, where the only way to observe the process is ion imaging, which can damage the sample. In a successful physical localization of IC fails, Zimmermann et al presented a completely in-situ two-beam technique that combines SEM monitored ion milling, XeF_2 staining and SEM imaging [20].

When dealing with charge neutralization, the two-beam system once more eliminates a problem encountered in the FIB. In general, when a sample is flooded with positively charged ions, the surface becomes charged, hurting the resolution of the image. With the availability of the electrons from the SEM, this problem is trivial. Similarly, a negative surface charge could be combated using the positively charged ions.

Another capability of the FIB machine, deposition of metal or insulating layers, can also be improved using the two-beam system. In some cases, using the ion beam for insulator deposition may lead to a material with poor insulating properties because of the high amount of gallium incorporated into the layer; when this is a problem, SEM induced deposition can be used, ensuring high insulating quality. In the case of

Figure 12. EBIC "hot spot" signal superimposed on secondary electron image showing one leakage site on this test structure [22].

SiO_x pads deposited by the SEM, it has been found that the resistance of the layer is two orders of magnitude higher than when using the ion beam [21]. In addition, while attempting to localize IC fails, this time electrically, Zimmermann *et al.* used the electron beam to deposit a highly insulating SiO_x layer in-situ in the two-beam system. A Pt probe pad was then deposited on the insulating SiO_x layer with ion beam assistance [22]. As before, the combination of FIB and SEM systems has proved both practical and valuable.

Finally, the SEM brings with it to the two-beam system an electron beam induced current (EBIC), a feature not present in the individual FIB system [23]. This is a useful technique for locating defects in diodes, transistors, and capacitors. The SEM beam generated a characteristic electron flow signal in the sample, which is then collected and amplified, providing an EBIC "hot spot" image (figure 12). This is powerful in pinpointing the bad site in a device that may contain as many as a few hundred thousand structures. Since the FIB can be used to prepare the sample for EBIC characterization and then to further analyze the located failure site through precision cross-sectioning, the two-beam has an advantage above the SEM alone.

We have seen now that the two-beam FIB system is an excellent multifunctional tool for sample imaging, modification and analysis. It offers high resolution 3D non-destructive imaging with the electron beam, as well as useful contrast imaging and SIMS chemical mapping with the ion beam. It can carry out precise, highly controlled an automated sample modification including ion milling, lithography, micro-machining, gas-assisted etching, and deposition. With all these and other capabilities, it is no question that the two-beam system will have an even longer list of applications. The remainder of this review is devoted to highlighting some of the most important of these applications.

III. APPLICATION OF FOCUSED ION BEAM INSTRUMENTATION

Thus far, we have discussed in depth the tremendous power of the FIB machine, with its wide-ranging functions and capabilities. Yet the capabilities by themselves are not the direct reason for the great industrial interest in FIB technology; instead, it is the abundance of applications yielded by these capabilities that have given rise to excitement. These applications seem to come with a promise of commercial success, and will therefore be discussed in detail in this review. We will begin with applications related to surface structure modifications, then look at the popular application of TEM sample preparation, and finally go over the imaging applications found in the FIB machine.

3.1. Surface Structure Modification

As discussed in earlier sections, the focused ion beam has the capabilities of a milling tool at micro and nano scales, making it an attractive new technique for micro-machining among other tasks. Three industrial fields particularly stand out as ideal candidates for the application of this ability in the FIB system. It is an almost perfect fit for the integrated circuit industry, and its need for more precise failure analysis and modification of its products. The FIB provides engineers with a tool for defect location and investigation, defect sample prep, circuit rewiring, and surface modification, all vital in circuit design and fabrication. In addition, the FIB holds lithographic capabilities that are becoming more and more effective. Critics have questioned the speed of the FIB system for patterning and lithography because of its serial nature, but studies have shown not only that FIB is up to par with lithographic standards, but also that FIB lithography and patterning have proven effective with numerous materials and specific applications.

Of course, as with any new technology, it is important to study its limitations. With the FIB, there is interest in knowing what surface modification at low ion dosage will do. At the Oxford University Department of Materials, a recent study was done to characterize low-dose ion beam surface manipulation. A Si wafer was bombarded with Ga^+ ions at different doses between 10^{13} and 10^{19} ions cm^{-2} (where the dose = $(I_{ion} * t_{exposure})/(A_{pattern} * 1.602\ E^{-15})$). Rectangles with depths varying around 4 nm and dimensions of 4×1 μm were created and investigated, along with grids consisting of lines 5 μm by 20 nm. Atomic force microscopy was then used to characterize the topology of the altered surface. It was discovered that the results differed for varying doses of ions. In almost all cases, edge protrusions were noted, ranging from 0 nm (10^{13} ions cm^{-2}) to a maximum height of approximately 1 nm, when below a dosage of 10^{17} ions cm^{-2} (figure 13). Edge effects were blamed mainly on Ga^+ implantation from stray ions that would unintentionally deposit in nearby regions, as well as redeposition of secondary and backscattered material [24].

With the knowledge we have from this and other research, we can understand the limitations of the FIB system and work within these limits to achieve successful applications of this valuable technology.

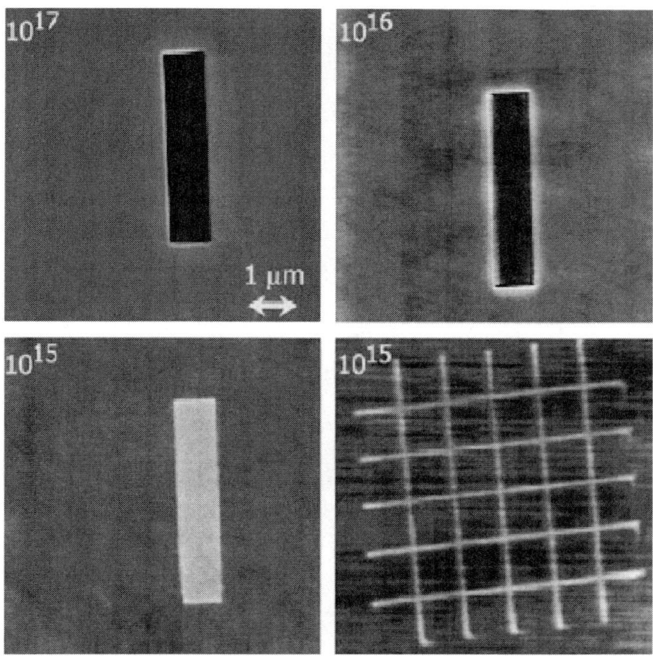

Figure 13. AFM image of features produced by low does Ga^{+} ions on Si. Note the protruding edge effect on 10^{16} ions cm^{-2} dosage (top right) [23].

3.1.1. Integrated Circuit Analysis and Modification

The semiconductor industry has made tremendous use of the FIB machine for its applications in defect analysis and modification of integrated circuits during the prototype stages. The benefits of being able to both remove and deposit material can be directly observed, as the ion beam is capable of milling away enough to cut a wire while depositing conducting material in another area to connect two pieces of wire (figure 14). All this is done with minimal damage to the wafer itself, allowing for the test product of be continually improved. The FIB can easily be used to create samples of the wafer for further analysis.

Unfortunately, when investigating integrated circuits for defects, it is often difficult to find the specific area of interest because of the lack of visible signs on the surface of the material. As good of a repair tool as the FIB may be, the failure cannot be repaired if it cannot be found. This problem has been analyzed at FEI Europe, Ltd. and a technique has been developed to effectively locate the area of concern utilizing the FIB machine for the entire process. The group examined a transistor containing a certain gate (approximately 1 μm^2) under which a thin gate oxide layer had broken down due to excessive voltage application (figure 15). The source and drain lay on opposite ends of the gate, which sat on top of the gate oxide layer, all of which resided on a silicon substrate. TEM samples of the failure site (<100 nm in diameter) were

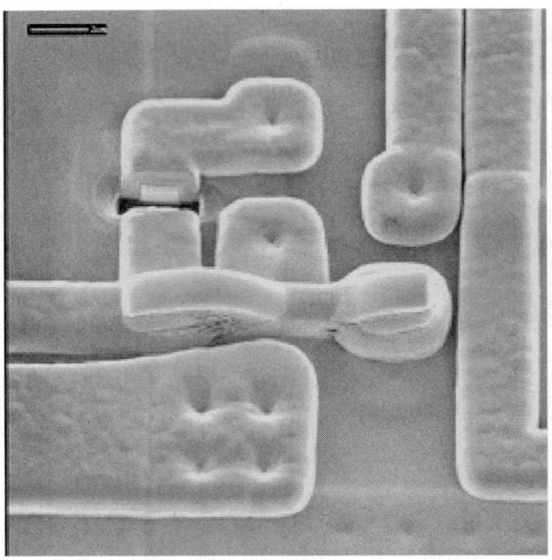

Figure 14. Image of rewired circuit. (*Courtesy of Integrated Reliability*)

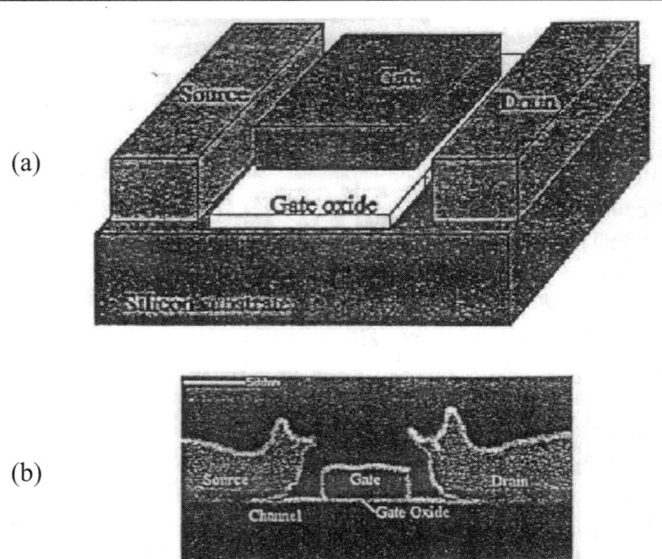

Figure 15. (a) Transistor Structure (b) Image of defect, where the break is clearly visible [24].

Figure 16. Combination SEM and FIB passive voltage contrast images. The FIB portion (at the right) shows an area of memory cells after exposing the wings of the floating gates. One floating gate appears bright, indicating that it is grounded and therefore the tunnel or gate oxide has failed [25].

desired for investigation on the mechanics behind the problem, but the area of failure had to first be located. By starting on high beam currents, an outline and subsequent trenches were dug around the region thought to be of focus. Removal of thin slices together with fast imaging was performed until the structure became apparent. Once achieved, low beam currents were used to uncover the region, noting that all focusing and stigmation adjustment were done away from the region. It was found that these basic steps succeeded in efficiently locating an area of defect [24]. Again, once the defect is located, the FIB's capabilities make it an ideal tool for use in integrated circuit modification.

Other work, reported by Haythornthwaite *et al.*, shows the versatility of the FIB system and its wide range of circuitry applications. The group worked with electrically erasable programmable read only memory, which after extensive use experiences failures. The FIB machine proved useful in three stages of the failure analysis process. Initially, using the FIB in passive voltage contrast mode helped confirm the occurrence and location of the failures (figure 16). Then, as an added benefit, the FIB was able to identify weak storage transistors on the memory device, which if not taken care of could soon lead to other failures. Finally, TEM samples were created using the FIB system, a process which will be discussed in depth in a later section [25].

3.1.2. Lithography and Patterning

Another application of the FIB machine related to surface modification involves its lithographic capabilities in nanofabrication. It holds clear advantages over other methods in its ability for high resolution patterning and depth of focus, and therefore should

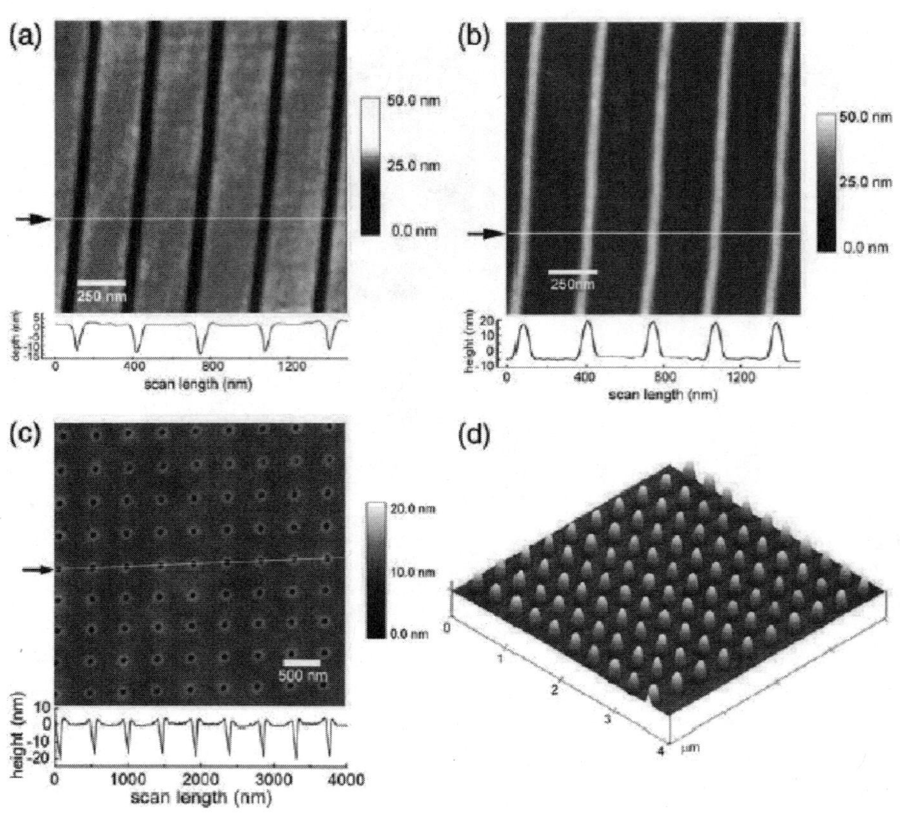

Figure 17. AFM images of FIB patterning. On the left are the actual FIB cuts, and on the right are their PDMS replicas [26].

be studied as a practical alternative. Unfortunately, because of the precision of the FIB system and the detail of lithography, this application of the FIB has one major drawback; although producing high quality results, the FIB machine appears unable to perform high throughput production. This problem was studied in depth at the University of Cambridge, where Li *et al.* replicated FIB created structures with nanocontact imprinting. Silicon wafers were patterned (10 μm single pixel line and 20 nm diameter dots) with a 30 kV Ga^+ ion beam at currents of 1, 4, 11, and 70 pA. Analysis was done using the AFM and replicas were made by pouring polydimethylsiloxane (PDMS) over the print master, then peeling it off (figure 17). Dwell time, ion dosage, and beam current were characterized, and satisfactory molds were created on almost all accounts [26].

As a result of its slow rate of progress, the FIB system, though proven capable of lithography, is still often criticized. With patterning rates generally around 0.1–1 $\mu m^3/nC$ incident ion current, its limited ability has been deemed as a restriction in the applicability of FIB. However, Liu *et al.* have reported the ability to rapidly

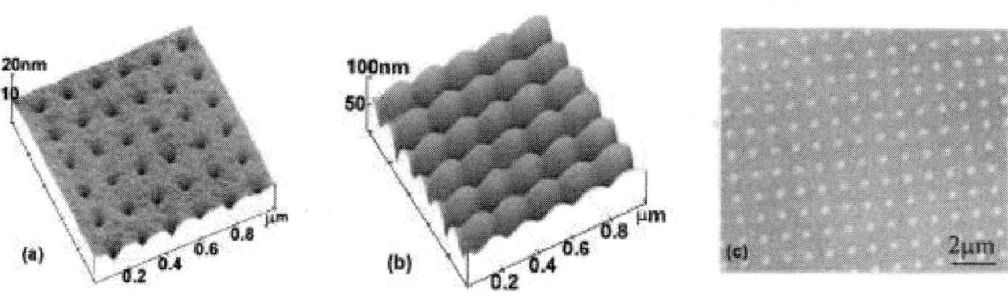

Figure 18. AFM images of topography in PMMA following FIB exposure at (a) 1 pA beam current and a total irradiation time of 20 µs/feature, (b) 11 pA beam current and a total irradiation time of 500 µs/feature. (c) TEM image (recorded at 200 kV, using mass–thickness contrast with an objective aperture, including just the transmitted beam of an array of sputtered features created using an 11 pA beam and a total irradiation time of 5 ms per feature). Samples were prepared by direct spinning of PMMA films onto TEM grids coated with thin (5 nm) amorphous C films [27].

create features, at rates exceeding those of past methods. Patterns were created on polymethylmethacrylate (PMMA) that was spun onto Si wafers, producing a 120 nm thin film (figure 18). Using a Ga^+ ion beam with beam currents varying from 1–70 pA, features were created with sizes ranging from 60–200 nm in diameter and 5–30 nm in depth. At diameters of 60 nm and depth of 5 nm, the paper reports milling times at 20 µs per feature and a material removal rate of 1000 $\mu m^3/nC$. Other features were also created at rates orders of magnitude faster than previously reported. One point of particular interest was the fact that all sputtering yields were consistently unusually high. The group proposed the possibility that this was the result of a depolymerization of the polymer being etched that was aided by the ion beam, a phenomena generally seen at higher temperatures [27]. This is only one of several cases of successful high speed lithography using the FIB. Although this method may not be the best in all situations, there are clearly cases when the FIB is the best tool, both in speed and quality, for lithographic and micro-printing applications.

Recently, a group at the University of Limerick in Ireland developed a two step negative resist image by dry etching (NERIME) process for lithography. The method combines exposure to Ga^+ ions from the FIB machine with reactive ion etching (RIE). By doing so, the group has further strengthened their abilities beyond those of conventional lithography, as the process eliminates some of the limitations of basic FIB lithography, such as low penetration depth and sample damage [28].

3.1.3. Materials Characterization and Alteration

We have already discussed in depth the characteristics of the FIB system which make it excel at material characterization and alteration. Again, its milling and deposition abilities allow the creation of almost any three-dimensional microstructure (figure 19), and its cross-sectioning capability proves to be very useful in the study and characterization of materials. As a result, there is a seemingly endless list of applications that can be derived from the FIB system. We discuss here just some of the many studies done

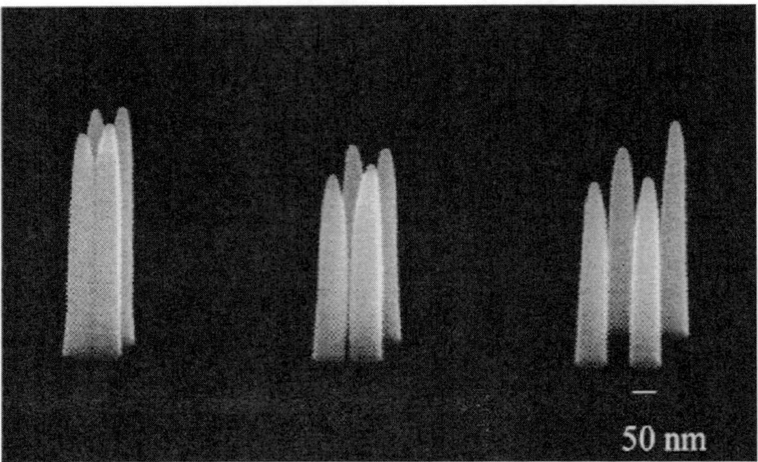

Figure 19. Creation of nanometer 3D structures. (*Courtesy of FEI Company*)

with the FIB into materials characterization and alteration. As with any new machine, there is an interest in finding out new processes and techniques that improve the old ones, as well as an interest in new information that can be gathered. This section serves to highlight some of these studies.

Gallium Nitride (GaN) is a widely applicable semiconductor material that has a high melting point and strong light sensitivity properties, which make it well suited for both high temperature devices and light detecting or emitting devices [29, 30]. Unlike traditional methods of etching, the FIB does not need an etch mask for manipulation because of its local specificity and precision; this is a most valuable feature in semiconductor fabrication. In a study done at the University of Bristol, FIB etching onto GaN structures was studied with both AFM and SEM techniques. Using an FIB machine with a gallium ion gun as well as a magnetic sector mass analyzer, a 20 nm Ga^+ ion beam was used to etch square patterns of 49 μm^2 onto a 1.2 μm thick GaN specimen at doses of 500, 1000, 1500, and 2000 $pC/\mu m^2$ (all done without an etch mask). Images were obtained from the SEM using a secondary electron detector, as well as performing a SIMS analysis. The etch depths and surface roughness were closely studied, and it was found that the etch rate increased linearly with ion dose, which in turn resulted in an increase in etch depth. The edge sections were found to have a roughness below 0.1 μm, a quite favorable number. All in all, it was found that although slower than traditional methods, FIB etching on GaN proved to be more advantageous in its high quality production and versatility [30].

In other experiments, it has been shown that ion bombardment on hydrogenated silicon-carbon alloy films (a-SiC:H) creates optical contrast between crystal layers (differentiating between those bombarded and those not) (figure 20). This is of great interest because having a large optical contrast in a-SiC:H will result in a good material for opto-electronic devices as well as for difficult environmental conditions. A

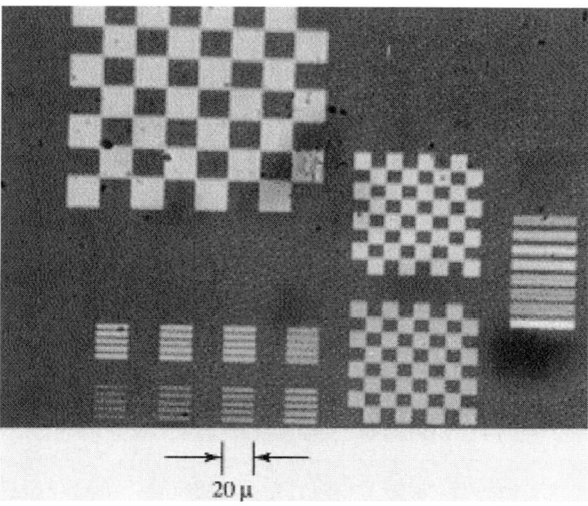

Figure 20. Optical contrast pattern written into a GD a-$Si_{0.85}C_{0.15}$:H film deposited on Corning 7059 glass substrate. The transparent and opaque regions represent unimplanted and implanted parts of the film, respectively. The patterning has been performed with the help of a program-controlled Ga^+-focused ion beam. The minimum feature size is 2.5 μm.

recent study compared Ga^+ and Sn^+ bombardment with previous experiments using As^+. Films were prepared with two methods, one being a glow discharge technique (GD), and the other reactive magnetron sputtering (SP). Samples were bombarded at 50 keV (Ga^+) and 60 keV (Sn^+), with an ion-beam intensity of 2 $\mu A/cm^2$ and doses between 10^{15} and 10^{17} ions cm^{-2}. In both cases, huge increases in optical absorption coefficients were noted as compared to previous experiments with As^+, indicating that the formation of optical contrast was much more prevalent. Although strong in both cases, this effect was even greater in the GD films as compared to the SP films [31].

Furthermore, the extent of the FIB's material altering capabilities, particularly when part of an integrated two-beam system, was demonstrated in work done by Anzalone *et al.* in creating complex 3D structures. They show that using the milling and deposition techniques discussed earlier, it is possible to model structures based on software defined inputs or bitmap files using a digital patterning generator. Computer Aided Design (CAD) files were used in conjunction with ion beam assisted CVD to create 3D helical structures (figure 21) [32].

In addition to alteration, many experiments have shown that the FIB machine is a great tool for the characterization of nanosized objects. For example, we expanded the applicability of the FIB by using it in combination with high-resolution strain mapping software. These two techniques, when used together, provide a new method for in situ measurement of the residual stresses in thin films. First, the FIB system is used to create narrow slots having precise location. These slots serve to relieve residual stress, causing the surrounding film to displace. The strain mapping software is then

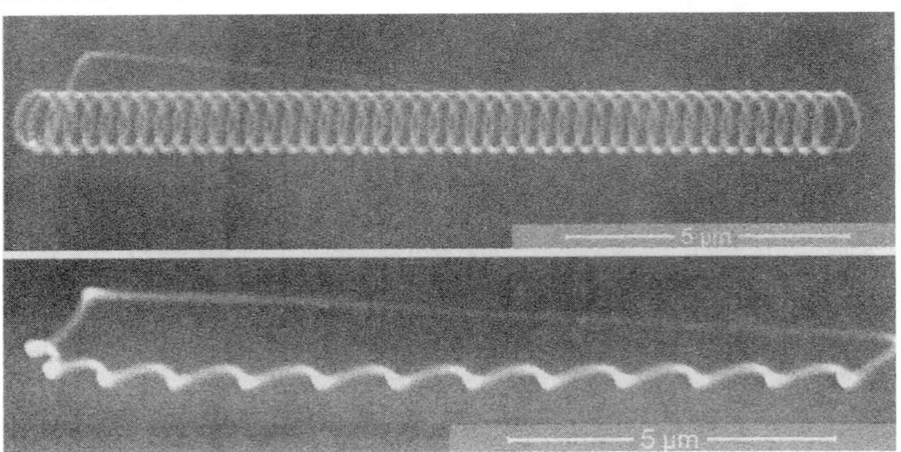

Figure 21. SEM images of an electron beam deposited Pt helical feature obtained with the use of the digital patterning generator [32].

used to measure these displacements and relate them to the residual stress (figure 22) [33]. Clearly, the FIB system is not only a useful stand-alone machine, but that it also complements other techniques, making them more efficient and of a higher quality.

In a different experiment, we studied the lubricated wear of steel couples coated with W-DLC. Among the methods used to characterize the samples, the two-beam FIB system was used to reveal the sub-surface condition of the coating and substrate. The ion beam was used to make small sections at specific locations, which were then observed using the electron beam (figure 23) [34]. Similarly, the FIB was used to characterize sub-surface damage in a study of foreign object damage (FOD) in a thermal barrier system. The conditions were set to simulate those of a turbine engine, and the FIB observations indicated damage in the thermal barrier coating (figure 24). The study demonstrated that these changes were caused by particle impact, confirming the presence of FOD [35]

These are, of course, only a few examples of the tremendous applicability of the FIB system for the alteration and characterization of materials. With all its uses, the FIB has become an invaluable tool for nanotechnology research.

3.2. TEM Sample Preparation for Imaging and Analysis

Perhaps the most studied and commonly used application of the FIB is for transmission electron microscopy (TEM), specifically in sample preparation, where the FIB provides revolutionary methods of creating electron transparent samples [36–41]. The TEM is very useful for finding information on the atomic structure and composition of solid materials. As its name suggests, the TEM collects its data from electrons that are transmitted (or diffracted) through the sample. Therefore, TEM samples must be extremely thin, with optimal samples less than 50 nm thick and an maximum near

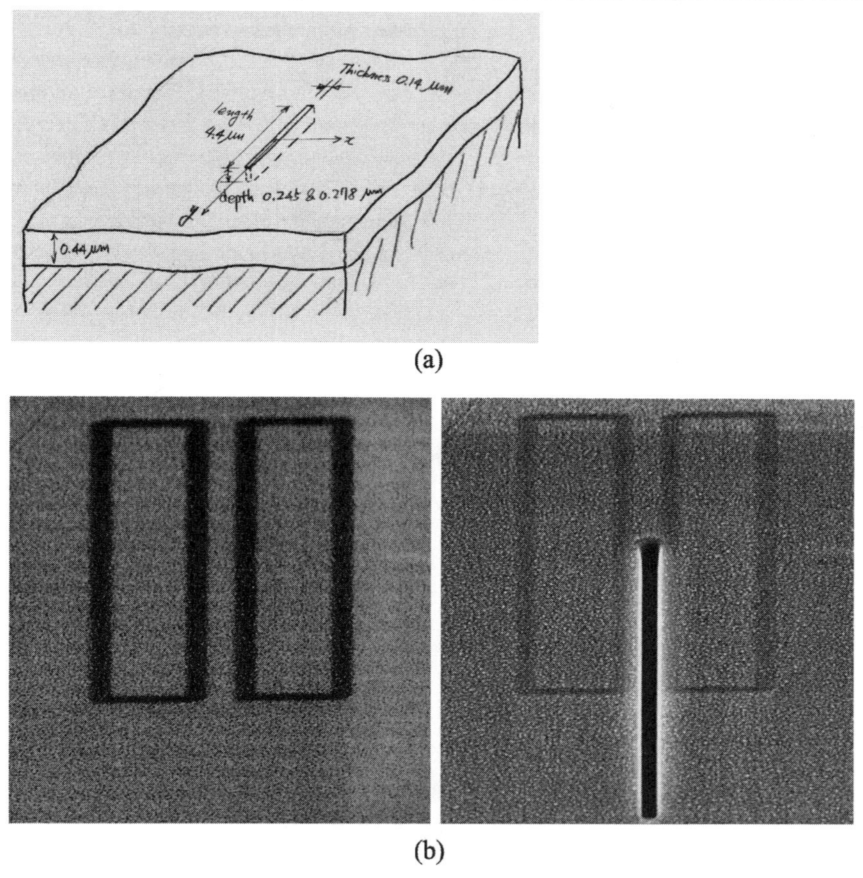

Figure 22. (a) A schematic of a FIB slot introduced into a film, defining the coordinates. (b) SE images of a region of a diamond-like carbon film before and after the introduction of the FIB slot [33].

200 nm. In addition, it is quite necessary to have a sample which has been cleanly extracted from the bulk specimen because of the fact that defect analysis (a major use of TEM) is so location specific. In the past tedious time consuming techniques for producing samples such as grinding and polishing were constantly a limiting factor in the progression of research. In recent years however, the focused ion beam machine has shown vast improvement in the time and quality of sample preparation.

The FIB holds a number of advantages over previous methods for TEM sampling. For example, since the sample can be rotated and oriented without ever having to take it out of the chamber, much time is saved while digging or etching the sample. The defective area can be found by thinning away at different areas until the defect is located, and then the milling process can begin. Also, samples can be much more precise in there overall shape and size due to the exactness of the FIB. Lastly, the FIB lets the user watch as he or she is sectioning a sample, allowing for much better

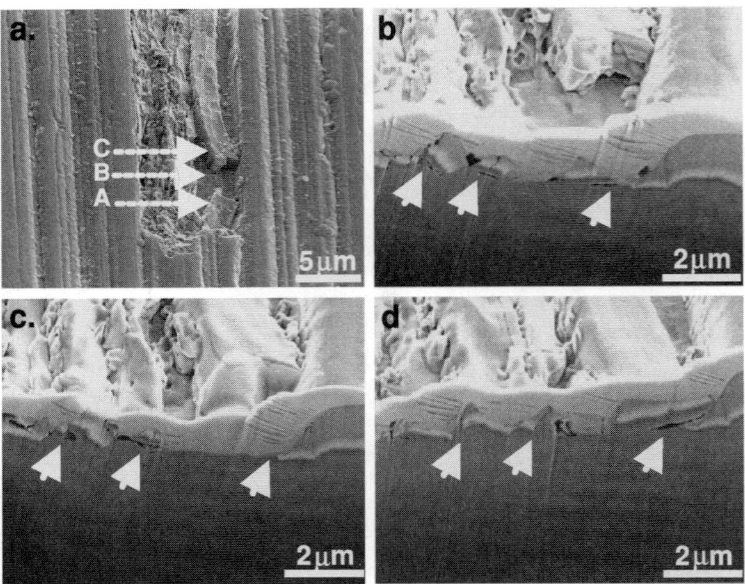

Figure 23. (a) SE image of surface morphology of damaged area. (b–d) SE images of FIB cross-section of W-DLC coating at 'A,' 'B,' and 'C,' respectively. The arrows identify regions where the retained W-DLC coating segments have either translated or rotated [34].

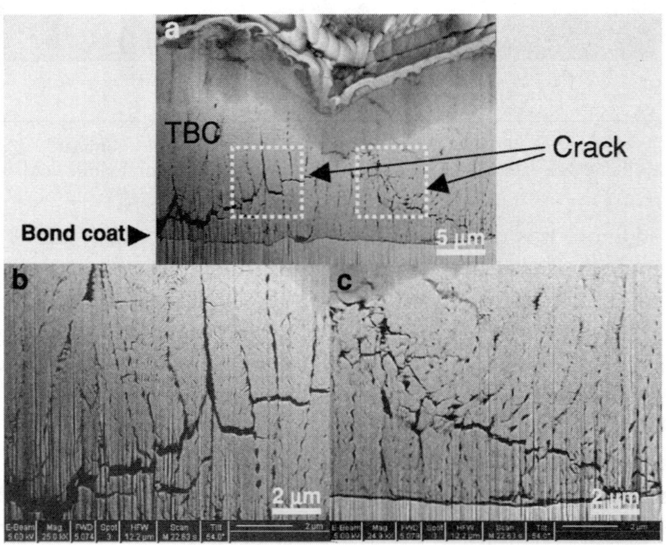

Figure 24. SE image of FIB cross-section showing the kink band beneath the impact site. The cracks, magnified in (b) and (c), extend along the kink band from the densified zone to the TGO layer [35].

decision making. This feature is greatly enhanced with a two-beam FIB system, where the SEM can be used for high-resolution imaging. In general, the FIB's capabilities of precise cutting and polishing give it a unique ability to quickly create excellent TEM samples, a process which once needed numerous machines to take care of. Cairney, *et al.* have taken advantage of this, using the FIB to create sample of "large, uniformly areas with relative ease" for their TEM work studying TiN and TiAlN thin films [37]. Similarly, Volkert *et al.* used the FIB to prepare TEM samples for their work on synthetic fluorapatite-gelatine composite particles. They report that the FIB-prepared samples make "high-quality, crack-free specimens with no apparent ion beam-induced damage" [38].

The most practical method for sample creation is the so called lift-out technique, where the focused ion beam, with its massive density, is able to perform site specific milling on bulk samples. The samples are generally coated ahead of time with some conducting metal so as to eliminate charging, and are then placed in the FIB machine for alteration. Once secured in the chamber, the region of investigation is found and a trench is dug around the area in a step-like fashion. It should be noted that the X and Y dimensions of the rectangular trenches must be at a ratio of 2:1 respectively so that one is able to image the entire trench. When the sample is thin enough, it is basically cut along the attached edges and lifted out with a glass rod, then placed in the TEM grid for analysis (figure 25). The lift-out technique was used by Wang *et al.* to prepare CMOS cross-sectional specimens starting with integrated circuit wafers. They report success in producing a "large and uniform sectional specimen in a very short time." On the other hand, they mention a disadvantage of the lift-out technique, namely that a finished specimen cannot be refabricated if it is too thick; newer techniques using the FIB have been able to overcome this problem [39, 40].

Another significant problem which has been observed to occur during thinning in the lift-out method is a warping effect in the sample. As thinning occurs (to roughly less than 200 nm), stressed samples fall into a bent shape to alleviate the pressure (figure 26). Stress can be caused by the bulk sample because of poor mounting or can be the result of the internal structure of the material. In either case, a study done at FEI Europe Ltd. found that cutting either one or both of the edges (depending of the seriousness of the warping) of the near fully thinned sample would alleviate the stress by giving it room for relaxation. This technique has proved to solve almost all cases of warping problems [41].

An alternative to the lift-out method of TEM sample preparation is called micro-pillar sampling. This technique is particularly useful in preparing specimens in situ for the scanning transmission electron microscope (STEM). The FIB micro-sampling method is used to extract a pillar shaped sample from the specimen, which is then mounted on top of a conical stage developed specifically for this method. The sample is then transferred to the STEM, where various imaging techniques are used [40].

In addition to material science, the FIB machine's success in TEM sample preparation has been used for environmental science applications. Copper run off from the roofs of many buildings is becoming a serious environmental concern, and strategies to reduce infiltration are being explored. One such idea has been to place filter systems in major

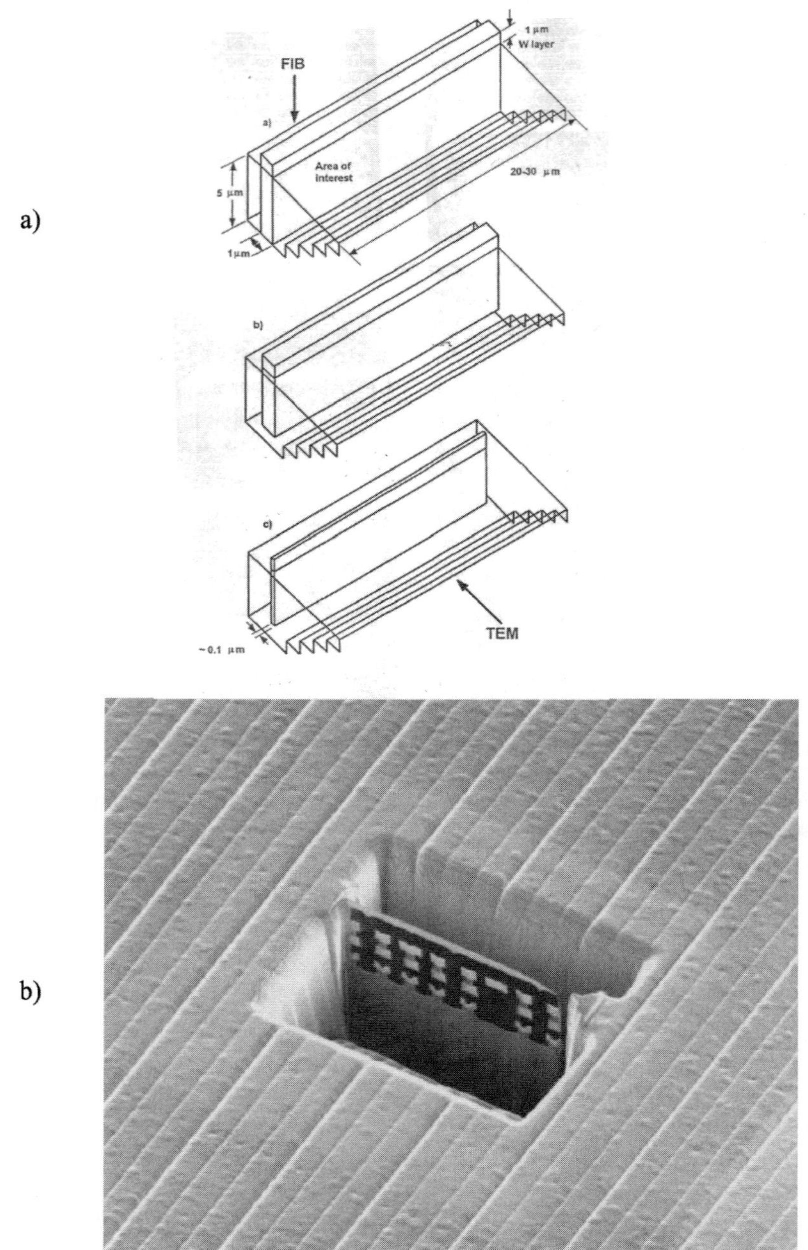

Figure 25. (a) Schematic of FIB cutting for TEM sample prep [39]. (b) SE image of sample directly before actual lift out. (*Courtesy of FEI Company*)

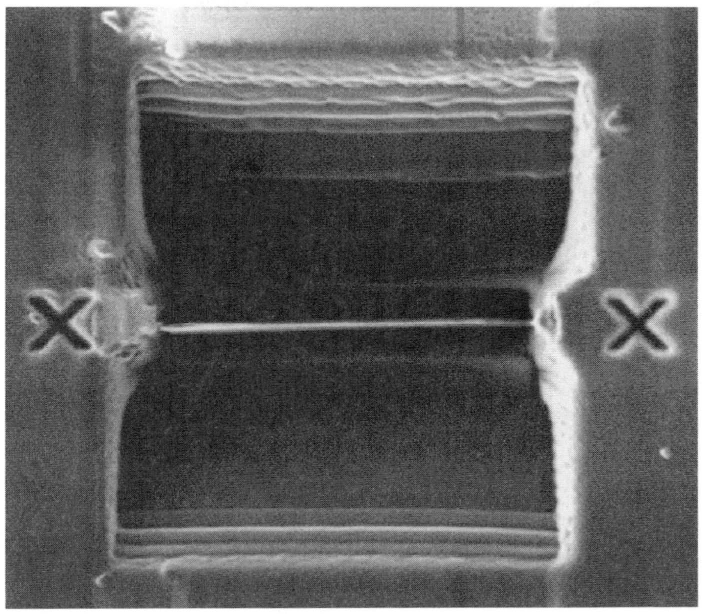

Figure 26. Image of warping effect. Cuts will be made around two endpoints to decrease warping.

runoff areas, the focus of a study done by the Swiss Federal Institute for Environmental Study and Technology. Iron hydroxide was found to be the most effective Cu filter. Testing focused on Cu adsorption on suspended specimens in the water as well in the iron hydroxide. TEM was used with iron hydroxide samples prepared using the FIB lift out technique, and it was found that indeed the iron hydroxide was effective in decreasing the copper runoff [43].

With TEM sample preparation, as with most other FIB applications, the two-beam system can provide certain advantages beyond those already discussed. Sivel *et al.* have studied two cases in which the two-beam FIB system has succeeded where other methods have failed. In the first, the two-beam was used to make samples for TEM imaging of an interface in a metal/organic coating sample (figure 27). The samples were composed of an 80 μm thick epoxy coating applied onto a 1 mm thick aluminum alloy. Sample preparation of interfaces is generally difficult because of the different mechanical properties of the two layers, but while conventional ion milling damaged the sample, preparation using the two-beam was successful. Similary, the two-beam was needed to successfully prepare a TEM sample of a grain boundary in a Cu_3Au alloy [44].

The TEM itself has a wide-range of uses, which can all now be added to the list of applications for which the FIB system is helpful. Undoubtedly, using the FIB has made improvements in both speed and quality of TEM sample production, and is a most important tool to have available.

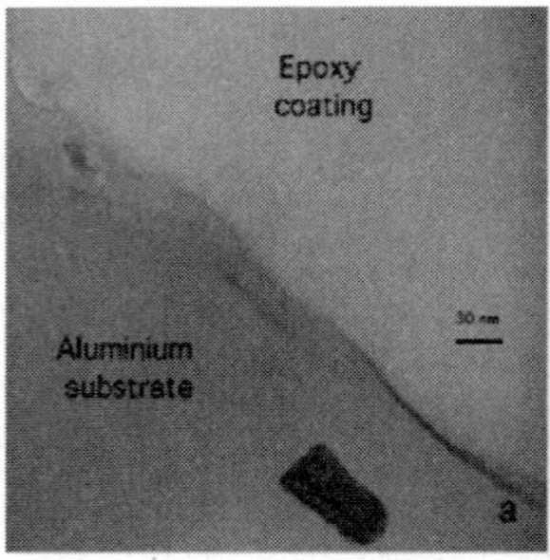

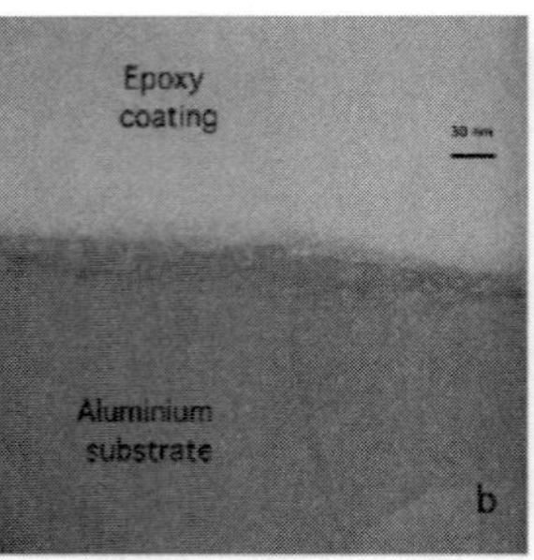

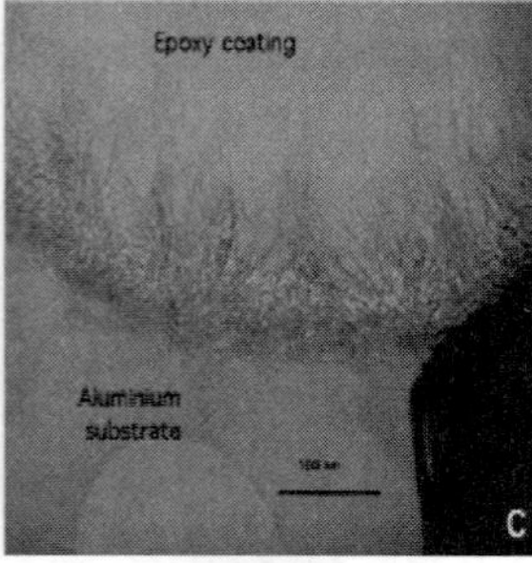

Figure 27. (a) Bright-field TEM image of the aluminum-epoxy interface of sample 1. No interface layer is visible. (b) TEM image of sample 2. A regular 30-nm-thick hydroxide layer is visible at the interface. (c) TEM image of the interface of sample 3. A 150-nm-thick pseudoboehmite layer can be seen. The epoxy coating has fully penetrated the rough pseudoboehmite layer [44].

Figure 28. High grain boundary definition in Nickel Polycrystals ion beam image [45].

3.3. Sample Imaging—Defining the Third Dimension

As discussed earlier, the FIB machine has the capability to image a surface, and the two-beam system in particular can obtain images using either an electron beam or and ion beam. Using the ion beam is advantageous for obtaining images that illustrate high grain boundary definition, as compared to the electron beam which gives crisp resolute surface images. This feature of ion beam images can be observed in work done on modeling the plasticity in LIGA nickel MEMS structures (figure 28) [45]. Though there have been marked advancements in FIB imaging such as adding oxides or carbides for enhancement, one must be aware that using the focused ion beam to image a surface will damage the surface to some extent. This is where the advantages

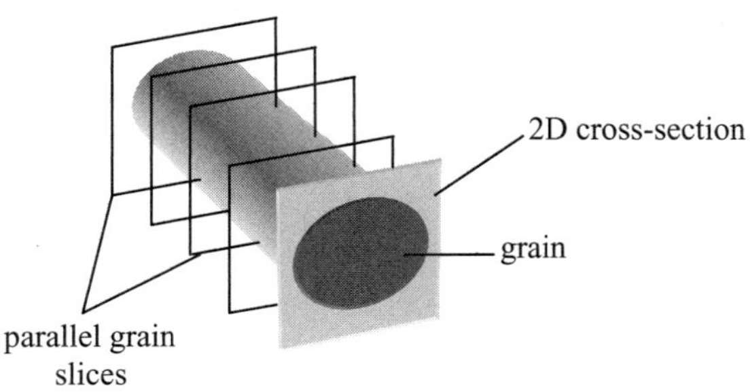

Figure 29. Schematic of cross sectioning for grain shape determination.

of the two-beam system come into play by allowing the user to switch back and forth between etching with the FIB and imaging with the SEM [7, 18, 46].

What is so unique about the FIB machine is its ability to uncover the third dimension of a material for imaging. Revealing the internal structure is of the utmost importance to many researchers because it yields valuable information regarding the properties of the material and predictable behavior under certain conditions. In particular, studying the grain structure or grain boundary of a material can provide good insight. In this respect, the two-beam FIB machine can be of great aid. Three-dimensional images are put together using a group of two dimensional images, obtained by shearing off sections of a sample using the FIB, then imaging with the SEM (figure 29). Interpolation of the graphs gives a three-dimensional representation, and important data about the grain structure and other features can then be attained. Internal elemental distribution can also be obtained using SIMS [18, 46]. Dunn and Hull recently demonstrated this ability of the FIB to create three-dimensional volume reconstructions. Their method used FIB serial sectioning and linear interpolation, and was able to produce well defined images of the sample's 3d structure (figure 30) [47].

Cross-sectioning with data analysis using the FIB machine has also been shown to be useful in investigating the effect of additive exposure. A study done at the University of Tokyo used the FIB machine to create cross sections in non-woven fibers with additives for spatial and distribution analysis. Researchers took non-woven polyester fiber samples and added Chimassorb 944, an additive known to improve the functionality of the fiber. Elemental distribution was of particular interest, and a new method of gathering this data was performed using FIB/SIMS technology. Samples of altered fiber were cross sectioned throughout the interior and surface in 3 different manners: perpendicular to the fibers, at a 45° angle, and parallel to the length of the fibers. SIMS mapping was then performed on the cross sections yielding the desired chemical distribution. Signals for C^-, O_2^-, AlO^-, Na^+, K^+, Ca^+, CaO^+, and C^+ ions were detected, analyzed and mapped for three-dimensional spatial analysis. Voids were found in the material where the additive seemed to concentrate, but most importantly a method of three-dimensional analysis using FIB/SIMS/SEM technology was proven successful (figure 31) [48].

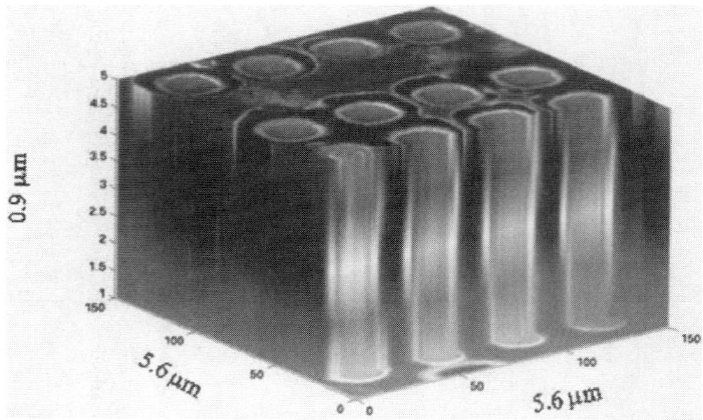

Figure 30. An array of vias reconstructed from a series of secondary electron images collected as a function of depth [47]

Figure 31. Focused Ion beam-induced secondary electron images. The cross-section is approximately perpendicular (a), parallel (b) and 45°. (c) to the longitudinal of the fiber, respectively, as shown in the diagram [48].

As usual, these are just a minute sample of the ways the FIB has been applied to 3D imaging purposes. Nonetheless, they demonstrate the great success and practicality of the FIB system, characteristics that make it an ideal tool.

3.4. Damage to the Sample Induced by the FIB

Having discussed many of the advantages of using the FIB system, this review would not be complete without a warning of the possibility of causing damage to the sample. Typically, the focused ion beam has several detrimental effects which result from its use. Firstly, bombardment with Ga^+ ions will likely result in some level of gallium

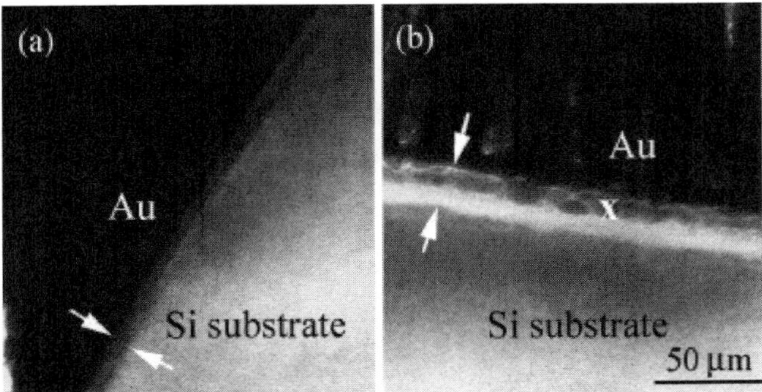

Figure 32. The structure of (a) the side-wall and (b) bottom-wall of the trench milled using a 10 keV beam energy and a 250 pA ion beam current [49].

implantation into the surface layers of the sample. Additionally, this bombardment can cause the formation of an amorphous layer, as atoms are knocked out of place and vacancies are created. Finally, some heterogeneity in thickness can occur by means of preferential sputtering or redeposition. It is very important that we gain an understanding of the nature and cause of this sample damage, so that we are both able to recognize it and develop techniques to minimize its occurrence.

Many studies of FIB-induced damage have been undertaken. For example, Rubanov *et al.* looked at cross-sections of trenches milled using the FIB to observe the types and range of damage incurred. They discovered that side-wall damage layers were amorphous, a direct result of using the gallium beam. As for the bottom-wall damage, the group found that there was a layer of material rich in gallium, and that local recrystalization had occurred (figure 32). They experimented with different beam energies, and noted that a reduction from 30 keV to 10 keV cut the thickness of the damage layer in half. In more complex milling patterns, redeposition often occurred [49].

Of course, these limitations of the FIB should not discourage the use of the FIB machine for research. It is the very ability to cause damage to the sample that gives the FIB its remarkable range of capabilities and applications. In addition, several methods to reduce the damaging effect are being developed, each of which has both benefits and weaknesses. Sutton *et al.* looked at the affect of the sample tilt, and noted that the depth of damage could be decreased by tilting the specimen by angles between 4 and 8 degrees at the end of the thinning process. The downside of this technique is its inability to create parallel sidewalls for chemical analysis [50]. Alternatively, using the GAE techniques discussed earlier has been shown to not only increase the etching rate, but also reduce the rate of re-deposition and gallium implantation [51]. The optimal method depends on the needs of the specific user and experiment, but in almost all cases, the FIB is considered as a versatile tool that combines both nanofabrication and microscopy capabilities for nanotechnology research.

ACKNOWLEDGEMENTS

The author is grateful to his students: Nathan Etessami, Gabriel Mas, Fei Wang, who have partially contributed in preparation of the manuscript; to the National Science Foundation-MRSEC program and New Jersey Commission of Science and Technology for their support.

REFERENCES

1. Krohn, V. E. Progr. Astronautics R (1961) 73.
2. Krohn, V. E. and Ringo, G. R., "Ion source of high brightness using liquid metal", *Appl. Phys. Lett.* **27** (1975) 479.
3. Orloff, J., Utlaut, M., and Swanson, L., *High Resolution Focused Ion Beams: FIB and Its Applications,* Kluwer Academic/Plenum Publishers, New York, NY (2003).
4. Phaneuf, M.W., Applications of focused ion beam microscopy to materials science specimens, *Micron,* **30** (1999) 277.
5. Van Doorselaer, K., Van den Reeck, M., Van den Bempt, L., Young, R., and Whitney, J., How to Prepare Golden Devices Using Lesser Materials in, *Proc. 19th International Symposium for Testing and Failure Analysis* (1993).
6. Russell, P. E., Stark, T. J., Griffis, D. P., and Gonzales, J. C., Chemically Assisted Focused Ion Beam Micromachining: Overview, Recent Developments and Current Needs, *Microsc. Microanal.,* **7S2** (2001) 928.
7. Phaneuf, M. W. and Li, J., FIB Techniques for Analysis of Metallurgical Specimens, *Microsc. Microanal.,* **6S2** (2000) 524.
8. Gerlach, R. and Utlaut, M., Focused ion beam methods of nanofabrication: room at the bottom, *Proc. SPIE Int. Soc. Opt. Eng.,* **4510** (2001) 96.
9. Mitsuishi, K., Shimojo, M., Han, M., and Furuya, K., Electron-beam-induced deposition using a subnanometer-sized probe of high-energy electrons, *Appl. Phys. Lett.,* **83** (2003) 2064.
10. Khizroev, S., Bain, J. A., and Litvinov, D., Fabrication of nanomagnetic probes via focused ion beam etching and deposition, *Nanotechnology,* **13** (2002) 619.
11. Allameh, S. M., Yao, N., and Soboyejo, W. O., Synthesis of self-assembled nanoscale structures by focused ion-beam induced deposition, *Scripta Mater.,* **50** (2004) 915.
12. Strobel, M., Heinig, K.-H., and Möller, W., Can core/shell nanocrystals be formed by sequential ion implantation? Predictions from kinetic lattice Monte Carlo simulations, *Nucl. Instr. Meth. B,* **148** (1999) 104.
13. Young, R. J., Application of the focused ion beam in materials characterization and failure analysis, *Microstuctural Science,* **25** (1997) 20.
14. Bender, H., Voltage contrast in the focused ion beam, *Scope,* **13** (2001).
15. Shih, W. C., Ghiti, A., Low, K. S., Greer, A. L., O'Neill, A. G., and Walker, J. F., Analysis of Geometrical and Microstructural Effects on Void Formation in Metallization: Observation and Modeling, *Mater. Res. Soc. Symp. Proc.,* **428** (1996) 243.
16. Crow, G. A., Christman, L., and Utlaut, M., A focused ion beam secondary ion mass spectroscopy system, *J. Vac. Sci. Technol. B,* **13** (1995) 2607.
17. Dong, L. F., Jiao, J., Tuggle, D. W., and Foxley, S., Synthesis and Characterization of Carbon Nanotubes on Porous Silicon Substrates, *Microsc. Microanal.,* **7S2** (2001) 398.
18. Hull, R., Dunn, D., and Kubis, A., Nanoscale Tomographic Imaging Using Focused Ion Beam Sputtering, Secondary Electron Imaging and Secondary Ion Mass Spectrometry, *Microsc. Microanal.,* **7S2** (2001) 34.
19. Gnauck, P., Zeile, U., Rau, W., and Schumann, M., Real time SEM imaging of FIB milling processes for extended accuracy in Cross Sectioning and TEM Preparation, *Microsc. Microanal.,* **9S3** (2003) 524.
20. Zimmermann, G. and Chapman, R., In-Situ Dual Beam (FIBSEM) Techniques for Probe Pad Deposition and Dielectric Integrity Inspection on 0.2 mm Technology DRAM Single Cells in, *Proc. 25th International Symposium for Testing and Failure Analysis* (1999).

21. Lipp, S., Frey, L., Lehrer, C., Frank, B., Demm, E., Pauthner, S., and Ryssel, H., Tetramethoxysilane as a precursor for focused ion beam and electron beam assisted insulator (SiO_x) deposition, *J. Vac. Sci. Technol. B*, **14** (1996) 3920.
22. Rice, L., Semiconductor Failure Analysis Using EBIC and XFIB, *Microsc. Microanal*, **7S2** (2001) 514.
23. Huey, B. D. and Langford, R. M., Low-dose focused ion beam nanofabrication and characterization by atomic force microscopy, *Nanotechnology*, **14** (2003) 409.
24. Walker, J. F., TEM sample preparation: site specific search strategy, FEI Europe Ltd.
25. Haythornthwaite, R., Nxumalo, J., and Phaneuf, MW., Use of the focused ion beam to locate failure sites within electrically erasable read only memory microcircuits, *J. Vac. Sci. Technol. A*, **22** (2004) 902.
26. Li, H. W., Kang, D. J., Blamire, M. G., and Huck, W. T. S., Focused ion beam fabrication of silicon print masters, *Nanotechnology*, **14** (2003) 220.
27. Liu, Y., Longo, D. M., and Hull, R., Ultrarapid nanostructuring of poly(methylmethacrylate) films using Ga^+, *Appl. Phys. Lett.*, **82** (2003) 346.
28. Arshak, K., Mihov, M., Arshak, A., McDonagh, D., and Sutton, D., Novel dry-developed focused ion beam lithography scheme for nanostructure, *Microelectron. Eng.*, **73–73** (2004) 144.
29. Harris, G. and Zhou, P., The Growth and Characterization Processes of Gallium Nitride (GaN) Nanowires, *National Nanofabrication User Network*, **2001** (2001) 34.
30. Flierl, C., White, I. H., Kuball, M., Heard, P. J., Allen, G. C., Marinelli, C., Rorison, J. M., Penty, R. V., Chen, Y., and Wang, S. Y., Focused ion beam etching of GaN, *MRS Internet J. N. S. R.*, **4S1** (1999) G6. 75.
31. Bischoff, L., Teichert, J., Kitova, S., and Tsvetkova, T., Optical pattern formation in a-SiC : H films by Ga+ ion implantation, *Vacuum*, **69** (2002) 73.
32. Anzalone, P. A., Mansfield, J. F., and Giannuzzi, L. A., DualBeam Milling and Deposition of Complex Structures Using Bitmap Files and Digital Patterning, *Microsc. Microanal.*, **10S2**(2004) 1154.
33. Kang, K. J., Yao, N., He, M. Y., and Evans, A. G., A Method for *in-situ* Measurement of the Residual Stress in Thin Films by Using the Focused Ion Beam, *Thin Solid Films*, **443** (2003) 71.
34. Yao, N., Evans, A. G., and Cooper, C. V., Wear mechanisms operating in diamond like carbon coatings in contact with machined steel surfaces, *Surface and Coatings Technology*, **179** (2004) 306.
35. Chen, X., Wang, R., Yao, N., Evans, A. G., Hutchinson, J. W., and Bruce, R. W., Foreign Object Damage in a Thermal Barrier System: Mechanisms and Simulations, *Materials Science and Engineering A*, **352** (2003) 221.
36. Giannuzzi, L. A., Drown, J. L, Brown, S. R., Irwin, R. B., and Stevie, F. A., Focused ion beam milling and micromanipulation lift-out for site specific cross-section TEM specimen preparation, *Mat. Res. Soc. Symp. Proc.*, **480** (1997) 19.
37. Cairney, J. M., Harris, S. G., Munroe, P. R., and Doyle, E. D., Transmission electron microscopy of TiN and TiAlN thin films using specimens prepared by focused ion beam milling, *Surf. Coatings Technol.*, **183** (2004) 239.
38. Volkert, C. A., Busch, S., Heiland, B., and Dehm, G., Transmission electron microscopy of fluorapatite-gelatine composite particles prepared using focused ion beam milling, *J. Microsc.-Oxford*, **214** (2004) 208.
39. Rossie, B. B., Shofner, T. L., Brown, S. R., Anderson, S. D., Jamison, M. M., and Stevie, F. A., A Method for Thinning FIB Prepared TEM Specimens After Lift-Out, *Microsc. Microanal.*, **7S2** (2001) 940.
40. Yaguchi, T., Konno, M., Kamino, T., Hashimoto, T., Onishi, T., and Umemura, K., FIB Micro-pillar Sampling Technique for 3D Stem Observatin and its Application, *Microsc. Microanal.* 9S2 (2003) 118.
41. Wang, Z. G., Kato, N., Sasaki, K., Hirayama, T., and Saka, H., Electron holographic mapping of two-dimensional doping areas in cross-sectional device specimens prepared by the lift-out technique based on a focused ion beam, *J. Electron Microsc.*, **53** (2004) 115.
42. Walker, J. F., Preparing TEM Sections by FIB: Stress Relief to Straighten Warping Membranes, *Inst. Phys. Conf. Ser.*, **157** (1997) 469.
43. Mavrocordatos, D., Steiner, M., and Boller, M., Analytical electron microscopy and focused ion beam: complementary tool for the imaging of copper sorption onto iron oxide aggregates, *J. Microsc.-Oxford*, **210** (2003) 45.
44. Sivel, V. G. M., Van den Brand, J., Wang, W. R., Mohdadi, H., Tichelaar, F. D., Alkemade, P. F. A., and Zandbergen, H. W., Application of the two-beam FIB/SEM to metals research, *J. Mircrosc.-Oxford*, **214** (2004) 237.
45. Lou, J., Shrotriya, P., Allameh, S., Yao, N., Buchheit, T., and Soboyejo, W. O., Plasticity Length Scale in LIGA Nickel MEMS Structure, *Mat. Res. Soc. Symp. Proc.*, **687** (2002) 41.

46. Inkson, B. J. and Möbus, G., 3D determination of grain shape in FeAl by focused ion beam (FIB) tomography, *Microsc. Microanal.*, **7S2** (2001) 936.
47. Dunn, D. N., and Hull., R., Three-dimensional volume Reconstructions Using Focused Ion Beam Serial Sectioning, *Microsc. Today*, **12(4)** (2004) 52.
48. Takanashi, K., Shibata, K., Sakamoto, T., Owari, M., and Nihei, Y., Analysis of non-woven fabric fibre using an ion and electron multibeam microanalyser, *Surf. Interface Anal.*, **35** (2003) 437.
49. Rubanov, S. and Munroe, P. R., FIB-induced damage in silicon, *J. Microsc.-Oxford*, **214** (2004) 213.
50. Sutton, D., Parle, S. M., and Newcomb, S. B., Focused ion beam damage: its characterisation and minimization, *Inst. Phys. Conf. Ser.*, **168** (2001) 377.
51. Russell, P. E., Stark, T. J., Griffis, D. P., Phillips, J. R., and Jarausch, K. F., Chemically and geometrically enhanced focused ion beam micromachining, *J. Vac. Sci. Technol. B*, **16** (1998) 2494.

10. ELECTRON BEAM LITHOGRAPHY

ZHIPING (JAMES) ZHOU

1. ELECTRON BEAM LITHOGRAPHY AND NANOTECHNOLOGY

The field of nanotechnology covers nanoscale science, engineering, and technology that create functional materials, devices, and systems with novel properties and functions that are achieved through the control of matter, atom by atom, molecule by molecule or at the macromolecular level. The domain of nanoscale structures, typically less than 100 nm in size, lies dimensionally between that of ordinary, macroscopic or mesoscale products and microdevices on the one hand, and single atoms or molecules on the other. There are two approaches to making building block artifacts such as quantum dots, nanotubes and nanofibers, ultrathin films and nanocrystals, nanodevices: bottom-up synthesis and top-down miniaturization.

The bottom-up approach ingeniously controls the building of nanoscale structures. This approach shapes the vital functional structures by building atom by atom and molecule by molecule. Researchers are working to find the mechanism of "self-assembly". "Self-assembly" involves the most basic ingredients of a human body self-reproducing the most basic structures. "Self-assembly" covers the creation of the functional unit by building things using atoms and molecules, growing crystals and creating nanotubes.

"Top down" is an approach that downsizes things from large-scale structures into small-scale structures. For example, vacuum tubes yielded to transistors before giving way to ICs (integrated circuits) and eventually LSIs (large scale integrated circuits). This method of creating things by downsizing from centimeter size to micrometer size is called "microelectronics".

It is a well-known fact that microelectronics has advanced at exponential rates during the past four decades. Due to its rich functionality in applications, low energy consumption in operations, and low cost in fabrication, microelectronics has entered into almost all aspects of our lives through the invention of novel small electronic devices. The most important advancement is the extension of microelectronics and its fabrication methodology into many non-electronic areas such as micro-actuators, micro-jet, micro-sensors, and micro DNA probes.

As this technology continues to advance, it has been extended from micrometer to nanometer scale, hence the existence of "nanotechnology" or "nanofabrication". Using nanotechnology, the narrowest line pattern on massive produced semiconductor devices is now approaching the 50-nanometer level. In research labs, horizontal dimensions of the device feature sizes have been further scaled down from 130 nanometers to 6 nanometers [1] and its vertical dimensions have been reduced to less than 1.5 nanometers or a couple of atoms [2]. These nanoscale devices known as "nanodevices" are obtained through the top-down miniaturization approach.

The heart of the top-down approach of miniaturization processing is the nanolithography technique, such as Electron Beam Lithography (EBL), Nanoimprint Lithography (NIL), X-ray Lithography (XRL), and Extreme Ultraviolet Lithography (EUVL). Among the four techniques of nanolithography, the EBL approach is the front-runner in the quest for ultimate nanostructure due to its ability to precisely focus and control electron beams onto various substrates. It has been demonstrated that electron beams can be focused down to less than 1 nm. This will extend the resolution of EBL to the sub-nanometer region provided that appropriate resistant material is available.

Electron Beam Lithography is a method of fabricating sub-micron and nanoscale features by exposing electrically sensitive surfaces to an electron beam. It utilizes the fact that certain chemicals change their properties when irradiated with electrons just as photographic film changes its properties when irradiated with light. With computer control of the position of the electron beam it is possible to write arbitrary structures onto a surface, thereby allowing the original digital image to be transferred directly to the substrate of interest. EBL followed soon after the development of the scanning electron microscope [3]. Almost from the very beginning, sub-100 nm resolution was reported. As early as 1964, Broers [4] reported 50 nm lines ion milled into metal films using a contamination resist patterned with a 10 nm wide electron beam. Later in 1976, with improved electron optics, 8 nm lines in Au-Pd were reported using a 0.5 nm probe [5]. In 1984, a functioning Aharonov-Bohm interference device was fabricated with EBL [6]. Muray *et al.* [7] reported 1 to 2 nm features in metal halide resists. Until recently, EBL was used almost exclusively for fabricating research and prototype nanoelectronic devices. Currently, its precision and nanolithographic capabilities make it the tool of choice for making masks for other advanced lithography methods.

In EBL nanofabrication, working conditions at which electron scattering causes minimal resist exposure is required. To achieve this goal, either very high energy or very low energy [8] electrons are used. In high-energy case, the beam broadening in the resist through elastic scattering is minimal [5] and the beam penetrates deeply into

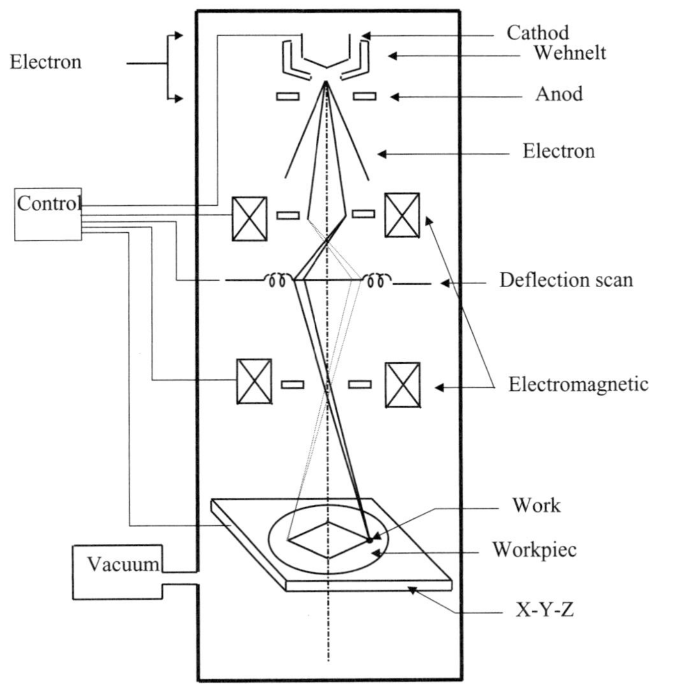

Figure 1. Electron beam lithography instrument. Electrons are generated and accelerated by the electron gun, and guided through the column by the electromagnetic lenses and the deflection scan coil. Both the scanning system and the X-Y-Z stage are used to define the working point on the workpiece.

the substrate. Low energy electron approaches are effective because the electrons have too low an energy to scatter over large distances in the resist.

To implement electronic beam nanolithography into a manufacturing process, speed and precision are required as well as control and yield in the nanofabrication processes. At nanoscale, the fundamental limits of e-beam resist interactions are also important issues, which concern electron scattering and the sensitivity of particular classes of resists to low-voltage in elastically scattered electrons. The issues of throughput, precision, and yield are relevant to instrument design, resist speed, and process control.

2. INSTRUMENTATION OF ELECTRON BEAM LITHOGRAPHY

2.1. Principle

An EBL instrument is a result of working a scanning electron microscope (SEM) in reverse, that is, using it for writing instead of reading. Its view field and throughput are, therefore, limited by the nature of this working principle. Similar as in the SEM, an EBL instrument consists of three essential parts: an electron gun, a vacuum system, and a control system. Figure 1 shows the diagram of an EBL instrument.

An electron gun is a device that generates, accelerates, focuses, and projects a beam of electrons onto a substrate. Electrons are first produced by cathodes or electron emitters. They are then accelerated by electrostatic fields to obtain higher kinetic energy and shaped into an energetic beam. Finally, the guidance system, consisting of the electric and magnetic focusing lenses and deflecting system, transmits the beam to a work point on the substrate.

The electron beam can only be properly generated and unrestrictedly propagated to the substrate in high vacuums. Depending on the material used for the electron gun and the application of the electron beam processing, the vacuum level requirement can usually range from 10^{-3} to 10^{-8} mm Hg. Therefore, the vacuum system, which creates a vacuum environment in the electron gun column and the working chamber, is considered one of the most important parts in the electron beam processing instrument.

The control system provides the manipulation capability for the electron beam generation, propagation, and timing. It also provides control over substrate translation and other functions. The coordination between translating substrates and blinking the electron beam on and off makes it possible to transfer the AutoCAD design onto a thin layer of electron beam resist.

There are two ways to generate actual patterns using an EBL instrument: raster-scanning and vector-scanning. A raster-scanning system patterns a substrate by scanning the exposing beam in one direction at a fixed rate while the substrate is moved under the beam by a controlled stage. In order to compose a designed pattern the electron beam is blanked on and off thousands of times during each scan. It is much like the raster scanning of a television. The vector-scanning scheme attempts to improve throughput by deflecting the exposure beam only to those regions of the substrate that require exposure. In this way, significant time can be saved since the beam skips over the areas that have no pattern.

2.2. Electron Optics

The theory of electron beam lithography can be understood through the electron motion in electric and magnetic fields and the basic Electron Optical Elements. Generally speaking, the electron motion in electric and magnetic fields can be described by Maxwell's equations. However, it is very difficult to solve the practical design problem of an electron beam system by simply applying the boundary conditions to the Maxwell's equations. Therefore, only the basic electron dynamics will be given in this section.

2.2.1. Basic Electron Dynamics

Assuming the velocity of the electrons during the processing is very small compared to the speed of light, assuming the applied electric and magnetic fields are static or varying slowly so that they can be treated as constants, and assuming electrode shapes, potentials, and magnetic field configuration are known, the general equation of motion

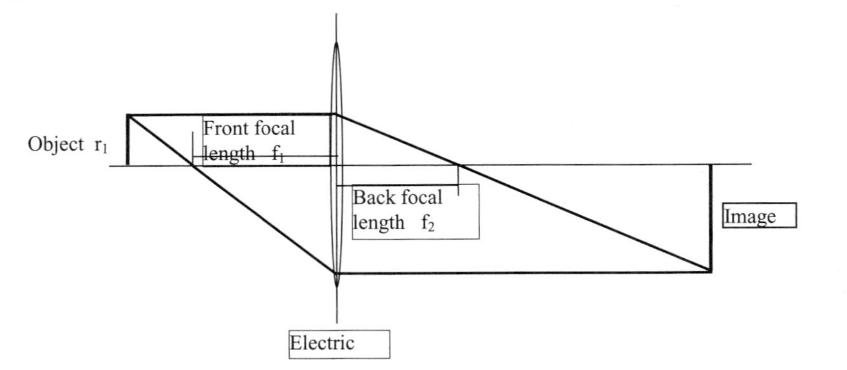

Figure 2. Ray diagram of electric lens. It is used for deriving the focal lengths of a thin lens.

for an electron in electric and magnetic fields can be written as:

$$\frac{d^2\mathbf{r}}{dt^2} = \frac{q}{m}(\mathbf{E} + \mathbf{v} \times \mathbf{B}) \tag{1}$$

where q is the charge of the electron, m the mass of the electrons, and $\mathbf{r}$ a position vector locating the electron with respect to any origin. **E** and **B** denote electric and magnetic field, respectively. **v** is the velocity of the electron moving in the fields.

ELECTRIC LENS Considering an axially symmetric field system of the electron beam generating column, the electron beam passes through a common point near the axis can be made to pass through another common point by a relatively limited region of the field variation. In analogy to the light optics, it is appropriate to call the first common point the object, the second the image, and the region of the fields the electric lens. Properties and parameters of the electric lens can be derived from the following paraxial ray equation:

$$\frac{d^2r}{dz^2} + \frac{dr}{dz}\left(\frac{V_0'}{2V_0}\right) + \frac{r}{4}\frac{V_0''}{V_0} = 0 \tag{2}$$

where V_0 is the potential on the axis. For examples, since the derivatives V_0' and V_0'' are normalized with respect to V_0, it is understandable that the field distribution rather than the intensity of the potential will determine the electron trajectories. The equation is unchanged in form even if a scale factor is applied to the location r. This indicates that all trajectories parallel to the axis will have the same focus regardless of their initial radius. It should be noticed that the electron charge q and electron mass m are absent from the ray equation. This implies that the equation is also applicable to other particles such as ions. Using the ray diagram in Figure 2, two focal lengths of a thin electric lens can be obtained from equation (2) as

$$\frac{1}{f_i} = (-1)^i \frac{1}{4\sqrt{V_i}} \int_1^2 \frac{V_0''}{\sqrt{V_0}} dz \quad i = 1, 2 \tag{3}$$

and the relationship between two focal lengths is similar to that in optics

$$\frac{f_2}{f_1} = -\frac{\sqrt{V_2}}{\sqrt{V_1}}, \tag{4}$$

where $\sqrt{V_i}$ is equivalent to the optical index of refraction, N_i.

Magnetic lens As in the electric case, an axially symmetric magnetic field also has lens characteristics and is called a magnetic lens. The paraxial ray equation for a magnetic lens is written as:

$$\frac{d^2r}{dz^2} = \frac{\frac{q}{m} r B_0^2}{8V} \tag{5}$$

where B_0 is the magnetic field on the axis. Clearly, the magnetic lens effect will depend on the charge and mass of the electrons involved. The magnetic lens is symmetrical because the equation is unchanged if B_0 is reversed in sign. The spatial invariance of the magnetic lens ensures that electronic imaging can be performed without distortion near the axis. Similar to the case of the electric lens, the focal lengths are given as

$$\frac{1}{f_2} = -\frac{\frac{q}{m}}{8V} \int_1^2 B_0^2 \, dz \tag{6}$$

and

$$f_1 = -f_2. \tag{7}$$

The symmetry in equation (5) has been applied to obtain the equation (7). Since electrons have a negative charge of q, the back focal length of f_2 is always positive. Therefore, the magnetic lens is always convex.

Bipole element Electron beam deflection could be achieved by using electrostatic and magnetic bipole elements. An electrostatic field will bend the passing electron beam towards the positive pole, while a magnetic field deflects the beam to the direction perpendicularly to the direction of the field. A pure magnetic field will change the direction of the electron's motion, but not the speed. The relationship between the magnetic field and the curvature of the electron path can be written as

$$R = \frac{mv}{qB\sin\theta} \tag{8}$$

where R is the instantaneous center of curvature and θ is the angle between the magnetic field and the velocity vectors.

Equations (1)–(8) describe the focusing imaging, and deflection of the electron beam and provide the basic electron dynamics needed in simple electron beam system design.

2.2.2. Electron Optical Elements

Compared with other lithography instrument, the use of electron guns is the core characteristic of the electron beam lithography technique. The principles of electron guns can be understood through the way of optical pass to rays of light. The electron optical elements are simply their optical counterparts. Therefore, this section will be mainly focused on the electron guns, including source generation, beam shaping, and the beam guiding system.

Based on the physical laws of electron emission and the desired energy conversion at the work point, almost all guns are of similar design, although they might differ widely with respect to beam power, acceleration voltage, and electron current. In the gun, free electrons are first generated from emitters, or cathodes, and are then shaped into a well-defined beam, which is ultimately projected onto the work point. The common concerns of the source generation and beam shaping systems are described below.

Source generation

Emission There are two kinds of electron emission. The first kind, called thermionic emission, happens when the emissive materials are heated up to a high enough temperature. The second type is field emission, in which electrons are produced due to an intense applied electric field. Since thermionic emission has higher efficiency in producing electrons at lower cost, it is widely used in industry and will be our primary concern here.

According to quantum dynamics, electrons are at rest in the ground state at 0°K and their energy levels and bands are well defined. As the temperature of the material increases, some electrons obtain more energy and jump to higher energy levels. Therefore the width of the energy bands increases. When the temperature is high enough, the electrons obtain sufficient energy to overcome the natural barrier, the work function that prevents them from escaping. In particular, as the temperature increases, the upper limit on the conduction band of metals smears and stretches out. Some of the conduction electrons obtain enough energy to overcome the potential barrier at the surface of the metal. These electrons may then be drawn off by the application of a suitable field. If the field is sufficiently high to draw all the available electrons from a cathode of work function Φ, the saturation current density J obtained at temperature T

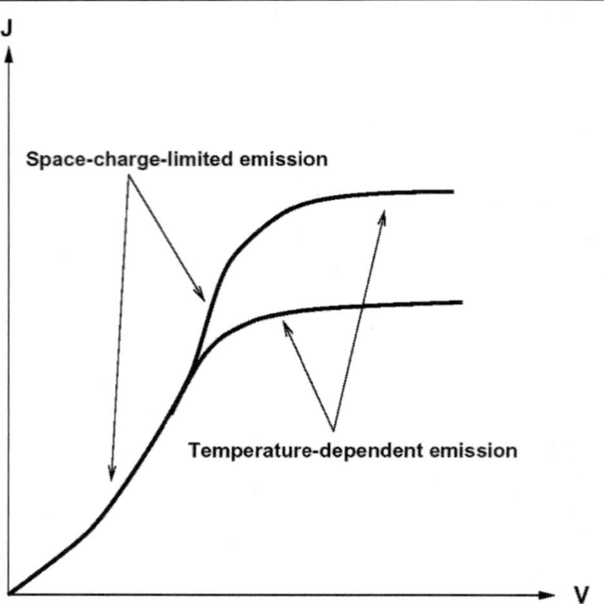

Figure 3. Space-charge-limited emission and temperature-dependent emission. Most cathodes in electron guns operate in the transition range between the space-charge and saturation regimes so that the desired emission current density can be obtained at the lowest cathode temperature.

is given by the well-known Richardson-Dushman Law

$$J = AT^2 \exp\left(-\frac{q\Phi}{kT}\right) \tag{9}$$

where A is a constant determined by the material and k is the Boltzmann constant.

In practical electron gun design, less than the saturation current is usually drawn from the gun. In this situation, the field is not strong enough to draw off all the available free electrons from the cathode. Therefore, the residual electrons are accumulated near the surface of the cathode, forming an electron cloud layer. Such operation, termed space-charge-limited emission, has the advantage that a smaller virtual cathode is formed slightly in front of the cathode that has a stable charge density, essentially independent of cathode temperature. The current that flows between parallel electrodes is given by Child's law [9]

$$J = \frac{4\sqrt{2}\varepsilon_0}{9}\sqrt{e/m}\frac{V^{3/2}}{L^2} = 0.0233\frac{V^{3/2}}{L^2} \tag{10}$$

where V is the acceleration voltage and L is the distance between the cathode and anode.

Figure 3 shows the current density in the range of space-charge-limited emission and temperature-dependent emission. Most cathodes in electron guns operate in the transition range between the space-charge and saturation regimes so that the desired emission current density can be obtained at the lowest cathode temperature.

Equations (9) and (10) give the conditions for obtaining the required emission from a given cathode material. As long as the material is specified, the preliminary cathode design can be completed.

Materials Free electrons may be obtained from the cathodes of many kinds of materials. However, the primary gun design requirements are that the cathode has a low work function and good thermal efficiency, supplies an adequate emission current, and is simple to construct. Among all the constraints, the vacuum condition of the electron gun puts strong limits on the choice of cathode materials.

At low vacuum level (less than 1×10^{-5} mm Hg), materials with low work functions and high bulk evaporation rates, such as barium, are frequently used. The material is first contained within the body of another material which provides structure and shape for the cathode, and then migrated to the surface by a diffusion process. This kind of cathode is called a dispenser cathode. The dispenser cathode generates and maintains an excess of barium metal at its surface and relies on that excess for its emission properties. In this configuration, the evaporation of the materials can be slowed down and easily controlled.

At vacuum levels higher than 1×10^{-5} mm Hg, the choice of cathode material is restricted to the refractory metals, which have higher work functions and operate at higher temperatures. The most attractive refractory metals are tungsten and tantalum with work functions of 4.55 and 4.1 electron volts, respectively. The melting point of tungsten is 3410°C, while that of tantalum is 2996°C. At temperatures below 2500°C, tantalum will emit 10 times the current of tungsten. Tantalum is also easy to work with and can be formed into a sheet to produce special cathode shapes.

If the vacuum is to be recycled to atmosphere but not operated above 5×10^{-6} mm Hg, a cathode of lanthanum hexaboride (LaB_6), with work function of 2.4 electron volts, may be used [10]. This arises from the need for relatively high emission current densities at lower emission temperatures. Among other activated cathodes, LaB_6 is much less sensitive to problems such as cathode contamination and lifetime but its long-time stability and thermal cycling stability are still unsolved problems.

Among all these cathode materials, tungsten may not be the best in most respects, but for normal applications it is a cheap, robust, and reliable emissive source. As of today, tungsten remains the most important cathode material in the field of electron beam processing, even though tantalum, LaB_6, as well as tungsten with emission-increasing alloying elements are also widely used.

Beam shaping and guidance After the free electrons are emitted from the cathode, they are first shaped into a well-defined beam with the desired beam diameter and focal length and then guided to the work point on the workpiece. This is achieved through different gun design and via focusing and deflection by using the principles of electron optics.

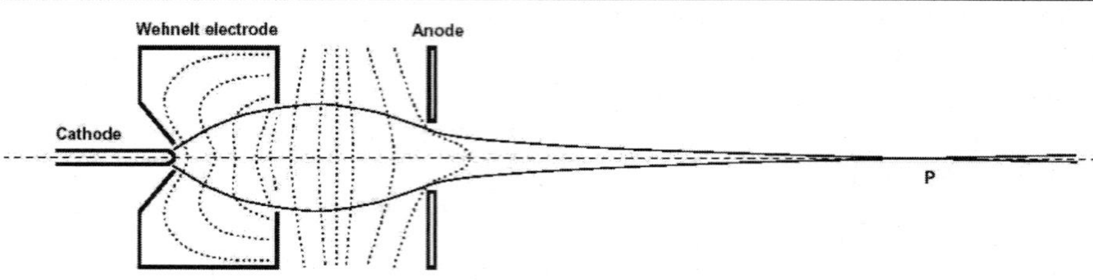

Figure 4. Three-electrode telefocus gun. Its long focal length is primarily due to the hollow shape and negative bias of the Wehnelt electrode, which acts as a simple electrostatic lens.

Gun type A basic electron gun consists of a cathode, a focusing electrode, and an anode. It is called a two-electrode gun if the focusing electrode has the same potential as that of the cathode. A design with different potentials of the cathode and the focusing electrode is called a three-electrode gun. Multielectrode guns have several focusing electrodes or control electrodes at different potentials.

Analogous to the terminology in light optics, it is called an axial gun if the elements of the beam-generating system, the electrostatic field, and the beam itself are rotationally symmetrical. There are three basic axial gun types for general use: the Telefocus gun, the Gradient gun and the Pierce gun.

The Telefocus gun is a three-electrode gun, see Figure 4. It is primarily designed to produce a relatively long focal length. The long focus effect is due to the hollow shape and negative bias of the Wehnelt electrode, which acts as a simple electrostatic lens. It operates as follows. First, the electrons near the cathode are pushed outwards along the diverging electric field. Due to the special design, the equipotentials between the Wehnelt electrode and the anode then become flat, and finally converge toward the anode (shown as dotted lines). At this final step, the electron beam obtains a net radial velocity inward. The magnitude of the net radial inward velocity is smaller than the initial outward velocity because the electrons now have higher energy. Consequently the electron beam converges quite slowly and has a long focal length. If the bias on the Wehnelt electrode increases, the field curvature in the cathode region also increases. Therefore, the focal length will be longer because the starting electron beam is more divergent. The ray traces are shown in solid lines. Position P is the focal point.

The Gradient gun [11] shown in Figure 5 is a post-acceleration gun. Similar to the conventional triode, the relatively high voltages and large currents are controlled by a small "grid" voltage V_1. Thus the total beam power may be varied over a wide range with a small variation in spot size. In order to take full advantage of the gun's capabilities, the total accelerating voltage must be much larger than the controlling voltage V_1. Also, V_1 must be high enough to draw adequate emission from the cathode.

In many applications in semiconductor manufacturing, uniform high intensity electron beams are required. It was suggested by Pierce that such a uniform electron beam could be obtained over a limited region if the region is considered a segment of extensive beam flow, and the electrodes, including cathode and anode, are shaped to maintain

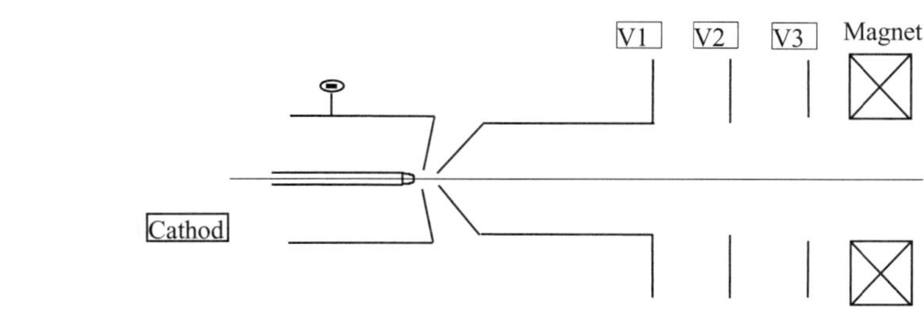

Figure 5. Triode gradient gun. Similar to the conventional triode, the relatively high voltages and large currents are controlled by a small "grid" voltage V_1. Thus the total beam power may be varied over a wide range with a small variation in spot size.

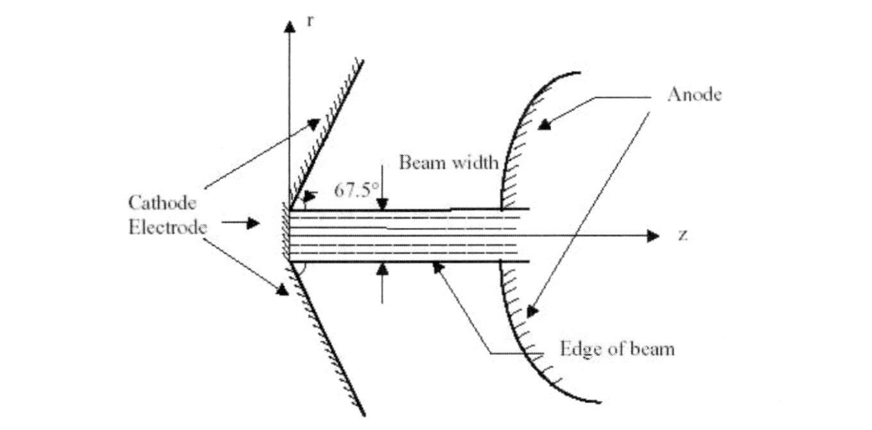

Figure 6. Two-electrode pierce gun. It is designed to produce, under the space-charge-limited emission, a parallel or slightly divergent uniform high intensity electron beams.

the same voltage along the edge of the segment. The so-called Pierce gun is designed to produce, under the space-charge-limited emission, a parallel or slightly divergent beam, see Figure 6. In this design, a broad electron beam is emitted by a flat cathode and propagates as a parallel laminar flow with a sharp planar or cylindrical surface. To keep this beam propagating as a parallel beam, the shape of the electrodes outside the beam must be carefully considered. The simplest solution is to have a 67.5° angle at the cathode and the curved anode surface, which coincides with an equipotential. A spherically curved cathode will converge the beam. However, the resultant focus point will be relatively large due to the outward directed force of the space charge. The Pierce gun is a two-electrode gun and is easy to design. The beam can be parallel, divergent or convergent. The efficiency of the gun can be as high as 99.9 % or more.

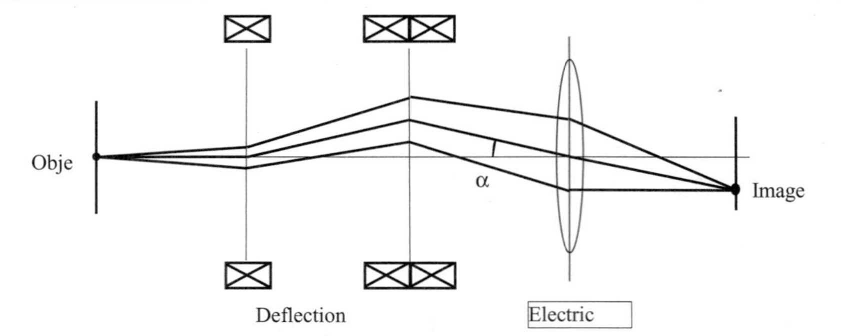

Figure 7. A double deflection system in which a focal point at the object plane is deflected by α degree at the image plane.

Beam guidance Shaped in the gun, the beam is characterized by the parameters of the focal spot. The most important focal spot parameters are the diameter and location of the focal spot on the axis, the current density and current density distribution on the focal plan, and the aperture. The object of the beam guidance system is to transform these parameters into parameters required by the particular application process on the workpiece. Figure 7 shows a simple beam guiding system. In this system, a focal point at the object plane is first deflected by a double deflection system, and then imaged and refocused onto the image plane. For some applications, the beam diameter formed inside the gun must be imaged either on an enlarged or reduced scale to obtain a beam with a defined diameter, a particular current density, and a specified power density on the workpiece. The beam current at the working point may be lower than the beam current in the gun through aperture limiting. Other applications may require that the beam be guided into the working chamber without any noticeable loss in beam current.

Like all other electron beam applications, beam guidance for electron beam processing is achieved via imaging, focusing and deflection under the principles of electron optics. In general, rotationally symmetrical magnetic fields produced by magnetic-lens systems are used for imaging and focusing; either plain or crossed magnetic bipole elements are often used for beam deflection. In the case of turning the beam over wide angles, magnetic sector fields may be added for additional deflection.

The magnetic lenses can be generated by permanent magnets. It can also be created by electrical coils. The simplest magnetic lenses are iron-clad coils, as shown in Figure 8. In this configuration, the magnetic induction is proportional to the excitation NI, where N is the number of turns and I is the coil current. The magnetic field profile and the electron optical features of the lens are totally dependent on the gap width, w, and the bore diameter of the pole pieces, D. In practice, the aberration and astigmatism should also be considered in lens design.

It can be seen from Equations (6) and (7) that all magnetic lenses are convex lenses. These lenses can be used for either producing a magnified image of the object or

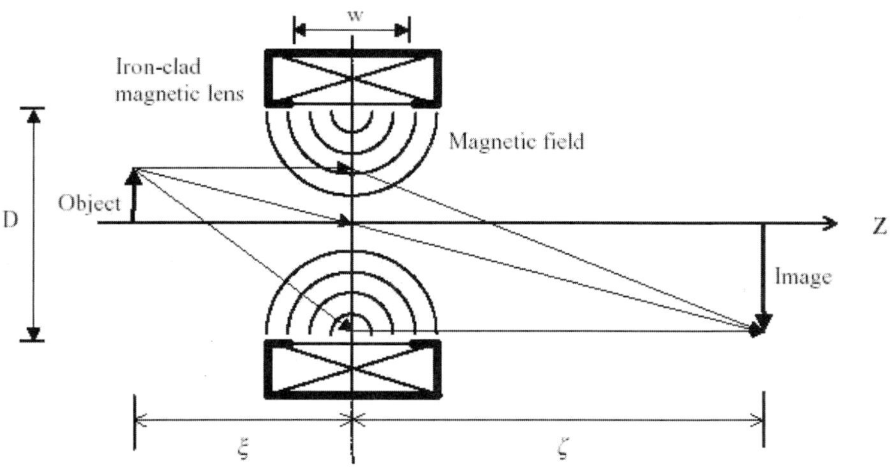

Figure 8. A magnetic lens generated by iron-clad coils, which is a basic element of the electron beam processing equipment to project an electron beam pattern to the workpiece.

focusing a parallel electron beam to a fine point. Assuming the front and back focal lengths of the convex "thin" lens are same, Newton's lens equation can be applied for the electron beam formation:

$$\frac{1}{\xi} + \frac{1}{\zeta} = \frac{1}{f} \tag{11}$$

where u is the distance between the object and the lens, v the distance from the lens to the image, and f the focal length of the lens. It can be seen that to obtain a real magnified image, both ξ and ζ should be greater than f. The magnification is defined as:

$$M = \frac{\zeta}{\xi}. \tag{12}$$

As in a well-designed optical imaging system, it is often necessary to change the magnification while operating the electron beam system. The magnification of the electron beam system is varied by changing the strength of its electric or magnetic lens. This is totally different from the light beam system, in which the magnification is changed by moving the optical lens or the objective back and forth.

Both electrostatic and magnetic bipole elements can be used for beam turning and deflection. They are created by electrical fields between two plates or by magnetic fields between the opposite poles of a permanent magnet and inside current-carrying coils. In electron beam processing, the electrostatic bipole element is usually employed for beam blanking or some other special purposes.

There are many designs of magnetic bipole elements. In the simplest case, the field between the poles of a permanent magnet is used for deflection. The pole-piece spacing w and their width b are usually much larger than the electron beam diameter. In most of cases, the magnetic fields for deflection elements are produced electromagnetically. The magnetic induction B is directly proportional to the excitation NI, and indirectly proportional to the pole-piece spacing w. In order to obtain the highest possible induction at a given excitation, the magnetic circuit must have fairly large dimensions.

Figure 9(a) and (b) shows narrow and wide angle deflection in a uniform magnetic field normal to the electron beam direction. Based on electron dynamics, the radius of the electron trajectory is given by

$$R = \left(\frac{2m}{q}\right)^{1/2} \frac{V^{1/2}}{B} = 3.37 \times 10^{-6} \frac{V^{1/2}}{B}. \tag{13}$$

When the electron beam vertically enters a limited magnetic field [12], the beam deflection over a narrow angle is expressed as:

$$\sin\theta = 2.97 \times 10^{5} \frac{LB}{V^{1/2}} \tag{14}$$

where L is the field length. In a magnetic sector field, the deflection angle can be found through the following equation:

$$\theta = \alpha - \beta_1 + \beta_2 \tag{15}$$

Clearly, the deflection angle can be enlarged by increasing the sector angle α.

3. ELECTRON-SOLID INTERACTIONS

3.1. Electron Scattering in Solid

As the electrons penetrate the solid materials, such as the electron beam resist, the interaction can be characterized by scatterings. There are two kinds of scatterings during the electron-solid interaction: small angle scattering (forward scattering), which tends to broaden the initial beam diameter, and the large angle scattering (backscattering), which causes the proximity effect [13], where the dose that a pattern feature receives is affected by electrons scattering from other nearby features. The backscattering happens as the electrons penetrate through the resist into the substrate.

During the interaction, the primary electrons slow down, producing a cascade of electrons called secondary electrons with energies from 2 to 50 eV. These are responsible

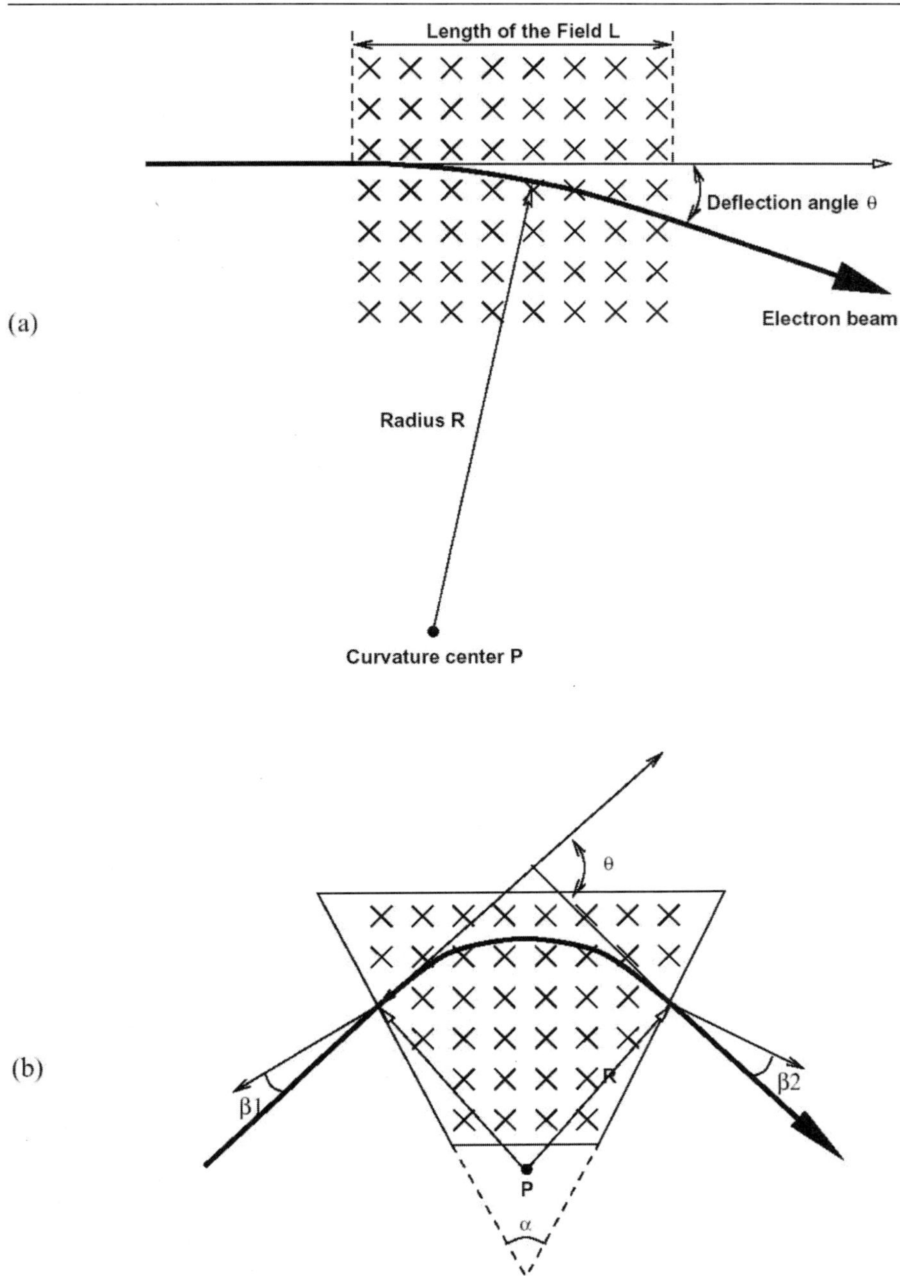

Figure 9. Electron beam deflection in a homogeneous magnetic field: (a) deflection in a limited field; (b) bending in a sector field. The sector field produces larger deflection angle.

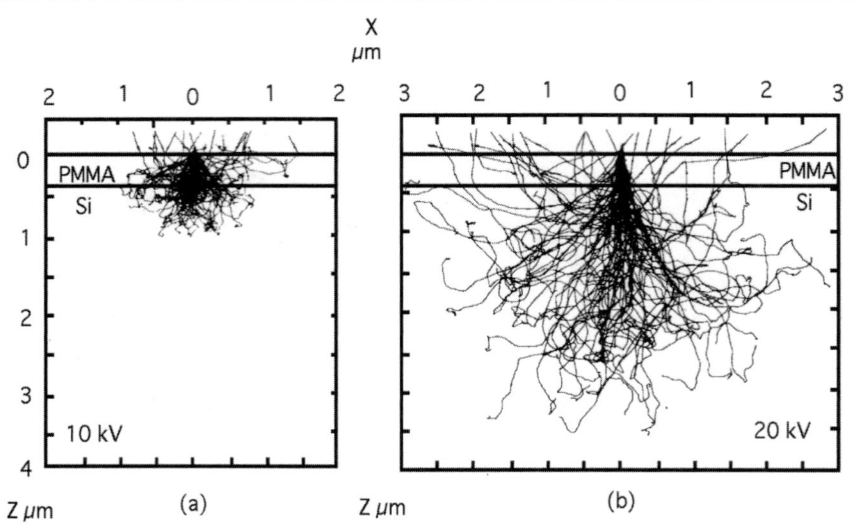

Figure 10. Monte Carlo simulation of electron scattering in resist on a silicon substrate at a) 10 kV and b) 20 kV (Kyser and Viswanathan, 1975).

for the bulk of the actual resist exposure process. Since their range in resist is only a few nanometers, they contribute little to the proximity effect. However, a small fraction of secondary electrons may have significant energies, on the order of 1 keV. These so-called "fast electrons" can contribute to the proximity effect in the range of a few tenths of a micron.

Figure 10 shows a computer simulation of electron scattering in typical samples [14]. The combination of forward and backscattered electrons results in an energy deposition profile in the resist that is typically modeled as a sum of two Gaussian distributions.

3.2. Proximity Effect

The result of the electron scattering is that the dose delivered by the EBL system is not confined to the shapes that the system writes. This results in pattern specific linewidth variations known as the proximity effect. Due to the proximity effect, a narrow line between two large exposed areas may receive so many scattered electrons that it can actually be developed away in positive resist.

Higher beam voltages, from 50 kV to 100 kV, are used to minimize both forward and backward scatterings when writing on very thin membranes such as those used for x-ray masks [15]. On the other hand, by using very low beam energies, where the electron range is smaller than the minimum feature size, the proximity effect can also be eliminated [8]. In this case, the thickness of a single layer resist must also be less than the minimum feature size so that the electrons can expose the entire film thickness.

3.3. Electron Beam Resists

EBL is classified as a reactive processing in terms of Electron Beam Processing. In this process, ionization and excitation of constituent molecules of the material occur during the scattering of the incident electrons. Some excited molecules lose their energy by collisions with other molecules and change into radicals. All these ions, excited molecules, radicals, and the secondary electrons are called active species that induce chemical reactions inside the material. Electron beam resists are used as the recording and transfer media for EBL process. The electron beam exposure alters the nature of the resist, through the breaking of chemical bonds, with the result that a subsequent immersion of the sample in a chemical solution removes the exposed parts, or the unexposed parts, of the resist film.

Therefore, the selected resist must be sensitive to the active species, which means that it must be altered by the beam in such a way that, after development, the portions exposed to the beam are removed (positive resist), or remain (negative resist). These materials are usually polymeric solutions that are applied to the surface of the substrate by a spin coater and dried by baking to form a uniform thin layer. To reduce proximity effects, positive resist is usually exposed and/or developed lightly while still adequately clearing the resist down to the substrate for all features.

In the EBL process, it is possible to make resist structures with very high aspect ratios (>5:1). But, when the aspect ratio exceeds about 5:1, the tall features may fall over during development, due primarily to surface tension in the rinse portion of the development sequence. For typical applications, the resist thickness should not exceed the minimum feature size required.

When exposing resist on insulating substrates, substrate charging causes considerable distortion [16]. A simple solution for exposure at higher energies (>10 kV) is to evaporate a thin (10 nm) layer of gold or chrome on top of the resist. Electrons travel through the metal with minimal scatter, exposing the resist. The metal film should be removed before developing the resist [17–18].

3.3.1. PMMA Electron Beam Resists

Polymethyl methacrylate (PMMA) is the standard high-resolution polymeric electron beam resist. PMMA at lower doses is a positive-acting resist and at higher doses is a negative-acting resist. For line doses the sensitivity difference between the positive resist and the negative resist is a factor of 20–30X. The best resolution obtained in PMMA was 10-nm lines in its positive mode. In negative mode, the resolution is about 50 nm. It was suggested [19] that the limitation was due to secondary electrons generated in the resist, although the effect of molecule size and development could also play a role. By exposing PMMA on a thin membrane, the exposure due to secondary electrons can be greatly reduced and the process latitude thereby increased.

The sensitivity of PMMA is roughly proportional with electron acceleration voltage, with the critical dose at 50 kV being roughly twice that of exposures at 25 kV. When using 50 kV electrons and 1:3 MIBK:IPA developer, the critical dose is around 350 $\mu C/cm^2$. PMMA has poor selectivity in plasma etching, compared to novolac-based photoresists. It has an approximately 1:1 etch selectivity for silicon nitride [20] and silicon dioxide [21]. PMMA makes a very effective mask for chemically assisted ion beam etching of GaAs and AlGaAs [22].

3.3.2. Positive Resists

In positive resists, electron beam breaks polymer backbone bonds and transfer the exposed polymer into fragments of lower molecular weight. A solvent developer selectively washes away the lower molecular weight fragments and leaves the unexposed portion of the resist film intact. Three most commonly used positive resists are described below.

EBR-9 EBR-9 is an acrylate-based resist, poly(2, 2, 2-trifluoroethyl-chloroacrylate) [23]. It is 10 times faster than PMMA. But, its resolution is more than 10 times worse than that of PMMA. EBR-9 is perfect for mask-writing applications, not because of its speed but because of its long shelf life, lack of swelling in developer, and large process latitude.

PBS Poly(butene-1-sulfone) (PBS) is a common high-speed positive resist used for mask making due to its very high sensitivity ($\sim$1 to 2 $\mu C/cm^2$). However, the processing of PBS is difficult: masks must be spray developed at a tightly controlled temperature and humidity [24]. For small to medium scale mask production, the time required for mask processing can make PBS slower than some photoresists [25].

ZEP ZEP-520 from Nippon Zeon Co. [26] is a relative new resist for EBL. It consists of a copolymer of -chloromethacrylate and -methylstyrene, with sensitivity of an order of magnitude faster than PMMA and similar to the speed of EBR-9. Comparing with EBR-9, the resolution of ZEP is very high and close to that of PMMA. ZEP has about the same contrast as PMMA. The etch resistance of ZEP in CF_4 RIE is around 2.5 times better than that of PMMA but is still less than that of novolac-based photoresists.

3.3.3. Negative Resists

In negative resists, electron beam cross-links the polymer chains together so that they are less soluble in the developer. Negative resists have less bias (the difference between a hole in the resist and the actual electron beam size) than positive resists. However, they tend to have the problems of insoluble residue in exposed areas, swelling during development, and bridging between features. Two commercially available negative electron beam resist are follows.

COP COP is an epoxy copolymer of glycidyl methacrylate and ethyl acrylate, P(GMA-co-EA). The speed of this resist is very high: 0.3 $\mu C/cm^2$ at 10 kV. But the resolution is only about 1 um [27]. The plasma etch resistance of COP is relatively poor. The resist requires spray development to avoid swelling. Because cross-linking occurs by cationic initiation and chain reaction, the cross-linking continues after exposure. Therefore, the size of the features depends on the time between exposure and development. COP is commonly used for making mask plates [28–29].

SAL The popular SAL resist [30] consists of three components: a base polymer, an acid generator, and a crosslinking agent. After exposure, a baking cycle enhances reaction and diffusion of the acid catalyst, leading to resist hardening by cross-linking. Common alkaline photoresist developers will dissolve the unexposed regions. The acid reaction and diffusion processes are important factors in determining the resolution [31], and a tightly controlled postexposure baking process is required. The extent of the cross-linking reaction is therefore affected by the thermal conductivity of the sample and by the cooling rate after the bake. The sensitivity of SAL is about 7 to 9 $\mu C/cm^2$ at either 20 or 40 kV, and is therefore suitable for mask making. A resolution of 30 nm has been demonstrated at very low voltage [32], and 50 nm wide lines have been fabricated using high voltage [33]. This novolac base polymer has etching properties similar to those of positive photoresists. It is interesting to note that, unlike PMMA, the critical dose of SAL does not scale proportionately with accelerating voltage. Although it is not as sensitive as other negative resists, such as COP, SAL has far better process latitude and resolution.

3.3.4. Multilayer Systems

Multilayer resist systems are needed in the following cases: when an enhanced undercut is needed for lifting off metal, when rough surface structure requires planarization, and when a thin imaging (top) layer is needed for high resolution.

Bilayer systems The simplest bilayer technique is to spin a high molecular weight PMMA on top of a low molecular weight PMMA. The low weight PMMA is more sensitive than the top layer, so the resist develops with an enhanced undercut. The two-layer PMMA technique was patented in 1976 by Moreau and Ting [34] and was later improved by Mackie and Beaumont [35] by the use of a weak solvent (xylene) for the top layer of PMMA.

Trilayer systems Virtually any two polymers can be combined in a trilayer resist if a barrier such as Ti, SiO_2, aluminum, or germanium separates them [36–37]. This trilayer system has been applied for fabricating dense and high aspect ratio resist profiles as described below. First, the top layer is exposed and developed and the pattern is transferred to the interlayer by RIE in CF_4 (or by Cl_2 in the case of aluminum). Then, using the interlayer which serves as an excellent mask, the straight etch profile is obtained by oxygen RIE. Such high aspect ratio of resist profiles can then be used

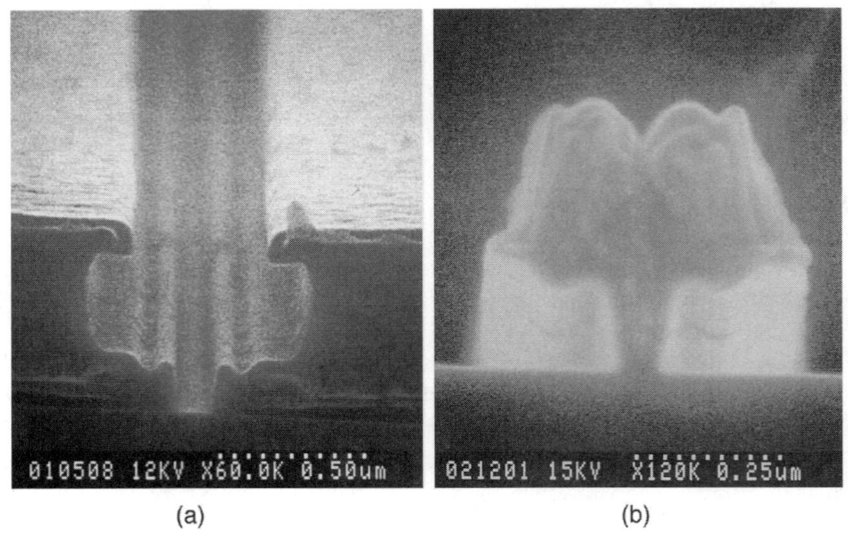

Figure 11. (a) Resist cross-section (PMMA on P(MMA-MAA) on PMMA) for the lift-off of a "T" shaped gate. (b) Metal gate lifted off on GaAs (R. C. Tiberio *et al.*, 1989).

for liftoff or for further etching into the substrate. Figure 11 shows two gate structures obtained by using the multiplayer system [38].

4. PATTERN TRANSFER PROCESS

After the resist is patterned by EBL, it is necessary to transfer the pattern onto the underlying substrate. There are two basic pattern transfer methods: additive process or subtractive process.

4.1. Additive Processes

Lift-Off and Plating are two basic additive processes as shown in Figure 12 and Figure 14.

4.1.1. Lift-Off Process

The lift-off process is based on the following requirements: there is an undercut profile in developed resist and the resist remains soluble after other material has been deposited onto it. The undercut profile is used as a "stencil" during the additive deposition process. Since the resist remains soluble after the deposition, soaking the substrate in a good resist solvent lifts the unwanted material together with the resist while the desired pattern structure remains on the substrate, hence the name "Lift-off", (see figure 12). It was predicted and observed that undercut profiles can be obtained in PMMA resist under a certain exposure dose and beam voltage [39–40]. Figure 13 shows that about

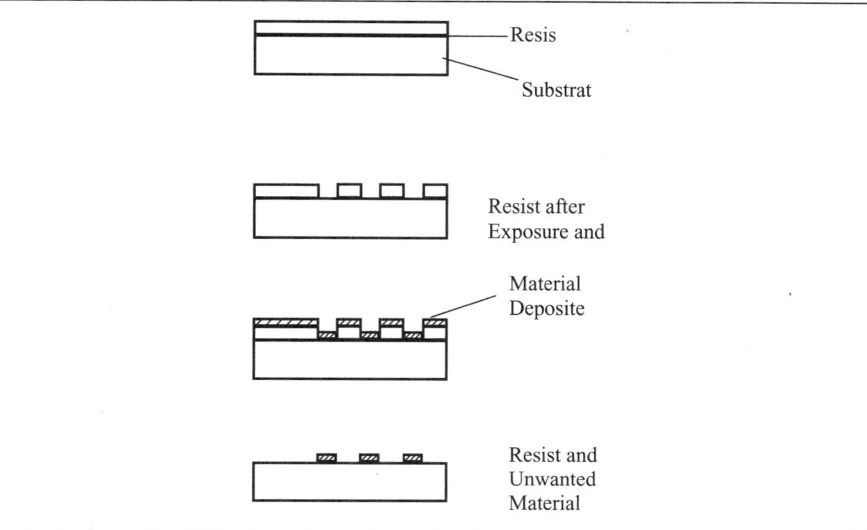

Figure 12. The lift-off process flow.

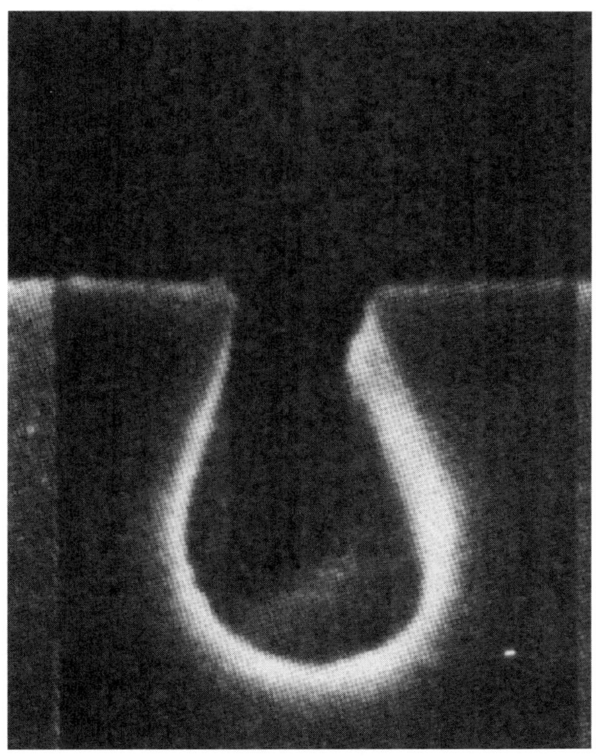

Figure 13. An undercut profile in PMMA resist.

100 μC/cm^2 at 15 kV produces a good undercut profiles in one micron thick PMMA resist.

In order to apply lift-off successfully, however, certain requirements must be met. 1). The resist undercut profile angle to the surface normal must always be larger than the deposition beam angle. 2). The temperature stability of the resist, which is determined by the glass transition temperature, Tg, should not be exceeded during the material deposition or substrate cleaning. Above Tg, the resist will flow and cause distortion of developed image and loss of the undercut profile. For PMMA resist, it will flow above 110°, therefore, care should be taken not to exceed 100°C temperature on the substrate during deposition. 3). Care also should be taken to remove the resist and clean the substrate after the material deposition. 4). The resist should remain soluble in some solvent or liquid after deposition, otherwise, lift-off cannot be completed. In some cases, ultrasonic agitation may be necessary in order to lift-off the unwanted metal, although this should be used only as a last resort, since the deposited pattern on the substrate may also be damaged especially if the material adhesion to the substrate is not very good.

4.1.2. Plating Processes

The second additive technique, which has found uses in circuit board fabrication and also in nanofabrication of such devices as zone plates for x-ray imaging, is plating of the metal in the areas where the resist has been removed after development. More recently this technique has also been used in the fabrication of x-ray masks where gold is used as the absorber. The most commonly used method is electroplating and this requires that a thin, electrically conducting layer is used as a plating base under the resist that is continuous over the entire surface of the substrate so that electrical contact can be made to it during electroplating. Figure 14 shows a schematic of the plating process.

4.2. Subtractive Processes

Subtractive patterning processes comprise all processes in which the layer to be patterned is deposited first, as a uniform layer, on the substrate followed by the resist, Figure 15. After exposure and development, the parts of the underlayer not protected by the resist, are removed either by immersion in an acid or other liquid, (wet etching), or by placement in a plasma reactor in which a chemically active gas has been added, (dry etching). Although wet etching is very fast and inexpensive, this process presents a major disadvantage that prohibits its use in high-resolution lithography. The resolution limitation of wet etching is a direct result of undercutting or metal etching under the resist mask due to the isotropic nature. Subtractive patterning in micro and nanostructure fabrication became possible only after the development of plasma or dry etching and specifically reactive ion etching, (RIE) in which the reactive gas ions are accelerated on to the surface of the resist covered substrate striking it

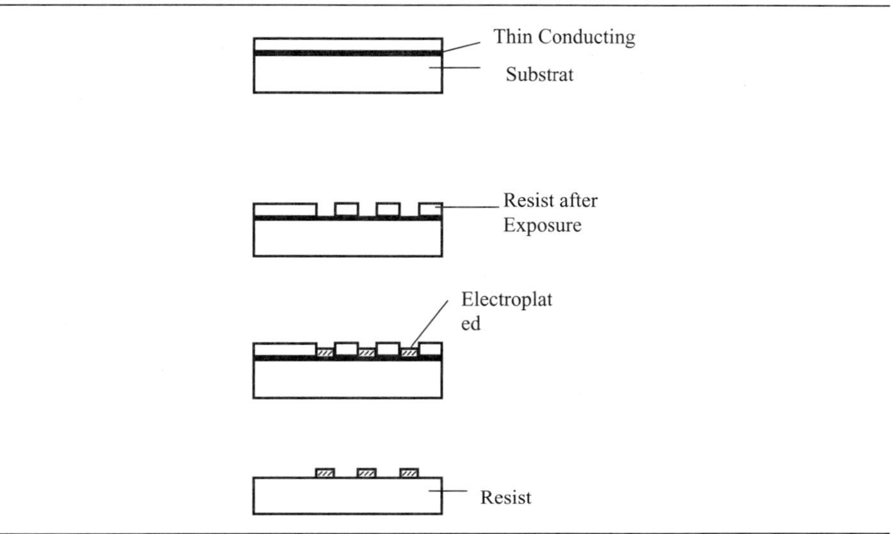

Figure 14. The electroplating process flow.

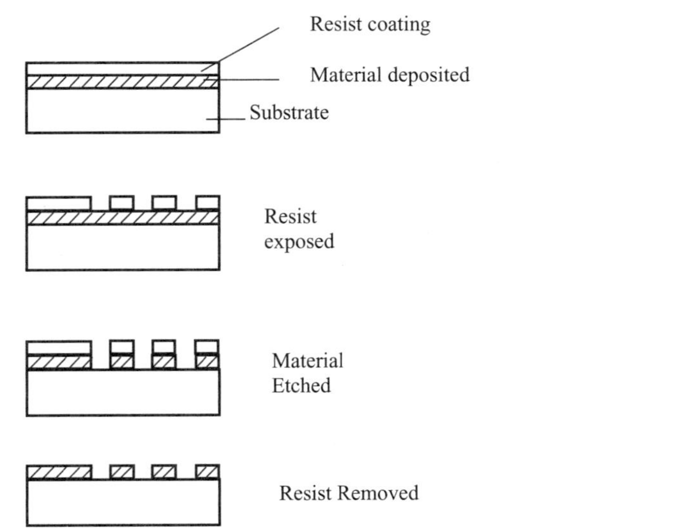

Figure 15. Subtractive pattern transfer process.

in a direction perpendicular to the substrate. This way, the etching is anisotropic or only in the vertical direction thereby eliminating undercutting effects. This process is used today, almost exclusively, for the patterning of most layers, including metals, in memory and logic circuit production with a few exceptions where lift-off is still used.

It should be mentioned that for additive processes positive resists are generally used because it is not possible to obtain the required undercut profiles from negative resists due to electron beam scattering effects, as seen in Figure 13, which force negative resists to develop sloping profiles, opposite to undercut. In subtractive processes, however, any resist can be used. In most subtractive cases the resist profile is replicated on the substrate after etching since the resist is also etched during substrate etching.

The etching of silicon, silicon oxides or silicon nitrides is done with gas mixtures that contain fluorine as the active gas while metals such as aluminum or chromium require chlorine gas for etching. Reactive ion etching systems use, in general, two parallel plates with the top plate grounded and the bottom plate, on to which the samples or silicon wafers to be etched are placed, connected to the RF power source, typically set at 13.5 MHz. Since the resist is also etched during the process, the plasma parameters, such as power, pressure, gas flow and gas composition, have to be optimized in order to increase the etching selectivity between resist and other materials.

It has been found that acrylic-type polymers, such as PMMA or its copolymers, are not very stable in the plasma and, therefore not very useful in RIE transfer processes, while phenolic-type polymers of which most AZ-type photo-resists are made, are much more stable and more widely used. For this reason and because phenolic resins can resist higher temperatures than acrylic polymers, all new resists, including the new, acid-catalyzed ones use as their base resin phenolic or aromatic polymers. Also, the choice between positive or negative resists is independent of the RIE process, as long as both types are made with aromatic polymers, and depends only on the density of the circuit pattern. This is especially true with electron beam lithography where the pattern polarity (and the resist) is chosen to minimize the beam writing time.

The resolution of the new, very sensitive acid-catalyzed resists does not approach that of PMMA. At this moment, it can only resolve line widths in periodic patterns of about 0.2 μm in 0.4-μm thick layers. While intensive investigation is been performed in many labs, PMMA is still in use for sub-100 nm structures.

5. APPLICATIONS IN NANOTECHNOLOGY

5.1. Mask Making

5.1.1. IC Fabrication Mask

As microelectronics continues to advance, the field has been extended from micrometer to nanometer scale, hence the development of nanotechnology and nanoelectronics. However, nanoscale mask making remains the key technique in nanoelectronics technology, as it has been in microelectronics technology. Due to its inherently high spatial resolution (less than 1 nm) and wide process margins, EBL is still the technological choice for mask making.

The most popular and well-established mask making system is the MEBES [41]. The MEBES uses a focused Gaussian spot to write a pattern in stripes while moving the stage continuously. The beam deflection is primarily in one direction, perpendicular to the motion of the stage. The MEBES is designed for high-throughput mask making, with a minimum feature size of 0.25 m.

Figure 16. (a) Imprint mold with 10 nm diameter pillars (b) 10 nm diameter holes imprinted in PMMA (Stephen Y. Chou, Princeton).

The EBES4 mask making system [42] is another system using a Gaussian spot. In this system the coarse/fine DAC beam placement is augmented with an extra (third) deflection stage, and the mask plate is moved continuously, using the laser stage controller to provide continuous correction to the stage position. The EBES4 mask making system has a spot size of 0.12 um, uniformity to 50 nm (3), stitching error of 40 nm, and repeatability of 30 nm over a 6 in. reticle.

5.1.2. Nanoimprint Mask

Nanoimprint lithography [43] patterns a resist by deforming the resist shape through embossing (with a mold), rather than by altering resist chemical structures through radiation (with particle beams). After imprinting the resist, an anisotropic etching is used to remove the residue resist in the compressed area to expose the substrate underneath. It is a major breakthrough in nanotechnology because it can produce sub-10 nm feature size over a large area with a high throughput and low cost.

One of the key elements for nanoimprint lithography is the mold, which relies on EBL technology to be produced. Figures 16 shows an imprint mold with 10 nm diameter pillars and the 10 nm diameter holes imprinted in PMMA material.

5.1.3. X-ray Lithography Mask

X-ray lithography's (XRL) penetrative power and scatterings free of X-rays are two primary reasons for its popularity in nanotechnology and nanofabrication. But, the mask fabrication has been the most difficult challenge to XRL. The XRL mask consists of a thin layer (200–250 nm) of X-ray absorbent material (e.g. Au and W), supported

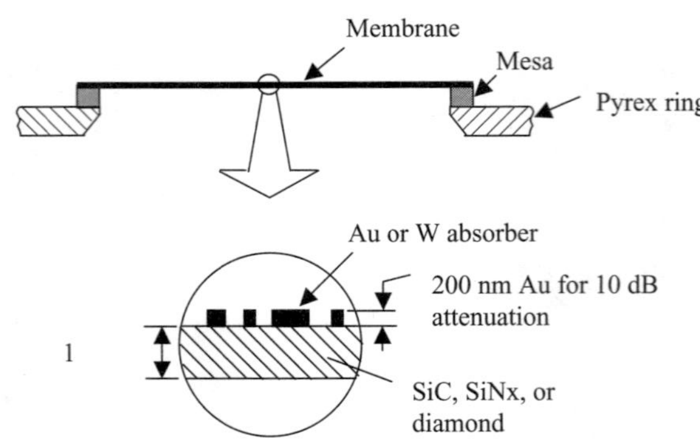

Figure 17. An x-ray mask configuration suitable for a 10–dB attenuation at the wavelength of 1.32 nm (Henry I. Smith and M. L. Schattenburg, MIT).

on a thin membrane (~1 μm) that is 20–30 mm in diameter. Figure 17 shows a mask configuration suitable for a 10-dB attenuation at the 1.32 nm radiation. The primary concern for the mask is distortion, which can be introduced by the following four sources: the original mask patterning; the mask frames; radiation damage; or absorber stress. To have a reasonable yield of the desired nanostructure, mask distortion at any point in a pattern should not exceed 1/5 to 1/10 of the minimum feature size. For a 10 nm feature, the distortion should be no more than a couple of nanometers.

Many techniques, such as e-beam lithography, photolithography, holographic lithography, X-ray lithography, and ion-beam lithography, have been developed to make XRL masks. But, e-beam lithography is the most frequently used method. Either additive or subtractive process has been used for pattern formation on the mask.

Examples of x-ray mask fabrication schemes for an Au additive process are shown in Figure 18. The process involves plating x-ray absorber on the resist-patterned membrane. It includes deposition of membrane film on a silicon wafer, back-etching the silicon to the membrane film, glass frame attachment, deposition of chrome for plating base, resist coating, pattern formation by electron beam lithography, Au plating (additive process), and finally resist removal. Figure 19 is an actual X-ray mask with 75 nm features fabricated by EBL.

5.2. Direct Writing

5.2.1. Self-Assembly

Carbon nanotubes (NTs) [44] have opened a promising path in nanotechnology. They provide insulating, semiconducting or truly conducting nanoscale wires for electronic applications. Devices such as a junction [45–47] and a field-effect transistor [48–51] have been demonstrated. But, up to now, all the demonstrated NT electrical devices have been fabricated either by randomly depositing NTs on a multi-electrode array or by patterning contacts onto randomly deposited NTs, after their observation.

K. H. Choi *et al.* [52] demonstrated a method for achieving controlled fabrication with the help of the EBL. This method is based on the electrostatic anchoring of

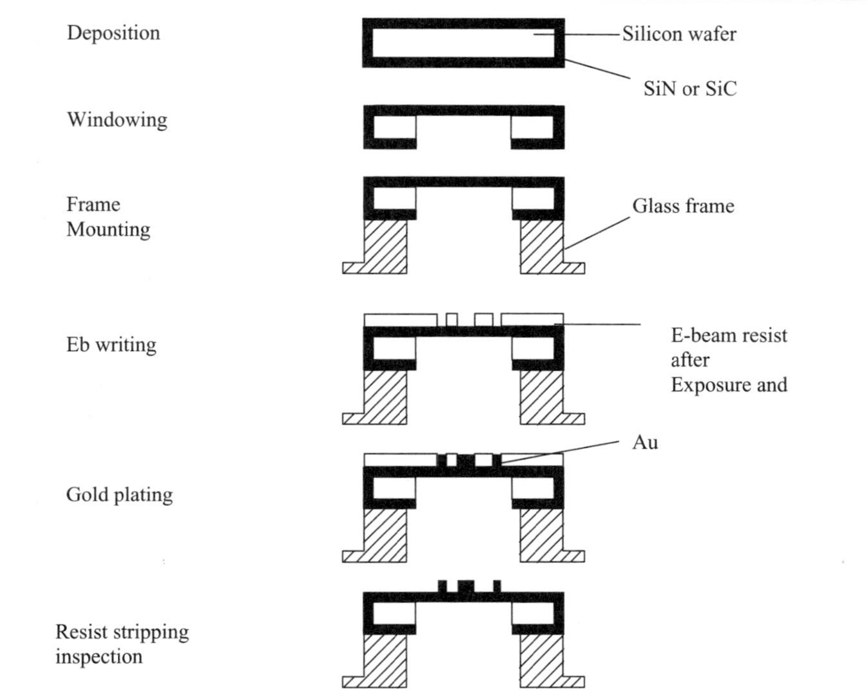

Figure 18. A gold additive process for x-ray mask fabrication.

Figure 19. 75 nm features of plated gold absorber on an X-ray membrane.

surfactant covered NTs onto amino-silane functionalized surfaces [53]: first, a reactive amino-silane template is prepared using chemical vapour deposition of silane molecules through a PMMA mask patterned by conventional electron-beam lithography. Surfactant covered NTs are then selectively deposited onto the template. Finally, the PMMA mask is lifted-off, leaving the tubes on the template. Figure 20 shows AFM images of NTs onto the patterned silane monolayer. The typical thickness of the NTs of (a) (measured relative to the silane surface) is 1.6 ± 0.2 nm.

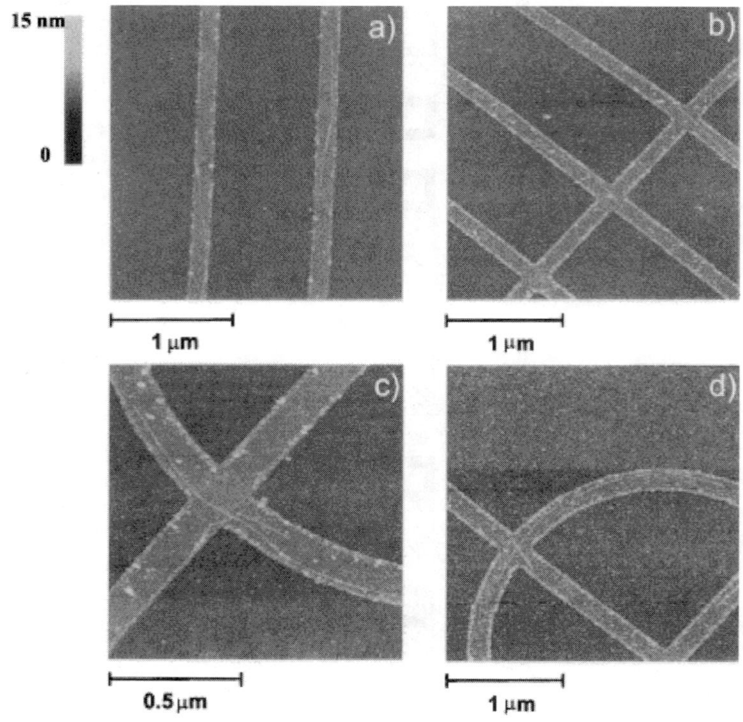

Figure 20. NTs onto the patterned silane monolayer. The silane stripes appear brighter than the bare silica surface. The NTs appear brighter than the silane stripes onto which they are adsorbed (K. H. Choi *et al.*, 2000).

Similarly, patterned amine-functionalized self-assembled monolayers have potential as a template for the deposition and patterning of a wide variety of materials on silicon surfaces, including biomolecules. C. K. Harnett *et al.* [54] obtained results for low-energy electron-beam patterning of 2-aminopropyltriethoxysilane and (aminoethylaminomethyl) phenethyltrimethoxysilane self-assembled monolayers on silicon substrates. They demonstrated that, on the ultrathin (1–2 nm) monolayers, lower electron beam energies (<5 keV) produce higher resolution patterns than high-energy beams. At 1 keV, a dose of 40 $\mu C/cm^2$ is required to make the patterns observable by lateral force microscopy. Features as small as 80 nm were exposed at 2 keV on these monolayers. After exposure, palladium colloids and aldehyde- and protein-coated polystyrene fluorescent spheres adhered only to unexposed areas of the monolayers.

5.2.2. *Nanoscale Device Fabrication*

The steps taken to produce a nanoscale device by EBL are shown in Figure 21: the sample is covered with a thin layer of PMMA, then the desired structure or pattern is exposed with a certain dose of electrons. The exposed PMMA changes its solubility

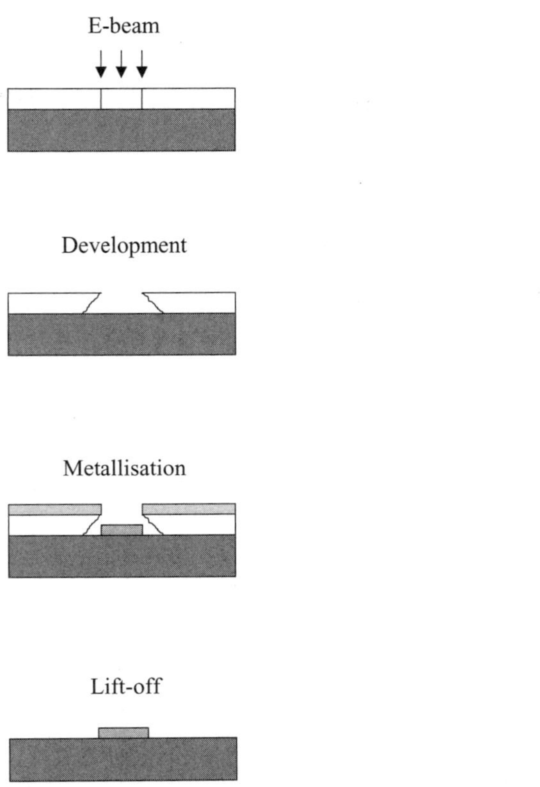

Figure 21. EBL steps for fabricating nanoscale devices.

towards certain chemicals. This can be used to produce a trench in the thin layer. If one wants to produce a metallic structure, a metal film is evaporated onto the sample and after dissolving the unexposed PMMA with its cover (lift-off) the desired metallic nanostructure remains on the substrate.

Until recently, the EBL system was used almost exclusively for fabricating research and prototype nanoelectronic devices. It is still the only tool used for this kind of device scale since a lithography system for mass production is not available for deep nanometer features [55–57]. Devices fabricated by using EBL include quantum dots, single electron transistors [58], nanotube transistors [59], and other nanoscale structures.

The following four examples of nanoscale structures are fabricated by using EBL. Figure 22 is a pattern for "Binary position-modulated sub-wavelength grating (BPMSG)" [60] fabricated using a "lift-off" process. BPMSG is a very important building block for many micro scale optical instruments such as spectrometers and multiplexors.

In Figure 23, the hexagon array was written with the beam making a single pass over each line. First, the beam moved back and forth across the entire structure writing the non-vertical lines. Then, each short vertical line was written to complete the pattern.

Figure 22. EBL generated 150 nm chromium lines with variable spaces ranging from 150 nm to 300 nm on a silicon substrate. The thickness of the lines is about 100 nm. [Zhou, 1993].

Figures 24–25 were fabricated at the Nanoscale Science Laboratory of the University of Cambridge headed by Professor Mark Welland [61]. Figure 24 is a field-emitting device fabricated on silicon. Its feature size is in the 100 nm regime. Source, drain and gate are made from tungsten.

Figure 25 shows a line of 5 nm wide written into PMMA. This line has a homogeneous width over more than 100 nm and is an example of the smallest features that can be produced with electron beam lithography.

5.2.3. Electron Beam Processing

Another important application of electron beams is their use as energy carriers for the local heating of a work point in a vacuum. Electron beams are specially used to carry high power densities and to generate a steep temperature rise on the work point. Furthermore, the energy input at the work point can be accurately controlled

Figure 23. EBL generated Hexagons. Copyright (c) 1996 JC Nabity Lithography Systems.

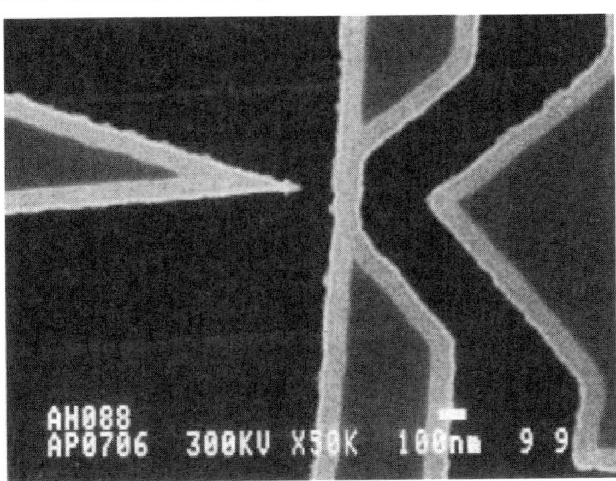

Figure 24. A field-emitting device fabricated on silicon using EBL.

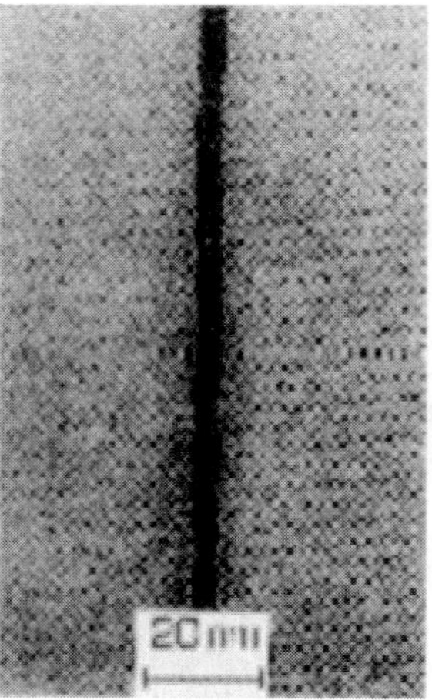

Figure 25. A line of 5 nm in width written into PMMA by EBL.

with respect to time and space. With such desirable characteristics, electron beam processing has been widely used in semiconductor manufacturing for the last three decades.

More recently, patterning of porous silicon (PS) at the nanometer length scale has been obtained by electron-beam-induced carbon masking [62]. This technique is able to locally modify or mask the Si substrate before the PS formation. It provides great possibilities in the field of micromachining [63] and photonics [64–65].

On the other hand, direct electron-beam irradiation of semiconductor surfaces has been attempted in recent years to remove H passivation [66] or to favor O desorption [67]. In many cases, it has been demonstrated that, as a consequence for this treatment(s), the surface shows a selective reactivity to subsequent treatments. The same effect, even larger as a consequence for the higher reactivity, can be expected for nanoporous silicon (NPS). This direct electron-beam treatment on PS opens the possibility of defining nanometer-sized structures on a nanosized material.

The availability of EBL apparatus allows for the structuring of several semiconducting and metallic materials down to few tens of nm. For these reasons, electron irradiation of submicrometric areas of PS can have interesting applications in micromachining and could allow for the realization of two- or three-dimensional photonic structures.

6. SUMMARY AND FUTURE PERSPECTIVES

One of the methods to create functional materials, devices, and systems through the control of matter at the atomic or molecular level is the top-down miniaturization.

The top down approach downsizes things from large-scale structures into small-scale structures. It is the major method used for the microelectronics development. As the microelectronics technology continues to advance, it has been extended from micrometer to nanometer scale, hence the "nanotechnology" or "nanofabrication". Using nanotechnology, the narrowest line pattern on mass produced semiconductor devices is now approaching the 50-nanometer level. In research labs, horizontal dimensions of the device feature sizes have been further scaled down from 130 nanometers to 6 nanometers and its vertical dimensions have been reduced to less than 1.5 nanometers or a couple of atoms.

The heart of the top-down approach of miniaturization processing is the nanolithography technique. Among many techniques of nanolithography, the EBL technique is, at least for now, the best choice for ultimate nanoscale structures due to its ability to precisely focus and control electron beams onto various substrates, and therefore, create virtually any kind of nanostructures. It has been demonstrated that electron beams can be focused down to less than 1 nm. This will extend the resolution of EBL to the sub-nanometer region provided that appropriate resistant material is available.

However, parallel writing schemes of EBL have to be developed so that higher throughput can be achieved.

REFERENCES

1. Bruce Doris, Meikei Ieong, Thomas Kanarsky, Ying Zhang, Ronnen A. Roy, Omer Dokumaci, Zhibin Ren, Fen-Fen Jamin, Leathen Shi, Wesley Natzle, Hsiang-Jen Huang, Joseph Mezzapelle, Anda Mocuta, Sherry Womack, Michael Gribelyuk, Erin C. Jones, Robert J. Miller, H-S Philip Wong, and Wilfried Haensch, IEDM Tech. Dig. (2002), pp. 267–270.
2. Timp, G., Bude, J., Bourdelle, K. K., Garno, J., Ghetti, A., Gossmann, H., Green, M., Forsyth, G., Kim, Y., Kleiman, R., Klemens, F., Kornblit, A., Lochstampfor, C., Mansfield, W., Moccio, S., Sorsch, T., Tennant, D. M., Timp, W., and Tung, R., IEDM Tech. Dig. (1999).
3. Smith, K. C. A., and Oatley, C. W., Br. J. Appl. Phys., 6, 391 (1955).
4. Broers, A. N., in *Proceedings of the First International Conference on Electron and Ion Beam Technology* (R. Bakish, ed.) Wiley, New York, p. 181, 1964.
5. Broers, A. N., Molten, W. W., Cuomo, J. J., and Wittels, N. D., Appl. Phys. Lett., 29, 596 (1976).
6. Umbach, C. P, Washburn, S., Laibowitz, R. B., and Webb, R. A., Phys. Rev. B, 30, 4048 (1984).
7. Muray, A., Scheinfein, M., Isaacson, M., and Adesida, I., J. Vac. Sci. Technol. B, 3, 367 (1985).
8. Y. Yau, R. F. W. Pease, A. Iranmanesh, and K. Polasko, J. Vac. Sci. Technol., 19, 1048 (1981).
9. Child, D. C., *Phys. Rev.*, 32, p. 492 (1911).
10. Lafferty, J. M. Boride cathode. *J. Appl. Phys.* 22(3), pp. 299–309 (1951).
11. Morley, J. R., *Proc. Third Symp. Electron Beam Tech.* p. 26 (1961).
12. Sayegh, G., Dumonte, P., Theoretical and experimental techniques for design of electron beam welding guns, effect of welding conditions on electron beam characteristics. In Salva, R. M. *Third electron Beam Processing Seminar*, Stratford-upon-Avon, England, 1974. Dayton, Ohio: Universal Technology (1974), pp. 3a1–3a87.
13. T. H. P. Chang, "Proximity effect in electron beam lithography," *J. Vac. Sci. Technol.* **12**, 1271–1275 (1975).
14. D. F. Kyser and N. S. Viswanathan, "Monte Carlo simulation of spatially distributed beams in electron-beam lithography," *J. Vac. Sci. Technol.* **12**(6), 1305–1308 (1975).
15. K. K. Christenson, R. G. Viswanathan, and F. J. Hohn, "X-ray mask fogging by electrons backscattered beneath the membrane," *J. Vac. Sci. Technol.* **B8**(6), 1618–1623 (1990).
16. J. Ingino, G. Owen, C. N. Berglund, R. Browning, R. F. W. Pease, "Workpiece charging in electron beam lithography," J. Vac. Sci. Technol. **B12** (3) 1367 (1994).
17. Gold etch solution type TFA from Transene Co., Rowley MA.
18. Chrome etch type CR-14 from Cyantek Corp., 3055 Osgood Ct., Fremont CA 94538.
19. Broers, A. N., J. Electrochem. Sec., 128, 166 (1981).
20. D. W. Keith, R. J. Soave, M. J. Rooks, "Free-standing gratings and lenses for atom optics," J. Vac. Sci. Technol. **B9** (6) 2846 (1991).

21. W. C. B. Peatman, P. A. D. Wood, D. Porterfield, T. W. Crowe, M. J. Rooks, "Quarter-micrometer GaAs Schottky barrier diode with high video responsivity at 118 m," Appl. Phys. Lett. **61** 294 (1992).
22. R. C. Tiberio, G. A. Porkolab, M. J. Rooks, E. D. Wolf, R. J. Lang, A. D. G. Hall, "Facetless Bragg reflector surface-emitting AlGaAs/GaAs lasers fabricated by electron-beam lithography and chemically assisted ion-beam etching", J. Vac. Sci. Technol. **B9** 2842 (1991).
23. T. Tada, "Highly sensitive positive electron resists consisting of halogenated alkyl -chloroacrylate series polymer materials," J. Electrochem. Soc. **130** 912 (1983).
24. K. Nakamura, S. L. Shy, C. C. Tuo, C. C. Huang, "Critical dimension control of poly-butene-sulfone resist in electron beam lithography," Jpn. J. Appl. Phys. **33**, 6989 (1994).
25. M. Widat-alla, A. Wong, D. Dameron, C. Fu, "Submicron e-beam process control," Semiconductor International (May 1988), p. 252.
26. Nippon Zeon is represented in the US by Nagase California Corp., 710 Lakeway, Suite 135, Sunnyvale, CA 94086. 408-773-0700.
27. E. Reichmanis, L. F. Thompson, "Polymer materials for microlithography," in *Annual Review of Materials Science*, v.17, R. A. Huggins, J. A. Giordmaine, J. B. Wachtman Jr., eds. (Annual Reviews, Palo Alto, 1987) p.238.
28. Mead Chemical Co., 10750 County Rd. 2000, PO Box 748, Rolla, MO 65401. 314-364-8844.
29. C. G. Willson, "Organic Resist Materials", and L. F. Thompson, "Resist Processing", in *Introduction to Microlithography, Second Edition,* L. F. Thompson, C. G. Willson, M. J. Bowden, eds. (American Chemical Society, Washington DC, 1994).
30. Shipley Inc., 455 Forest St., Marlboro, MA 01752. 800-343-3013.
31. T. Yoshimura, Y. Nakayama, S. Okazaki, "Acid-diffusion effect on nanofabrication in chemical amplification resist," J. Vac. Sci. Technol. **B10**(6) 2615 (1992).
32. E. A. Dobisz, C. R. K. Marrian, "Sub-30 nm lithography in a negative electron beam resist with a vacuum scanning tunneling microscope," Appl. Phys. Lett. **58**(22) 2526 (1991).
33. A. Claßen, S. Kuhn, J. Straka, A. Forchel, "High voltage electron beam lithography of the resolution limits of SAL601 negative resist," Microelectronic Engineering **17** 21 (1992).
34. W. Moreau, C. H. Ting, "High sensitivity positive electron resisit," US Patent 3934057, 1976.
35. S. Mackie, S. P. Beaumont, Solid State Technology **28** 117 (1985).
36. R. E. Howard, E. L. Hu, L. D. Jackel, "Multilevel resist for lithography below 100 nm," IEEE Trans. Electron. Dev. **ED-28**(11) 1378 (1981).
37. R. E. Howard, D. E. Prober, "Nanometer-scale fabrication techniques," in *VLSI Electronics: Microstructure Science* vol. 5 (Academic Press, New York, 1982).
38. R. C. Tiberio, J. M. Limber, G. J. Galvin, E. D. Wolf, "Electron beam lithography and resist processing for the fabrication of T-gate structures," Proc. SPIE **1089**, 124 (1989).
39. M. Hatzakis, C. H. Ting, and N. Viswanathan, in *Proceedings of the Symposium on Electron and Ion Beam Science and Technol.*, ECS Inc., Princeton, NJ, 542–579 (1974).
40. A. R. Neureuther, D. F. Kyser, K. Murata, and C. H. Ting, in *Proceedings of the Symposium on Electron and Ion Beam Science and Technol.*, ECS Inc., Princeton, NJ, 265–275 (1978).
41. M. Gesley, F. Abboud, D. Colby, F. Raymond, S. Watson, "Electron beam column developments for submicron- and nanolithography," Jpn. J. Appl. Phys. **32** 5993 (1993).
42. Lepton Inc., Murray Hill NJ 07974, 908-771-9490.
43. Stephen Y. Chou, Peter R. Krauss, and Preston J. Renstrom, J. Vac. Sci. Technol. B 14 (1996) 4129.
44. S. Iijima. *Nature* **354** (1991), p. 56.
45. R. D. Antonov and A. T. Johnson. *Phys. Rev. Lett.* **83** (1999), p. 3274.
46. Z. Yao, H. W. C. Postma, L. Balents and C. Dekker. *Nature* **402** (1999), p. 273.
47. M. S. Fuhrer, J. Nygard, L. Shih, M. Forero, Y. G. Yoon, M. S. C. Mazzoni, H. J. Choi, J. Ihm, S. G. Louie, A. Zettl and P. L. McEuen. *Science* **288** (2000), p. 494.
48. S. J. Tans, A. R. M. Vershueren and C. Dekker. *Nature* **39** (1998), p. 49.
49. R. Martel, T. Schmidt, T. Hertel and P. Avouris. *Appl. Phys. Lett.* **73** (1998), p. 2447.
50. L. Roschier, J. Penttilae, M. Martin, P. Hakonen, M. Paalanen, U. Tapper, E. Kauppinen, C. Journet and P. Bernier. *Appl. Phys. Lett.* **75** (1999), p. 728.
51. H. T. Soh, A. F. Morpurgo, J. Kong, C. M. Marcus, C. F. Quate and H. Dai. *Appl. Phys. Lett.* **75** (1999), p. 627.
52. K. H. Choi, J. P. Bourgoin, S. Auvray, D. Esteve, G. S. Duesberg, S. Roth, and M. Burghard, Surface Science, 462 (2000), pp. 195–202.
53. M. Burghard, G. S. Duesberg, G. Philipp, J. Muster and S. Roth. *Adv. Mater.* **10** (1998), p. 584.
54. C. K. Harnett, K. M. Satyalakshmi, and H. G. Craighead, App. Phys. Lett., 76 (2000), pp. 2466–2468.

55. Beaumont S. P., Bower P. G., Tamamura T. and Wilkinson C. D. W., Appl. Pliys. Lett., 38 (1981) 436.
56. Craighead H. G., Howard R. E., Jackel L. D, and Mankiewich P. M., Appl. Phys. Lett., 42 (1983) 38.
57. Ochiai Y., Baba M., Watanabe H. and Matsui S., Jpn. J. Appl. Phvs., 30 (1991) 3266.
58. Claes Thelander, Martin H. Magnusson, Knut Deppert, Lars Samuelson, Per Rugaard Poulsen, Jesper Nygård, and Jørn Borggreen, Appl. Phys. Lett.79 (2001) pp 2106–2108.
59. S. J. Tans *et al.*, Nature, 7 May 1998.
60. Zhou, Z. J., Ph.D. Thesis, Georgia Institute of Technology (1993).
61. http://www-g.eng.cam.ac.uk/nano/emlitho.htm last visited on August 30, 2003.
62. T. Djenizian, L. Santinacci, and P. Schmuki, Appl. Phys. Lett. **78**, 2940 (2001).
63. W. Lang, P. Steiner, and H. Sandmaier, Sens. Actuators A **51**, 31 (1995).
64. P. Ferrand, D. Loi, and R. Romestain, Appl. Phys. Lett. **79**, 3017 (2001).
65. A. M. Rossi, G. Amato, V. Camarchia, L. Boarino, and S. Borini., Appl. Phys. Lett. **78**, 3003 (2001).
66. F. Y. C. Huy, G. Eres, and D. C. Joy, Appl. Phys. Lett. **72**, 341 (1998).
67. T. Yasuda, D. S. Hwang, J. W. Park, K. Ikuta, S. Yamasaki, and K. Tanaka, Appl. Phys. Lett. **74**, 653 (1999).

II. ELECTRON MICROSCOPY

11. HIGH-RESOLUTION SCANNING ELECTRON MICROSCOPY

JINGYUE LIU

1. INTRODUCTION: SCANNING ELECTRON MICROSCOPY AND NANOTECHNOLOGY

Ultimately, all materials, organic or inorganic, have their origins in the collective assembly of a small number of atoms or molecules. Human beings have been fascinated by the interior world: the secrets of cells and the building blocks of matter. Ever since various types of microscopes were discovered, they have been the primary instruments for helping us to directly observe, understand, and manipulate matter or cells on an ever-decreasing scale. By understanding what the building blocks are and how they are arranged together to form architectures that possess unique properties or display specific functions, we hope to modify, manipulate, and sculpt matter at nanoscopic dimensions for desired purposes. To understand the fundamental properties of various nanosystems, it is necessary to characterize their structures at a nanometer or atomic level and integrate the nanoscale components that provide macroscale functions or properties. This, in turn, allows us to understand the synthesis-structure-property relationships of nanosystems, thus achieving the ultimate goal of molecular engineering of functional systems.

The scanning electron microscope (SEM) is undoubtedly the most widely used of all electron beam instruments. The popularity of the SEM can be attributed to many factors: the versatility of its various modes of imaging, the excellent spatial resolution now achievable, the very modest requirement on sample preparation and condition, the relatively straightforward interpretation of the acquired images, the accessibility of associated spectroscopy and diffraction techniques. And most importantly its

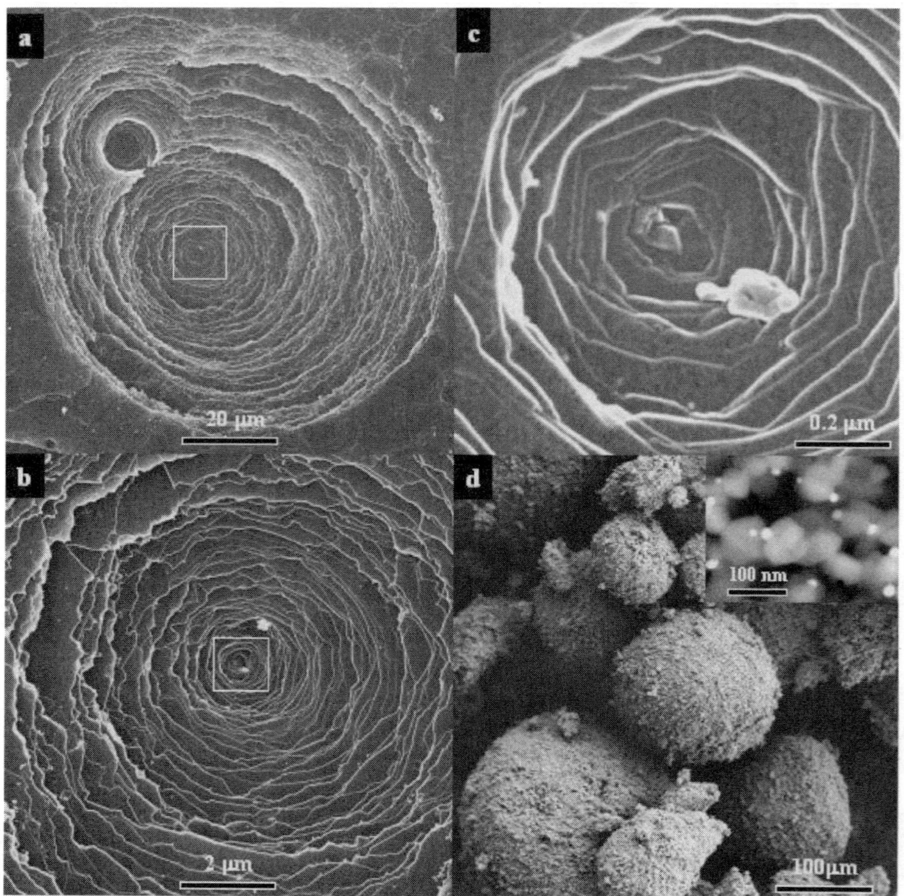

Figure 1. Secondary electron images of a Pt/graphite system showing the surface morphology at low magnification (a), medium magnification (b) and high magnification (c); low magnification secondary electron image (d) of a Pd/TiO_2 catalyst showing the general morphology of the catalyst powders and high-resolution backscattered electron image (inset) showing the size and morphology of the Pd and TiO_2 nanoparticles.

user-friendliness, high levels of automation, and high throughput make it accessible to most research scientists. With the recent generation of SEM instruments, high-quality images can be obtained with an image magnification as low as about 5× and as high as >1,000,000×; this wide range of image magnifications bridges our visualization ability from naked eyes to nanometer dimensions. Image resolution of about 0.5 nm can now be achieved in the most recent generation field-emission-gun SEM (FEG-SEM), clearly rivaling that of a transmission electron microscope (TEM); the sample size, however, can be as large as production-scale silicon wafers.

As an example, Fig. 1 shows a set of images that demonstrate the usefulness of correlating low magnification images to high-resolution images. Figure 1a shows a low magnification secondary electron (SE) image of an annealed Pt/graphite sample. The sample was prepared by sputtering Pt nanoparticles onto freshly cleaved graphite surface and then the Pt/graphite composite was annealed at 900°C in a flowing mixture of N_2 and H_2 for 10 hours. The purpose was to study the dependence of Pt-graphite interaction processes on the gas environment, annealing temperature, and residence time. Figure 1a and many other images clearly showed that many large pits had been generated on the graphite surface during the annealing process. Medium magnification images such as the one shown in Fig. 1b revealed the presence of many surface steps, terraces, and the presence of some small particles within the pits. High-resolution images such as the one shown in Fig. 1c not only showed the size and shape of some nanoparticles residing on the graphite terraces and steps but also showed the step structure and the fine details on the graphite terraces. By analyzing these SE images and the corresponding X-ray energy dispersive spectroscopy (XEDS) data, information about the interaction processes between Pt nanoparticles and the graphite substrate and how the gas environment affects these processes can be extracted. Figure 1d shows another example of extracting useful information on nanostructured Pd/TiO_2 catalyst, which has important industrial applications. Nanoscale features on the catalyst beads can be easily obtained by zooming the magnification of the region of interest; and Pd nanoparticles can be easily identified in high-resolution backscattered electron (BE) images such as the one shown in the inset of Fig. 1d. Not only the size and spatial distribution of the Pd nanoparticles can be determined but also their spatial relationship to the TiO_2 nanocrystals can be extracted.

The SEM had its origins in the work of Knoll [1] and von Andenne [2, 3]; images of surfaces with material, topographic, and crystallographic contrast were obtained [4]. The first modern SEM, however, was described by Zworykin *et al.* [5]; the instrument incorporated most of the features of a current SEM such as a secondary electron detector and a cathode-ray-tube display, achieving an image resolution of about 5 nm on solid specimens. The aggressive development of the various components of a modern SEM by the Oatley group at Cambridge University led to the first commercial production of the SEM [6].

The introduction of the use of a field-emission gun (FEG) in a scanning electron microscope made it possible to image individual heavy atoms in the transmission mode by collecting scattered electrons with an annular detector [7–9]. The wide use of the field-emission guns in the commercial SEM during the 1980s and 1990s, the significant improvement in the design of the probe-forming lens, the development of new detection schemes and detectors, and the revival of the development of high-resolution SE imaging in this time frame established the current practice and understanding of the high-resolution SEM techniques [10–20]. The continuous improvement in the field-emission guns, the probe forming lenses, the efficiency of various detectors, the automation of sample stages, and the digital image acquisition systems has made the modern high-resolution FEG-SEM a powerful and high-throughput tool for examining the physicochemical properties of a plethora of inorganic materials and biological systems on a nanometer scale.

The development of low-voltage SEM (LV-SEM) clearly broadened the application of the SEM techniques to non-conducting or delicate specimens; LV-SEM is now the preferred mode of operation [20–23]. Although low-voltage SEM was not a new concept [1, 5] nanometer scale resolution is achievable only in the modern FEG-SEM; nanometer scale surface features of bulk samples is now obtainable with primary electron energies <1 keV. The significant reduction in the electron-specimen interaction volume clearly makes low-voltage SEM more surface-sensitive; and the charge balance at the specimen surface makes it possible to directly observe non-conducting materials without a conductive coating layer.

A very recent development is the use of a retarding field to modulate the landing energy of the primary electrons. By simply applying a negative potential to the specimen, a retarding field can be generated between the specimen and a grounded electrode just above the specimen; with this arrangement the specimen itself is part of a "cathode lens" system and high-resolution images can be obtained with ultra-low-energy electrons [24–25]. The use of primary electrons with a landing energy below 100 eV significantly enhances the surface sensitivity and greatly reduces the electron-beam-induced irradiation effects. With further improvement of low-voltage FEG-SEM optical systems the ultra-low-voltage SEM (ULV-SEM) technique will become a powerful tool for studying nanoscale systems.

The combination of a high-performance thermal FEG with improved condenser lens designs can produce electron nanoprobes with high stability and high probe-current, thus enabling various signals generated from the specimen to be collected with good signal-to-noise ratio. The incorporation of a high-sensitivity detector for collecting electron backscatter diffraction patterns (EBSP) allows lattice orientations of crystalline materials to be measured at each pixel and thus orientation images of individual grains can be formed. The incorporation of high-efficiency spectrometers makes it possible to perform high spatial resolution chemical microanalysis by XEDS [22–23] and wavelength dispersive spectroscopy (WDS) techniques.

Among many characterization techniques, microscopy should and will play an important role in understanding the nature of nanosystems. Microscopy and the associated analytical techniques can provide information on the physicochemical properties of individual nanocomponents as well as the spatial relationships among these nanocomponents. Among many microscopy techniques that are discussed in this volume, SEM techniques are unique in the sense that almost any sample can be examined in a modern, variable pressure or environmental FEG-SEM [26–29]; powders, films, pellets, conducting, semiconduing, non-conducting, "dirty", hydrated, and even wet samples can be examined with high spatial resolution and high image quality. Another advantage of the SEM is the amount of materials that can be examined: from macroscale such as pellets, powders, films, or even whole silicon wafers to nanoscale such as carbon nanotubes, metal nanoparticles, ceramic nanopowders, or nanoscale networks in polymers and fibers. The ability to quickly correlate the observations of nanostructures to macroscale properties is invaluable for providing statistically meaningful data in developing nanostructured systems.

The modern high-resolution FEG-SEM and associated techniques are suitable for applications to both the materials and life sciences. For example, it can be used in the automotive industry for metallurgy applications, in the chemical industry for

From Illuminating Laser
Objective
Lens
To Detector
Sample
in x-y
plane
Optic z axis
Dichroic
Mirror
Pinhole

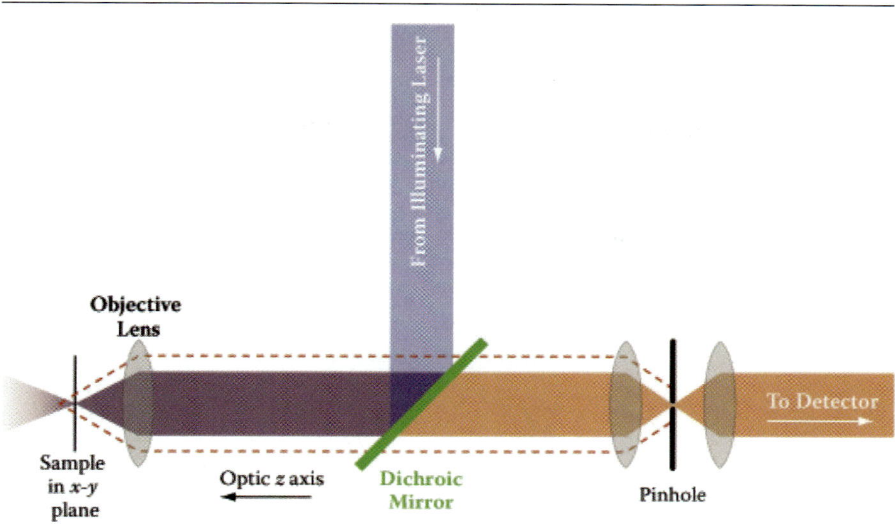

Plate 1.

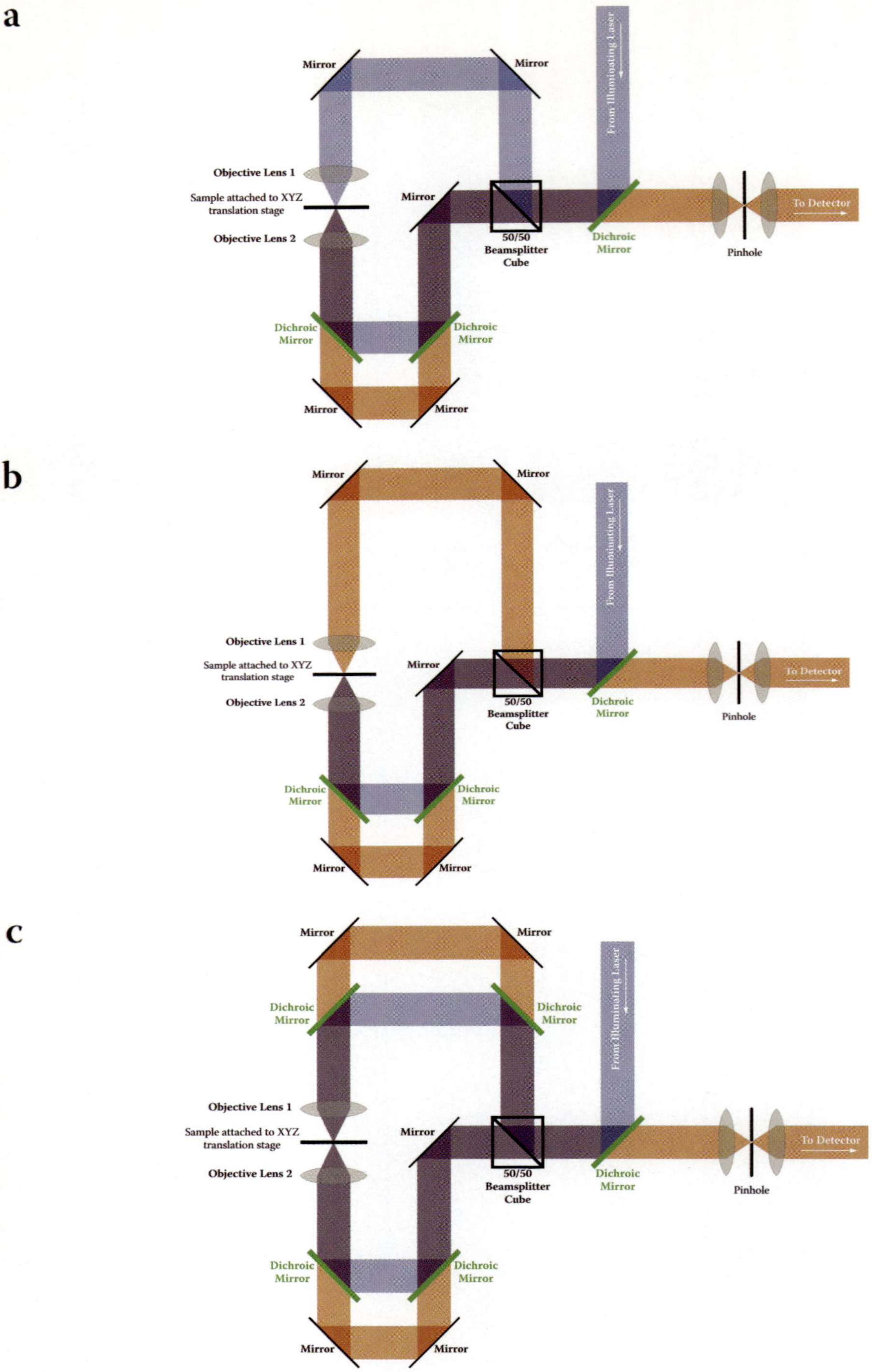
a
Mirror
Mirror
From Illuminating Laser
Objective Lens 1
Sample attached to XYZ translation stage
Objective Lens 2
Mirror
50/50 Beamsplitter Cube
Dichroic Mirror
Pinhole
To Detector
Dichroic Mirror
Dichroic Mirror
Mirror
Mirror
b
Mirror
Mirror
From Illuminating Laser
Objective Lens 1
Sample attached to XYZ translation stage
Objective Lens 2
Mirror
50/50 Beamsplitter Cube
Dichroic Mirror
Pinhole
To Detector
Dichroic Mirror
Dichroic Mirror
Mirror
Mirror
c
Mirror
Mirror
Dichroic Mirror
Dichroic Mirror
From Illuminating Laser
Objective Lens 1
Sample attached to XYZ translation stage
Objective Lens 2
Mirror
50/50 Beamsplitter Cube
Dichroic Mirror
Pinhole
To Detector
Dichroic Mirror
Dichroic Mirror
Mirror
Mirror

Plate 2.

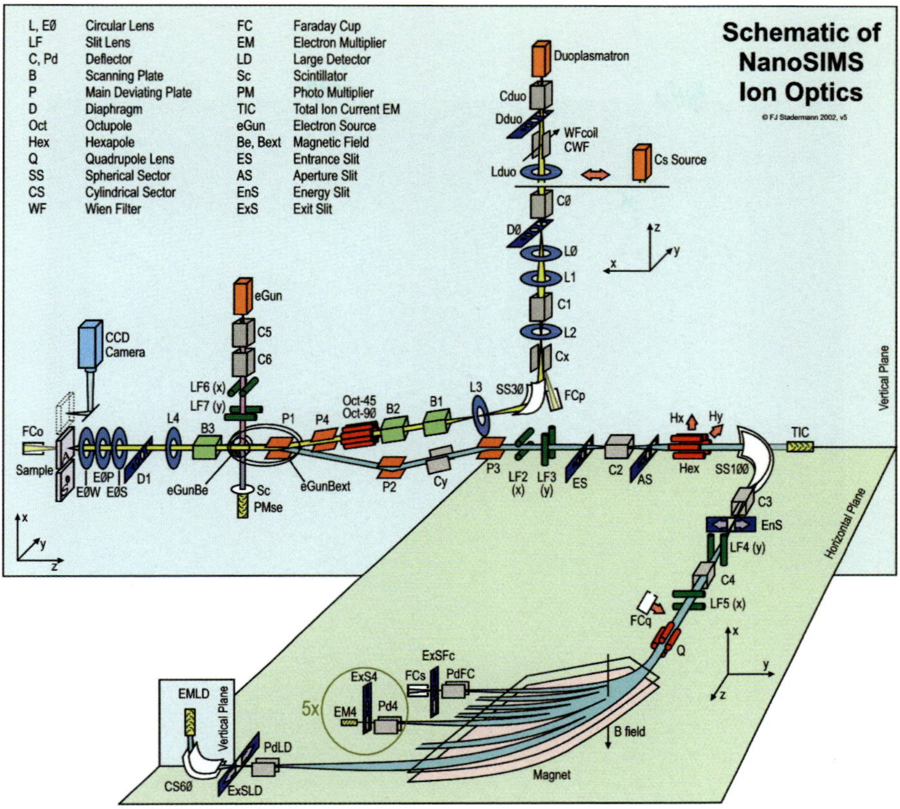

Schematic of NanoSIMS Ion Optics
© FJ Stadermann 2002, v5
L, EØ Circular Lens
LF Slit Lens
C, Pd Deflector
B Scanning Plate
P Main Deviating Plate
D Diaphragm
Oct Octupole
Hex Hexapole
Q Quadrupole Lens
SS Spherical Sector
CS Cylindrical Sector
WF Wien Filter
FC Faraday Cup
EM Electron Multiplier
LD Large Detector
Sc Scintillator
PM Photo Multiplier
TIC Total Ion Current EM
eGun Electron Source
Be, Bext Magnetic Field
ES Entrance Slit
AS Aperture Slit
EnS Energy Slit
ExS Exit Slit
Duoplasmatron
Cduo
Dduo
WFcoil
CWF
Lduo
Cs Source
CØ
DØ
LØ
L1
C1
L2
Cx
SS3Ø
FCp
eGun
C5
C6
CCD Camera
LF6 (x)
LF7 (y)
L4
B3
P1
P4
Oct-45
Oct-9Ø
B2
B1
L3
FCo
Sample
EØP
EØW
EØS
D1
eGunBe
Sc
PMse
eGunBext
P2
Cy
P3
LF2 (x)
LF3 (y)
ES
C2
AS
Hx
Hy
Hex
SS1ØØ
TIC
Vertical Plane
Horizontal Plane
C3
EnS
LF4 (y)
C4
LF5 (x)
FCq
Q
ExSFc
FCs
PdFC
ExS4
5x
EM4
Pd4
B field
EMLD
Vertical Plane
PdLD
CS6Ø
ExSLD
Magnet

Plate 3.

~50 nm
1.00 µm
fcc_9.im - 28Si
Lin[0..91]

(e)

1.00 µm
28Si-
27Al16O-

(f)

Plate 4.

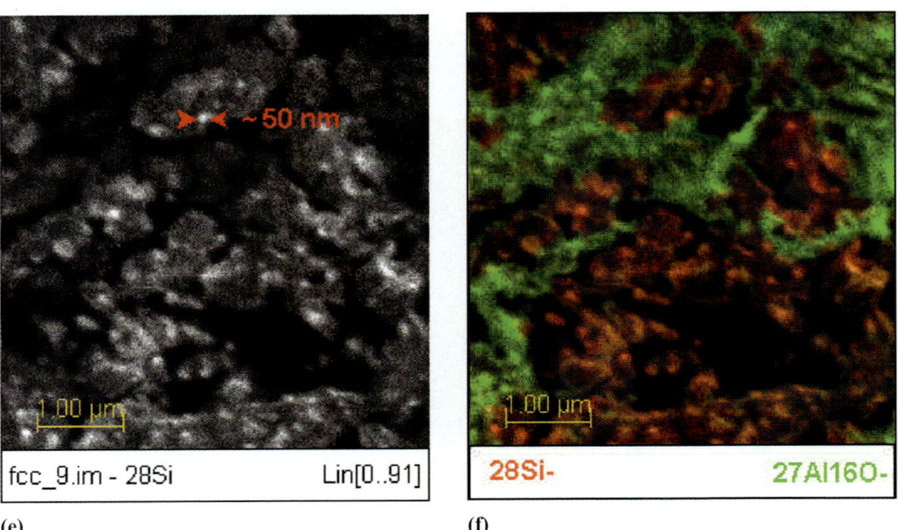

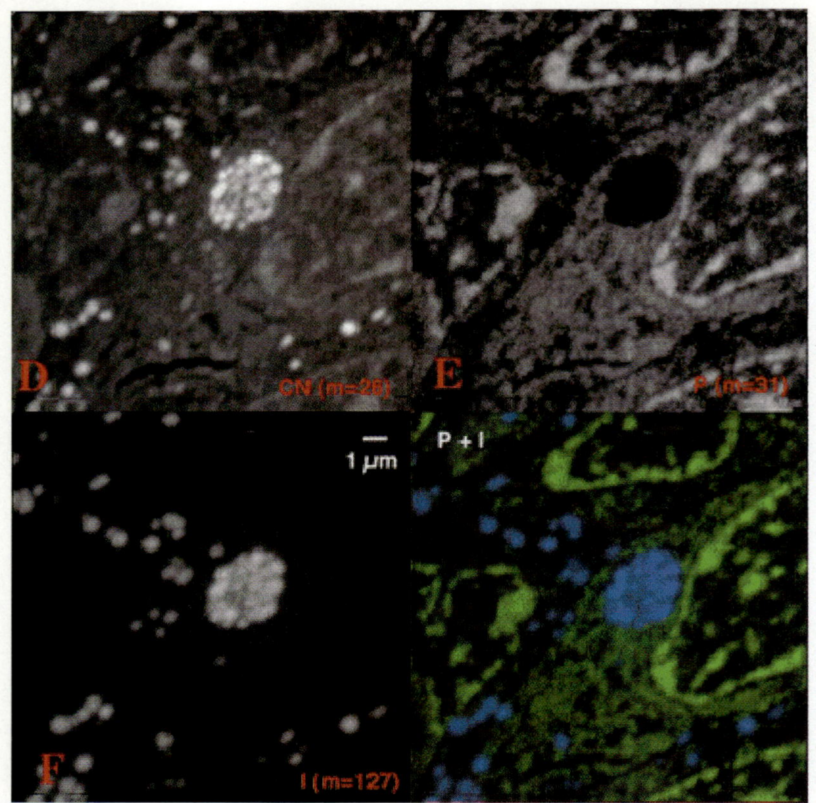

Plate 5.

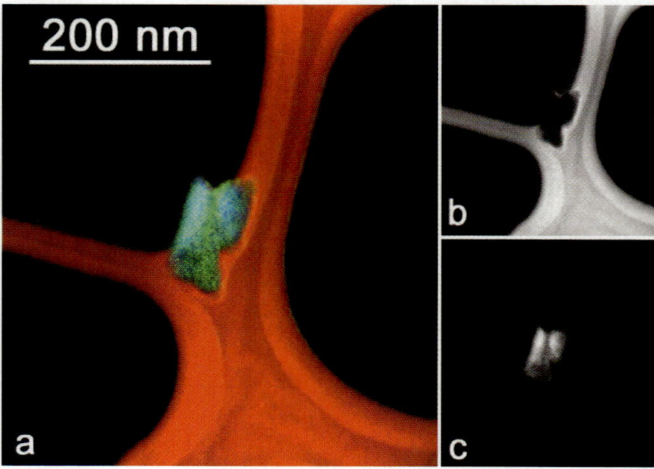

Plate 6.

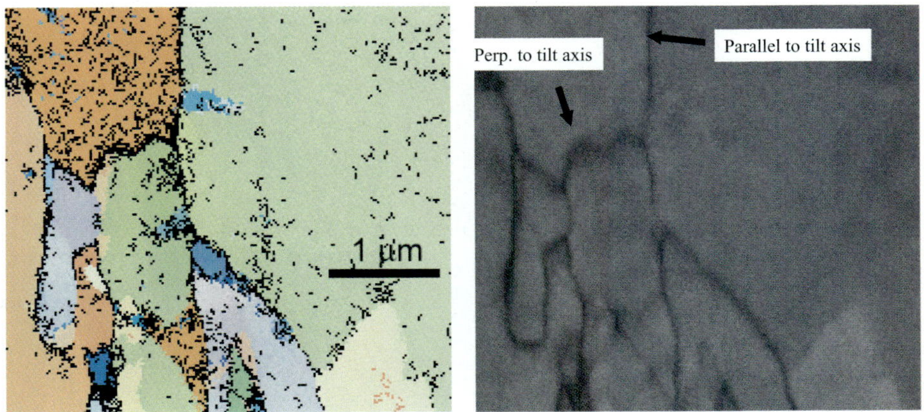

Plate 7.

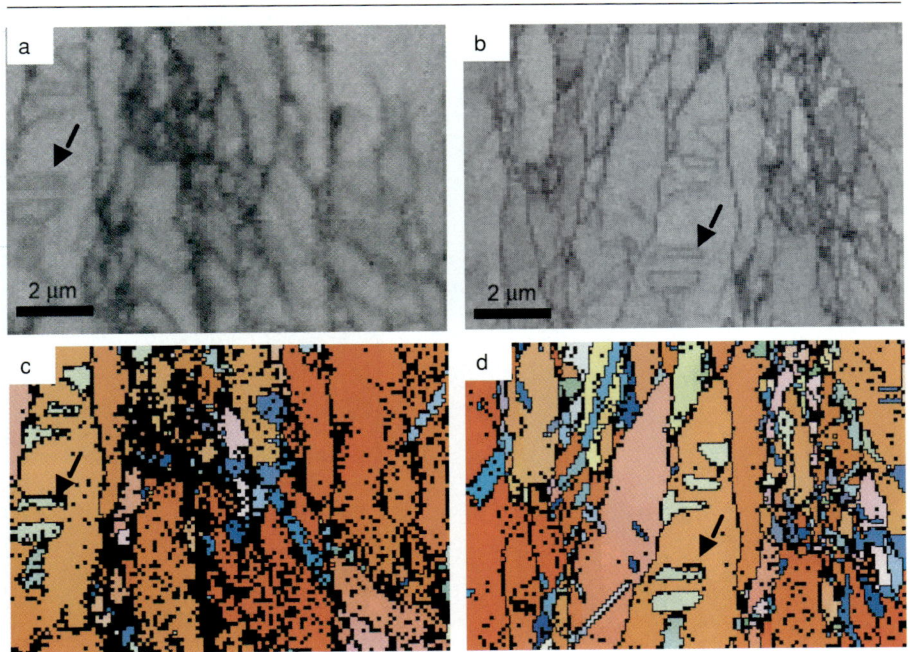

Plate 8.

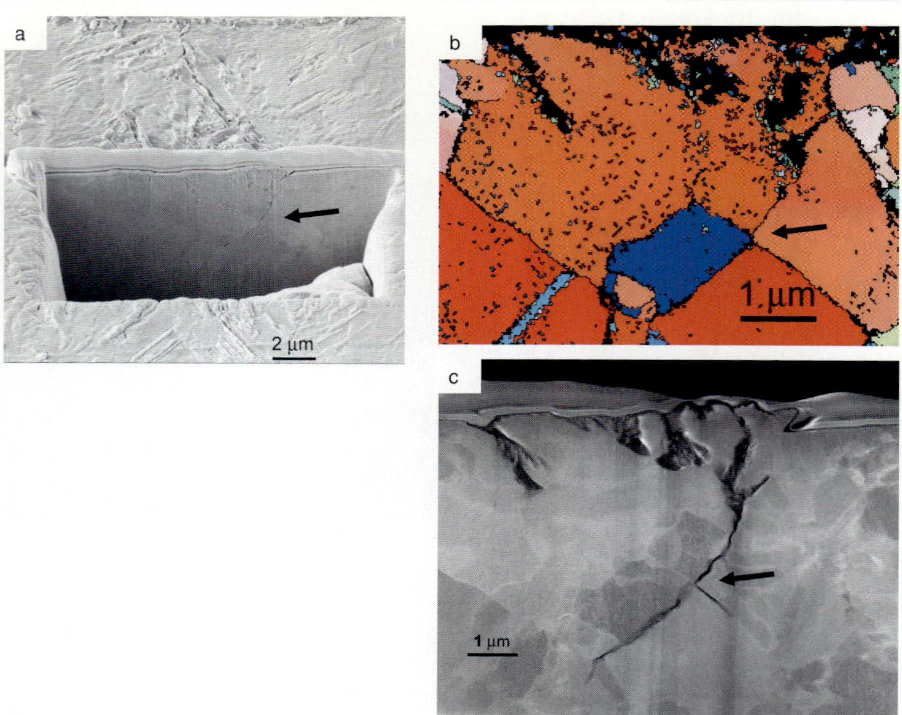

Plate 9.

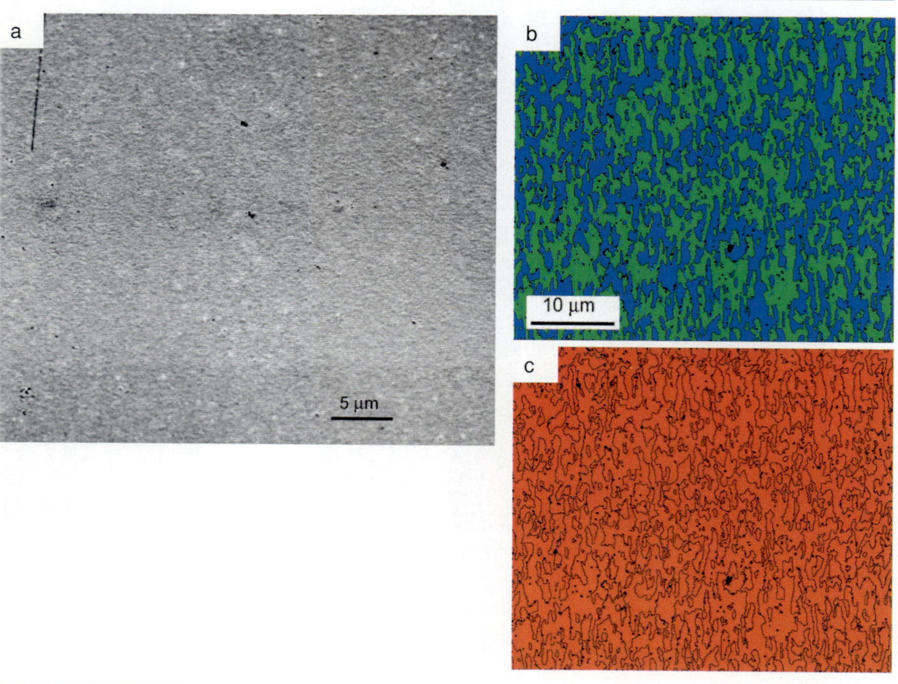

Plate 10.

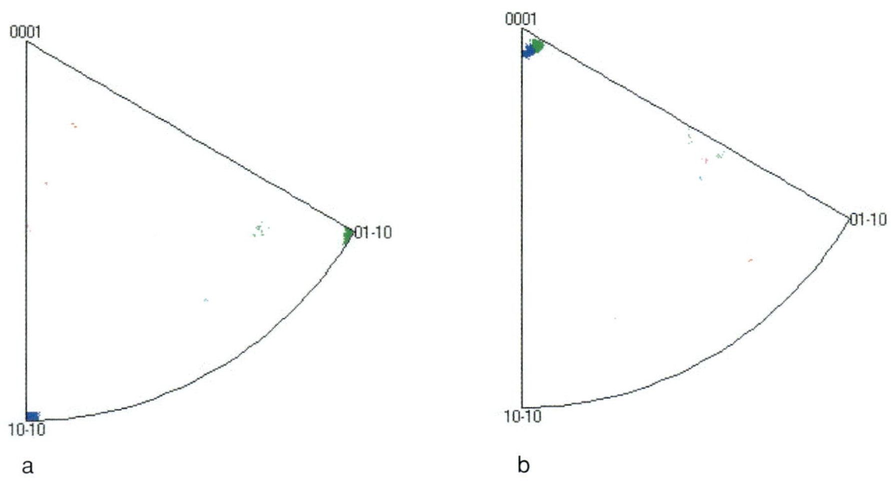

Plate 11.

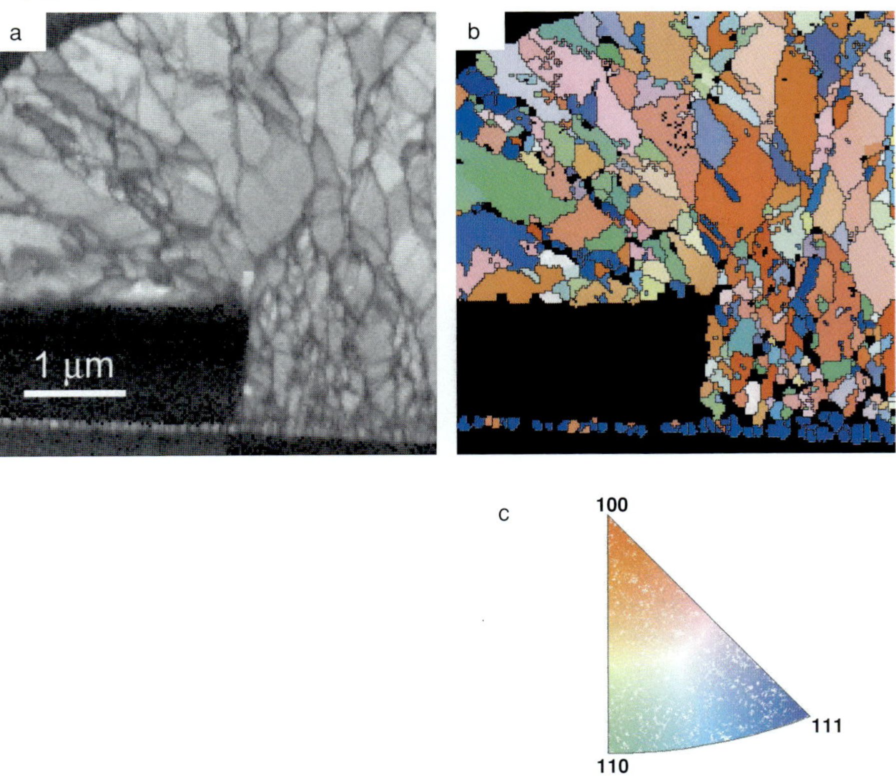

Plate 12.

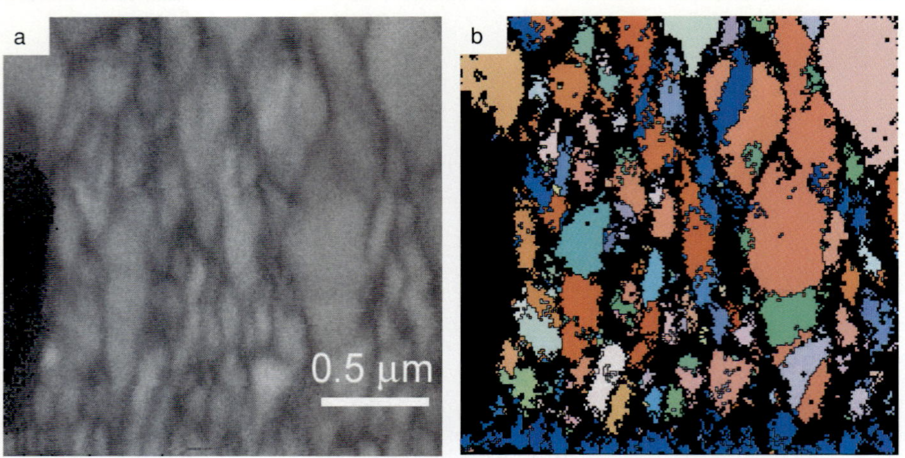

Plate 13.

Plate 14.

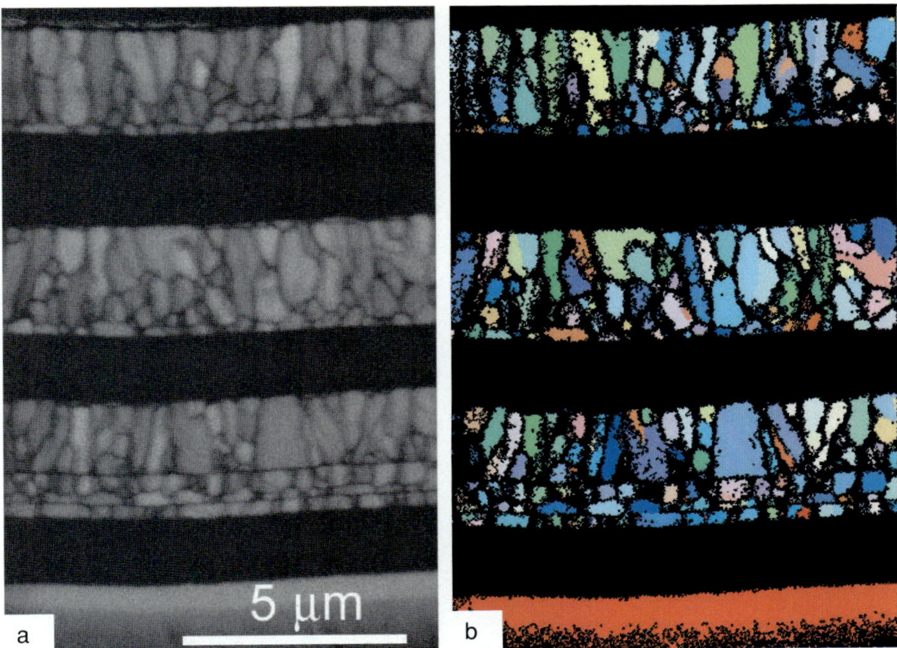

Plate 15.

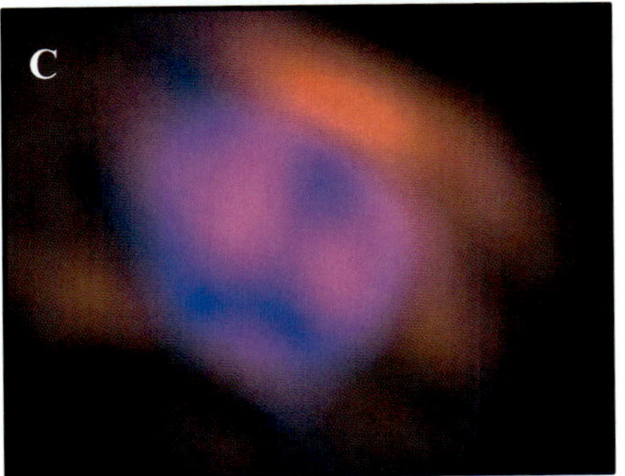

Plate 16.

Bonding electrons in Cuprite
Copper
Atom
Oxygen
Atom
Nature, 1999
Zuo et. al.,

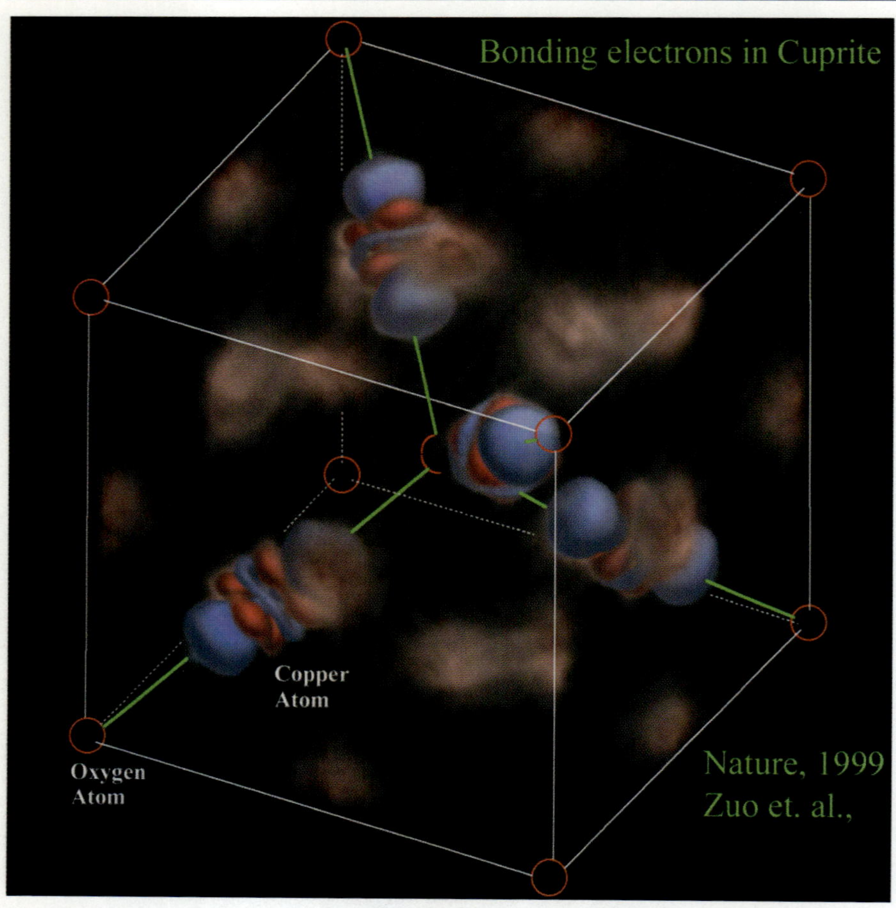

Plate 17.

a)
b)
2-D simulation
y (um)
x (um)
1.10
1.05
1.00
0.95
0.90
0.85
0.20
0.25
0.30
0.35
0.40
0.45
0.50
0.55
0.60
0.65
0.70
0.75
0.80
0.85
0.90

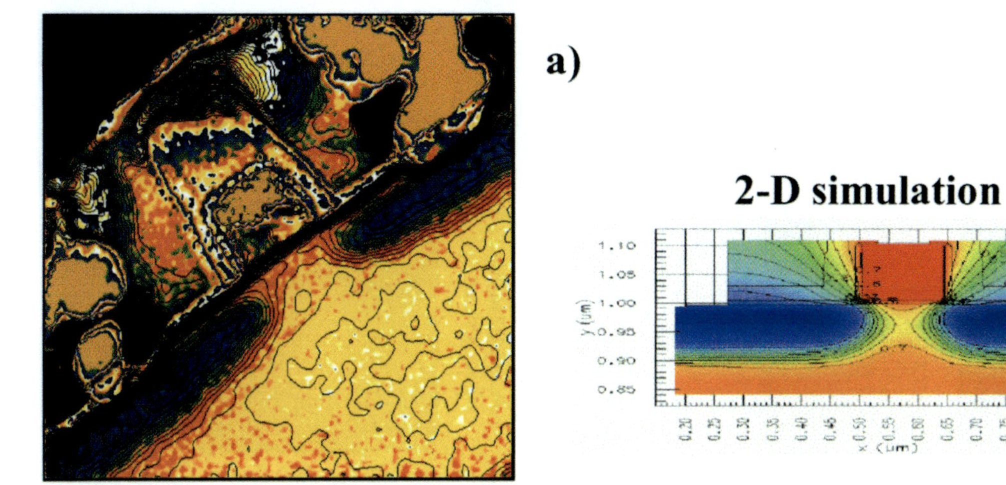

Plate 18.

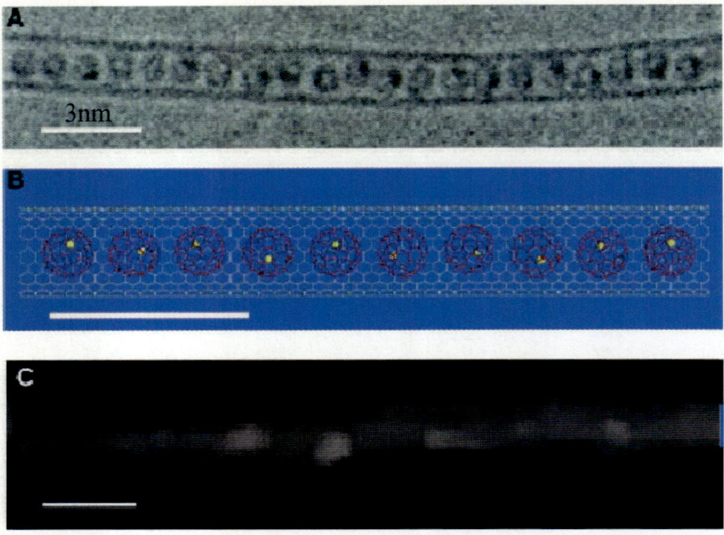

Plate 19.

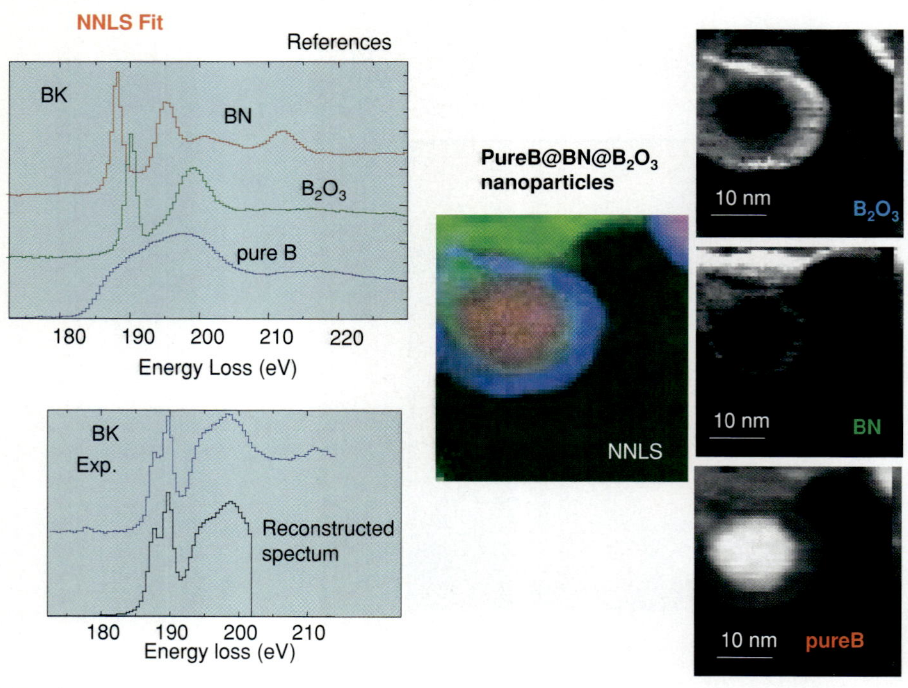

Plate 20.

polymer and catalyst applications, in the semiconductor industry for device applications, in the food industry for particle and morphological applications, in the consumer health industry for nano-particulate applications, and in the pharmaceutical industry for drug and drug delivery applications. With the use of a low temperature system, fully hydrated biological systems can be examined at high spatial resolution providing morphological and ultrastructural information. With the rapid development of the computer technology and automated operations, live SEM images are now used in classroom lectures or are observed by scientists located at remote sites; telemicroscopy is now a reality.

Although there are many imaging, diffraction, and spectroscopic techniques available in the modern FEG-SEM, we review, in this chapter, only the recent development of high-resolution imaging techniques and the applications of these techniques to studying various types of nanophase materials. Interested readers should consult the existing textbooks for more detailed discussions on the fundamentals of SEM and related techniques [30–31]. This chapter is closely related to the chapters in this volume by D. Newberry on high-resolution quantitative chemical microanalysis, by J. Michael on EBSP and orientation imaging microscopy, and by J. M. Cowley on scanning transmission electron microscopy.

2. ELECTRON-SPECIMEN INTERACTIONS

When an electron beam interacts with a bulk specimen, a variety of electron, photon, phonon, and other signals can be generated (Fig. 2). There are three types of electrons that can be emitted from the electron-entrance surface of the specimen: secondary electrons with energies <50 eV, Auger electrons produced by the decay of the excited atoms, and backscattered electrons that have energies close to those of the incident electrons. All these signals can be used to form images or diffraction patterns of the specimen or can be analyzed to provide spectroscopic information. The de-excitation of atoms that are excited by the primary electrons also produces continuous and characteristic X-rays as well as visible light. These signals can be utilized to provide qualitative, semi-quantitative, or quantitative information on the elements or phases present in the regions of interest. All these signals are the product of strong electron-specimen interactions, which depends on the energy of the incident electrons and the nature of the specimen.

2.1. Electron-Specimen Interactions in Homogeneous Materials

Both elastic and inelastic scattering processes critically depend on the energy of the incident electrons. In a SEM instrument, the elastic and inelastic mean-free-paths of the primary electrons rapidly decrease with decreasing energy. Both the electron range, which is a measure of the penetration depth of the incident electrons, and the interaction volume, which is a measure of the diffusion of the incident electrons, are significantly reduced at low energies. By performing Monte Carlo simulations of the electron cascading processes inside a specimen, the energy dependence of the change in

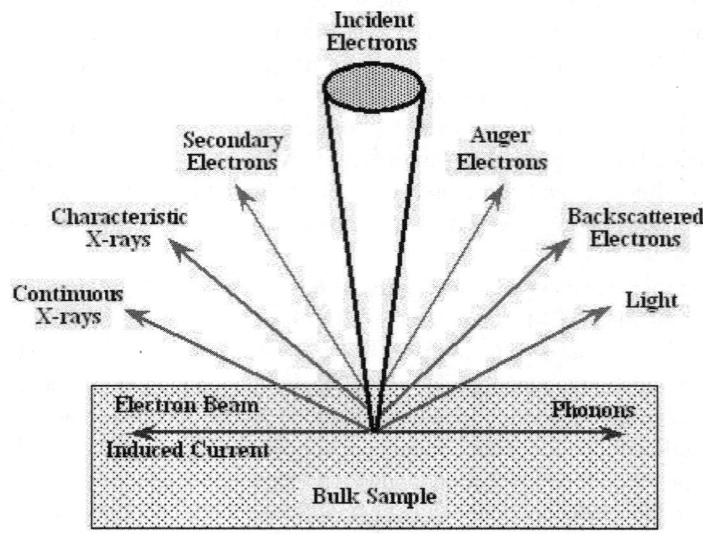

Figure 2. Schematic drawing illustrates the signals generated inside a scanning electron microscope when an electron beam interacts with a specimen.

the interaction volume, the yields of secondary electrons and backscattered electrons, and the penetration depth can be demonstrated [32].

As an example, Fig. 3 shows trajectories of incident electrons in bulk carbon and platinum for electrons with energies of 1 keV and 15 keV, respectively. In low atomic-number (Z) materials such as carbon, a large fraction of the high-energy electrons is scattered through small angles and subsequent electron diffusion results in a pear-shaped spatial distribution (Fig. 3b). In high-Z materials such as platinum, large-angle scattering events are enhanced and thus the electron trajectories are more strongly coiled up (Fig. 3d), reducing the penetration of the primary electrons into the sample. At low voltages, the differences in electron range between low-Z and high-Z materials, although still appreciable, are considerably reduced. For low-Z materials such as carbon, alumina, or silica, the electron interaction volume is reduced by five to six orders of magnitude from 15 keV to 1 keV. The corresponding electron range decreases by about 100 times. For high-Z materials such as platinum, gold, or tungsten, however, the decrease in the electron range and the interaction volume is not as drastic as that for low-Z materials. It is interesting to note that the range of 1-keV electrons in bulk platinum is less than about 10 nm.

Figure 3 also shows that at higher energies there are significant differences in the electron range and interaction volume between low-Z and high-Z materials. At lower energies, however, these differences rapidly diminish. At electron energies <0.5 keV, all materials have approximately the same range and interaction volume. At ultra-low electron energies (<100 eV), surface scattering may play a significant role in determining the nature of the electron-specimen interactions. The imaging theory of

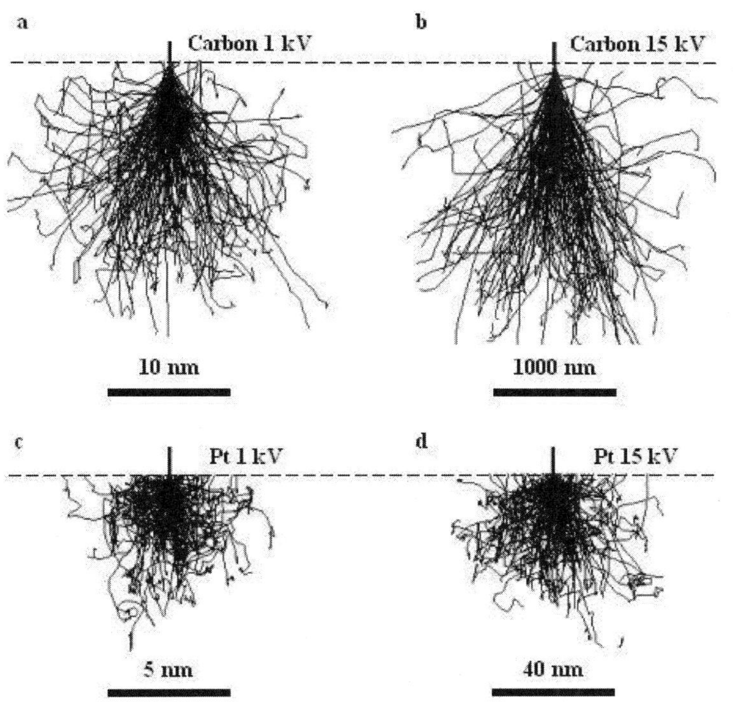

Figure 3. Monte Carlo simulations of the electron beam interactions with bulk carbon (a, b) and platinum (c, d) for 1-keV (a, c) and 15-keV (b, d) electrons, respectively.

low energy electron microscopy (LEEM) may be applicable here if clean surfaces are examined in ultra-high vacuum SEM [33]. In most practical applications, however, this condition will not be met.

Although the electron backscattering coefficient of a bulk sample is almost constant for electrons with energies >10 keV, it changes significantly between 1 keV and 10 keV because of the stronger electron-specimen interactions at low energies. For incident electrons with energies >10 keV, the electron backscattering coefficient increases monotonically with increasing Z. At low energies, however, with decreasing electron energy the electron backscattering coefficient increases for low-Z materials and decreases for high-Z materials. Therefore, we have to be cautious in interpreting low-voltage backscattered electron images. At ultra-low voltages, both the SE and BE signals are surface sensitive and it is possible that the BE signal may have an average escape depth smaller than that of the corresponding SE signal. Strongest surface interactions occur for electrons with energies in the range of 20 eV to 100 eV.

The drastic reduction in the interaction volume provides many advantages of using low-energy electron beam techniques to characterize a plethora of materials including semiconductor devices and nanoparticles. With low-energy electrons, we can perform high spatial resolution microanalysis of bulk samples; we can reduce the electron-induced damaging of delicate specimens; we can tune the surface-sensitivity of the detected signals; and we can examine non-conducting or less conducting materials with high spatial image resolution.

2.2. Electron-Specimen Interactions in Composite Samples

When the sample of interest is chemically non-homogeneous, the electron-specimen interaction processes are more complex. We use here a model system consisting of metallic nanoparticles supported on the surface of, or embedded in, a carbon substrate to illustrate the complexity of electron-specimen interactions in composite systems. The fundamental scattering processes we discussed here should apply to other nanoscale, non-homogeneous systems as well.

To demonstrate the effect of small metal particles on the scattering of incident electrons, we have performed Monte Carlo simulations of the trajectories of incident electrons and the corresponding backscattering coefficients from small particles supported on, or embedded in, a bulk substrate. Backscattered electrons are defined here as those incident electrons that are scattered out of the target after suffering single or multiple deflections through such an angle that they leave the target on the side by which they entered. A plural scattering model with appropriate modified scattering cross-sections was used for all the simulations [32].

Figures 4a and 4b show electron trajectories in bulk carbon for electron energies of 3 keV and 30 keV, respectively. In the plots shown in Fig. 4, the trajectories of all incident electrons are traced until they rest in the sample or escape out of the sample. It is clearly shown that a large number of the 3-keV electrons are multiply scattered close to the electron entrance surface while the 30-keV electrons penetrate deep inside the carbon without encountering many scattering events. When a small, heavy-element particle is located on, or inside, a light-element substrate, the electron-specimen interaction with the composite sample is more complicated. Depending on the size and the location of the particle as well as the energy of the incident electrons, the electron-particle interaction may play a dominant role in determining the scattering processes of the incident electrons. Figures 4c and 4d show trajectories of 3-keV and 30-keV electrons, respectively, for normal incidence at the center of a Pt particle with a size of 20-nm in diameter, positioned on the external surface of a carbon substrate. Most of the 3-keV electrons interact strongly with the 20-nm Pt particle and a large fraction of the incident electrons can escape at the particle surface as backscattered or as high-angle scattered electrons (The trajectories of the backscattered electrons are not easily discernible in the plot. But careful examination of Figure 4c should show that many of the incident electrons are backscattered by the small Pt particle). In contrast, most of the 30-keV electrons penetrate through the particle and diffuse deep into the carbon substrate although a few of the incident electrons can escape the particle surface. Note that some of the high-angle scattered electrons re-enter the carbon substrate and can travel along the carbon surface for a distance from several to tens of microns. These "grazing" electrons can generate appreciable amount of secondary electrons from the carbon substrate. When the 20-nm Pt particle is positioned 20 nm below the

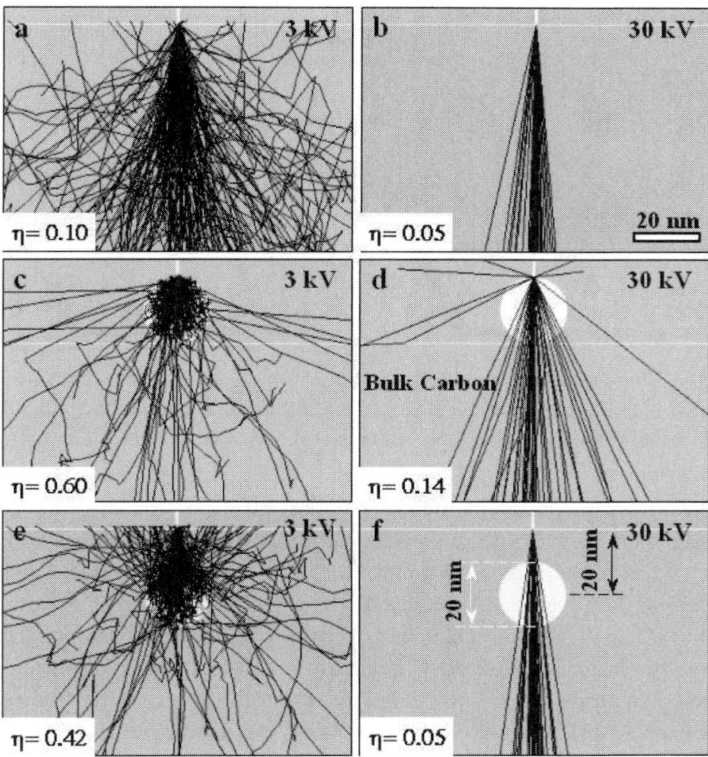

Figure 4. Monte Carlo simulations of the electron beam interactions with bulk carbon (a, b), a 20 nm Pt particle on the surface of the carbon support (c, d), and a 20 nm Pt particle located 20 nm inside the carbon support (e, f) for 3-keV (left hand side) and 30-keV (right hand side) electrons, respectively. The backscattering coefficients are also shown.

carbon surface, a large fraction of the 3-keV electrons are first backscattered by the Pt particle, then travel through the carbon substrate, and finally exit the carbon surface as backscattered electrons (Fig. 4e). The 30-keV electrons, however, do not experience significant high-angle scattering events (Fig. 4f); most of the high-energy electrons pass through the Pt particle without being backscattered.

The value of η for a particle-substrate composite strongly depends on the primary electron energy E and the size and location of the particle. A 20-nm Pt particle located on the surface of a carbon support may give a very high contrast in BE images obtained with 3-keV electrons since $\eta_P = 0.6$ for the platinum-carbon composite sample and $\eta_C = 0.1$ for the carbon support. A contrast

$$C = 100\% \times (\eta_P - \eta_C)/(\eta_P + \eta_C) = 71\% \quad (1)$$

is then obtained for this Pt particle in high-resolution BE images. On the other hand, a 20-nm Pt particle embedded 20 nm below the surface of the carbon support may not be

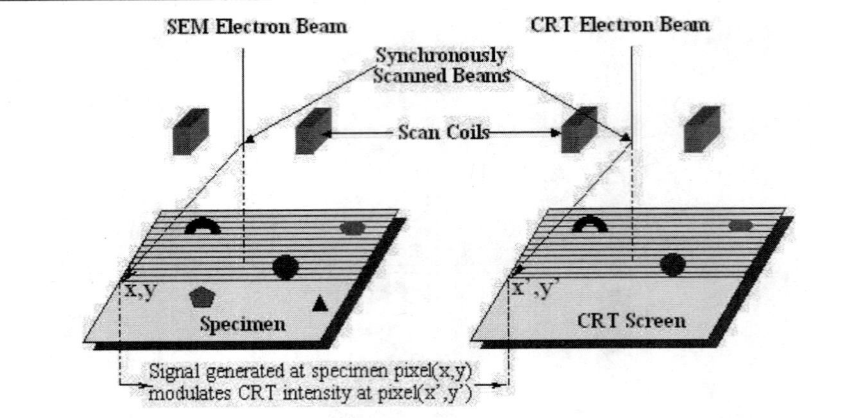

Figure 5. Schematic diagram illustrates the formation of SEM images.

easily observed in BE images obtained with 30-keV electrons (Fig. 4f) since the electron backscattering coefficient for the particle-carbon composite sample ($\eta_P = 0.05$) does not change appreciably from that for pure carbon. If 3-keV electrons are used (Fig. 4e), this Pt particle, however, can be easily observed with a contrast of about 62% since now $\eta_P = 0.42$, much larger than that of pure carbon. Electron interactions with composite samples are complicated, especially when low-energy electrons are used. The interaction processes of low-energy electrons also critically depend on the size of the Pt particle. For example, although 3-keV electrons cannot penetrate through a Pt particle with a size of 50 nm in diameter, electrons scattered to high angles by the Pt particle can exit the particle surface and strongly interact with the carbon substrate.

3. INSTRUMENTATION OF THE SCANNING ELECTRON MICROSCOPE

3.1. General Description

Unlike in a TEM where a stationary, parallel electron beam is used to form images, the SEM, similar to that of a fax machine or a scanning probe microscope, is a mapping device. In a SEM instrument, a fine electron probe, formed by using a strong objective lens to de-magnify a small electron source, is scanned over a specimen in a two-dimensional raster. Signals generated from the specimen are detected, amplified, and used to modulate the brightness of a second electron beam that is scanned synchronously with the SEM electron probe across a cathode-ray-tube (CRT) display. Therefore, a specimen image is mapped onto the CRT display for observation (Fig. 5). If the area scanned on the sample is A_s and the corresponding area on the CRT display is A_d, then the magnification (M) of a SEM image is simply given by $M = A_d/A_s$. The SEM magnification is purely geometric in origin and can be easily changed by varying the scanned area on the sample. These operating principles are the same as those of the STEM instrument.

Since SEM is a serial recording instead of a parallel recording system, the whole process of generating a SEM image could be slower than that in the TEM. A high-quality SEM image usually builds up over several seconds to several minutes, depending on the types of signals; thus, high probe-current within a small electron nanoprobe is desirable and microscope and sample stability is critical to obtaining high-quality and high-resolution SEM images. Unlike in TEM, there is no rotation between the object and image planes, and the microscope magnification can be changed without refocusing the electron beam to obtain an optimum focused image. The resolution of SEM images at high magnifications is primarily determined by the size of the incident electron probe, the stability of the microscope and the sample, and the inherent properties of the signal-generation processes.

Multiple imaging and analytical detectors have been developed to simultaneously collect several signals that can be displayed individually or combined in perfect register with each other. This unique capability makes the SEM a powerful microanalytical tool since multiple views of a sample, in different imaging, diffraction, or spectroscopy modes, can be collected, analyzed, and compared in a single pass of the incident electron beam. When a thin specimen is used, both bright-field (BF) and annular dark-field (ADF) detectors can be used to form STEM images [34]; the image formation theory, the contrast mechanisms, and the applications of the STEM are fully discussed by Cowley in this volume. Figure 6 illustrates the available detectors commonly used for collecting imaging and analytical signals in a modern high-resolution FEG-SEM instrument.

Because of the intrinsic nature of a mapping device, the SEM is ideal for digital imaging and for on-line or off-line processing of images, spectra, and EBSP. Signals from several detectors can be digitally acquired either simultaneously or independently; different signals can also be combined together by addition, subtraction, multiplication, or other mathematical manipulations to gain insight about the structure of the sample. These digital images or spectra can be electronically transferred to remote locations through the intranet/internet or the world-wide-web for live image observation or for fast dissemination of vital information.

3.2. Performance of a Scanning Electron Microscope

The performance of an electron microscope is generally defined by its achievable spatial resolution. In the case of the high spatial resolution FEG-SEM the attainable resolution is determined by many parameters including the diameter d of the effective electron probe, the total beam current I_b contained within that probe, the magnification of the image (or pixel resolution), and the type of the imaging mode and the corresponding electron-specimen interaction processes.

The diameter of the effective electron-probe at the specimen surface is by far the most important parameter that determines the performance of the FEG-SEM. The current distribution in an electron nanoprobe depends on the beam parameters (brightness β, energy E, and energy-spread ΔE), the lens system parameters (spherical and chromatic aberration coefficients C_s and C_c), the application settings (probe current I_b and the aperture angle α), and the focus value f [30].

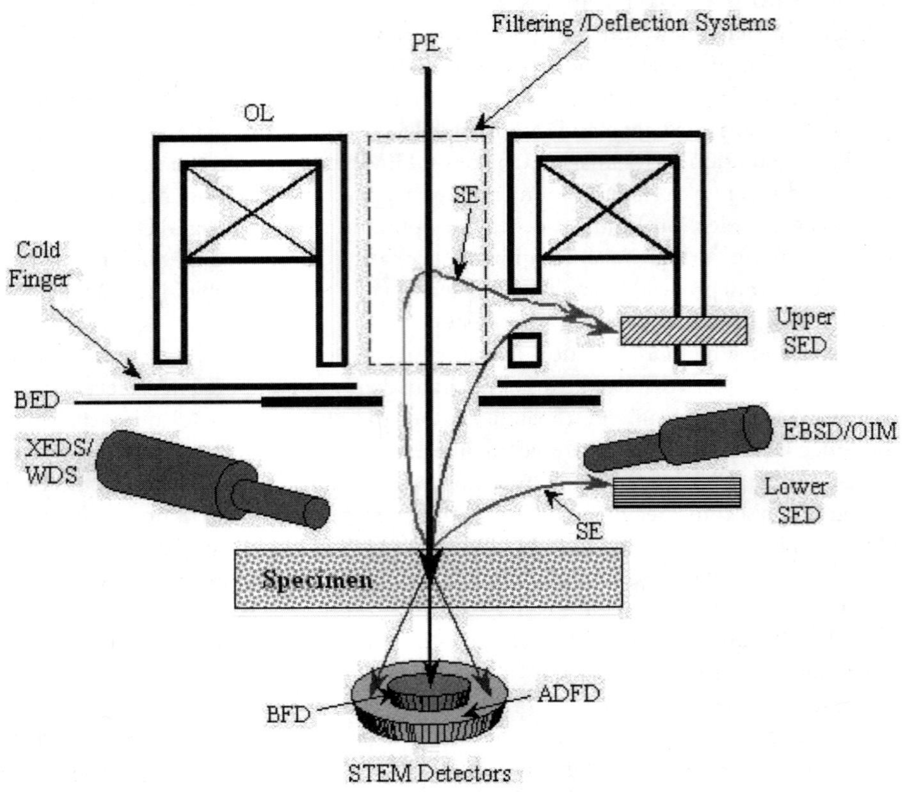

Figure 6. Schematic diagram illustrates the available detectors in a modern FEG-SEM: primary electron (PE), objective lens (OL), secondary electron detector (SED), backscattered electron detector (BED), X-ray energy dispersive spectrometer (XEDS), wavelength dispersive spectrometer (WDS), electron backscattering diffraction (EBSD), orientation imaging microscopy (OIM), secondary electron (SE), bright-field detector (BFD), annular dark-field detector (ADFD), scanning transmission electron microscopy (STEM).

The current density within an electron nanoprobe is not uniform; thus, measurement of the width of the current density distribution is not a good measure of the characteristics of the nanoprobe. The commonly used FWHM (full width at half maximum) method is too simplistic and can give misleading information about the characteristics of the electron probe; electron probes having the same FWHM value can have drastically different current density distributions as shown in Fig. 7 of three very different probes but they all have the same FWHM value. More than 98% of the electrons are contained within the FWHM of the top-hat probe (#1); about 78% of all electrons are contained within the FWHM of the Gaussian probe (#2); and only about 57% of the total number of electrons are contained within the FWHM of the spike probe (#3). Probe #3 may yield higher image resolution but probe #1 can provide a better "analytical" resolution.

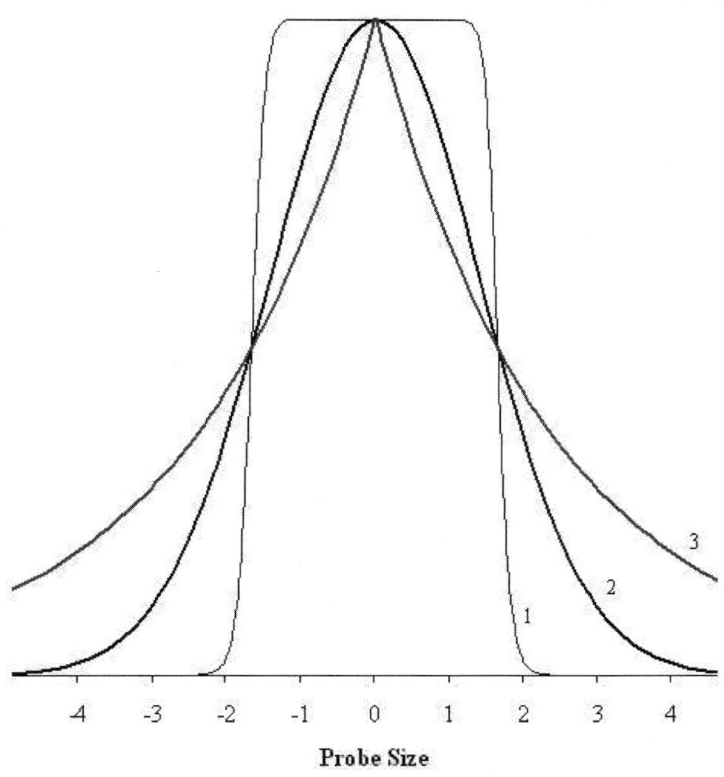

Figure 7. Calculated one-dimensional curves illustrate that different current density distributions within a probe can give the same value of full width at half maximum (FWHM) but may give very different image resolution or contrast.

To most of the practitioners, the full wave optical description of the current density distributions of electron nanoprobes and how these distributions vary with the beam and lens parameters is too complicated to provide intuitive guidance for understanding the electron probe-forming system in a FEG-SEM. Although the probe current distribution in an electron nanoprobe needs to be calculated from the wave-optical theory, geometrical optical theory of the electron-probe formation processes can provide some insights and can provide practical guides to control and to manipulate attainable sizes of the electron nanoprobes.

Considering the geometric probe-diameter d_0 and its broadening due to the action of the lens aberrations of the objective lens, the effective electron-probe diameter d_p can be estimated by [30, 30–36]

$$d_p = \left(d_0^2 + d_d^2 + d_s^2 + d_c^2\right)^{1/2}$$
$$= \left\{\left[C_0^2 + (0.6\lambda)^2\right]\alpha_p^{-2} + 0.25C_s^2\alpha_p^6 + \left(C_c\frac{\Delta E}{E}\right)^2\alpha_p^2\right\}^{1/2} \quad (2)$$

where d_d is the probe broadening due to the diffraction effect, d_s due to the effect of spherical aberration, and d_c due to the effect of chromatic aberration. The parameter α_p is the electron probe aperture and is defined by the size of the objective aperture. The parameters C_s and C_c are the spherical and chromatic coefficients of the final probe-forming lens, respectively. The parameters λ and E are the wavelength and the energy of the primary electrons, respectively; and ΔE is the energy spread of the electron beam. The parameter C_0 contains the probe current I_p and the gun brightness β and is given by

$$C_0 = \left(\frac{4I_p}{\pi^2\beta}\right)^{1/2} \tag{3}$$

The final probe size is therefore determined by the size of the objective aperture, the gun brightness and the total probe current, the spherical and chromatic aberration coefficients, and the energy and energy spread of the primary electron beam.

Figure 8a shows each of the various factors affecting the effective probe size and their dependence on the probe-aperture angle for two electron energies. The logarithmic plot is used here to show the various dependencies and for easy discussion and visualization. The diffraction lines for both the 30-keV and the 1-keV plots have a slope of −1; the probe size can never reach the regions below these lines due to the uncertainty principle. The plots for chromatic aberration give a line of slope +1; the effect of chromatic aberration on the final probe size becomes dominant with decreasing incident electron energy. The effect of spherical aberration does not depend on the incident electron energy; with a line of slope +3, it quickly becomes, however, dominant when larger probe apertures are used.

In a modern FEG-SEM, when high-energy electrons (10–30 keV range) are used, the contribution of d_c to the final probe size is negligible; with a high-brightness FEG, the probe size of a high-energy electron beam is primarily determined by the C_s value, the diffraction limit, and the total beam current. When low-energy electrons (<5 keV) are used, however, the effect of the chromatic aberration becomes increasingly dominant and the achievable resolution is primarily limited by the C_c value.

Figure 8a also shows the dependence of the effective probe sizes $d_{1\text{keV}}$ and $d_{30\text{keV}}$ on the probe-aperture angle (ref equation (2)). There exists an optimum probe aperture which provides the minimum probe size. The value of the optimum probe aperture depends on the values of C_c, C_s, C_0, ΔE, and the incident beam energy E. Figure 8b shows the dependence of the minimum probe size on the energy of the incident electrons (RMS curve); and Fig. 8c shows, for 30-keV electrons, how the total probe current affects the obtainable minimum probe size and the value of the optimum probe aperture.

Figure 8d shows how the optimum probe aperture and minimum probe size vary with C_c and C_s values for 1-keV electrons. To achieve a 1-nm probe size with 1-keV electrons both the chromatic and spherical aberrations of the objective lens have to be

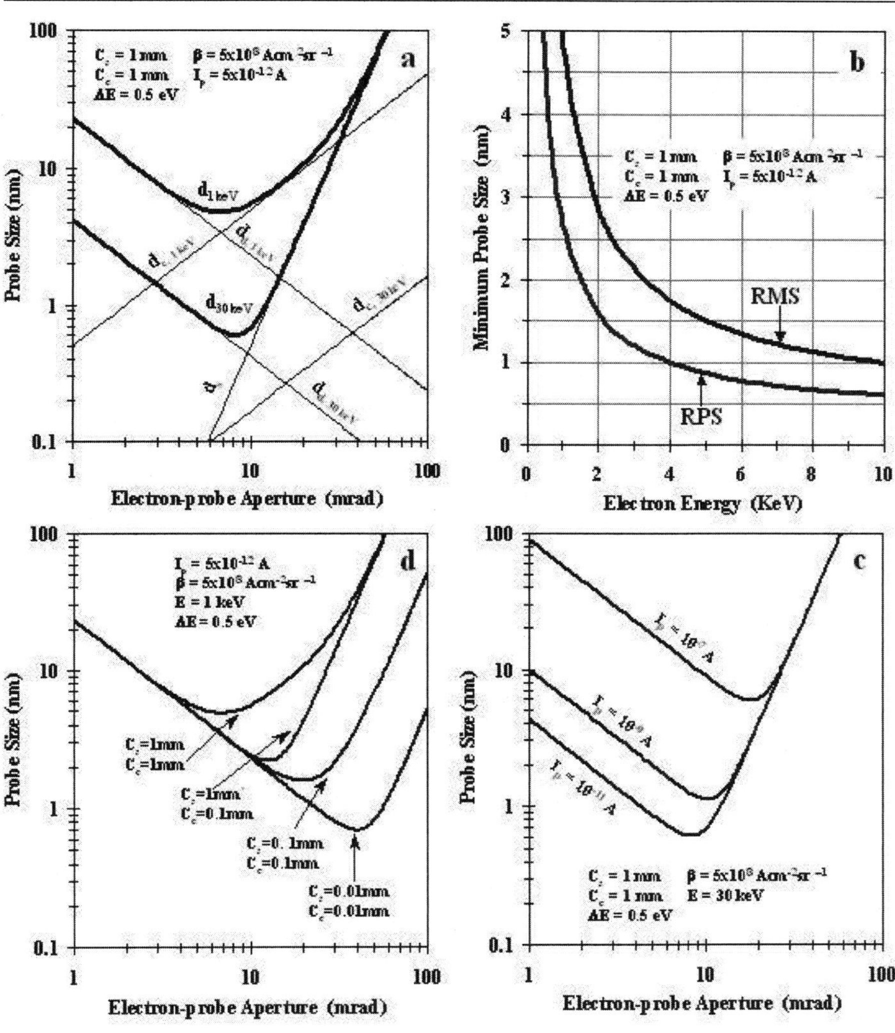

Figure 8. Plots (a) show the dependence of the probe size on the probe-aperture angle for diffraction-limited probe, spherical aberration limited probe, chromatic aberration limited probe, and the root mean square sum of these contributions for 1-keV and 30-keV electrons, respectively. Plots (b) show the dependence of minimum probe size on the primary electron energy based on the root mean square (RMS) rule and the root power sum (RPS) rule. Plots show (c) the dependence of the probe size on the probe-aperture angle for different total beam current values. Plots (d) show the dependence of the obtainable minimum probe sizes on the spherical and chromatic aberration coefficients.

dramatically reduced. This has been recently achieved by using C_c and C_s correctors in a low voltage FEG-SEM [39]. Besides achieving ultrahigh image resolution, other advantages of using aberration-correctors include increase of the total beam current at a given probe size and the increase of the effective working distance without sacrificing the image resolution. For ultra-low-energy electrons, the use of a cathode lens in combination with the electron-focusing lens may be a better alternative for achieving high spatial resolution imaging [40].

While the plots in Fig. 8 allow us to gain some insight into the various contributions to the probe size formation and how the electron probe changes with the probe aperture, they do not allow us to calculate the smallest probe size achievable. The reason, as pointed out in [37], is that we cannot simply form the root mean square of the various contributions to find the minimum size of the electron probe since different components have very different characters and may not even occur at the same position along the optical axis of the electron microscope. A root-power-sum algorithm is recently proposed to find the optimized probe size for given parameters [38]; and the authors showed that the values obtained from this new algorithm is very close to those given by the full wave optical calculations. The dependence of the optimum probe size (the probe size here is defined as the width of the probe current distribution that contains 50% of the total probe current) calculated from the root-power-sum algorithm is plotted in Fig. 8b (the RPS curve). It clearly shows that the obtainable minimum probe sizes are much smaller than those estimated from equation (2), especially at low energies; the trend, however, is similar.

When high-current electron probes are desired, for example, for applications in electron beam nanolithography, electrons within the finite-sized nanoprobe interact with each other through Coulomb force; the electron-electron repulsive interaction can perturb the axial (along the electron optic axis) and transverse velocities of the emitted electrons, especially at crossovers of intense electron beams. The electron-electron interactions can affect the property of an electron beam in three ways [41]: 1) the space charge effect, 2) the trajectory displacement effect, and 3) the energy broadening effect. The space charge effect refers to the deflection of an electron, traveling along the optic axis, by the effective average charge of all other electrons within the electron beam; the degree of deflection is proportional to the distance of the electron from the optic axis, thus causing a defocus of the electron beam. The space charge effect increases linearly with the total current of the electron beam; it becomes dominant in high current applications.

The effect of defocus on the probe size is a simple first order effect originating from a shift along the optic axis by an amount, say, Δz; this will produce a uniformly illuminated circle of diameter given by $\Delta d = \Delta z \tan \alpha_p$, where α_p is the probe-aperture angle. The effect of defocus on the probe size determines the depth of focus in SEM images. If we assume the resolution on a cathode ray tube display is δ_{CRT}, then the corresponding resolution required at the specimen is given by $\delta = \delta_{CRT}/M$ and specimen details at a depth

$$T = \frac{\delta}{\tan \alpha_p} = \frac{\delta_{CRT}}{M \tan \alpha_p} \tag{4}$$

will appear sharp in the SEM images. If we further define a field width of the image on the CRT display as W_{CRT}, then the width of the scanned area at the specimen is

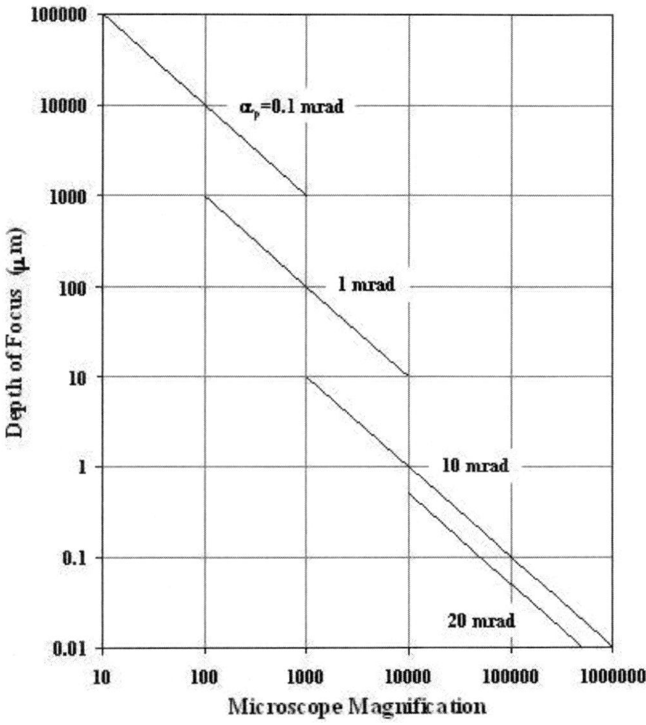

Figure 9. Plots show the dependence of the depth of field in a SE image on the image magnification and the probe-aperture angle. Higher image resolution, which requires larger probe-aperture angles, results in lower depth of field.

given by $W = W_{CRT}/M$. The normalized depth of field t, which is more meaningful in practice, is given by

$$t = \frac{T}{W} = \frac{\delta_{CRT}}{W_{CRT} \tan \alpha_p} \sim \frac{10^{-3}}{\tan \alpha_p} \tag{5}$$

In equation (5), we approximated δ_{CRT} to one pixel and W_{CRT} to 1024 pixels. The normalized depth of focus clearly drops dramatically with the increase of the probe-aperture angle. Figure 9 shows plots of the dependence of depth of focus on the microscope magnification and the probe-aperture angles. In practical applications of a modern FEG-SEM, one needs to effectively utilize the interrelated relationships among probe-aperture angle, magnification, depth of focus, and image resolution.

Figure 8d shows that probe-aperture angles >20 mrad have to be used to obtain a probe size $\leq$1 nm with low-energy electrons; then, the depth of field in the high-resolution SEM image is only about 5% of the viewing screen. The use of large probe-aperture angles significantly reduces the depth of focus in SEM images; and thus makes it possible to observe only those features that are located on the focal plane; out of focus features will be highly blurred and contribute only to the image background. This feature can be effectively utilized to obtain confocal SEM images that provide not only high spatial resolution in the x-y plane but also useful information on the z-axis direction. Three-dimensional images of non-flat surfaces may be obtainable with high spatial resolution confocal SEM images.

To effectively employ the power of a modern high-resolution FEG-SEM, the environmental conditions of the microscope become a critical issue. For example, presence of the stray magnetic field near the microscope can severely interfere its performance, especially at low electron energies. Thermal disturbances, mechanical vibrations, and acoustic disturbances can all profoundly affect the performance of a modern-day low voltage SEM.

4. THE RESOLUTION OF SECONDARY AND BACKSCATTERED ELECTRON IMAGES

Probe size and resolution are interrelated in the SEM but they are not the same. The image resolution can be worse but not better than the probe size of the electron beam (Digital processing, via deconvolution processes, may provide a resolution better than the probe size; but these computationally labor-extensive processes are not routinely used for two-dimensional images and deconvolution procedures have their intrinsic limitations.) At low magnifications, the image resolution is usually determined by the pixel size since features smaller than the pixel size cannot be observed. At high magnifications, however, the pixel size may be much smaller than the probe size so resolvable features can be displayed in several pixels. The information source, which we will discuss below, becomes the resolution-limiting factor at high magnifications.

When a SEM image is formed, the specimen is illuminated by a primary beam of a total current I_p with a current density distribution function $i(\boldsymbol{r})$. Upon entering the specimen, the electron probe interacts with a thin slice of material from which various signals can be generated and some of these signals escape from the specimen surface. If we assume that for a point source the spatial distribution of the generated signals can be described by a function $S(\boldsymbol{r})$, then the information source generated by a finite-sized electron probe at position $\boldsymbol{r}$ is given by

$$\int i(\mathbf{r}')\,S(\mathbf{r}-\mathbf{r}')\,\mathrm{d}\mathbf{r}' \tag{6}$$

When a primary electron beam interacts with a specimen, secondary, Auger, and backscattered electrons can all be generated. On their way traveling to the specimen surface, these electrons will generate more secondary electrons with lower energies; therefore, a cascade process occurs within the specimen. If we designate the yield of the secondary electrons that are directly generated by the primary beam as δ, the BE yield as η, and the yield of the secondary electrons that are generated by the backscattered electrons as $\beta\delta$, then for each probe position the amplitude distribution of the information source in the specimen surface plane is given by

$$\frac{I(\boldsymbol{r})}{I_p} = \delta\int i(\boldsymbol{r}')S_\delta(\boldsymbol{r}-\boldsymbol{r}')\mathrm{d}\boldsymbol{r}' + \eta\int i(\boldsymbol{r}')S_\eta(\boldsymbol{r}-\boldsymbol{r}')\mathrm{d}\boldsymbol{r}' + \eta\beta\delta\int i(\boldsymbol{r}')S_{\eta\delta}(\boldsymbol{r}-\boldsymbol{r}')\mathrm{d}\boldsymbol{r}' \tag{7}$$

The first term in equation (7) represents the spatial distribution of the secondary electrons that are directly generated by the primary electrons; these secondary electrons are usually called SE1. The second term in equation (7) represents the spatial distribution of the backscattered electrons at the specimen surface. The third term in equation (7) represents the spatial distribution of the secondary electrons that are generated by the backscattered electrons; these secondary electrons are usually called SE2.

Detailed calculations of equation (7) is not possible at the present time since the $S(r)$ functions are not known although many approximations and models have been proposed to gain insights into the generation, diffusion, and escape processes of the backscattered and secondary electrons. Monte Carlo simulations, however, have provided useful knowledge on the electron-specimen interaction and SE generation processes (see reference [42] for a recent review). When high-energy electrons are employed, the SE1 and SE2 can be spatially separated since the SE1 signal is localized within a very close range of the impact position of the incident electron beam, but the SE2 signal can originate from a region similar to that of the electron range of the primary electron beam. Although the strength of the SE2 signal can be higher than that of the SE1 due to the cascading process, the current density distribution of the SE1 signal is much more shaper than that of the SE2 signal. Therefore, at high magnifications, the resolution of the SE image is determined by the spatial distribution of the SE1 signal; the SE2 signal contributes to the background of the high magnification SE images. Since the strength of the SE1 signal may be only a small fraction of that of the SE2 signal, the high-resolution details revealed in a SE image has poor signal-to-noise and signal-to-background ratios.

At middle range magnifications, the pixel size may be close to the range of the SE2 signal and the image resolution and contrast are now determined by the properties of the SE2 since SE2 contributes to most of the detected signal. We should note that the energy distribution of the SE1 signal could not be distinguished from that of the SE2 signal. Only if a very thin specimen is used can SE2 be significantly reduced or eliminated. The SE2 signal should carry information similar to that of the BE signal.

When the energy of the primary electron probe is reduced, the range of the SE2 signal significantly decreases as demonstrated in Figure 3. The spatial distribution of the SE2 approaches that of the SE1 signal; thus, the SE1 and SE2 signals can no longer be differentiated even at high magnifications. In this case, both components contribute to high-resolution information; thus, high-resolution SEM is significantly more efficient at low electron energies although the probe size may be larger as demonstrated in Fig. 8. With the use of aberration correctors, the resolution of an SEM will become less dependent on the primary beam energy; the accelerating voltage of the SEM can therefore be used to modulate the interaction volume, to optimize the image contrast, or to probe the subsurface information.

The ultimate resolution limit of SE imaging is determined by the first term in equation (7). When a point probe or a very small electron probe is used, the image resolution is determined by the source function $S_\delta(r)$. The ultimate resolution is determined by the spatial distribution of the secondary electrons initially excited by the primary beam; the subsequent transportation of these hot electrons toward the specimen surface may not deteriorate the image resolution. The values of the inelastic mean free paths of the generated secondary electrons are not related to the ultimate achievable resolution of SE images at all; very high-resolution SE images of insulators such as MgO smoke crystals have been obtained [43].

To determine the source function $S_\delta(r)$ we need to know the specific processes for generating secondary electrons inside solid specimens; this is a non-trivial problem and the detailed discussions are beyond the scope of this chapter. Recent data from the electron coincidence spectroscopy experiments [44–45] confirmed the early proposal that the high-resolution SE signals are generated via large momentum transfer excitation events and are thus highly localized in the initial generation process [15]. The detailed generation and escape processes of SE1 signals are still not clearly understood. Inner-shell and single electron valence excitations may account for the SE production, especially for the high spatial resolution signals [46].

The resolution and the contrast of a SE image depend on the rate at which the collected signal changes as the electron probe is moved across a non-homogeneous region; and this depends primarily on the change in the number of the collected signals. Thus, even if the escape distance is not very small and the acquired signals actually originate from some other region of the specimen, variations in composition, for example, can still be detected over small distances because of the associated change in the elastic or inelastic scattering events of the primary electrons.

In view of the above discussion, BE signals can also provide high spatial resolution information; and the high sensitivity of the BE signal to the atomic number Z of solid materials makes the high-resolution BE imaging the method of choice in the modern FEG-SEM for providing compositional information on a nanometer scale. Nanometer-scale metal nanoparticles in supported catalysts can be easily detected by collecting the BE signal; information on the spatial and size distributions of metal nanoparticles can be obtained on bulk samples and even the depth distribution of the metal nanoparticles can be inferred [20]. At lower primary beam energies, the BE signal becomes more surface sensitive and its source of generation becomes much smaller, approaching that of the SE1. The recent development of high-sensitivity BE detectors will make the BE imaging or the combination of BE and SE imaging the most powerful tool for studying nanophase materials or other nanosystems.

The current achievable image resolution in a modern FEG-SEM is still primarily limited by the achievable probe size and poor signal-to-noise ratio. An image resolution of about 0.5 nm or better, however, can be achieved in both the in-lens type FEG-SEM or in STEM instruments. Atomic resolution imaging, for example, of lattice fringes of bulk crystals that are oriented at the exact channeling conditions, has not been achieved yet.

5. CONTRAST MECHANISMS OF SE AND BE IMAGES OF NANOPARTICLES AND OTHER SYSTEMS

The contrast mechanisms of SE images have been of extensive study ever since the SEM was invented. Depending on the materials of interest many factors can contribute to the observed image contrast: material contrast, topographic contrast, voltage contrast, magnetic contrast, electron channeling contrast, charging contrast, etc. Some or most of these contrast mechanisms may play a role in the SE images of interest. Detailed discussions on the origin of these contrast mechanisms have been well documented in literature and readers of interest should consult the relevant textbooks, for example, references [30–31]. We will briefly discuss below, however, some recent developments that are relevant to the study of nanosystems and surfaces.

5.1. Small Particle Contrast in High-Resolution BE Images

The small particle contrast is of interest to correctly interpreting high-resolution SE and BE images of supported catalysts, nanoparticle systems, or other nanoscale systems. Figure 10 shows a set of SE and BE images of a carbon supported Pt catalyst obtained with different primary electron energies; all the images were obtained from the same sample region. Because of its high Z-contrast, the high-resolution BE images clearly revealed the Pt nanoparticles located either on the carbon surface or within the highly porous carbon support. A better image resolution is achieved at higher incident electron energies because of the smaller size of the incident probe and less beam broadening in the sample. The image intensity of a Pt nanoparticle does not necessarily increase monotonically with its size in BE images.

In Fig. 10a, for example, some smaller Pt particles (e.g., the particle indicated by the letter B) clearly have higher visibility than that of some bigger Pt particles (e.g., the Pt particle indicated by the letter A). In Fig. 10b, however, the particle A shows a much higher intensity than that of particle B. The interpretation of high-resolution BE images of supported nanoparticles is complicated and many factors may contribute to the observed image contrast. To understand the effect of the incident electron energy and the size and location of Pt nanoparticles on the particle contrast, Fig. 11a shows the dependence of $\eta(E)$ on incident electron energy for various sizes of Pt particles located on the surface of a bulk carbon support. The values of $\eta(E)$ for bulk carbon and bulk Pt are also shown for comparison. With increasing electron energy, $\eta(E)$ first increases, reaches a maximum, and then slowly decreases. This trend is true for all sizes of the particles although the rate of change depends on the particle size. For each particle size, there exists an optimum electron energy, E_{op}, which gives a maximum value of $\eta_{max}(E_{op})$. This maximum value is reached when the range $(R(E_{op}))$ of the incident electrons is about the diameter d of the Pt particle. The E_{op} value increases with the size of the Pt nanoparticle. Note that the maximum value of the backscattering coefficient η_{max} from a small Pt nanoparticle is about the same for different sizes of particles. The

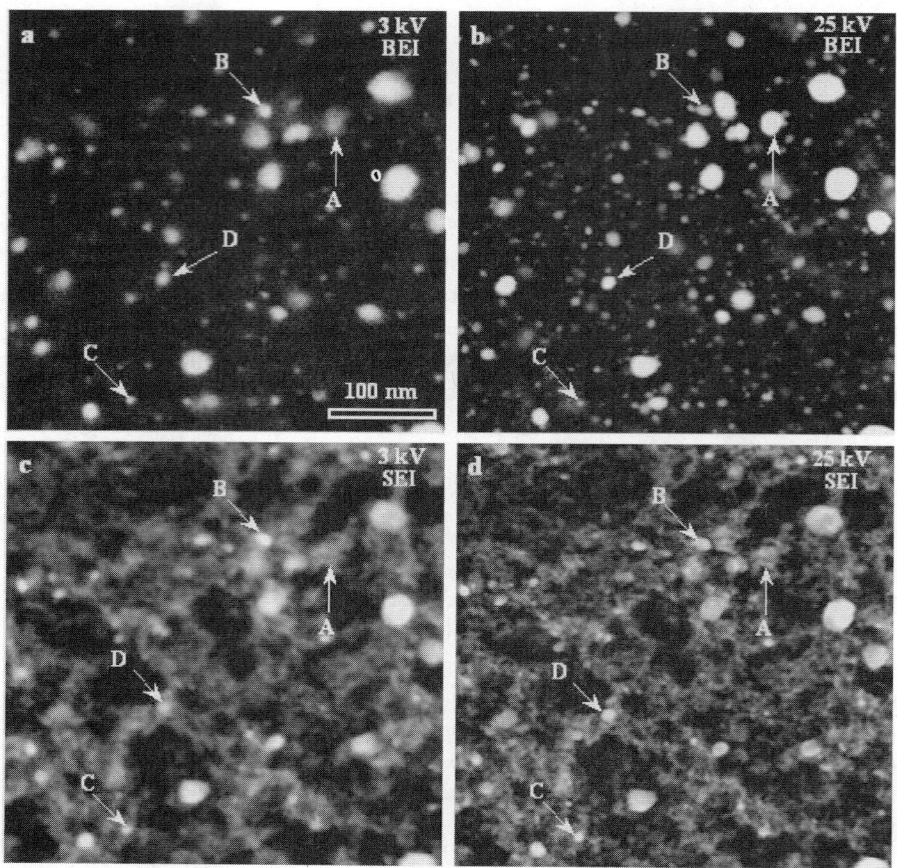

Figure 10. High-resolution SE and BE images of a Pt/C catalyst obtained with the primary electron energies of 3 keV and 20 keV, respectively, clearly showing the spatial and size distribution of the Pt nanoparticles as well as the detailed morphology of the carbon support. See text for detailed discussion.

value of η_{max} may be related to the scaling parameter $\alpha = d/R(E_{op})$ which is about unity for all particle sizes. For smaller Pt particles, although η decreases rapidly with increasing electron energy it does not approach the value of bulk carbon. Thus, small Pt particles positioned on the surface of a carbon substrate will always give an observable bright contrast in high-resolution BE images regardless of the incident electron energy.

Figure 11b shows the dependence of $\eta(d)$ on the size of a Pt particle at different electron energies. With low-energy electrons, a smaller Pt particle may have a higher intensity at the center of the particle than that of a larger Pt particle. For example, a 20 nm Pt particle will give a maximum contrast in BE images obtained with 4 keV electrons while both smaller and larger particles will give a lower image contrast. With

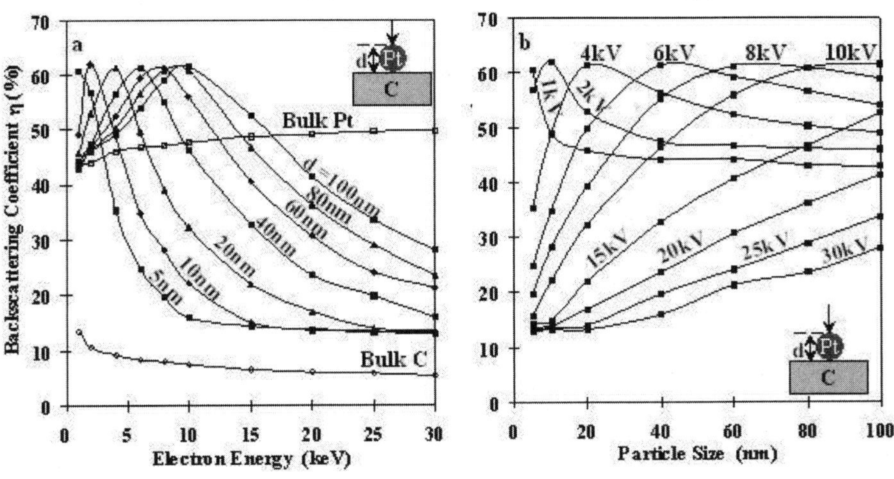

Figure 11. Monte Carlo calculations of backscattering coefficients (η) of Pt particles positioned on a carbon substrate: (a) as a function of electron energy for various sizes of Pt particles and (b) as a function of the size of Pt particles for various energies of the incident electrons. The values of η for bulk carbon and bulk platinum are also shown for comparison. Number of trajectories used: 30,000.

higher-energy electrons, however, the image intensity of a small Pt particle increases monotonically with its size. With further increases of the particle size, the η values for all electron energies will gradually approach their corresponding values of bulk Pt. Figure 11 indicates that the interpretation of high-voltage BE images of small Pt particles located on the surface of bulk substrates is straight forward while that of low-voltage BE images is more complicated.

Another complicating factor in interpreting BE images of supported metal catalysts or other inhomogeneous systems is that some small particles are often located below the substrate surface. For example, some of the small Pt particles with a diffuse contrast in Fig. 10a (e.g., particle A) are definitely located at various depths below the carbon surface. Monte Carlo simulations of the contrast of nanoparticles embedded within matrix materials have been performed [20]. Regardless of the incident electron energy, the backscattering coefficient is the highest when the Pt particle is located on the surface of the carbon substrate. The backscattering coefficient or the contrast of BE images, however, is sensitive to the geometry of the particle-matrix composite as well as the location of the nanoparticle. When a nanoparticle is located below the substrate surface, the incident electrons are first scattered by a layer of the substrate material and then are scattered by the nanoparticle. The electrons backscattered by the nanoparticle will be scattered again by a layer of carbon before they can finally exit the substrate surface as backscattered electrons. Since the energy of the incident electrons is continuously decreasing during these scattering processes and the electron backscattering coefficients from small nanoparticles strongly depend on the energy of the electrons (see Fig. 11a) the electron backscattering processes from the particle-matrix composite can be very complex. The particle-substrate interface may also play an important role in contributing to high-angle scattering events.

The Monte Carlo calculations clearly show the complex nature of the contrast of small Pt particles in high-resolution BE images of carbon supported Pt catalysts, especially at low voltages. However, by examining high-resolution BE images obtained with different incident electron energies and by comparing these BE images with the corresponding SE images, information about the lateral distribution as well as the depth distribution of metal nanoparticles in supported metal catalysts can be extracted [20].

5.2. Small Particle Contrast in High-Resolution SE Images

High-resolution SE imaging of supported metal catalyst samples was first demonstrated with 100-keV electrons in the STEM [13, 47–48]. The increased emission of secondary electrons from small, heavy-element particles is partially attributed to the local increase in the stopping power of high-energy electrons. Topographical contrast is also a main factor that contributes to the observed bright contrast of small particles. With high-energy incident electrons, the integrated SE intensity of a small metal particle is proportional to the total number of atoms within that particle [49], suggesting that the number of secondary electrons generated is proportional to the total stopping power of a small particle. Small metal particles embedded inside supporting materials may not be revealed in high-resolution SE images [43]. Only those metal particles that lay on, or very close to, the sample surface can give observable image contrast.

When the energy of the incident electrons decreases from high energies, the electron backscattering from a small, heavy-element particle increases. The backscattered electrons generate additional secondary electrons when they pass through the specimen surface (SE2 signal) or when they impinge onto the pole pieces and other parts of the optical system and specimen chamber (SE3 signal). Both the SE2 and SE3 signals carry information similar to that of backscattered electrons. Therefore, if the SE2 and SE3 signals significantly contribute to the collected SE signal then metal particles located deep inside a light-element support may also be revealed in high-resolution SE images.

To understand the contrast of small Pt particles in high-resolution SE images, both the BE and SE signals were collected to form images of the same area of a Pt/carbon catalyst (see Fig. 10). The relatively poorer image resolution shown in Fig. 10a and 10c is partially attributed to the larger working distance (WD = 7 mm) used to accommodate the retractable BE detector. The visibility of the smaller Pt particles in SE images is clearly low when compared to that in the corresponding BE images. Furthermore, only some of the Pt particles which are visible in the BE images are revealed in the SE images. Because of the deeper penetration of electrons at higher energies, more Pt particles are revealed in the higher energy BE image. In contrast, Fig. 10d shows almost the same number of Pt particles, with an improvement in image resolution though, as those revealed in Fig. 10c. Some smaller Pt particles (e.g., the particle indicated by the letter C) have a surprisingly high contrast compared to that of some bigger particles (e.g., the particle indicated by the letter D). While some Pt

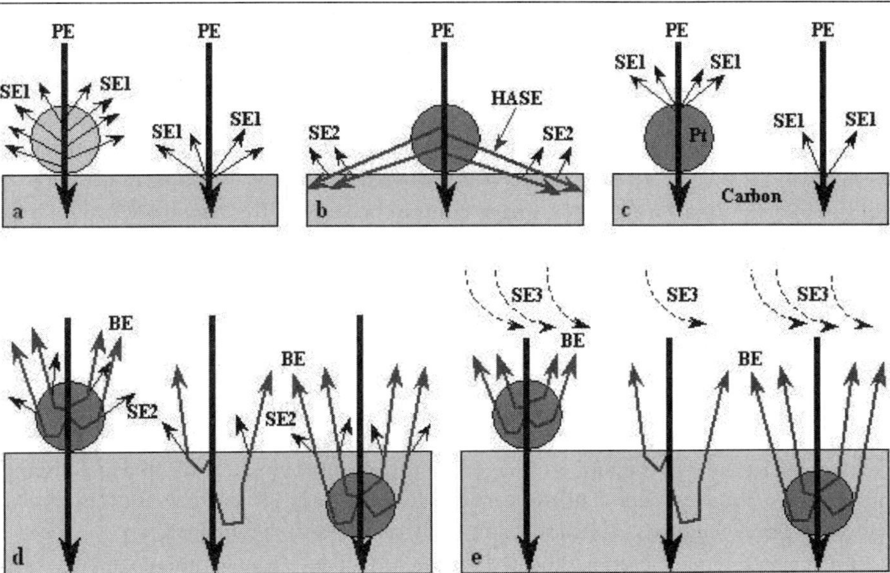

Figure 12. Schematic diagrams illustrate the contributions of various SE signals to the small-particle contrast in high-resolution SE images. PE—primary electrons; SE1—type 1 secondary electrons; SE2—type 2 secondary electrons; SE3—type 3 secondary electrons; HASE—high-angle scattered electrons; BE—backscattered electrons.

particles (e.g., the particle indicated by the letter B) are clearly visible in both the BE and SE images other Pt particles (e.g., the particle indicated by the letter A) are only revealed in the BE images. If we assume that the particle A was embedded inside the carbon support then the variations of the visibility of the particle A in both the BE and SE images of different primary electron energies can be explained. The visibility of small Pt particles in the SE images is also affected by the dominating topographic contrast of the carbon mesopores. A small positive voltage applied to the specimen can, however, greatly suppress the topographic contrast of the carbon support and thus enhance the visibility of metal nanoparticles [50].

The visibility of small metal particles in high-resolution SE images of supported metal catalysts is determined by the following factors: a) topographic contrast; b) diffusion contrast; c) material contrast due to the SE1 signal; d) material contrast due to the SE2 signal; and e) material contrast due to the SE3 signal. Experimental results showed that only those metal nanoparticles that are located on, or very close to, the surface of the substrate can be observable in high-resolution SE images.

The topographic contrast may play a dominant role for nanoparticles located on the surface of a substrate as illustrated in Fig. 12a. When the radius of a small particle becomes comparable to, or smaller than, the escape depths of the secondary electrons, most of the secondary electrons, generated inside the particle, with energies higher than the surface barrier may escape from the particle. In contrast, because of the total internal reflection of low-energy secondary electrons only about 10% of the total internal secondary electrons that are generated within the escape depth region, can escape from a flat surface. Because of this geometric effect in SE emission from spherical particles, small particles located on a flat substrate surface, even if the substrate and the particle are the same material, are often observed with a bright contrast in high-resolution SE images. In fact, small particles of low-Z materials supported on high-Z substrates can often give a bright image contrast because of this dominant topographic effect; this is in striking contrast to BE images in which high-Z materials always give bright image contrast.

The topographic contrast of small particles can be delineated by the ratio of the particle radius (R) to the average escape depth (L) of the collected secondary electrons. If $R/L \ll 1$, the image intensity is proportional to the total number of atoms within that particle. If $R/L < 1$, the image intensity increases with the particle size and has a maximum at the center of the particle. If $R/L > 1$, the image intensity slowly increases with particle size and the highest image intensity is approximately located at a distance $d = (R\text{–}L)$ from the center of the particle. If $R/L \gg 1$, the particle contrast evolves into the edge-brightness contrast, commonly observed in SE images.

For heavy-element particles located on the surface of a light-element support, electrons scattered to high angles (<90 degrees) by a small particle can re-enter the substrate at a glancing angle as schematically illustrated in figure 12b. Monte Carlo simulations show that these high-angle scattered electrons (HASE) can travel a long distance inside the substrate, generating additional secondary electrons that lie within the escape depth of the substrate material. These "extra" secondary electrons will add to the bright, topographic contrast of small particles. This electron diffusion effect is enhanced at low electron energies as demonstrated by the Monte Carlo simulations shown in Fig. 4. This electron diffusion contrast is significantly reduced for particles located just below the substrate surface. The fraction of high-angle scattered electrons by a small, heavy-element particle depends on the size and shape of the particle, the incident electron energy, the average atomic number of the particle, and the location of the particle relative to the substrate surface.

The increased stopping power in a small particle of heavy-element materials can generate more secondary electrons (SE1) per unit volume than that in the light-element supports (schematically illustrated in Fig. 12c). Taking into consideration of the effects of work function and escape depth on the emission of secondary electrons, heavy-element materials may have a higher SE yield than that of light-element materials. Thus, a material contrast may contribute to the visibility of nanoparticles in SE images. There is, however, no direct correlation between the SE yield of a material and its atomic number. So it cannot be generalized that high-Z elements will have a higher SE yield even though they may have a higher electron stopping power. For example, some non-conducting materials may have a much higher SE yield than that of heavy-element metals.

The SE2 signal due to a small, heavy-element particle can be generated within, and escape from the surface of, a small particle (if the particle is located on the surface of the substrate). They can also be generated in the substrate (if the particle is embedded in the substrate) and escape from the sample surface (schematically illustrated in Fig. 12d). Because of the higher BE yield from small, heavy-element particles, especially at low electron energies (see Fig. 4), more SE2 signals are generated whenever the incident electrons encounter a small particle. The small-particle contrast due to the SE2 signal is directly related to the image contrast of the corresponding BE images.

Similar to the discussion for the SE2 signal, the SE3 signal also carries information directly related to that of the BE signal (Fig. 12e). The SE3 signal should also give a bright contrast in SE images of small, heavy-element particles. Without special arrangements of the SE detection system, the effects of the SE2 and SE3 signals on the small-particle contrast cannot be distinguished or eliminated. Experimental results showed, however, that the contribution of the SE2 and SE3 signals from small, heavy-element particles to the collected SE signal may not be significant; the contribution of these signals from larger particles, however, may be appreciable [20].

5.3. Other Contrast Mechanisms

The contrast in both SE and BE images is a number contrast; any change (above the statistical noise level) in the number of the collected signal will give an image contrast. Thus, change in work function, which affects the emission step of the internal secondary electrons, can be visualized in biased SE images [51–52]. Submonolayer, monolayer, or multilayer deposits can be readily visualized and distinguished.

By using a directionally sensitive detection system, it is also possible to reveal, in strong contrast, atomic steps and the crystallographically equivalent but differently oriented 1×2 and 2×1 domains at the reconstructed Si (100) surface [53]. The origin of this contrast could come from a Bragg reflection effect in the reconstructed layer with a strong dependence on the azimuthal direction of the electron crossing the surface barrier [46]. Since the energy distribution of the internal secondary electrons near the surface of the specimen is highly skewed toward low energies, any change of the work function or even the shape of the surface potential may intricately modify the SE emission probability, especially when an external field is applied.

Local potential variations on a non-homogeneous specimen surface generate patch fields, which may influence the emission or movement of the secondary electrons. Depending on the polarity of the patch fields, these regions may give a reduced or enhanced SE emission, especially in the low energy part of the SE spectrum. There is not much experimental or theoretical work done on the effect of local patch fields, which may be becoming increasingly important in understanding nanosystems, on the emission of secondary electrons or on the image contrast of high-resolution SE images. Recent interest in this topic [46] may provide much needed impetus for further exploration on the possibility of high-resolution imaging of local variations in work function, contact potential, or even charge transfer in nanoscale systems.

Changes in surface barrier height or subsurface band bending effects have recently been shown to provide good image contrast of p-n junctions in high-resolution images of doped semiconductor devices [54]. The authors observed that the p-type regions always appear brighter relative to the n-type regions in a doped GaAs sample. The observed contrast was attributed to band pinning effects associated with the chemisorption of oxygen at the sample surface, which enhances the yield variations between the p- and n-type regions; band bending and the modification of the shape of the surface potential may as well play a role in the observed image contrast. The effect of energy filtering by the in-lens type FEG-SEM, which was used to obtain these results, on the observed image contrast was also proposed [21].

6. APPLICATIONS TO CHARACTERIZING NANOPHASE MATERIALS

High-resolution SEM imaging can be conveniently applied to extract nanoscale information from many different types of nanophase materials or systems. Any solid materials, even soft materials, can be imaged in a modern FEG-SEM; with the use of a cryo system or by using an environmental SEM, almost all the organic and inorganic materials can be examined and useful information about the morphology, ultrastructure, and composition of the region of interest can be extracted. Not only nanophase materials but also intact micro- or nano-devices can be imaged in the modern day, large chamber FEG-SEM instruments.

High-resolution SEM imaging and associated techniques are now routinely used in the semiconductor industry for failure analyses of semiconductor devices, for example, analyses of cell-blocks, trench capacitors, bit lines, gate conductors, or cross-sections of the DRAM cells. Specialized electron beam lithography and wafer processing inspection systems have been in extensive use in the semiconductor industry. Critical dimension measurements and nano-metrology will become more critical to the successful development of next generations of micro and nanodevices. Creation of micro electro mechanical systems, nanostructures, and nanodevices is certainly the most important frontier of nanoscience research and nanotechnology. To know what we have created or how they work or how to make improvements, we need to use high-resolution FEG-SEM techniques to examine these devices or materials. It is expected that the high-resolution FEG-SEM will become the most important tool in developing nanotechnology products, especially nanodevices and bio-nanodevices.

As an example of applications, Fig. 13a shows a low voltage (1 KV), high-resolution SE image of a freshly fractured IC device clearly showing the individual components. The cross-section device could not be examined at high voltages due to severe sample charging effect [25]. In this particular case, low-voltage SE imaging proves to be more useful and reliable. High-resolution BE images, especially at low voltages, also provided useful information on the identification and measurement of the individual components. Rapid and reliable identification of defects and high throughput are of essence here.

On the other hand, Fig. 13b shows a high voltage (100 KV), high resolution SE image of a MgO smoke crystal, clearly revealing surface steps, facets, and the various faces of the incomplete MgO cube. Individual (111) planes are not charge neutral; the presence of a net dipole moment normal to the surface leads to a divergence in the surface energy. Therefore, the (111) planes are not stable and facet into (100) planes as clearly shown in Fig. 13b. Figure 13b also shows that the MgO (110) planes are

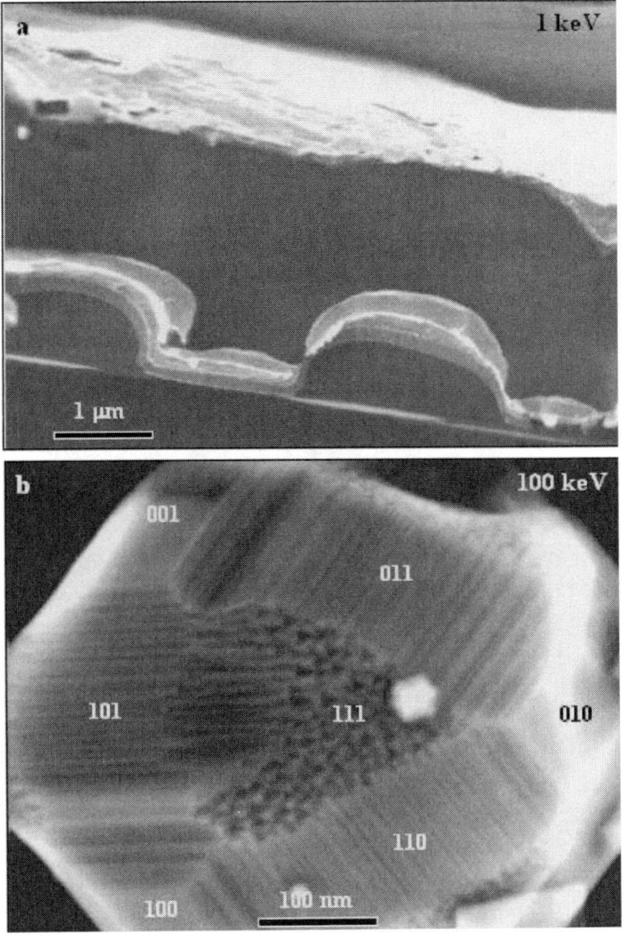

Figure 13. High-resolution, low-voltage SE image of a freshly fractured IC device clearly showing the different components of the device (a); high-resolution, high-voltage SE image of a MgO smoke crystal revealing steps, facets, and the different planes of the incomplete MgO cube (b).

also not stable although these surfaces are non-polar with equal numbers of cations and anions in each atomic plane parallel to the surface. The surface energy of MgO (110), however, is reported to be much higher than that for the (100) surface [55]; thus, the MgO (110) may relax or exhibit rumpling. The high-resolution SE image in Fig. 13b clearly shows that the (110) surfaces of the MgO smoke are not smooth and are faceted toward (100) planes, at least along the direction connecting the different (110) planes. These studies are important for understanding the formation mechanisms of nanoparticles or nanocrystals and for the interactions between metallic nanoparticles and metal oxide nanocrystals.

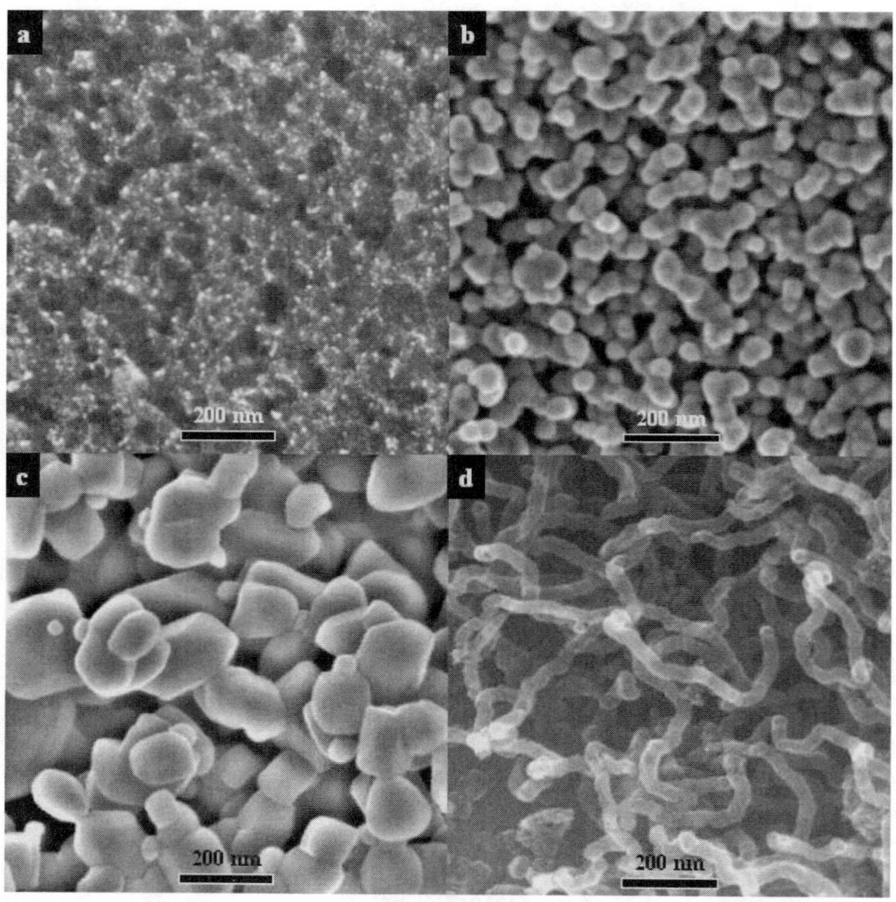

Figure 14. High-resolution SE images of (a) carbon supported Pt nanoparticle catalyst, (b) antimony-doped SnO_2 nanocrystals, (c) TiO_2 nanoparticles, and (d) carbon nanorods growing from a carbon substrate. These examples illustrate the powerful applications of high-resolution FEG-SEM to the characterization of various types of nanophase materials.

Other examples of applying high-resolution SEM imaging to the study of nanophase materials are shown in Fig. 14. Figure 14a shows a SE image of a carbon-supported Pt catalyst used in many catalytic chemical reactions. The image clearly provides information on the size and spatial distribution of the Pt nanoparticles; it also provides information on the pore structure of the carbon support as well as the spatial relationship between the Pt nanoparticles and the carbon pores. By varying the incident electron energy and by combining high-resolution SE imaging with BE imaging the three-dimensional distribution of the Pt nanoparticles may also be extracted. This type of information is invaluable for understanding the catalytic performance of the supported catalyst, and for developing better nanostructured heterogeneous catalysts.

Figure 14b shows a high-resolution SE image of antimony-doped SnO_2 (ATO) nanoparticles. The ATO nanoparticles were developed for applications in sensors and smart devices, for example, electrochromic displays and windows. The information on the size distribution, the connectivity among, and the aggregation behavior of, the ATO nanoparticles is critical to optimizing the performance of the desired devices; the detailed pore structure of the ATO nano-powders is also critical to the optimization of electrochemical devices.

Figure 14c shows a high-resolution SE image of TiO_2 nanoparticles. These nanoparticles have important applications as pigments, as photo-catalysts, or as catalyst support materials; again, the size and shape distributions as well as the aggregation behavior of the TiO_2 nanoparticles are important parameters in determining their properties and applications. Tuning of the size distribution and the electronic structure of the TiO_2 nanoparticles determines the specific applications of these important semiconduting metal oxide nanoparticles.

Figure 14d shows clearly the shape and morphology of carbon nanorods as well as some extremely small metallic nanoparticles on the surfaces of these nanorods. This type of information can be effectively used to correlate the structure of the fabricated carbon nanophase materials to the synthesis processes. Carbon nanotubes and nanotube-based devices are in the forefront of nanoelectronics; and the ability to be able to statistically analyze large numbers of nanotubes or to be able to examine how these nanotubes interact with various substrates or other nanosystems is critically important.

In addition to the wide applications of high-resolution SEM to the semiconductor and nanoelectronics industries, the above examples clearly demonstrate the power of high-resolution SEM in solving challenging nanoscale materials problems.

7. SUMMARY AND PERSPECTIVES

In this chapter, we illustrated by using nanoparticles as a specific example the recent developments in SEM instrumentation and applications of high-resolution SEM to characterizing nanophase materials. High-resolution SEM techniques are now widely used in R&D laboratories, in manufacturing facilities, or in areas such as art, archeology, forensic investigations, environment preservation, etc.

At high electron energies, nanometer or subnanometer resolution SE images can now be routinely obtained in the new generation FEG-SEM instruments. To achieve the same image resolution at low electron energies, however, is still very challenging. As we discussed in section 3, both C_c and C_s aberration correctors have to be used to significantly reduce the effect of spherical and chromatic aberrations on the formation of low-energy electron nanoprobes. Such an attempt has already resulted in significant resolution improvement [39]. The use of aberration correctors not only makes the obtainable probe size much smaller but it makes possible to obtain high beam current contained within a small nanoprobe; high total beam current provides adequate characteristic X-ray signals for chemical microanalysis or BE signals for high-quality, high-resolution imaging.

When the lens aberrations are corrected by using the aberration correctors, large probe-aperture angles have to be used, due to the diffraction limit, to obtain smaller probe sizes; the use of large probe-aperture angles significantly reduces the depth of field in SE images. However, we can effectively utilize this property to form confocal SEM images, similar to that of confocal laser scanning microscopy. The development of confocal SEM techniques will certainly make the high-resolution FEG-SEM technique much more powerful since true three-dimensional images of various samples can be obtained to provide not only the lateral but also the depth information on a nanometer scale.

For some samples, high probe current may be detrimental due to electron-beam induced irradiation damage of the sample; low probe current, however, usually results in poor signal-to-noise ratio and thus poor image or spectrum quality. Development of highly efficient detectors for both high-energy and low-energy electrons will prove to be fruitful in imaging and analyzing delicate samples. High-sensitivity BE detectors, especially at low-electron energies, are needed to differentiate phases that have small compositional differences. Use of SE and BE spectrometers in a modern FEG-SEM can further improve the image resolution and provide additional contrast. Energy-filtered SE and BE imaging will provide more detailed information on the samples of interest. It is possible that lattice fringes of bulk crystalline materials can be obtained in energy-filtered high-resolution BE images due to electron bean channeling effect.

The recent wide adoption of the environmental or variable pressure SEM instruments undoubtedly demonstrated the power of these microscopes. These microscopes can be operated as high-vacuum instruments or various gases can be allowed into the sample chamber to reduce sample-charging effects, to maintain low vapor pressure, or to study sample-gas interactions. It is anticipated that within a few years all the SEM instruments will be environmental or variable pressure SEM microscopes.

The recent development of ultra-low-voltage (ULV) imaging in the modern FEG-SEM is clearly interesting since this technique bridges the gap between SEM and low energy electron microscopy (LEEM) techniques. It is expected that different lens configurations will have to be used in order to achieve nanometer scale resolution with primary electron energies as low as a few eV. High-resolution ULV imaging makes it possible to reduce sample charging effect, to significantly reduce electron-induced damage of delicate samples, and to significantly increases surface sensitivity. For example, Fig. 15 shows a set of images obtained from the same region of a partially metallized polymer film. With the use of 500 eV electrons (Fig. 15a), the polymer film charged up and some of the metal particles appear dark in the image. With 100 eV electrons (Fig. 15b), the charging of the polymer film is significantly reduced and the metal particles appear bright now; but the contaminant in the center of the image was still charged up. When the electron energy was reduced to 50 eV or below, the image (Fig. 15c) clearly shows the sizes and morphology of the metal particles, the contaminant particle, and the surface details of the metal/polymer composites. Note

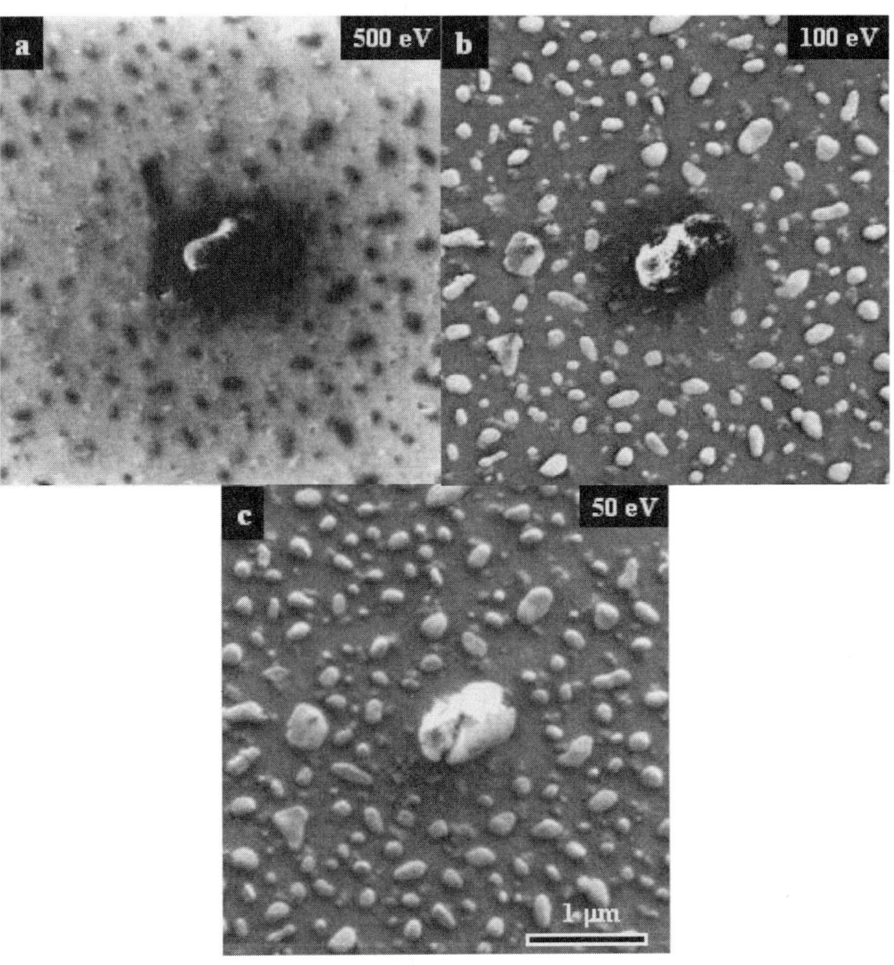

Figure 15. Low voltage (a, b) and ultra-low-voltage (c) SE images of a partially metallized polymer film illustrate the advantage of using low energy electrons. The Au particles are better revealed in images (b) and (c).

that the image resolution did not change much at all in all the images; high-resolution images can be obtained with primary electron energies <50 eV [23].

On the other hand, the recent development of ultra-high-voltage (~100 KV or above) SEM integrates the traditional SEM techniques to those of the STEM techniques; thus, both bulk samples and thin specimens can be examined by SEM techniques and TEM/STEM techniques can be applied to thin specimens. Subsurface structures can be revealed in high voltage SE or BE images. These microscopes are now broadly used in the semiconductor industry to characterize micro- or nano-devices.

The future generation electron microscopes may integrate all the functions that are now provided by SEM, TEM, and STEM techniques.

REFERENCES

1. M. Knoll, *Z. Tech. Phys.* 16 (1935) 497.
2. M. Von Ardenne, *Z. Phys.* 109 (1938) 553.
3. M. Von Ardenne, *Z. Tech. Phys.* 19 (1938) 407.
4. M. Knoll and R. Theile, *Z. Phys.* 113 (1939) 260.
5. V. K. Zworykin, J. Hillier, R. L. Snyder, *ASTM Bulletin* 117 (1942) 15.
6. C. W. Oatley, *J. Appl. Phys.* 53 (1982) R1.
7. A. V. Crew and J. Wall, *J. Mol. Biol.* 48 (1970) 373.
8. A. V. Crew, J. Wall and J. Langmore, *Science* 168 (1970) 1333.
9. A. V. Crew, *Rep. Progr. Phys.* 43 (1980) 621.
10. K.-R. Peters, *Scanning Electron Microscopy* 1982 (1982) 1359.
11. T. Nagatani, S. Saito, M. Sato, M.Yamada, *Scanning Microscopy* 1 (1987) 901.
12. D. Imeson, R. H. Milne, S. D. Berger, and D. McMullan, *Ultramicroscopy* 17 (1985) 243.
13. J. Liu and J. M. Colwey, *Ultramicroscopy* 23 (1987) 463.
14. J. Liu and J. M. Cowley, *Scanning Microscopy* 2 (1988) 63.
15. J. Liu and J. M. Cowley, *Scanning Microscopy* 2 (1988) 1957.
16. A. L. Bleloch, A. Howie and R. H. Milne, *Ultramicroscopy* 31 (1989) 99.
17. D. C. Joy and J. B. Pawley, *Ultramicroscopy* 47 (1992) 80.
18. A. Howie, *J. Microscopy* 180 (1995) 192.
19. R. Darji and A. Howie, *Micron* 28 (1997) 95.
20. J. Liu, *Microsc. Microanal.* 6 (2000) 388.
21. D. C. Joy and C. S. Joy, *Micron* 27 (1996) 247.
22. E. D. Boyes, *Adv. Mater.* 10 (1998) 1277.
23. J. Liu, *Mater. Characterization* 44 (2000) 353.
24. I. Mullerova and L. Frank, *Scanning* 15 (1993) 193.
25. J. Liu, *International Journal of Modern Physics* B16 (2002) 4387.
26. G. D. Danilatos, *Adv. Electron. El. Phys.* 71 (1988) 109.
27. P. J. R. Uwins, *Mater. Forum* 18 (1994) 51.
28. R. L. Schalek and L. T. Drzal, *Adv. Mater.* 32 (2000) 32.
29. A. M. Donald, *Nature Materials* 2 (2003) 511.
30. L. Reimer, *Scanning Electron Microscopy: Physics of Image Formation and Microanalysis*, Springer-Verlag, New York (1998).
31. J. I. Goldstein, D. E. Newbury, P. Echlin, D. C. Joy, A. D. Romig, Jr., C. E. Lyman, C. Fiori and E. Lifshin, *Scanning Electron Microscopy and X-Ray Microanalysis: A Text for Biologists, Materials Scientists, and Geologists*, Plenum, New York (1992).
32. D. C. Joy, *Monte Carlo Modeling for Electron Microscopy and Microanalysis*, Oxford University Press, New York (1995).
33. E. Bauer, *Rep. Prog. Phys.* 57 (1994) 895.
34. A. V. Crewe, J. Wall and L. M. Welter, *J. Appl. Phys.* 30 (1968) 5861.
35. V. E. Cosslett, *Optik* 36 (1972) 85.
36. J. Ximen, Z. Shao and P. S. D. Lin, *J. Microsc.* 170 (1993) 119.
37. A. V. Crewe, *Ultramicroscopy* 23 (1987) 159.
38. J. E. Barth and P. Kruit, *Optik* 101 (1996) 101.
39. J. Zach and M. Haider, *Nuclear Instruments and Methods in Physics Research* A363 (1995) 316.
40. M. Lenc and I. Mullerova, *Ultramicroscopy* 41 (1992) 411.
41. P. Kruit and G. H. Jansen, in *Handbook of Charged Particle Optics*, J. Orloff (Ed.), pp. 275–318, CRC Press, New York (1997).
42. Z. J. Ding and R. Shimizu, Scanning 18 (1996) 92.
43. J. Liu and J. M. Colwey, *Ultramicroscopy* 37 (1990) 50.
44. J. Drucker, M. R. Scheinfein, J. Liu and J. K. Weiss, *J. Appl. Phys.* 74 (1993) 7329.
45. H. Mullejans, A. L. Bleloch, A. Howie and M. Tomita, *Ultramicroscopy* 52 (1993) 360.
46. A. Howie, *J. Microscopy* 180 (1995) 192.
47. D. Imeson, *J. Microscopy* 147 (1987) 65.

48. J. Liu and J. M. Cowley, *Ultramicroscopy* 31 (1990) 119.
49. J. Liu and G. E. Spinnler, in: *Proceedings 50th Annual Meeting of EMSA* (1992), San Francisco Press, San Francisco, pp. 1288–1289.
50. J. Liu, R. L. Ornberg and J. R. Ebner, *Microscopy and Microanalysis, Vol. 3, Suppl. 2* (1997) 1123.
51. G. W. Jones and J. A. Venables (1985) *Ultramicroscopy* 18 (1985) 439.
52. J. A. Venables and J. Liu, in: *the Encyclopedia of Surface and Colloid Science*, A. P. Hubbard (Ed.), Marcel Dekker (2002).
53. Y. Homma, M. Suzuki, and M. Tomita, *Appl. Phys. Lett.* 62 (1993) 3276.
54. D. D. Perovic, M. R. Castell, A. Howie, C. Lavoie, T. Tiedje, and J. S. W. Cole, *Ultramicroscopy* 58 (1995) 104.
55. P. W. Tasker, *Adv. In Ceramics* 10 (1984) 176.

12. HIGH SPATIAL RESOLUTION QUANTITATIVE ELECTRON BEAM MICROANALYSIS FOR NANOSCALE MATERIALS

DALE E. NEWBURY, JOHN HENRY J. SCOTT, SCOTT WIGHT, AND JOHN A. SMALL

I. INTRODUCTION

A dominant theme in modern materials science is the determination of the relationships between material microstructure and macroscopic physical, chemical, and engineering properties. Because microstructures of materials consisting of several elements are typically segregated into two or more different chemical phases, a critical component of this task has been the development of tools that can characterize the compositional microstructure of materials on the micrometer to nanometer spatial scales, both laterally and in-depth. Electron beam excitation has provided an important class of these tools because of the powerful combination of high resolution morphological imaging through transmission and scanning electron microscopy, the elemental/chemical analysis made possible by associated x-ray and electron spectrometries, and the determination of crystal structure by electron diffraction methods [Newbury and Williams, 2000]. The emergence of nanoscale science and technology has increased the challenges facing these characterization methods.

Measuring elemental composition at high spatial resolution with electron microscopy is the theme of this contribution. For this discussion, the following arbitrary definitions will be adopted when referring to broad classes of concentration levels:

Major constituent (concentration, $C > 0.1$ mass fraction)
Minor constituent ($0.01 \leq C \leq 0.1$)
Trace constituent ($C < 0.01$)

By this convention, the lower limit of the trace category is not defined and is eventually determined when the single atom limit within the sampled volume is reached.

II. THE NANOMATERIALS CHARACTERIZATION CHALLENGE: BULK NANOSTRUCTURES AND DISCRETE NANOPARTICLES

Materials with nanometer-scale features can be divided into two broad classes for the development of appropriate microanalytical strategies: bulk nanostructures and discrete nanoparticles.

A. Bulk Nanostructures

Bulk nanomaterials include those "natural" systems in which rapid solidification from the melt and/or solid state reactions create nanoscale microstructural features in certain metal alloys and ceramics. A second class of bulk nanomaterials consists of "materials by design" that are built up from the atomic scale with processes such as ion sputtering deposition where indivual atoms, molecules, and clusters are manipulated. A third class of bulk nanomaterials includes engineered materials, such as complex layered electronic devices, that are synthesized by "building down" with massively parallel processing such as photolithography to create the discrete nanoscale features that compose advanced electronics.

Characterizing the fine structure of bulk materials with nanoscale features has been a long standing theme in materials science. Imaging and crystallographic studies with nanometer scale resolution have been performed for more than 40 years with the transmission electron microscope (TEM) [Williams and Carter, 1996]. For the past 25 years, these studies have been augmented by compositional measurements with energy dispersive x-ray spectrometry (EDS) and electron energy loss spectrometry (EELS) performed in the modified TEM known as the analytical electron microscope (AEM) [Joy *et al.*, 1986]. Examination of matter in the TEM/AEM places great demands on the form of the specimen. The range in nanometers of the primary electron beam in a target is given by equations of the form [Kanaya and Okayama, 1972]:

$$R(\text{nm}) = [(27.6A)/(\rho Z^{0.89})]E_0^{1.67} \tag{1}$$

where A is the atomic weight (g/mole), ρ is the density (g/cm^3), Z is the atomic number, and E_0 is the incident beam energy (keV). The range of high energy ($\geq$100 keV) electrons (see Table 1) in solid matter is so great that in order to retain high resolution imaging, diffraction, and analytical information, the beam electron interactions must be restricted to single or plural (at most) scattering events. To achieve this restriction, the specimen thickness must be limited to only tens to hundreds of nanometers, depending on composition. TEM/AEM examination of bulk specimens thus requires preparing a suitably thin cross section, and this step has often been the greatest impediment to a successful study, especially when a specific sub-micrometer feature of a complex device must be located and thinned for examination. When the dimensions of specimen features are less than the foil thickness, as is likely to be the case for fine scale nanostructured materials, multiple features are likely to be encountered

Table 1. Electron Range in Bulk Materials (Kanaya-Okayama Range)

E_0 (keV)	C	Si	Fe	Ag	Au
100	73.5 μm	70.4 μm	23.6 μm	20.2 μm	12.6 μm
30	9.9 μm	9.3 μm	3.2 μm	2.7 μm	1.7 μm
20	5.0 μm	4.7 μm	1.6 μm	1.4 μm	860 nm
10	1.6 μm	1.5 μm	500 nm	430 nm	270 nm
5	490 nm	470 nm	160 nm	130 nm	85 nm
2.5	160 nm	140 nm	50 nm	43 nm	27 nm
1	34 nm	32 nm	11 nm	9 nm	6 nm

through the thickness. In viewing/analyzing such a complex thin section, information from the different structures through the thickness will be superimposed in transmission images and integrated into composite x-ray and electron spectra. To truly isolate such nanostructures, exceptionally thin sections must be prepared, with the desired thickness depending on the scale of the features. The traditional methods of thin section preparation, which have utilized mechanical abrasion, chemical polishing, and ion beam sputtering, have been greatly advanced with the evolution of systems that combine an ion beam sputtering system with a scanning electron microscope (SEM) to achieve exquisite control in locating the target area of interest and preparing the thin section [e.g., Phillips *et al.*, 2000]. With successful thin foil preparation, typical characterization problems include determining the morphology and composition of single phase regions, information that includes both the elemental concentrations derived from spectrometry and the crystallographic parameters derived from diffraction patterns. More complex problems involve the nature of interfaces between different phases in natural materials or the engineered interfaces in technological structures that often control the macroscopic behavior of a device.

The development of the SEM has provided a powerful alternative tool for probing the first surface of massively thick bulk specimens, thus simplifying the specimen preparation problem to that of a single surface, and indeed permitting examination of the surface of rough materials in the as-received condition [Goldstein *et al.*, 2003]. In the "conventional" beam energy range for the SEM, 10–30 keV, the range is greater than a micrometer for materials of low and intermediate atomic numbers (e.g., $Z < 40$), as listed in Table 1, which is not satisfactory for characterizing the individual components of bulk nanomaterials. However, the emergence of the high performance SEM based upon the high brightness field emission gun (FEG-SEM) has enabled practical operation of the SEM in the "low voltage" regime, $E_0 \leq 5$ keV, where the range is typically reduced to 100 nm or less, even in materials of low atomic number, as also listed Table 1. Low voltage SEM thus provides access to nanoscale structures in bulk materials with the possibility of minimum or even no specimen modification.

B. Nanoparticles

The second broad class of nanomaterials consists of discrete particles with nanometer dimensions which will be referred to as "nanoparticles" [e.g., Komarneni, *et al.*, 1997]. Discrete nanoparticles with dimensions $\leq$100 nm are sufficiently small that they are already compatible with high beam energy transmission electron microscope techniques without further specimen preparation. Following dispersion and mounting on a suitable thin foil substrate, such as 10-nm thick amorphous carbon film supported on a Cu, Ni or C grid, nanoparticles can be examined directly in the as-received condition in the TEM/AEM or SEM. More complex preparation problems are encountered when nanoparticles are aggregated, either as a result of the fabrication scheme or after subsequent processing, which may require redispersion or thinning for AEM/TEM examination. Alternatively, the SEM can be applied to study the outer layer of aggregated nanoparticles regardless of the total thickness of the aggregate without the need for thinning methods which might alter the structure or composition.

The typical characterization problem required for discrete nanoparticles is the determination of the morphology, composition and crystalline phase of individual particles and the variation of these properties within particle populations. While individual nanoparticles may be found that are homogeneous, populations of particles with different compositions may be found intermixed. More complex nanoparticles may have a surface coating with a composition different from the interior, and the nature of this coating and its interface to the underlying substrate particle can provide a range of increasingly challenging characterization problems. Finally, nanoparticles may have internal compositional heterogeneities on an even finer scale, down to clusters of atoms.

III. PHYSICAL BASIS OF THE ELECTRON-EXCITED ANALYTICAL SPECTROMETRIES

The physical basis for spectrometric analysis under electron bombardment is illustrated in Figure 1 for a carbon atom. The initial interaction is an inner shell ionization event created by inelastic scattering of the energetic beam with a bound inner shell electron. In this inelastic event, the beam electron transfers an amount of energy at least equal to the binding energy (critical ionization energy), E_c, of the bound atomic electron, which is then ejected from the atom, leaving a vacancy in the shell. If the incident beam electron has a sharply defined energy, E_0, then because the ionization edge energy E_c is also sharply defined, such an inelastic interaction, a so-called "core loss" event, will cause a well-defined change in the energy of the beam electron. Following such a core loss event, the incident beam electron leaves the sample with a reduced energy given by $E_0–E_c$. Because the amount of energy lost is defined by the binding energy of the inner shell atomic electron, a spectrum of energy losses experienced by the incident beam forms the basis for elemental analysis of the sample; this techniques is known as electron energy-loss spectrometry (EELS) (Egerton, 1986).

After the primary ionization of a K-shell of carbon ($E_c = 240$ eV) in Figure 1, the excited atom resides in the excited state for a few picoseconds and then undergoes electron transitions between the L and K shells to lower its energy back toward ground state. The lower section of Figure 1 illustrates the first stage of this decay process. In one case (radiative transition, referred to as fluorescence), the excess energy difference after the electron transition between the shells is expressed in the form of electromagnetic energy, as an x-ray photon. Because the energy levels of the atomic shells are sharply defined and specific to each element, the transition of an electron from one shell to

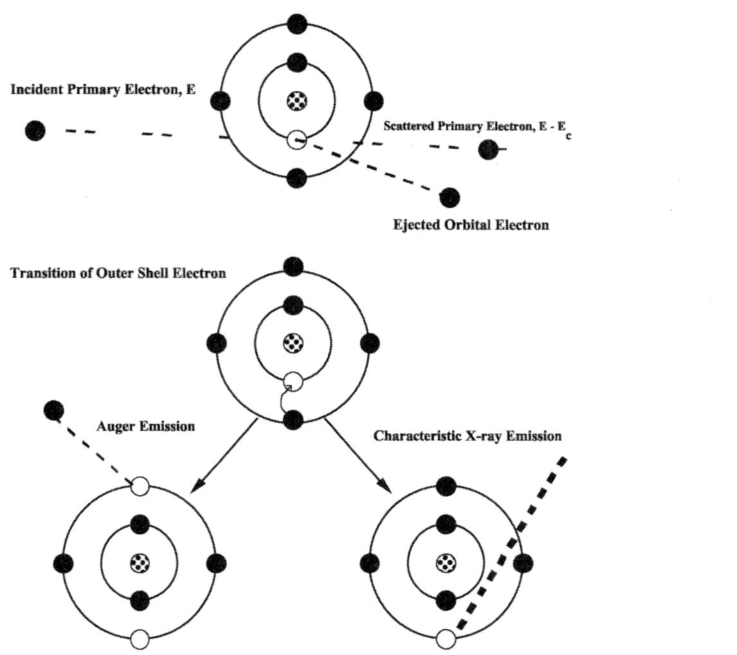

Figure 1. Inner shell ionization and de-excitation tree for carbon (Goldstein *et al.*, 2003).

another produces a sharply defined difference in energy that manifests as the energy of the x-ray photon, making it characteristic of the particular atom species. As a result of the transition process, the vacancy moves out to an outer atomic shell, where the transition process can repeat for a sufficiently complex (intermediate to high atomic number atom), resulting in a family of characteristic x-ray photons. Measuring the energy of the photons forms the basis of x-ray spectrometry [Goldstein *et al.*, 2003].

In the second case (the non-radiative transition), as the excited atom undergoes electron transitions, the energy difference between the shells is imparted to another bound outer shell electron, known as the Auger effect (Auger, 1923). This atomic electron is ejected with a kinetic energy equal to the energy difference of the initial transition minus the binding energy of the ejected electron. Because all of the energy levels are sharply defined, the as-ejected Auger electron also has a characteristic kinetic energy that identifies the atomic species responsible for the emission. Measuring the energy of the electrons ejected from the target forms the basis of Auger electron spectrometry (AES) (Bishop, 1989).

Each of the analytical spectrometries derived from the physics of inner shell ionization has its strengths and weaknesses for analysis of nanoscale materials, whether prepared in thin section or in the form of discrete nanoparticles. The following sections describe key factors that must be considered when approaching nanoscale analysis problems.

The electron microscopy platforms upon which these analytical spectrometries can be applied to nanoscale characterization problems can be divided into two broad classes depending upon beam energy. The high energy ($\geq$100 keV) class is represented by the transmission electron microscope (TEM)/scanning transmission electron microscope (STEM) equipped for analysis as the analytical electron microscope (AEM), while the intermediate ($10 \leq E_0 \leq 30$ keV) and low energy ($\leq$5 keV) classes are based upon the scanning electron microscope (SEM) (Sarikaya *et al.*, 1994).

IV. NANOSCALE ELEMENTAL CHARACTERIZATION WITH HIGH ELECTRON BEAM ENERGY

A. Electron Energy Loss Spectrometry

Electron energy loss spectrometry (EELS) in the AEM involves the measurement of the kinetic energy of an electron that has passed through the specimen (Egerton, 1986). In order for this kinetic energy to be analytically meaningful, the incident electron energy must be sharply defined, so that the energy loss after the interaction results in a value that is characteristic of that particular inelastic scattering process, such as inner shell ionization. The beam electrons emitted from the gun of the electron microscope satisfy the requirement to be sharply defined in energy, with a kinetic energy range that depends upon the type of source: thermionic, thermal field emission, or cold field emission. Even when using a source dependent on thermionic emission (which yields the broadest distribution of energies), the incident energy for a high voltage transmission electron microscope operating at 100 keV is typically defined within 3 eV. This represents a relative energy spread of 3 eV/100,000 eV, or 3×10^{-5}, which is sufficient for useful EELS spectrometry of inner shell ionization features, which are typically spaced in energy by tens or hundreds of electron volts.

The measurement of the EELS spectrum is accomplished with a parallel detection spectrometer consisting of electromagnetic lenses which efficiently couple the beam that exits the specimen to a magnetic prism, and the prism to an imaging detector that can simultaneously view a large energy loss band, e.g., 0–1 keV. Generally, the limit on the energy losses is approximately 2.5 keV.

1. Analytical Aspects of EELS

The characteristic core loss edge structures provide the critical information for elemental analysis by EELS. Inevitably, EELS is sensitive to the complete range of inelastic scattering processes encountered by the energetic beam electrons. These phenomena range in energy loss from low energy transfer processes such as phonon scattering at a fraction of an eV energy loss, to ionization of valence and molecular-bonding electrons in the 1 eV–10 eV range, to scattering with bulk and surface plasmons in the 5 eV–50 eV range, and finally to the ionization of core-level electrons that can be used to identify and quantify the elements present. The core loss peaks useful for elemental analysis are illustrated for NIST Standard Reference Material 2063a (Thin Glass Film) in Figures 2 and 3. While all of these interactions enrich the information available, they also complicate the EELS spectrum of a thick specimen through multiple inelastic scattering. When multiple scattering occurs, the energy losses of different events are

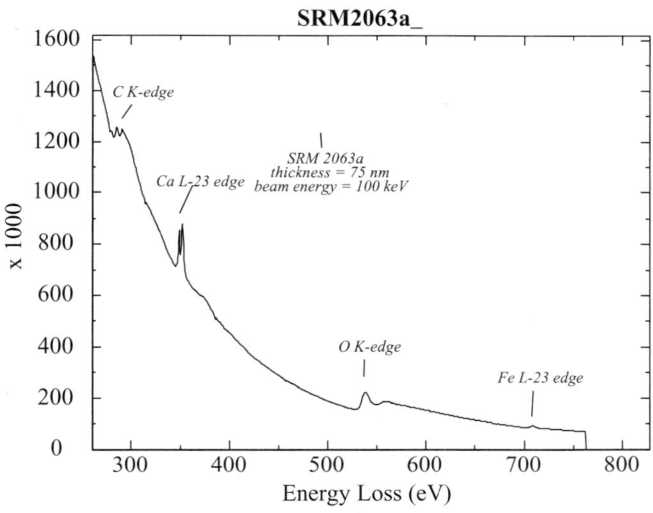

Figure 2. EELS spectrum of NIST Standard Reference Material 2063a (Thin Glass Film) showing core edges for Ca L, O K, and FeL edges (Leapman and Newbury, 1993).

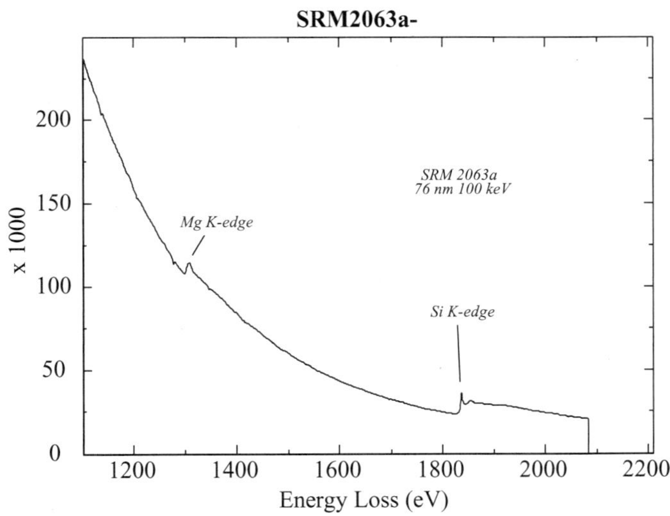

Figure 3. EELS spectrum of NIST Standard Reference Material 2063a (Thin Glass Film) showing core edges for Mg K and Si K edges edges (Leapman and Newbury, 1993).

superimposed upon the same beam electron, which has the effect of degrading the specific detail at an excitation edge or other feature. Multiple inelastic scattering increases with thickness. With sufficiently thin specimens where the inelastic scattering consists of, at most, single events per beam electron, the excitation edges permit, in principle, the detection of any element in the entire periodic table from a K, L, M, or N edge in the energy loss range from 0–2.5 keV. Calculation procedures have been developed to deconvolve the effects of multiple scattering, extending the practical thickness range for EELS (Egerton, 1986). Since inelastic cross sections generally decrease with increasing beam energy, the EELS technique is applied in the high voltage analytical electron microscope, where 100 keV $\leq E_0 \leq$ 300 keV. Nevertheless, specimen thickness is the chief limitation to the application of EELS to nanoscale materials, especially bulk nanoscale and aggregated nanoparticles which must be thinned prior to examination.

At 100 keV in Si, the average rate of energy loss with distance from all inelastic processes is estimated to be 0.6 eV/nm from the Bethe energy loss model. Because of the stochastic nature of inelastic scattering, an individual beam electron will have lost an indeterminate amount of energy after passing through a portion of the specimen. While such an energetic electron can still initiate the ionization tree shown in Figure 1 further along its trajectory in the specimen, the measured kinetic energy of the beam electron after the event is no longer characteristic of a single inner shell ionization event that identifies the atom species present but will be transferred to the background continuum below the edge energy. The usefulness of EELS is thus restricted to specimens whose thickness is of the order of 10–100 nm, depending on the atomic number of the most abundant atom species. For Si, a thickness of 50 nm would result in a dispersion of approximately 30 eV in the beam energy through the foil.

The fraction of beam electrons that will undergo a specific event while traversing a particle or foil of thickness t is given approximately by (t/λ), where λ is the mean free path for the event:

$$\lambda = A/(N_0 \rho Q) \tag{2}$$

A is the atomic weight (g/mol), N_0 is Avogadro's number, ρ is the density (g/cm^3), and Q is the cross section for the process of interest. For inner shell ionization, the cross section can be described by:

$$Q = 6.51 \times 10^{-20} (n_s b_s) / \left(U E_c^2\right) \log_e (c_s U) \tag{3}$$

where n_s is the number of electrons in the shell being ionized, b_s and c_s are constants, E_c is the shell binding energy (keV), and U is the overvoltage $= E_0/E_c$ and the dimensions of Q are [ionizations/e/(atom/cm^2)]. Brown (1974) took $c_K \sim c_L \sim 1$, with the following expressions for b_s:

$$b_K = 0.52 + 0.0029 Z \tag{4a}$$

$$b_{L23} = 0.44 + 0.0020 Z \tag{4b}$$

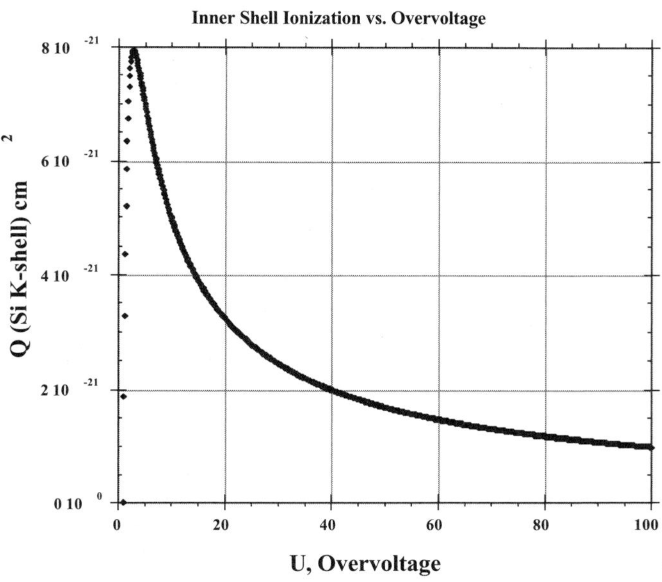

Figure 4. Plot of the inner shell ionization cross section Q of Jakoby versus overvoltage, U.

where Z is atomic number. The inner shell ionization cross section for the Si K-shell is plotted versus U in Figure 4, where the value of Q initially rises from $U = 1$, reaches a peak at a value of $U = 3$, and then decreases asymptotically for further increases in U.

For silicon, the ionization energy $E_K = 1.838$ keV, so that for $E_0 = 100$ keV, $U = 54.4$ which gives $Q = 1.59 \times 10^{-21}$ cm^2. This cross section results in a mean free path for Si K-shell ionization as $\lambda = 0.0128$ cm. In traversing a 50 nm Si foil, $(t/\lambda) = 3.9 \times 10^{-4}$, which means only about 4 in 10,000 of the incident electrons will undergo Si inner shell ionization while traversing 50 nm of Si. Thus, inner shell ionization is expected to be a weak feature in an EELS spectrum compared to higher probability inelastic events, such as plasmon scattering. Fortunately, these rare core-loss electrons are still traveling in much the same direction after interaction so that they can be efficiently collected. The mean scattering angle, θ_E, for a given process is given by:

$$\theta_E = \Delta E / 2E_0 \tag{5}$$

where ΔE is the energy loss. At $E_0 = 100$ keV and for Si ionization where $\Delta E = 1.838$ keV, $\theta_E = 9.2 \times 10^{-3}$ radians or $\sim$0.5 degrees, which is within the maximum angle that can be accommodated with an EELS spectrometer. Moreover, the EELS measurement does not depend on the subsequent decay of the excited core edge state and partitioning of the emission, as is the case for x-ray or Auger spectrometry. Finally, although there are certainly beam–sensitive materials such as biological and organic specimens, the EELS measurement is usually non-destructive, so the collection of atoms being examined can be repeatedly measured. As a result of these considerations, EELS can achieve extraordinary absolute mass sensitivity despite the relative paucity of ionization events, reaching the attogram to zeptogram level. Under some circumstances where the core edge displays features that are sharply defined in energy loss (e.g., "white lines"), nanoscale analysis with trace sensitivity as low as 10^{-5} mass fraction has been demonstrated by accumulating high count spectra combined with special mathematical methods of spectrum processing ("first difference" processing) (Leapman and Newbury, 1993).

The typical maximum practical energy loss utilized in a parallel collection EELS spectrometer is approximately $\Delta E = 2.5$ keV, making the S K-edge at 2.47 keV the highest energy K-edge before L-edges, M-edges, etc. must be utilized. For some elements, the L-, M- and N-edges are not as sharply defied in the onset energy as the K-shell of similar energy, which restricts the sensitivity for these elements.

2. EELS QUANTIFICATION

Quantification of elemental constituents with EELS is based upon integrating the energy loss region beginning at the ionization edge to a defined limit above the edge, typically a window 50 eV wide (Egerton, 1986). This range represents atomic electrons scattered to various final kinetic energies. For thin specimens where multiple scattering is minimized, the intensity I_K measured above the K-ionization edge for an atomic species A, is given by:

$$I_{K,A} = N_A Q_{K,A} I_0 \tag{6a}$$

where N_i is the number of atoms of element "i" in the excited volume of the specimen, Q_K is the K-ionization cross section, and I_0 is the incident beam current. Similar relations can be written for L-, M-, N- etc. edges. Because the thickness of a specimen is often locally variable and difficult to measure, absolute analysis is not typically performed, but rather where multiple elements coexist in the excited region of the specimen, one element is typically determined relative to another. For relative elemental analysis of two species "A" and "B", equation (6) can be written for each species, and by separating the parameters equal to the beam current, which is identical for both "A" and "B" since the two quantities are simultaneously measured with a parallel spectrometer, the following relation is obtained:

$$I_{K,A}/(N_A Q_{K,A}) = I_0 = I_{K,B}/(N_B Q_{K,B}) \tag{6b}$$

$$N_A/N_B = (I_{K,A}/I_{K,B})(Q_{K,B}/Q_{K,A}) \tag{6c}$$

The relative error budget of this EELS quantification procedure consists of the measurement errors associated with the intensities I (the variation in I_A/I_B is likely to be less than 5% relative with parallel EELS spectrometry) and uncertainties in the theoretical ionization cross sections (partial in energy range ΔE and scattering angle ΔQ for which errors are more difficult to estimate). If a standard of known composition

Table 2. Sigma-K, Sigma-L EELS Quantitative Analysis of SRM 2063a Leapman and Newbury (Analytical Chemistry, vol. 65, 1993, p. 2409)

Element/Edge		Certified atom fraction	Measured by EELS atom fraction	Relative error, %
O	K	60.8 ± 2.2	61 ± 3	0.3%
Mg	K	7.5 ± 0.3	7.6 ± 0.4	1.3
Si	$L_{2,3}$	20.3 ± 0.9	19.2 ± 1.0	−5.4
Ca	$L_{2,3}$	6.6 ± 0.3	7.9 ± 0.4	19.7
Fe	$L_{2,3}$	4.5 ± 0.4	4.6 ± 0.2	2.2

(i.e., the ratios N_A/N_B for all elements) is available, then the overall error budget can be estimated by comparing the calculated ratios of N_A/N_B to the known values. Such studies have suggested that quantitative EELS measurements of K-shells can determine simple stoichiometric binaries, such as BN or SiC, within 5% of the ideal value, and more complex systems within 20% relative when the specimens are sufficiently thin that multiple scattering effects are negligible. An example that demonstrates this performance is given in Table 2, where the Sigma-K and Sigma-L cross sections of Egerton (1986) were used for quantification.

3. SPATIAL SAMPLING OF THE TARGET WITH EELS

The spatial resolution of any of the techniques based upon the ionization tree in Figure 1 is determined by two major factors, the diameter of the incident probe that contains sufficient beam current for the measurement and the scattering of the beam electrons and the emitted radiation within the target. The smallest volume of analysis would be the ideal cylinder defined by the footprint of the beam on the entrance and exit surfaces and the specimen thickness, with the exit diameter only increased by the angle of divergence of the beam, as shown in Figure 5. Compared to the spatial resolution situation for x-ray spectrometry described below, the action of elastic scattering to remove beam electrons from the ideal cylinder and degrade the lateral spatial resolution of analysis is *not* a significant effect for EELS. The reason is that any significant elastic scattering will cause the electron trajectory to deviate outside the input acceptance angle of the EELS spectrometer so that these scattered and spatially degraded electrons cannot contribute to the energy loss spectrum. Thus, although EELS must be restricted to thin specimens to reduce multiple scattering so as to preserve the integrity of the core loss information, a positive aspect is that the lateral spatial resolution can actually correspond to the practical beam diameter, e.g., 1 nm or smaller in a field emission AEM.

B. X-ray Spectrometry

The x-ray spectrum is almost universally measured with the semiconductor energy dispersive x-ray spectrometer, usually of the silicon (lithium compensated) type, designated Si(Li) or Si-EDS, which operates at a temperature near 77 K. Such Si-EDS spectrometers are generally capable of an optimum energy resolution of approximately 130 eV at the energy of MnKα radiation (5890 eV). The resolution is dependent

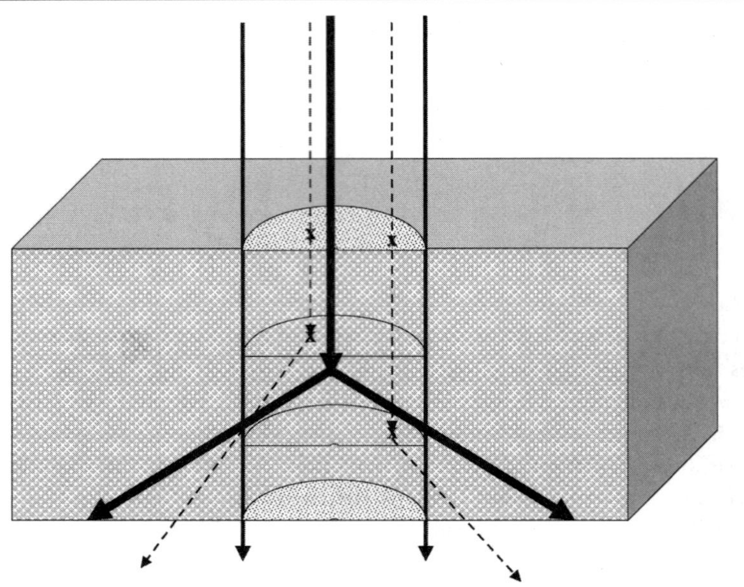

Figure 5. Schematic of the excitation cylinder plus the x-ray excitation cone calculated with the mid-specimen beam broadening criterion (thick lines).

on the energy of the x-ray photon, with peak widths of 65 eV at C K (282 eV) and 175 eV at CuKα(8040 eV). The limiting count rate is approximately 3 kHz at optimum resolution (long pulse integration time constant) and 25 kHz with degraded resolution at short time constant [Goldstein *et al.*, 1993].

Two recent and ongoing developments in x-ray spectrometry may soon radically change this x-ray measurement performance. (1) The "silicon drift detector" (SDD) is a new design for the Si-EDS that operates at 250 K and achieves similar energy resolution to the Si-EDS when long pulse time constants are used [Struder *et al.*, 1998; Iwanczyk *et al.*, 2001]. However, the SDD is capable of operating with a much shorter time constant, with some compromise in resolution, that enables spectrum counting rates of 400 kHz or higher. (2) The microcalorimeter EDS determines the energy of an x-ray photon by measuring the temperature rise when the photon is absorbed in a metal target held at a temperature of 100 mK [Wollman *et al.*, 1997]. The resolution of the microcalorimeter EDS has been demonstrated to be 4.5 eV at MnKα, with a resolution of 2 eV at AlKα (1487 eV). The limiting count rate is approximately 1 kHz.

1. Analytical Aspects of X-ray Spectrometry

Once the x-ray photon is emitted from an atom inside a solid, Figure 1, it must propagate through the other atoms of the material to escape. X-rays are subject to photoelectric absorption (i.e., the x-ray is annihilated during an interaction with a bound atomic electron which gains the entire energy of the x-ray and which is subsequently ejected from the atom as a photoelectron), incoherent (inelastic) scattering, and coherent scattering. Incoherent or inelastic scattering, which alters the x-ray energy, is much less probable, by a factor of 10^{-6} to 10^{-2}, than photoelectric absorption over the range of photon energies of analytical interest (0.1–15 keV). For the range of electron excitation through which the emitted x-rays subsequently must propagate, inelastic scattering is so much lower in probability than absorption that for practical measurements, it can be ignored. An x-ray photon will either be annihilated by absorption or else it will escape carrying its original energy. Thus, a characteristic x-ray will remain capable of identifying its source atom even after it has passed through the solid.

2. X-ray Spectrometry Quantification: Thin Specimens (High Beam Energy AEM Case)

When x-ray spectrometry is employed in the high beam energy (100–300 keV) analytical electron microscope (AEM), the specimens are in the form of thin foils or discrete nanoparticles. This simple specimen geometry and limited mass thickness through which the x-rays must propagate permits a straightforward empirical approach to quantitative analysis through the application of sensitivity factor analysis and the use of calibration standards. A sensitivity factor k_{AB} for element A relative to element B is defined as follows (Cliff and Lorimer, 1975):

$$k_{AB} = \left[\frac{C_A}{C_B}\right] * \left[\frac{I_B}{I_A}\right] \tag{7a}$$

where C is the weight (mass) concentration of the element in the standard, I is the measured x-ray intensity, and A and B represent any two elements in the specimen. B is the reference element, generally chosen to be a major constituent of the standard whose concentration is known with good accuracy. Once values of k_{AB} are measured experimentally from known standards, they can be applied to the analysis of unknowns. Any measured relative x-ray intensity ratio I_A/I_B from an unknown can be converted to a relative concentration ratio C_A/C_B by multiplying the intensity ratio by k_{AB}:

$$C_A/C_B = (I_A/I_B)k_{AB} \tag{7b}$$

The concentrations of the individual constituents in terms of the concentration of the reference element B can be obtained by rearranging equation (7b):

$$C_A = C_B * k_{AB} * [I_A/I_B] \tag{7c}$$

By measuring all elemental constituents, the absolute concentrations can be calculated from the following relationship:

$$C_B + \sum_i C_i = 1 \tag{8a}$$

where C_B is the concentration of the reference element and C_i represents all other elemental constituents. Equation 7(c) can be substituted in equation 8(a) for each element i, so that equation 8(a) is reduced to an equation in one unknown, C_B:

$$C_B + \sum_i C_B k_{iB}(I_i/I_B) = 1 \tag{8b}$$

because the k_{iB} terms are known from standards and the (I_i/I_B) terms are the experimental measurements on the unknown. Equation (8b) can then be solved to give an absolute value of the concentration for the reference element in the unknown, C_B. This C_B value can be substituted into equation (7c) to give absolute values of the concentrations of all other constituents. The error budget of such an empirical procedure is such that the errors are in the range 5–10% relative, typically limited by the accuracy with which the standard concentrations are known.

When suitable multiple element standards of known composition that are homogeneous on the nanometer scale are not available to determine the values of the k_{AB} sensitivity factors, an alternate approach based upon physical theory is possible. The x-ray intensity, I_A^*, generated for an element A by an electron beam that passes through a solid of thickness t, where t is small compared to the mean free path for elastic scattering, $t \ll \lambda_{\text{elastic}}$, is given by:

$$I_A^* = \text{constant}\, C_A\, Q_A\, \omega_A\, a_A\, \rho t / A_A \tag{9}$$

where C is the weight (mass) fraction, Q is the ionization cross section, ω is the fluorescence yield, a is the x-ray family abundance for the peak measured (e.g., $K\alpha$ or $K\beta$), ρ is the density (g/cm^3), and t is the thickness (cm). The fraction of this radiation that is actually measured depends upon absorption or penetration through in the components of the x-ray spectrometer:

$$I_A = I_A^* \varepsilon_A \tag{10}$$

where ε_A is the efficiency of the spectrometer. The spectrometer efficiency can be generally described as (Goldstein *et al.*, 1993):

$$\varepsilon_A = \left\langle \exp\left[-\left(\frac{\mu}{\rho}\right)_{win}^{A} \rho_{win} s_{win} - \left(\frac{\mu}{\rho}\right)_{ice}^{A} \rho_{ice} s_{ice} - \left(\frac{\mu}{\rho}\right)_{Au}^{A} \rho_{Au} s_{Au} - \left(\frac{\mu}{\rho}\right)_{SiDL}^{A} \rho_{Si} s_{SiDL}\right]\right\rangle * \left\langle 1 - \exp\left[-\left(\frac{\mu}{\rho}\right)_{Si}^{A} \rho_{Si} s_{Si}\right]\right\rangle \tag{11}$$

In equation (11), the terms of the form (μ/ρ) represent the mass absorption coefficients for x-rays of element A and the terms "s" the thicknesses of the various spectrometer components including the window(s), pathological ice buildup on the detector, gold

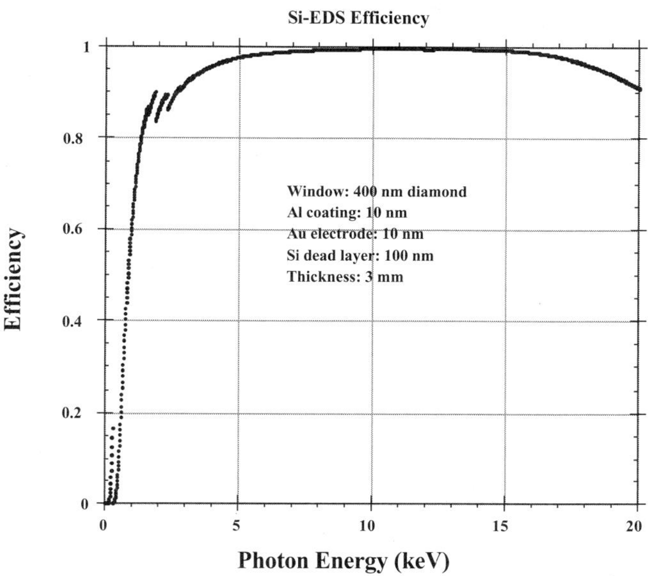

Figure 6. EDS detector efficiency versus photon energy for a silicon detector with a 10 nm Al light reflection coating, a 400 nm diamond window, a 10 nm gold surface electrode, 100 nm silicon dead layer, and an overall Si detector thickness of 3 mm.

front surface electrode, the "dead" or partially active silicon layer just below the Au electrode, and the overall silicon detector thickness. The terms in the first set of brackets <> describe the absorption of x-rays passing through the window and various components of the EDS to reach the active volume of the detector. The second term in brackets <> represents the loss of x-rays through the thickness of this active volume, which becomes significant for photon energies above about 10 keV depending on detector thickness. An example of the detector efficiency response calculated for a detector with a 400 nm diamond window, 100 nm of ice, 10 nm gold surface electrode, 100 nm silicon dead layer, and a 3 mm detector thickness is shown in Figure 6.

If equations (9) and (10) are written for two elements "A" and "B", then the ratio I_A/I_B can be expressed as:

$$\frac{I_A}{I_B} = \frac{C_A}{C_B}\left[\frac{(Q\omega a/A)_A\,\varepsilon_A}{(Q\omega a/A)_B\,\varepsilon_B}\right] \tag{12}$$

Note that in this ratio, the mass thickness term, ρt, divides out quantitatively so that knowledge of the local specimen thickness is not required. Matching terms in

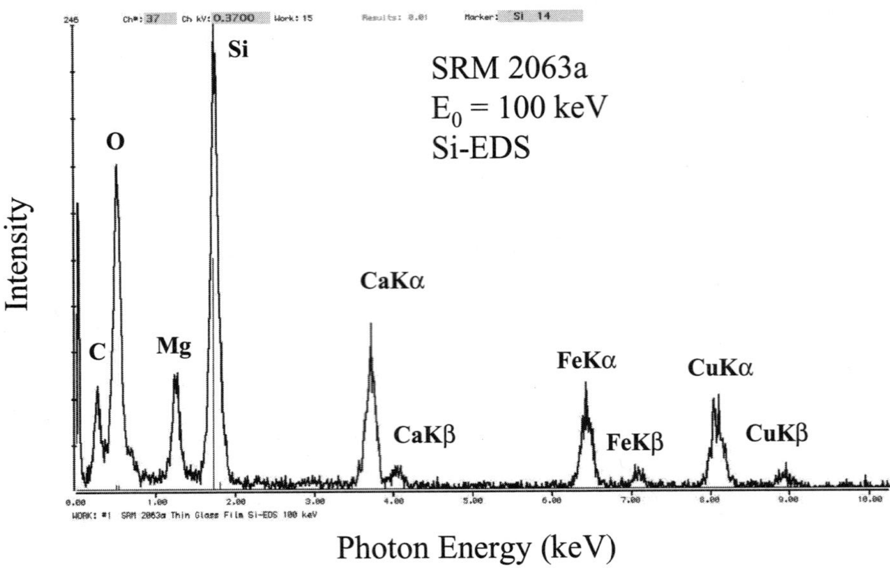

Figure 7. EDS spectrum of SRM 2063a recorded in an AEM with $E_0 = 100$ keV.

equation 12 with equation (7b), the sensitivity factor k_{AB} is explicitly calculated as:

$$\frac{1}{k_{AB}} = \left[\frac{(Q\omega a/A)_A\,\varepsilon_A}{(Q\omega a/A)_B\,\varepsilon_B}\right] \tag{13}$$

Equation 12 enables the analyst to convert measured intensity ratios to ratios of concentration from "first principles" with suitable expressions for Q, ω, and a for the elements of interest and with a model for the efficiency of the EDS detector. The efficiency of the EDS detector can be estimated from the known window and detector parameters, and then "fine tuned" by comparing the x-ray continuum from a known target. Following this procedure for the EDS x-ray spectrum of the SRM 2063 thin glass film shown in Figure 7, yields the results given in Table 3a (100 keV) and 3b (300 keV). In general, the concentration values obtained from the first principles calculation fall within ±25% of the known standard reference values for elements measured with K-shell x-rays.

The error budget has significant contributions associated with the detector efficiency and the physical calculations. The detector efficiency is modeled from knowledge of the materials of construction (window, detector electrode, silicon dead layer, etc.) and tested by comparing measured thin film continuum spectra with calculated continuum spectra. Such a procedure is likely to have significant errors for low energy photons, e.g., $E < 2$ keV and especially for light elements like oxygen, where the efficiency is low and sensitive to contamination. The magnitude of the contributions to the error budget associated with calculating sensitivity factors with equation (12) depend

Table 3a. "First Principles" Quantitative EDS Analysis of SRM 2063a at 100 keV (All elements measured with K-shell x-rays)

Element	Certified atom fraction	Measured by EDS* atom fraction	Relative error, %
O	60.8 ± 2.2	62.3 ± 2.5**	+2.5%
Mg	7.5 ± 0.3	5.8 ± 4.1	−23
Si	20.3 ± 0.9	20.4 ± 2.0	+0.5
Ca	6.6 ± 0.3	6.6 ± 3.1	0
Fe	4.5 ± 0.4	5.0 ± 3.6	+11

* Quantification with Jakoby cross section for K-shell ionization
** relative standard deviation of the count (1 σ)

Table 3b. "First Principles" Quantitative EDS Analysis of SRM 2063a at 300 keV (All elements measured with K-shell x-rays)

Element	Certified atom fraction	Measured by EDS* atom fraction	Relative error, %
O	60.8 ± 2.2	55.9 ± 1.1**	−8.0%
Mg	7.5 ± 0.3	7.8 ± 1.4	+4.0
Si	20.3 ± 0.9	23.6 ± 0.7	+16
Ca	6.6 ± 0.3	7.4 ± 1.1	+12
Fe	4.5 ± 0.4	5.3 ± 1.3	+18

* Quantification with Jakoby cross section for K-shell ionization
** relative standard deviation of the count (1 σ)

on the atomic shells involved. Thus the ionization cross section, fluorescence yield, and family abundance factor are generally known within ±5% to 10% relative for the K-shell, but are more poorly known for the L- and M-shells. The error budget can be especially large when calculation of k_{AB} must involve different shells, e.g., K vs. L or L vs. M, for elements "*A*" and "*B*". Relative errors of 20% to 50% or more can be encountered with such raw theoretical calculations for L and M shells. To reduce the errors contributed by the imperfect knowledge of the physical parameters, a good strategy is to use even a limited suite of known standards, ideally with K, L, and M-shell elements represented, to provide an empirical measurement base with which to constrain the theoretical calculations. The theoretical calculations are then used to fill in the k_{AB} values that are not represented in the standards suite but which are required for the analysis of the unknown. Such a combined procedure is capable of achieving errors within 10–20% relative when known stoichiometric compounds are tested.

From the point of view of quantitative x-ray microanalysis, a specimen is considered 'thin" as long as the error contribution of the differential self-absorption of the characteristic x-rays remains below a specific value, e.g., 5%. As the specimen thickness increases, the effects of specimen self-absorption will eventually influence the relative x-ray intensities. The thickness at which this happens depends on the photon energies of interest and the composition. This relative absorption effect is most pronounced if a range of photon energies from low (<1 keV) to high (>5 keV) is represented in the elements to be measured. Goldstein *et al.* (1977) have described a criterion for estimating the limit of specimen thickness for which the thin film approximation is valid for any pair of elements, based upon the differential x-ray absorption:

$$\left| \frac{\mu^B}{\rho_{spec}} - \frac{\mu^A}{\rho_{spec}} \right| \rho t \csc\varphi < 0.2 \tag{14}$$

where μ/ρ is the mass absorption coefficient for the respective element in the specimen, ψ is the spectrometer take-off angle, ρ is the density, and t is the thickness.

For specimens that exceed this thickness criterion, an absorption correction is necessary. Goldstein *et al.* (1986) described a model for a multiplicative absorption correction factor (ACF) that considered a constant rate of x-ray production through the thickness, which is a good approximation for the AEM case and nanoscale materials:

$$\frac{C_A}{C_B} = \frac{I_A}{I_B} k_{AB} ACF \tag{15a}$$

where the absorption correction factor is given by:

$$ACF = \left[\frac{(\mu/\rho)^A_{spec}}{(\mu/\rho)^B_{spec}} \right] \left[\frac{1 - \exp\left[-(\mu/\rho)^B_{spec}\, \rho t \csc\varphi\right]}{1 - \exp\left[-(\mu/\rho)^A_{spec}\, \rho t \csc\varphi\right]} \right] \tag{15b}$$

To apply the absorption correction factor, a specific estimate of the thickness, t, must be provided.

Additionally, a correction for fluorescence induced by the characteristic radiation may be needed for thicker specimens when the characteristic x-ray energy of a major element "A" is within 1 keV above the critical ionization energy of a second element "B", which leads to extra emission of "B" beyond that directly excited by the electron beam (Goldstein *et al.*, 1986). Such a correction is needed when a wedge-shaped specimen is used, such that x-rays created by the beam impact in the thin area can propagate laterally into the thick rim of "bulk" material. However, because of the much greater range of x-rays in matter, compared to electrons, such a fluorescence correction is almost never significant for nanoscale particles and ultrathin material sections.

3. Spatial Resolution of X-ray Microanalysis by AEM

The fundamental spatial resolution of x-ray microanalysis in the AEM is determined by the same interaction volume as that for AEM-EELS described above: a cylinder defined by the beam footprint on the entrance and exit surfaces and the specimen thickness, Figure 5. However, for x-ray microanalysis, the effects of elastic scattering of a fraction of the beam electrons cannot be neglected because x-ray events produced by these elastically scattered electrons outside of the beam cylinder are collected by the EDS with equal efficiency as those x-rays produced inside the cylinder. Goldstein *et al.* (1977) developed a formula that describes the degradation of the spatial resolution of x-ray microanalysis for the single elastic scattering regime by considering that beam electrons scattered from the fundamental cylinder are distributed in a cone shaped volume centered at the middle of the object, Figure 5. The diameter, b (cm), of the base of the cone at the exit surface that contains 90% of the electrons is given by (Goldstein *et al.*, 1977):

$$b = 6.25 \times 10^5 (Z/E_0)(\rho/A)^{1/2} t^{3/2} \quad (16)$$

where Z is the atomic number, E_0 is the incident beam energy in eV, ρ is the density (g/cm^3), A is the atomic weight (g/mole), and t is the specimen thickness (cm).

V. NANOSCALE ELEMENTAL CHARACTERIZATION WITH LOW AND INTERMEDIATE ELECTRON BEAM ENERGY

The SEM becomes the instrument of choice when nanomaterials and nanoparticles must be examined on a bulk substrate or within the interior of bulk material that cannot be thinned for the AEM but which may be directly viewed at the naturally exposed surface or which can be revealed by fracture or another procedure (Goldstein *et al.*, 2003). Because image spatial resolution is inevitably a critical performance factor for such nanoscale problems, the field emission gun SEMs, which have a high source brightness and therefore optimum probe diameter/current characteristics, are the best choice. In applying the FEG-SEM to nanomaterial/nanoparticle problems, the choice of the beam energy depends strongly on the type of problem to be solved. The electron optical performance, as measured by the source brightness, is proportional to the beam energy. Thus, a 20 keV beam is 20 times brighter than 1 keV beam, with proportional improvement in the focused beam characteristics. This high brightness at the upper end of the operational energy scale can be of great value in nanoscale particle investigations. When the specimen consists of nanoscale particles dispersed on a thin support foil which minimizes scattering, a good strategy is to operate the FEG-SEM at the extreme upper end of the intermediate beam energy regime, e.g., 30 keV or more, to achieve the high spatial imaging and analytical resolution of the finely focused beam. However, when the specimen has "bulk" characteristics (i.e., thickness greater than the electron range), the critical factor that controls the spatial resolution for the SEM is the elastic scattering of the beam electrons regardless of how small the beam is initially focused (Goldstein *et al.*, 2003).

A. Intermediate Beam Energy X-ray Microanalysis

1. *X-ray Range in Bulk Materials*

The range of the production of characteristic x-rays by beam electrons in a bulk target can be described through a modification of equation 1 to account for the limit of x-ray production at the critical excitation energy, E_c (Goldstein *et al.*, 2003):

$$R(\text{nm}) = [(27.6\,A)/(\rho Z^{0.89})]\left(E_0^{1.67} - E_c^{1.67}\right) \quad (17)$$

For a flat specimen placed normal to the beam, the range of x-ray production can be thought of as the radius of a hemisphere whose origin is centered on the beam impact

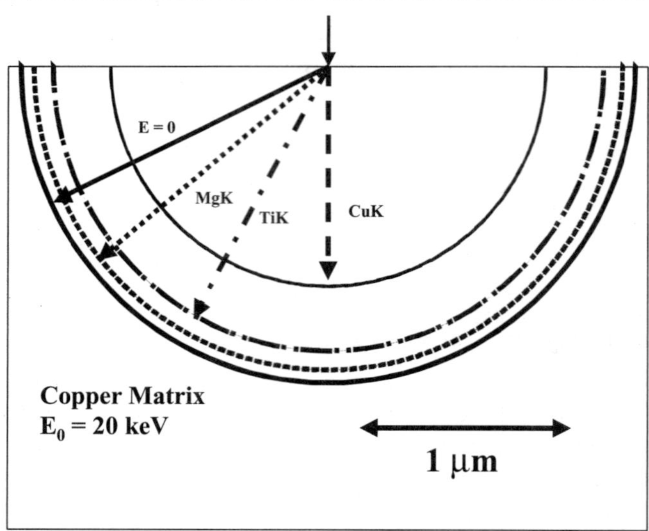

Figure 8a. Range given by equation 17 for E_0 = 20 keV in Cu and various x-ray edges: CuK; TiK, and MgK;

footprint at the specimen surface, as shown in Figure 8(a). This x-ray production range is only a crude description of the interaction, which has a strongly varying volume density as a function of location within the interaction volume, as shown in the Monte Carlo simulation of the depth distribution of x-ray production shown in Figure 8(b).

Equation 17 and Figure 8(a) reveal that the sampling of the target is inevitably different depending on the characteristic photon energies being measured. The difference in sampling volumes can be quite large, an order of magnitude or more, when there is a large range in photon energies measured, e.g., Mg Kα (1.25 keV) vs. CuKα (8.04 keV). Thus, when bulk nanostructures are being examined, it may not be possible to confine the excitation for x-ray photons of different energies within a nanoscale structure of interest using a fixed beam energy in the intermediate range (10 keV to 30 keV).

2. *Matrix Effects in Quantitative X-ray Microanalysis*

Furthermore, quantitative x-ray microanalysis of bulk materials with x-ray excitation in the intermediate beam energy range is subject to "matrix or interelement effects" (Goldstein *et al.*, 2003). Matrix effects arise because of the elastic and inelastic scattering of the beam electrons to form the interaction volume and the subsequent propagation of x-rays through the specimen. All of these physical effects are dependent upon the particular atomic species present. That is, the generation and propagation of the characteristic x-rays of a particular element are modified by the presence of the other elements that make up the specimen at the location sampled by the beam. Thus, the "atomic number effect" arises from the combined effects of electron backscattering, which reduces the ionization power of the beam, and inelastic scattering, which determines

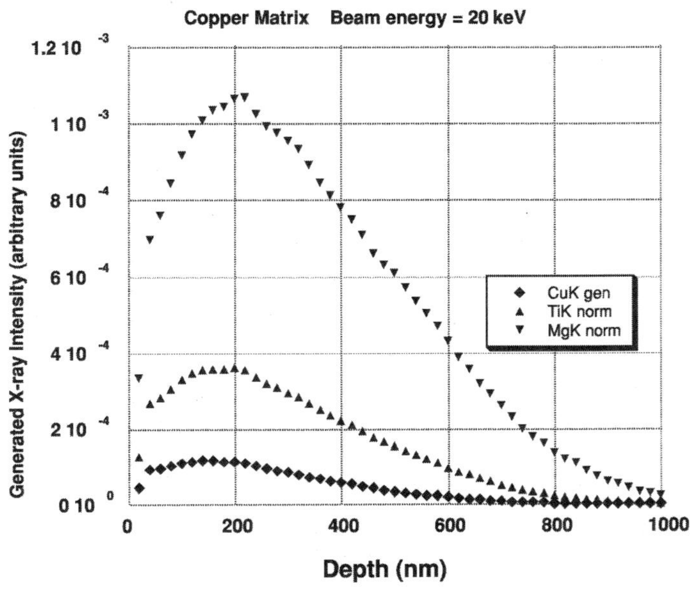

Figure 8b. Monte Carlo simulation of the depth distribution of x-ray production in Cu at $E_0 = 20$ keV.

the ionization power with depth. The "absorption effect" results from the photoelectric absorption of x-rays while propagating through the specimen, and the "secondary fluorescence effect" is a consequence of photoelectric absorption, in which the absorbing atom can subsequently emit its own characteristic x-ray, in addition to those x-rays generated directly by the beam electrons. For intermediate beam energy analysis of thick, bulk specimens, these matrix effects can change the measured intensity relative to the generated intensity, which directly depends on the concentration, by a factor of 10 or more, especially for x-rays of low energy ($E < 2$ keV), and for situations involving high atomic number elements ($Z > 40$) which both strongly scatter electrons and have high absorption for x-rays. Detailed methods based upon a combination of empiricism and physical theory exist for calculating the matrix effects to derive quantitative results with an acceptable error budget of $\pm 5\%$ relative or less (Goldstein *et al.*, 2003). However, intermediate beam energy x-ray microanalysis is not generally appropriate to bulk nanoscale materials because the sampling volume is too large to isolate individual nanoscale features of interest.

3. Analysis of Nanoparticles

Intermediate beam energy x-ray microanalysis does have significant advantages when the specimen consists of discrete, well dispersed nanoscale particles. Beam electrons only undergo modest to negligible scattering when passing through nanoscale particles, and if the particles are supported upon a thin (e.g., 10 nm thick) carbon film, the beam electrons that are transmitted through the particle/support film can be eliminated by trapping in a Faraday cup. More usefully, such transmitted electrons can be collected

Table 4. Quantitative EDS Analysis of SRM 2063a at 30 keV (All elements measured with K-shell x-rays)

Element	Certified atom fraction	Measured by EDS* atom fraction	Relative error, %
O	60.8 ± 2.2	58.3 ± 0.4**	−4.1%
Mg	7.5 ± 0.3	8.1 ± 0.3	+8.0
Si	20.3 ± 0.9	22.3 ± 0.2	+9.9
Ca	6.6 ± 0.3	6.7 ± 0.3	1.5
Fe	4.5 ± 0.4	4.7 ± 0.3	4.4

* Quantification with Jakoby cross section for K-shell ionization
** relative standard deviation of the count (1 s)

with an appropriate detector to provide the signal for scanning transmission electron microscopy (STEM) images, which are highly sensitive to small amounts of particle mass.

When x-ray microanalysis of nanoscale particles is considered under these conditions, the reduction in electron scattering and retardation and the extremely short x-ray absorption path lengths result in matrix corrections that tend to unity. For nanoscale particles measured with intermediate beam energies (e.g., 20–40 keV), the Cliff-Lorimer sensitivity factor method from the AEM can be applied. When appropriate standards are available to determine the Cliff-Lorimer factors, the error budget consists mainly of contributions from the uncertainty in the standard compositions and the counting statistics for the standards and the unknowns. First principles analysis with calculated corrections for the ionization cross section and the detector efficiency, illustrated in Table 4, can produce relative errors for the K-shell that are generally 10% or less, but L- and M-shell measurements are subject to much larger error uncertainties.

B. Low Beam Energy X-ray Microanalysis: Bulk Nanostructures

1. Range at Low Beam Energy

The strong exponential dependence on the beam energy indicates that improved spatial resolution can be obtained at low beam energies for bulk materials. As shown in the Monte Carlo simulation for Si in Figure 9, the interaction volume dimensions decrease into the nanometer scale when the beam energy is lowered. At $E_0 =$ 2 keV, the interaction volume is less than 90 nm in radius. This dramatic reduction in excitation volume, coupled with the high brightness FEG-SEM to achieve useful beam size/current performance, has led to the development of the "low voltage microscopy" regime, arbitrarily defined as $E_0 \leq 5$ keV. This mode when applied to nanoscale materials/particles offers the important advantage that the specimen can be examined without thinning, thus minimizing or avoiding sample preparation artifacts. Additionally, the matrix effects that operate strongly in the intermediate beam energy x-ray microanalysis of bulk materials are much reduced at low beam energy. The x-ray path length through the specimen is sufficiently short such that, except for x-rays of the lowest energy, $E < 500$ eV, absorption is effectively eliminated. The

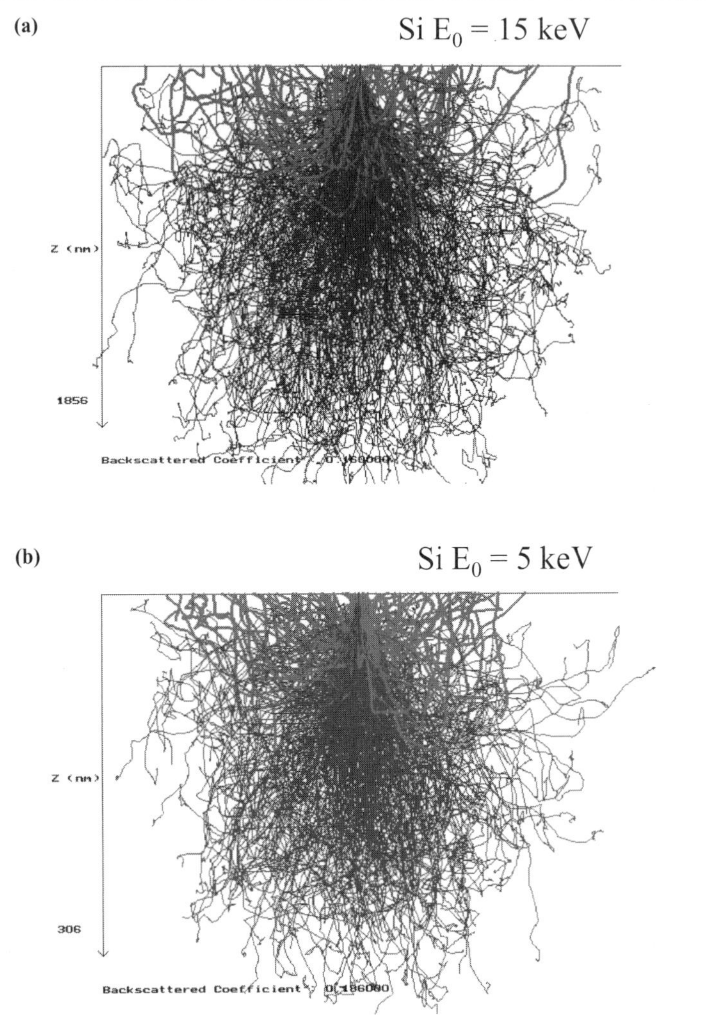

Figure 9. Monte Carlo simulation of electron interactions in silicon at (a) $E_0 = 15$ keV. (b) 5 keV; (c) 2 keV. Depth scales: (a) 1856 nm; (b) 306 nm; (c) 75 nm

electron scattering effects are also much reduced because of the limited energy lost to backscattering. Matrix correction factors thus tend to unity at low beam energy.

2. Limits Imposed by X-ray Spectrometry and Specimen Condition

The negative consequences of performing x-ray spectrometry in the low voltage microscopy regime are the limited photon energies that can be excited with the low beam energy, $E_0 \leq 5$ keV and the low overvoltage that is available for those x-rays that can be excited. The production of x-rays depends upon the beam electron having

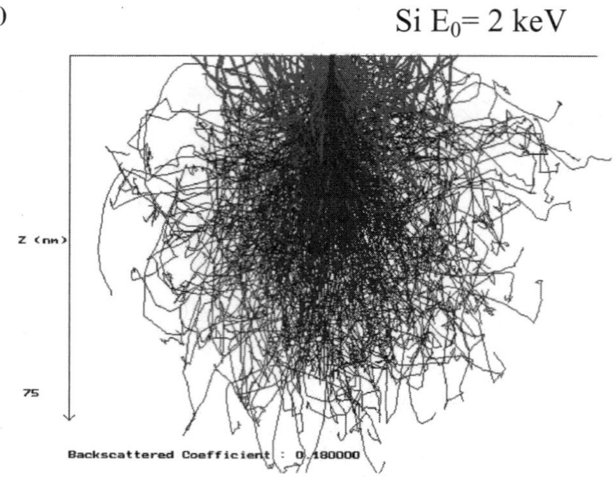

Figure 9. (continued)

sufficient energy, E_0, to ionize the inner shell with an ionization energy, E_c, expressed as the overvoltage:

$$U = E_0/E_c \tag{18a}$$

For a solid specimen, the x-ray intensity generated depends upon the overvoltage:

$$I \approx (U-1)^n \tag{18b}$$

where n is an exponent with a value in the range 1.3 to 1.7, depending on the element and shell. For practical x-ray spectrometry, U must at least have a value of 1.1 to detect major constituents ($C > 0.1$ mass fraction) in a reasonable measurement interval, e.g., 100 seconds. Figure 10(a,b,c) shows EDS spectra of silicon with overvoltage values decreasing toward unity. The Si K-peak decreases relative to the continuum background as U decreases. The limit of detection, C_{DL}, can be estimated from pure element spectra, such as those shown for silicon in Figure 10, by using the formula of Ziebold (1967):

$$C_{DL} \geq 3.3a/[n\tau\ P(P/B)]^{0.5} \tag{19}$$

where a, which generally has a value near unity, is the coefficient in the Ziebold-Ogilvie hyberbolic expression relating concentration and measured x-ray intensity ratio (unknown/pure standard), n is the number of replicate measurements, t is the integration time, P is the peak count rate on a pure element, and (P/B) is the spectral peak-to-background.measured on the same pure element. Figure 11 shows the calculated value of C_{DL} as a function of U for $\tau = 100$s and $n = 1$. The value of P was defined for a beam current that produced a significant EDS detector deadtime of 25%

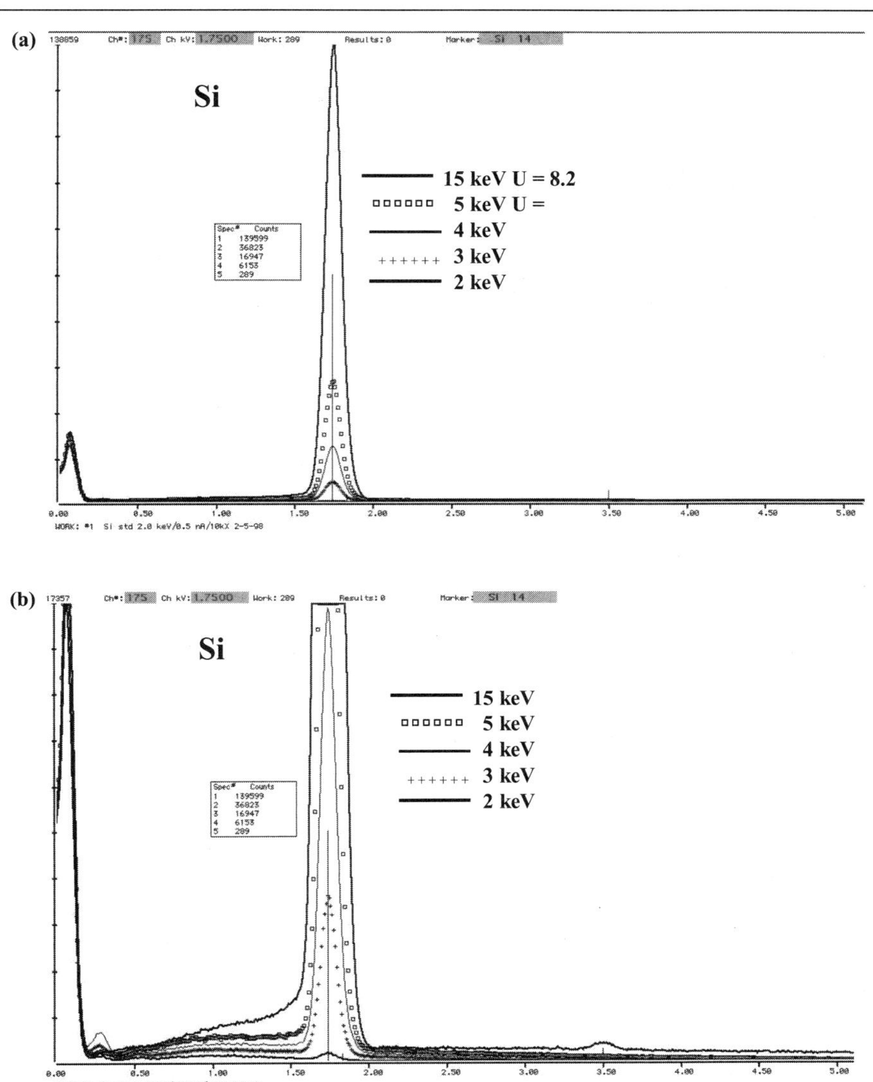

Figure 10. EDS spectra of Si as a function of E_0. (a) Linear; (b) expanded linear; (c) logarithmic.

at $U = 1.5$. Generally to achieve at least minor element detectability under low voltage conditions requires a U of at least 1.25 and greater than 100s integration time.

However, a complicating factor is the often complex surface layer structure that occurs on most materials. "Native oxide" layers often form due to the inherent reactivity of the material with oxygen from the atmosphere. A bare aluminum surface, for example, when exposed to the atmosphere rapidly forms a native oxide layer that is about 4 nm thick, and most elements are similarly reactive. Thus, the measurement of

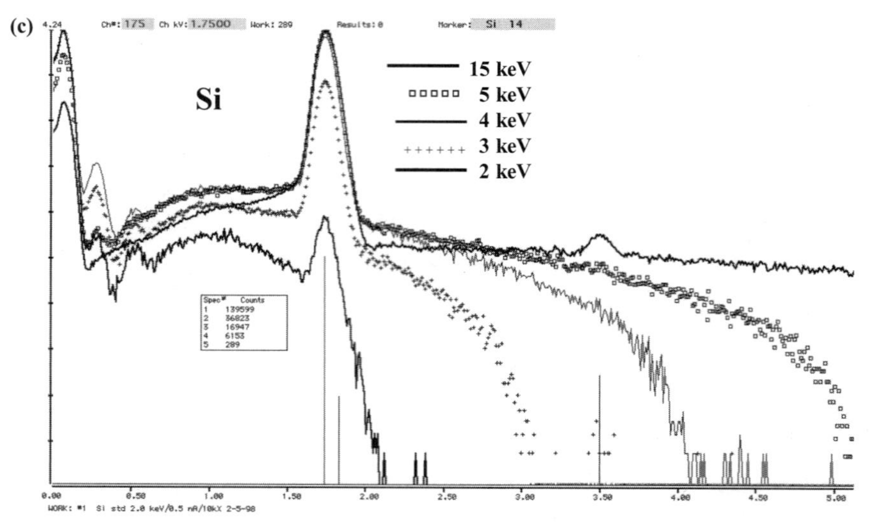

Figure 10. (continued)

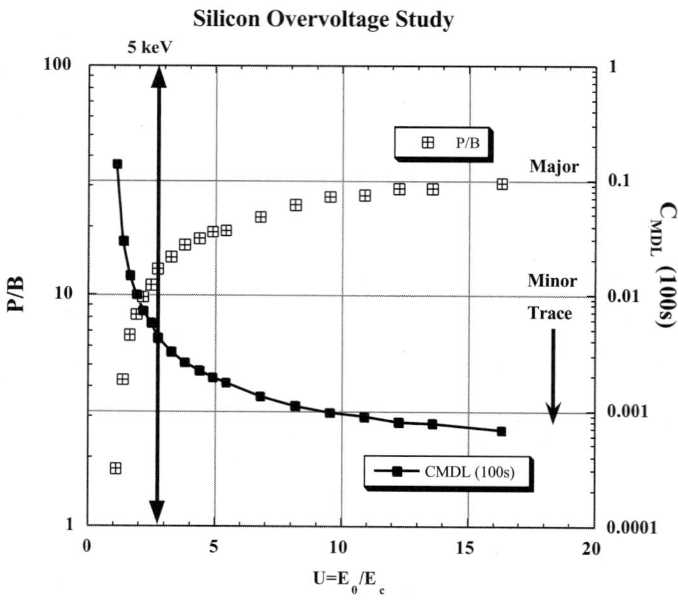

Figure 11. EDS concentration limit of detection calculated from data of figure 10.

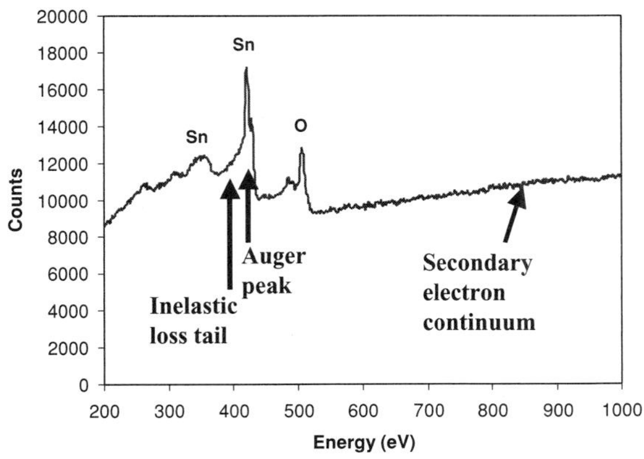

Figure 12. Auger spectrum of tin oxide, showing the characteristic peaks, inelastic loss tail, and the continuous secondary electron background.

a standard is likely to yield a complex spectrum unless special efforts are undertaken to clean the surface and preserve it with very high vacuum conditions.

C. Auger Spectrometry

1. Sampling Depth

The ionization and de-excitation tree shown in Figure 1 shows that Auger electrons and characteristic x-rays come from the same primary ionization events, so that the spatial distribution of Auger electron emission sites within the beam interaction volume must be exactly the same as the emission sites for characteristic x-rays. The only difference between x-ray production and Auger production is the relative abundance of the emission products, with the Auger effect strongly favored for low energy shell ionizations, with the x-ray emission increasing for higher ionization energies, $E_c > 4$ keV (Bishop, 1989). An example of an Auger spectrum from a small particle of tin oxide is shown in Figure 12. The characteristic features from the major constituents are observed as relatively small peaks with an asymmetric low energy tail on a high, continuous background. The low energy tail consists of Auger electrons of the peak energy that have lost energy due to inelastic scattering while escaping from the sub-surface region.

There is a marked difference in the spatial distribution of the *detected* Auger electrons and characteristic x-rays (Bishop, 1989; Goldstein *et al.*, 2003). As noted above, x-rays passing through matter are subject to photoelectric absorption, which completely consumes the x-ray, but those x-rays which are not absorbed do not suffer significant inelastic scattering, so that they arrive at the detector bearing their original energy, thus remaining characteristic of the atom that emitted the x-ray. Auger electrons, however, are subject to inelastic scattering so that after traveling a few nanometers in the target, depending on the initial kinetic energy, the Auger electron will lose energy due to a variety of inelastic events, so that its energy will no longer be characteristic

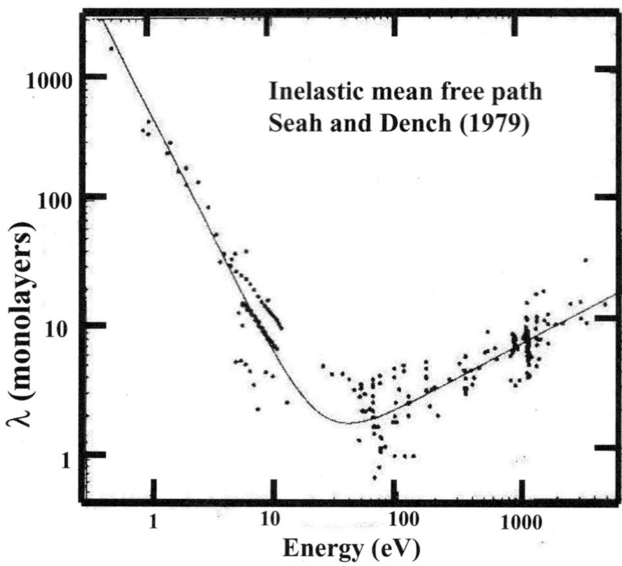

Figure 13. Inelastic mean free path as a function of initial electron energy (Seah and Dench, 1979).

of the initial emission. Such degraded electrons contribute to the background under the characteristic peaks, which also includes beam electrons that have backscattered and have lost energy due to inelastic scattering (Bishop, 1989; Seah, 1989). The classic approach to estimating the Auger sampling depth is shown in Figure 13, which shows the mean free path for inelastic scattering as a function of the inital Auger kinetic energy. Generally, the mean free path is below 10 nm for all but the lowest Auger energies. A more advanced approach is to model the complete Auger scattering history, including elastic scattering, to interpret the measured spectrum. The Auger signal is thus confined to surface/near-surface sensitivity, while the x-ray signal generated with the same incident beam energy samples more of the "bulk" of the material. For intermediate beam energies, $E_0 \geq 10$ keV, the spatial sampling difference can easily be a factor of 1,000, or nanometers for Auger versus micrometers for x-rays. Under low voltage microscopy conditions, the range of the primary beam is greatly restricted, so that the Auger and x-ray sampling begin to approach a common value.

2. Lateral Sampling

While the sampling depth of the Auger signal is limited by inelastic scattering, the lateral resolution of the Auger signal for a bulk target is degraded by the Auger electrons produced by the backscattered electrons (Bishop, 1989). The contributions of the focused beam and the backscattering can be modeled as a pair of Gaussian distributions with different values of the standard deviation, σ_B and σ_{BSE}, that scale each distribution. With a high brightness field emission gun, σ_B can be 1 nm or less. For intermediate beam energies, $E_0 \geq 10$ keV, the value of σ_{BSE} determined by the backscattering distribution can be a micrometer or more, while at low beam energies, $E_0 < 5$ keV, σ_{BSE} can be a few tens of nanometers or less, depending on the matrix. The backscattering coefficient is atomic number sensitive, so that for a copper target, 30% of the total Auger signal may come from the backscattered electron contribution, with lower fractions for lower atomic number matrices.

3. Quantification of Auger Signals

The development of "first principles" quantification methods is rapidly proceeding, including advanced methods that model the inelastic and elastic scattering that the Auger electrons undergo while propagating through complex layered structures (Werner, 2001; Powell and Jablonski, 2002). However, most practical quantitative analysis is currently performed by means of instrumental sensitivity factors measured on pure elements or simple stoichiometric compounds using the same instrument under identical conditions as used for analysis of the unknowns (Bishop, 1989). Because of the action of inelastic scattering, which can be strongly dependent on local composition and structure, to transfer electrons from the characteristic peak energy into the "shoulder" on the low energy side of the peak, careful attention must be paid to establishing consistent sampling of the band of energy used to define the Auger peak for the intensity measurement. With careful attention to the spectral intensity measurement, the error budget is such that measurements within ±10% relative can be achieved.

D. Elemental Mapping

An especially effective way to study materials that are laterally heterogeneous is to prepare an elemental map that shows the distribution of one or more atomic species (Goldstein *et al.*, 2003). Elemental maps are generated by scanning the focused primary beam over the specimen in a regular raster pattern, usually controlled digitally from a computer. The beam is addressed to a location (x, y) and the signal intensity corresponding to one or more regions of the spectrum is measured for a defined dwell time and recorded as a data matrix (x, y, I). In the simplest case, the resulting intensity maps provide qualitative information that enables a user to determine the spatial relationships of the constituents. In more advanced systems, the entire spectrum of interest will be recorded at each location, thus constructing a "data cube" (x, y, N(E)), where N(E) is the intensity versus channel number (or other calibrated value). This "spectrum imaging" form of mapping is the most efficient procedure because it maximizes the information per unit radiation dose to the specimen and it does not require the analyst to presuppose anything about the specimen composition. By recording all of the available data, post-collection processing can be used to recover any desired compositional information that is resident in the region of the spectrum that has been collected. This spectrum imaging approach is especially powerful because sufficient data is available to perform peak fitting, background corrections, and full physical matrix corrections to

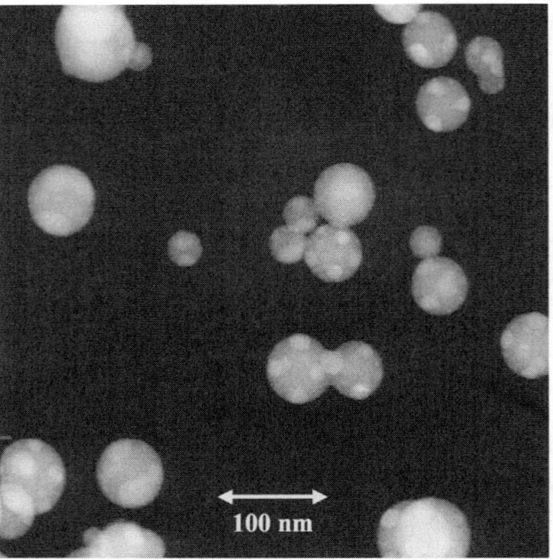

Figure 14. Fe_2O_3-SiO_2 particles as viewed in dark field scanning transmission electron microscopy, where higher intensity indicates stronger scattering power.

achieve quantitative compositional mapping. Such maps are vital to eliminate artifacts when constituents at minor or trace levels are pursued.

For EELS, an alternative mapping mode to scanning the primary beam and recording the entire spectrum exists, that of transmission electron microscope imaging with energy filtering. In this case all points in the image are simultaneously illuminated and recorded, but only a narrow energy slice ΔE is transmitted through an appropriate electron optical prism. By successively changing the average energy transmitted by the prism and recording a series of images, the information to construct a comprehensive data cube is again obtained. Both scanning and direct EELS imaging have their respective merits. Generally, direct imaging can record a higher density of discrete picture elements, but scanning permits full use of every channel of the recorded spectrum.

VI. EXAMPLES OF APPLICATIONS TO NANOSCALE MATERIALS

A. Analytical Electron Microscopy

1. STEM Imaging/EELS Spectrometry of Nanoparticles

One of the most surprising aspects of nanoscale particles is that, as small as they are, such particles frequently display an even finer scale internal structure. An interesting system that forms such ultrafine structure is iron oxide—silicon dioxide, where the insolubility of iron oxide in silicon dioxide is such that when the two substances are forced to co-exist by a flame synthesis process, particles with a distinctive substructure are formed, as shown in Figure 14. Particles with distinct Fe-rich inclusions located

Table 5. Fe/O (atom ratio)

Location	Inclusion	Adjacent Matrix
Particle 1	0.92 (0.7%)*	0.12 (1.5%)
Particle 2	0.74 (1.0%)	0.11 (1.9%)
Particle 3	0.70 (0.5%)	0.043 (1.1%)
Particle 4	0.60 (1.5%)	0.090 (2.8%)
Particle 5	0.41 (0.7%)	0.11 (0.8%)
Particle 6	0.39 (0.4%)	0.079 (0.6%)
Particle 7	0.35 (0.5%)	0.052 (1.0%)
Particle 8	0.28 (1.4%)	0.065 (2.3%)
Particle 9	0.25 (0.5%)	0.055 (0.8%)
Particle 10	0.24 (1.5%)	0.094 (1.7%)

* Relative standard deviation based on counting statistics only.

at the particle periphery were selected for EELS analysis. Table 5 contains the results for 10 randomly chosen particles in which the inclusion and the matrix immediately adjacent were measured. For measurements made when an inclusion was located at an extreme edge of a particle, and therefore the minimum amount of matrix signal should be observed, the maximum Fe/O atom ratio was found to exceed 0.9, which suggests as a possible phase the compound FeO. Three other edge locations yielded Fe/O values in the range 0.6–0.74, which is consistent with Fe_2O_3. The remaining Fe/O values were much lower (0.24–0.41), most likely due to the contribution of O-signal from the underlying SiO_2 matrix. The measurements of the Fe/O ratio in the matrix gave a lowest value of 0.043 with a range up to 0.12. Close examination of Figures 14 and 15 reveals that there is a size spectrum of ultra fine-scale inclusions, recognizable by the enhanced scattering detected in the dark field image, within the SiO_2 matrix that may be responsible for the apparent iron signal detected there.

2. STEM/EELS Compositional Measurements of Nanoparticles at High Fractional Sensitivity

Equation (19) for the instrumental limit of detection, C_{MMF}, in x-ray microanalysis can be generalized for other spectrometries around the two critical spectrometry terms: P, the peak counting rate (characteristic x-ray peak, counts/second) and P/B, the spectral peak-to-background [Ziebold, 1967]:

$$C_{MMF} \sim 1/[n\tau P(P/B)]^{0.5} \tag{20}$$

Electron energy loss spectrometry is capable of high absolute mass sensitivity, but EELS is not generally considered to be capable of achieving high fractional sensitivity because the P/B term is generally low, as can be seen for the major constituent peaks visible in Figures 2 and 3. However, a situation does exist for EELS whereby high fractional sensitivity can be achieved, namely the existence of "white lines", i.e. sharp threshold peaks with an inherently high P/B that occur at the critical ionization edges due to resonance effects (Egerton, 1986). With advanced EELS spectrum acquisition and processing based upon "second difference" techniques, such white lines have been demonstrated to provide fractional limits of detection in the range 10^{-5} to 10^{-4}

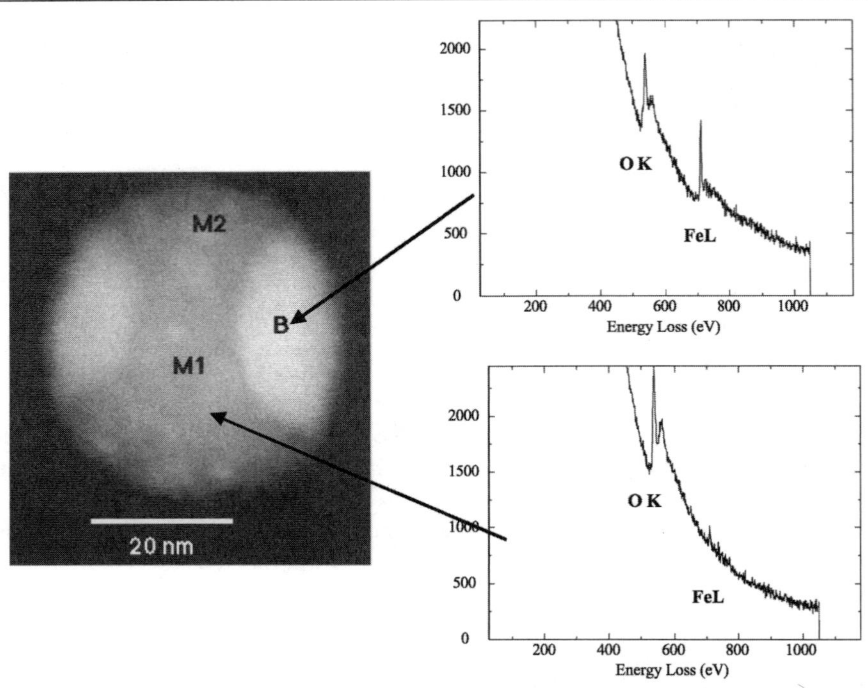

Figure 15. Fe_2O_3-SiO_2 particle as viewed in dark field STEM and electron energy loss spectra recorded in the matrix and in the bright (strong scattering) phase.

mass fraction (10–100 ppm) (Leapman and Newbury, 1993) depending on the specific element. An example of the simultaneous detection of several elements at low fractional levels below 10^{-4} mass fraction is shown in Figure 16 for a single nanoscale particle (thickness estimated to be less than 100 nm) of a multi-constituent glass standard reference material. The ultimate limits of detection under these conditions (beam energy 100 keV; beam diameter 1 nm, and beam current 1 nA) have been estimated to be approximately 10^{-6} to 10^{-5} mass fraction (1–10 ppm). This combination of analysis with trace sensitivity and nanoscale spatial specificity has been termed "trace nanoanalysis." For an EELS analytical volume in a silicon target defined by the scanning the beam over an area of 10 nm × 10 nm with a thickness of 100 nm, the total number of atoms contained is approximately 500,000. For an EELS limit of detection of 10^{-5} mass fraction, this implies a detection of fewer than 5 atoms of similar mass to silicon.

3. EELS Elemental Mapping

One of the most powerful techniques for analyzing nanoparticles in the AEM is energy-filtered transmission electron microscopy (EFTEM). As with all AEM characterization techniques that are based on EELS, in EFTEM the electrons are transmitted through

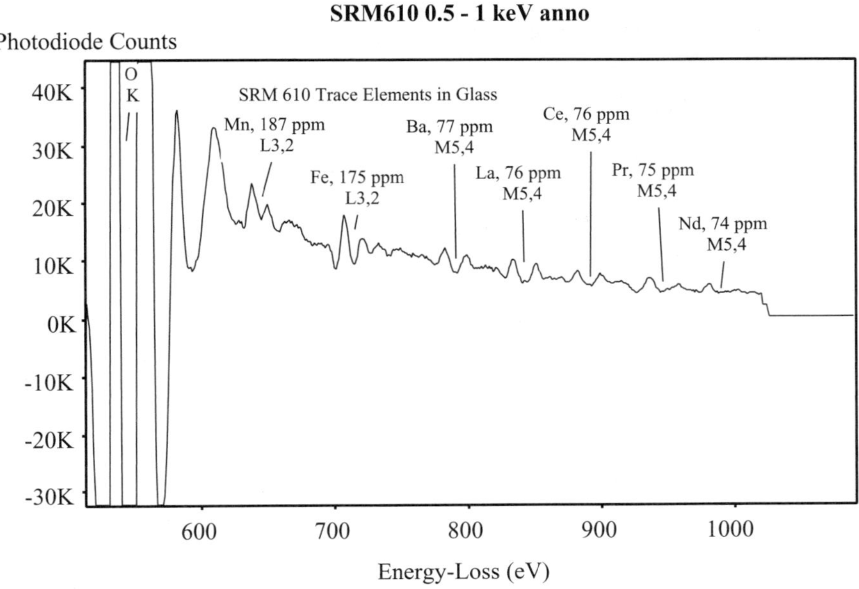

Figure 16. EELS of a single particle of NIST SRM 610 (Trace Elements in Glass) with a thickness below 100 nm demonstrating detection of several constituents at trace levels below 10^{-4} mass fraction (Leapman and Newbury, 1993).

the sample, many experiencing energy losses from inner shell scattering events that are characteristic of the elements in the sample. Unlike conventional point-analysis EELS, in EFTEM the entire TEM image is passed through the electron spectrometer in a way that preserves the spatial relationship of the pixels with a minimum of aberration. Instead of producing a spectrum from these electrons, in EFTEM the spectrometer is used as an imaging energy filter. A slit is placed in the electron optical plane containing the EELS spectrum to filter out all electrons except those within a user-defined pass band of energies. This pass band is usually 10 eV or 20 eV wide and can be centered precisely in the energy spectrum, either before, after, or on top of an element-specific feature in the EELS spectrum such as a core-loss ionization edge. A charge-coupled device (CCD) detector after the spectrometer and post-slit electron optics captures the filtered TEM data, producing an energy-selected image. When many such images are combined into a multispectral or hyperspectral data cube, quantitative chemical compositions can be determined for each pixel in the field of view.

An example of this capability is shown in Figure 17, using a manganese oxide nanoparticle supported on a holey carbon support grid as a test specimen. Figure 17(b) is an example of a map of the carbon content (brighter pixels mean more carbon at that location). The map was made using the carbon K ionization edge in the electron energy-loss spectrum (EELS) of the sample. Figure 17(c) is an analogous map of the

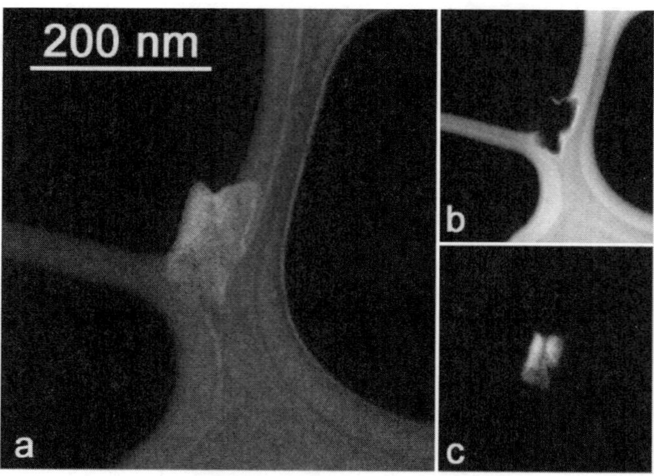

Figure 17. EFTEM images of a nanoparticle of manganese oxide: (a) color superposition of carbon (red), oxygen (green), and manganese (blue). (b) Carbon content map (brighter pixels mean more carbon at that location). The map was made using the carbon K-edge. (c) Analogous map of the manganese using the Mn L-edge. (See color plate 6.)

manganese in the sample. These two maps were combined with an oxygen map (not shown) to produce a three-color elemental map showing carbon (red), oxygen (green), and manganese (blue) all in one image. This combined, multicolor elemental map is shown in Figure 17(a).

Although this figure is composed only of the actual elemental maps themselves, the level of spatial detail and the signal-to-noise ratio in the image is so high it rivals a conventional TEM image in quality. This is in contrast to scanned-beam techniques which typically provide much more spectral detail than EFTEM, but at the expense of spatial detail. While the spatial resolution of scanned-beam techniques is comparable to that achievable in EFTEM, the time needed to acquire scanned elemental maps often limits the image sizes to 256×256 pixels, and sometimes much smaller than this. Because EFTEM images acquire all the spatial pixels in parallel, the acquisition time is independent of the number of pixels in the image. EFTEM maps of 1024×1204 and 2048×2048 pixels are common, presenting the analyst with nearly two orders of magnitude more spatial information than scanned-beam elemental maps.

B. Low Voltage SEM

1. Low voltage x-ray mapping for characterization of nanoscale structures in an electronic device.

The relentless push to making engineered devices with ever smaller features has resulted in electronic and microelectromechanical systems with features below 300 nm

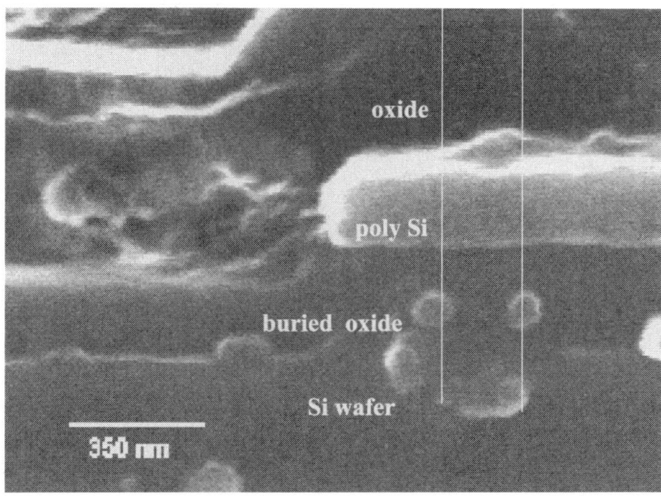

Figure 18. FEG-SEM secondary electron image (through the lens detector) of the poly Si gate region of a CMOS device is shown in Figure 1. Elemental x-ray maps for Si-K at 1.74 keV, and 0–K at 0.53 keV, were collected at a nominal magnification of 70,000 diameters at beam energies of 15 keV and 5 keV.

in dimension. Such a scale is well below the resolution of x-ray mapping with the SEM/EPMA in the conventional beam energy range. For example, equation (17) gives a range of 2860 nm for Al K-shell x-ray production in silicon at 15 keV. X-ray mapping with the low voltage FEG-SEM can provide a sufficient improvement in spatial resolution to distinguish nanoscale structures. For $E_0 = 5$ keV, equation (17) gives a range of 400 nm for Al K x-rays in Si, but the x-ray production is very much more concentrated near the beam impact point because of the more rapid loss of energy with distance traveled in the target. Figure 18 shows an SEM image (secondary electron signal collected with a through-the-lens detector) of a cross section of a CMOS device with a complex structure. The Si K x-ray maps are shown in Figure 19. In the 15 keV Si map, Figure 19(a), the variation in intensity over the mapped area is so low that only the position of the Si wafer base can be recognized, so that only minimal information can be deduced about the structure of the gate. In comparison the 5 keV Si map, Figure 19(b), while noisy, clearly shows the four separate zones of the device corresponding to the top oxide layer, the polycrystalline Si, the buried oxide, and the Si wafer. The corresponding O K x-ray maps are shown in Figure 20. Similar to the results from the Si maps, the 15 keV O map, Figure 20(a), shows the oxidized regions versus the underlying Si wafer, but again this map provides minimal information on the structure of the gate. The 5 keV map, Figure 20(b), clearly shows the structure of the device. While the 5 keV maps are noisy due to the low efficiency of x-ray generation and the low primary beam current, the observer's eye can integrate the area information sufficiently to recognize the regions of interest. With data integration

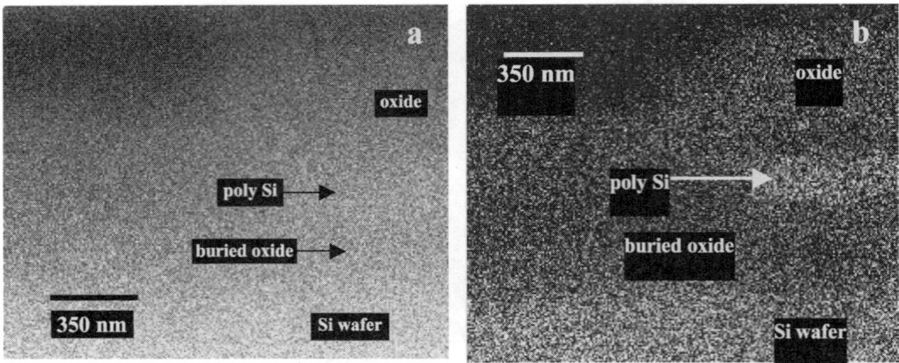

Figure 19. Si K x-ray maps. (a) 15 keV electron beam energy. (b) 5 keV electron beam energy.

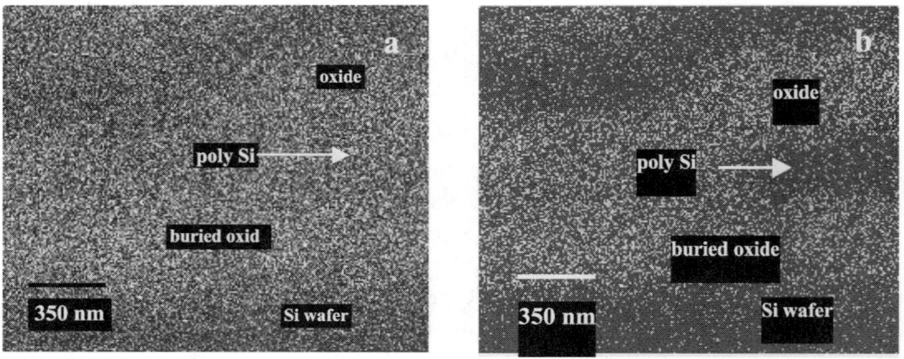

Figure 20. O K x-ray maps. (a) 15 keV electron beam energy. (b) 5 keV electron beam energy.

across the window shown superimposed in Figure 18, the various parts of the structure become more obvious, and the feature interfaces can be distinguished in the intensity versus position plots shown in Figure 21.

C. Auger/X-ray SEM

1. Simultaneous Auger and EDS X-ray Microanalysis of Particles

Often the surface of a material differs in composition from the interior, either intentionally as a result of the fabrication or accidentally as a result of corrosion effects in service. Simultaneous x-ray and Auger analyses provide a powerful combination to distinguish between elements partitioned between the surface and interior regions because of the differences in relative sampling depths of the characteristic signals. As the dimensions of objects enter the nanoscale domain, the fraction of an object that

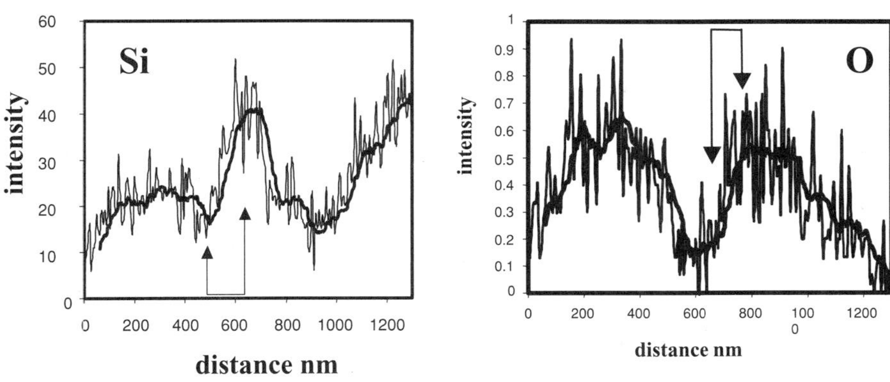

Figure 21. Line scans for the Si K and O K x-rays at $E_0 = 5$ KeV. The line scans were determined from the x-ray maps along the white lines shown in Figure 18 with integration across the width of the region defined by the lines.

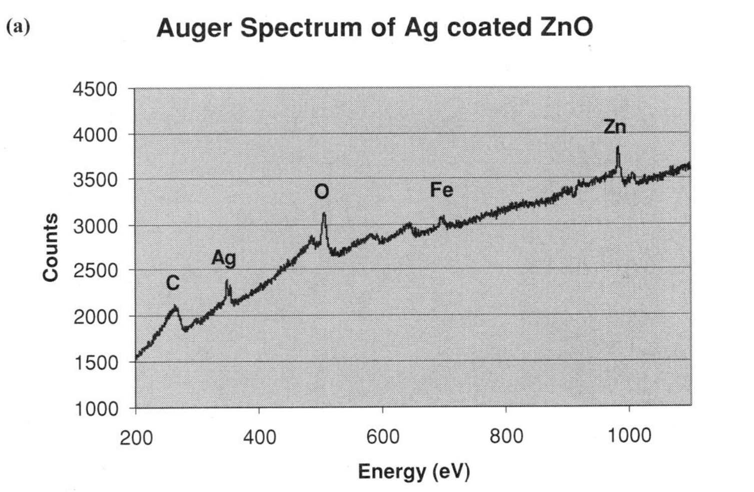

Figure 22. Combined x-ray and Auger electron analysis of nanoscale particles (ZnO coated with silver). Beam energy 5 keV. (a) Auger electron spectrum. (b) EDS x-ray spectrum. (c) FEGSEM image.

lies within the "surface" region sampled by the Auger signal increases until, for sufficiently small particles, even the most remote interior portions fall within the Auger sampling depth. Because of the effects of excitation and propagation of both x-rays and Auger electrons of different energies, each analytical situation must be carefully evaluated to determine if sensible differences between the interior and the surface can be distinguished. Figure 22(c) shows an aggregate of zinc oxide particles coated with silver as imaged with the FEG-SEM. The individual particles comprising the aggregate have lateral dimensions that are generally below 100 nm while the thicknesses of

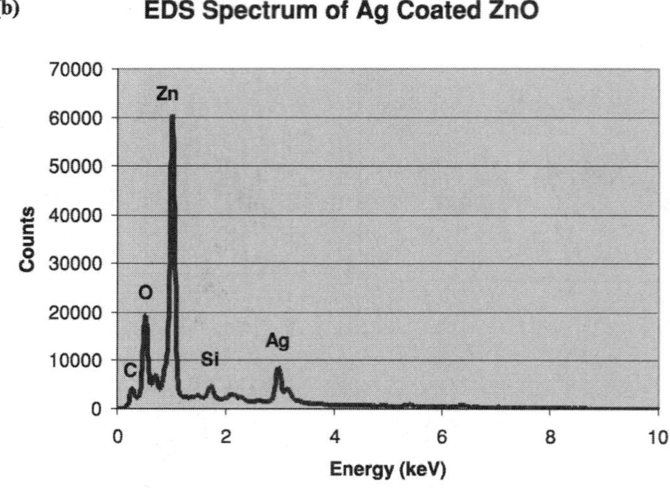

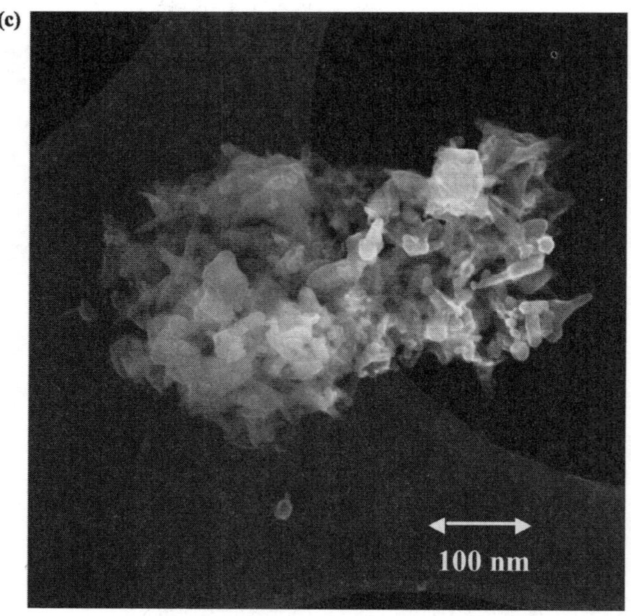

Figure 22. (continued)

the flake-like particles are substantially less than 100 nm. The corresponding Auger electron and EDS x-ray spectra measured simultaneously in the same instrument are shown in Figures 22(a) and (b). Examination of these spectra reveal the same principal elements in both, which may arise because of the nature of the specimen, e.g., incomplete surface coverage of Ag could permit the underlying ZnO to be detected in the Auger spectrum, or because of similarities in the relative sampling depths for both types of characteristic radiation due to the thin dimensions of the particles. Differences in depth may exist in the distribution of minor and trace elements, although the spectrum interpretation is problematic. The Auger spectrum shows a small peak for iron that does not appear in the EDS spectrum, but the severe interference from O K on the FeL peak in EDS prevents detection of the low level of iron. A small peak for silicon is seen in the EDS spectrum, but the energy range of the Auger spectrum does not include an appropriate Si Auger peak. Developing careful analytical strategy is clearly critical for drawing meaningful conclusions from simultaneous X-ray and Auger spectrometries.

VII. CONCLUSIONS

Transmission and scanning electron microscopes provide platforms for a powerful arsenal of electron and x-ray spectrometries that yield chemical characterization of nanoscale particles and nanostructured bulk materials. Combined with the imaging and crystallographic measurement functions of TEMs and SEMs, comprehensive characterization of morphology, crystal structure, and composition becomes possible. The advent of more efficient electron optical systems, spectrometers, and digital imaging devices promises to increase the throughput of these instruments, many of which are already capable of automatic, unattended operation, at least in some operational modes, to vastly increase the accumulation of data. Thus, techniques that are capable of isolating a single nanoparticle or nanostructure feature are also becoming capable of accumulating a great deal of information about a particle array or complex bulk nanostructure. Such large scale databases of information themselves pose a challenge to extract relevant information. Parallel developments in "data mining" techniques will increasingly come into play to solve the challenges raised by nanoscale materials.

REFERENCES

Auger, P. (1923) Compt. Rend. **177**, 169.

Bishop, H. E. (1989) "Auger Electron Spectroscopy" in *Methods of Surface Analysis*, J. Walls, ed., (Cambridge) 87.

Cliff, G. and Lorimer, G. W. (1975) J. Micros. **103**, 203.

Egerton, R. F., Electron Energy-Loss Spectroscopy in the Electron Microscope (Plenum Press, New York, 1986).

Goldstein, J. I., Newbury, D. E., Joy, D. C., Lyman, C. E., Echlin, P., Lifshin, E., Sawyer, L., and Michael, J., *Scanning Electron Microscopy and X-ray Microanalysis*, 3rd edition (Kluwer Academic Plenum Press, New York, 2003).

Goldstein, J. I., Costley, J. L., Lorimer, G. W., and Reed, S. J. B. (1977) SEM/77, O. Johari, ed., (Chicago, IITRI) 315.

Goldstein, J. I., Williams, D. B., and Cliff, G. (1986) "Quantitative X-ray Analysis" in *Principles of Analytical Electron Microscopy*, D. C. Joy, A. D. Romig, Jr., and J. I. Goldstein, eds. (Plenum, New York) 155.

Iwanczyk, J. S., B. E. Patt, C. R. Tull, and S. Barkan (2001) "High-Throughput, Large Area Silicon X-ray Detectors for High-Resolution Spectroscopy Applications," Micros. Microanal. (Suppl 2) **7**, 1052.

Joy, D. C., Romig, A. D., Jr., and Goldstein, J. I., eds., *Principles of Analytical Electron Microscopy* (Plenum Press, New York, 1986).

Kanaya, K. amd Okayama, S. (1972) J. Phys. D.: Appl. Phys. **5**, 43.

Komarneni, S., Parker, J. C. and Wollenberger, H. J., eds., *Nanophase and Nanocomposite Materials II, vol 457* (Materials Research Society, Warrendale, PA, 1997).

Leapman, R. D. and Newbury, D. E. (1993) "Trace Elemental Analysis at Nanometer Spatial Resolution by Parallel-Detection Electron Energy Loss Spectroscopy", Analytical Chemistry, **65**, 2409.

Newbury, D. E. and Williams, D. B. (2000) "The Electron Microscope: The Materials Characterization Tool of the Millenium," Acta mater. **48**, 323.

Phillips, J. R., Griffis, D. P., and Russell, P. E. (2000) J. Vac. Sci. Tech. A, **18**, 1061.

Powell, C. J. and Jablonski, A. (2002) Surf. Interface Anal., **33**, 211.

Seah, M. P. (1989) "Electron and Ion Energy Analysis" in *Methods of Surface Analysis*, J. Walls, ed., (Cambridge) 57.

Seah, M. P. and Dench, W. A. (1979) Surf. Interface Anal., **1**, 2.

Sarikaya, M., Wickramasinghe, H. K., and Isaacson, M., eds. (1994) Determining Nanoscale Physical Properties of Materials by Microscopy and Spectroscopy, *vol 332* (Materials Research Society, Warrendale, PA).

Small, J. A. & Bright D. S., "Comparison of High- and Low-Voltage X-ray Mapping of An Electronic Device", Proceedings of the 2000 International Conference on Characterization and Metrology for ULSI Technology, eds Seiler *et al.* AIP Conference Proceedings 550, 2000, pp. 596–600.

Struder, L., C. Fiorini, E. Gatti, R. Hartmann, P. Holl, N. Krause, P. Lechner, A. Longoni, G. Lutz, J. Kemmer, N. Meidinger, M. Popp, H. Soltau, and C. von Zanthier (1998) "High Resolution Non Dispersive X-ray Spectroscopy with State of the Art Silicon Detectors," Mikrochim. Acta. Suppl. **15**, 11.

Werner, W. S. M. (2001) Surf. Interface Anal, **31**, 141.

Williams, D. B. and Carter, C. B., *Transmission Electron Microscopy* (Plenum Press, New York, 1996).

Wollman, D. A., K. D. Irwin, G. C. Hilton, L. L. Dulcie, D. E. Newbury, J. M. Martinis (1997) High-Resolution, Energy-Dispersive Microcalorimeter Spectrometer for X-Ray Microanalysis, J. Microscopy **188**, 196.

13. CHARACTERIZATION OF NANO-CRYSTALLINE MATERIALS USING ELECTRON BACKSCATTER DIFFRACTION IN THE SCANNING ELECTRON MICROSCOPE

J. R. MICHAEL

1. INTRODUCTION

Nano-crystalline and ultra-fine-grained materials provide unique challenges in characterization. Until recently, the study of crystallography at the sub-micrometer scale has been the exclusive domain of the transmission electron microscope (TEM), or if microstructural observations are not needed x-ray diffraction (XRD). Electron backscatter diffraction (EBSD) now allows the crystallography of bulk samples to be studied in the SEM with sub-micron spatial resolution. The obvious advantage to the study of crystallography in the SEM over the TEM is that the bulk specimen used in the SEM can reduce the need for thin sample preparation and may permit larger, more representative areas of the sample to be analyzed.

EBSD in the SEM has been developed for two different purposes. The oldest application of EBSD is for the measurement of texture on a pixel-by-pixel basis. Texture determined in this way has been termed the microtexture of the sample. The other use of EBSD is for the identification of micrometer or sub-micrometer crystalline phases. However, phase identification has not yet succeeded in addressing features smaller than 100 nm [1, 2]. Both applications of EBSD add an important new tool to the SEM which now has the capability to study the morphology of a sample through either secondary or backscattered electron imaging, the chemistry through energy dispersive spectrometry (EDS) or wave-length dispersive spectrometry (WDS) and the addition of EBSD permits the crystallography of the sample to be studied.

Important structure-property relationships may depend strongly on the crystallographic texture of the sample. Properties of polycrystalline materials are determined by the individual properties of the grains and the statistical characteristics of the polycrystal as a whole. The information that may be obtained through EBSD are the distributions of crystallographic orientations of the individual grains and the distribution of the orientations between the grain or the grain boundary misorientations. The development of electron backscattering diffraction (EBSD) into commercial tools for phase identification and orientation determination has provided a new insight into the links between microstructure, crystallography and materials physical properties that was not previously attainable. The combination of field emission electron sourced scanning electron microscopes and EBSD has now allowed EBSD to be applied to nano-crystalline materials (grain size < 100 nm) and ultra-fine grained materials (grain size < 500 nm) [3].

Microtexture is a term that means a population of individual orientations that are usually related to some feature of the sample microstructure. A simple example of this is the relationship of the individual grain size to grain orientation. The concept of microtexture may also be extended to include the misorientation between grains, often termed the mesotexture. Thus, the misorientation of all the grain boundaries in a given area may be determined. It is now possible using EBSD to collect and analyze tens of thousands of orientations per hour thus allowing excellent statistics in various distributions to be obtained. The ability to link texture and microstructure has lead to significant progress in the understanding of recrystallization, gain boundary structure and properties, grain growth and many other physical phenomena and properties important in nanocrystalline materials [4, 5].

This chapter will first describe the origin of the EBSD pattern in the SEM. The hardware required to collect these patterns is then discussed along with the instrumental operating conditions. The chapter will then proceed to discuss the details of microtexture determination using EBSD.

2. HISTORICAL DEVELOPMENT OF EBSD

This technique has had many different names over the past 40 years. The original developers called them high angle Kikuchi patterns (HAKP) [6]. Others have used electron backscatter patterns (EBSP), electron backscatter diffraction (EBSD), and backscattered electron Kikuchi diffraction (BEKP or BKD). We will use the terms electron backscatter diffraction to describe this technique. The first EBSD patterns were observed over 40 years ago and were termed high angle Kikuchi patterns. This was almost ten years before the first SEM was built so a special experimental apparatus had to be constructed. The apparatus consisted of a cylindrical chamber that was lined with a strip of photographic film. An electron beam was focused on to a tilted sample of LiF exposing the film. This produced patterns that are every bit as good as those collected today on modern SEMs with modern EBSD acquisition hardware [6].

The addition of an appropriate camera system to an SEM for the observation of EBSD patterns was demonstrated in the early 1970's [7]. These authors demonstrated that sample orientations could easily be measured from EBSD patterns and coined the term electron backscatter patterns (EBSP). At about that same time others demonstrated some of the significant advantages held over selected area electron channeling patterns (SAECP) [8]. Subsequently, in the late 1980's and early 1990s, the automated indexing of EBSD patterns was developed followed by systems that could scan the sample point-by-point and determine the orientation at each point automatically [9]. This is the basis for determining the microtexure of a sample region. There are now many commercial tools that represent variations of these original developments. Also in the early 1990's, EBSD for phase identification was perfected, which has also served as the basis for a commercial phase identification system [10, 11]. A more detailed review of the historical development of EBSD may be found elsewhere [12].

3. ORIGIN OF EBSD PATTERNS

EBSD patterns are obtained in the SEM by illuminating a highly tilted specimen with a stationary electron beam. Currently two mechanisms may describe the formation of EBSD patterns. In one description, the elastic scattering of previously inelastically scattered electrons forms the patterns. These backscattered electrons appear to originate from a virtual point below the surface of the specimen. Some of the backscattered electrons will satisfy the Bragg condition ($+\theta$ and $-\theta$) and are diffracted into cones of intensity with a semi-angle of ($90°-\theta$), with the cone axis normal to the diffracting plane. Since the wavelength of the electron is small, the Bragg angle is typically small, less than 2°. These pairs of flat cones intercept the imaging plane and are imaged as two nearly straight Kikuchi lines separated by an angle of 2θ. An alternative description of EBSD pattern formation is the single event model. In this model, it is argued that the inelastic and elastic scattering events are intimately related and may be thought of as one event. In this case the electron channels out of the sample and forms the EBSD pattern [13].

A typical electron backscatter diffraction pattern is shown in Figure 1. The pattern consists of a large number of parallel lines of intensity. These lines may appear as discreet lines or may appear as a band of increased intensity in the patterns. Frequently, as shown in figure 1, one of the pair of a set of Kikuchi lines will appear dark relative to the other. This is a result of the unequal angles between the scattered beam and the primary beam that leads to an unequal intensity transfer to one of the diffracted cones of intensity. There are many places where a number of line pairs intersect. The intersections are zones axes and are related to specific crystallographic directions in the crystal. The lines represent planes with in the crystal structure and the zone axes represent crystallographic directions within the crystal. The actual trace of the plane for a given set of diffracted lines lies exactly in between the two lines or at the center of a band. The EBSD pattern is essentially a map of the angular relationships that exist within the crystal.

The spacing between a pair of lines is inversely related to the atomic spacing of those planes within the crystal. The EBSD pattern is an angular map of the crystal

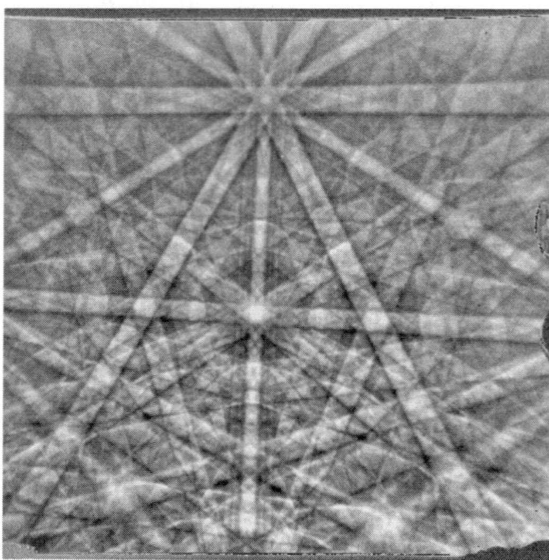

Figure 1. A typical EBSD pattern from Fe_2O_3 (hematite, rhombohedral) recorded at 20 kV with a CCD-based slow scan camera.

so we should relate the distance between any pair of lines to an angle. The angle for diffraction from a given set of planes is called the Bragg angle and is given by:

$$\lambda = 2d \sin\theta \tag{1}$$

Where d is the spacing of the atomic planes, λ is the wavelength of the electron and θ is the Bragg angle for diffraction. From this equation, it can be clearly seen that as the interatomic spacing increases the Bragg angle must decrease. Thus, smaller atomic plane spacings will result in wider line pairs.

3.1. Collection of EBSD Patterns

Acquisition of EBSD patterns in the SEM is relatively easy. Some of the earliest images of these patterns were obtained by directly exposing photographic emulsions inside the specimen chamber of an SEM [6]. Although this is still an option, all commercial systems now use a phosphor screen inside the sample chamber of the SEM that is imaged by an external camera through a leaded glass window. A schematic of a typical experimental setup is shown in Figure 2. The tilt angle of the sample with respect to the normal to the electron beam is typically set to about 70°. The exact value is not critical, as patterns have been obtained from tilt angles as small as 40°. Figure 3 shows an actual SEM chamber fitted with an EBSD camera.

There are no specific requirements on the electron beam parameters. The only requirement is that the diffracted electrons reaching the recording medium (either a phosphor screen or a photographic emulsion) must have a sufficient kinetic energy

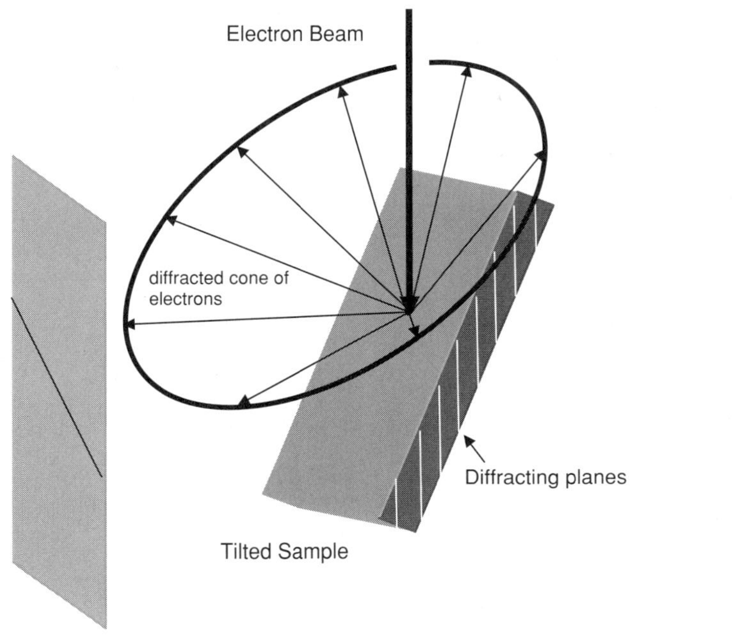

Figure 2. Typical EBSD experimental configuration. Note the relationship between the electron beam, the tilted sample and the phosphor screen.

to produce an appropriate response (generation of light) in the recording medium and must have a large enough current to produce a measurable signal. Generally, this means that a beam current of greater than 0.1 nA and an accelerating voltage of greater than 2.5 kV must be used. More typical values are an accelerating voltage of 10 kV and a beam current of 0.5 nA or larger. Due to the response of the commonly used phosphor medium, it is usually not possible to work at very low beam currents and low accelerating voltages.

The detection of the patterns has been accomplished using many different types of cameras and photographic film. The most commonly used camera is either a TV rate low light level camera or a CCD-based imaging system. The CCD-based systems are most versatile as a variety of parameters, such as binning and gain levels, can be set to provide excellent EBSD patterns over a range of microscope operating conditions. Figure 4 shows a series of EBSD patterns collected from Ni on a CCD-based camera system using a range of accelerating voltages.

3.2. Automated Orientation Mapping

The study of the orientation of individual grains within a polycrystalline assemblage has become indispensable in materials science. It has only been recently that EBSD in the SEM and automated pattern indexing and orientation measurement have been

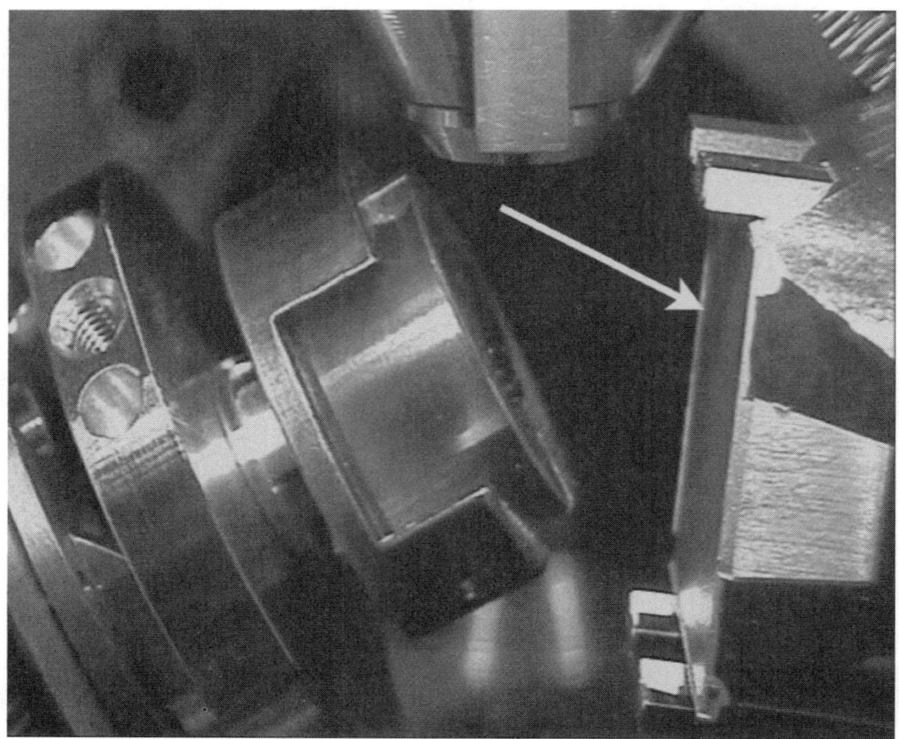

Figure 3. Arrangement of the EBSD camera (arrowed) with respect to the specimen in a SEM.

combined. A use of EBSD in the SEM is to determine the local relationships between the microstructure of the sample and crystallography. This may simply involve the direct correlation between the orientations of the grains in a polycrystalline material in which some interesting event is occurring or determining the orientation relationship between some precipitate and the matrix phase or the relationship between the fracture path and crystallography. These cases require only a few measurements at discreet points and do not need fully automatic orientation measurement, although automation makes this work quite easy. Other cases, for example, the measurement of texture, require many thousands of measurements so that the distribution of orientations may be obtained.

There are two ways that automated orientation mapping can be accomplished. Stage scanning requires the electron beam to be held stationary and the sample is rastered by moving the sample stage. Beam scanning utilizes the scan coils in the SEM column to scan the beam over a stationary sample. In the early days of EBSD, the preferred method of scanning the beam over the sample surface was stage scanning. Stage scanning has a number of important advantages. The first advantage is that there

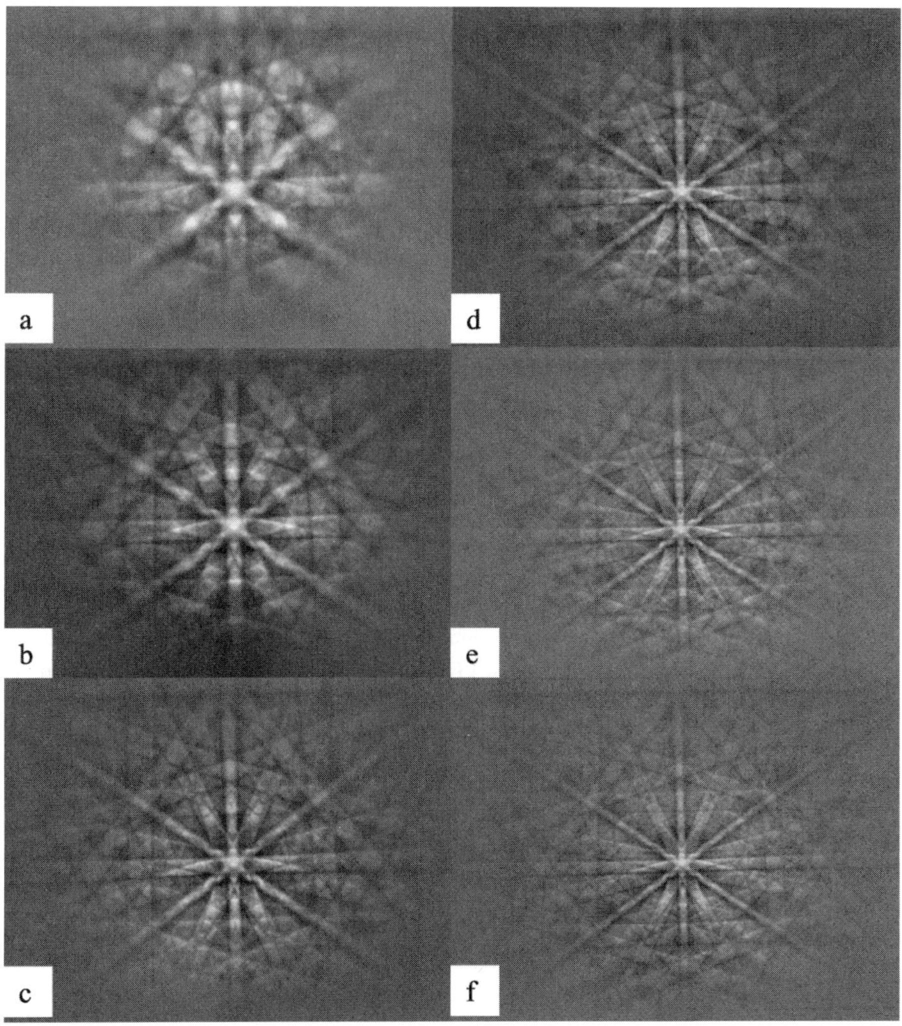

Figure 4. EBSD patterns of Ni collected at a) 5 kv, b) 10 kV, c) 15 kV and d) 20 kV e) 25 kV and f) 30 kV.

is no defocus associated with moving the beam over the highly tilted sample. This is only possible if the x and y motor drives for the stage are in the plane of the tilted sample. The second advantage is related in that the beam position with respect to the EBSD phosphor screen is fixed and thus the calibration (the pattern center and the distance between the sample and the phosphor screen) does not vary. The disadvantage of stage scanning is that it is much slower than beam scanning. Generally, the fastest stage scan requires about 1 second per pixel. The other disadvantage is that the accuracy of movement is not much better than 1 μm and is therefore not of much use to the study of nano-crystalline materials.

Beam scanning is the more commonly used technique for automated orientation mapping. For beam scanning, the scan generator of the SEM is disabled and the electron beam is rastered by an external scan generator controlled by the EBSD computer system. The advantage of this system is that the time required to reposition the beam is quite short when compared to the time required to collect and index the EBSD pattern. Thus, beam scanning is much more rapid when compared to stage scanning. The disadvantages of stage scanning are related to the motion of the electron beam with respect to the SEM column axis and the EBSD phosphor screen. The tilted sample results in an out of focus condition at the highest and lowest positions on the specimen. This can be partially compensated for by using dynamic focus corrections and small objective apertures, but the defocus at low magnifications cannot be eliminated. The other related disadvantage is that the calibration of the EBSD system changes with the position of the beam on the sample. This effect may be reduced by automatically correcting the calibration of the system as a function of beam position on the sample. These difficulties are primarily noted at lower magnification where there may also be some distortion of the microscope scans anyway. At higher magnification these disadvantages are not significant and beam scanning is recommended over stage scanning for the study of fine-grained or nano-crystalline materials. One very useful mode of operation is a mix of beam scanning and stage scanning. In this mode the beam is scanned over an area of interest. After the scan is complete the stage is moved and then another beam scan is commenced. Accurate positioning of the areas allows the maps to be knitted together to produce larger area maps.

At each pixel the EBSD pattern is acquired using a low-light level video or CCD-based camera that images the phosphor screen. The bands or lines in the pattern are then detected through the use of the Hough or Radon transform of the image [14]. Once the lines are found, the angles between the bands can be calculated and compared to a look up table of angles that is constructed from the crystallography of the sample. Once a consistent set of indices can be assigned to the bands, the orientation of the pixel can be described with respect to some external set of reference axes. This process is then repeated at every pixel throughout the area of interest [15].

4. RESOLUTION OF EBSD

4.1. Lateral Resolution

In order to use EBSD for the study of nano-crystalline materials we must understand the spatial resolution of the measurement. The spatial resolution of EBSD is strongly influenced by the atomic number of the material to be studied, the accelerating voltage of the SEM, the focused probe size and the sample tilt. For EBSD to be useful for nano-crystalline materials, each of these parameters must be carefully set to achieve high spatial resolution required for the study of nano-crystalline materials.

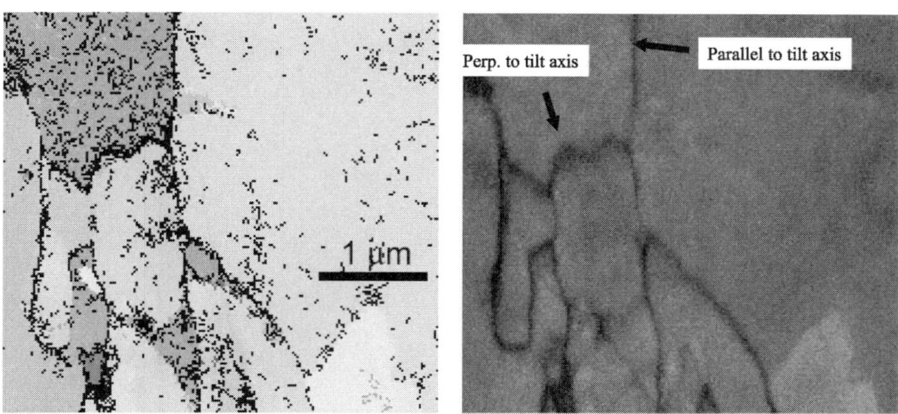

Figure 5. Resolution differences resulting from the use of a highly tilted sample. Note that boundaries parallel to the tilt axis are sharper than those perpendicular to the tilt axis. a) Orientation map with respect to an in-plane direction. Pixels not indexed are shown in black, indexed pixels in color. b) Pattern quality image of same area. (See color plate 7.)

The spatial resolution of EBSD is strongly influenced by the sample tilt. The high sample tilts required for EBSD results in an asymmetric spatial resolution parallel and perpendicular to the axis of tilt. This asymmetric spatial resolution also influences the depth from which the EBSD pattern is generated. The resolution parallel to the tilt axis is much better than the resolution perpendicular to the tilt axis due to the high sample tilt angles used to acquire EBSD patterns. The resolution perpendicular to the tilt axis is related to the resolution parallel to the tilt axis by:

$$L_{perp} = L_{para}(1/\cos\theta) \tag{2}$$

Where θ is the sample tilt with respect to the horizontal, L_{perp} is the resolution perpendicular to the tilt axis and L_{para} is the resolution parallel to the tilt axis. The resolution perpendicular to the tilt axis is roughly three times the resolution parallel to the tilt axis for a tilt angle of 70° and increases to 5.75 times for a tilt angle of 80°. Thus, it is best to work at the lowest sample tilt angles possible consistent with obtaining good EBSD patterns. The difference in resolution parallel to and perpendicular to the tilt axis is apparent in Figure 5. The image in this figure is generated by measuring the pattern sharpness or quality at each pixel within the map, and then displayed as a gray scale. The pattern sharpness or contrast is reduced when the electron beam is very close to a grain boundary due to contributions to the patterns from the grains on either side of the boundary resulting in an EBSD pattern containing overlapping patterns. The grain boundary parallel to the tilt axis is wider than the boundary perpendicular to the tilt axis demonstrating the reduced resolution perpendicular to the tilt axis.

The spatial resolution of EBSD is directly related to the electron beam size or focused probe size at the sample. Smaller beam sizes result in higher spatial resolution

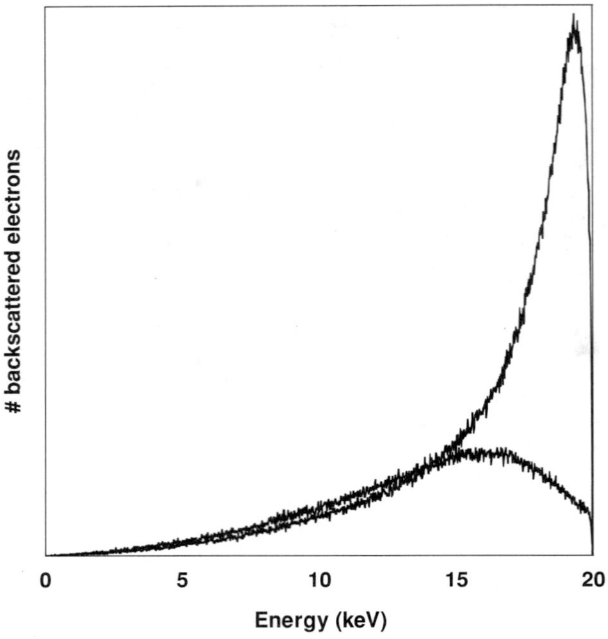

Figure 6. Energy distribution of backscattered electrons for normal incidence and a 70° sample tilt.

for EBSD measurements. For this reason, when studying nano-crystalline materials, an SEM with a field emission source must be considered mandatory for producing EBSD patterns. Although EBSD is a technique that utilizes backscattered electrons, the resolution of the technique is generally much better than can be achieved with standard backscattered electron imaging in the SEM. This is a direct result of the way in which the patterns are formed. Monte Carlo electron trajectory simulations are used to study the energy distribution of the electrons that exit the sample and an example of this is shown in Figure 6, where the energy distribution of the backscattered electrons is plotted for an untilted sample and one tilted 70° from the horizontal position. This plot is drawn for an initial electron beam energy of 20 kV interacting with a Ni specimen. The number of initial trajectories was the same for each sample orientation. There are two important points to note from this figure. First, it is very clear that tilting the sample to a high angle with respect to the normal results in a much higher backscatter yield. This is important as the increased signal improves the quality of the EBSD patterns and decreases the acquisition time. The other important point is the shape of the backscattered electron energy distribution. It is apparent from Figure 6 that many of the backscattered electrons from the tilted sample have nearly the initial electron beam energy. The large peak in the tilted sample backscattered energy distribution is only a few hundred eV less than the initial beam energy. The electrons in the sharp energy peak contribute to the crystallographic information in the EBSD pattern, while the electrons in the remainder of the distribution contribute to the

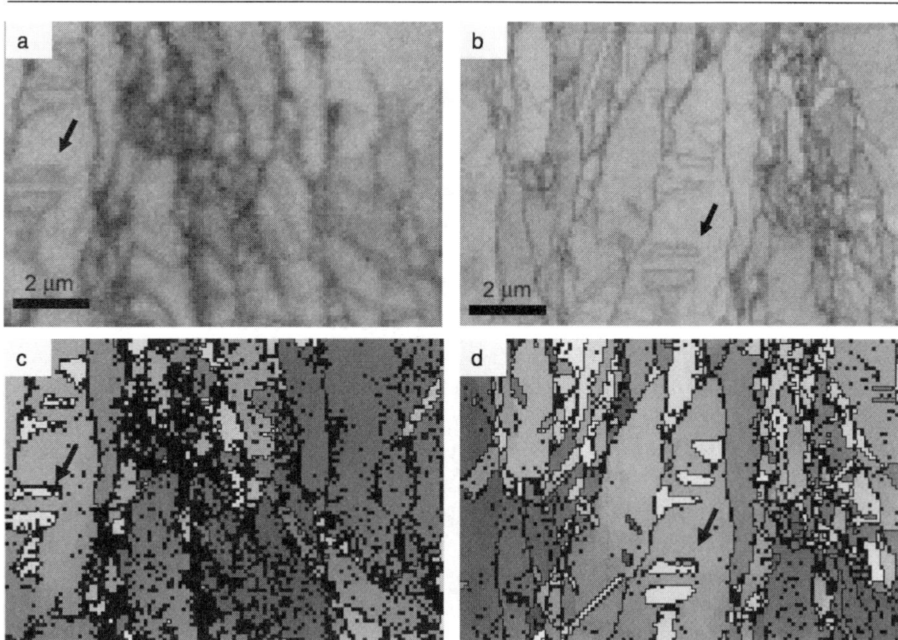

Figure 7. Comparison of EBSD resolution with large and small electron beam sizes. a) Pattern quality map for a large electron beam size, b) pattern quality map for a much smaller electron beam size, c) orientation map corresponding to beam size used in a and d) orientation map corresponding to beam size used in b. Black pixels in the orientation maps are where the EBSD patterns could not be indexed. Arrows indicate the same areas in each image. All maps are 120 × 80 pixels with a step size of 100 nm and were collected in 10 minutes. (See color plate 8.)

overall background intensity of the EBSD pattern. The sample that is normal to the electron beam has a broad spectrum of energies associated with the backscattered electrons which mainly contribute to producing the background signal in an EBSD pattern with no visible crystallographic information. Under certain special conditions it has been shown that EBSD patterns may be generated at normal incidence, but there are significant challenges to utilizing this geometry [16].

Monte Carlo electron trajectory simulations have shown that the low loss backscattered electrons emerge from the specimen surface very near the initial beam impact point [17]. This causes the resolution of EBSD to depend mainly on the electron probe size. The spatial resolution is also a function of the accelerating voltage and the atomic number of the sample. Lower accelerating voltages will improve the spatial resolution due to the decreased range of the backscattered electrons within the sample as long as the electron beam size does not increase (may be difficult to achieve in practice). Higher atomic number samples will have improved spatial resolution when compared to lower atomic number specimens. This is a result in the decreased range of the backscattered electrons in the higher atomic number sample. Figure 7 can be used to illustrate the effect of electron beam size. Figure 7a and b are gray scale maps of the EBSD pattern sharpness at each pixel. Figure 7a was acquired using

Table 1. Lateral Spatial Resolution in nm for Al and Cu vs. SEM Operating Voltage

	10 kV	20 kV	30 kV
Al	60 nm	200 nm	300 nm
Cu	30 nm	50 nm	80 nm

non-optimum beam parameters to produce a rather larger electron beam size. Note the lack of sharpness of the grain boundaries and the overall low image quality. Figure 7b was generated using parameters that provided a much smaller beam diameter than in 7a with adequate beam current. Note the obvious blurring of the grain boundary regions in the patterns quality map of Figure 7a. The blurring of the image near the grain boundaries is due to overlap of the beam onto two grains and results in two superposed patterns. Figure 7c and 7d are orientation maps of the same region. Black is used to show positions on the sample from which the EBSD patterns could not be indexed. It is immediately obvious that the smaller electron beam size used in Figure 7b and 7d has resulted in a higher percentage of indexed patterns. This demonstrates the importance of utilizing the smallest electron beam size that has sufficient current to generate EBSD patterns in fine-grained or nano-crystalline materials.

There have been some studies that have measured the spatial resolution of EBSD. We will discuss only those studies that have used field emission electron sourced SEM instruments. Generally, the spatial resolution of the technique has been determined by placing the electron probe near a grain boundary and noting the minimum distance from the grain boundary that allows EBSD patterns to be obtained with no overlap from the opposite grain. This technique has produced results that vary, but can be used as a guideline for the use of EBSD in fine-grained materials. One study of Ni found that the lateral spatial resolution was 150 nm, 100 nm and 50 nm at 30 kV, 20 kV and 10 kV respectively [18]. Another study has shown that for Al the lateral spatial resolution of about 20 nm [19]. A more comprehensive study that utilized experimental measurements and electron trajectory simulations has determined the improved spatial resolutions resulting from lower accelerating voltage SEM conditions, as shown in Table 1 [20]. Table 1 shows that the spatial resolution of the technique is quite high and suitable for studying nano-crystalline materials and can be better than 50 nm. It is clear that the best spatial resolutions are achieved by operating at lower accelerating voltages and in higher atomic number materials. A purely experimental study of EBSD spatial resolution has shown a resolution of about 20 nm in Al and as good as 9 nm in brass. These results are calculated based on the fraction of indexed EBSD patterns as a function of grain size and not on a physical measurement of the EBSD interaction volume [21].

4.2. Depth Resolution

The depth resolution of EBSD is also a strong function of the specimen atomic number and the operating voltage of the SEM and is more difficult to measure experimentally. The maximum distance that a Bloch wave can travel unscattered is about 2 to 3 extinction distances for the given reflection. Thus the depth resolution has been described as 2 or 3 times the many-beam extinction distance for the accelerating voltage and the material [22]. This leads to a worst-case depth resolution at 20 kV of about 100 nm for Al and about 50 nm for Ni. As in the lateral resolution, the depth resolution is improved by operating at lower operating voltages or for higher atomic number samples. Monte Carlo electron trajectory simulations have been used to predict the depth resolution of EBSD based on the use of a 90% energy loss cut off with a resulting depth resolution at 20 kV of 100 nm for Al and 20–30 nm for Ni [20]. These results represent a worst-case value due to the generous energy cut off. In general the depth resolution appears to be similar to the lateral resolution discussed previously. An example of thin film analysis is discussed in the section 6.1.

5. SAMPLE PREPARATION OF NANO-MATERIALS FOR EBSD

The main sample preparation requirement for EBSD studies is that the sample surface shall be clean, representative of the bulk of the material and free from damage or deformation resulting from the preparation process. Some materials, like epitaxial or heteroepitaxial layers may require no additional sample preparation steps and EBSD patterns may be obtained from the sample in the as-deposited condition. Other samples may require more complex procedures to produce good samples for EBSD. For nano-crytalline materials it is imperative that proper sample preparation procedures are followed as the grain size of the material approaches the resolution of the EBSD technique. Improper sample preparation can result in degraded patterns or the total loss of the EBSD patterns due to surface deformation or damage induced by the preparation technique.

For larger samples of nano-crystalline materials, it may be possible to employ conventional specimen preparation techniques. Standard metallographic mounting and polishing can produce samples suitable for EBSD. In some cases it may be necessary to follow the mechanical polishing with a chemical attack etch to remove the last remnants of damaged material caused by the mechanical polishing. This technique has been shown to work well for many ceramics as well as metal samples. One of the best techniques for EBSD preparation is electropolishing, similar to the techniques used for TEM sample preparation. Unfortunately, the small sample size often associated with nano-crystalline materials, makes electropolishing difficult or impossible.

The small size of many nano-crystalline samples can make conventional sample preparation difficult or impossible. One very useful tool for the preparation of nano-crystalline materials for EBSD is the focused ion beam (FIB) tool. FIB tools have been used for many years to produce thin samples for transmission electron microscopy (TEM) where the ion beam is used to micromachine a thin membrane from the sample. The advantage to FIB preparation is that the sample can be prepared from a very specific location. Once the thin sample is prepared using FIB the sample is mounted on a suitable substrate, usually a carbon coated TEM grid using the same procedures used for TEM sample preparation [23]. This method does not allow any

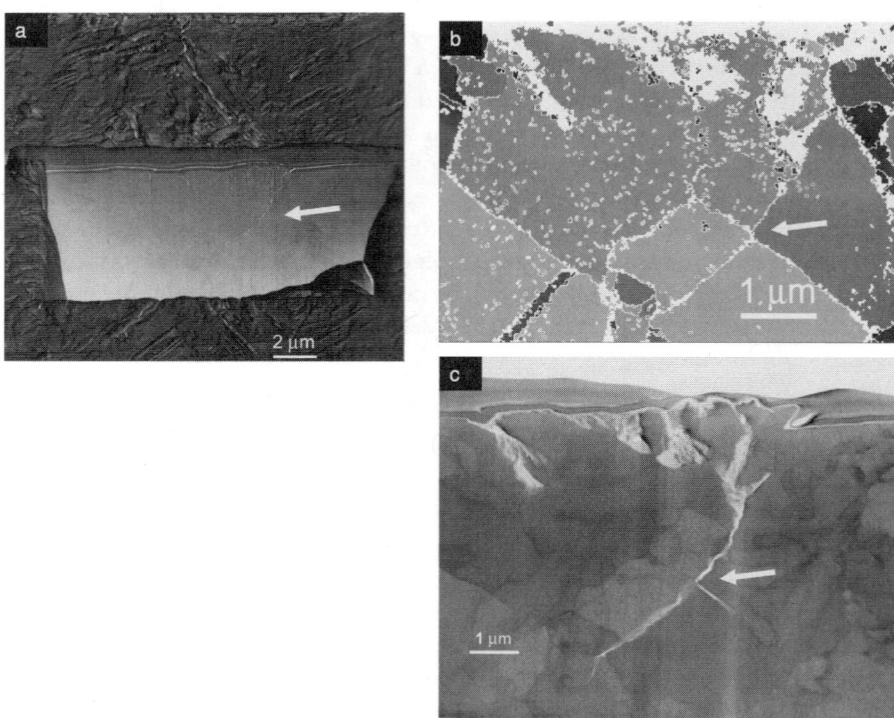

Figure 8. Combined EBSD and TEM of a FIB prepared cross section through a fatigue crack in a Ni MEMS device. a) FIB cross section that shows the location of the section with respect to the fatigue crack, b) EBSD map of area around fatigue crack, c) annular dark field STEM image of same sample as in b. Arrows indicate the same point in all three images. The orientation map is 300 × 200 pixels with a step size of 25 nm and required 30 minutes to collect. (See color plate 9.)

subsequent treatment of the sample. Alternatively, the sample can be attached to the side of the grid using the metal deposition capability of the FIB so that subsequent low energy ion milling may be performed [24]. The main disadvantage to FIB sample preparation is the limited sample size (about 50 μm wide by 20 μm deep). Thus, the area or feature of interest must be small and near the surface of the sample, although this is not often a problem with nano-materials. Even with the sample size restriction, FIB has been used to prepare useful samples for EBSD analysis [25, 26].

Figure 8 shows an example of a FIB section prepared from the vicinity of a crack in a fatigued electrodeposited Ni sample. The sample was mounted onto a carbon film and an orientation map was obtained with no further preparation. We have found this to work for many transition metals and many ceramic materials. Other materials, like Si and compound semiconductors, require addition preparation due to damage to the crystal structure caused by the ion beam. These materials require further low energy ion polishing in order to obtain good EBSD patterns. Low voltage (2–3 kV), low-angle (3–4°) argon ion polishing following FIB micromachining, for a very short time is all that is needed to produce good EBSD patterns from these materials.

One additional advantage of the FIB technique is that the sample may be made sufficiently thin for TEM analysis that could follow the EBSD analysis. Thus, correlation of the EBSD orientation maps with TEM images of the material is relatively easy and can be a powerful route to nano-materials characterization. An example of this is shown in Figure 8b and 8c, which are an orientation map of the fatigue crack region of electrodeposited Ni sample and a STEM annular dark field image of the same area. This ability to perform correlative imaging is very powerful and allows the sample to be more fully characterized. There are a few texts that discuss specimen preparation for EBSD, but additional sample preparation details have been published [27].

6. APPLICATIONS OF EBSD TO NANO-MATERIALS

6.1. Heteroepitaxy of Boron Arsenide on [0001] 6H-SiC

Icosohedral boron arsenide (B_{12} As_2, rhombohedral, hexagonal parameters, a = 0.615 nm, c = 1.191 nm) is a wide-band gap semiconductor exhibiting exception radiation hardness and is potentially useful for the fabrication of beta-cells for direct nuclear to electrical energy conversion. Chemical vapor deposition has been shown to produce crystalline films of As_2B_{12} on 6H-SiC [28]. Characterization of the films is important to understand the properties and the performance of energy conversion devices. Figure 9a is a SEM image of the surface of the boron arsenide film. The films are quite smooth and were shown to have useful electrical properties. Figure 9b and 9c are orientation maps of a similar area. The thin black lines superposed on both images in 9b and 9c represent high angle grain boundaries. Figure 9b is a map that is colored with respect to an in-plane direction and shows that two rotational variants are present. It is important to note that when utilizing orientation maps one must usually prepare maps that represent orientations in two orthogonal directions to get a complete picture of the sample texture as these are equivalent to inverse pole figures and suffer the same disadvantages. Figure 9c is an orientation map that is colored with respect to the surface normal. The red color over the entire sample demonstrates that all the rotational variants have the same crystallographic direction parallel to the sample surface or the film growth direction. Figure 9b and 9c show that all areas of the film have grown with [0001]B_2As_{12}//[0001]6H-SiC as shown by the red color indicating that the film has a [0001] plane parallel to the surface. Figure 10 a and b, are inverse pole figures from the map shown in Figure 9b and c. Figure 10a is the in-plane inverse pole figure that clearly shows the two rotational domains consisting of 60° rotations about the <0001> axis as shown in the inverse pole figure of 10b. Bright and dark-field TEM images are shown of the boron arsenide film in Figure 11a and b. Note that the film is 100 nm thick and as shown in the bright field images consists of domains of different rotational orientations as well as translational variants. There are also many smaller translational variants that are not detected using using EBSD mapping, making TEM imaging an important component of understanding these thin films [29].

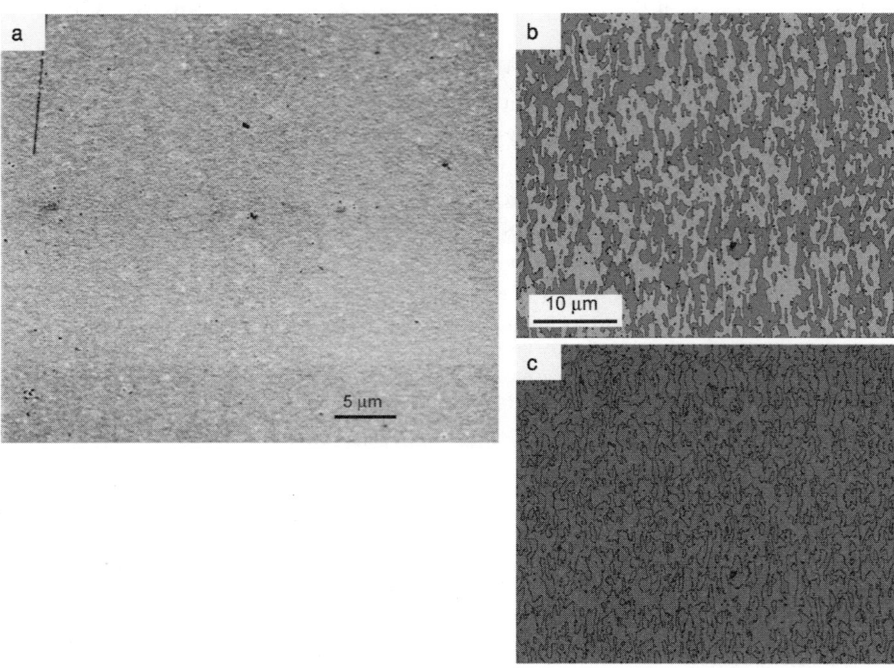

Figure 9. Heteroepitaxy of Boron arsenide thin-film deposited on 6H-SiC. a)SEM image of sample tilted for orientation mapping, b) Orientation map of the in-plane texture, c) Orientation map of the out-of-plane, or surface normal orientation. Lines on the orientation maps represent boundaries between rotational variants. The map is 450 × 350 pixels with a step size of 100 nm and was collected in 6 hours. (See color plate 10.)

6.2. Electrodeposited Ni for MEMS Applications

Electodeposition of metals has now become quite common process in many areas of micro-electronics and micro-electro mechanical systems (MEMS). Metals are needed for some MEMS devices due to the need to have thicker structures that are capable of transmitting large sustained loads or torques [30, 31]. The microstructural scale of electrodeposited metals can be quite small. In order to optimize the process and to understand the resulting properties of the electrodeposit it is important to understand the crystallographic texture of electrodeposited metals for MEMS applications. There are two examples shown in this section, electrodeposition of Ni onto an Au seed layer and the electrodeposition of Ni for the fabrication of small mechanical parts using the LIGA process.

EBSD orientation studies were carried out in order to understand the electro-decomposition of Ni over patterned lines on a Au seed layer. Due to the small size of the features involved it was necessary to use the FIB to prepare cross sections through the electrodeposits. These sections were mounted on Cu TEM grids and then mounted in the SEM for orientation mapping. Figure 12 shows an overall view of one of the electrodeposited Ni lines. Figure 12a is a pattern quality map of the sample area. These

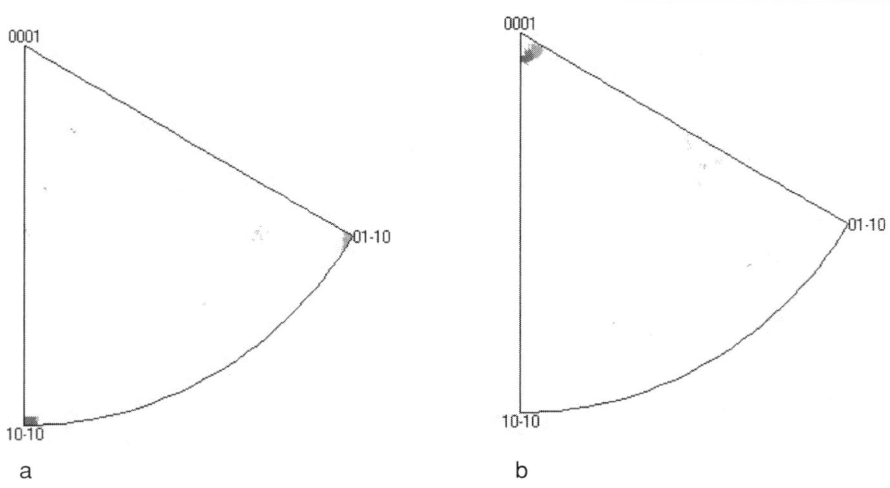

Figure 10. Inverse pole figures formed from the data shown in Figure 9b. a) Inverse pole figure with respect to the x direction that shows the two 60° rotational variants b) Inverse pole figure with respect to the surface normal demonstrating the slight misalignment of the basal planes from each rotational variant. (See color plate 11.)

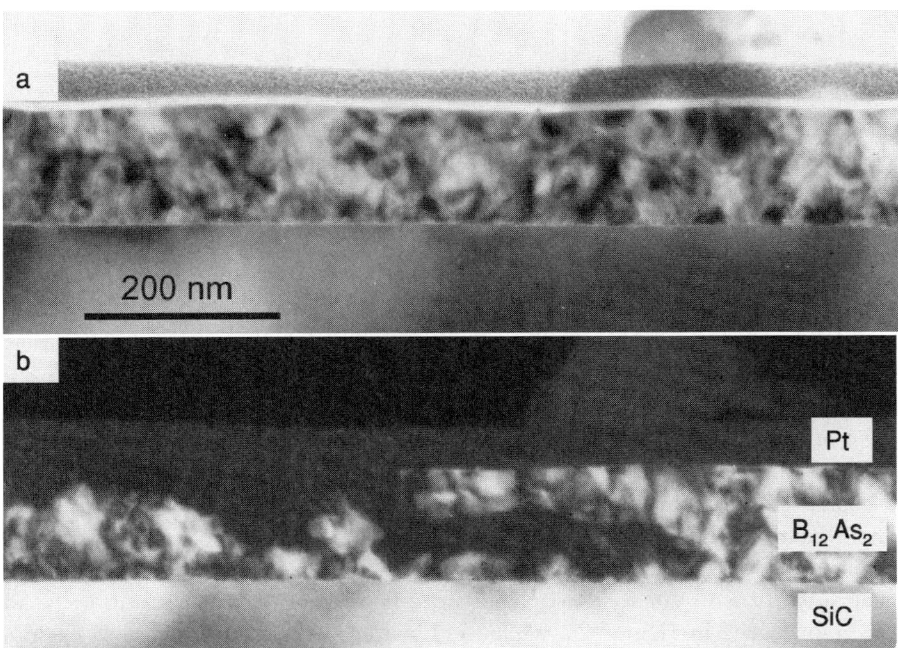

Figure 11. TEM cross sectional images of the boron arsenide films of Figure 9. a) Bright field TEM image that shows the large number of small translational domains in the sample that are not detected by EBSD, b) Dark field TEM that shows one of the rotational domains in contrast. Surface layers are due to FIB deposited Pt to aid in sample preparation.

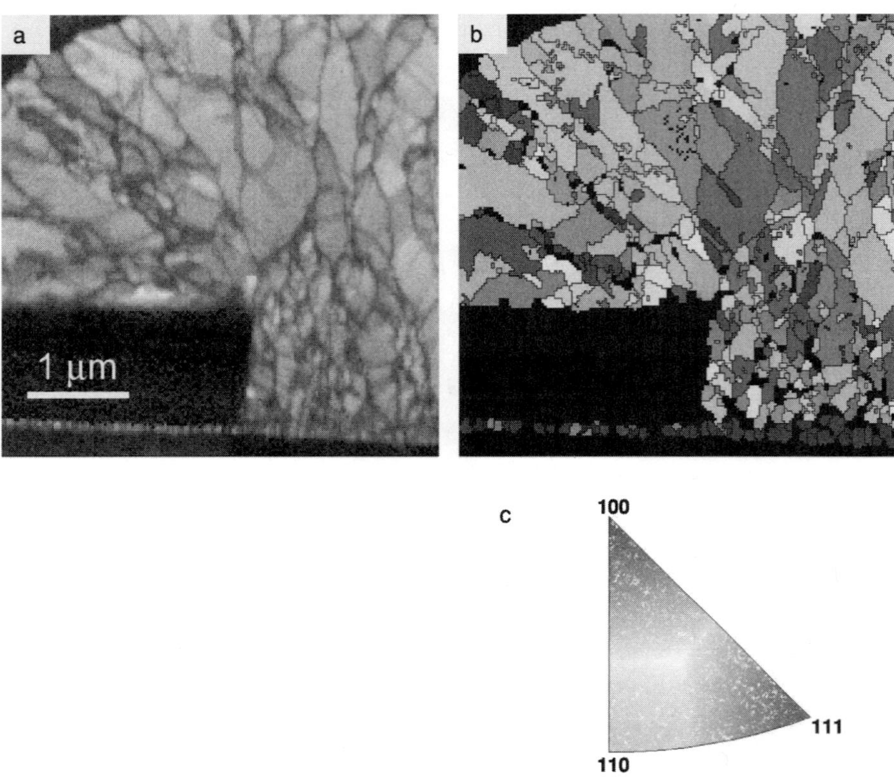

Figure 12. Ni electrodeposited lines on patterned substrate. The thin layer at the base of the deposit is a thin Au layer. a) Pattern quality map. b) Orientation map with respect to the deposition direction. c) Stereographic triangle color legend for the orientation map in b. The map is 175 × 175 pixels with a step size of 25 nm and was collected in 50 minutes. (See color plate 12.)

images provide excellent detailed information about the general microstructure of the sample. The microstructural details are clearly visible in the pattern quality image. Figure 12b is an orientation map with respect to the growth direction. This map is colored with respect to the stereographic triangle shown in Figure 12c. The narrow band of blue colored grains (indicated with the arrow) at the bottom of the image is the Au seed layer. These grains have formed with a <111> fiber texture. Figure 13 is a higher resolution image that shows the details of the electrodeposit at the Au seed layer. Figure 13a is a pattern quality map that shows the microstructure of the deposit. Figure 13b is an orientation map with respect to the growth direction of the Ni. It is apparent that the Au seed layer has formed with a <111> fiber texture (refer to the color legend in Figure 12c) while the Ni layer has grown with a more random texture. The largest of the Au grains is about 100 nm in diameter demonstrating that the resolution of the technique is quite good. From this work, the growth texture of the Ni deposit is not strongly influenced by the texture of the Au seed layer.

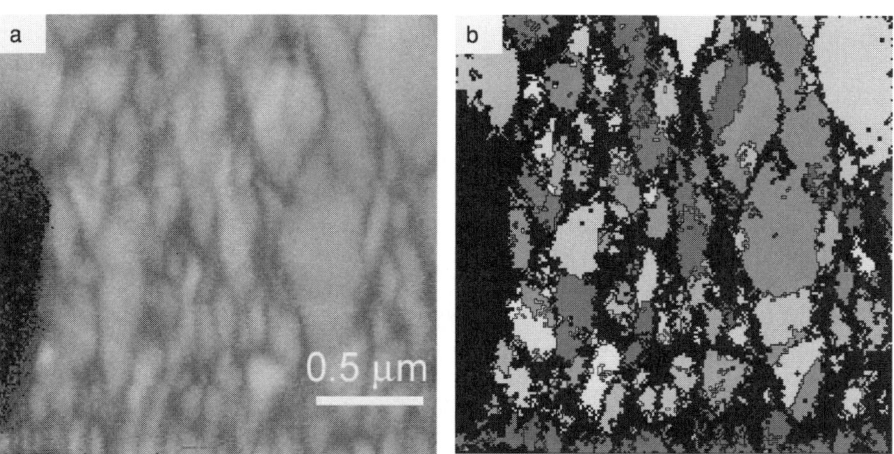

Figure 13. Higher magnification of the electrodeposit shown in Figure 12. a) Pattern quality map, b) Orientation map with respect to the growth direction using the color legend shown in 12c. The fiber oriented Au is at the bottom of the image. The map is 200 × 200 pixels with a step size of 10 nm and was collected in 1 hour. (See color plate 13.)

LIGA, an acronym of the German words "Lithographie, Galvanoformung, Abformung," is a microfabrication process in which structural material is electrodeposited into a photolithographical realized mold. LIGA permits the fabrication of parts with extremely small cross sections, on the order of a few micrometers with accuracy better than one micrometer [30]. The microstructure of LIGA formed parts depends primarily on the processing parameters during electrodeposition of the metal. The texture of the deposit is important in determining mechanical properties. Thus careful characterization and control of LIGA materials and processes is required to produce high quality, reliable components. The small physical size of many LIGA parts makes characterization by standard x-ray techniques difficult or impossible. EBSD orientation mapping provides the needed resolution and the ability to sample a larger area than TEM analysis makes EBSD orientation mapping ideal for the study of these materials.

X-ray diffraction of these materials has been used to determine their macrotexture, but this does not give any indication of the relationship between texture and microstructure. EBSD orientation mapping was carried out to determine the relationship between the texture of an electroformed deposit and the grain size and shape. Also, the distributions of grain boundary orientations was of interest. For this analysis the sample was mounted and polished using standard metallographic procedures so that the deposit could be studied in cross section. Figure 14a is a pattern quality map of an NiMn electroformed part that shows there is a duplex grain size with some large columnar grains separated by smaller areas of equiaxed grains. Figure 14b is an orientation map of the area colored with respect to the growth direction. Refer to the stereographic triangle in Figure 12c for the color legend. It is immediately obvious that the long columnar grains have a <110> direction parallel to the growth direction and

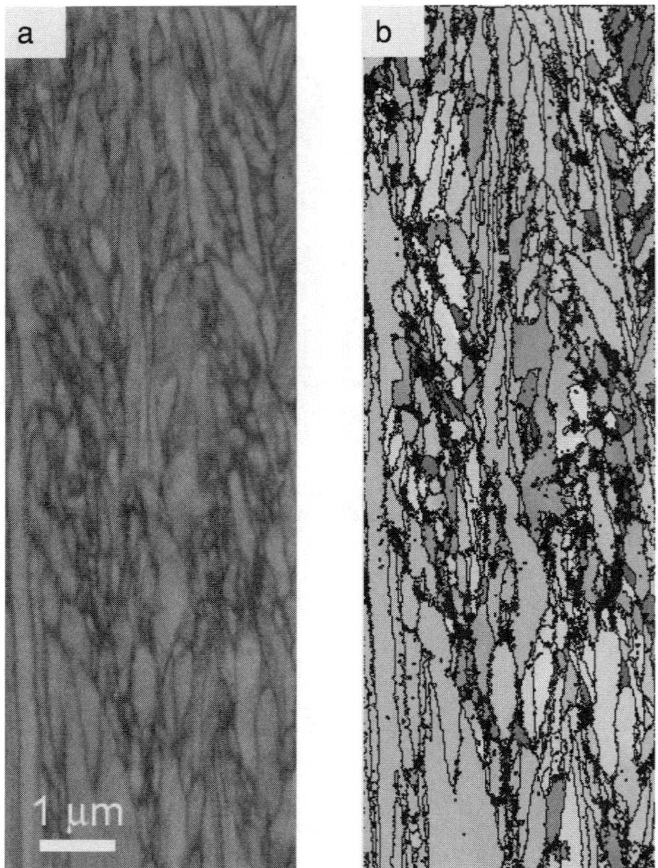

Figure 14. Electroformed NiMn for LIGA MEMS devices. a) Pattern quality map that shows the grain structure of the deposit b) Orientation map colored as shown in Figure 12c. The large green shaded grains represent a <110> fiber texture. Smaller grains are randomly oriented with respect to the growth direction. The map is 200 × 600 pixels with a step size of 20 nm and was collected in 2.5 hours. (See color plate 14.)

the smaller equiaxed regions are more randomly oriented. From this data the volume fraction of <110> fiber oriented grains can be determined by selecting only those pixels with orientations within 10° of a <110> fiber texture demonstrating that about 50% of the grains have a <110> fiber orientation with respect to the growth direction. This is important for a full understanding of the mechanical properties of this material.

There have been a number of recent studies that have shown that the distribution of grain boundary misorientations can influence the performance of polycrystalline materials. Early studies focused on the corrosion properties of a variety of metals [32]. Further studies have shown that by increasing the fraction of special boundaries (typically defined as grain boundaries near the low Σ misorientations) the corrosion

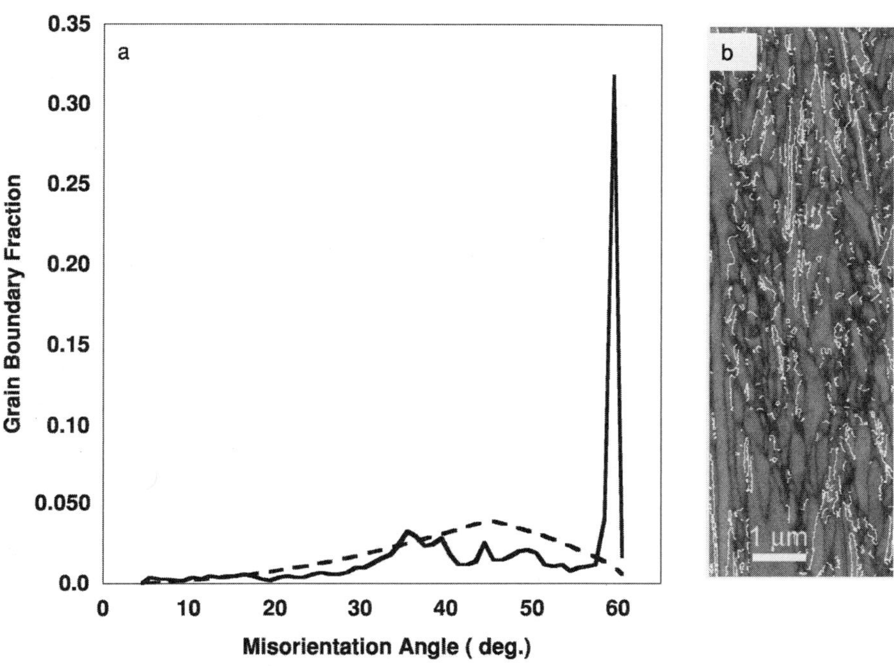

Figure 15. Representations of the distributions of the grain boundary misorientations in the NiMn electrodeposits. a) Grain boundary misorientation distributions, solid line is experimental data from map shown in Figure 12, dotted line is the distribution for a random arrangement of grains. b) Pattern quality map with twin boundaries shown in white.

behavior, creep resistance and weldability of the material can be enhanced [33]. One important concern with the electrodeposited NiMn material was the influence of grain boundary misorientations on the annealing behavior of the material. Previous work in Ni deposits has shown that the material recrystallized in a manner that maximized the number of low energy (twin type or $\Sigma 3$ or a 60° rotation about the <111>) grain boundaries [31]. Therefore, it is important to know the misorientation distribution of the as-deposited material. Figure 15a shows the grain boundary misorientation distribution for the NiMn sample. This information is calculated from the orientation data so no additional scanning needs to be performed. The data from the NiMn electrodeposit is compared with the expected distribution of grain boundary misorientations derived ffrom the random orientations of cubes [34]. It is clear that there is a large difference between the expected random distribution and the measured distribution. The most striking difference is the large number of 60° misorientations. These rotations can be shown to be about the <111> and are therefore $\Sigma 3$ twin type boundaries. Figure 15b is a pattern quality image with the twin type boundaries shown in white. The high density-of-twins in the microstructure are important to understanding the recrystallization behavior and the response to high temperature exposures of electrodeposited Ni.

Figure 16. shows a FIB prepared cross section through an unreleased multilayer polycrystalline Si MEMS device. The sample was micromachined using the FIB in a manner similar to that used to make TEM samples and then attached to the grid using Pt deposition.

6.3. Polycrystalline Si For MEMS Applications

Polycrystalline silicon thin films are an important component of many advanced MEMS structures. Reliability of these devices is closely related to the mechanical performance of the thin films of polycrystalline silicon used to fabricate the devices. The need for an understanding of these structure-property relationships in these films is made more important by the fact that the microstructure may consist of highly textured columnar grains [35, 36]. Many polycrystalline silicon MEMS devices are produced by a surface micromachining that uses silicon oxide coating as sacrificial spacers between the structural layers of polysilicon.

Characterization of these types of materials are complicated by the small size of the devices which in many cases may have features that do not exceed 1 μm. EBSD is ideal for determining the grain structure and texture of the multiple layers of polysilicon used to produce these devices. Preparation of samples is best performed with the FIB. In the case of silicon and the results shown here, subsequent low energy ion milling is required to produce EBSD patterns of sufficient quality for automated mapping.

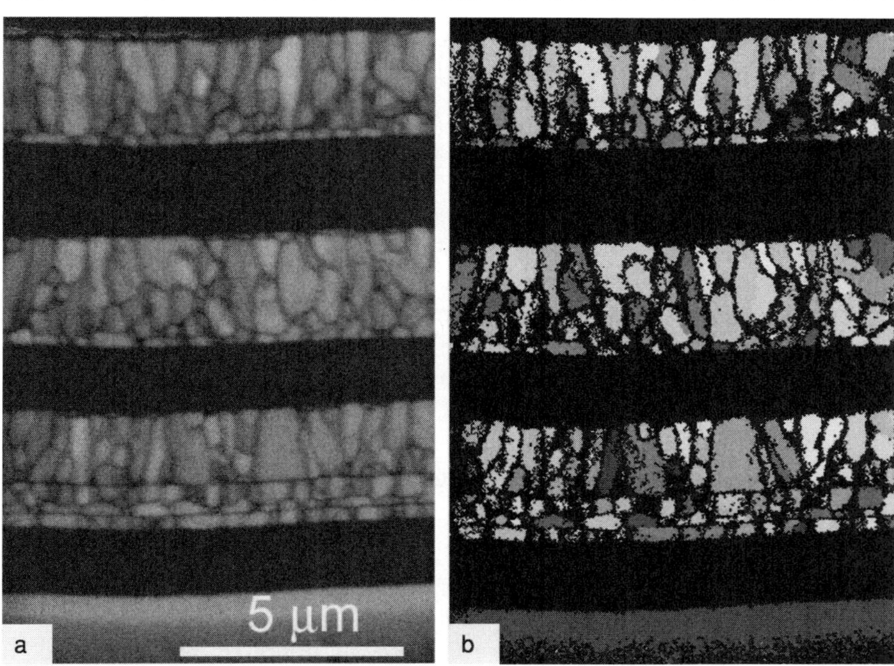

Figure 17. Polycrystalline structures for MEMS devices. a) Pattern quality map, the single crystal Si substrate is at the bottom of the image. The polycrystalline Si layers are separated by oxide. b) Orientation map with respect to the growth direction. The map is 600 × 500 pixels with a step size of 25 nm and was collected in 8.5 hours. (See color plate 15.)

Figure 15 shows a FIB prepared cross section through an unreleased (the oxide spacers have not been removed) multilayer MEMS device. The sample was micromachined using the FIB in a manner similar to that used to make TEM samples [23]. The cross section was attached to a support grid by depositing Pt at the sample grid junction, due to the need for subsequent low energy ion milling to allow EBSD patterns to be acquired. This arrangement is robust and allows the sample to be handled with little danger of damage. The sample was low-energy argon ion milled for 30 seconds at 3 keV. After ion milling excellent EBSD patterns were obtained from the silicon. Figure 16a is a pattern quality map of the multilayer device. Note that the individual levels of polycrystalline silicon appear to be made up of multiple layers. Figure 16b is a orientation map with respect to the growth direction. Note that the substrate silicon is red indicating a <100> direction normal to the sample surface (please refer to Figure 12c for the complete color legend). Each layer has a slightly different microstructure due to the accumulated time at temperature for the first layer as compared to the subsequent layers. The layer furthest from the substrate is more columnar when compared to the first layer. In addition there are slight textural differences from the first layer to the last. The first layer represents annealed material and exhibits a tendency for a <111>

fiber texture as compared to the last layer which has not been at temperature for as long and exhibits a tendency to a <110> fiber. These differences in texture result in different mechanical properties for each layer, which assists in the proper design of MEMS components.

7. SUMMARY

EBSD should now be considered a standard analytical accessory for the SEM. The spatial resolution of EBSD is quite high (as high as a few nm, depending upon material) and is suitable for the study of many nano-crystalline materials. For applications of EBSD where the highest spatial resolution is required, an SEM with a field emission electron source is mandatory as the resolution of the technique is most directly determined by the electron beam size. Further developments the electron optics of the SEM should allow even higher spatial resolutions. For nano-crystalline materials, EBSD is most useful for the mapping of grain orientations and grain boundary misorientations as demonstrated in the examples shown in this chapter.

ACKNOWLEDGEMENTS

Special thanks to my colleagues at Sandia National Laboratories for bring their special materials problems to my attention. In particular, I would like to thank Terry Aselage for the boron arsenide work, Sean Hearne for the nucleation of electrodeposited Ni study, Brad Boyce for the fatigue samples and Tom Buchheit and Steve Goods for the LIGA NiMn material.

Sandia is a multiprogram laboratory operated by Sandia Corporation, a Lockheed Martin Company, for the United States Department of Energy under contract DE-AC04-94AL85000.

REFERENCES

1. J. A. Small and J. R. Michael, *J. of Microsc.* 201 (2001) 59.
2. J. A. Small, J. R. Michael and D. S. Bright, *J. of Microsc.* 206 (2002) 170.
3. E. Ma, *Scripta Mat.* 49 (2003) 663.
4. A. J. Schwartz, M. Kumar and B. L. Adams (Eds.), *Electron Backscatter Diffraction in Materials Science*, Kluwer Academic/Plenum Publishers, New York (2000).
5. V. Randle, *Microtexture Determination and its Application*, The Institute of Materials, London (1992).
6. M. N. Alam, M. Blackman and D. W. Pashley, *Proc. Royal Society of London*, A221 (1954) 224.
7. J. A. Venable and A. Bin-Jaya, *Phil. Mag.*, 35 (1977) 1317.
8. D. J. Dingley, *Scanning Electron Microscopy IV*, 4 (1981) 273.
9. S. I. Wright and B. L. Adams, *Met. Trans. A.*, 23 (1992) 759.
10. J. R. Michael and R. P. Goehner, *MSA Bulletin*, 23 (1993) 168.
11. R. P. Goehner and J. R. Michael, *J. of Res. of the National Institute of Standards and Technology*, 101 (1996) 301.
12. D. J. Dingley, in: *Electron Backscatter Diffraction in Materials Science*, A. J. Schwartz, M. Kumar and B. L. Adams (Eds.), pp. 1–19, Kluwer Academic/Plenum Publishers, New York (2000).
13. O. L. Wells, *Scanning*, 21 (1999) 368.
14. V. F. Leavers, *Shape Detection in Computer Vision Using the Hough Transform*, Springer-Verlag, New York (1992).
15. S. I. Wright, in: *Electron Backscatter Diffraction in Materials Science*, A. J. Schwartz, M. Kumar and B. L. Adams (Eds.), pp. 51–64, Kluwer Academic/Plenum Publishers, New York (2000).

16. J. K. Farrer, M. M. Nowell, D. J. Dingley and D. J. Dingley, Proceedings of Microscopy and Microanalysis 2003, pp. 80–81, Cambridge University Press, New York (2003).
17. K. Murata, *J. Appl. Phys.*, 45 (1974) 4110.
18. T. C. Isabell and V. P. Dravid, *Ultramicroscopy*, 67 (1997) 59.
19. F. J. Humphreys, Y. Huang, I. Brough and C. Harris, *J. of Microsc., 195 (1999) 212.*
20. S. X. Ren, E. A. Kenik, K. B. Alexander and A. Goyal, *Microscopy and Microanalysis*, 4 (1998) 15.
21. F. J. Humphreys and I. Brough, *J. of Microsc*, 195 (1999) 6.
22. D. C. Joy, Proceedings of Microscopy Society of America 1994, pp. 592–593, San Francisco Press, San Francisco, California (1994).
23. M. W. Phaneuf, *Micron*, 30 (1999) 277.
24. M. W. Nowell, *J. of Electronic Materials*, 31 (2002) 23.
25. T. L. Matteson, S. W. Schwartz, E. C. Houge, B. W. Kempshall and L. A. Giannuzzi, *J. of Electronic Materials*, 31 (2002) 33.
26. S. V. Prasad, J. R. Michael and T. R. Christenson, *Scripta Mat.*, 48 (2003) 255.
27. J. Goldstein, D. Newbury, D. Joy, C. Lyman, P. Echlin, E. Lifshin, L. Sawyer and J. Michael, *Scanning Electron Microscopy and X-ray Microanalysis*, pp. 537–564, Kluwer Academic/Plenum Publishers, New York (2003).
28. R. H. Wang, D. Zubia, T. O'Neil, D. Emin, T. Aselage, W. Zhang and S. D. Hersee, *J. of Electronic Materials*, 29 (2000) 1304.
29. C. P. Flynn and J. A. Eades, *Thin Solid Films*, 389 (2001) 116.
30. T. E. Buchheit, D. A. Lavan, J. R. Michael, T. R. Christenson and S. D. Leith, *Metallurgical and Materials Trans. A.*, 33A (2002) 539.
31. K. J. Hemker and H. Last, *Mat. Sci. and Eng. A.*, A319 (2000) 882.
32. P. Lin, G. Palumbo, U. Erb and K. T. Aust, *Script Met.*, 33 (1995) 1387.
33. M. Kumar, W. E. King and A. J. Schwartz, *Acta Mat.*, 48 (2000) 2081.
34. J. K. MacKenzie, *Biometrica*, 45 (1958) 229.
35. S. Jararaman, R. L. Edwards and K. J. Hemker, *J. of Materials Science*, 14 (1999) 688.
36. M. Furtsch, M. Offenberg, A. Vila, A. Cornet and J. R. Morante, *Thin Solid Films*, 296 (1997) 177.

14. HIGH RESOLUTION TRANSMISSION ELECTRON MICROSCOPY

DAVID J. SMITH

1. HRTEM AND NANOTECHNOLOGY

With the rapidly escalating attention recently being given to nanoscale science and technology, techniques capable of structural characterization on the nanometer scale have central importance. The high-resolution transmission electron microscope (HRTEM) has evolved over many years to such an extent that resolving powers at or close to the one Ångstrom (0.1 nm) level can nowadays be attained almost on a routine basis. Using correct operating conditions and well-prepared samples, high-resolution image characteristics are interpretable directly in terms of projections of individual atomic-column positions. With quantitative recording and suitable image processing, atomic arrangements at defects and other inhomogeneities can be reliably and accurately determined. Since nanoscale irregularities have a marked influence on bulk behavior, the HRTEM has become a powerful and indispensable tool for characterizing nanostructured materials. This chapter first outlines some of the basic theoretical principles and practical aspects of the HRTEM. Representative applications to problems involving nanostructured materials are then described. Ongoing trends are identified and some underlying problems associated with use of the HRTEM are briefly discussed. The interested reader is referred to review articles [1–4], conference proceedings [5–10] and research monographs [11–13] for further information about HRTEM and additional applications.

2. PRINCIPLES AND PRACTICE OF HRTEM

2.1. Basis of Image Formation

The process of image formation in the HRTEM can be considered as occurring in two stages. Incoming or incident electrons undergo interactions with atoms of the specimen, involving both elastic and inelastic scattering processes. The electron wavefunction which leaves the exit surface of the specimen is then transmitted through the objective lens and subsequent magnifying lenses of the electron microscope to form the final enlarged image. Our attention here will be focused on those electrons that are elastically scattered since these contribute to formation of the high-resolution bright-field image. Note, however, that the inelastically scattered electrons can also be used to provide invaluable information about the sample composition using the technique of electron-energy-loss spectroscopy (EELS). Moreover, electrons scattered to very large angles are utilized for Z-contrast annular-dark-field imaging in the scanning transmission electron microscope (STEM). These possibilities are explored elsewhere in other chapters.

Electron scattering is a strongly dynamical process, unlike X-ray or neutron scattering, so that a simple kinematical scattering approximation is insufficient for understanding image formation for all but the thinnest of samples. Multiple electron scattering and large phase changes are typical for most samples, which means that the relative heights of the different atoms in the specimen must be taken into account when attempting quantitative image interpretation. Image simulations, which must also take into account additional effects caused by the objective lens, are essential before detailed information about atomic arrangements at crystal defects such as dislocations and interfaces can be confirmed. Several alternative approaches to image simulation have been developed [14, 15]. The most widespread is based on n-beam dynamical theory [16, 17], and is termed the multislice approach. In this theory, atoms in the specimen are considered as located on narrowly-separated planes (or slices), normal to the beam direction. The electron wavefunction is then propagated slice-by-slice through the sample to eventually form the exit-surface wavefunction. This iterative process lends itself to convenient computer algorithms which enable rapid computations to be carried out, which are especially useful during the refinement of unknown defect structures [18, 19]. Further information about other equivalent approaches to electron scattering can be found in the monograph by Cowley [20].

The electron wavefunction at the exit surface of the specimen must still be transferred to the final viewing screen or recording medium. This process is dominated by the transfer characteristics of the objective lens of the microscope. The effect of this lens on image formation is conveniently understood by reference to what is termed the phase contrast transfer function (TF), as described by Hanszen [21]. The TF is both specimen- and microscope-independent so that a single set of universal curves enables the transfer characteristics of all objective lenses to be described. Electron microscopes that have different objective lenses, or operate at different accelerating voltages, are easily compared by using suitable scaling factors. Figure 1 shows TFs computed for the optimum defocus of the objective lens for a typical 400-kiloVolt HRTEM. Two

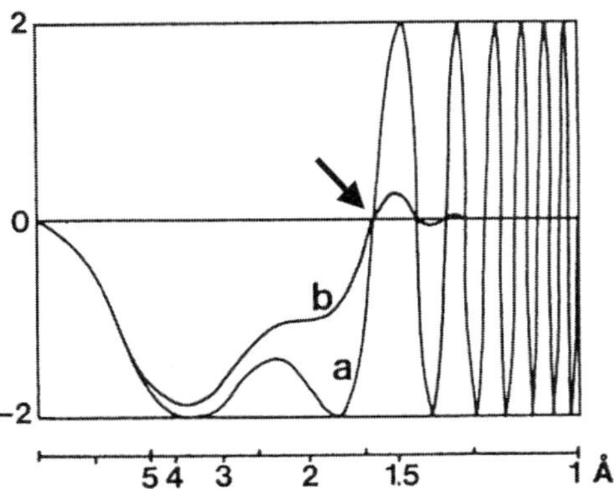

Figure 1. Transfer functions for high-resolution phase-contrast imaging at optimum defocus for objective lens (C_s = 1.0 mm) of 400-kV HREM: (a) coherent illumination; and (b) partially coherent illumination, with focal spread of 8 nm and incident beam convergence half-angle of 0.5 mrad. Arrow indicates interpretable resolution limit.

cases are shown, corresponding to: (a) coherent, and (b) partially coherent, incident electron illumination. Notice that the TF has an oscillatory nature, meaning that electrons scattered to different angles undergo relative phase reversals, which would lead directly to artefactual detail in the final image. Note also that the shape of the TF is further affected by defocus changes, which result in additional phase changes that will alter the image appearance. It is thus essential to have an accurate knowledge of the lens defocus, since much of the detail contained in the recorded high-resolution image will otherwise be uninterpretable. In the ideal case, the incident electron beam will be a coherent, monochromatic plane wave. In practice, some inevitable loss of coherence will result from focal spread (temporal coherence) and finite beam divergence (spatial coherence) [22, 23]. Mathematically, the effects of partial coherence can be represented by envelope functions. These will cause dampening of the TF for spatial frequencies corresponding to larger scattering angles, as illustrated, for example, by curve (b) in Fig. 1. Note also that the positions of the TF crossovers, or "zeroes", are not affected by the envelope functions. However, specimen information that is scattered to higher spatial frequencies, which would be equivalent to higher image resolution, is liable to be lost during image formation. Lanthanum hexaboride has been used as the electron source for calculation of the curve (b) with partial coherence. As discussed in the following section, additional specimen information can possibly become available by using a high-coherence electron source such as the field-emission gun (FEG).

2.2. Definitions of Resolution

Resolution is traditionally discussed in terms of the ability of an imaging system to discriminate between two discrete objects, and is usually closely coupled to the wavelength of the incident illumination. Since high-energy electrons have picometer wavelengths, resolution limits on the same scale might be anticipated. In practice, electron lenses have unavoidable aberrations, and perfectly coherent electron sources are unattainable. It then becomes necessary to make a compromise between the traditional diffraction limit, which varies inversely as the aperture angle, and spherical aberration, which varies rapidly with the cube power of the angle. The end result is an expression of the form

$$d = AC_s^{1/4}\lambda^{3/4} \tag{1}$$

where C_s is the spherical aberration coefficient of the objective lens, λ is the electron wavelength, and the value of the constant A depends on certain assumptions that are made about the imaging conditions.

In practice, there are several resolution limits that are applicable to any specific HRTEM [24]. These are conveniently understood by reference to the TF of the objective lens. The *interpretable* resolution, which is sometimes known as the *structural* or *point* resolution, is defined by the position of the first zero crossover of the TF at the optimum defocus. This resolution is indicated by the arrow in Fig. 1. This specific defocus gives the widest possible band of spatial frequencies without a phase reversal, and the corresponding interpretable resolution is then given by $\delta \sim 0.66(C_s\lambda^3)^{1/4}$. Values of C_s inevitably increase as the electron energy is increased due to limits on the saturation of the pole-piece material. However, overall improvements in δ should be obtained due to reductions in λ as the accelerating voltage is increased. Typical resolution figures using realistic C_s values (which range from 0.3 mm at 100 keV to about 1.5 mm at 1.0 MeV) are in the range of 0.25 to 0.12 nm. Other practical factors such as the size and cost of higher-voltage electron microscopes, as well as the increasing likelihood of electron irradiation damage due to the higher energy electron beam, become important considerations. Intermediate-voltage HRTEMs which operate in the range of 200 to 400 kV have become widespread due to the impact of these factors.

The *instrumental* resolution or *information limit* of the HRTEM is usually defined in terms of the damping produced by the envelope functions. A value of approximately 15% [i.e., exp(-2)] is commonly taken as the resolution cutoff since this level is regarded as the minimum acceptable for image processing requirements [25]. In recent 200- or 300-kV HRTEMs equipped with an FEG electron source, this resolution limit extends well beyond the interpretable resolution. Because of the oscillatory nature of the TF, the very fine detail present in the image at such resolution levels will, however, not be easily related to specimen features. Higher-order diffracted beams with inverted phase can be prevented from contributing to the image by using an objective aperture of suitable diameter located in the back focal plane of the objective lens. Conversely, provided that the defocus and C_s values are known with sufficient accuracy, then the phase-modulating effects of the TF can be removed by *a posteriori* image processing. Image interpretability with improved resolution can be achieved, as demonstrated by the pioneering work of Coene *et al.* [26] who used a focal-series reconstruction approach to resolve oxygen atoms for the first time in a high-temperature superconductor.

The term *lattice-fringe* resolution refers to the finest spacings of the lattice fringes that are visible in high-resolution images from a particular HRTEM. These lattice fringes result from the interference between two or more diffracted beams from a crystalline material. This resolution is determined by the overall instrumental stability of the HRTEM, and is often a direct reflection of freedom of the microscope environment from adverse external factors such as acoustic noise, mechanical vibrations, and stray magnetic fields. It must be noted, however, that lattice fringes with very fine spacings do not usually provide any useful information about local atomic arrangements. The interfering diffracted beams typically originate from comparatively large specimen areas, and the lattice fringe images may be recorded at significant underfocus conditions when there are many TF oscillations. For many years, the lattice-fringe resolution was widely regarded as the ultimate figure of merit for an HRTEM. It should be appreciated that the interpretable and instrumental resolution limits are nowadays considered far more useful for comparison purposes.

2.3. Lattice Imaging or Atomic Imaging

Most elemental and compound materials have unit cells that are relatively large in comparison with the resolution limits of commercially available HRTEMs. High-resolution lattice-fringe images are thus relatively straightforward to obtain when these materials are viewed in major low-index projections. However, it is important to appreciate that only a very small subset of these lattice images can be interpreted in terms of atomic arrangements. The basic problem is that the phase reversals that result from TF oscillations when the defocus of the objective lens is changed make it difficult to recognize or select the appropriate optimum defocus. The characteristic image of a crystal defect or the Fresnel fringe along the edge of the sample is often indispensable for recognizing a particular defocus setting. Some prior knowledge or calibration of the focal step size corresponding to the finest focus control is thus essential. The situation is even more complicated for materials with small unit cells which have comparatively few diffracted beams contributing to the final image. For these materials, identical images, referred to as *Fourier* or *self-images*, recur periodically with changes in defocus, with the period given by $2d^2/\lambda$, where d is the corresponding lattice spacing [27, 28]. With a half-period change in defocus, all image features reverse in contrast, with black spots becoming white and *vice versa*. As an example, a series of image simulations for [001] tin dioxide is shown in Fig. 2, where the major reversals in contrast of the image features from white to black to white to black are readily apparent [29].

Further misleading complications are likely to arise in thicker crystals because of dynamical multiple scattering. As the sample thickness is increased, intensity will be progressively lost from the directly transmitted beam, and it will instead build up in the diffracted beams. Eventually, a thickness is reached where the direct beam will be relatively low in intensity (in the vicinity of what is termed the thickness extinction contour), and the resulting image will be dominated by (second-order) interference processes between the various diffracted beams. In some small-unit-cell materials, these interferences can often lead to pairs of white spots, graphically referred to as *dumbbells*, which appear to portray faithfully all atom positions in the unit cell [30]. Except under highly specific thickness and defocus values, it has been found that the separation

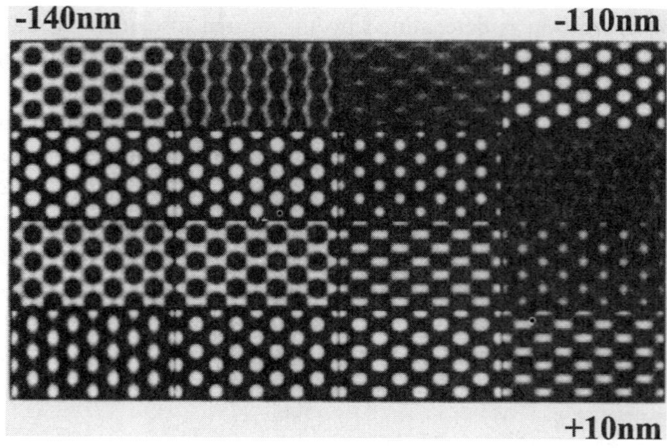

Figure 2. Through-focal series of image simulations for crystal of tin dioxide in [100] projection: 500 keV, $C_s = 3.5$ mm, crystal thickness -2.8 nm. Focus extends in 10 nm steps from -140 nm (top left) to $+10$ nm (bottom right) [29].

between these spots will invariably be slightly different from the correct projected atomic separations for the specific material [31]. Moreover, erroneous image features can also be anticipated to occur at structural discontinuities such as interfaces [32]. Such interference lattice-fringe images from thicker specimen regions should thus be considered as nothing more than interesting coincidental effects and they should never ever be interpreted in terms of atomic positions.

2.4. Instrumental Parameters

In common with other sophisticated pieces of equipment, the HRTEM has an array of adjustable parameters that must often seem almost overwhelming to the newcomer. However, only a small subset must be known with any degree of accuracy. Experimental determination of these instrumental parameters is well documented elsewhere [24], so these details are not reproduced here. For atomic-resolution imaging, it is essential that the accelerating voltage be fixed and highly stable, preferably to within one part per million (ppm), but the exact value is relatively unimportant. A similar situation holds for the current of the objective lens, which will determine its focal length and several other imaging parameters, including C_s, the spherical aberration coefficient. In practice, it is highly desirable to operate with a fixed objective lens current, due to the extreme sensitivity of several adjustable parameters such as the incident-beam tilt alignment and the objective lens astigmatism to the exact current setting. The objective lens current should thus be monitored continuously, and when the sample is tilted or another field of view is selected, the sample height should be adjusted rather than altering the objective lens current. Knowledge of the objective lens defocus is critical to image interpretation because of the extreme sensitivity of the image appearance to the defocus value. The highly characteristic image appearance of complicated structures at the optimum defocus can be recognized in some special cases, but for unknown aperiodic features such as a dislocation or grain boundary, some other means for selecting defocus is required. It has become increasingly common to generate a through-thickness-through-focus tableau of images, at least of the perfect crystal structure, before commencing microscopy [33]. Some electron microscopists have recommended recording a focal series of images with pre-calibrated focal step sizes, which thereby reduces somewhat the subjective element of the matching process [34]. Methods based on cross-correlation [35], and non-linear least squares [36] have been developed for matching experimental and simulated images, and these can provide determination of both defocus and local specimen thickness on a local scale.

At extreme resolution limits, several higher-order objective-lens aberrations have a dominant role in determining the overall image integrity. Two-fold image astigmatism is well-known to result in focal length differences in orthogonal image directions, with a corresponding effect on the image appearance, but its presence can usually be detected by reference to the image of a thin amorphous material, such as carbon or germanium: Fourier transformation reveals a set of concentric rings, that are more or less elliptical depending on the amount of astigmatism present [37]. The successful implementation of image reconstruction schemes involving either focal series [26] or off-axis electron holography [38] requires knowledge of the spherical aberration coefficient with an error of no greater than 1%. The graphical method of optical diffractogram analysis [37] is no longer sufficiently accurate, and alternative methods are required in order to reach the desired level of accuracy [39]. The third-order aberrations of axial coma and three-fold astigmatism are more difficult to detect and quantify because their influence is only clearly visible in images of amorphous materials recorded with tilted incident illumination [40, 41]. Detection and correction of third-order aberrations is discussed further in Section 3.4.

2.5. Further Requirements

There are additional requirements that must be satisfied before the atomic-scale resolution of the HRTEM can be fully utilized for characterizing nanostructured materials. The sample region being imaged must be sufficiently thin, typically meaning that its thickness should be on the order of 15 nm or less depending on its composition. Otherwise, multiple electron scattering effects become significant, and the very fine details present in the image are not likely to be interpretable in terms of atomic arrangements, even with the assistance of image simulations. Since the final recorded image represents a two-dimensional projection of the crystal structure, the incident electron beam must be aligned closely with a major zone axis of the crystal. The tolerance for adjustment of the crystal alignment to avoid overlap of projected atomic columns obviously becomes more demanding for thicker crystals and smaller column separations [42]. The examination of crystalline defects must be confined to planar faults, such as twin boundaries and stacking faults, and linear defects, such as dislocations, which are aligned so that they are parallel with the incident beam direction. The defects should also be periodic through the entire projected structure. Inclined faults and curved grain boundaries will not in general be amenable to fruitful study by high-resolution imaging (nor any other imaging technique). Despite these various constraints, there have been many, many successful studies where the HREM has proven indispensable in evaluating nanostructured materials. The highly selective group of examples briefly described below is chosen to be illustrative of the myriad possibilities. The need for complementary information from macroscopic measurements, including electrical and optical properties, as well as other types of microscopy, should not be overlooked.

2.6. Milestones

The performance of the "high" resolution electron microscope first exceeded that of the optical microscope in the mid-1930s [43], but direct correspondence between lattice images and projected crystal structure of large-unit-cell block oxides was not achieved for many years [44, 45]. Improved instrumentation eventually led to directly interpretable information about atomic arrangements for metals, ceramics and semiconductors. Individual atomic columns in a small gold particle were resolved [46], and several high-voltage HRTEMs surpassed the 0.2 nm interpretable resolution limit on a routine basis in the early 1980s [47–50]. Intermediate-voltage HRTEMs that could attain this performance level regularly also became available commercially [31, 51, 52]. The highly coherent illumination available with 300-kV FEG TEMs facilitated the achievement of instrumental resolutions closely approaching 0.1 nm [53], and image interpretation to the same level was achieved by through-focal series reconstruction [26] and off-axis electron holography [38]. The latest generation of high-voltage (1–1.5 MV) HRTEMs have succeeded in closely approaching the long-sought-after goal of 0.1 nm without reversals in the contrast transfer function [54–56], enabling direct image interpretation without the need for *a posteriori* image processing. Significant information transfer beyond the first zero crossover of 0.105 nm was clearly demonstrated by the 1.25-MeV HRTEM in Stuttgart [57]. Sub-Ångstrom electron microscopy to resolutions of better than 0.09 nm has since been achieved using exit-wave retrieval [58, 59] and also using aberration-corrected annular-dark-field imaging with a scanning transmission electron microscope [60].

3. APPLICATIONS OF HRTEM

Imaging with the HRTEM enables individual atomic columns to be resolved in most inorganic materials, making it possible to determine the atomic-scale microstructure of lattice defects and other inhomogeneities. Structural features of interest include planar faults such as grain boundaries, interfaces and crystallographic shear planes, linear faults such as dislocations and nanowires, as well as point defects, nanosized particles and local surface morphology. Additional information can be extracted from high-resolution studies, including unique insights into the controlling influence of structural discontinuities on a range of physical and chemical processes such as phase transformations, oxidation reactions, epitaxial growth and catalysis. The HRTEM has impacted many scientific disciplines, and the technique has generated a vast scientific literature, far too extensive to be reviewed in any detail. Our intent here is to describe,

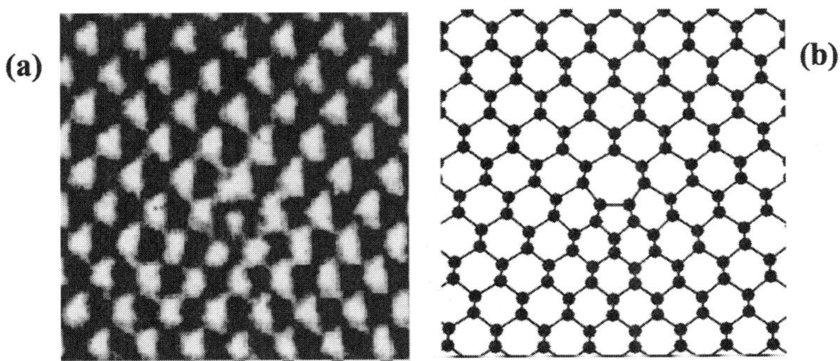

Figure 3. (a) HRTEM micrograph showing core structure of symmetrical Lomer edge dislocation at Ge/Si(001) heterointerface, as recorded with 1.25-MeV atomic-resolution electron microscope; (b) corresponding structural model [65].

albeit very briefly, some representative examples that illustrate the range of applications, while additional information and many more examples can be found elsewhere [1–13].

3.1. Semiconductors

The characterization of semiconductors has been a highly active and fruitful area of research for HRTEM. Because of the relatively large unit-cells of elemental and compound semiconductors, ranging from 0.54 nm (Si) to 0.65 nm (CdTe), high-resolution lattice-fringe images are easily obtained in the <110> orientation. However, individual atomic columns are not separately resolved unless the microscope resolution allows {004}-type reflections to contribute to the imaging process [61]. True atomic imaging for elemental Si and Ge was demonstrated in <100>, <111> and <013> orientations under carefully chosen imaging conditions using a 400-keV HRTEM [62]. Discrimination between atomic species on the basis of image spots of different intensity can be achieved for compound semiconductors with sufficiently different atomic numbers, which is particularly useful for analysing interfaces between dissimilar materials.

Dislocations: Individual atomic columns have not actually been resolved in almost all HRTEM structural studies so far reported. Extensive comparisons with simulated images for various models of Lomer edge dislocation cores in Si and Ge enabled the locations of atomic columns to be deduced to within ~0.025 nm [63]. HRTEM analysis of dissociated 60° dislocations in CdTe showed that alternative structural models (glide set and shuffle set) could be clearly differentiated, but again without resolution of individual atomic columns [64]. By taking advantage of the improved microscope resolution that has recently become available, more accurate modelling of dislocation core structures in semiconductors should nowadays become possible. Figure 3 shows an example, recorded with the 1.25-MeV HRTEM in Stuttgart, of a symmetrical Lomer edge dislocation at a Ge/Si(001) heterointerface, together with the structural model derived directly from the HRTEM image appearance [65].

Figure 4. Cross-section of $CoSi_2$ nanowire grown by reactive epitaxy on Si(100) surface at 750°C [74].

Interfaces: Many semiconductors have grain boundaries (GBs) and heteroepitaxial interfaces that are almost atomically flat and can be aligned edge-on to the incident beam direction, with the crystals on both sides of the boundary oriented to a low-index zone axis. The Ge (130) $\Sigma = 5$ GB was studied in [001] and [013] projections, so that a complete three-dimensional crystallographic analysis could be performed [66]. The Ge (112) $\Sigma = 3$ GB was shown to have a $c(2 \times 2)$ periodic supercell of the geometrical coicidence lattice of the boundary [67], and the rotation-twin nature of a GaP $\Sigma = 3$ $\{111\}$ twin boundary was unambiguously identified from the characteristic image features [57].

Heterointerfaces are of immense practical importance. Interface roughness can be made apparent by choosing appropriate thickness and defocus values to emphasize contrast differences between materials, and composition gradients can be assessed from the abruptness with which the two characteristic contrast motifs terminate at the interface. For materials with large lattice mismatch, significant image features can usually be interpreted without resorting to special defocus/thickness combinations. Examples of large misfit systems where the atomic structure of interfacial misfit dislocations have been successfully determined include the GaAs/Si(001) interface [68], the CdTe(001)/GaAs(001) interface [69], and the CdTe(111)/GaAs(001) interface [70]. For silicide-silicon interfaces, the atomic facetting of asymmetrical twins and asymmetrical "hetero"-twins was investigated [71], and structural studies have been reported for various $CoSi_2$ (A, B)/Si(111) interfaces [72, 73]. Figure 4 shows an interesting example of a $CoSi_2$ nanowire formed by self-assembled epitaxial growth on Si(100) substrate at 750°C [74].

The Group III-nitrides of AlN, GaN and InN have potential applications in short wavelength optoelectronic devices, based on their wide bandgaps which range from 1.9 eV (InN), to 3.4 eV (GaN), to 6.2 eV (AlN), and their excellent thermal properties make them ideal candidates for high-temperature and high-power devices [75]. However, the lack of substrate materials that are both lattice- and thermally-matched

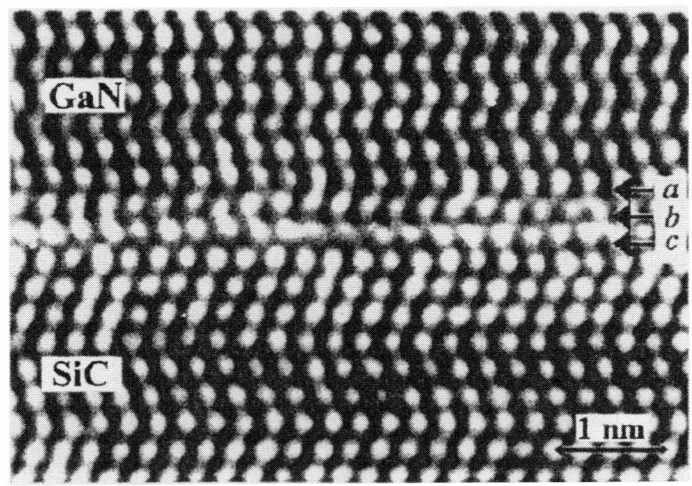

Figure 5. High-resolution electron micrograph showing cross-section of GaN/SiC interface. Analysis of lattice spacings along planes labeled *a*, *b*, and *c* confirmed interface abruptness [76].

represents a serious obstacle to ongoing research. Thus, there is much interest in understanding the atomic structure at the substrate-nitride interface. Figure 5 shows a high-resolution image of an GaN/SiC interface, recorded at the optimum defocus so that atomic-column projections have black contrast. Image analysis established that the GaN/SiC interface was atomically abrupt (between the planes labelled *a* and *b*), and that atomic arrangements across the interface primarily consisted of N bonded with Si, but with some Ga bonded with C in order to maintain charge balance [76].

3.2. Metals

Defects in metals pose greater challenges for characterization using HRTEM because of reduced unit-cell dimensions. Atomic imaging is restricted to comparatively few low-index zone axes even with the latest generation of instruments. High-resolution imaging requires very thin foils but substantial atomic rearrangements, especially relaxation in the vicinity of lattice defects and crystal surfaces, are liable to alter the defect structure so that it becomes atypical.

Grain boundaries and interfaces: High-resolution observations of symmetrical Au <110> tilt GBs established the presence of different recurring structural units [77]. Structural characterization of an Al [100] 45° twist plus 17.5° tilt GB revealed that this asymmetrical GB was composed of a mixture of two basic structural units [78]. Observations of a Mo bicrystal revealed the presence of both $\Sigma = 25$ and $\Sigma = 41$ structural units along the GB [79]. Bi segregation was reported to modify the structure of grain boundaries in Cu [80], and GBs became facetted and an ordered Bi layer was observed at a $\Sigma = 3$ twin boundary [81]. Figure 6 shows an example of an Al 6° [001]

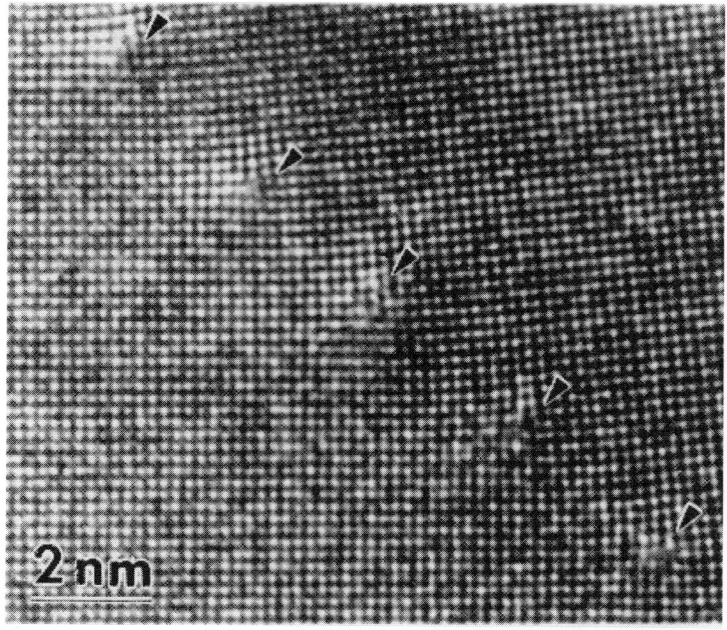

Figure 6. Atomic-resolution electron micrograph of Al 6° [001] symmetric tilt grain boundary with misfit accommodation by [110]/2 edge dislocations (arrowed). Each black spot corresponds to projection of individual Al atomic column [82].

symmetric tilt GB with prominent 1/2<110> edge dislocations that accommodate the tilt misalignment [82].

Structural modelling based upon atomistic simulations has accompanied many HRTEM GB studies. The incoherent Al {112} twin GB was observed [83], and comparisons were made with structures calculated using the Embedded Atom Method (EAM). For the Al $\Sigma = 9$ (221)[110] GB, glide-plane and mirror-plane symmetric structures were observed to alternate periodically along the boundary: the predictions of the various atomistic modelling approaches could only be distinguished for one of these two structures but not the other [84]. The atomic structure of the Nb (310) twin boundary was generated using interatomic potentials derived from several approaches but only one, the so-called model-generalized pseudopotential theory, successfully predicted the mirror symmetry later observed in experimental images [85]. For the Nb $\Sigma = 25$ (710)/[001] twin, a multiplicity of stable, low-energy structures were predicted by EAM: four possibilities out of 13 remained after careful comparison with experimental micrographs [86]. Molecular dynamics simulations of the Ag $\Sigma = 3$<110> (211) twin boundary predicted a thin boundary phase having the rhombohedral 9R structure, and this prediction was confirmed by experimental HRTEM micrographs [87]. Figure 7 is an experimental image showing the existence of the same rhombohedral phase at an incoherent Cu $\Sigma = 3$ (211) boundary [88].

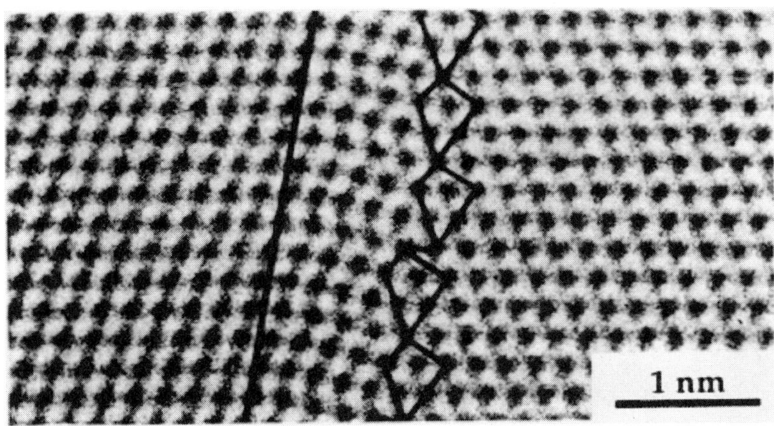

Figure 7. Atomic-resolution electron micrograph of Cu $\Sigma = 84°$ grain boundary showing presence of predicted rhombohedral 9R phase located between large-angle grain boundary at right (structural units outlined) and small-angle boundary at left. Black spots represent positions of individual Cu atomic columns [88].

Atomic-level structural investigations of intermetallic alloys and several metal/metal systems have been reported. For example, the NiAl $\Sigma = 5$ [001](310) GB was imaged by high-resolution electron microscopy and analyzed with the assistance of image simulations [89]: changes in local stoichiometry and a rigid body translation along but not normal to the boundary were inferred. A quantitative study of the NiAl $\Sigma = 3$ (111) GB enabled atomic positions at the boundary core to be determined with an accuracy of ~0.015 nm [90].

Dislocations: Determination of the atomic structure of dislocation cores is simplified by the presence of only one atomic species but the core may not be stable in metal foils thin enough for atomic-resolution viewing due to the possibility of core spreading or even defect motion in the form of glide to the thin foil edge [91]. Moreover, structural rearrangements during high-resolution observations of thin metal foils have been observed by many workers (see, for example, [92]). The 1/3<1120>{1010} edge dislocation in α-Ti was determined to have a planar elongated core structure [93]. Atomic modelling of the core structure of **a**<100> and **a**<110> dislocations in the intermetallic alloy NiAl revealed that the former had large elastic strain fields but were usually undissociated, whereas the latter either decomposed into other types of dislocations or climb-dissociated into two partial dislocations [94]. An investigation of dislocation core structures in the ordered intermetallic alloy TiAl has also been reported [95].

3.3. Oxides and Ceramics

Most ceramic materials consist of two or more atomic species and only rarely can the separate elements be discriminated in HRTEM micrographs. The image of the 6H polytype of SiC in the (1120) orientation shown in Fig. 8 is a particularly noteworthy

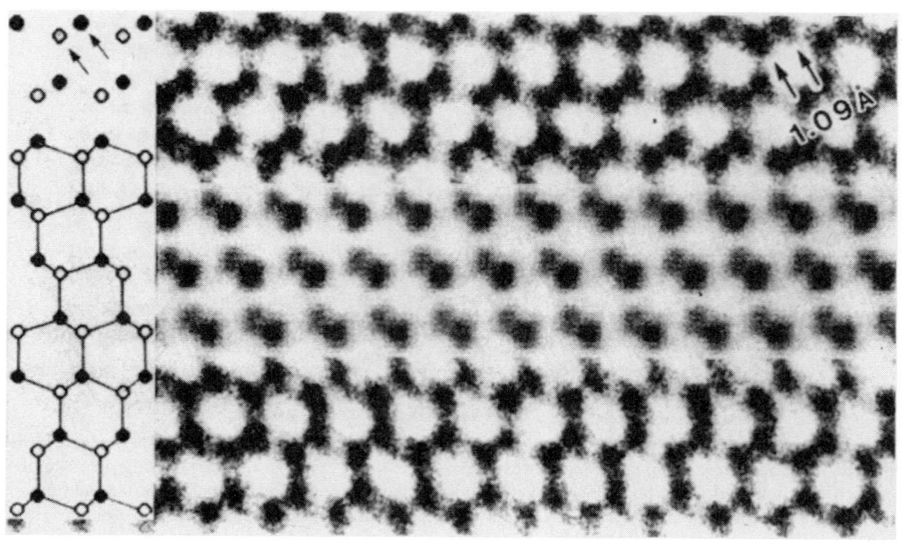

Figure 8. Atomic-resolution electron micrograph of 6H SiC polytype recorded with 1250-keV HRTEM installed in Tokyo. Image simulation (inset) confirms separation of individual Si and C atomic columns [55].

example since the closest separation of the clearly resolved Si and C columns is only 0.108 nm [55]. Structure imaging in the electron microscope originated with large-unit-cell block oxides, although the image interpretation was initially based on contrast features corresponding to the location of tunnels [45]. Improvements in HRTEM resolution to better than 0.2 nm led to oxide images in which the black spots visible at optimum defocus were interpretable in terms of atomic column positions. It is then possible to deduce directly the detailed atomic structure of complicated shear defects and precipitation phenomena observed in doped and slightly reduced oxides, as represented by the example of a pentagonal bipyramidal defect as shown in Fig. 9 [96].

Grain boundaries. Symmetric tilt GBs in NiO bicrystals had a multiplicity of distinct structural units [97], and asymmetric structural models based on experimental micrographs, differed from symmetric models derived from theoretical considerations. Observations of the $\Sigma = 5$ (210)/[001] symmetric tilt GB in yttrium aluminum garnet were compared with model atomic structures [98]. Structural investigations of a near-$\Sigma = 5$ (210) GB in TiO_2, rutile, revealed a stepped boundary with well-defined lattice dislocations at the steps, and extended, flat terraces that consisted of $\Sigma = 5$ (210) segments with mirror glide symmetry [99]. Observation of an undoped $SrTiO_3$ $\Sigma = 5$ (130) symmetrical tilt GB revealed that it was composed of repeating structural units [100]. Examination of a 25° [001] tilt boundary in $SrTiO_3$ provided an initial structural model, and bond-valence sum calculations based on electron-energy-loss spectroscopy at the boundary were then used to refine the O atom positions [101]. A combined structural

Figure 9. (a) Atomic-resolution electron micrograph of nonstoichiometric (W, Nb)$O_{2.93}$ showing pairs of pentagonal bipyramidal columnar defects. (b) Corresponding structural model. Occupied tunnel sites are located by direct visual inspection [96].

and spectroscopic investigation of coherent (111) twins in $BaTiO_3$ enabled a modified structural model of the boundary to be proposed [102]. The atomistic structure of 90° domain walls in ferroelectric $PbTiO_3$ thin films was investigated by using digital processing to determine lattice parameter variations across the walls [103].

Interfaces. Knowledge of metal-ceramic interfaces promotes a better understanding of bulk mechanical properties. For thin layers of Nb and Mo deposited on R-plane sapphire, misfit dislocations were offset from the interface for Nb films whereas they were localized very close to the interface for Mo films [104]. Interactive digital image matching was used in a comprehensive study of the Nb/sapphire interface: and translation vectors were determined with a precision of ~0.01 nm [105]. The atomic core structure of the dislocations was determined in another study of the same interface [106].

Ceramics. Most ceramics are close-packed materials with tetrahedral bonding so that atomic resolution is very difficult to obtain. Careful examination of {100} nitrogen platelets in diamond led to the development of a novel "nitrogen-fretwork" model [107], while later observations led to a more refined model [108]. Interfacial structures and the defects that occurred during heteroepitaxial growth of β-SiC films on TiC substrates have been characterized [109]. The atomic structure of Ti(C,N)-TiB_2 interfaces was investigated using a combined microscopy-simulation study [110].

3.4. Surfaces

Several different TEM configurations provide atomic-scale information about surfaces [111]. In surface profile imaging [112], the electron microscope is operated in the normal HRTEM mode and the optimum defocus image displays the surface profile at the structural resolution limit. Gold received much early attention, mainly because its surface was relatively inert and electron irradiation readily caused desorption of surface

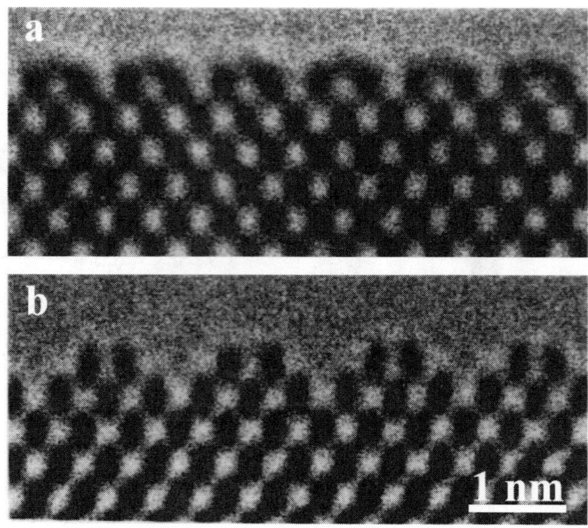

Figure 10. Surface profile images showing reconstructed CdTe (001) surface at different temperatures: (a) 2 × 1 at 140°C; (b) 3 × 1 at 240°C [127].

contamination overlayers, and also because its large interatomic spacings and high atomic-column visibility made it easy to characterize atomic rearrangements [113]. Thus, profile imaging was first applied to observations of a 2 × 1 reconstruction of the gold (110) surface [114]. Adsorbed Bi atoms have been identified on reconstructed Si(111) surfaces using an ultrahigh-vacuum (UHV) transmission electron microscope [115], and the overgrowth of Au on ZnTe has been investigated [116]. In later experiments involving *in situ* Au evaporation under UHV conditions, the ×5 superperiod associated with the corrugated (5 × 28) reconstructed Au surface was observed [117].

Oxide surfaces are generally more straightforward to prepare for profile imaging [118], but surface modification may occur under intense electron irradiation due to electron-stimulated desorption of oxygen from near-surface regions [119, 120]. A complex spinel catalyst developed surface rafts, identified as ZnO, following prolonged use as an oxidation catalyst [121]. Surface profile images were central to studies of terbium oxide [122], Eu_2O_3 [123], and $\beta-PbO_2$ [124]. A direct correlation was made between the exposed surface structure of V_2O_5 oxide catalysts and the catalyst selectivity [125].

Profile images of semiconductor surfaces can only really be considered as valid when the surface has been cleaned inside the microscope. A novel 1 × 1 dimer reconstruction of a Si (111) surface was reported after the sample had been heated *in situ* to 1000°C [126]. As shown by the surface profile images in Fig. 10, the CdTe (001) surface undergoes a reversible phase transformation from a 2 × 1 structure at temperatures below about 200°C to a 3 × 1 structure at higher temperatures [127]. The atomic columns at the surface were located to within ∼0.01 nm, and structural models were developed: the 2×1 was determined as being Cd-stabilized whereas the 3×1 was found to be Te-stabilized.

3.5. Dynamic Events

Dynamic events such as phase transitions, defect motion and interface dynamics can be documented directly at the atomic-scale by means of a low-light-level TV camera attached to the base of the microscope without any loss of resolution. However, image recording is not useful for quantitative image analysis due to reduced dynamic response of the camera, as well as image distortions. Further information about dynamic *in situ* studies can be found in the chapter by Sharma and Crozier.

Surfaces. Surface profile imaging revealed movement of Au atomic columns across extended Au (110) surfaces [113], and rapid structural changes and the existence of "atom clouds" extending out from Au surfaces were reported [128]. Hopping of Au columns between surface sites on small Au particles was recorded at TV rates [129, 130], and similar hopping of Pt columns was also studied [131].

Small particles. Under strong electron irradiation, small metal particles (<8–10 nm) rapidly change their shape and orientation. Structural rearrangements from single crystal twinned to multiply-twinned were observed in small Au particles [130, 132]. Structural rearrangements as well as surface hopping were also documented in small particles of Pt [131] and Rh [133]. In the case of small Ru particles (~2.5 nm), the internal stacking changed between cubic-close-packed (ccp) and hexagonal-close-packed (hcp) which is the stable bulk form of Ru [134]. The term "quasi-melting" has been used to describe the structural fluctuations which have been reported for Au clusters supported on pillars of MgO [135].

4. CURRENT TRENDS

4.1. Image Viewing and Recording

Image viewing or recording should not normally be expected to affect resolution limits of the HRTEM but the recording media must still be properly optimized to ensure efficient operation. Ideally, every incident electron should be detected but readout noise and shot noise could affect the overall recording efficiency. The low-light-level TV camera has steadily evolved to the point where it has replaced the fluorescent screen for most image viewing. A camera is easily attached beneath the viewing chamber without affecting microscope performance, and high brightness images can be obtained even for very low exposure levels. Enlarged specimen detail is easily visible on a TV monitor, and permits image focussing and rapid astigmatism correction, while dynamic events within the sample can be viewed and recorded at TV rates. Nevertheless, high dark current and amplifier noise limit dynamic range, the input-output linearity is poor, and the number of resolvable picture elements is restrictive. The intensified TV camera is seriously inadequate as a recording medium for quantitative HRTEM studies.

The development of the slow-scan CCD camera has led to a revolution across the entire field of electron microscopy, and it has particular value for quantitative

HRTEM applications. The potential of the CCD camera had long been recognized [136], especially its sensitivity, wide dynamic range and overall usefulness for the electron microscopist [137]. However, dynamic viewing at TV rates seems unlikely to be achieved without sacrificing image quality. The imaging properties of the slow-scan CCD camera, the intensified TV camera and the photographic plate have been compared [138], and a comprehensive overview of characterization methods and important detection parameters, such as modulation transfer function and input-output linearity, for CCD cameras has been published [139]. The fixed location of the CCD camera enables geometric distortions of the imaging system to be accurately compensated [140], which is advantageous for extracting quantitative phase information during off-axis electron holography. On-line acquisition of a digital signal also enables automated microscope control or "autotuning" [137], as described in the following section.

4.2. On-Line Microscope Control

As resolution limits improve, it becomes progressively more difficult for an operator to adjust focus, to correct the objective lens astigmatism and to align the incident beam direction (coma-free alignment) with the accuracy that is required to ensure that image interpretation is not compromised. Accordingly, attention has been directed towards on-line computer control or "autotuning" of the microscope [137, 141]. The desired end-product of routine, top-quality micrographs should then leave the microscopist free to concentrate on solving the particular materials problem at hand.

Several criteria have been proposed as a basis for autotuning, including diffractograms, beam-tilt-induced image displacement, and contrast analysis for amorphous materials [24]. These methods rely upon signals that are fed into a computer, which analyses the data and then makes appropriate on-line adjustments to the microscope controls. Contrast analysis, the first autotuning method to be implemented successfully [141], locates a global minimum of the image variance as the computer successively iterates through the focus, beam tilt controls (in two orthogonal directions) and two objective lens stigmator controls. High accuracy for beam alignment and astigmatism can be achieved but the method is inefficient in terms of dose and inapplicable in the absence of amorphous material.

The latest variant of autotuning to be implemented [137], which is based on automated diffractogram analysis (ADA), utilizes diffractograms computed from slow-scan CCD images of amorphous material. The method works best at high image magnification, but is again inapplicable in the absence of an amorphous specimen region. The set of diffractograms in Figure 11, recorded (a) before, (b) after one cycle, and (c) after two cycles, of autotuning, illustrate the improvements that can be obtained with a 200-keV FEG-TEM. Astigmatism correction and focus adjustment to within 1 nm can be achieved, and the beam-tilt alignment is better than 0.1 mrad, which is beyond the adjustment/detection limit of the available instrumentation. Autotuning should be regarded as an essential procedure during preparation of the microscope for HRTEM observation.

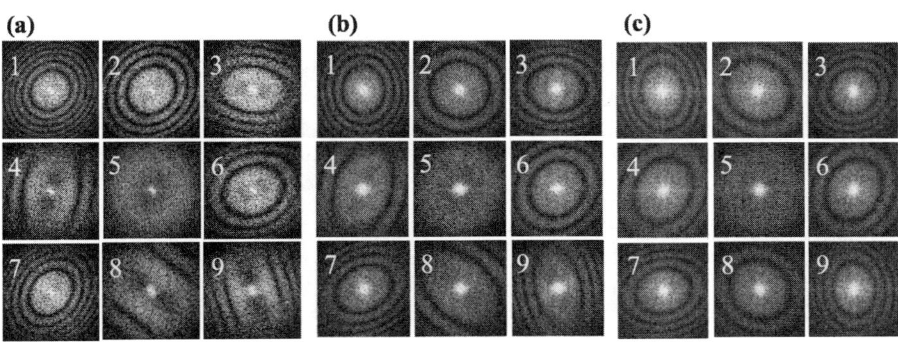

Figure 11. Diffractogram tableaus from thin amorphous carbon film recorded (a) before; (b) after one cycle; (c) after two cycles, autotuning sequence based on automated diffractogram analysis (ADA). Initial tilt misalignment reduced successively from 4 mrad (a) to 0.4 mrad (b) and <0.1 mrad (c).

4.3. Detection and Correction of Third-Order Aberrations

All electron lenses suffer from performance-limiting aberrations that must be corrected when possible, or at least quantified and accounted for during image interpretation. Two-fold astigmatism and third-order axial coma can be corrected during autotuning as discussed in the previous section. Detection and correction of spherical and chromatic aberration, as well as three-fold objective-lens astigmatism become much more critical as microscope resolution limits extend towards and beyond the 0.1-nm barrier [142–144].

Spherical aberration is unavoidable in rotationally symmetric electron lenses [145, 146]. Elimination of C_s (and C_c) by a suitable combination of multipole elements attracted much attention over many years but all early correction attempts failed, due primarily to insufficient electrical stability and lack of alignment precision [146]. In recent years, a double-hexapole C_s-corrector system attached to a 200-keV FEG-TEM has enabled the normal 0.23 nm interpretable resolution limit of the instrument to be surpassed, with a level of ~0.13 nm eventually being achieved after adventitious instabilities were removed [147]. Concurrently, a corrector system, incorporating multiple quadropole-octopole elements, has been applied to the probe-forming lens of a 100-keV scanning transmission electron microscope, and probe sizes of less than 0.1 nm can be achieved [148]. Note that both approaches to aberration correction are completely dependent on computer analysis of the imaging conditions and high-precision feedback to the numerous deflector and corrector power supplies. Some initial experiences with aberration-corrected HRTEM are described later in Section #4.5.

Knowledge of the chromatic aberration coefficient of the objective lens is not critical for high-resolution imaging since C_c does not affect the interpretable resolution, and the temporal coherence envelope is determined by an effective focal spread, which can be estimated empirically if required [23]. Nevertheless, C_c does impact temporal coherence, which is performance-limiting for an LaB_6 electron source, so that reduction or correction of chromatic aberration is desirable and worth pursuing. Correction of C_c in a low-voltage scanning electron microscope was achieved [149] but no substantial success in the energy range applicable for HREM (upward from about 100 kV) has so far been reported. An alternative approach to reducing chromatic effects is to locate a monochromator immediately following the electron gun, which could serve to reduce the energy spread to about 0.1 eV [150], but this improvement can only be achieved at the expense of reduced beam current [151].

Three-fold astigmatism was quantified during the first comprehensive study of coma-free alignment [40]. However, it was not until the implementation of the ADA method for autotuning that the disturbing implications of three-fold astigmatism for high-resolution imaging at the 0.10 nm level came to be fully appreciated [41]. Like axial coma, three-fold astigmatism is effectively invisible in an axial bright-field image or the corresponding diffractogram since its basic effect is an asymmetrical shift of phase information. However, image simulations have shown that the impact on very-high-resolution images of crystalline materials can be highly detrimental, depending on the relative orientation of the three-fold astigmatism and any crystal symmetry directions [143, 144]. The magnitude of three-fold astigmatism can be estimated using diffractograms from four mutually orthogonal beam-tilt directions, and correction can be achieved using a pair of sextupole stigmator coils located near the back focal plane of the objective lens: Reasonable adjustments can be reached using fixed correction currents in existing objective stigmator coils provided that the coils are energized separately rather than being wired in pairs [152]. Microscopists purchasing new microscopes should insist that this highly desirable correction of three-fold astigmatism be done during installation and commissioning of their instrument.

4.4. Quantitative HRTEM

A major attraction of HRTEM is the possibility that atomic arrangements at local irregularities such as dislocations and interfaces can be determined to very high accuracy, some times closely approaching 0.01 nm. However, the refinement process is heavily demanding both of the microscopist and the instrument, as well as being computing-intensive and often extremely time-consuming with much trial-and-error parameter fitting. Prior knowledge (or elimination) of essential experimental parameters facilitates the goal of interactive structure refinement. Correction of the three-fold astigmatism at the time of microscope installation should reduce its effect to the level where it has almost negligible influence on the image. Routine application of autotuning, which implies the availability of an online CCD camera as well as computer control of relevant power supplies, would remove two-fold astigmatism (two parameters) and coma/beam tilt (two more parameters) from consideration. Thus, additional resources must be committed in order to make progress towards the ultimate goal of real-time, interactive structure refinement. For aperiodic structural features, it appears that this goal will not be easily attained since some parameters can only be optimized iteratively [105], although it is possible to automate considerably parts of the iteration process [153].

Historically, the determination of defect structures has relied upon qualitative comparisons between experimental micrographs and image simulations that were based on various alternative structural models. The acceptability of a specific structural model was generally considered as being enhanced when an image match was achieved for multiple members of a focal series (see, for example, refs. 85, 107). An alternative approach has been to overlay projected atomic column positions on the experimental and/or simulated images [66], and other studies have utilized superposition [79] or subtraction [154] of simulated and experimental images during refinement. A non-linear least-squares optimization approach was used to refine atomic positions at Nb GBs [155].

Attention has been given to the issue of quantifying the "goodness of fit" between experimental micrographs and the corresponding simulated images, as derived from postulated structural model(s). The so-called reliability or R-factors of X-ray or neutron diffraction studies refer to the contents of the entire unit cell, unlike aperiodic defects such as grain boundaries or dislocation cores studied by HRTEM where such well-defined discrete entities do not generally exist. Thus, alternative image agreement factors (IAFs) have been proposed and utilized at different times, and a useful summary of those most commonly used can be found in the Appendix of ref. [105]. However, it should be noted that slightly different results can be obtained depending on which IAF is used, possibly because different IAFs weight bright and dark contrast areas differently. It would be helpful if defocus and other microscope parameters were removed from the actual refinement process by instead using the complex exit-surface wavefunction of the specimen as the basis for comparison between simulation and experiment. At that stage, a χ^2 goodness-of-fit criterion would be an appropriate test of overall convergence [156]. An additional benefit of determining the exit-surface wavefunction would be the availability of both phase and amplitude information about electron scattering by the sample. The possibilities for determining unknown object structures would be enhanced because this wavefunction directly reflects electron scattering by the object.

4.5. Aberration-Corrected HRTEM

Compensation of the spherical aberration of the objective lens offers the exciting prospect of directly interpretable image detail extending out to the HRTEM information limit without the need to unscramble the artefactual detail normally caused by TF oscillations. One additional benefit of the aberration-corrected HRTEM is that image delocalization, which is a major source of imaging artefacts at discontinuities such as interfaces and surfaces, is markedly reduced [147, 157]. Another benefit is that other imaging aberrations are also substantially reduced, which should simplify the process of exit-wave retrieval using through-focal reconstruction [158], and also alleviate the accuracy needed for sample tilting. However, it is relevant to reiterate here that conventional TF theory for the HRTEM refers to phase contrast imaging, so that when the C_s value is reduced exactly to zero it would be necessary to use a projected charge density approach for image interpretation [159]. In practice, having an adjustable spherical aberration can provide additional flexibility to the microscopist interested in solving particular materials problems. For example, a small negative C_s with a slightly overfocus condition enabled imaging of oxygen atom columns in a perovskite ceramic [160]. It is clear that much experimentation is still required to examine the full range of possibilities for aberration-corrected HRTEM, and possibly reach some consensus about standard imaging conditions.

5. ONGOING PROBLEMS

5.1. The Stobbs' Factor

It has become increasingly obvious, and highly disconcerting, that there are substantial, seemingly inexplicable, discrepancies between the contrast levels of experimental and simulated images as well as diffracted beam intensities [161]. These differences were not apparent in earlier qualitative studies using photographic film when there was no simple measure of absolute intensity, and the contrast range in image simulations could easily be scaled to match that of the experimental micrographs. Initial quantitative studies revealed major differences in image contrast, sometimes by factors as large as 6 to 8 [161], although factors of about three are reported to be more typical [162] Obvious sources of error, such as contributions from inelastic scattering and surface contamination overlayers, have not adequately accounted for these contrast differences [163], stimulating further concerted efforts to identify the origin(s) of what has come to be called the Stobbs' factor [164]. Thermal diffuse scattering has been at least partly implicated by recent experiments combining energy-filtered imaging with off-axis electron holography [165], leading to suggestions for further experiments that might finally be definitive [164].

5.2. Radiation Damage

Interactions between the highly energetic electron beam and the sample within the electron microscope are always likely to result in permanent structural modification. There are two basic types of electron beam damage [166]: radiolytic processes (sometimes known as "ionization damage") involve electron-electron interactions and affect most covalent and ionic solids; and direct atomic "knock-on" displacements, which occur above characteristic energy thresholds. An electron energy of 400 keV is sufficient to cause bulk displacements in elements as heavy as copper ($Z = 29$). Moreover, because of reduced binding energy, the energy needed for the activation of surface sputtering or for diffusion at defects or interfaces is considerably less than bulk values. For example, under continuous electron irradiation (400-keV, 5–15 A/cm^2), electron-stimulated desorption causes depletion of oxygen from the near-surface region of maximally valent, transition-metal oxides, leaving thin layers of reduced oxide covering the surface [119]. In a quantitative comparison of difference images, preferential damage was documented to occur at a Cu/sapphire interface [167], which limited the useful viewing time to a total of ~10 mins (1250 keV, magnification of 600,000×, specimen current density of ~1.6 A/cm^2).

Thus, the electron microscopist must be continually alert to the likelihood that the sample morphology has altered, probably irreversibly, during HRTEM imaging and especially microanalysis. Higher viewing magnifications as a means to improve signal statistics result in higher current densities, which will mean higher damage rates since the current density at the sample increases with the square of the magnification when the screen brightness is maintained at a constant level. For quantitative studies, the image magnification and the beam current density should be limited whenever possible, and the region of sample studied must be periodically checked for signs of structural change. A cautious microscopist will monitor the sample appearance during observation and thus erroneous results can be discounted once changes start to become apparent [168]. Digital micrographs of an NiAl $\Sigma = 3$ (111) twin boundary, recorded with a slow-scan CCD camera in a high-voltage HREM, were compared at regular intervals as a means of establishing the useful observation time [90]. Finally, it should be apparent that structural change is even more likely to occur when the highly intense, focussed probe of the STEM is used for nanoscale microanalysis, and this possibility should always be monitored.

5.3. Inversion of Crystal Scattering

Inversion of crystal scattering to retrieve the crystal potential in the presence of dynamical scattering is a major unresolved problem. Several methods exist for retrieving the exit-surface wavefunction, and Fourier inversion can be used to extract the crystal potential for very thin samples when the kinematical or weak phase object approximations are valid. However, *ab initio* inversion of crystal scattering to retrieve the crystal potential has received comparatively scant attention over the years. An iterative method based on inversion of the multislice algorithm has been proposed [169], but further work is still needed to explore the applicability of the method, and for extending thickness limits in the case of non-periodic wavefields [170]. An alternative approach based on a simulated annealing algorithm has been explored for a perfect GaAs crystal with zincblende structure [171]. Successful reconstruction was achieved for a thickness of 5.6 nm but not for a thickness of 11.2 nm. Overall, it still holds true that because of the likelihood of multiple solutions to the inverse scattering problem for almost any sample of reasonable thickness, the uniqueness of the inversion process for unknown structures remains an unresolved issue.

6. SUMMARY AND FUTURE PERSPECTIVE

This chapter has provided an overview of HRTEM, with the objective of highlighting some of its applications and achievements, as well as identifying areas of ongoing research and development. The HRTEM enables the atomic structure of interfaces and defects to be determined routinely, reliably and with very high positional accuracy, thus providing better insights into the physical behavior of many nanostructured materials. Experts in the HRTEM field should find ways to ensure that their colleagues who are interested in the properties of these types of materials are given the opportunity and assistance required to capitalize on its attractive possibilities. Meanwhile, further developments in quantification and many novel applications can be anticipated. Online microscope control ("autotuning"), digital recording and computer processing, (almost) real-time structure refinement, and *in situ* environmental electron microscopy are likely areas of concentrated activity. Many challenges remain. The differences between simulated and experimental contrast levels need to be fully explained. Better approaches to inversion of crystal scattering are needed. Operating conditions for aberration-corrected imaging need to be further explored. These issues will surely receive much attention over the next several years.

REFERENCES

1. D. J. Smith, *Rep. Prog. Phys.* 60 (1997) 1513.
2. F. Ernst and M. Rühle, *Current Opin. Solid State & Mats. Sci.* 2 (1997) 469.
3. J. C. H. Spence, *Mater. Sci. Eng.* R26 (1999) 1.
4. Z. L. Wang, *Adv. Maters.* 15 (2003) 1497.
5. W. Krakow, F. A. Ponce, and D. J. Smith (Eds.), *High Resolution Microscopy of Materials,* MRS Symp. Proc. Vol. 139, Materials Research Society, Pittsburgh (1989).
6. R. Sinclair, D. J. Smith, and U. Dahmen (Eds.), *High Resolution Electron Microscopy of Defects in Materials*, MRS Symp. Proc. Vol. 183, Materials Research Society, Pittsburgh (1990).
7. D. Biegelsen, D. J. Smith, and S.-Y. Tong (Eds.), *Atomic Scale Imaging of Surfaces and Interfaces*, MRS Symp. Proc. Vol. 295, Materials Research Society, Pittsburgh (1993).
8. D. J. Smith and J. C. H. Spence (Eds.), *Aspects of Electron Microscopy, Diffraction, Crystallography and Spectroscopy* Ultramicroscopy 52, Nos. 3/4, December 1993.
9. M. Rühle, F. Phillipp, A. Seeger, and J. Heydenreich (Eds.), *High-Voltage and High-Resolution Electron Microscopy* Ultramicroscopy 56, Nos. 1–3, November 1994.
10. M. Sarikaya, H. K. Wickramasinghe, and M. Isaacson (Eds.), *Determining Nanoscale Physical Properties of Materials by Microscopy and Spectroscopy*, MRS Symp. Proc. Vol. 332, Materials Research Society, Pittsburgh (1995).
11. J. C. H. Spence, *Experimental High Resolution Electron Microscopy*, Clarendon, Oxford (1988).
12. S. Horiuchi, *Fundamentals of High Resolution Transmission Electron Microscopy*, North Holland, Amsterdam (1994).
13. D. Shindo and K. Hiraga, *High Resolution Electron Microscopy for Materials Science*, Springer, Tokyo (1998).
14. P. G. Self and M. A. O'Keefe, In: *High-Resolution Transmission Electron Microscopy and Associated Techniques*, P. R. Buseck, J. M. Cowley, and L. Eyring (Eds.), Chapter 8, Oxford University Press, New York (1988).
15. D. Van Dyck, *J. Microscopy* 132 (1983) 31.
16. J. M. Cowley and A. F. Moodie, *Acta Cryst.* 10 (1957) 609.
17. P. Goodman and A. F. Moodie, *Acta Cryst.* 30 (1974) 280.
18. P. A. Stadelmann, *Ultramicroscopy* 21 (1987) 131.
19. P. G. Self, M. A. O'Keefe, P. R. Buseck, and A. E. C. Spargo, *Ultramicroscopy* 11 (1983) 35.
20. J. M. Cowley, *Diffraction Physics*, Third Revised Edition, North Holland, Amsterdam (1995).
21. K.-H. Hanszen, *Adv. Opt. Electron Microsc.* 4 (1971) 1.
22. J. Frank, *Optik* 38 (1973) 519.
23. W. O. Saxton, *Optik* 49 (1977) 51.
24. D. J. Smith, *Adv. Opt. Electron Microsc.* 11 (1988) 1.
25. M. A. O'Keefe, *Ultramicroscopy* 47 (1992) 282.
26. W. Coene, G. Janssen, M. Op de Beeck, and D. van Dyck, *Phys. Rev. Lett.* 69 (1992) 3743.
27. J. M. Cowley and A. F. Moodie, *Proc. Phys. Soc.* 70 (1957) 486.
28. S. Iijima and M. A. O'Keefe, *J. Microscopy* 17 (1979) 347.
29. D. J. Smith, L. A. Bursill, and G. J. Wood, *J. Solid State Chem.* 50 (1983) 51.
30. K. Izui, S. Furuno, T. Nishida, H. Otsu, and S. Kuwabara, *J. Electron Microscopy* 27 (1978) 171.
31. J. L. Hutchison, T. Honda, and E. D. Boyes, *JEOL News* 24E (1986) 9.
32. D. J. Smith, R. A. Camps, and L. A. Freeman, *Inst. Phys. Conf. Ser.* 61 (1982) 381.
33. M. A. O; Keefe, U. Dahmen, and C. J. D. Hetherington, *Mater. Res. Soc. Symp. Proc.* 159 (1990) 453.
34. A. R. Wilson, A. E. C. Spargo, and D. J. Smith, *Optik* 61 (1982) 63.
35. A. Thüst and K. Urban, *Ultramicroscopy* 45 (1992) 23.
36. W. E. King and G. H. Campbell, *Ultramicroscopy* 56 (1994) 46.
37. O. L. Krivanek, *Optik* 45 (1976) 97.
38. A. Orchowski, W. D. Rau, and H. Lichte, *Phys. Rev. Lett.* 74 (1995) 399.

39. W. M. J. Coene and T. J. J. Denteneer, *Ultramicroscopy* 38 (1991) 235.
40. F. Zemlin, K. Weiss, P. Schiske, W. Kunath, and K.-H. Herrmann, *Ultramicroscopy* 3 (1978) 49.
41. O. L. Krivanek, *Ultramicroscopy* 55 (1994) 419.
42. D. J. Smith, W. O. Saxton, M. A. O'Keefe, G. J. Wood, and W. M. Stobbs, *Ultramicroscopy* 11 (1983) 263.
43. E. Ruska, *Zeit. f. Physik* 85 (1934) 580.
44. J. G. Allpress and J. V. Sanders, *J. Appl. Cryst.* 6 (1969) 165.
45. S. Iijima, *J. Appl. Phys.* 42 (1971) 5891.
46. V. E. Cosslett, R. A. Camps, L. A. Freeman, W. O. Saxton, D. J. Smith, W. C. Nixon, H. Ahmed, C. J. D. Catto, J. R. A. Cleaver, P. Ross, K. C. A. Smith, and A. E. Timbs, *Nature* 281 (1970) 49.
47. M. Hirabayashi, K. Hiraga, and D. Shindo, *Ultramicroscopy* 9 (1982) 197.
48. D. J. Smith, R. A. Camps, V. E. Cosslett, L. A. Freeman, W. O. Saxton, W. C. Nixon, C. J. D. Catto, J. R. A. Cleaver, K. C. A. Smith, and A. E. Timbs, *Ultramicroscopy* 9 (1982) 203.
49. K. Yagi, K. Takayanagi, K. Kobayashi, and S. Nagakura, In: *Proc. Seventh Int. Conf. High Voltage Electron Microscopy,* R. M. Fisher, R. Gronsky, and K. H. Westmacott (Eds.), Lawrence Berkeley Laboratory, Berkeley (1983) p. 11.
50. R. Gronsky and G. Thomas, In: *Proc. 41st. Ann. Meet. Electron Microscopy Society of America,* G. W. Bailey (Ed.), San Francisco Press, San Francisco (1983) p. 310.
51. D. J. Smith, J. C. Barry, A. K. Petford, and J. C. Wheatley, *JEOL News* 22E (1985) 2.
52. A. Bourret and J. M. Penisson, *JEOL News* 25E (1987) 2.
53. D. van Dyck, H. Lichte, and K. D. vander Mast, *Ultramicroscopy* 64 (1997) 1.
54. S. Horiuchi, Y. Matsui, Y. Kitami, M. Yokoyama, S. Suehara, X. J. Wu, I. Matsui, and T. Katsuta, *Ultramicroscopy* 39 (1991) 231.
55. H. Ichinose, H. Sawada, E. Takano, and M. Osaki, *J. Electron Microscopy* 48 (1999) 887.
56. F. Phillipp, R. Höschen, Osaki, G. Möbus, and M. Rühle, *Ultramicroscopy* 56 (1994) 1.
57. F. Phillipp, *Adv. Solid State Physics* 35 (1996) 257.
58. M. A. O'Keefe, C. J. D. Hetherington, Y. C. Wang, E. C. Nelson, J. H. Turner, C. Kisielowski, J.-O. Malm, R. Mueller, J. Ringnalda, M. Pan, and A. Thust, *Ultramicroscopy* 89 (2001) 215.
59. C. Kisielowski, C. J. D. Hetherington, Y. C. Wang, R. Kilaas, M. A. O'Keefe, and A. Thust, *Ultramicroscopy* 89 (2001) 243.
60. P. E. Batson, N. Dellby, and O. L. Krivanek, *Nature* 418 (2002) 617.
61. R. W. Glaisher, A. E. C. Spargo, and D. J. Smith, *Ultramicroscopy* 27 (1989) 131.
62. A. Bourret, J. L. Rouviere, J. Spendeler, *phys. stat. sol.* (a) 107 (1988) 481.
63. A. Bourret, J. Desseaux, and A. Renault, *Phil. Mag.* A45 (1982) 1.
64. P. Lu and D. J. Smith, *Phil Mag.* B62 (1990) 435.
65. J. N. Stirman, P. A. Crozier, D. J. Smith, F. Phillipp, G. Brill, and S. Sivananthan, *Appl. Phys. Lett.* (2004) in press.
66. A. Bourret, J. L. Rouviere, and J. M. Penisson, *Acta Cryst.* A44 (1988) 838.
67. A. Bourret and J. J. Bacmann, *Surf. Sci.* 162 (1985) 495.
68. D. Gerthsen, *Phil. Mag.* A67 (1993) 1365.
69. J. E. Angelo and M. J. Mills, *Phil. Mag.* A72 (1995) 635.
70. A. Bourret, P., G. Feuillet, and S. Tatarenko, *Phys. Rev. Lett.* 70 (1993) 311.
71. W.-J. Chen and F.-R. Chen, *Ultramicroscopy* 51 (1993) 316.
72. C. W. T. Bulle-Lieuwma, A. F. de Jong, A. H. van Ommen, J. F. van der Veen, and J. Vrijmoeth, *Appl. Phys. Lett.* 55 (1989) 648.
73. A. Catana, P. E. Schmid, P. Lu, and D. J. Smith, *Phil. Mag.* A66 (1992) 933.
74. Z. He, D. J. Smith, and P. A. Bennett, *Phys. Rev. Lett.* (2004) in press.
75. H. Morkoç, S. Strite, G. B. Gao, M. E. Lin, B. Sverdlov, and M. Burns, *J. Appl. Phys.* 76 (1994) 1363.
76. J. N. Stirman, F. A. Ponce, A. Pavlovska, I. S. T. Tsong, and D. J. Smith, *Appl. Phys. Lett.* 76 (2000) 822.
77. H. Ichinose, Y. Ishida, N. Baba, and K. Kanaya, *Phil. Mag.* A52 (1985) 51.
78. M. Shamsuzzoha, D. J. Smith, and P. Deymier, *Phil. Mag.* A64 (1991) 719.
79. J. M. Penisson, T. Nowicki, and M. Biscondi, *Phil. Mag.* A58 (1988) 947.
80. D. E. Luzzi, *Ultramicroscopy* 37 (1991) 180.
81. D. E. Luzzi, M. Yan, M. Sob, and V. Vitek, *Phys. Rev. Lett.* 67 (1991) 1894.
82. M. Shamsuzzoha, D. J. Smith, and P. Deymier, *Scripta Met. et Mater.* 24 (1990) 1611.
83. J. M. Penisson, M. J. Mills, and U. Dahmen, *Phil. Mag. Lett.* 64 (1991) 327.
84. M. J. Mills, M. S. Daw, G. J. Thomas, and F. Cosandey, *Ultramicroscopy* 40 (1992) 247.
85. G. H. Campbell, W. L. Wien, W. E. King, S. M. Foiles, and M. Rühle, *Ultramicroscopy* 51 (1993) 247.

86. G. H. Campbell, S. M. Foiles, P. Gumbsch, M. Rühle, and W. E. King, *Phys. Rev. Lett.* 70 (1993) 449.
87. F. Ernst, M. W. Finnis, D. Hofmann, T. Muschik, U. Schönberger, U. Wolf, and M. Methfessel, *Phys. Rev. Lett.* 69 (1992) 620.
88. D. Hofmann and F. Ernst, *Interface Sci.* 2 (1994) 201.
89. R. W. Fonda and D. E. Luzzi, *Phil. Mag.* A68 (1993) 1151.
90. K. Nadarzinski and F. Ernst, *Phil. Mag.* A74 (1996) 641.
91. M. J. Mills, M. S. Daw, and S. M. Foiles, *Ultramicroscopy* 56 (1994) 79.
92. D. L. Medlin, C. B. Carter, J. E. Angelo, and M. J. Mills, *Phil. Mag.* A75 (1997) 733.
93. A. de Crecy, A. Bourret, S. Naka, and A. Lasalmonie, *Phil. Mag.* A47 (1983) 245.
94. M. J. Mills and D. B. Miracle, *Acta metall. mater.* 41 (1993) 85.
95. K. J. Hemker, B. Viguier, and M. J. Mills, *Mats. Sci. Eng.* A164 (1993) 391.
96. L. A. Bursill and D. J. Smith, *Nature* 309 (1984) 319.
97. K. L. Merkle and D. J. Smith, *Phys. Rev. Lett.* 59 (1987) 2887.
98. G. H. Campbell, *J. Amer. Ceram. Soc.* 79 (1996) 2883.
99. U. Dahmen, S. Paciornik, I. G. Solorzano, and J. Vandersande, *Interface Sci.* **2** (1994) 125.
100. V. Ravikumar and V. P. Dravid, *Ultramicroscopy* 52 (1993) 557.
101. M. M. McGibbon, N. D. Browning, M. F. Chisholm, A. J. McGibbon, S. J. Pennycook, V. Ravikumar, and V. P. Dravid, *Science* 266 (1994) 102.
102. A. Recnik, J. Bruley, W. Mader, D. Kolar, and M. Rühle, *Phil. Mag.* B70 (1994) 1021.
103. S. Stemmer, S. K. Streiffer, F. Ernst, and M. Rühle, *Phil. Mag.* A71 (1995) 713.
104. D. M. Tricker and W. M. Stobbs, *Phil. Mag.* A71 (1995) 1037.
105. G. Möbus and M. Rühle, *Ultramicroscopy* 56 (1994) 54.
106. G. Gutenkunst, J. Mayer, and M. Rühle, *Scripta metall. mater.* 31 (1994) 1097.
107. J. C. Barry, *Phil. Mag.* A64 (1991) 111.
108. P. J. Fallon, L. M. Brown, J. C. Barry, and J. Bruley, *Phil. Mag.* A72 (1995) 21.
109. F.-R. Chien, S. R. Nutt, J. M. Carulli, N. Buchan,C. P. Beetz, and W. S. Yoo, *J. Mater. Res.* 9 (1994) 2086.
110. J. Y. Dai, Y. G. Wang, D. X. Li, and H. Q. Ye, *Phil. Mag.* A70 (1994) 905.
111. D. J. Smith, In: *Chemistry and Physics of Solid Surfaces VI*, R. Vanselow and R. Howe (Eds.), Springer, Berlin (1986) Chapter 15.
112. D. J. Smith, R. W. Glaisher, P. Lu, and M. R. McCartney, *Ultramicroscopy* 29 (1989) 123.
113. D. J. Smith and L. D. Marks, *Ultramicroscopy* 16 (1985) 101.
114. L. D. Marks and D. J. Smith, *Nature* 303 (1983) 316.
115. Y. Haga and K. Takayanagi, *Ultramicroscopy* 45 (1992) 95.
116. M. Shiojiri, N. Nakamura, C. Kaito, and T. Miyano, *Surf. Sci.* 126 (1983) 719.
117. T. Hasegawa, K. Kobayashi, N. Ikarashi, K. Takayanagi, and K. Yagi, *Japan. J. Appl. Phys.* 25 (1986) L333.
118. D J. Smith, L. A. Bursill, and D. A. Jefferson, *Surf. Sci.* 175 (1986) 673.
119. D. J. Smith, L. A. Bursill, and M. R. McCartney, *Ultramicroscopy* 23 (1987) 299.
120. M. I. Buckett, J. Strane, D. E. Luzzi, J. P. Zhang, B. W. Wessels, and L. D. Marks, *Ultramicroscopy* 29 (1989) 217.
121. J. L. Hutchison and N. A. Briscoe, *Ultramicroscopy* 18 (1985) 435.
122. Z. C. Kang, D. J. Smith, and L. Eyring, *Surf. Sci.* 175 (1986) 684.
123. X.-G. Ning and H.-Q. Ye, *Phil Mag.* A62 (1990) 431.
124. Z. C. Kang and L. Eyring, *Ultramicroscopy* 23 (1987) 275.
125. A. Andersson, J.-O. Bovin, and P. Walter, *J. Catal.* 98 (1986) 204.
126. J. M. Gibson, M. L. McDonald, and F. C. Unterwald, *Phys. Rev. Lett.* 55 (1985) 1765.
127. P. Lu and D. J. Smith, *Surf. Sci.* 254 (1991) 119.
128. J.-O. Bovin, L. R. Wallenberg, and D. J. Smith, *Nature* 317 (1985) 47.
129. L. R. Wallenberg, J.-O. Bovin, and D. J. Smith, *Naturwiss.* 72 (1985) 539.
130. S. Iijima and T. Ichihashi, *Japan. J. Appl. Phys.* 24 (1985) 1125.
131. L. R. Wallenberg, J.-O. Bovin, A. K. Petford-Long, and D. J. Smith, *Ultramicroscopy* 20 (1986) 71.
132. D. J. Smith, A. K. Petford-Long, L. R. Wallenberg, and J.-O. Bovin, *Science* 233 (1986) 872.
133. A. K. Petford-Long, N. J. Long, D. J. Smith, L. R. Wallenberg, and J.-O. Bovin, In: *Physics and Chemistry of Small Clusters,* P. Jena, B. K. Rao, and S. N. Khanna (Eds.), Plenum, New York (1987) p. 127.
134. J.-O. Bovin, J.-O. Malm, A. K. Petford-Long, D. J. Smith, and G. Schmid, *Z. Angew. Chem.* 27 (1988) 555.

135. P. M. Ajayan and L. D. Marks, *Phys. Rev. Lett.* 60 (1988) 585.
136. K.-H. Herrmann, D. Krahl, H.-P. Rust and O. Ulrichs, *Optik* 44 (1976) 393.
137. O. L. Krivanek and P. E. Mooney, *Ultramicroscopy* 49 (1993) 95.
138. W. J. de Ruijter and J. K. Weiss , *Rev. Sci. Instr.* 63 (1992) 4314.
139. W. J. de Ruijter, *Micron* 26 (1995) 247.
140. W. J. de Ruijter and J. K. Weiss, Ultramicroscopy 50 (1993) 269.
141. W. O. Saxton, D. J. Smith, and S. J. Erasmus, *J. Micros.* 130 (1983) 187.
142. O. L. Krivanek and M. L. Leber, In: *Proc. 51st. Ann. Meet. Microscope Society of America*, G. W. Bailey and C. L. Rieder (Eds.), San Francisco Press, San Francisco (1993) p. 972.
143. W. O. Saxton, *Ultramicroscopy* 58 (1995) 239.
144. O. L. Krivanek and P. A. Stadelmann, *Ultramicroscopy* 60 (1995) 103.
145. O. Scherzer, *J. Appl. Phys.* 20 (1949) 20.
146. H. Rose, *Ultramicroscopy* 56 (1994) 11.
147. M. Haider, H. Rose, S. Uhlemann, E. Schwan, B. Kabius, and K. Urban, *Ultramicroscopy* 75 (1998) 53.
148. O. L. Krivanek, N. Dellby, and A. R. Lupini, *Ultramicroscopy* 78 (1999) 1.
149. M. Haider and J. Zach, In: *Proc. Microscopy and Microanalysis 1995*, G. W. Bailey, M. H. Ellisman, R. A. Henninger and N. Zaluzec (Eds.), Jones and Begall, New York (1995) p. 596.
150. H. Rose, *Optik* 85 (1990) 95.
151. H. W. Mook and P. Kruit, *Ultramicroscopy* 81 (2000) 129.
152. M. H. F. Overwijk, A. J. Bleeker, and A. Thust, *Ultramicroscopy* 67 (1997) 163.
153. G. Möbus, R. Schweinfest, T. Gemming, T. Wagner, and M. Rühle, *J. Microscopy* 190 (1998) 109.
154. T. Hoche, P. R. Kenway, H.-J. Kleebe, and M. Rühle, In: *Atomic Scale Imaging of Surfaces and Interfaces*, D. Biegelsen, D. J. Smith, and S.-Y. Tong (Eds.), MRS Symp. Proc. Vol. 295, Materials Research Society, Pittsburgh (1993) p. 115.
155. W. E. King and B. S. Lamver, In: *Microbeam Analysis*, D. G. Howitt (Ed.), San Francisco Press, San Francisco (1991) p. 217.
156. W. E. King and G. H. Campbell, *Ultramicroscopy* 51 (1993) 128.
157. K. Urban, B. Kabius, M. Haider, and H. Rose, *J. Electron Microscopy*, 48 (1999) 821.
158. J. H. Chen, H. W. Zandbergen, and D. van Dyck, *Ultramicroscopy* (2004) in press.
159. M. A. O'Keefe, Microsc. & Microanal. 4 (1998) 382.
160. C. L. Jia, M. Lentzen, and K. Urban, *Science* 299 (2003) 870.
161. M. J. Hÿtch and W. M. Stobbs, *Ultramicroscopy* 53 (1994) 191.
162. C. B. Boothroyd, *J. Microscopy* 190 (1998) 99.
163. C. B. Boothroyd, *Ultramicroscopy* 83 (2000) 159.
164. A. Howie, *Ultramicrosopy* (2004) in press.
165. C. B. Boothroyd and R. E. Dunin-Borkowski, *Ultramicroscopy* (2004) in press.
166. L. W. Hobbs, In: *Quantitative Electron Microscopy*, J. N. Chapman and A. J. Craven (Eds.), Scottish Universities Summer School in Physics, Edinburgh (1984) p. 399.
167. G. Dehm, K. Nadarzinski, F. Ernst, and M. Rühle, *Ultramicroscopy* 63 (1996) 49.
168. D. Hofmann and F. Ernst, *Ultramicroscopy* 53 (1994) 205.
169. M. J. Beeching and A. E. C. Spargo, *Ultramicroscopy* 52 (1993) 243.
170. M. J. Beeching and A. E. C. Spargo, *J. Microscopy* 190 (1998) 262.
171. M. Lentzen M and K. Urban, *Ultramicroscopy* 62 (1996) 89.

15. SCANNING TRANSMISSION ELECTRON MICROSCOPY

J. M. COWLEY

1. INTRODUCTION

It was realized in the early days of electron microscopy [1] that the same electromagnetic, or electrostatic, lenses used in the conventional transmission electron microscopy (TEM) instruments could also be used to form a small electron probe by demagnifying a small bright source, and that probe could be scanned over the specimen in a two-dimensional raster. Scanning Electron Microscopy (or, Secondary Electron Microscopy) for the study of specimen morphologies and surface structure by the detection of low-energy secondary electrons, was widely developed in the 1950s, as was the microanalysis of specimens, with associated imaging, by detection of emitted X-ray. Scanning Transmission Electron Microscopy (STEM), making use of the electrons transmitted through a thin specimen, was developed more slowly. It was not until the 1970s that it was shown that STEM could compete with TEM in resolution and could have some important advantages (along with some disadvantages).

Albert Crewe realized that the way to produce an electron probe of very small diameter was to make use of a field emission gun (FEG) in which electrons are extracted from a cold, sharply-pointed metal tip by use of a high electrical field gradient. He and his associates built a small microscope column using electrons accelerated by 30 kilovolts and focused with a strong, short-focus electron lens to form a probe of diameter of approximately 0.4 nm [2]. The electrons scattered out of the incident beam cone were collected by use of an annular detector to form a dark-field image. With this arrangement, the first transmission electron microscope images of individual

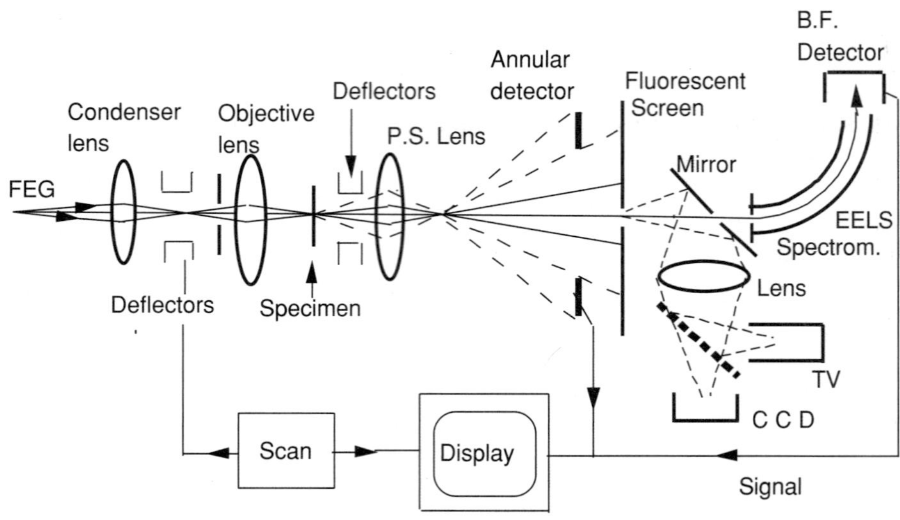

Figure 1. Diagram of a STEM instrument, modified for the convenient display and recording of shadow images and nanodiffraction patterns. A condenser lens and an objective lens produce the incident beam probe on the specimen and one (or more) post-specimen (P.S.) lenses govern the display of the diffraction pattern on a transmission phosphor screen which may be viewed using a TV-VCR system or a CCD camera with digital recording.

single heavy atoms, thorium atoms on a very thin amorphous carbon support, were recorded [3].

Successful applications to many areas of non-biological and biological science then followed rapidly. Several companies attempted to produce commercial STEM instruments. The only company to succeed in this was VG Microscopes, Ltd., of England which produced a succession of increasingly complicated instruments from about 1974 until their demise in 1997.

For STEM, the recording is essentially of a serial nature as the image signals are derived as the incident beam is scanned over the specimen, whereas in TEM the recording is in parallel with the whole image recorded simultaneously. One big advantage of STEM arises because, with the serial mode of recording, images formed with several different detectors may be recorded simultaneously. Bright-field and dark-field images and images with secondary radiations can be obtained during the same single scan of the incident beam over the specimen.

The principal components of a modified STEM instrument [4] are illustrated, diagrammatically, in Fig. 1. The source of electrons is a field-emission electron gun (FEG), preferably operating with a cold (i.e., room temperature) sharpened tungsten wire tip in ultra-high vacuum. A positive voltage, of about 3,000 volts with respect to the tip, creates a sufficiently high voltage gradient to draw the electrons out of the tip. The effective diameter of this very bright electron source is as small as 4 or 5 nm.

The field emission tip is held at a negative voltage of 100 to 300 kV. An earthed anode forms a weak electrostatic lens, which may be supplemented by the field of a weak electromagnetic lens, to form a narrow electron beam in the microscope column. The main lens producing the demagnification of the effective source, to form the small-diameter probe at the specimen level, is the objective lens, similar to the objective lenses used in TEM, with a high magnetic field and a focal length as small as 1 mm. One or more condenser lenses are placed before the objective lens in order to provide the flexibility in the beam parameters that are desirable for various modes of operation. Deflector coils, built into the bore of the objective lens, serve to scan the electron probe over the specimen.

For each position of the incident electron beam on the specimen, a diffraction pattern of the illuminated area is formed on a distant fluorescent screen. Since the beam focused by the objective lens is necessarily convergent, the pattern is a convergent-beam electron diffraction (CBED) pattern. Since the illuminated region of the specimen has a diameter approximately equal to the resolution limit of the images formed, usually less than 1 nm, the pattern is often referred to as an electron nano-diffraction (END) pattern. The scale of the diffraction pattern on the screen is affected by the weak post-specimen field of the objective lens and may be conveniently adjusted by means of one or more weak, post-specimen, magnetic lenses. The diffraction pattern may be observed and/or recorded by the use of a light-optical system consisting, for example, of a wide-aperture optical lens forming an image of the pattern in a low-light-level TV camera, coupled to a video-tape recorder (VCR), or else on a CCD camera for digital recording.

An aperture in the observation screen allows some part of the diffraction pattern, usually the central spot, to be transmitted into an electron energy-loss spectrometer (EELS), so that the EELS spectrum may be recorded or else an image may be made with electrons having any desired energy loss. The normal bright-field STEM image is obtained with the zero-loss electrons from the central beam of the pattern. Dark-field images may be obtained by selecting electrons that have lost the few electron volts associated with plasmon excitation, or with the many-electron-volt losses due to the inner-shell electron excitations characteristic of the various elements present. Dark-field images are also formed when any electrons scattered out of the incident beam cone are detected, with or without energy losses. Post-specimen deflection coils are used to direct any chosen part of the diffraction pattern to the entrance aperture of the spectrometer.

The most common form of dark-field imaging in STEM is given when all, or some selected part of the diffraction pattern, outside of the central beam spot, is detected, without energy analysis, by use of an annular detector. Crewe *et al.* [3] showed that the collection of the image signal in this way is much more efficient than for any of the dark-field TEM modes and led to interesting possibilities for signal manipulation.

Detectors may be placed close to the specimen to record the intensities of secondary radiations produced by the fast incident electron beam. Energy-sensitive X-ray detectors allow for the microanalysis of the illuminated specimen areas, or for the imaging of the specimen with the characteristic X-rays from particular elements. Low-energy

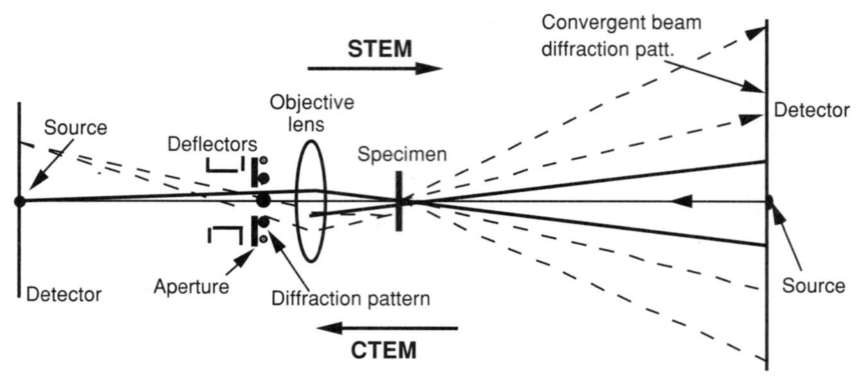

Figure 2. Diagram of the principal components of a STEM and a conventional TEM suggesting that, if the beam direction is reversed, the two imaging systems are equivalent, in accordance with the Reciprocity Relationship.

secondary electrons, or Auger electrons from specimen surfaces, may be detected and used to form secondary-electron (SEM) images of either the top side or the bottom side of the specimen.

Because the cold FEG must be operated in ultra-high vacuum, the whole microscope column, connected to the gun through a small aperture, should preferably be maintained at a high- (but not necessarily ultra-high-) vacuum. In the VG instruments, the column vacuum is normally better than 10^{-8} torr, an order of magnitude better than for most TEM instruments. This level of vacuum is attained only through the use of special techniques of construction of the column components, but has the advantage that the cleanliness of specimen surfaces may be maintained and the contamination of specimens under electron radiation may be kept to a minimum.

Initially, it was assumed that the contrast of STEM images could be interpreted in terms of an incoherent imaging theory, with the intensity at each point of the image given by the intensity of scattering from each irradiated point of the specimen. However, it was shown that the image intensities could be related to those of TEM by application of the Principle of Reciprocity [5, 6]. In the idealized case, it is stated that, for a scalar wave and elastic scattering, the amplitude of irradiation at a point B due to a point source at A, is the same as the amplitude at A due to a point source at B. Extending this to consider each point of a finite source and each point of a detector separately, one can confirm that the image given in a STEM system is the same as that for a TEM system, given the same components of lenses and apertures, but with the direction of propagation of the electrons reversed, as suggested in Fig. 2. Thus it has been demonstrated that the same variety of phase-contrast and amplitude-contrast modes as in TEM is possible in STEM, and the well-developed theory for TEM image contrast can be applied immediately for STEM imaging.

The commercially-provided STEM instruments used electrons accelerated by either 100 or 300 kV. In various laboratories, special instruments have been built in order to exploit the advantages of using much higher or lower electron energies. Of the several projects to build million-volt instruments, the only one to produce useful results was that of Strojnik [7] who demonstrated the attainment of resolutions of about 1 nm at about 600 kV. As predicted, the high-voltage STEM gave much better imaging of very thick specimens than the contemporary high-voltage TEM instruments.

Medium-energy STEM instruments, operating at voltages of 5 to 20 kV, and used mostly in the reflection mode, have been built especially for surface studies in ultra-high vacuum [8–10]. Within the last few years, new TEM instruments with field-emission guns have become available commercially, and a flexibility of the electron optics allows them to be used for STEM imaging also. These so-called TEM/STEM instruments, operating at 200 or 300 kV, have been shown to produce excellent STEM images using the high-angle annular dark-field (HAADF) mode and also spectrometry and imaging with energy-loss electrons (EELS, and energy-loss imaging) [11–13]. It has not yet been demonstrated that such instruments have the same versatility in allowing the imaging, diffraction and microanalysis with the variety of modes possible with a dedicated STEM instrument, but their advent broadens the range of STEM instrumentation available at a time when the commercial manufacture of dedicated STEM instruments has temporarily ceased.

New experimental dedicated STEM instruments are now being tested. Some, incorporating aberration-corrector systems for the objective lenses and special monochromators, aim to provide resolutions of better than 0.1 nm [14, 15].

The STEM techniques clearly have many potential applications in the rapidly expanding field of nanotechnology. Many of the present and anticipated advances in this field involve the production and application of particles, tubes or wires, and interfacial assemblies for which the critical dimensions are of the order of one nanometer. The STEM instruments are ideally designed for the observation, characterization and analysis of materials having such dimensions. Adequate image contrast can be given by thicknesses of 1 nm with lateral resolutions of better than 0.2 nm. Compositions can be derived by microanalysis, and crystal structures can be recognized or determined from nanodiffraction patterns from regions having diameters of 1 nm or less. Special STEM techniques may be applied for the study of the structures of surfaces and the analysis of amorphous materials. Applications to the study of biological systems on a comparable scale are necessarily limited by radiation damage caused by the incident electron beam, but it has been shown that valuable data can be derived concerning the inorganic components of such systems. In this review, an account of the technical and theoretical bases for understanding STEM capabilities will be followed by examples of various applications that have been made to the study of samples relevant for future nanotechnology.

2. STEM IMAGING

The simplest form of imaging in a STEM instrument is shadow-imaging (or, point-projection imaging) in which a stationary incident beam is greatly defocused so that a small cross-over is formed well before, or after, the specimen, as suggested in the

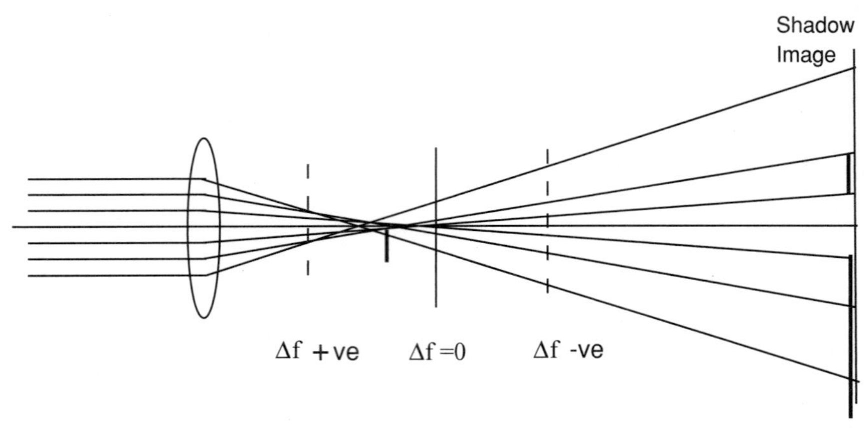

Figure 3. Diagram for the formation of shadow (point-projection) images for a lens having spherical aberration. If the cross-over comes before the specimen (over-focus) the image has positive magnification. If the cross-over comes after the specimen, the magnification is negative. For an object close to the paraxial focal point, there is infinite magnification for a particular radius in the image.

simple, geometric-optics diagram of Fig. 3. Images formed in this way are often useful for the initial survey of a specimen at low magnification, although it can be shown theoretically that the resolution attainable in this mode is given by the dimensions of the cross-over and is equal to that of the scanned STEM images. The magnification is positive or negative depending on whether the cross-over comes before or after the specimen [16].

When the cross-over is close to the specimen and the magnification becomes large, the image is distorted by the aberrations of the lens. When third-order spherical aberration is important, as for electromagnetic electron lenses, the image for a specimen placed close to the position of minimum diameter of the cross-over, as in Fig. 3, is greatly distorted. For incident paraxial rays, the cross-over is after the specimen so that for the central part of the image the magnification is high but negative. For rays making a large angle with the axis, the cross-over is before the specimen so that for the outer part of the image the magnification is high but positive. For one particular angle of incidence, the rays cross over at the specimen level so that the magnification becomes infinite. Taking into account the three-dimensional configuration of the rays, it can be seen that there is one radius in the image for infinite radial magnification and another radius for infinite circumferential magnification [17]. If the objective lens has astigmatism, the circular symmetery in the variations of magnification of the image is distorted. The image of a straight edge becomes a loop. This is in direct analogy with the "knife-edge test" of classical light optics.

More dramatic and useful effects appear if the specimen transmission is periodic. For a thin crystalline specimen, the shadow image of a set of parallel lattice planes, near focus, is distorted by the lens aberrations to give a set of loops, known as Ronchi-fringes in honor of Ronchi [18] who observed that such fringes are given when a

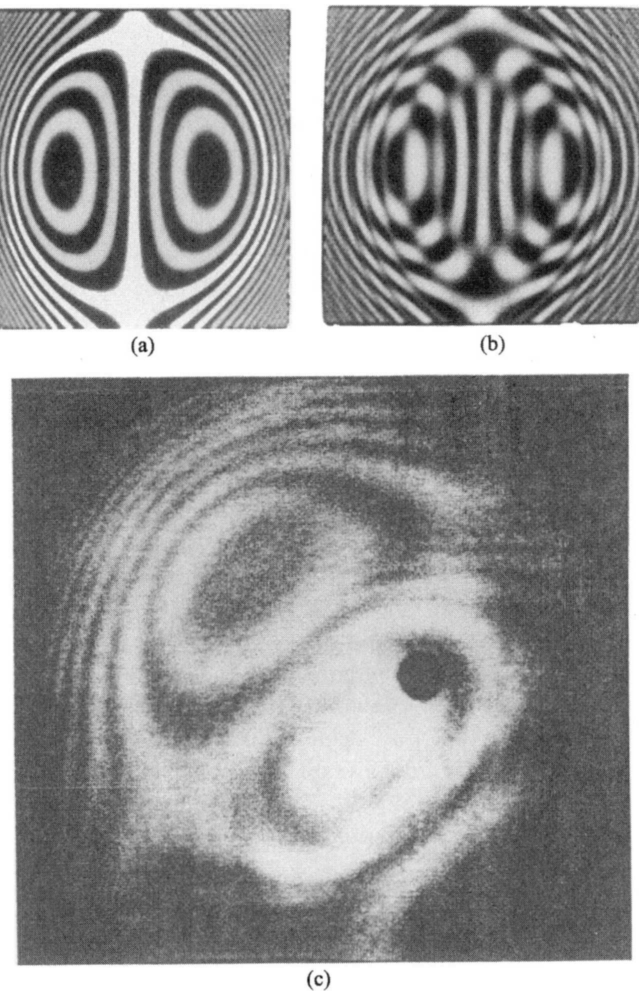

Figure 4. Ronchi fringes formed in the shadow image of a set of lattice planes in a crystal. (a,b) Simulated fringes for particular defocus values. (c) Observed fringes from the 0.34 nm periodicity of one side of a multi-wall carbon nanotube [136].

diffraction grating is placed near the focus of a large telescope mirror and showed that the fringes could be used to measure the lens aberrations. Typical Ronchi fringes in the electron shadow image of a crystal are shown in Fig. 4 [19, 12]. After the loops have been made symmetrical by careful correction of the astigmatism, measurement of their dimensions allows the accurate determination of the spherical aberration constant from a set of images obtained with known differences of defocus [20].

The circles of infinite magnification for general specimens, or the Ronchi fringes from crystalline specimens, also provide a convenient means for alignment of the electron-optics of the microscope. These features of the shadow image allow the axis of the objective lens to be aligned with the detector aperture and provide a guide for the insertion of other apertures in the column.

The theoretical basis for STEM imaging may conveniently be referred to that for TEM imaging, making use of the Principle of Reciprocity. We show the equivalence for the simplest case.

The usual theory of electron microscope imaging is based on the Abbe theory for coherent radiation. For plane-parallel incident radiation, the transmission function for a thin object may be written, in one dimension for simplicity, $q(x)$. Then the bright-field image wave function for TEM is given by

$$\psi(x) = q(x) \otimes t(x) \equiv \int q(X) \cdot t(x - X) \cdot dX \tag{1}$$

and $t(x)$, is known as the Spread Function, which describes the loss of resolution due to the lens aberrations [21]. It is given by the Fourier transform of the lens transmission function, $T(u)$, where u is the angular variable, equal to $(2/\lambda) \cdot \sin(\phi/2)$, and ϕ is the angle of scattering. The convolution integral in (1), denoted by the $\otimes$ symbol, has the effect of smearing-out the $q(x)$ function. The image intensity is then given by the modulus squared of (1).

For STEM imaging, it is considered that a lens having a transfer function $T(u)$ forms a small probe for which the amplitude distribution is given by $t(x)$. The wave transmitted through the specimen when the probe is displaced by an amount X is then $q(x) \cdot t(x - X)$, and the intensity on the detector plane is given by the modulus squared of the Fourier transform, $Q(u) \otimes [T(u) \cdot \exp(2\pi i u X)]$. The signal collected to form the image is given by multiplying this intensity distribution by some detector function, $D(u)$, and integrating the transmitted intensity to give

$$I(X) = \int |Q(u) \otimes [T(u). \exp\{2\pi i u X\}]|^2 \cdot D(u).du \tag{2}$$

The form of $D(u)$, is related by Reciprocity to the TEM bright-field illumination. The equivalent of TEM with a plane-parallel wave incident, is the delta function detector, $\delta(u)$. For this case, (2) reduces to

$$I(X) = \left| \int Q(U) \cdot T(U) \cdot \exp\{2\pi i UX\} \cdot dU \right|^2 = |q(X) \otimes t(X)|^2 \tag{3}$$

which is the same result as for bright-field TEM, given above, since the STEM image is formed by recording the observed intensity as a function of the probe displacement, X.

The equation (2) may be used to determine the detected signal and the image contrast for other detector configurations, such as the cases where the detector is an aperture of finite diameter, or, with convenient approximations, for the dark-field images for which the detector records the electrons scattered outside the incident beam cone. For this latter case, with an annular detector with a central aperture having a diameter slightly greater than the incident beam cone, a simple result is obtained if it is assumed that the intensity of the radiation scattered within the incident beam cone is proportional to the intensity scattered outside this cone. This assumption is good for the imaging of single isolated heavy atoms, but may fail if the specimen contains structure on a scale that will give diffraction maxima for scattering angles approximately equal to the half-angle of the incident cone, i.e., for structural details near the resolution limit of the microscope [22].

For this annular dark-field (ADF) mode, it is convenient to consider weakly-scattering objects and to write the transmission function as $q(x) = 1 + p(x)$, where $p(x)$ is much smaller than unity. Then it can be shown that, by taking $D(u) = 1$, and neglecting the scattering within the incident beam cone, the intensity of the image is given as $I(X) = |p(X)|^2 \otimes |t(X)|^2$. Thus the square of the scattering function is imaged with a resolution defined by the intensity distribution of the incident beam This result is interpreted as showing that the ADF contrast is given by "incoherent" imaging, in that the signal is proportional to the scattering power of the object with a spread function equal to the intensity distribution of the incident beam.

It may be noted that, if the amplitude distribution of the incident beam is assumed to be gaussian, the squaring of this gaussian gives an intensity distribution which is a gaussian with a width smaller by a factor of the square root of 2. Hence it may be concluded that dark-filed STEM gives a resolution better than that of bright-field TEM or STEM by a factor of about 1.4.

Application of the principle of reciprocity suggests that, with a finite detector diameter, rather than a delta-function detector as assumed in the derivation of (3), the resolution of the STEM bright-field image is degraded in the same way that the resolution in bright-field TEM is affected by applying a "damping function", or : "envelope function" to the phase-contrast transfer function, to take account of the illumination from a finite, incoherent source [21]. However, a simple argument indicates that, if the STEM detector has a diameter exactly equal to the diameter of the incident beam cone, the image signal, in the absence of absorption, is given by the total incident intensity minus the intensity in the annular dark-field image, and the image resolution of the bright-filed image should be the same as for the ADF image, although the contrast is necessarily lower [23].

On the basis of the simple incoherent imaging approximation to the contrast for the ADF STEM images, the contrast for single atoms should be approximately proportional to the square of Z, the atomic number. Hence heavy atoms could be distinguished clearly when held on a light-atom support. Also it was noted that for inelastic scattering the dependence on the atomic number is different so that, if the inelastically scattered electrons could be detected by use of an energy-loss spectrometer, the ratio of the signals recorded simultaneously from the elastically and inelastically scattered electrons would give image contrast proportional to some power of the atomic number and independent of the number of atoms present. This provided a further means for separating the signals from heavy and light atoms.

A further valuable form of dark-field STEM imaging makes use of an annular detector having a large inner diameter so that it detects only those electrons scattered to relatively high angles, usually of 50mrad or more for 100keV electrons. It was pointed out by Howie [24], that in ADF images of heavy-atoms on light-atom materials, such as metal particles in light-atoms catalyst supports, confusion could arise from strong signals given by small crystalline regions of the supports. This trouble could be avoided by noting that the diffraction spots from crystals extend to only limited scattering angles. By use of a high-angle annular dark-field (HAADF) detector, these contributions could be avoided. It was realized that the scattering to higher angles was mostly thermal diffuse scattering, due to the atomic vibrations. While the elastic crystalline reflection intensities fall off with scattering angle approximately in proportion to $f^2(u)$, the square of the atomic scattering factor, the intensity for first-order thermal diffuse scattering is proportional to $u^2f^2(u)$, and so peaks at higher angles [25]. For thick specimens, the higher-order thermal diffuse scattering can become important and this peaks at even higher angles [26].

As a first approximation, it may be assumed that, as in the case of first-order thermal diffuse scattering, the HAADF signal may be assumed to be proportional to the square of the atomic number of the elements present. More complicated dependencies are appropriate when multiple diffuse scattering is present or in the case of crystalline specimens, to be considered later.

A further useful form of dark-field STEM imaging is given by the use of a thin annular detector, in which the outer and inner diameters of the detector differ by as little as 10 percent, so that only those electrons scattered within a limited range of angles are detected. The effective mean radius of the thin annular detector may be selected by varying the strengths of the post-specimen lenses to vary the magnification of the diffraction pattern on the detector plane. Then this thin-annular-detector dark-field (TADDF) mode may be useful in some cases, for imaging particular components of a compound specimen [27, 28] as in the case illustrated in Fig. 5.

In particular, it has been shown that the TADDF mode can be useful in the preferential imaging of nanocrystalline or amorphous phases in a composite specimen when the components have maxima of their scattering intensities for different scattering angles. For example, nanocrystals of carbon in a layer 1 nm thick could be readily distinguished from a supporting film of amorphous silica, 6 nm thick, when the thin annular detector was set to collect the scattering to the strong carbon peak, corresponding to a d-spacing of about 0.34 nm, which is well separated from the main peaks of the scattering intensity from the amorphous silica [29].

Further interesting imaging modes are given when the post-specimen lenses are used to enlarge the diffraction pattern until the central beam spot is comparable in diameter with the inner diameter of the annular detector. If the incident beam spot is just slightly smaller the inner detector diameter, bright image contrast is given when a rapid variation of the specimen thickness or scattering strength acts like a prism to deflect the incident beam. This so-called "marginal" imaging mode then shows the locations of edges, or any sudden variation of the scattering power of the specimen.

If the central beam spot is made slightly larger than the inner diameter of the annular detector, a bright-field image is produced. An analysis of this mode suggests that the

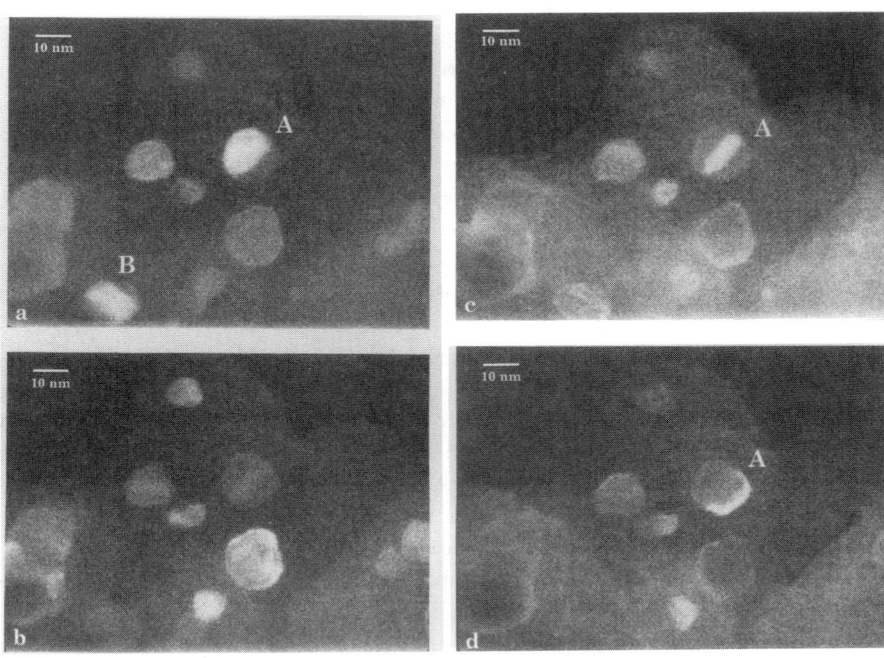

Figure 5. STEM images of a specimen of nanocrystals of Pt in amorphous carbon using a thin annular detector with average detection angles corresponding to d-values of (a) 0.056 nm, (b) 0.076 nm, (c) 0.098 nm and (d) 0.120 nm. For (c), the scattering from the a–C is strong [27].

resolution of this bright-field image should be better than that of the normal, axial bright-field STEM image by a factor of about 1.7 [28]. Application of the reciprocity principle suggests that this mode of bright-field STEM imaging with a thin annular detector should be equivalent to the bright-field TEM imaging mode in which the specimen is illuminated by a hollow cone of incident electrons, or else, more conveniently, the incident beam is tilted to a fixed angle and then gyrated so that the integration over time is equivalent to the hollow-cone illumination [30]. For similar geometries, this hollow-cone TEM mode should show the same properties as the TADDF STEM in dark-field and the same improvement of resolution for bright-field images. For practical applications, the TADDF STEM mode is probably more simple, experimentally, and more convenient.

Detectors of non-circular symmetry, including those with two separate halves or four individual quadrants, have been employed for high-resolution, differential phase contrast and the detection of weak magnetic fields [31].

3. STEM IMAGING OF CRYSTALS

3.1. Very Thin Crystals

When the specimen is a very thin crystal with the incident beam direction close to a principal crystal axis, the diffraction pattern on the detector plane is a regular array of diffraction spots with separations inversely proportional to the dimensions of the projected unit cell of the crystal. Because, in STEM, the incident beam is convergent, each diffraction spot is spread out to form a circular disk of the same diameter as the incident beam disk. Provided that the crystal region illuminated is very thin, the intensity distribution across each diffracted beam disk is uniform. For increasing thicknesses, the diffraction disks are crossed by dark or bright lines, corresponding the variations of diffracted beam intensities as the incident beam orientations varies, as in the convergent-beam electron diffraction (CBED) patterns commonly observed in TEM instruments. For relatively very thick crystals, as in CBED, a diffuse background, arising from inelastic scattering processes, accumulates, crossed by dark and light Kikuchi lines.

In STEM instruments, the illuminated area of the specimen is normally 1 nm or less in diameter and the incident beam cone may be considered as completely coherent. For CBED in TEM instruments, the area of the specimen illuminated normally has a diameter of about 100 nm, and, to a good approximation, the incident beam may be assumed to be completely incoherent. It has been shown that, for the elastic scattering from a perfect crystal, without defects or boundaries, the diffraction intensities are exactly the same for the coherent and incoherent cases [32]. Hence, the diffraction patterns observed in a STEM instrument may be used for all of the many applications developed for the CBED technique using TEM instruments [32, 33].

When the cone angle for the incident beam in STEM is so large that the diffraction spots overlap, or when there are defects or boundaries in the crystal region illuminated, giving rise to scattering between the perfect-crystal diffraction spots, there can be coherent interaction between electron waves scattered into the same direction, but coming from different directions within the incident beam cone. These coherent interactions give rise to some striking interference effects. Interpretation of the "coherent CBED" observations can provide important additional information about the specimen structure and allow interpretation of STEM images of crystals and their defects.

The condition that diffraction spot disks should overlap appreciably is the condition that the incident beam should have a diameter smaller than the periodicity in the crystal corresponding to the diffraction spot separation. Then electron waves coming from different directions within the incident beam cone can give diffracted beams in the same direction for the two different reflections, as suggested in Fig. 6. These two coincident diffracted beams then interfere, and a set of fringes appears in the region of overlap of the spots. The positions of the fringes depend on the relative phases of the two reflections. This is the basis of the suggestion that the coherent CBED patterns should give the solution to the "phase problem" of X-ray diffraction (and kinematical electron diffraction) in that the observation of the regions of overlap of all the pairs of diffraction disks in a two-dimensional diffraction pattern would supply the information on the relative phases of the reflections, needed for the unambiguous determination of the projected crystal structure [34–36]. This is the basis for the proposed technique of ptychography for crystal structure analysis.

However, the relative phases of the diffracted beams depend on the position of the incident beam relative to the origin of the unit cell. As a beam is translated across

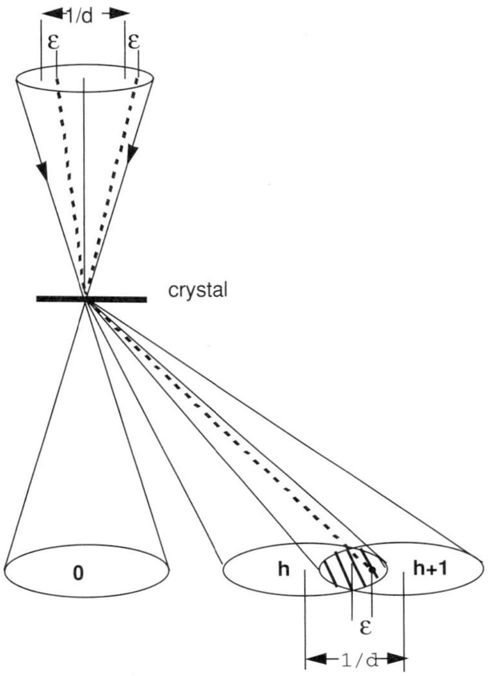

Figure 6. Diagram to suggest the interference of rays from different parts of a coherent source to give interference fringes in the region of overlap of adjacent orders of reflection from a thin crystal.

the crystal, the interference fringes in the areas of overlap of the diffraction spots are seen to move. These movements of the fringes provide a basis for interpreting lattice fringes in STEM images. If a small detector is placed in the region of overlap of two diffraction spots, it will record a sinusoidally-varying signal as the incident beam is translated, so that the image will show fringes having the periodicity of the crystal lattice [36].

If the incident beam cone is further enlarged, until the diffraction spot disks from many reflections overlap at the origin of the diffraction pattern, the image formed with a small axial detector is given by the coherent sum of all the diffraction beam amplitudes, and the image shows the periodic structure revealing the distribution of the scattering matter within the unit cell. Such an incident beam cone size corresponds to an incident beam diameter much smaller than the crystal periodicity, and the image resembles that for HRTEM with an objective aperture large enough to include the corresponding number of diffracted beams [37].

The reciprocity relationship allows the STEM images of thin single crystals to be understood and calculated by the standard methods used for TEM. For larger crystal thicknesses, especially when the inelastic contributions to the image begin to become important, the differences between the STEM and TEM cases may become significant.

3.2. Dynamical Diffraction Effects

For all crystals except those composed of light atoms and no more than a few nanometers thick, dynamical diffraction effects (coherent multiple scattering) are significant and dominate the scattering for most experimental cases, especially when heavy atoms are present [21]. For a crystal of gold in [100] orientation with 100 keV electrons, for example, the diffracted beams attain the same intensity as the incident beam for a thickness of about 2 nm, and the incident and diffracted beam intensities oscillate with thickness increase, almost sinusoidally, with a periodicity of about 4 nm.

In the presence of strong dynamical scattering, there is no intuitively obvious connection between the diffraction pattern or image contrast with the projection of the structure of the crystal. Various schemes have been proposed whereby the atom arrangement can be derived from the observed intensities of series of diffraction patterns or images [38, 39] but these have not led to practical routine procedures. The only practical procedure at present is the established trial-and-error method of postulating a structure and then calculating the image or diffraction pattern by use of one of the computer programs, based on either the multislice approach [40–42] or on the Bloch-wave method, or scattering matrix multiplication method, derived from the Bethe theory [43–45]. Accounts of these computing techniques are given, for example, in references [32, 46, 47].

In order to calculate the dynamical diffraction effects in STEM diffraction patterns or bright-field images, one could, in principle, take the square of the sum of the amplitudes calculated for each incident beam direction within the incident beam cone, added coherently with the correct phase relations, for each incident beam position. Alternatively, one could take, as the input to a multislice calculation, the amplitude distribution of the focused incident beam. Since this distribution is essentially non-periodic, it is then necessary to use the periodic-continuation approximation, assuming a periodic array of incident beams at the origin points of a very large unit cell, with many times the dimensions of the projected crystal unit cell. However, the large amount of computing involved in these approaches may often be avoided by making the calculation for the reciprocity-related TEM conditions, for which a single multislice or Bethe-type calculation may suffice. Calculations that involve inelastic scattering into a continuous diffuse background of the diffraction pattern are necessarily more complicated and tedious [48].

For ADF STEM images, the simple incoherent imaging approximation given above, with image intensity proportional to the square of the projected potential function, fails in the presence of dynamical scattering. The image intensities may be calculated by summing the computed intensities of the diffraction pattern, outside of the central beam spot, obtained for each position of the incident beam. For HAADF imaging it may usually be assumed that the image intensity is proportional to the thermal diffuse scattering (TDS) intensity. Since the integration is made over a wide range of scattering angles and all scattering directions, the assumption of incoherent scattering from individual atoms may be made. For single-scattering TDS, the assumption may be made of a simple TDS form factor for each atom, dependent on the atomic number, the temperature and the atom environment. An extension of such an assumption to include the possibility of multiple TDS scattering, is feasible [49].

3.3. Channeling

For the observation of electron diffraction patterns and the high-resolution imaging of crystals, it is often of interest to consider the special case in which the incident beam is parallel to one of the principle axes of the crystal structure. Then the incident electrons are directed along the rows of closely-spaced atoms, and the channeling phenomenon is observed. The projected potential of the crystal may be considered to have deep wells at the atomic positions and the electrons may be thought of as trapped within the potential wells, giving rise to special transmission effects. As in the case of electrons bound to the three-dimensional potential wells of atoms in normal quantum theory, it may be considered that the electrons in the two-dimensional fields of the atom rows may have bound *1s*, *2s*, *2p*, etc states, with the *1s* states usually predominating.

Treatments along these lines were first introduced with plane waves incident, to describe the diffraction and imaging in HRTEM [50]. Related treatments for the STEM case of incident focused beam probes followed [51]. It has been shown that for an incident conical STEM beam with its axis aligned with a row of atoms through a crystal, in the absence of absorption, the intensity along a line through the centers of the atoms varies periodically with distance into the crystal. The width of the beam is a minimum when the intensity on the central line is a maximum. For a crystal of gold in [100] orientation, for example, the beam diameter may periodically become as small as 0.03 nm, or as large as 0.1 nm. The periodicity of this variation in the beam direction is about 5 nm [52]. For graphitic carbon, this periodicity is as great as 60 nm and the minimum beam diameter is about 0.06 nm [53]. In the presence of absorption, the oscillations of the beam intensity are damped for increasing beam paths.

This is the "atomic focuser" effect. A single heavy atom, or a row of atoms extending through a thin crystal, may be thought of as constituting an electrostatic lens for electrons, providing a focused beam with a diameter of 0.05 nm or less at the exit face. The effective focal length of such a lens is about 2 nm [54]. It has been suggested that such fine probes may form the basis for a STEM imaging scheme giving resolutions of 0.05 nm or better. Alternatively, the periodic arrays of such probes created when a plane wave is transmitted through a thin crystal in an axial direction, may form the basis for new methods for STEM or TEM imaging with comparable resolution [55, 56]. Preliminary experiments with specially favorable configurations [53] and computer simulations [57] have confirmed these possibilities.

4. DIFFRACTION IN STEM INSTRUMENTS

4.1. Scanning Mode Electron Diffraction

In the early commercial STEM instruments, provision was made for the serial recording of diffraction patterns. Post-specimen deflector coils, the so-called Grigson coils, are used to scan the diffraction pattern over a detector aperture, which may be the entrance aperture to the energy-loss spectrometer. The signal from the elastically or inelastically scattered electrons can be recorded or can give a display of the diffraction pattern intensities on the cathode ray screen. An important advantage over the TEM SAED mode is that the diffraction pattern may come from a much smaller specimen area, of diameter equal to that of the incident beam probe.

The principle practical limitations of this mode arise because of the inefficiency in the detection system. Only a very small fraction of the diffraction pattern is recorded at a time. The small area of the specimen giving the pattern is being irradiated by the incident beam during the rather extended recording time and no image of the specimen is available during this time. The effects of specimen drift and irradiation damage may be important.

4.2. Two-Dimensional Recording Systems

The diffraction pattern formed for any position of the incident beam on the specimen may be observed on a fluorescent screen, as suggested in Figure 1. The pattern may be viewed and recorded using a low-light-level television camera, coupled with a video-cassette recorder (VCR), or else using a CCD camera with an associated digital image recording system. With the field-emission gun and normal lens settings, sufficient intensity is available to allow diffraction patterns to be recorded with the TV-VCR system, at the TV rates of 30 patterns per second. With the CCD, the digital recording is more quantitative and more sensitive to small signals, but the recording is slower, with normally only a few frames per second.

In a configuration equivalent, by reciprocity, to the SAED mode in TEM, the condenser lens may be focused to form a cross-over just before the objective lens so that the objective lens forms a parallel beam at the specimen, giving a sharply-focused diffraction pattern on the observation screen. The diameter of the illuminated region of the specimen can be as small as a few tens of nm, and is given by the reduced image of the aperture placed in what, for TEM, would be the image plane of the objective lens. This configuration has been used effectively for quantitative diffraction by Zuo *et al.* [58].

When the incident beam is focused on the specimen, the diffraction patterns are necessarily convergent beam patterns, with spot sizes inversely proportional to the width of the incident beam. For the minimum beam dimensions at the specimen level of about 0.2 nm, the diffraction patterns spots are so large that they overlap for most crystals. These conditions are valuable for special purposes when coherent interference effects are being observed, but for most practical investigations of local structure, it is convenient to use an incident beam of larger diameter, of the order of 1 nm, and the higher intensity obtained by suitable settings of the condenser lenses. Figure 7, for example, shows some diffraction patterns obtained with a beam probe of diameter about 0.7 nm in diameter, for which the diffraction spots are well separated for most crystals having unit cell dimensions no greater than about 1 nm. Such electron nanodiffraction (END) patterns may usually be interpreted in

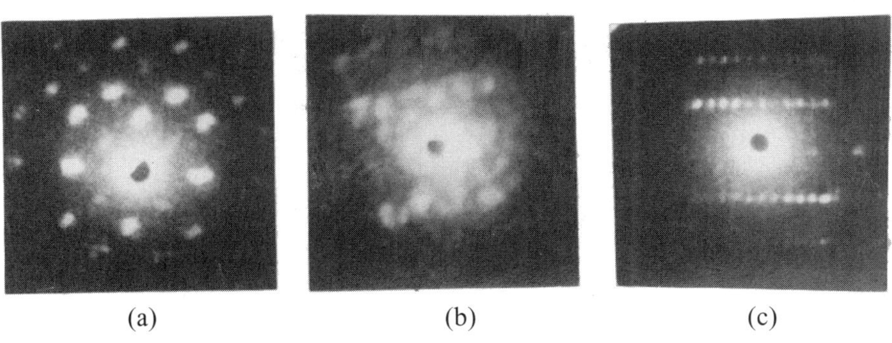

Figure 7. Nanodiffraction patterns obtained with a focussed incident beam of diameter 0.7 nm from the iron-containing cores of ferritin molecules. (a) A magnetite-like phase in [100] orientation. (b) The hexagonal phase in [110] orientation showing the lines of spots corresponding to the 0.94 nm spacing. (c) As for (b), but with the beam defocused to give smaller spot sizes [59].

the same way as parallel-beam diffraction patterns except that the diffraction spots are larger. They have proven useful for the study of many nanocrystalline materials, the localized defects in larger crystals, and the local ordering in quasi-amorphous materials.

For those specimens having lattice plane spacings so great that the diffraction spots in the END patterns overlap, it is often convenient to make the beam diameter at the specimen larger by defocusing the objective lens. Then the spot sizes for small crystals are governed more by the crystallite sizes than by the beam convergence, and the close rows of spots may be separated, as shown in Figure 7, (b) and (c) [59].

4.3. Convergent-Beam Electron Diffraction

In TEM instruments, the CBED method has many valuable applications, depending, for the most part on the observations and measurements of dynamical diffraction effects. The applications include the determination of lattice parameters and thicknesses of crystals, the determinations of space-group symmetries without the normal limitations of kinematical diffraction, the highly accurate measurement of electron scattering amplitudes and the consequent determination of the potential, and electron-density, distributions in relatively simple crystals [33, 60].

For CBED in a normal TEM instrument, without a field-emission gun (FEG), the diameter of the region giving the CBED pattern is usually 20 to 100 nm. In a STEM instrument, or in a TEM/STEM instrument, with a field emission gun as a source, CBED patterns may be recorded with a beam cross-over at the specimen having a diameter approximately equal to the resolution limit for dark-field images, which may be as small as 0.2 nm.

The CBED patterns obtained with a STEM instrument may be used for all the same purposes as CBED in TEM instruments, but with the added advantage that the perfect-crystal regions studied may be much smaller. However, to date, very little has been done to take advantage of these possibilities, presumably because the equipment for making accurate intensity measurements, including CCD detectors or image plates, has not been installed in STEM instruments and other instrumental factors have not received the same amount of attention.

4.4. Coherent Nanodiffraction

A number of interesting and potentially useful effects have been observed in nanodiffraction patterns obtained from crystal specimens containing discontinuities. A planar discontinuity, parallel to the incident beam, for example, gives a continuous streak, perpendicular to the discontinuity, in kinematical diffraction patterns. But for a coherent convergent incident beam, there can be interference between the electrons scattered in neighboring directions. The circular disks of the spots given by the perfect crystal may be modulated so that bright arcs appear on the edges of the disks separated along the line perpendicular to the discontinuity. For example, with the incident beam illuminating the edge of a small gold crystal or a cube-shaped crystallite of MgO, the diffraction spots appear to be split into two arcs [61]. It was shown by Cowley and Spence [61] that such a splitting is to be expected from the interference of the electron waves from the two sides of the discontinuity.

For diffraction from the edges of a small crystal, the amount and form of the modulation of the diffraction spots is dependent on the lateral extent of the edge and the form of the edge; whether there is a flat face parallel to the beam or whether the beam cuts across the top of a pyramid or wedge of crystal. In some cases, the complete circumference of the circular diffraction spot may have enhanced intensity, so that the spots become rings [62]. In principle, it should be possible to derive the complete three-dimensional geometry of a crystallite by observing the form of the diffraction spots obtained from each of its edges.

Similar spot splittings or distortions may be observed for some of the diffraction spots when there is an internal boundary between two regions of different structure within a crystal. For crystals of binary alloys having long-range ordering, nanodiffraction from the regions of out-of-phase domain boundaries, show a splitting of the "superlattice reflections", but not of the "fundamental" reflections. By observing the form of the splitting it is possible to derive the form of the boundary [63, 64].

Similarly, in face-centered cubic metals, where twinning occurs on $\{111\}$ planes, if the incident beam is parallel to a twin plane, some reflections are not affected by the twinning and so the diffraction spots are not affected, but for other reflections the crystal periodicity shows a discontinuity so that the diffraction spots are split [65]. From such observations, the nature of any discontinuity can be deduced.

When a large aperture, or no aperture, is used in the STEM instrument, a perfect thin crystal gives the pattern of Ronchi fringes, as described above. The perturbation of the Ronchi pattern in the presence of a discontinuity in the crystal structure is profound and may, in principle, be analysed to deduce the nature of the discontinuity. However, the possibility of such analyses has been very little explored.

5. MICROANALYSIS IN STEM

5.1. Electron Energy Loss Spectroscopy and Imaging

It was realized, from the beginning [2], that it is easy and useful to attach electron spectrometers to STEM instruments so that images may be formed with electrons that have lost energy through inelastic scattering processes, and energy-loss spectra may be obtained to allow compositional analysis of very small regions of the specimen. Use of the field emission guns allowed ample signal strength for the analytical data and imaging by EELS or X-ray emission with spatial resolutions small as 1 nm [66]. It may be noted that these techniques form valuable adjuncts for other applications of STEM instruments especially because they are readily available and their access does not hinder the use of any of the STEM imaging or diffraction modes.

Bulk plasmon losses, due to the collective excitations of nearly-free electrons in the interior of solids, ranging from a few volts up to about 50 eV, may give strong characteristic peaks in the energy-loss spectrum, and provide information of modifications of nearly-free electron states in solids due to variations of composition or bonding [67]. These excitations are generally considered not to be highly localized, but experimental results have indicated that they can show changes in the electron energy states over distances as small as 1 nm [68].

5.2. Secondary Emissions

In addition to the emission of X-rays, characteristic of the elements present in the sample, the incident high-energy electron beam gives rise to the emission of low-energy secondary electrons with energies of a few eV, and the Auger electrons, with energies of tens or hundreds of eV, produced when the energy from the excited inner shells of atomic electrons is transferred to produce the emission of outer-shell electrons. Because these low-energy electrons have very limited path-lengths in solids, the images created by detecting them comes from the thin surface layers of solids. Dedicated instruments designed for secondary electron microscopy (SEM) or Auger electron microscopy (AEM) have long been used for the study of surface structure and morphology. The spectroscopy of Auger electrons (AES), provides the means for the analysis of the chemical composition of surface layers. The advantage of STEM instruments for SEM or AEM imaging or for AES is that the spatial resolution may be high, of the order of 1 nm or better. Several authors have determined that the spatial resolution is not degraded by delocalization of the process of exciting the electron emission [68].

The specimen in a STEM instrument is usually immersed in the strong magnetic field of the objective lens. The low-energy electrons emitted from the specimen travel in spirals around the lines of force of the magnetic field. When they exit the field, they may be deflected into a detector or else into a spectrometer so that their energy distribution may be analysed. For the thin specimens studied in STEM instruments, low energy electrons are emitted from both the top and bottom surfaces and, in specially-designed instruments, images or analyses of both surfaces may be obtained simultaneously, and compared with the images or analyses produced by the transmission of the incident fast electrons [69].

6. STUDIES OF NANOPARTICLES AND NANOTUBES

6.1. Nanoparticles

The STEM techniques have obvious relevance for the rapid increase of scientific and industrial interest in the formation and properties of small particles having diameters of the order of one, or a few, nanometers. Such particles can be imaged preferentially by use of special STEM imaging modes. Their crystalline structures can be determined by nanodiffraction and their compositions can be determined by microanalysis with high spatial resolution.

Studies of supported metal catalysts have proved valuable. For dispersions of nanometer-size platinum particles on supports of alumina or other oxides, nanodiffraction from individual particles chosen from STEM images, has revealed their structures and the presence of twinning. In some cases it has been shown that the oxides of platinum may be present [70].

For gold-ruthenium catalysts particles on a support of magnesia, one unexpected finding was the presence of ruthenium in a body-centered cubic modification instead of the usual hexagonal structure [71]. This finding is consistent with the increasing evidence that nanometer-size metal particles may often have structures other than those stable in bulk.

Small particles, particularly of noble metals, are frequently found to be twinned, or multiply twinned [72]. Gold particles with diameters from 10 to 100 nm take the form of decahedra, with five tetrahedrally-shaped regions of perfect crystal related by twinning on (111) planes, or of icosahedra with twenty mutually twinned regions. The evidence from TEM is inconclusive for particles smaller than 10 nm. Nanodiffraction from a sample of gold particles on a polyester support showed that some particles of sizes 3 to 5 nm had multiple twinning of this type, but smaller particles were mostly untwinned [73].

Some minerals are found only in microcrystalline form and the determinations of their structures by conventional X-ray or electron diffraction methods are difficult. The mineral ferrihydrite, for example, occurs in two forms. The "six-line" form, 6LFh, gives X-ray powder patterns with only six diffuse lines. Patterns from "two-line" form, 2LFh, contain only two very diffuse lines. A hexagonal crystal structure has been proposed for the 6LFh, [74]. but only guesses are available for the 2LFh. Nanodiffraction has provided single-crystal patterns from nanosized particles of both types of material, allowing their crystal structures to be determined and refined. In both cases, the material was found to consist of a mixture of crystallites of several phases [75, 76]. The 6LFh samples, for example, contain about sixty percent of a hexagonal phase similar to that previously proposed for the mineral, but also contain cubic structures similar to those of magnetite, Fe_3O_4, and a disordered form of wüstite, FeO.

The molecules of the protein ferritin, important for the transport of iron in the bodies of most living entities, from bacteria to humans, consist of a spherical shell of protein with an iron-rich core. The iron-rich cores were said to have the same structure as the ferrihydrite mineral, and high-resolution electron micrographs appeared to confirm the presence of the hexagonal 6LFh phase [77]. Nanodiffraction in a STEM instrument, however, showed that, as in the ferrihydrite mineral, a variety of phases is present (see Fig. 7) with much the same phases as in the mineral, but in appreciably different proportions, varying with the origins of the ferritin [59].

6.2. Nanotubes and Nanoshells

Since the time of their discovery by Iijima [78] carbon nanotubes have been studied extensively for both their intrinsic interest as representing a new form of matter, and their potential industrial applications. The determinations of their detailed structures have been made mostly by use of high resolution TEM and by selected-area electron diffraction (SAED). However, the use of STEM instruments offers advantages for these purposes The SAED patterns are weak because they come from areas of much larger diameter than the tube diameters and usually depend on having long straight tubes of uniform structure, but STEM nanodiffraction patterns, given with incident beam diameters comparable with, or less than, the tube diameters, can be used to determine the structures of any small region of a tube, or of attachments or inclusions in a tube (see Fig. 8).

In a recent review [79], an account has been given of the results of bright-field and dark-field STEM imaging and individual nanodiffraction patterns or series of nanodiffraction patterns recorded in transit across a variety of nanotube and related systems. These techniques have contributed to our knowledge as follows.

Some multi-walled nanotubes do not have the circular cross-sections, universally assumed, but have cross-sections which are polygonal, and usually pentagonal.

The helix angles of the individual layers of carbon atoms in a multi-walled tube are not uniform, but tend to change at intervals of four or five layers.

The determination of the helix angles for single-walled nanotubes is quick and simple when nanodiffraction is used, so that statistics for the distributions of helix angles in various samples, or in local regions of any one sample may be accumulated readily.

In ropes of single-walled nanotubes, the helix angles may be uniform over regions 10 nm in diameter, but may vary slightly between such regions, giving rise to twists and bends of the ropes.

Multi-walled nanotubes and the associated near-spherical nanoshells, containing inclusions of metals or, usually, metal carbides, the crystallographic relationship between the nanotube walls and the inclusions gives evidence of the process of formation of these systems.

Similar structures have been found for nanotubes composed of the tungsten and molybdenum sulfides. Further applications of the nanodiffraction and STEM techniques have included studies of nanobelts such as those formed of ZnO and other oxides [80] and of boron and its carbides [81].

7. STUDIES OF CRYSTAL DEFECTS AND INTERFACES

The defects of crystals and the interfaces between differing crystal structures are of prime importance for the development of semiconductor and related components for

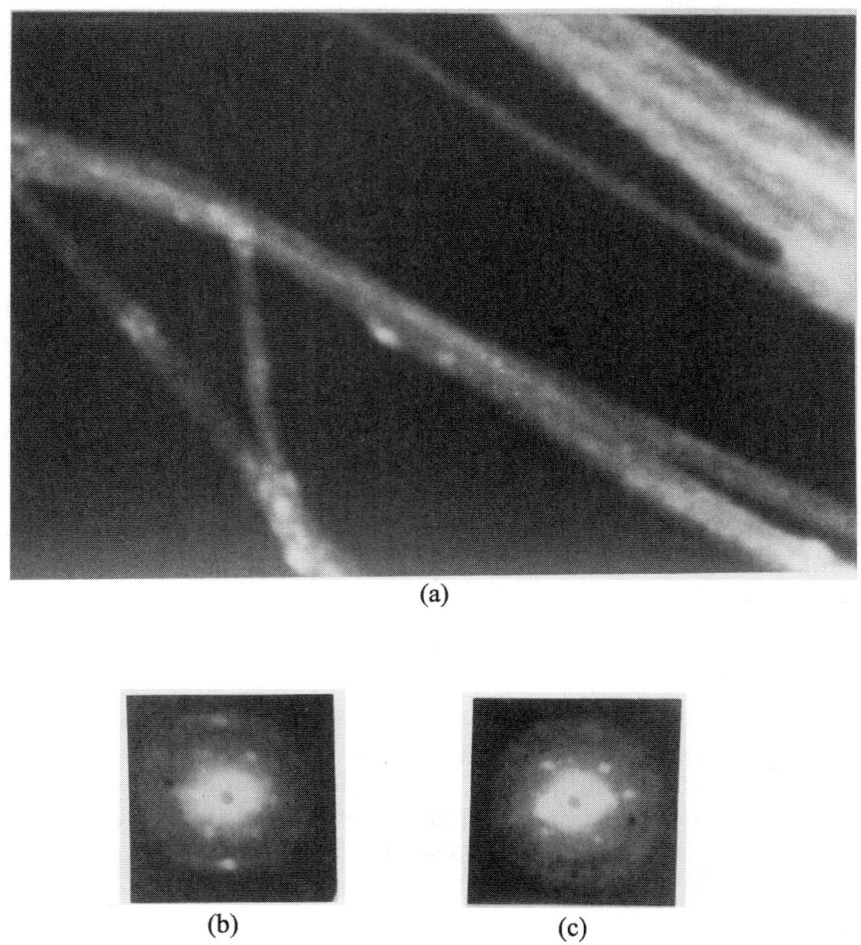

Figure 8. (a) A dark-field STEM image showing single-walled carbon nanotubes (SWnT) and bundles of SWnT, with nanocrystals of lead oxide attached. (b) Nanodiffraction pattern of SWnT, diameter 2 nm, helix angle 30 degrees and tilt 0. (c) Nanodiffraction pattern of SWnT, diameter 2 nm, helix angle 0, tilt 30 degrees.

the burgeoning micro-electronics industry. STEM imaging, particularly in the HAADF mode, is proving to be an important tool for their study. In common practice, a thin slice is cut to show a cross-section of the critical region of a device and the high-resolution image is obtained by viewing the slice along an axial direction of the crystal lattices involved.

For thin crystals, a detailed analysis of a defect may be possible. Planar defects have been seen in TEM images in diamond crystals [82]. Series of nanodiffraction patterns were obtained as a beam of 0.3 nm diameter was translated across such a defect at intervals of 0.02 nm. Comparison of the nanodiffraction pattern intensities with dynamical diffraction calculations for various models showed best agreement with the Humble model [83].

It was shown by Pennycook *et al.* [84] that the high-resolution HAADF images obtained from reasonably thick crystals, and interfaces with their planes parallel to the electron beam, are in many case superior to those of HRTEM in that they do not suffer from the same complications of dynamical scattering effects which give strong variations of TEM contrast for variations of crystal thickness and orientation. Because the HAADF contrast arises from the incoherent addition of thermal diffuse scattering from all depths in the crystal, it may be considered, to a reasonable approximation, as being incoherent imaging, depending only on the number of atoms aligned in the beam direction.

The quasi-incoherent nature of the HAADF imaging of crystals aligned with the incident beam almost parallel with a principle crystal axis, has been confirmed by experiment and by theoretical analyses [85]. For such orientations, the channeling effect gives rise to an oscillation of the incident electron beam intensities and widths along the rows of atoms. As a consequence, there is some oscillatory component in the integrated intensity from the thermal diffuse scattering [26], but such oscillatory effects are relatively small and are rapidly damped with increasing thickness.

Because the electrons are channeled along the lines close to the atom rows, by excitation of the *1s* states of the potential wells of the rows of atoms, there is little spreading of the beams within the crystal and the intensity distribution at the exit face, and in the corresponding image, has clear maxima at the positions of the atomic rows, provided that the atomic rows are well separated. Figure 9 shows an image of the multiple interfaces of crystals of Si and Ge. The images of the separate atomic rows are clear within the individual crystal layers and at the boundaries [86].

Because the incident electron beam in a STEM instrument is not sharply defined, but has weak "tails" extending beyond the central intensity maximum, it is possible that some artefacts will arise. For example, weak maxima may appear in an image at positions close to atomic rows, where no atoms exist [87]. Also, when the rows of atoms are close together, the wavefields in the two potential wells may overlap, and the image intensities are affected. Recently, computer simulations have been made to evaluate these effects [88–91]. Similar, even more prominent and complicated effects of 'non-localization' appear in the corresponding HRTEM images. These effects are reduced when the resolution of the instrument is improved, for example by the use of systems for the correction of the objective lens aberrations.

8. THE STRUCTURE AND COMPOSITION OF SURFACES

8.1. Ultra-High Vacuum Instruments

Although the vacuum in the column of most dedicated STEM instruments is normally an order-of-magnitude better than that in most TEM instruments, it does not usually approach the ultra-high vacuum (UHV) level (better than10^{-10} torr) considered necessary for the studies of surfaces which have been made for many years using the techniques of LEED (low-energy electron diffraction) or Auger analysis [92]. For some

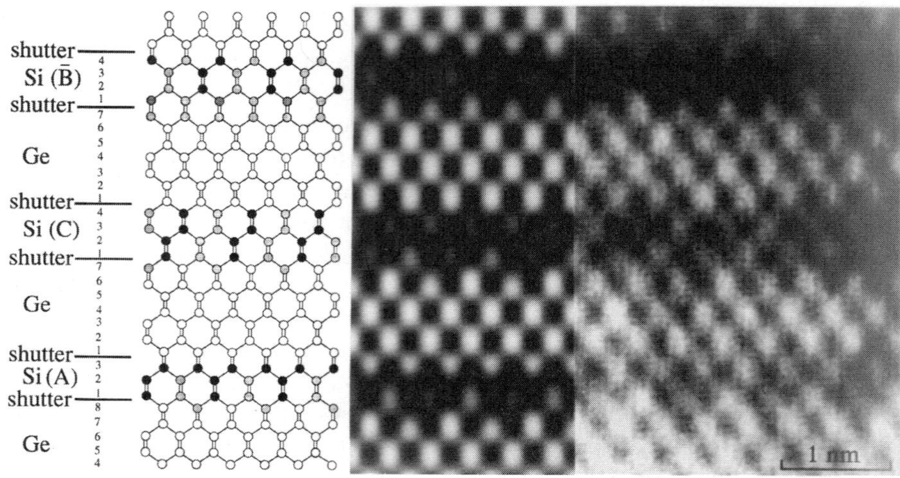

Figure 9. HAADF STEM image of a cross-section of a Si-Ge multilayer in <110> orientation, deposited in an MBE system with the shutter opened and closed at the indicated times to give the alternating Si and Ge layers, showing (left to right) a diagram of the structure, a simulated image and the observed image. [Courtesy of S. J. Pennycook,] c.f. [84].

materials, such as noble metals and oxides, having non-reactive surfaces, the vacuum of a normal STEM instrument may suffice, but for more general surface studies, special UHV instruments must be built [93, 94].

In the MIDAS instrument (a Microscope for the Imaging Diffraction and Analysis of Surfaces) the whole column is constructed of UHV-compatible components, a UHV specimen preparation and treatment chamber is attached and UHV is maintained throughout by the use of suitable ion pumps [93]. This instrument is equipped with the special devices needed for the imaging and analysis with low-energy secondary electrons and Auger electrons. The low-energy electrons emitted from the surfaces of a suitably biased specimen are detected or energy-analysed after they have been collected in the field-free spaces above or below the lens. In this way, high-resolution SEM or AEM images of the top and bottom surfaces of the specimen, and Auger analyses of the composition of the surfaces may be made and compared with the bright-field transmission image.

This instrument has been used for various studies of the initial stages of crystal growth on surfaces, including the nucleation and growth of crystallites of Ge on Si [94]. The image resolution is superior to that of existing UHV SEM instruments or AEM instruments and the possibility of correlating the images with those from STEM imaging modes and nanodiffraction adds greatly to its power and versatility.

8.2. Reflection Electron Microscopy

The technique of reflection electron microscopy (REM) has been widely developed and applied in many studies of the structure of crystals surfaces and of thin layers or deposits on surfaces [95, 96]. Applications of the technique using TEM instruments

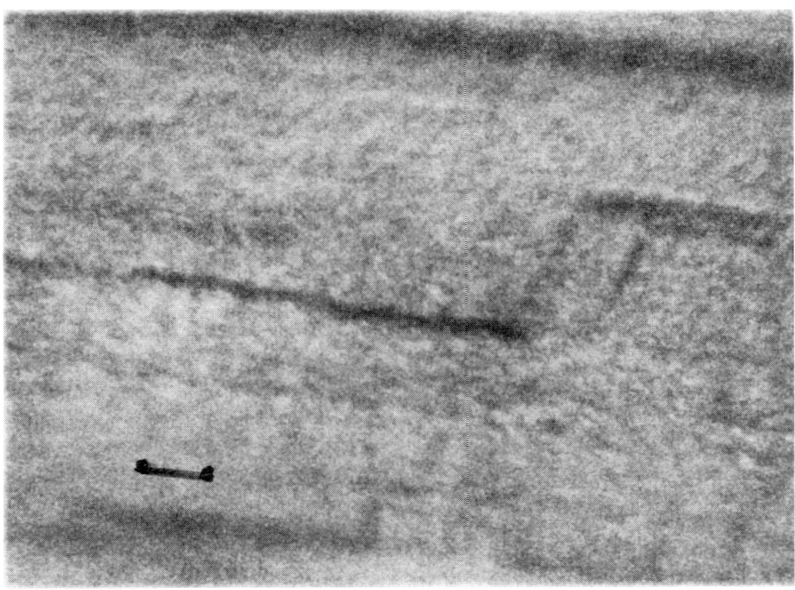

Figure 10. Scanning reflection electron microscopy (SREM) image of the surface of a crystal of MgO, obtained with the specular (400) reflection, showing surface steps [98]. Marker = 10 nm.

with grazing angles of incidence are normally limited by the relatively poor vacuum levels, but REM in special UHV instruments has allowed important studies of the structures, growth and interactions of steps and superlattices on semiconductor and other surfaces [97].

The equivalent scanning reflection electron microscopy (SREM) technique is possible with STEM instruments, using equivalent geometries. Figure 10, for example, is a SREM image showing atom-high steps on the surface of a crystal of MgO [98]. Also superlattice structures on crystal surfaces may be directly imaged [99], and surface reactions, such as the oxidation of copper, have been studied [100].

The advantage of SREM over REM is that the imaging may readily be combined with the observation of nanodiffraction patterns or EELS or X-ray microanalysis of small surface regions or of small protrusions from the surface. For example, nanodiffraction patterns from crystallites of palladium, a few nm in diameter, sitting on the surface of MgO crystals showed their partial and progressive oxidation to PdO [101]. The special imaging modes developed for STEM may be applied for SREM. When a HAADF detector is used, the image of a surface depends very little on the crystal structure, but shows more clearly the surface topography [101].

The intensities of RHEED patterns and the contrast in REM or SREM images may be calculated by use of modified computer programs adapted from those originally developed for LEED [92, 102] or for TEM [21]. The multislice programs, used for transmission diffraction or for HRTEM, may be applied if the slices are taken perpendicular to the surface. Since the slice content is essentially non-periodic, a periodic-continuation approximation is made with the vacuum and topmost crystal atomic layers repeated at large intervals [103, 104]. For SREM, the localized amplitude of the incident beam is used for the input, and the progressive penetration of the wave into the surface and its interaction with any perturbation of the surface structure can be followed.

8.3. Surface Channeling Effects

Calculations made using the multislice computer programs have revealed that, for certain diffraction geometries involved in REM or SREM, the penetration of the incident beam into the surface of a perfect crystal may be very small [105].

A surface-channeling phenomenon takes place, in which the electron wave is confined to just the first one or two atomic layers. The necessary condition is that a strong diffraction beam is generated in a direction parallel to the plane of the surface. Under these conditions, the specular reflection in the RHEED pattern becomes very strong, and the whole RHEED pattern, including the background of Kikuchi lines, is enhanced [106, 107].

With surface channeling, the REM or SREM image contrast is highly sensitive to small perturbations of the surface structure and the EELS analysis of the diffracted beams can be highly sensitive to the composition of the top surface layers [108].

8.4. MEED and MEEM

The severe fore-shortening of REM and SREM images may be avoided, to some extent, if the angles of diffraction by the surface planes are increased by decreasing the beam voltage and so increasing the wavelength. A number of medium-energy instruments operating in the range of 5 to 20 keV have been constructed for diffraction (MEED) and scanning microscopy (MEEM) [8–10]. With the simpler electron optics and the more compact design possible for this energy range, it was easier to produce systems operating in UHV with adequate provision for the preparation and treatment of clean surfaces.

With a cold field-emission electron source, operating in the same UHV environment, a relatively simple electron lens system can give image resolutions of a few nm. The diffraction pattern can be observed on a spherically curved fluorescent screen with energy-filtering meshes, as used for LEED, and a hole in the screen allows selected diffracted beams to be transmitted to a detector.

Such instruments have been applied to studies of the oxidation of copper surfaces [109] and for numerous studies of the surface structure and modifications of semiconductors and related materials.

9. AMORPHOUS MATERIALS

9.1. Thin Quasi-Amorphous Films

For a wide range of materials, including most biological materials, there is no periodic structure. Some local ordering often occurs on a nanometer scale, but in the absence of long-range order, the diffraction patterns obtained by X-ray diffraction or SAED techniques show only a few diffuse haloes, and the materials are considered to be "amorphous". However, with STEM imaging and nanodiffraction, the local ordering may usefully be detected and measured. For the organic and biological materials, such studies are usually limited by their great susceptibility to radiation damage by the incident electron beam, but important progress has been made with investigations of thin films of many inorganic materials.

When a STEM beam of nanometer diameter passes through a thin film of quasi-amorphous material, the nanodiffraction pattern does not contain the diffuse haloes of normal SAED patterns, but shows a two-dimensional array of diffuse maxima and minima, corresponding to the local ordering of the atoms within the illuminated area of the specimen. The patterns change as the beam is moved over distances of the order of 1 nm, and different arrays of atoms are illuminated. Assuming that, for very thin films, the scattering can be interpreted in terms of the simple kinematical approximation, it should, in principle be possible to derive some information about the arrangement of the atoms within each illuminated area. It has been considered that, from the observation of many such patterns, it should be possible to derive not only the average pair-wise, atom-to-atom distances, as in the case of SAED patterns [110], but also the presence of many-atom configurations, or medium-range order [111].

It has been shown by Treacy and Gibson [112] that the technique of TEM with a hollow-cone (or gyrating beam) illumination may form the basis for 'variable-coherence' microscopy, or 'fluctuation microscopy'. By observation of the variation of the speckle in the dark-field images of thin amorphous films, the degree of medium range order present may be investigated. In such a way, they derived information of the extended correlations in semiconductor films [113]. These authors suggested that the reciprocity principle implied that TADDF STEM images using a thin annular detector, could give the same information. An analysis of the equivalent STEM technique was given by Cowley [114], who pointed out that several possible, and experimentally simpler, approaches are possible if the STEM methods are combined with nanodiffraction observations.

9.2. Thick Amorphous Films

For many biological samples, such as those containing complete cells, and for specimens held in environmental cells with controlled atmospheres, it is desirable to obtain images of specimens which may be several micrometers thick. To achieve the required increase in penetration, high-voltage TEM instruments, operating at 1 MeV or more, have been employed, in spite of the large and expensive installations required. However, it has been shown that, for the same voltages, the effective penetration should be greater for STEM and several attempts have been made to build high-voltage STEM instruments operating at up to 1 MeV.

The loss of resolution for thick specimens derives from two causes. Because of multiple elastic scattering in the specimen, there is uncertainty in the position of scattering of any given electron so that the location of any scattering object becomes uncertain. Also, for a thick object there can be multiple energy losses of the electrons, with an energy loss of about 25 eV, on the average, for each event, and the chromatic aberration of the imaging system in TEM gives a spread of the image. For STEM, the first of these causes for resolution loss applies, but the second does not apply because the main imaging components come before the specimen. For typical specimens, the electron energy needed for a given penetration is reduced by a factor of at least 2 for STEM, or the resolution for a given electron energy may be improved by an equal factor [115, 116]. Observations made using the high-voltage STEM instrument built by Strojnik, [7] have confirmed these estimates.

For specimens held in environmental cells, the path length of the electrons in a scattering medium, including the gas in the cell, is greater and the first of the above resolution-impairing factors is more important. The advantage of the STEM system over the TEM is then not so great.

10. STEM HOLOGRAPHY

10.1. Gabor's in-line Holography

In his original proposal for holography, Gabor [117] suggested that the deleterious effects of lens aberrations on electron microscopy could be overcome by reconstructing the object wave from a hologram in which the wave scattered by the object is made to interfere with a reference wave. He proposed a scheme suitable for use with very thin specimens or specimens for which the object of interest covers only a small part of the image field. A small probe formed by imaging a small bright source with the strong objective lens, as in a STEM instrument, is focused close to the specimen, as suggested in Fig. 3. The greatly enlarged, out-of-focus shadow image is then the hologram, containing the interference effects of the transmitted wave, acting as a reference, and the waves scattered by the object. In the course of reconstruction of the object wave, compensation can be made for the phase distortions due to the aberrations of the objective lens by using either an optical system, as suggested by Gabor, or by the computerized manipulation of a digitized hologram [118].

A demonstration of this form of in-line holography has been given, with relatively low resolution, with reconstruction from a defocused shadow image in a STEM instrument [119]. However, because the phases of the waves are lost in the recording of the intensity distribution in the hologram, the reconstructed image is always accompanied by a defocused conjugate image. Various schemes have been proposed whereby this conjugate image may be made negligibly small or eliminated. By recording the hologram very far out-of-focus, the conjugate image may be made very diffuse so that it forms only a weakly-modulated background; or it may be removed, in principle, if a series of holograms is recorded with small translations of the incident beam [120].

A further limitation of in-line holography is that the reconstruction from the intensity distribution may contain terms of second and higher order in the scattering function of the object. These terms are avoided only if the scattering by the object is very weak compared with the reference wave so that only first-order interference terms

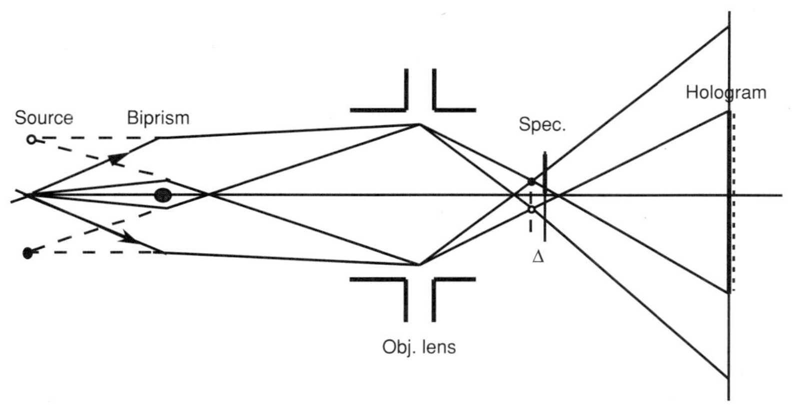

Figure 11. Diagram of the arrangement for off-axis electron holography in a STEM instrument. A biprism in the illuminating system gives two small probes near the specimen, one passing through the specimen and one through vacuum.

are significant. Thus in-line holography can, in practice, be applied only to very thin specimens of light-atom material.

10.2. Off-Axis Holography

The major advances in electron holography have come from the use of off-axis holography in which the reference wave has passed only through vacuum and is made to interfere with the wave passing through the specimen after the two waves have been deflected by an electrostatic biprism placed in the imaging system. This form of electron holography has been widely used for the enhancement of image resolution, for the mapping of weak electrical and magnetic fields, within and around solids and for other purposes [121, 122].

It has been pointed out that many forms of electron holography are possible, and that, as suggested by the reciprocity principle, for each form of holography applicable in a TEM instrument, there is a STEM equivalent [123]. In the STEM equivalent of the off-axis form illustrated in Fig. 11, an electrostatic biprism is placed in the illuminating system of the instrument so that two focused probes are formed by the objective lens in the specimen plane, and these are scanned in parallel, one through the specimen and one through vacuum, to record the hologram.

Little use has been made, as yet, of the off-axis form of STEM holography for the enhancement of resolution, but the technique has proved highly effective for the study of magnetic fields in and around ferromagnetic objects [124]. Figure 12, for example, shows a display of the contoured field strength in a thin film of Co. [125]. The spatial resolution for such images can be about 1 nm, which is adequate for most purposes. Applications of this technique have been made to the study of the configurations of magnetic domains within small crystals and the fields within multi-layer composites of ferromagnetic and non-ferromagnetic materials [126].

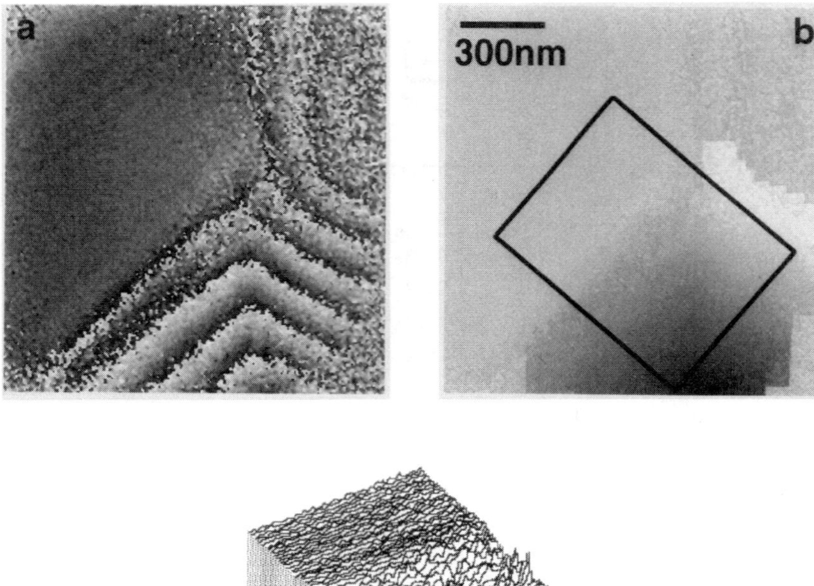

Figure 12. Determination of the magnetization in a 20 nm Co film using STEM holography. (a) Wrapped phase image. (b) Partially unwrapped phase image. (c) Three-dimensional view of the region marked in (b); the gradient of the phase difference taken perpendicular to the edge determines the absolute value of the magnetization [125].

11. ULTRA-HIGH-RESOLUTION STEM

11.1. Atomic Focusers

As pointed out in our discussion of the STEM imaging of crystals, above, the passage of an incident STEM beam along a row of atoms in a thin crystal can produce a cross-over near the exit face having a diameter of 0.05 nm or less. If this fine probe can be scanned over a thin specimen, STEM imaging with a correspondingly high resolution should be possible [54, 55]. Because the effective focal length of such an "atomic-focuser" lens is of the order of 2 nm, there are severe practical difficulties in realizing this STEM mode with the atomic-focuser crystal and the specimen so close together.

If the distance between the atomic-focuser crystal and the specimen is increased to, perhaps, 100 nm, the arrangement is that for in-line electron holography. The area of the specimen illuminated at any time has a diameter of about 10 nm and no scanning of the beam is needed to produce a useful image. Computer simulations have shown that the resolution attainable in the image reconstructions from such an arrangement may be better than 0.05 nm when a thin gold crystal is used as the focuser [127]. The conjugate image is defocused by an amount equal to twice the separation of the crystals and so forms only a very diffuse background.

Further schemes for attaining ultra-high resolution in a STEM instrument depend on the formation of a periodic, coherent array of fine cross-overs at the exit face of a thin 'atomic-focuser' crystal, and the formation of Fourier images at regular intervals in the subsequent space [21]. An experimental realization of one such scheme has been shown with a STEM beam focused on one wall of a carbon nanoshell to image the opposite wall [53]. The resolution observed in these experiments was approximately equal to 0.06 nm, the diameter of the nanoprobe formed by a graphite crystal.

11.2. Aberration Correction

In recent years, considerable progress has been made in the attempts to correct the aberrations of the objective lens used for TEM and STEM by the addition of systems of multipole auxiliary lenses, following suggestions by Scherzer [128], Rose [129] and Crewe [130]. For STEM, a quadrupole-octupole aberration corrector has been shown to be effective [15]. The third-order spherical aberration constant may be reduced to zero, or to some controlled small value. The resolution limit possible for STEM imaging is thereby reduced from about 0.2 nm to better than 0.1 nm. As shown in Fig. 13, the correction of aberrations of the objective lens of a STEM instrument gives considerable improvement in definition, contrast and signal-to-noise ration for the imaging of individual heavy atoms on a light-atom support [131].

Such resolution improvement is often demonstrated by images of thin crystals of silicon in [110] orientation. In this projection, there are pairs of atom rows with a separation of 0.136 nm, giving the so-called "silicon dumb-bells" in images with sufficient resolution. A more important feature of the improved resolution attainable with the aberration correction is the improvement of the localization in the images of crystals. There is less cross-talk between images of adjacent rows of atoms, and the positions of atom rows in asymmetrical surroundings, as at boundaries or interfaces are more accurately indicated [16].

11.3. Combining Nanodiffraction and Imaging

For normal STEM imaging, only one signal is derived, either from the incident beam spot or from all or parts of the diffraction pattern, to form the image. It has long been realized that it should be possible to take advantage of the fact that for each incident beam position a complete two-dimensional nanodiffraction pattern can, in principle, be recorded, and this should allow one to derive much more information, and possibly much better resolution, than that from the standard techniques.

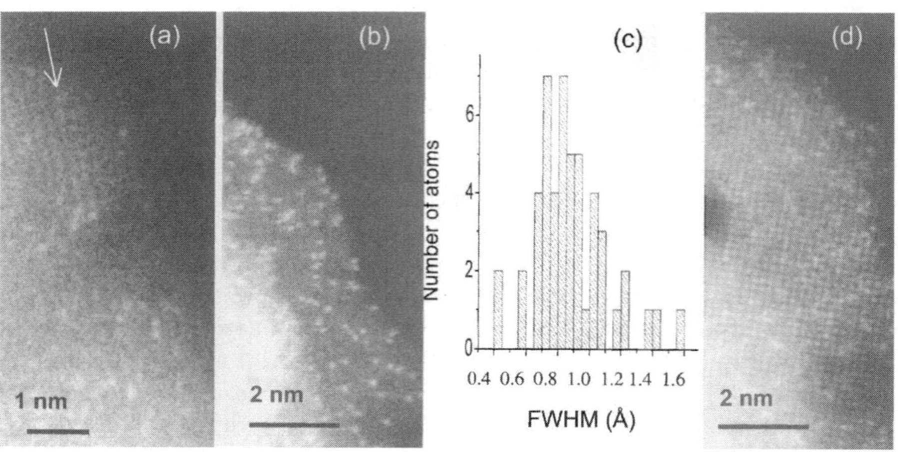

Figure 13. Images of individual La atoms on γ-alumina obtained with a 300 kV STEM, before (a) and after (b, d) aberration correction, showing the improvement in contrast, resolution and signal-to-noise ratio. In (d), the La atoms are visible at the same time as the γ-Al_2O_3 lattice in [100] orientation. The histogram (c) shows the FWHM of intensity profiles. (Courtesy of Dr. S. J. Pennycook], [131]

In a scheme proposed by Konnert *et al.* [132], the nanodiffraction patterns are recorded for a series of closely-spaced positions of the incident beam. Fourier transform of the nanodiffraction pattern intensity distributions gives a series of correlated real-space autocorrelation functions from which the structure of the specimen can be derived. This process was carried out for a thin crystal of silicon in [110] orientation using a STEM instrument having a resolution of about 0.3 nm and gave a calculated image having a resolution which was clearly better than 0.1 nm.

In a formulation of the problem by Rodenburg and Bates [133], a four-dimensional intensity function is derived with the two dimensions of the image and two dimensions of the nanodiffraction intensity distributions. It was shown that, by taking a particular two-dimensional section of this four-dimensional function, an image may be derived with an improvement of resolution by at least a factor of two. The validity of this approach was demonstrated by light-optical analogue experiments and by STEM experiments with moderate resolutions [134]. The theoretical treatment of this procedure shows that the perturbations of the phase distributions in the transfer function of the objective lens due to the lens aberrations are cancelled out, and the resolution obtainable should be twice as good as that corresponding to the objective aperture size, which can, in principle, be arbitrarily large [135]. The practical limitations on resolution enhancement then come from the incoherent limiting factors of microscope instabilities. In experimental tests of the method, made for simplicity on a one-dimensional object, the side of a multi-wall carbon nanotube, a resolution of better than 0.1 nm was demonstrated using a STEM instrument for which the resolution for normal STEM imaging is about 0.3 nm [136].

One limitation of the Rodenburg scheme is that it is applicable only to thin, weakly scattering objects since there are background contributions to the derived image arising from a defocused conjugate image and second- and higher-order terms in the scattering function, as in in-line holography. Also there is the practical problem that an enormous amount of data-recording and computation is involved in the derivation and treatment of the four-dimensional intensity function, requiring long exposure times for reasonable specimen areas. It has been suggested [137] that the first of these difficulties may be avoided if the scheme is combined with off-axis holography, making use of an electrostatic biprism to produce two focussed probes in the STEM instrument (Fig. 11). The second difficulty may be avoided if the required two-dimensional section of the four-dimensional function is automatically produced during the recording process by scanning the two probes in opposite directions, which is possible if the voltages on two perpendicular biprisms are scanned [138].

12. CONCLUSIONS

In this review, it has been shown that STEM imaging, particularly when accompanied by the associated techniques of nanodiffraction and the microanalysis of nanometer-sized regions, has the potential for being of increasing value for studies of fundamental importance for nanotechnology. The dedicated STEM instrument with a cold-field-emission gun offers the highest available intensity in a probe of nanometer diameter or less and thus ensures the greatest capability for studies of the shapes, sizes, crystal structures, surface structures and compositions of nanometer-size regions of small particles, thin films and nanocrystals.

As in TEM, the specimens needed for STEM imaging or nanodiffraction or microanalysis must usually be very thin with thicknesses less than few hundred nm. Nanoparticles may be supported on very thin supporting films, usually of amorphous carbon, or else, in favorable cases, they may be self-supporting as is usually true for nanotubes and nanowires. For the studies of three dimensional structures such as the components of microcircuitry, it is necessary to cut thin sections. Efficient procedures have been developed for this purpose [139]. Adaptation of the STEM instrumentation for surface studies allows the imaging, diffraction and analysis of the surfaces of bulk materials, with high-resolution SEM and AEM for general morphological studies and, in the case of flat crystals surfaces, SREM and associated nanodiffraction and microanalysis.

Further developments of the STEM techniques are to be anticipated. Because the number of dedicated STEM instruments is much smaller than the number of TEM instruments, and the number of investigators concerned with the development of new methods for their use is correspondingly limited, many possibilities remain to be explored. The excitement associated with the recent development of the techniques for ultra-high resolution may well prompt new attempts to explore further the new and existing capabilities.

The measurement and quantitative interpretation of STEM image intensities is being pursued in some limited cases, such as for the HAADF images of crystals of simple structure and their interfaces, and work has commenced on the theoretical basis for detailed interpretations [140, 141].

There have been some measurements of image intensities from thin crystals as adjuncts to the structure analysis of crystals by electron diffraction methods [142], but these measurements have been confined to TEM and SAED techniques. The corresponding measurements with STEM and nanodiffraction, applicable to much smaller regions of crystallites and crystal defects, or to nanoparticles, are missing. There have been indications that nanodiffraction, combined with STEM imaging, may be used to explore the arrangements of atoms, and the local symmetries, within the unit cells of complex structures [143], but this possibility has not been pursued much further.

A few of the possibilities for coherent nanodiffraction in STEM instruments have been explored to a limited extent, as outlined above. There have been some limited observations but few quantitative measurements. Much more remains to be done before the wealth of potential applications in nanotechnology and nanoscience can be fully exploited.

ACKNOWLEDGEMENTS

The author is grateful to the many students, post-docs and collaborators who have contributed to the exploration of the STEM techniques over the past twenty five years or so in this laboratory; to the Center for High Resolution Electron Microscopy at ASU for provision of the instruments and facilities; and to the staff of the Center, particularly Al Higgs, John Wheatley and Karl Weiss, for their continued support and help in the essential task of keeping the instruments running. Many thanks to Dr. Stephen Pennycook for providing figures 9 and 13.

REFERENCES

1. M. von Ardenne, *Z. Phys.*, 109 (1938) 553.
2. A. V. Crewe and J. Wall, J. Mol. Biol., 48 (1970) 373.
3. A. V. Crewe, J. Wall and J. Langmore, *Science*, 168 (1970) 1333.
4. J. M. Cowley, *Micros. Res. Tech.*, 46 (1999) 75.
5. J. M. Cowley, *Appl. Phys. Letts.*, 15 (1969) 58.
6. E. Zeitler and M. G. R. Thompson, *Optik* 31(1970), 258 and 359.
7. A. Strojnik, *Scanning Electron Microscopy/1981/1*, SEM Inc., AMF O'Hare (Chicago) Om Johari, Ed., (1981), 117 and 122.
8. C. Elibol, H.-J. Ou, G. G. Hembree and J. M. Cowley, *Rev. Sci. Instrum.*, 56 (1985) 1215.
9. J. A. Venables, A. P. Janssen, P. Akhter, J. Derrien and C. J. Harland, *J. Micros.*, 118 (1980) 351.
10. M. Ichikawa, T. Doi, M. Ichihashi and K. Hayakawa, *Jpn. J. Appl. Phys.*, 23 (1984) 913.
11. E. M. James and N. D. Browning, *Ultramicroscopy*, 78 (1999) 125.
12. N. D. Browning, L. Arslan, P. Moeck and T. Topuria, *phys. stat. sol (b)*, 227 (2001) 229.
13. Y. Hirotsu, M. Ishimaru, T. Ohkubo, T. Hanada and M. Sugiyama, *J. Electron Micros.*, 50 (2001) 435.
14. P. E. Batson, N. Dellby and O. L. Krivanek, *Nature*, 418 (2002) 617.
15. O. L. Krivanek, N. Dellby, M. F. Murfit, P. D. Nellist and Z. Szilagyi, *Micros. Microanal.* (In Press).
16. J. M. Cowley and M. M. Disko, *Ultramicroscopy*, 5 (1980) 469.
17. R. W. Carpenter and J. C. H. Spence, *J. Microsc.*, 142 (1986) 211.
18. V. Ronchi, *Appl. Optics*, 3 (1964) 437.
19. J. M. Cowley, *Ultramicroscopy*, 7 (1981) 19.
20. J. A. Lin and J. M. Cowley, *Ultramicroscopy*, 19 (1986) 31.
21. J. M. Cowley, *Diffraction Physics (3rd revised edition)*, Elsevier Science B.V., Amsterdam, The Netherlands (1995).
22. J. M. Cowley and A. Y. Au, in *Scanning Electron Microscopy, 1978, Vol. 1*, Om Johari, Ed., SEM Inc., Illinois, (1978) 53.

23. J. Liu and J. M. Cowley, *Ultramicroscopy*, 52 (1993) 335.
24. A. Howie, *J. Microscopy*, 117 (1979) 11.
25. Z. L. Wang and J. M. Cowley, *Ultramicroscopy*, 31 (1989) 437.
26. S. Hillyard and J. Silcox, *Ultramicroscopy*, 52 (1993) 325.
27. R.-J. Liu and J. M. Cowley, *J. Micros. Soc. Amer.*, 2 (1996) 9.
28. J. M. Cowley, *J. Electron Micros.*, 50 (2001) 147.
29. J. M. Cowley, V. I. Merkulov and J. S. Lannin, *Ultramicroscopy*, 65 (1996) 61.
30. J. M. Gibson and A. Howie, *Chem. Scripta*, 14 (1978) 109.
31. J. M. Chapman, P. E. Batson, E. M. Waddell and R. P. Ferrier, *Ultramicroscopy*, 3 (1978) 203.
32. J. C. H. Spence and J. M. Zuo, *Electron Microdiffraction*, Plenum Press, New York and London (1992).
33. M. Tanaka, M. Terauchi, K. Tsuda and K. Saitoh, *Convergent beam electron diffraction, IV*, JEOL Ltd., (2002).
34. W. Hoppe, *Ultramicroscopy*, 10 (1982) 187.
35. B. C. McCallum and J. M. Rodenburg, *Ultramicroscopy*, 52 (1993) 85.
36. J. C. H. Spence and J. M. Cowley, *Optik*, 50 (1978) 129.
37. J. M. Cowley, *Chem. Scripta*, 14 (1978–79) 279.
38. J. C. H. Spence, *Acta Cryst. A*, 54 (1998) 7.
39. A. E. C. Spargo, M. Beeching and L. J. Allen, *Ultramicroscopy*, 55 (1994) 329.
40. J. M. Cowley and A. F. Moodie, *Acta Cryst.*, 10 (1957) 609.
41. P. Goodman and A. F. Moodie, *Acta Cryst.*, *A*, 30 (1974) 280.
42. K. Ishizuka and N. Uyeda, *Acta Cryst. A.*, 33 (1977) 740.
43. H. A. Bethe, *Ann. Physik.*, 87 (1928) 55.
44. P. B. Hirsch, A. Howie, R. B. Nicholson, D. W. Pashley and M. J. Whelan, *Electron Microscopy of Thin Crystals*, Butterworth and Co. London (1965).
45. C. J. Humphreys and E. G. Bithel, in *Electron Diffraction Techniques, Vol.1*, J. M. Cowley, Ed., Oxford Univ. Press, Oxford. (1992).
46. J. Barry, in *Electron Diffraction Techniques, Vol. 1*, J. M. Cowley, Ed., Oxford Univ. Press, Oxford (1992).
47. K. Ishizuka, *Ultramicroscopy*, 90 (2002) 71.
48. Z. L. Wang, *Elastic and Inelastic Scattering in Electron Diffraction and Imaging*, Plenum Press, New York, (1995).
49. P. Rez. *Ultramicroscopy*, 81 (2000) 195.
50. D. Van Dyck and M. Op de Beeck, *Ultramicroscopy*, 64 (1996) 99.
51. S. J. Pennycook, *Advances in Imaging and Electron Physics*, 123 (2002) 173.
52. M. Sanchez and J. M. Cowley, *Ultramicroscopy*, 72 (1998) 213.
53. J. M. Cowley and J. B. Hudis, *Micros. Microanal.*, 6 (2000) 429.
54. V. V. Smirnov, *J. Phys. D*, 31 (1998) 1548.
55. J. M. Cowley, J. C. H. Spence and V. V. Smirnov, *Ultramicroscopy*, 68 (1997) 135.
56. J. M. Cowley, R. E. Dunin-Borkowski and M. Hayward, *Ultramicroscopy*, 72 (1998) 223.
57. R. E. Dunin-Borkowski and J. M. Cowley, *Acta Cryst. A*, 55 (1999) 119.
58. J. M. Zuo, I. Vartanyants, M. Gao, R. Zhang and L. A. Nagahara, *Science* (2003) (In press).
59. J. M. Cowley, D. E. Janney, R. C. Gerkin and P. R. Buseck, *J. Struct. Biol*, 131 (2000) 210.
60. J. M. Zuo, M. Kim, M. O'Keeffe and J. C. H. Spence, *Nature*, 401 (1999) 49.
61. J. M. Cowley and J. C. H. Spence, *Ultramicroscopy*, 6 (1981) 359.
62. M. Pan, J. M. Cowley and J. C. Barry, *Ultramicrsocopy*, 30 (1989) 385.
63. J. Zhu and J. M. Cowley, *Acta Cryst. A*, 38 (1982) 718.
64. J. Zhu, H. Q. Ye and J. M. Cowley, *Ultramicroscopy*, 18 (1985) 111.
65. J. Zhu and J. M. Cowley, *J. Appl. Cryst.*, 16 (1983) 171.
66. R. F. Egerton, *Electron Energy Loss Spectroscopy*, Plenum Press, New York (1986).
67. H. Raether, *Excitation of Plasmons and Interband Transitions by Electrons*, Springer-Verlag, Berlin (1980).
68. M. R. Scheinfein, J. S. Drucker and J. K. Weiss, *Phys. Rev. B*, 47 (1993) 4068.
69. G. G. Hembree, P. A. Crozier, J. S. Drucker, M. Krishnamurthy, J. A. Venables and J. M. Cowley, *Ultramicroscopy*, 31 (1989) 111.
70. M. Pan, J. M. Cowley and I. Y. Chan, *J. Appl. Cryst.*, 20 (1987) 300.
71. J. M. Cowley and R. J. Plano, *J. Catalysis*, 108 (1987) 199.
72. J. A. Allpress and J. V. Sanders, *Surf. Sci.*, 7 (1965) 1.
73. J. M. Cowley and R. A. Roy, in *Scanning Electron Microscopy/1981*, Om Johari, Ed., SEM Inc., AMF O'Hare, (Chicago), (1982) 143.
74. V. A. Drits, B. A. Sakharov, A. L. Salyn and A, Manceau, *Clay Minerals*, 28 (1993) 185.

75. D. E. Janney, J. M. Cowley and P. R. Buseck, *Amer. Mineral.*, 85 (2000) 1180.
76. D. E. Janney, J. M. Cowley and P. R. Buseck, *Amer. Mineral.*, 86 (2001) 327.
77. W. H. Massover and J. M. Cowley, *Proc. Nat. Acad. Science*, 70 (1973) 3847.
78. S. Iijima, *Nature*, 354 (1991) 56.
79. J. M. Cowley, in *Electron Microscopy of Nanotubes and Nanowires*, Ed. Z. L. Wang, Kluwer Academic/Plenum Publishers (2003).
80. Z. W. Pan, Z. R. Dai and Z. L. Wang, *Science*, 291 (2001) 1947.
81. C. Jones Otten, O. R. Lourie, M.-F. Yu, J. M. Cowley, M. J. Dyer, R. F. Rouff and W. E. Buhro, *J. Amer. Chem. Soc.*, 124 (2002) 4564.
82. P. Humble, *Proc. Roy, Soc. London*, A381 (1982), 65.
83. Cowley, J. M., Osman, M. A. and Humble, P., *Ultramicroscopy*, 15 (1984) 311.
84. S. J. Pennycook and D. E. Jesson, *Ultramicroscopy*, 37 (1991) 14.
85. P. D. Nellist and S. J. Pennycook, *Ultramicroscopy*, 78 (1999) 111.
86. B. Rafferty, P. D. Nellist and S. J. Pennycook, *J. Electron Micros.*, 50 (2001) 227.
87. K. Watanabe, N. Nakanishi, T. Yamazaki, M. Kawasaki, I. Hashimoto and M. Shiojiri, *phys. stat. solidi, (b)*, 235 (2003) 179.
88. G. R. Anstis, D. Q. Cai and D. J. H. Cockayne, *Ultramicroscopy*, 94 (2003) 309.
89. C. Dwyer and J. Etheridge, *Ultramicroscopy* (In press).
90. P. M. Voyles, J. L. Grazul and D. A. Muller, *Ultramicroscopy* (In press).
91. Z. Yu, P. E. Batson and J. Silcox, *Ultramicroscopy* (In press).
92. J. B. Pendry, *Low Energy Electron Diffraction*, Academic Press, New York, London (1974).
93. J. A. Venables, *Introduction To Surface and Thin Film Processes*, Cambridge University Press, UK, (2000).
94. M. Krishnamurthy, J. S. Drucker and J. A. Venables, *J. Appl. Phys.*, 69 (1991) 6461.
95. J. M. Cowley, in *Handbook of Microscopy, Vol.1, Methods 1*, Eds., S Amelinckx, D. Van Dyck, J. F. Van Landuyt, and G. Van Tenderloo, VCH Verlag, Weinheim, Germany (1997).
96. K. Yagi in *Electron Diffraction Techniques, Vol. 2*, J. M. Cowley, Ed., Oxford Univ. Press, Oxford, (1993) 260.
97. Y. Tanishiro, K. Takayanagi and K. Yagi, *J. Microsc.*, 142 (1986) 211.
98. J. M. Cowley, *Ultramicroscopy*, 27 (1989) 319.
99. J. Liu and J. M. Cowley, *Ultramicroscopy*, 48 (1993) 381.
100. Milne, R. H., in *Reflection High Energy Electron Diffraction*, Eds. P. K. Larson, P. J. Dobson, Plenum Press, New York and London (1988), p. 317.
101. H.-J. Ou and J. M. Cowley, *phys. stat. solidi (a)*, 107 (1988) 719.
102. J. M. McCoy and P. A. Maksym, *Surface Sci.*, 310 (1994) 217.
103. L. M. Peng and J. M. Cowley, *Acta Crystallog., A*, 42 (1986) 552.
104. Y. Ma and L. D. Marks, *Micros. Res. Tech*, 20 (1992) 371.
105. Z. L. Wang, J. Liu, P. Lu and J. M. Cowley, *Ultramicroscopy*, 27 (1988) 101.
106. N. Yao and J. M. Cowley, *Ultramicroscopy*, 33 (1990) 237.
107. M. Gajdardziska-Josifoska and J. M. Cowley, *Acta Crystallog., A*, 47 (1991) 74.
108. Z. L. Wang and J. M. Cowley, *J. Micros. Spectros. Electroniques*, 13 (1988) 184.
109. J. D. Landry, G. G. Hembree, P. E. Hojlund-Nielsen and J. M. Cowley, in *Scanning Electron Microscopy, 1976/1.*, Om Johari, Ed., IITResearch Institute, Chicago, Ill., (1976), 239.
110. D. J. H. Cockayne, D. R. McKenzie, W. McBride, C. Goringe and D. McCullock, *Microsc. Microanal.* 6 (2000) 329.
111. J. M. Cowley in *Diffraction Studies of Non-Crystalline Substance*, I. Hargittai and W. J. Orville-Thomas, Eds., Elsevier Sci. Publ. (Amsterdam) (1981) 847.
112. M. M. J. Treacy and J. M. Gibson, *Acta Crystallog., A*, 52 (1996) 212.
113. J. M. Gibson, M. M. J. Treacy and P. M. Voyles, *Ultramicroscopy*, 83 (2000) 169.
114. J. M. Cowley, *Ultramicroscopy*, 90 (2002) 197.
115. H. T. Pearce-Percy and J. M. Cowley, *Optik*, 44 (1976) 273.
116. D. J. Smith and J. M. Cowley, *Ultramicroscopy*, 1 (1975) 127.
117. D. Gabor *Nature*, 161 (1948) 777.
118. J. M. Cowley and D. J. Walker, *Ultramicroscopy*, 6 (1981) 71.
119. J. A. Lin and J. M. Cowley, *Ultramicroscopy*, 19 (1986) 31.
120. J. A. Lin and J. M. Cowley. *Ultramicroscopy*, 19 (1986) 179.
121. A. Tonomura, L. F. Allard, G. Pozzi, D. C. Joy and Y. A. Ono, Eds. *Electron Holography*, Elsevier, Amsterdam (1995).

122. E. Völkl, L. F. Allard and D. C. Joy, Eds., *Introduction to Electron Holography*, Kluwer Academic/Plenum Publ., New York, etc. (1998).
123. J. M. Cowley, *Ultramicroscopy*, 41 (1992) 335.
124. M. Mankos, M. R. Scheinfein and J. M. Cowley, *J. Appl. Phys.*, 75 (1994) 7418.
125. M. Mankos, A. A. Higgs, M. R. Scheinfein and J. M. Cowley, *Ultramicroscopy*, 58 (1995) 87.
126. M. Mankos, M. R. Scheinfein and J. M. Cowley, in *Advances in Imaging and Electron Physics*, P. Hawkes, Ed., 98 (1996) 323.
127. V. V. Smirnov and J. M. Cowley, *Phys. Rev. B.*, 65 (2002), 064109.
128. O. Scherzer, *Optik*, 2 (1947) 114.
129. H. Rose, *Ultramicroscopy*, 56 (1994) 11.
130. A. V. Crewe and N. W. Parker, *Optik*, 46 (1976) 183.
131. S. J. Pennycook, A. R. Lupini, M. Varela, A. Borisevich, Y. Peng and P. D. Nellist., *in AIP Conf. Proceedings, From the Atomic to the Nanoscale* (In press).
132. J. Konnert, P. D'Antonio, J. M. Cowley, A. Higgs and H. J. Ou, *Ultramicroscopy*, 30 (1989) 371.
133. J. M. Rodenburg and R. H. T. Bates, *Philos. Trans. Roy. Soc. Lond.*, A339 (1992) 521.
134. J. D. Rodenburg, B. C. McCallum and P. D. Nellist, *Ultramicroscopy*, 48 (1993) 304.
135. J. M. Cowley, *Ultramicroscopy*, 87 (2001) 1.
136. J. M. Cowley and J. Winterton, *Phys. Rev. Letts.*, 87 (2001) 016101.
137. J. M. Cowley, *Ultramicroscopy*, 96 (2003) 163.
138. J. M. Cowley, *Microsc. Microanal.* (In press).
139. R. Anderson, B. Tracy and J. Bravman, *MRS Proceedings* (1991) 254.
140. L. J. Allen, S. D. Findlay, M. P. Oxley and C. J. Rossouw, *Ultramicroscopy*, 96 (2003) 47.
141. S. D. Findlay, L. J. Allen, M. P. Oxley and C. J. Rossouw. *Ultramicroscopy*, 96 (2003) 65.
142. D. L. Dorset, *Structural Electron Crystallography*, Plenum Press, New York and London (1995).
143. H.-J. Ou and J. M. Cowley, in *Proc. 46th. Annual Meeting, Electron Micros. Soc Amer.*, G. W. Bailey, Ed., San Francisco Press, San Francisco (1988), 882.

16. IN-SITU ELECTRON MICROSCOPY FOR NANOMEASUREMENTS

ZHONG LIN WANG

1. INTRODUCTION

In-situ microscopy refers to the techniques that allow a direct observation of the dynamic properties at nano-scale through imaging and diffraction. The most traditional in-situ technique is the thermal induced structural transformation and transition. More recent developments include in-situ property measurements of nanotubes/nanowires, electric transport through a nanotube, and environmental microscopy.

Due to the highly size and structure selectivity of nanomaterials, their physical properties could be quite diverse, depending on their atomic-scale structure, the size and chemistry [1]. Characterizing the mechanical properties of individual carbon nanotubes (NTs) is a challenge to many existing testing and measuring techniques because of the following constrains. First, the size (diameter and length) is rather small, prohibiting the applications of the well-established testing techniques. Tensile and creep testing require that the size of the sample be sufficiently large to be clamped rigidly by the sample holder without sliding. This is impossible for nanotubes using conventional means. Secondly, the small size of the nanotube makes their manipulation rather difficult, and specialized techniques are needed for picking up and installing individual nanostructure. Therefore, new methods and methodologies must be developed to quantify the properties of individual nanotubes [2].

In this chapter, we first show the temperature driven in-situ process, which is one of the most established application of TEM. Then we will described the theory and techniques that have been developed for characterizing the properties of quasi-one-dimensional nanostructures.

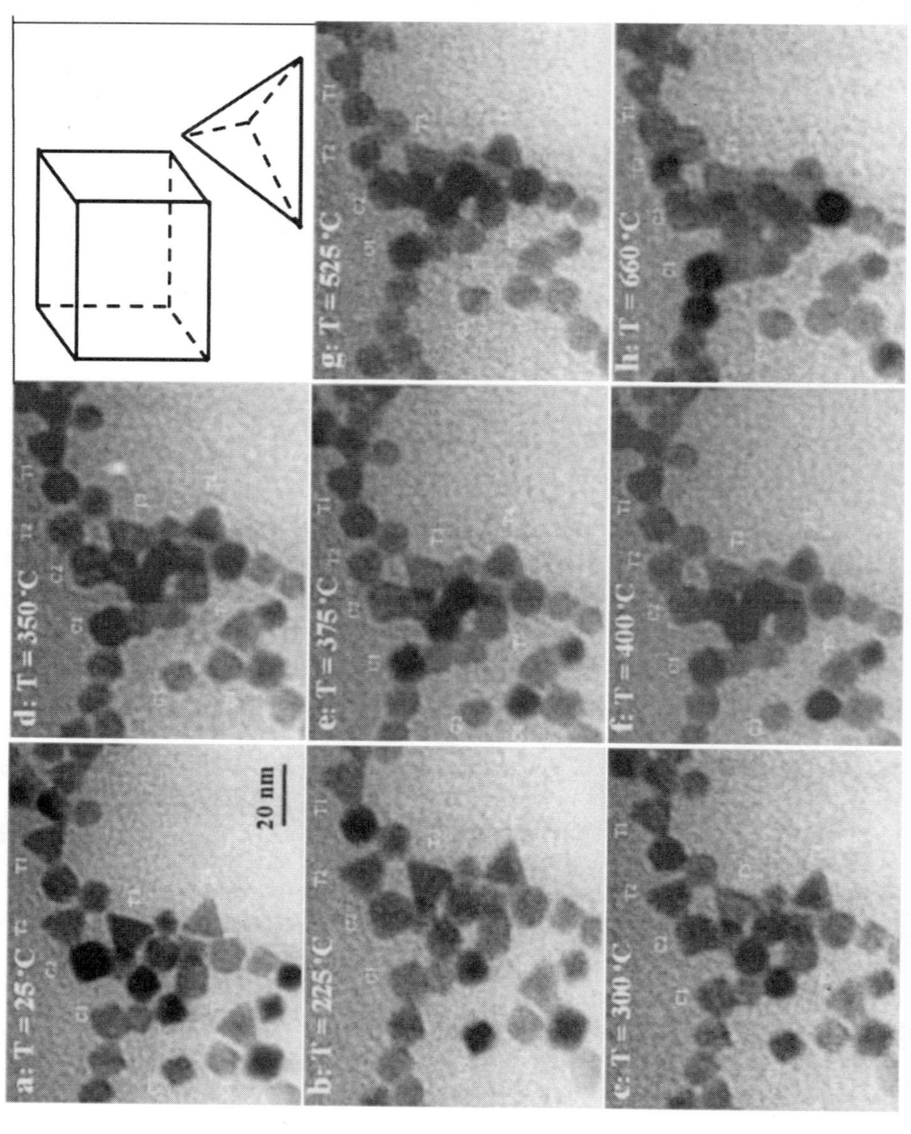
a: T = 25 °C
20 nm
b: T = 225 °C
c: T = 300 °C
d: T = 350 °C
e: T = 375 °C
f: T = 400 °C
g: T = 525 °C
h: T = 660 °C

2. THERMAL INDUCED SURFACE DYNAMIC PROCESSES OF NANOCRYSTALS

In-situ study of the temperature induced phase transformation, structural and chemical evolution of nanocrystals is important for understanding the structure and structural stability of nanomaterials. TEM is an ideal approach for conducting this type of experiments, in which a specimen can be cooled down to the liquid nitrogen or liquid helium temperatures or heated to 1200 °C. The in-situ process can be recorded at TV rate for exhibiting the time and temperature dependent phenomena.

The large percentage of surface atoms in nanocrystals is the origin of their unique properties. The melting point of a nanocrystal is much lower than the bulk melting temperature [3]. The melting of Pt particles is taken as an example here [4]. Platinum nanoparticles with a high percentage of cubic-, tetrahedral- and octahedral-like shapes, respectively, have been synthesized by changing the ratio of the concentration of polymer capping material (polyacrylate) to that of Pt^{2+} ions being reduced by H_2 from K_2PtCl_4 at room temperature [5]. Cubic and tetrahedral are the two most typical shapes for the Pt nanocrystals. We now apply in-situ TEM to determine the stability of particle shapes and the melting behavior.

Figure 1 shows a series of TEM images recorded from the same region when the specimen temperature was increased from 25 to 610 °C. These images were selected from a group of images to present the most significant changes in the particle shapes. For easy notation, particles are labeled as groups to track their shape transformation behavior. Most of the particle shapes showed no significant change when the specimen temperature was below 350 °C (Figures 1a–c). Truncated cubic and tetrahedral particles were formed when the temperature arrived 410 °C (Figure 1d). The corners and edges of the particles were disappearing because the local atoms have higher energy. The tetrahedral particles could still be identified even when the temperature reached 500 °C (Figure 1e), while the cubic particles became spherical when the temperature was above 500 °C. This indicates that the tetrahedral particles are more stable than cubic ones possibly because the {111} surfaces have lower surface energy than the {100}. A feature observed in Figure 1 is that the tetrahedral shape of a particles indicated by T_1 and T_3 was preserved when the specimen temperature was as high as 660 °C. This is possibly due to the contact of the apexes of the particle with the adjacent particles, so that the interparticle diffusion can still sustain the shape of the apexes. Our results indicate that the surface capping polymer is removed by annealing the specimen to a temperature of 180–250 °C, while the particle shape shows no change up to ~350 °C. In a temperature range of 350 to 450 °C, a small truncation occurs in the particle shape but no major shape transformation. The particle shape experiences a dramatic transformation into spherical-like shape when the temperature is higher than ~500 °C (the melting point of bulk Pt is 1773 °C).

←

Figure 1. A series of TEM images recorded in-situ from *Pt* nanocrystals dispersed on a carbon substrate and being heated to different temperatures, exhibiting the shape transformation as well as surface melting phenomenon of the particles at temperatures much lower than the melting point of the bulk. Models for cubic and tetragonal polyhedra are also shown.

2. MEASURING DYNAMIC BENDING MODULUS BY ELECTRIC FIELD INDUCED MECHANICAL RESONANCE

For the characterization of the mechanical properties of nanotubes/nanowires/nanobelts, the main challenge comes from the ultra small size of the object that prohibits the application of conventional techniques. We first have to see the nanotube and then measure it properties. There are two techniques having been developed for this application. The first technique relies on the thermal vibration of the carbon nanotube, and the second one depends on the electrically induced resonance of the nanotubes, both of them were performed in TEM. The details of these techniques will be described in following sections.

2.1. Young's Modulus Measured by Quantifying Thermal Vibration Amplitude

To characterize the mechanical properties of individual nanotubes and correlates them with the observed microstructure, one must do it using in-situ electron microscopy. A striking feature noticed in Figure 2a is that a few nanotubes, indicated by arrowheads, show blurring contrast towards the ends. A systematic measurement of the vibration amplitude as a function of temperature shows that the blurring contrast is due to the thermal induced vibration at the tip, and a quantitative analysis of the vibration amplitude gives the Young's modulus [6].

By assuming that a nanotube is equivalent to a clamped homogeneous cylindrical cantilever of length L, with outer and inner radii R and R_i, respectively, the vibration energy w_n is related to the horizontal vibration amplitude u, for a given mode n, is related to the Young's modulus

$$w_n = \frac{1}{2} C_n u_n^2 \tag{1a}$$

where the effective spring constant for the motion of the tip is

$$C_n = \pi \beta_n^4 Y \left(R^4 - R_i^4\right) / 16L^3 \tag{1b}$$

the values of $\beta_0 = 1.8751$ for the fundamental mode, $\beta_1 = 4.6941$ and $\beta_2 = 7.8548$ for the first two overtones, and R and R_i are the outer and inner diameters of the nanotube. Because there are both elastic and kinetic energy degrees of freedom in a vibrational mode, then $<w_n> = k_B T$ for each vibration mode, with w_n obeying the Boltzmann distribution, and k_B is the Boltzman constant. It can be proven that each mode of a stochastically-driven oscillator has a Gaussian vibration probability profile with a standard deviation of $\sigma_n = (k_B T/c_n)^{1/2}$. With consideration all of the possible modes, the standard deviation is given by

$$\sigma^2 = 16L^3 kT/\pi Y \left(R^4 - R_i^4\right) \sum_n \beta_n^{-4} \approx 0.4243 L^3 k_B T / Y \left(R^4 - R_i^4\right) \tag{1b}$$

This is the equation that can be used to fit the experimentally measured mean-square vibration amplitude $<u^2>$ of the nanotube (Fig. 2b). Thus, the Young's modulus

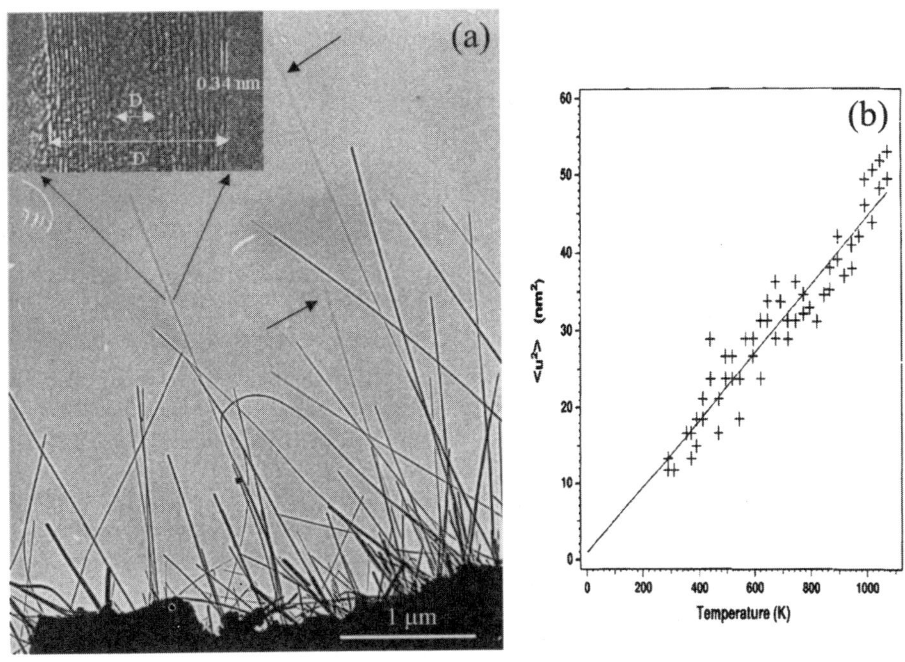

Figure 2. (a) A low magnification TEM image of carbon nanotubes grown by arc-discharging. The thermal vibration can be directly seen at the tips of the nanotubes with a large aspect ration, as indicated by an arrowhead. A high-resolution TEM image is inserted, which shows the inner and outer diameters of a nanotube. (b) Plot of the mean-square vibration amplitude of a carbon nanotube measured in TEM as a function of temperature. The length of the nanotube 5.1 μm, diameter 16.6 nm. The slope of the solid line is 0.044 nm^2K^{-1}, which gives an effective Young's modulus of 3.7 ± 0.2 TPa (data courtesy of Treacy *et al.*, 1996).

can be determined. The experimental data by Treacy *et al.* [6] for the multi-walled nanotube (MWNT) is in the range of 1 TPa.

This technique is likely to be appropriate for nanotubes with smaller diameters and longer lengths because the ones with larger diameters show no visible thermal vibration. The inaccuracy of this type of measurements arises from inaccuracy in determining the vibration amplitude experimentally, and in some cases, an error of more than 100% can be yielded.

This technique has been extended to measure the Young's modulus of individual single-walled nanotubes (SWNTs) [7]. By measuring the vibration amplitude from electron micrographs and assuming that the vibration modes are driven stochastically, an average Young's modulus of $E = 1.25$ TPa was found for SWNT, about 25% higher than the currently accepted value of the in-plane modulus of graphite.

An alternative has been developed by Osakabe *et al.* [8] to measure the thermal vibration frequency of a thin Pt wire. By placing the projected edge of a *Pt* wire partially blocking a small hole drilled at the entrance aperture prior to the detector, the intensity passing through the hole is modulated by the vibration of the wire. As a result of wire vibration, a Fourier transform of $I(t)$ gives the frequency-dependent power spectrum of the vibration, allowing the resonance frequency and the Ω factor to be measured. The advantage of this technique is the use of frequency information of the vibration rather than the vibration amplitude, which is likely to give a better accuracy for the measurements. The disadvantage is that this technique works well for low frequency vibrations due to the unavailability of electron counter for high-frequency vibration, technically limiting its application for smaller objects such as carbon nanotubes.

2.2. Bending Modulus by Electric Field Induced Mechanical Resonance

TEM is a powerful tool for characterizing the atomic-scale structures of solid materials. A modern TEM is a versatile machine that not only can provide a real space resolution better than 0.2 nm, but also can give a quantitative chemical and electronic analysis from a region as small as 1 nm. It is feasible to receive a full structure characterization from TEM. A powerful and unique approach could be developed if we can integrate the structural information of a nanostructure provided by TEM with the properties measured in-situ from the same nanostructure [9, 10]. This is a powerful technique that not only can provide the properties of an individual nanotube but also can give the structure of the nanotube through electron imaging and diffraction, providing an ideal technique for understanding the property-structure relationship. The objective of this section is to introduce this technique and its applications.

2.2.1. Experimental Method

To carry out the property measurement of a nanotube, a specimen holder for an 100 kV TEM was built for applying a voltage across a nanotube and its counter electrode (Figure 3) [10, 11]. In the area that is loading specimen in conventional TEM, an electromechanical system is built that allows not only the lateral movement of the tip, but also applying a voltage across the nanotube with the counter electrode. This set up is similar to the integration of scanning probe technique with TEM. The static and dynamic properties of the nanotubes can be obtained by applying a controllable static and alternating electric field.

The nanotubes used in the study is produced by an arc-discharge technique, and the as-prepared nanotubes are agglomerated into a fiber-like rod. An TEM image recorded from the vicinity of the tip is given in Fig. 4. The carbon nanotubes have diameters 5–50 nm and lengths of 1–20 μm and most of them are nearly defect-free. The fiber is glued using silver past onto a gold wire, through which the electric contact was made. The counter electrode is an Au/Pt ball of diameter ~0.25 mm.

2.2.2. Electrostatic Deflection and Elastic Limit

Figure 5 shows TEM images of carbon nanotubes prior and after applying a 60 V across the electrodes, showing static deflection due to electrostatic attraction. The bright contrast of the two tubes is due to the opposite charges built on the tubes. The nanotube can be bent reversibly over many cycles by turning on and off the

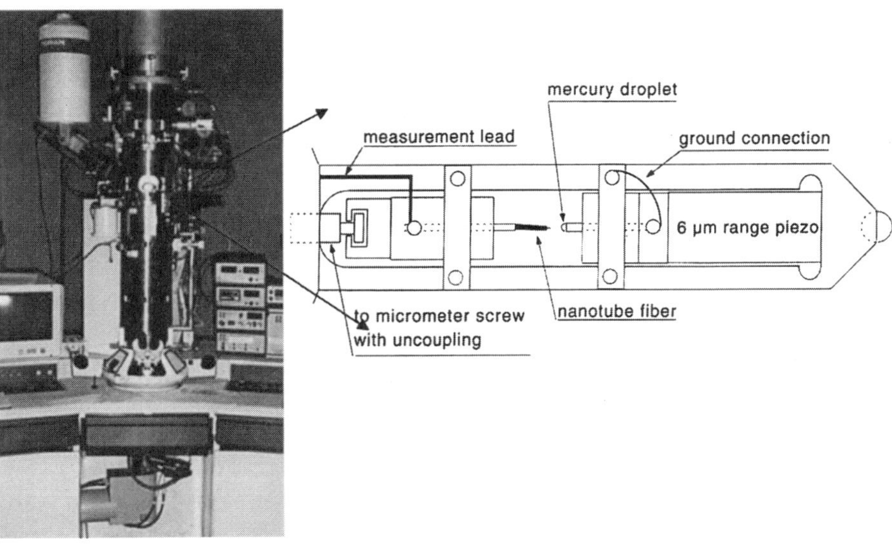

Figure 3. A transmission electron microscope and a schematic diagram of the newly built specimen holder for in-situ measurements.

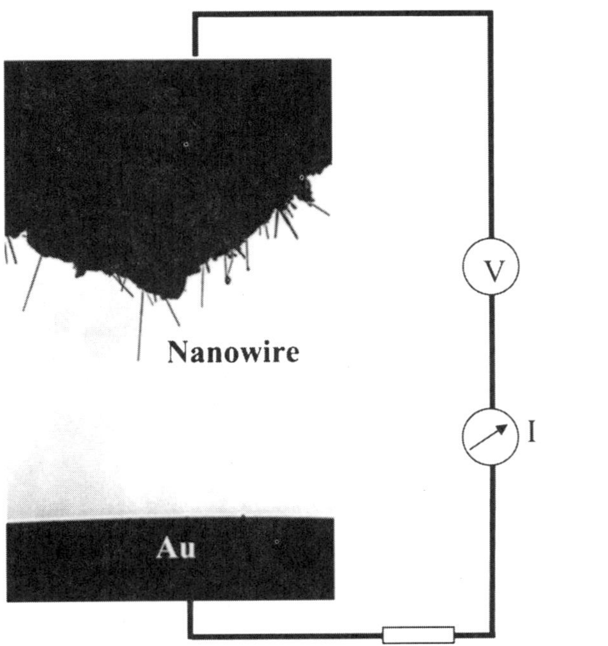

Figure 4. TEM image showing carbon nanotubes at the end of the electrode and the other counter electrode. A constant or alternating voltage can be applide to the two electrodes to induce electrostatic deflection or mechanical resonance.

Figure 5. Electrostatic deflection of carbon nanotubes induced by a constant field across the electrodes. The induced charges are mainly accumulated at the tip.

applied voltage. The nanotube is very flexible, tough and elastic to a large-degree of deformation, much higher than conventional materials.

2.2.3. The Fundamental Resonance Frequency and Nonlinear Effect

As demonstrated in Figure 5, a carbon nanotube can be charged by an externally applied voltage; the induced charge is distributed mostly at the tip of the carbon nanotube and the electrostatic force results in the deflection of the nanotube. Alternatively, if an applied voltage is an alternating voltage, the charge on the tip of the nanotube is also oscillating, so is the force. If the applied frequency matches the natural resonance frequency of the nanotube, mechanical resonance is induced. By tuning the applied frequency, the first and the second harmonic resonances can be observed (Figure 6). The analysis of the information provided by the resonance experiments relies on the theoretical model for the system. The most established theory for modeling mechanical system is the continuous elasticity theory, which is valid for large size object. For atomic scale mechanics, we may have to rely on molecular dynamics. The diameter of the

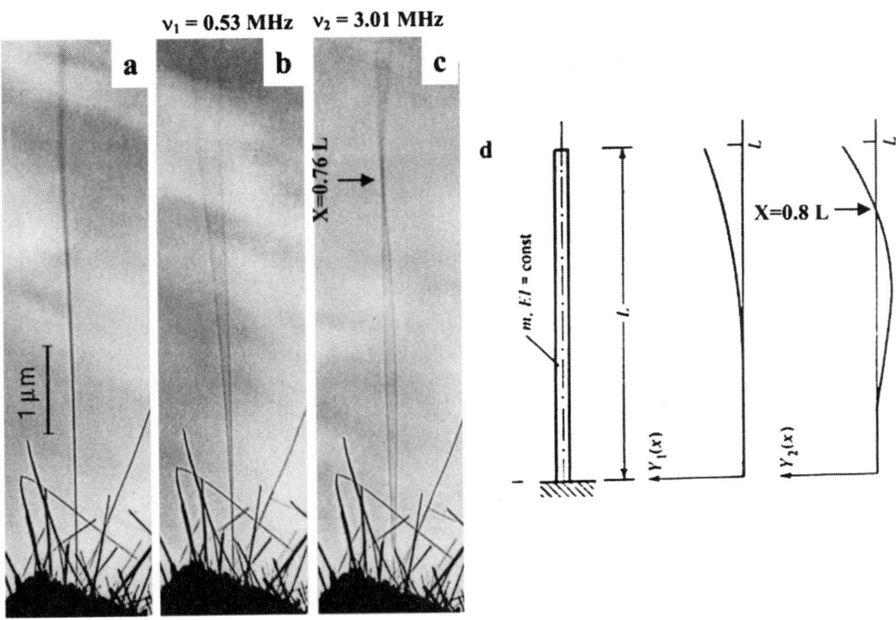

Figure 6. A selected carbon nanotube at (a) stationary, (b) the first harmonic resonance ($\nu_1 = 1.21$ MHz) and (c) the second harmonic resonance ($\nu_2 = 5.06$ MHz). The right-hand side shows the shape predicted based on elasticity theory for a uniform macroscopic beam. (d) Vibration traces of an one-end fixed beam predicted by elasticity theory.

nanotube is between the continuous model and the atomistic model, thus, we need to examine the validity of applying the classical elasticity theory for the data analysis.

We have compared the following three characteristics between the results predicted by the elasticity theory and the experimental results shown in Figure 6. First, the theoretical node for the second harmonic resonance occurs at 0.8 L, and the experiment showed ~0.76 L. Secondly, the frequency ratio between the second to the first mode is $\nu_2/\nu_1 = 6.27$ theoretically, while the observed one is $\nu_2/\nu_1 = 5.7$. The agreement is reasonably well if one looks into the assumptions made in the theoretical model: the nanotube is a uniform and homogeneous beam, and the root of the clamping side is rigid. The latter, however, may not be realistic in practical experiment. Finally, the shape of the nanotube during resonance has been compared quantitatively with the shape calculated by the elasticity theory, and the agreement is excellent. Therefore, we still can use the elasticity theory for the data analysis.

If the nanotube is approximated as a uniform solid bar with one end fixed on a substrate, from classical elasticity theory, the resonance frequency is given by [12]

$$\nu_i = \frac{\beta_i^2}{8\pi}\frac{1}{L^2}\sqrt{\frac{\left(D^2 + D_i^2\right)E_B}{\rho}} \tag{2}$$

where D is the tube outer diameter, D_i inner diameter, L the length, ρ the density, and E_b the bending modulus. The resonance frequency is nanotube selective and it is a specific number for a nanotube.

The correlation between the applied frequency and the resonance frequency of the nanotube is not trivial. From Figure 5 we know that there are some electrostatic charges built on the tip of the carbon nanotube. With consideration of the difference between the surface work functions between the carbon nanotube and the counter electrode (Au), a static charge exists even when the applied voltage is withdrawn. Therefore, under an applied field the induced charge on the carbon nanotube can be represented by $Q = Q_0 + \alpha V_0 \cos \omega t$, where Q_0 represents the charge on the tip to balance the difference in surface work functions, α is a geometrical factor, and V_0 is the amplitude of the applied voltage. The force acting on the carbon nanotube is

$$F = \beta(Q_0 + \alpha V_0 \cos \omega t) V_0 \cos \omega t \\ = \alpha\beta V_0^2/2 + \beta Q_0 V_0 \cos \omega t + \alpha\beta V_0^2/2 \cos 2\omega t, \quad (3)$$

where β is a proportional constant. Thus, resonance can be induced at ω and 2ω at vibration amplitudes proportional to V_0 and V_0^2, respectively. The former is a linear term in which the resonance frequency equals to the applied frequency, while the latter is a non-linear term and the resonance frequency is twice of the applied frequency. In practical experiments, the linear and non linear terms can be distinguished by observing the dependence of the vibration amplitude on the magnitude of the voltage V_0. This is an important process to ensure the detection of the linear term.

Another factor that one needs to consider is to identify the true fundamental resonance frequency. From Eq. (2), the frequency ratio between the first two modes is 6.27. In practice, if resonance occurs at ω, resonance could also occur at 2ω, which is the double harmonic. To identify the fundamental frequency, one needs to examine the resonance at a frequency that is half or close to half of the observed resonance frequency; if no resonance occurs, the observed frequency is the true fundamental frequency.

The diameters of the tube can be directly determined from TEM images at a high accuracy. The determination of length has to consider the 2-D projection effect of the tube. It is essential to tilt the tube and to catch its maximum length in TEM, which is likely to be the true length. This requires a TEM that gives a tilting angle as large as of $\pm$ 60°. Also the operation voltage of the TEM is important to minimize radiation damage. The 100 kV TEM used in our experiments showed almost no detectable damage to a carbon nanotube, while a 200 kV electrons could quickly damage a nanotube. The threshold for radiation damage of carbon nanotubes is ~150 kV.

2.2.4. *The "Rippling" and "Buckling" Effect*

After a systematic study of the multi-walled carbon nanotubes, the bending modulus of nanotubes was measured as a function of their diameters (Figure 7). The bending

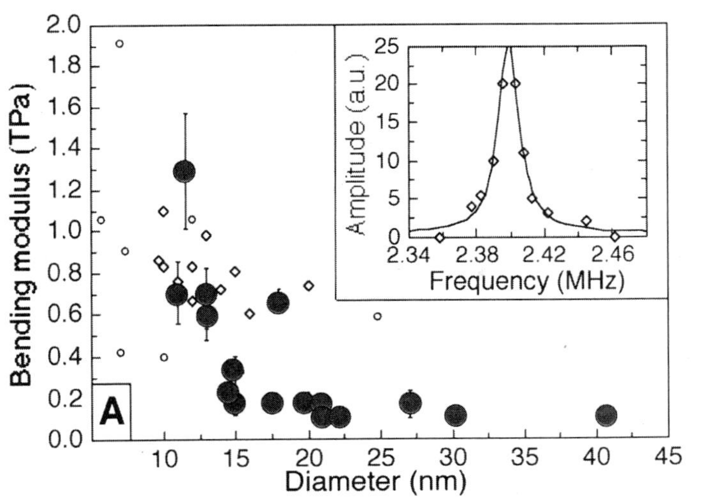

Figure 7. Bending modulus of the MWNT produced by arc-discharge as a function of the outer diameter of the nanotube. The inner diameter of the nanotubes is ~5 nm, independent of the outer diameter. The FWHM of the resonance peak is inserted.

modulus is as high as 1.2 TPa (as strong as diamond) for nanotubes with diameters smaller than 8 nm, and it drops to as low as 0.2 TPa for those with diameters larger than 30 nm. A decrease in bending modulus as the increase of the tube diameter is attributed to the wrinkling effect of the wall of the nanotube during small bending. For nanotubes bending to a large-degree of deformation, the rippling effect can be directly observed with TEM, as shown in Figure 8. The deformation is elastic and the nanotube recovers its shape after relieving. This effect is pronounced especially for larger size nanotubes, thus the geometrical shape makes significant contribution to the measured bending modulus (different from Young's modulus). However, with the decrease in nanotube diameter, the bending modulus approaches the Young's modulus.

The Young's modulus is a quantity that is defined to characterize the interatomic interaction force, and it is the double differentia of the bonding energy curve between the two atoms. The ideal case is that it is an intrinsic property at the atomic level and is independent of the sample geometry. For the nanotube case, the bending of a nanotube is determined not only by the Young's modulus, but also by the geometrical shape of the nanotube, such as the wall thickness and tube diameter. What we have measured by the in-situ TEM experiments is the bending modulus.

Theoretical investigation by Liu *et al.* [13] suggests that the equation (5) used for the analysis is based on the linear analysis, which is valid for small amplitude of vibration; for large vibration amplitude, the nonlinear analysis may have to be used. Based on the non-linear elasticity theory, they have successfully explained the rippling effect observed experimentally by Poncharal *et al.* [9]. In practical experiments, the resonance

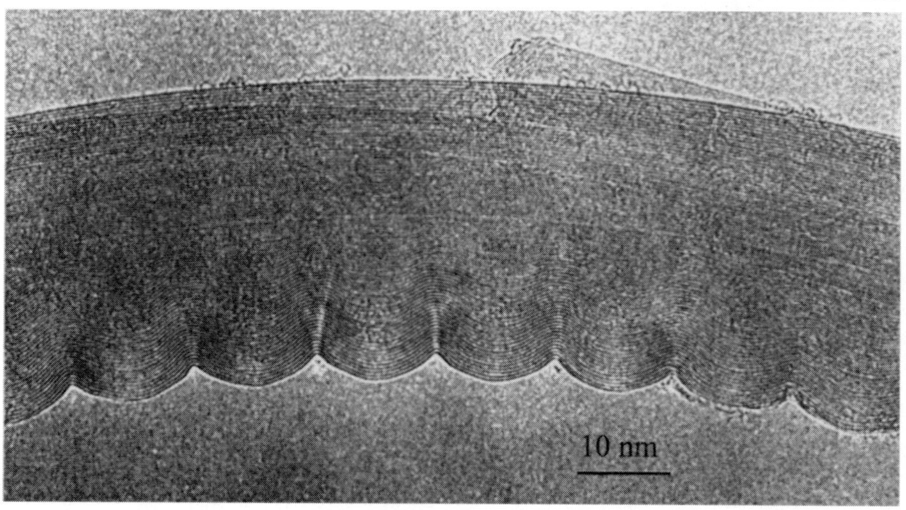

Figure 8. TEM image of a carbon nanotube after deformed to a large arc, showing the rippling effect of the graphitic layers to absorb the deformation energy and strain (Image courtesy of Dr. D. Ugarte).

frequency shows no drift as the vibration increases to as large as 30°, suggesting that the frequency measured can still be quantified using the linear analysis.

2.2.5. Effect from Defects on Bending Modulus

We know that the ultra-high mechanical strength of carbon nanotubes produced by arc discharge are due to their structural perfectness and uniformity. For applications in flat panel display and composite materials, aligned carbon nanotube arrays produced by catalyst assisted pyrolysis [14, 15, 16] are very attractive. From the structural point of view, carbon nanotubes produced by chemical synthesis are very different from those produced by arc-discharge, in that the former contain a higher density of point defects because of the introduction of pentagons and/or heptagons. We have applied the approach demonstrated in the last section for measurements the bending modulus of the carbon nanotubes with the presence of a high density of point defects [17], and the experimental results are summarized in Table 1. The bending modulus of nanotubes with point defects was ~30 GPa, which is about a factor of 7 smaller than the bending modulus of the same size of nanotubes that were grown by an arc-discharge technique. Therefore, the experimental approach demonstrated here provides solid data on structure and property for theoretical modeling of nano-scale wire systems.

The in-situ TEM technique demonstrated here provides a powerful approach towards nanomechanics of fiber-like nanomaterials with well-characterized structures. It can be applied to measure the mechanical properties of a wide range of nanowires, such as SiC, silica and Si nanowires, regardless their electrical conductivity. This is a universal approach for nanomechanics.

Table 1. Bending Modulus of Carbon Nanotubes Produced the Bending Modulus of Individual Carbon Nanotubes from Aligned Arrays Grown by Pyrolysis was Measured by in-situ Electromechanical Resonance in Transmission Electron Microscopy.

Nanotube	Outer diameter D(nm)(±1)	Inner diameter D_1 (nm)(±1)	Length L(μm) (±0.05)	Frequency ν(MHz)	E_b(GPa)
1	33	18.8	5.5	0.658	32 ± 3.6
2	39	19.4	5.7	0.644	26.5 ± 3.1
3	39	13.8	5	0.791	26.3 ± 3.1
4	45.8	16.7	5.3	0.908	31.5 ± 3.5
5	50	27.1	4.6	1.420	32.1 ± 3.5
6	64	27.8	5.7	0.968	23 ± 2.7

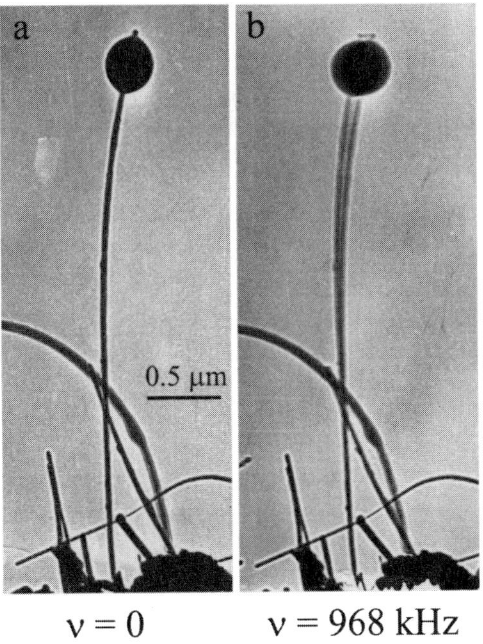

Figure 9. A small particle attached at the end of a carbon nanotube at (a) stationary and (b) first harmonic resonance ($\nu = 0.968$ MHz). The effective mass of the particle is measured to be ~22 fg ($1 f = 10^{-15}$).

2.2.6. The Nanobalance

In analogy to a spring pendulum, the mass of a particle attached at the end of the spring can be determined if the vibration frequency is measured, provided that the spring constant is calibrated. This principle can be adopted to determine a very tiny mass attached at the tip of the free end of the nanotube. The resonance frequency drops more than ~40% as a result of adding a small mass at its tip (Figure 9). The mass of the particle can be thus derived by a simple calculation using an effective mass in the calculation of the momentum of inertia. This newly discovered "nanobalance" has been shown to be able to measure the mass of a particle as small as 22 ± 6 fg (1f = 10^{-15}).

3. YOUNG'S MODULUS OF COMPOSITE NANOWIRES

Composite silicon carbide-silica nanowires have been synthesized by a solid-vapor process [18]. The as-synthesized materials are grouped into three basic nanowire structures: pure SiO_x nanowires, coaxially SiO_x sheathed β-SiC nanowires, and biaxial β-SiC-SiO_x nanowires. Figure 10 depicts a TEM image of the nanowires and their cross-section images, showing the coaxial and biaxial structures. The nanowires are uniform with diameter 50–80 nm, and a length that can be as long as 100 μm. The coaxial SiC-SiO_x nanowires have been extensively studied, and have a <111> growth direction with a high density of twins and stacking faults perpendicular to the growth direction.

The synthesized nanowires could be potentially useful for high-strength composites, in which the mechanical properties are critical. The technique demonstrated in section 2.2 can be applied to measure the Young's modulus of the composite nanowires. For a beam with one end hinged and the other free, the resonance frequency is given by:

$$\nu_0 = (\beta^2/2\pi)(EI/m)^{1/2}/L^2, \tag{4}$$

where ν_0 is the fundamental resonance frequency, $\beta = 1.875$, EI is the flexural rigidity (or bending stiffness), E is the Young's modulus, I is the moment of inertia about a particular axis of the rod, L is the length of the beam, and m is its mass per unit length. For a uniform solid beam with For a coaxial cable structured nanowire whose core material density is ρ_c and diameter is D_c and a sheath material density that is ρ_s with outer diameter D_s, the average density of the nanowire is given by

$$\rho_e = \rho_c\left(D_c^2/D_s^2\right) + \rho_s\left(1 - D_c^2/D_s^2\right). \tag{5}$$

The effective Young's modulus of the composite nanowire, E_{eff}, is

$$E_{eff} = \rho_e\left[8\pi f_0 L^2/\beta^2 D_s\right]^2 \tag{6}$$

The bending modulus for the coaxial cable structured SiC-SiO_x nanowires results in combination from SiC and SiO_x, where the contribution from the sheath layer of SiO_x is more than that from the SiC core because of its larger flexural rigidity (or bending stiffness). The bending modulus increases as the diameter of the nanowire increases (Table 2), consistent with the theoretically expected values of $E_{eff} = \alpha E_{SiC} + (1 - \alpha)E_{Silica}$, where $\alpha = (D_c/D_s)^4$. The data match well to the calculated values for larger diameter nanowires.

From the cross-sectional TEM image of a biaxially structured nanowire, the outermost contour of the cross-section of the nanowire can be approximated to be elliptical.

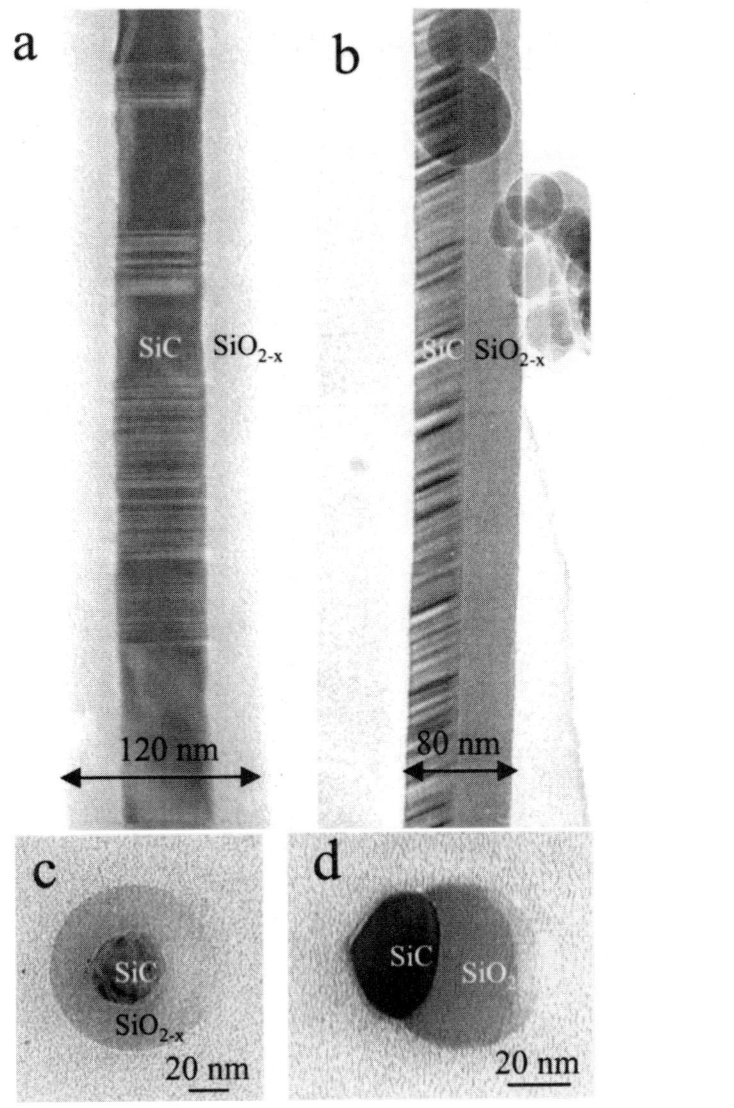

Figure 10. (a, b) TEM images of the coaxial and biaxial structured SiC-SiO_x nanowires, and (c, d) the cross-sectional TEM images, respectively.

Thus, the effective Young's modulus of the nanowire can be calculated using Eq. (6) with the introduction of an effective moment of inertia and density. For an elliptical cross-section of half long-axis a and half short-axis b, the moments of inertia are $I_x = \pi ab^3/4$, and $I_y = \pi ba^3/4$, where a and b can be calculated from the widths of the composite nanowire. With consideration of the equal probability of resonance with respect to the x and y axes, the effective moment of inertia introduced in the

Table 2. Measured Young's Modulus of Coaxial Cable Structured SiC-SiO_x Nanowires (SiC is the Core, and Silica is the Sheath) ($\rho_{Silica} = 2.2 \times 10^3$ kg/m^3; $\rho_{Sic} = 3.2 \times 10^3$ kg/m^3). The Young's Modulii of the Bulk Materials are $E_{SiC} = 466$ GPa and $E_{Silica} = 73$ GPa.

D_s (nm) (±2 nm)	D_c (nm) (±1 nm)	L (μm) (±0.2 μm)	f_0 (MHz)	E_{eff}(GPa) Exp.	E_{eff}(GPa) Theo.
51	12.5	6.8	0.693	46 ± 9.0	73
74	26	7.3	0.953	56 ± 9.2	78
83	33	7.2	1.044	52 ± 8.2	82
132	48	13.5	0.588	78 ± 7.0	79
190	105	19.0	0.419	81 ± 5.1	109

Table 3. Measured Young's Modulus of Biaxially Structured SiC-SiO_x Nanowires. D_{wire} and D_{SiC} are the Widths Across the Entire Nanowire and Across the SiC Sub-Nanowire, respectively.

D_{wire} (nm) (± 2 nm)	D_{SiC} (nm) (± 1 nm)	L (μm)(± 0.2 μm)	f_0 (MHz)	E_{eff} (GPa)Exp.
58	24	4.3	1.833	54 ± 24.1
70	36	7.9	0.629	53 ± 8.4
83	41	4.3	2.707	61 ± 13.8
92	47	5.7	1.750	64 ± 14.3

calculation is taken to be approximately $I = (I_x + I_y)/2$, and the density per unit length is $m_{eff} = A_{SiC}\,\rho_{SiC} + A_{Silica}\,\rho_{Silica}$, where A_{SiC} and A_{Silica} are the cross-sectional areas of the SiC and SiO_x sides, respectively. The experimentally measured Young's modulus is given in Table 3.

4. BENDING MODULUS OF OXIDE NANOBELTS

4.1. Nanobelts

In the literature, there are a few names being used for describing one-dimensionally elongated structures, such as nanorod, nanowire, nanoribbon, nanofiber and nanobelts. When we named the nanostructures to be "nanobelts" [19], we mean that the nanostructure has specific growth direction, the top/bottom surfaces and side surfaces are well defined crystallographic facets. The requirements for nanowires are less restrictive than for nanobelts because a wire has a specific growth direction but its side surfaces may not be well defined, and its cross-section may not be uniform nor specific shape. Therefore, *we believe that nanobelts are more structurally controlled objects than nanowires, or simply a nanobelt is a nanowire that has well-defined side surfaces.* It is well known that the physical property of a carbon nanotube is determined by the helical angle at which the graphite layer was rolled up. It is expected that, for thin nanobelts and nanowires, their physical and chemical properties will depend on the nature of the side surfaces.

The most typical nanobelt is ZnO (Figure 11a), which has a distinct in cross section from the nanotubes or nanowires [19]. Each nanobelt has a uniform width along its

Table 4. Crystallographic Geometry of Functional Oxide Nanobelts.

Nanobelt	Crystal Structure	Growth Direction	Top Surface	Side Surface
ZnO	Wurtzite	[0001] or [$01\bar{1}0$]	$\pm(2\bar{1}\bar{1}0)$ or $\pm(2\bar{1}\bar{1}0)$	$\pm(01\bar{1}0)$ or $\pm(0001)$
Ga_2O_3	Monoclinic	[001] or [010]	$\pm(100)$ or $\pm(100)$	$\pm(010)$ or $\pm(10\bar{1})$
t-SnO_2	Rutile	[101]	$\pm(10\bar{1})$	$\pm(010)$
o-SnO_2 wire	Orthorhombic	[010]	$\pm(100)$	$\pm$ (001)
In_2O_3	C-Rare earth	[001]	$\pm(100)$	$\pm(010)$
CdO	NaCl	[001]	$\pm(100)$	$\pm(010)$
PbO_2	Rutile	[010]	$\pm(201)$	$\pm(10\bar{1})$

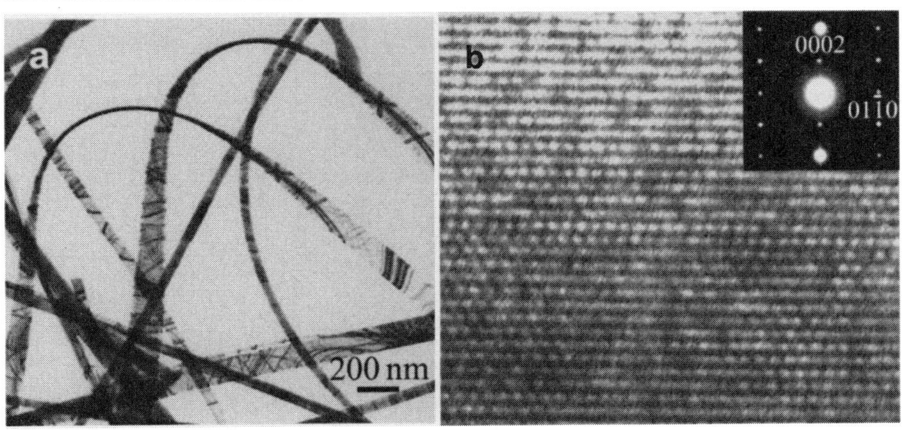

Figure 11. (a) TEM image of ZnO nanobelts synthesized by a solid-vapor phase technique. (b) High-resolution TEM image of a ZnO nanobelt with incident electron beam direction along [$2\bar{1}\bar{1}0$]. The nanobelt grows along [0001], with top/bottom surfaces ($2\bar{1}\bar{1}0$) and side surfaces ($01\bar{1}0$).

entire length, and the typical widths of the nanobelts are in the range of 50 to 300 nm. A ripple-like contrast appeared in the TEM image is due to strain resulted from the bending of the belt. High-resolution TEM (HRTEM) and electron diffraction studies show that the ZnO nanobelts are structurally uniform, single crystalline and dislocation free (Figure 11b).

There are a few kinds of nanobelts having been reported in the literature. Table 4 summarizes the nanobelt structures of function oxides [19, 20, 21, 22, 23, 24]. Each type of nanobelts is defined by its crystallographic structure, the growth direction, top surfaces and side surfaces. Some of the materials can grow along two directions, but they can be controlled experimentally. Although these materials belong to different crystallographic families, but they do have a common faceted structure, which is the nanobelt structure. In addition, nanobelts of $Cu(OH)_2$ [25], MoO_3 [26, 27], MgO [28, 29], and CuO [30].

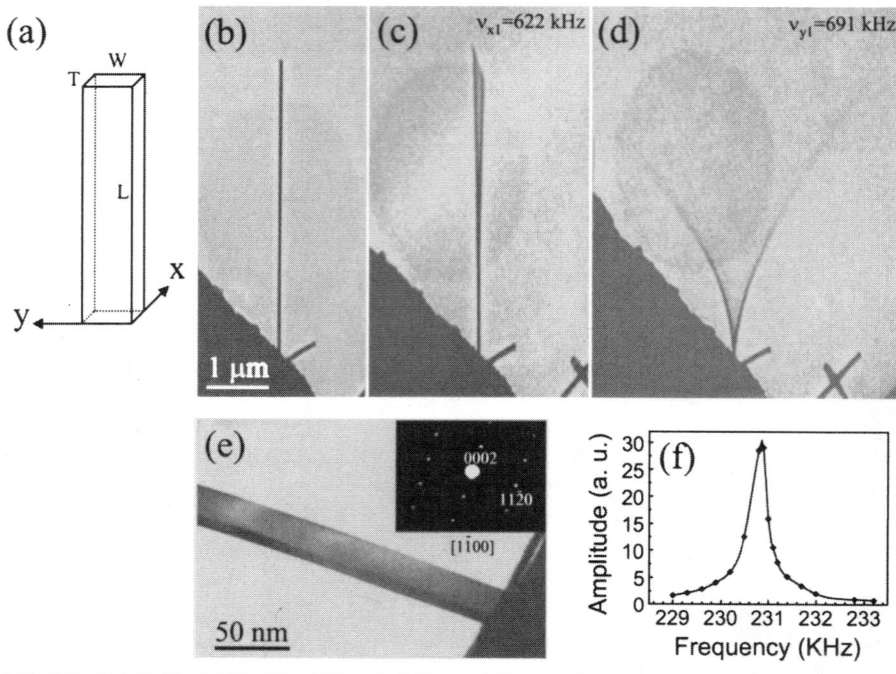

Figure 12. A selected ZnO nanobelt at (a, b) stationary, (c) the first harmonic resonance in x direction, $\nu_{x1} = 622$ kHz, and (d) the first harmonic resonance in y direction, $\nu_{y1} = 691$ kHz. (e) An enlarged image of the nanobelt and its electron diffraction pattern (inset). The projected shape of the nanobelt is apparent. (f) The FWHM of the resonance peak measured from another ZnO nanobelt. The resonance occurs at 230.9 kHz.

4.2. Dual-Mode Resonance of Nanobelts

Due to the mirror symmetry and its rectangular cross-section of the nanobelt (Fig. 12a), there are two distinct fundamental resonance frequencies corresponding to the vibration in the thickness and width directions, which are given from the classical elasticity theory as [31]

$$\nu_x = \frac{\beta_1^2 T}{4\pi L^2}\sqrt{\frac{E_x}{3\rho}}, \tag{7}$$

$$\nu_y = \frac{\beta_1^2 W}{4\pi L^2}\sqrt{\frac{E_y}{3\rho}}, \tag{8}$$

where $\beta_1 = 1.875$; E_x and E_y are the bending modulus if the vibration is along x-axis (thickness direction) and y direction (width direction), respectively; ρ is the density, L is the length, W is the width and T is the thickness of the nanobelt. The two modes are decoupled and they can be observed separately in experiments.

Table 5. Bending Modulus of the ZnO Nanobelts. E_x and E_y Represents the Bending Modulus Corresponding the Resonance Along the Thickness and Width Directions, respectively.

Nanobelt	Length L (μm) (±0.05)	Width W (nm) (±1)	Thickness T (nm) (±1)	W/T	Fundamental Fequency (kHz) ν_{x1}	ν_{y1}	ν_{y1}/ν_{x1}	Bending Modulus (GPa) E_x	E_y
1	8.25	55	33	1.7	232	373	1.6	46.6 ± 0.6	50.1 ± 0.6
2	4.73	28	19	1.5	396	576	1.4	44.3 ± 1.3	45.5 ± 2.9
3	4.07	31	20	1.6	662	958	1.4	56.3 ± 0.9	64.6 ± 2.3
4	8.90	44	39	1.1	210	231	1.1	37.9 ± 0.6	39.9 ± 1.2

Changing the frequency of the applied voltage, we found two fundamental frequencies in two orthogonal directions transverse to the nanobelt [31]. Figures 12c and 12d show the harmonic resonance with the vibration planes nearly perpendicular and parallel to the viewing direction, respectively. For calculating the bending modulus, it is critical to accurately measure the fundamental resonance frequency (ν_1) and the dimensional sizes (L and T *or* W) of the investigated ZnO nanobelts. To determine ν_1, we have checked the stability of resonance frequency to ensure one end of nanobelt is tightly fixed, and the resonant excitation have been carefully checked around the half value of the resonance frequency.

4.3. Bending Modulus of Nanobelt

The geometrical parameters are the key for derivation of the mechanical property from the measured resonance frequencies. The specimen holder is rotated about its axis so that the nanobelt is aligned perpendicular to the electron beam, thus, the real length (L) of the nanobelt can be obtained. The normal direction of the wide facet of the nanobelt could be firstly determined by electron diffraction pattern, which was $[2\bar{1}\bar{1}0]$ for the ZnO nanobelt. Then the nanobelt was tilted from its normal direction by rotating the specimen holder, and the tilting direction and angle were determined by the corresponding electron diffraction pattern. As shown in the inset of Fig. 12e, the electron beam direction is $[1\bar{1}00]$. The angle between $[1\bar{1}00]$ and $[2\bar{1}\bar{1}0]$ is 30°, i.e. the normal direction of the wide facet of this nanobelt is 30° tilted from the direction of the electron beam. Using the projected dimension measured from the TEM image (Fig. 12e), the geometrical parameters of this nanobelt are determined to be $W = 28$ nm and $T = 19$ nm. Based on the experimentally measured data, the bending modulus of the ZnO nanobelts is calculated using Eq. (7) and Eq. (8). The experimental results are summarized in Table 5 [31]. The bending modulus of the ZnO nanobelts was ~52 GPa. This value represents the modulus that includes the scaling effect and geometrical shape, and it cannot be directly compared to the Young's modulus of ZnO ($c_{33} = 210$ GPa, $c_{13} = 104$ GPa) [32], because the shape of the nanobelt and the anisotropic structure of ZnO are convoluted in the measurement.

The bending modulus measured by the resonance technique, however, has excellent agreement with the elastic modulus measured by nanoindentation for the same type of nanobelts [33].

Although nanobelts of different sizes may have slight difference in bending modulus, there is no obvious difference if the calculation was done using either Eq. (7) or Eq. (8). The ratio of two fundamental frequencies ν_{y1}/ν_{x1} is consistent to the aspect ratio W/T, as expected from Eqs. (7) and (8), because there is no significant difference between E_x and E_y.

5. NANOBELTS AS NANOCANTILEVERS

Cantilever based scanning probe microscopy (SPM) technique is one of the most powerful approaches in imaging, manipulating and measuring nanoscale properties and phenomena. The most conventional cantilever used for SPM is based on silicon, Si_3N_4 or SiC, which is fabricated by e-beam or optical lithography technique and has typically dimension of thickness of ~100 nm, width ~5 μm and length ~50 μm. Utilization of nanowire and nanotube based cantilever can have several advantages for SPM. Carbon nanotubes can be grown on the tip of a conventional cantilever and be used for imaging surfaces with a large degree of abrupt variation in surface morphology [34, 35]. We demonstrate here the manipulation of nanobelts by AFM and its potential for as nanocantilevers [36].

Combining MEMS technology with self-assembled nanobelts we are able to produce cost effective cantilevers with much heightened sensitivity for a range of devices and applications. Force, pressure, mass, thermal, biological, and chemical sensors are all prospective devices. Semiconducting nanobelts are ideal candidates for cantilever applications. Structurally they are defect free single crystals, providing excellent mechanical properties. The reduced dimensions of nanobelt cantilevers offer a significant increase in cantilever sensitivity. The cantilevers under consideration are simple in design and practice. Using a Dimension 3000 SPM in Tapping Mode, we have successfully lifted ZnO nanobelts from a silicon substrate. Capillary forces are responsible for the adhesion strength between the atomic force microscope probe and the ZnO nanobelts. Combining the aforementioned techniques with micromanipulation has led to the alignment of individual ZnO nanobelts onto silicon chips (Fig. 13). The aligned ZnO cantilevers were manipulated to have a range of lengths. This exemplifies our ability to tune the resonance frequency of each cantilever and thus modify cantilevers for different applications such as Tapping and Contact Mode AFM. The nanobelt based nanocantilever is ~50–1000 times smaller than the conventional cantilever. Decreased size in micro-optical mechanical devices corresponds to increased sensitivity. Combining the aforementioned techniques with micromanipulation has led to the horizontal alignment of individual ZnO nanobelts onto silicon chips (Fig. 13). This exemplifies our ability to tune the resonance frequency of each cantilever and thus modify cantilevers for different applications such as Contact, Non-Contact, and Tapping Mode AFM.

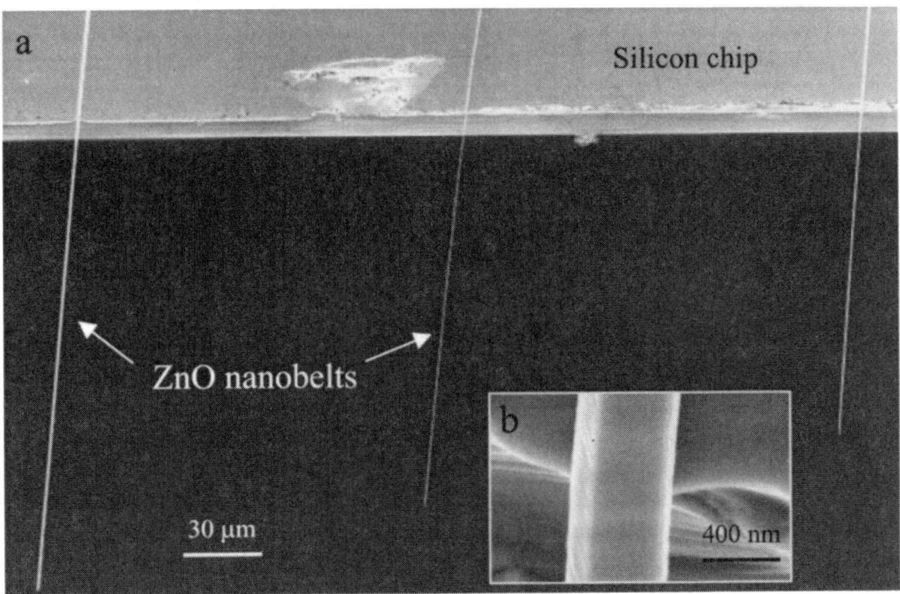

Figure 13. Site specific placement and alignment of ZnO nanobelts onto a silicon chip, forming nanocantilever arrays. The inset is an enlarged SEM image of the third nanocantilever showing its shape; the width of the cantilever was measured to be 525 nm.

6. IN-SITU FIELD EMISSION FROM NANOTUBE

Field emission, one of the most promising applications of carbon nanotubes, has been extensively studied. Several quantities have been defined to characterize the field emission performance of a material. Besides the work function, the turn-on field (E_{to}) and threshold field (E_{thr}) for electron emission, defined as the macroscopic fields needed to produce a current density of 10 $\mu A/cm^2$ and 10 mA/cm^2, respectively, are the two most typical parameters. The characterization of the field emission properties of carbon nanotubes uses the arrays of aligned carbon nanotubes. By specifying the distance between the tips of the nanotube to the cathode, the measured results are a statistical average of the nanotubes with a wide range distribution in diameters and lengths. The theory used for the data analysis still relies on the Fowler–Nordheim equation, which was derived for a semi-infinitive metal surface. For the carbon nanotubes, the geometry of the anode is an array of carbon nanotubes and the electrons are emitted from the tips. The success of growing highly aligned carbon nanotubes is a major advanced for field emission application [37, 38].

7. WORK FUNCTION AT THE TIPS OF NANOTUBES AND NANOBELTS

An important physical quantity in electron field emission is the surface work function, which is well documented for elemental materials. For the emitters such as carbon

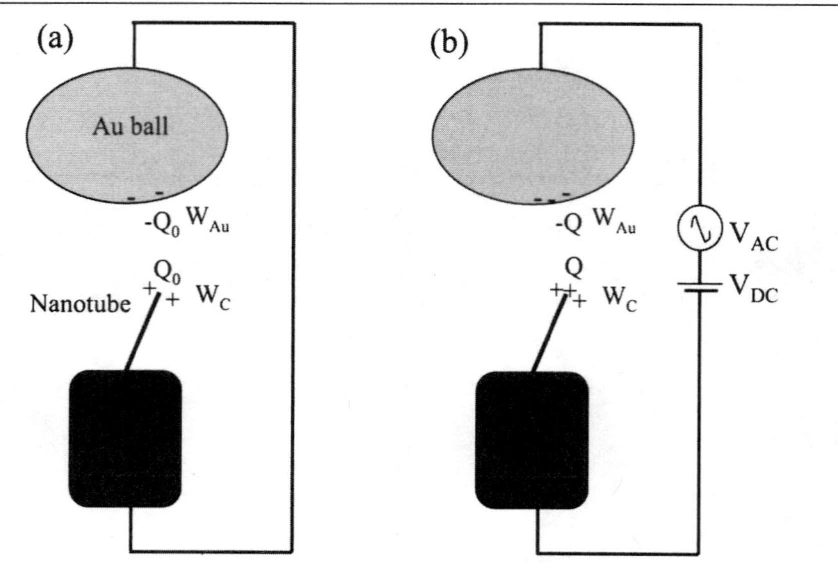

Figure 14. (a) Schematic diagram showing the static charge at the tip of carbon nanotube as a result of difference in work functions between the nanotube and the gold electrode. (b) Schematic experimental approach for measuring the work function at the tip of a carbon nanotube.

NTs, most of the electrons are emitted from the tips of the carbon NTs, and it is the local work function that matters to the properties of the NT field emission. The work function measured from the $\ln(J/E^2)$ versus $1/E$ characteristics curve, where E is the macroscopic applied electric field that is an average over all of the aligned carbon NTs that are structurally divers in diameters, lengths and helical angles. We have developed a technique for the measurement the work function at the tip of a single carbon nanotube [39].

Our measurement is based on the electric field induced mechanical resonance of carbon nanotubes. The principle for work function measurement is schematically shown in Figure 14a. We consider a simple case in which a carbon nanotube, partially soaked in a carbon fiber produced by arc-discharge, is electrically connected to a gold ball. Due to the difference in the surface work functions between the NT and the counter Au electrode, a static charge Q_0 exists at the tip of the NT to balance this potential difference even at zero applied voltage [40]. The magnitude of Q_0 is proportional to the difference between work functions of the Au electrode and the NT tip (NTT), $Q_0 = \alpha(W_{Au} - W_{NTT})$, where α is related to the geometry and distance between the NT and the electrode.

The measurement relies on the mechanical resonance of the carbon NT induced by an externally applied oscillating voltage with tunable frequency. In this case, a constant voltage V_{DC} and an oscillating voltage $V_{AC} \cos 2\pi ft$ are applied onto the NT, as shown in Figure 14b, where f is the frequency and V_{AC} is the amplitude. The total induced charge on the NT is

$$Q = Q_0 + \alpha e(V_{DC} + V_{AC}\cos 2\pi ft). \tag{9}$$

The force acting on the NT is proportional to the square of the total charge on the nanotube

$$\begin{aligned} F = \beta[Q_0 + \alpha e(V_{DC} + V_{AC}\cos 2\pi ft)]^2 \\ = \alpha^2\beta\left\{\left[(W_{Au} - W_{NTT} + e\,V_{DC})^2 + e^2 V_{AC}^2/2\right]\right. \\ \left. + 2e\,V_{AC}\,(W_{Au} - W_{NTT} + e\,V_{DC})\cos 2\pi ft + e^2 V_{AC}^2/2\cos 4\pi ft\right\} \end{aligned} \tag{10}$$

where β is a proportional constant. In Eq. (10), the first term is constant and it causes a static deflection of the carbon NT; the second term is a linear term, and the resonance occurs if the applied frequency f approaches the intrinsic mechanical resonance frequency f_0 of the carbon NT (Figure 15a). The last term in Eq. (10) is the second harmonics. The most important result of Eq. (10) is that, for the linear term, the resonance amplitude A of the NT is proportional to $V_{AC}(W_{Au} - W_{NTT} + eV_{DC})$. By fixing the V_{DC} and measuring the vibration amplitude as a function of $V_{AC,}$ a linear curve is received (Figure 15c).

Experimentally, we first set $V_{DC} = 0$ and tune the frequency f to find the mechanical resonance induced by the applied field. Secondly, under the resonance condition of keeping $f = f_0$ and V_{AC} constant, slowly change the magnitude of V_{DC} from zero to a value that satisfies $W_{Au} - W_{NTT} + eV_{DC0} = 0$ (Figure 15b); the resonance amplitude A should be zero although the oscillating voltage is still in effect. V_{DC0} is the x-axis interception in the $A \sim V_{DC}$ plot (Figure 15d). Thus, the tip work function of the NT is $W_{NTT} = W_{Au} + eV_{DC0}$ [39].

Several important factors must be carefully checked to ensure the accuracy of the measurements. The true fundamental resonance frequency must be examined to avoid higher order harmonic effects. The resonance stability and frequency-drift of the carbon nanotubes must be examined prior and post each measurement to ensure that the reduction of vibration amplitude is solely the result of V_{DC}. The NT structure suffers no radiation damage at 100 kV, and the beam dosage shows no effect on the stability of the resonance frequency. Figure 16 gives the plot of the experimentally measured V_{DC0} as a function of the outer diameter of the carbon NTs. The data show two distinct groups: −0.3 to −0.5 eV and ~ +0.5 eV. The work function shows no sensitive dependence on the diameters of the NTs at least in the range considered here. 75% of the data indicate that the tip work function of carbon NTs is 0.3 to 0.5 eV lower than the work function of gold ($W_{Au} = 5.1$ eV), while 25% of the data show that the tip work function is ~0.5 eV higher than that of gold. This discrepancy is likely due to the nature of some nanotubes being conductive and some being semiconductive, depending on their helical angles. In comparison to the work function of carbon ($W_C = 5.0$ eV), the work function at the tip of a conductive multiwalled carbon NT is 0.2–0.4 eV lower. This is important for electron field emission.

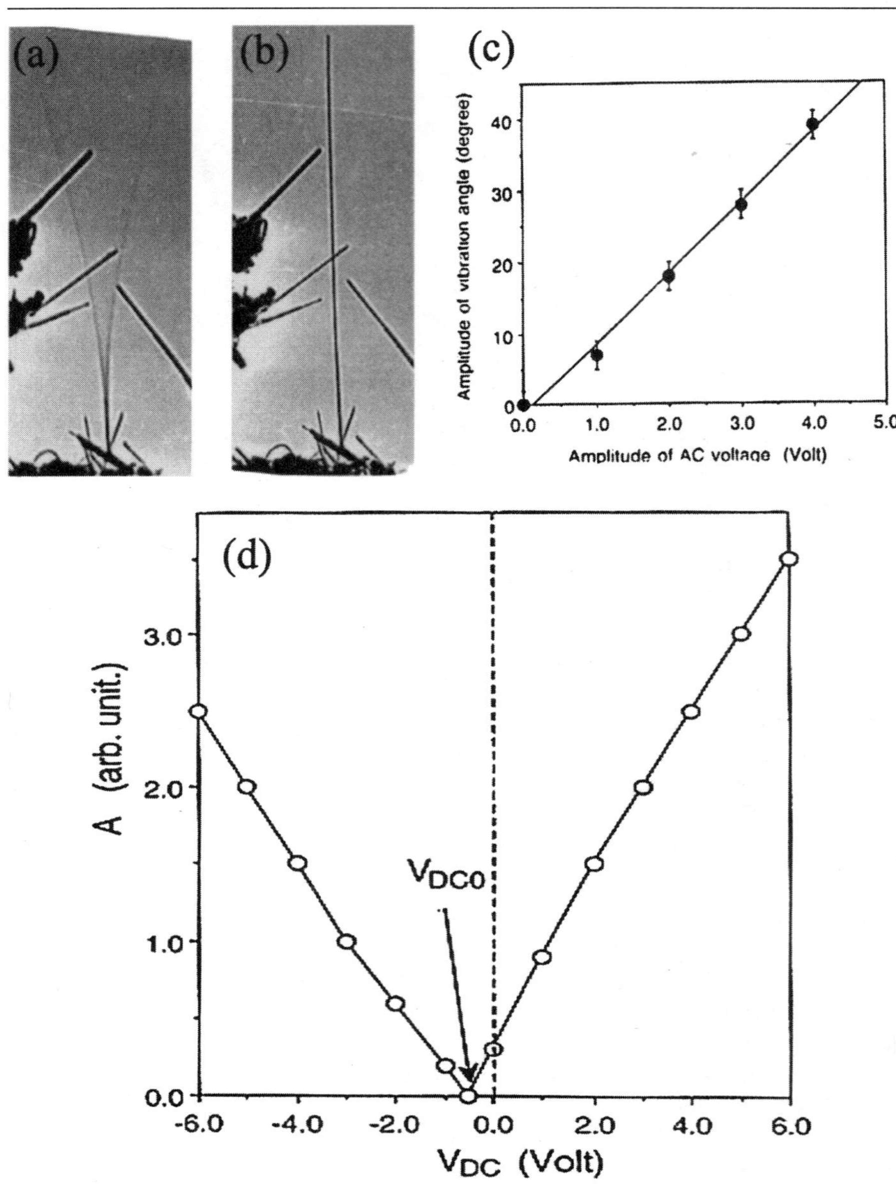

Figure 15. (a) Mechanical resonance of a carbon nanotube induced by an oscillating electric field; (b) Halting the resonance by meeting the condition of $W_{Au} - W_{NTT} + eV_{DC0} = 0$. (c) A plot of vibration amplitude of a carbon nanotube as a function of the amplitude of the applied alternating voltage V_{AC}. (d) A plot of vibration amplitude of a carbon nanotube as a function of the applied direct current voltage V_{DC}, while the applied frequency is 0.493 MHz and $V_{AC} = 5$ V.

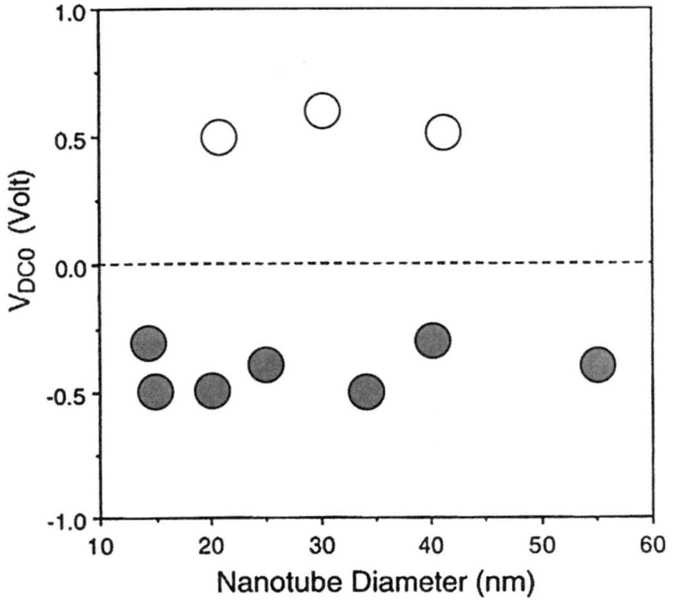

Figure 16. The experimentally measured V_{DC0} as a function of the outer diameter of the carbon nanotube.

The technique demonstrated here has also been applied to measure the work function at the tips of ZnO nanobelts [41].

8. MAPPING THE ELECTROSTATIC POTENTIAL AT THE NANOTUBE TIPS

It is known that the electrons are emitted from the tips of carbon nanotubes. The accumulated charges at the tips create intensive local electric field. The images shown in Figure 17 clearly show the potential distribution around the nanotube tip, but the quantification of the potential is not trivial, simply because that the image contrast shown in Figure 17 strongly depends on the size of the objective aperture and the defocus value. Quantitative mapping of the electric potential has to be done properly using the electron holography, through which the electron phase introduced by the potential can be retrieved.

An important application of electron holography is to retrieve the phase of the electron wave after interaction with a specimen. For simplicity, the non-relativistic approximation is made. When an electron enters a space with an electrostatic field distribution $V(x, y, z)$, the electron wave length is changed to conserve the total energy. The relative phase shift of the electron wave with respect to the reference wave is

$$\phi(x, y) = \sigma \int_{-\infty}^{\infty} dz\, V(x, y, z), \tag{11}$$

Figure 17. (a) Bright and (b, c) dark field images of carbon nanotubes recorded by selecting the electrons scattered off the center beam due to the electrostatic deflection produced by the nanotubes under an applied voltage of 40 V, showing the electric field distribution at the tips, from where the electrons are emitting.

which is proportional to the projected potential of the object along the incident beam direction. For a homogeneous specimen with constant thickness, in addition to the average potential introduced by the atoms in the crystal, charge barrier can be created at interfaces and defects. If the former contributes only a background, the latter can be retrieved experimentally. Electron holography has been applied to map the distribution of electrostatic charge in space [42].

The holographic technique has been applied to image the potential distribution around the tip of the nanotubes [43], as shown in Figure 18. The result shows that the electron emission and associated electric field is concentrated at the tip of the nanotube, as expected. The electric field magnitude and distribution are stable in time, even in cases where the nanotube field emission current exhibits extensive temporal fluctuations.

6. FIELD EMISSION INDUCED STRUCTURAL DAMAGE

Due to the large aspect ratio of carbon nanotubes, the electric field at the tips of the nanotubes can be rather large, thus, the local temperature could be very high at field emission. It is possible to induce structural damage at the tips. The classical definition of turn on field $E_t = V/d$, where V is the applied voltage and d is the distance from the tip of the field emitter to the surface of the counter electrode, may not be an adequate measurement on the local field at the tips of the carbon nanotube, due to its sharp needle geometry.

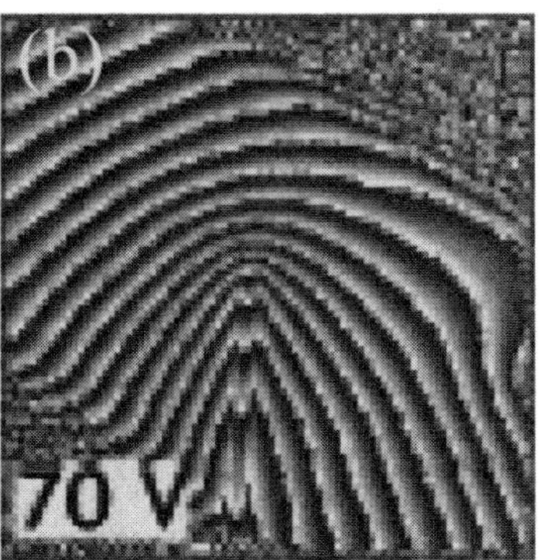

Figure 18. Electron phases image of a carbon nanotube prior and after applying 70 V of external electric voltage (Courtesy of Dr. A. Zettl).

An important phenomenon of our study is the observation of structural damage of a carbon nanotube during field emission under a higher voltage [44]. This study is useful in determining the structural stability of the nanotubes. Figure 19 shows a series of images of a nanotube that was being damaged by an applied voltage. The structural damage is apparent as the applied voltage increases. The damage occurs in such a way that the walls of the nanotubes are split patch-by-patch and segment-by-segment. A closer image of the splitting is shown in Fig. 19e. This damage process is different from the unraveling process proposed by Rinzler *et al.* [45], who believed that the nanotubes are damaged following a string-by-string removing of the carbon atoms along the circumference of the graphitic layer.

Figure 19 shows a "stripping" process of a carbon nanotube under the applied electric field. The diameter and length of the nanotube A decrease as it being damaged by the field. This is a sharpening process of the multiwalled nanotube. The structure of nanotube B is almost totally damaged by the field and finally becomes a graphitic structure.

The mechanisms of the field-induced damage are believed due to two processes. One, the electrostatic force acting on the tip of the nanotube can split the nanotube piece-by-piece and segment-by-segment, such as the one shown in Fig. 19. The second process is likely due to the local temperature created by the flow of emission current, which may "burn" the nanotube layer-by-layer, resulting in the sharpening at the nanotube tip. This process has recently been used for removing the walls of carbon nanotubes [46, 47].

It was reported by Rinzler *et al.* [45] that the current emitted by nanotubes fluctuates almost randomly as a function of time at the time scale of a couple of seconds, and this phenomenon was interpreted owing to a unraveling process of the carbon atom ring. Through in-situ TEM observation, we found that the fluctuation in emission current is due to a "head-shaking" effect of the nanotube while emitting electrons [44]. As shown in Fig. 5, the nanotube bends toward the counter electrode at an

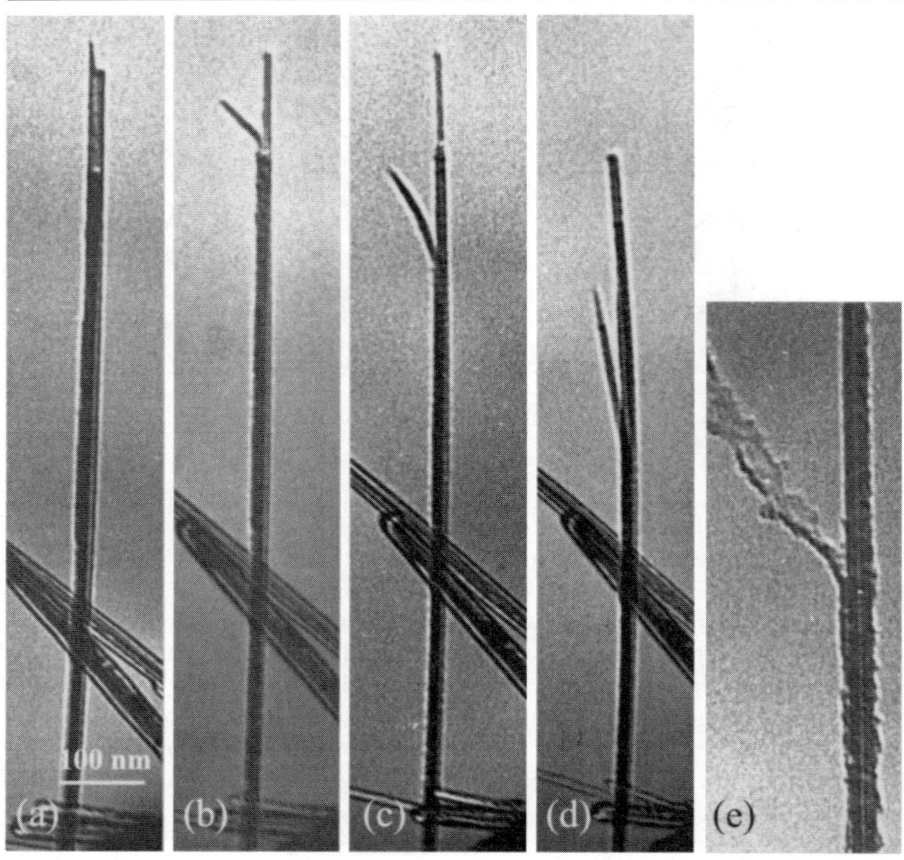

Figure 19. "Splitting" process in structural damage. (a–d) A series of TEM images showing the structural damage of a carbon nanotube during field emission, in which the applied voltage and the emission current are: (a) $V = 80$ V, $I = 10$ μA, (b) $V = 90$ V, $I = 40$ μA, (c) $V = 110$ V, $I = 100$ μA, and (d) $V = 130$ V, $I = 250$ μA. The distance from the tip of the nanotube to the counter electrode was ~2 μm. (e) A nanotube that is experiencing the splitting of its outer layers during the damage.

applied voltage. The emission of electrons from a nanotube is like to be a "ballistic" emission process in which the electrons are emitted as groups, although each emission can release many electrons. When the nanotube is fully charge prior to emission, the distance between the nanotube tip and the counter electrode is the smallest due to the strongest electrostatic attraction; as soon as the electrons are emitted as a group, the electrostatic force between the nanotube and the electrode drops slightly, resulting in the recovery of the nanotube shape and a larger distance from the electrode. The head-shaking of the nanotube due to "ballistic" emission results in a variation in the distance of its tip from the electrode, thus, leads to a fluctuation in the emission current. This may also account for the blinking of emission current from carbon nanotubes. The ballistic emission is possible because the small size of a nanotube that can only hold a small amount of electrons at its tip. A rough estimation indicates that loosing one electron at the tip can change the tip potential by ~0.15 V for a 20 nm diameter nanotube. The head-shaking is a result of its large aspect ratio that leads to body swing during field emission.

7. NANOTHERMOMETER AND NANOBEARING

There other interesting experiments having been carried out using in-situ TEM. Gao and Bando [48] have demonstrated a nanothermometer. By trapping liquid gallium in a closed carbon nanotube during the nanotube growth, a nano-thermometer is naturally formed. The gallium level increases linearly with temperature with repeatability.

MWNTs are composed of concentric graphitic walls. The bonding between the shells is van der Waals, thus the inter-wall linkage is rather weak. Thus, sliding between the walls is possible. By opening one end of a MWNT, Cummings and Zettl [49] have demonstrated the ultralow-friction nanoscale linear bearings and constant-force nanosprings. The inner walls can be repeatability reversibly pulled out by the electrostatic force and self-sucked in. Repeated extension and retraction of telescoping nanotube segments revealed no wear or fatigue on the atomic scale. Hence, these nanotubes may constitute near perfect, wear-free surfaces.

8. IN-SITU TRANSPORT MEASUREMENT OF NANOTUBES

Electrical transport in single-walled nanotubes (SWNTs) and multi-walled nanotubes (MWNTs) is of great importance for their applications in electronics [50]. The electronic band structure of SWNTs is well known: depending on the helicity and statistically in 1/3 of the cases, a tube has two one-dimensional subbands (channels) that intercept the Fermi level, giving rise to metallic conduction. More precisely, only armchair tubes are gapless: all others are often referred to as metallic although small gaps that are introduced by curvature effects of the order of 10 meV for 1.4 nm diameter SWNTs effect transport at low temperatures. The gap diminishes with increasing tube diameter. Measurements of nanotube conductance mainly use two techniques. Using lithographically made gold electrodes, a carbon nanotube is laid down across two or four electrodes, and the I–V characteristic is measured [51, 52]. The other technique takes the advantage of using liquid mercury as a soft contacting electrode, a nanotube is inserted into the mercury and the conductance is monitored as a function of the depth that the nanotube is inserted into the mercury [53]. The latter has been carried out in-situ in TEM. A comprehensive review about all of the existing literature and the comparison of data in electrical characterization can be found from ref. [54].

8.1. Ballistic Quantum Conductance at Room Temperature

Our original technique for measuring the conductance of a structurally perfect nanotube involved two contact measurements on free-standing MWNTs in-air [53]. An arc-produced fiber composed of MWNTs was attached to a conducting tip of a scanning probe microscope and dipped into various liquid metals. The conductance was recorded as a function of the depth Z that the nanotube penetrated into the liquid

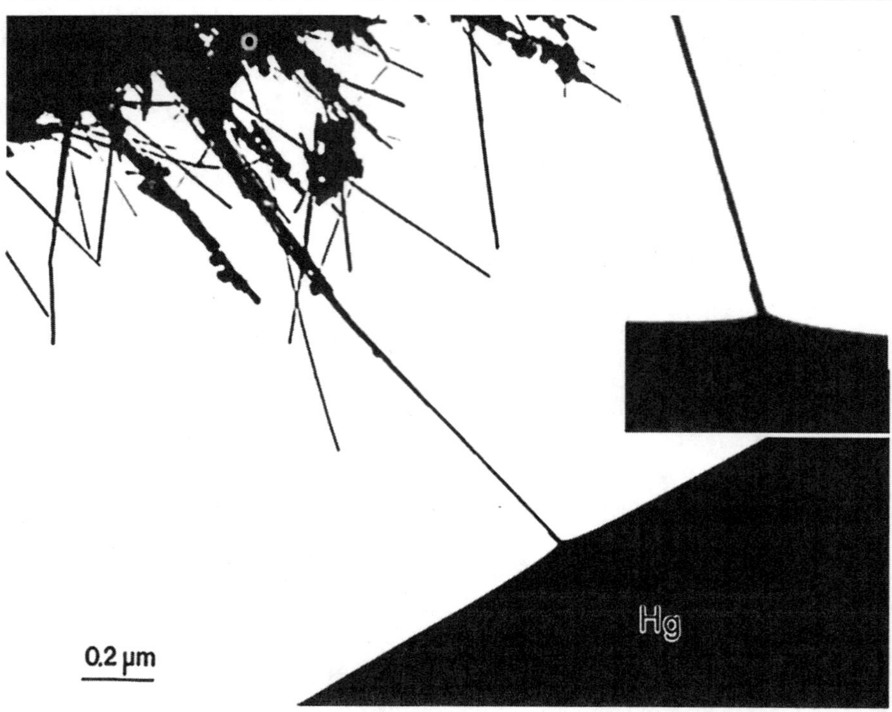

Figure 20. In-situ TEM image showing the conductance measurement through a single carbon nanotube. The inset is the contact area of the nanotube with the mercury surface.

metal surface. The experiment has three main advantages: 1) the carbon nanotube is as-synthesized without any contamination; 2) the soft metal contact reduces the contact resistance, allowing a sensitive measurement on the conductance of the nanotube; and 3) the length of the nanotube exposed outside the liquid metal surface is a variable that allows a direct measurement about the conductance of the nanotube as its length being reduced.

The conductance measurement of individual carbon nanotubes has been carried out using the in-situ technique introduced in Fig. 3. The same specimen holder used for the mechanical property measurements was used for the conductance measurement except that the counter electrode is replaced by a mercury droplet, which served as the soft contact for the measurement. Figure 20 shows the contact of a carbon nanotube with the mercury electrode, and the conductance of G_0 was observed (with $G_0 = 2e^2/h = 1/(12.9\ \text{k}\Omega)$). The conductance is quantized and it is independent of the length of the carbon nanotube. No heat dissipation was observed in the nanotube. This is the result of ballistic conductance, and it is believed to be a result of single graphite layer conductance. Recent observations using different techniques have confirmed our result [55, 56]. Theoretical calculations provide some interpretations about the

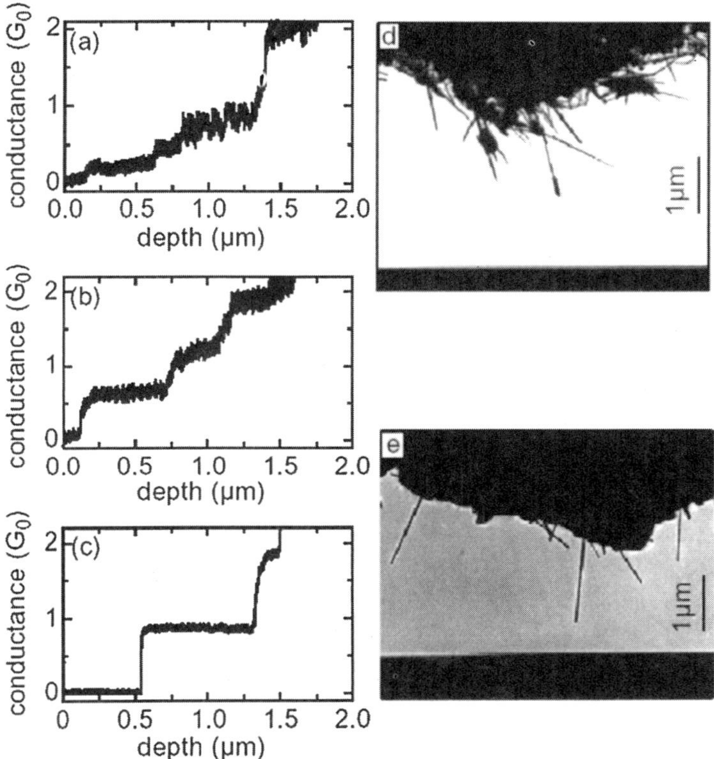

Figure 21. Cleaning of nanotubes and development of nanotube fiber properties by repeated dipping in Hg. Development of conduction properties (a, b, c) and the microstructure (d, e). (d) Electron micrograph of a virgin fiber tip opposing Hg surface (see also Fig. 2); note the contaminating graphitic particles and the loose structure of the tip. (e) Electron micrograph of a fiber tip that has previously been repeatedly dipped in Hg; the nanotubes are straight and free of particles and the fiber is compacted. (a) Conduction trace of the virgin fiber steps are barely discernable; (b) Steps have developed after a few hundred cycles but they still exhibit relatively large slopes; (c) After several thousand cycles, the steps are well developed and the pattern is stabile. The first step evolves from the shoulder seen in (a) (step: 0.2 G_0, slope: 36 kW/μm) to a rounded step in (b) (step: 0.62 G_0, slope: 4 kW/μm), to a sharp step in (c) (step 0.85 G_0, slope: -0.3 kW/μm). The second step is due to another tube and evolves analogously.

G_0 quantization [57] and fraction quamtum conductance when the tip contact the mercury surface [58].

It is also interesting to note that the contact area between the nanotube and the mercury surface is curved. This is likely due to the difference in surface work function between nanotube and mercury, thus, electrostatic attraction could distort the mercury surface. This effect effectively reduces the contact resistance between the nanotube and the Hg electrode.

In our original experiment nanotube fibers were conditioned by many (up to several thousand) dipping cycles after which the steps develop quantized (flat) plateaus. However the initial cycles do not show the effect. Figure 21a–c shows the evolution of the conductance versus depth measurements. The initial traces show little evidence for quantization. After many cycles a pattern of reproducible steps appears (each about 1 G_0 high), with cycle to cycle step height variations of the order of ± 0.05 G_0. The plateaus are remarkably flat (typically $|dG/dL| < 0.04$ $G_0/\mu m$), which implies a conductivity greater than for copper. Usually positive, but sometimes slightly negative plateau slopes are observed, as in Fig. 21c. This may be due to a slight reduction in the electronic transmission with increasing strain [59]. Flat plateaus (ideally $\rho = 0$) and plateau conductances G in the range 0.8 $G_0 < G < 1$ G_0, are consistent with conductance quantization and ballistic conduction allowing for a slight residual contact resistances at the fiber/nanotube contact.

The nanotubes (with typical diameters from about 10 to 30 nm) that protrude from a virgin fiber are initially covered with graphitic particles. In the process of contacting the Hg surface, some of these particles transfer to the metal surface causing the surface to become covered with particles while the tubes become cleaner (Fig. 21d). The dipping process is accompanied with relatively large stresses causing some tubes to become dislodged and displaced and occasionally even to break. We conclude that the tubes that remain in place after many conditioning cycles (as in the in-air experiment) are clean, well anchored to the fiber and relatively robust (i.e. defect free) (Fig. 21e).

8.2. Quantum Conductance and Surface Contamination

In-situ nanotube conductances are determined by applying a voltage difference of 100 mV between the Hg droplet and the nanotube fiber and measuring the current. Three types of behavior are found. Type 1: more than half of the nanotubes that clearly contact the Hg surface (as seen from a slight bowing of the nanotube, as for example in Fig. 22a, or a distortion of the Hg surface) have resistances R are out of our range i.e. >1 MΩ. For type 2 tubes, the conductance is 13 k$\Omega < R <$ 15 kΩ. For the remainder 15 k$\Omega < R <$ 100 kΩ (type 3). We have not observed nanotubes whose resistance is significantly less than 13 kΩ (i.e. 6.5 kΩ as expected for 2 G_0). This observation is consistent with the in-air experiments.

The nanotubes of type 2 are always straight and free of particles (Fig. 22a, b). Type 3 nanotubes are usually clearly covered to some degree with graphitic particles or have visible defects (Fig. 22c), suggesting that the increased resistance of type 3 compared to type 2 has its origin in these attributes. This is consistent with the observations carried out in air experiments where type 3 tubes convert to type 2 during the cleaning procedure as shown in Fig. 21.

We previously measured currents up to 1 mA through the tubes before they are damaged implying current densities at the outer layer up to 10^{10} A/cm^2 for a 10 nm diameter tube (assuming surface conduction). Higher currents (>1 mA) damage the nanotubes. We observe that defective nanotubes tend to break at the defects (Fig. 23) while type 2 nanotubes invariably break near the Hg contact at a higher applied voltage (such as 4 V). The presence of the point defect can destroy the ballistic conductance. In particular, type 2 nanotubes do not break halfway between the contacts, which is where the temperature would be greatest for freely suspended nanotubes if the

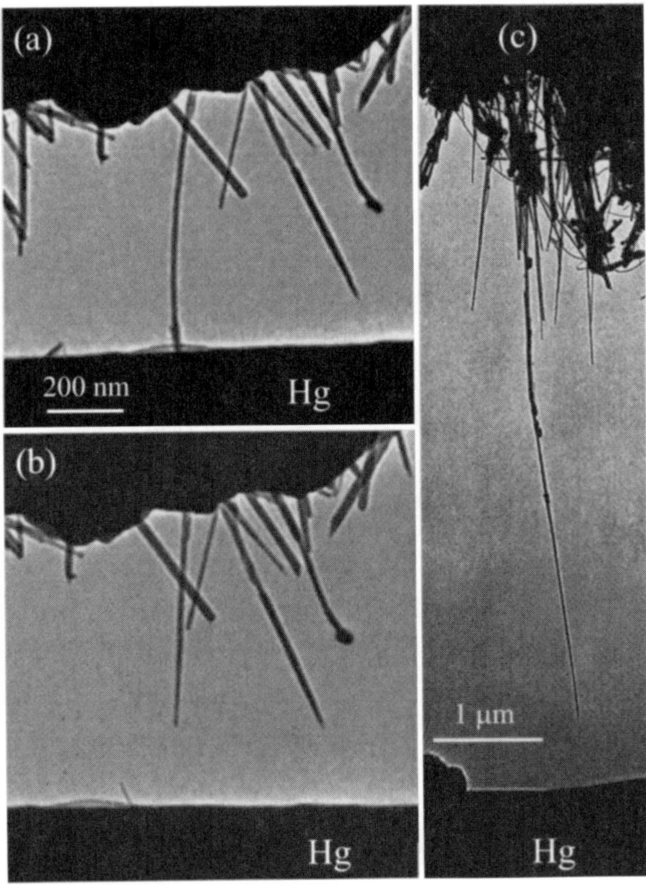

Figure 22. In-situ observation of electric transport through a single carbon nanotubes. (a) Applied voltage $V = 0.1$ V, current $I = 7.3$ μA, the resistance of the carbon $R = 12.7$ kΩ which corresponds to a conductance $G = (12.7\ \text{k}\Omega)^{-1} = 1.02\ G_0$. (b) After applying a 4 V voltage, the nanotube was broken and the break occurred at the contact of the nanotube and the Hg. (c) The conductance of this carbon nanotube contaminated with graphitic particles is 0.25 G_0.

nanotubes were dissipative conductors. Hence we conclude that the heat dissipation occurs primarily at the contacts, consistent with ballistic transport.

From our investigations, graphitic particles on the surfaces of freely suspended multiwalled carbon nanotubes as well as defects decrease their conductance. Extrapolating these results suggest that perfectly clean and defect free nanotubes will exhibit vanishing intrinsic resistivities, consistent with ballistic conduction. Since adhered particles can have such a pronounced effect on the transport properties of MWNTs, it is reasonable that related scattering effects reduce the conductance of substrate supported nanotubes. Often graphitic particles are found on are produced supported nanotubes. Surface

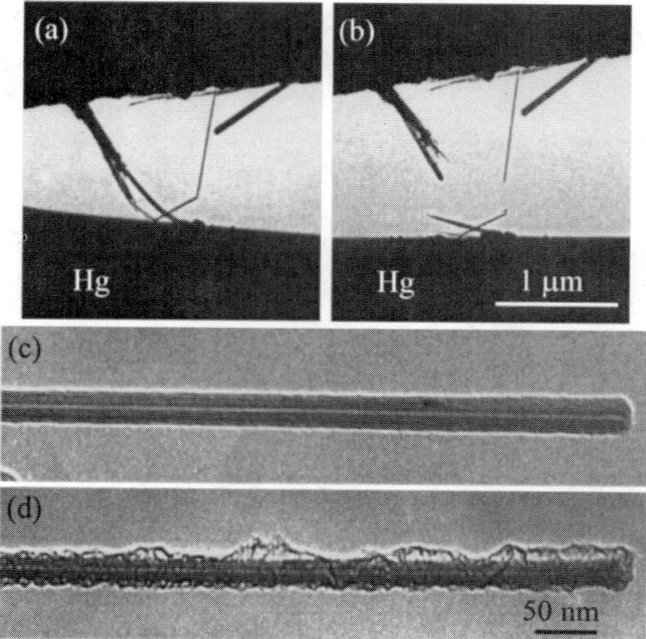

Figure 23. Examples of nanotube failure modes. MWNTs (a) before and (b) after passing a large current, showing the break points at the defect sites. (c, d) A MWNT before and after passing a large current through it. In (d) the tip of the MWNT had been contacted to a Au surface and current was passed trough it. Only the outer layer, starting at the contact point has been damaged, suggesting that the current flows over the surface.

contamination, such as oxygen adsorption [60], can drastically affect the conductance of carbon nanotubes.

8.3. Top Layer Transport in MWNT

We now examine the origin of quantum conductance in MWNT at room temperature. Graphite has a unique layer structure, in which the conductivity parallel to the graphitic plane (a–b axis plane) is about 50 times higher than that along the c-axis (Figure 24a). If the applied voltage is small, so that the interlayer tunneling effect can be ignored and the end of the nanotube is closed (Figure 24b), the electric current mainly flows along the top surface layer, provided the tube is perfect and there is no defect. Since the layer thickness of $\sim$0.34 nm is comparable to the wavelength of the conduction electron in the graphitic ($\sim$0.5 nm), quantum conductance is possible. The unique structure of the nanotube greatly reduces the phonon density of states, thus the mean-free-path length of the electron is extraordinarily long. Theoretical calculation shows that the mean-free-path length for SWNT is $>$10 μm [61], possibly resulting in room temperature quantum effect. Our recent experiments have shown a mean-free-path length of $>$50 μm. This is possible for a structurally perfect nanotube with closed ends (Figure 24c). If the applied voltage is large, the interlayer tunneling among the nanotube

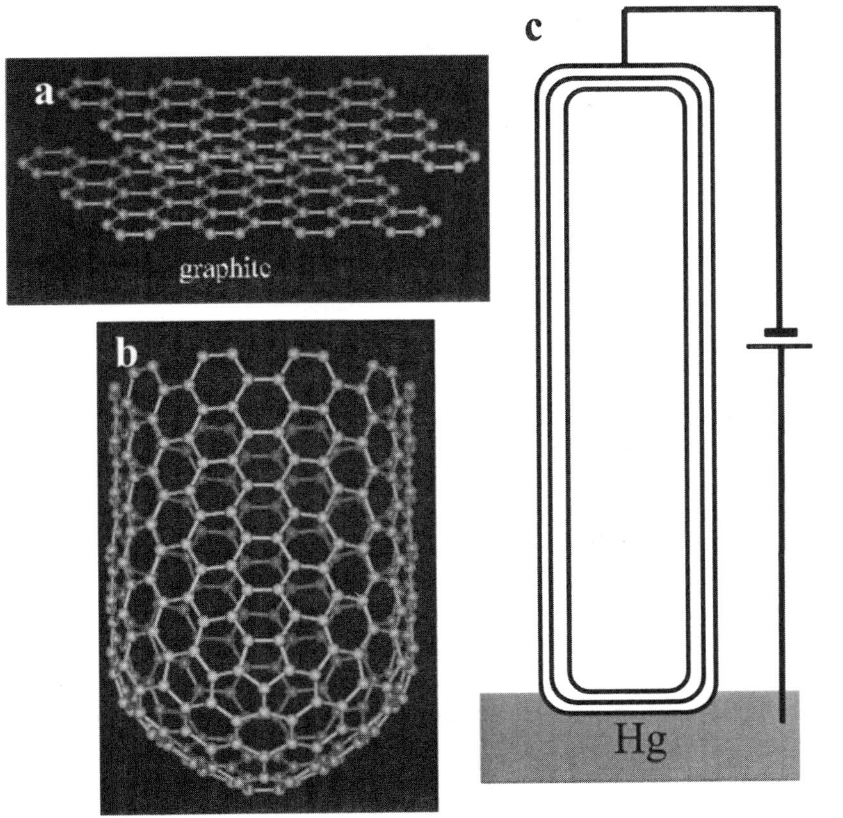

Figure 24. (a) Atomic structure of graphitic sheets; (b) A carbon nanotube model with a closed; (**c**) top layer conductance of a defect-free, close-end MWCT.

walls is inevitable, the effective size of the object that carries the current is large, finally destroying the quantum conductance effect. On the other hand, if there is a defect, the density of phonons near the defect is different from the rest of the nanotube, a reduced mean-free-path length at the defect site results in the disappearance of the quantum effect as well. Therefore, nanotubes grown by CVD process that contain a large density of defects are unlikely to exhibit the quantum conductance.

Figure 23d shows a MWNT after passing a large current, which was brought into contact with an Au surface (which replaced the Hg droplet in this experiment) during the field emission experiments. Due to a flow of a shock short-circuit current, the surface of the nanotube is disrupted starting at the contact point while the interior layers are not visibly affected. This observation provides further evidence that the electronic current passes over the surface layers of the nanotube, e.g., the single layer conductance. The observation of Aharonov-Bohm oscillations in carbon nanotubes also supports the top layer conductance [62].

9. SUMMARY

Property characterization of nanomaterials is challenged by the small size of the structure because of the difficulties in manipulation. In this chapter, we demonstrated a novel approach that allows a direct measurement of the mechanical properties, the electron field emission and the ballistic quantum conductance in individual nanotubes and nanowires by in-situ transmission electron microscopy (TEM). The technique is powerful in a way that it can directly correlate the atomic-scale microstructure of the carbon nanotube with its physical properties, providing a one-to-one correspondence in structure-property characterization.

To carry out the property measurement of a nanotube/nanobelt, a specimen holder for an TEM was built for applying a voltage across a nanotube and its counter electrode. Static and dynamic properties of the nanotubes can be obtained by applying controllable static and alternating electric fields. To measure the bending modulus of a carbon nanotube, an oscillating voltage is applied on the nanotube with ability to tune the frequency of the applied voltage. Resonance can be induced in carbon nanotubes by tuning the frequency, from which the bending modulus can be derived.

Due to the rectangular cross-section of the nanobelt, two fundamental resonance modes have been observed in corresponding to two orthogonal transverse vibration directions, showing the versatile applications of nanobelts as nanocantilevers and nanoresonators. The bending modulus of the ZnO nanobelts was measured to be ~52 GPa. Nanobelts have also been demonstrated as ultra-small nanocantilevers for sensor and possibly imaging applications in AFM.

For carbon nanotube emitters, most of the electrons are emitted from the tips of the tubes, and it is the local work function that matters to the properties of the tube field emission. Since the lack of suitable technical method to measure the work function of carbon nanotubes, the value of work function of carbon nanotubes used in the literatures is mainly from the well-studied carbon or graphite. Thus, it is necessary to experimentally measure the work function of carbon nanotubes. We presented experimental measurements of tip work functions of individual carbon nanotubes. Our results indicate that the tip work function show no significant dependence on the diameter of the nanotubes in the range of 14–55 nm. Majority of the nanotubes have a work function of 4.6–4.8 V at the tips, which is 0,2–0.4 V lower than that of carbon.

The conductance of a carbon nanotube was measured as a function of the depth with which the tube was inserted into the mercury. Surprisingly, the nanotube displays quantum conductance No heat dissipation was observed in the nanotube. This is the result of ballistic conductance, and it is believed to be a result of single graphite layer conductance.

ACKNOWLEDGEMENT

The results reviewed in this paper were partially contributed from my group members and collaborators: Ruiping Gao, Xuedong Bai, William Hughes, Enge Wang, P. Poncharal, James, Gole, and W. de Heer, to whom I am very grateful. Research supported by NSF, NASA and NSFC.

REFERENCES

[1] Z. L. Wang, Y. Liu and Z. Zhang (ed.) *Handbook of Nanophase and Nanostructured Materials*, Vol. I–IV, Tsinghua University Press (Beijing) and Kluwer Academic Publisher (New York) (2002).
[2] Z. L. Wang and C. Hui (ed.) *Electron Microscopy of Nanotubes*, Kluwer Academic Publisher (New York) (2003).
[3] Ph Buffat, J.-P. Borel, *Phys. Rev. A*, 13 (1976) 2287.
[4] Z. L. Wang, J. Petroski, T. Green, M. A. El-Sayed, *J. Phys. Chem. B*, 102 (1998) 6145.
[5] T. S. Ahmadi, Z. L. Wang, T. C. Green, A. Henglein, M. A. El-Sayed, *Science*, 28 (1996) 1924.
[6] M. M. J. Treacy, T. W. Ebbesen and J. M. Gibson, *Nature*, 381 (1996) 678.
[7] A. Krishnan, E. Dujardin, T. W. Ebbesen, P. N. Yianilos, M. M. J. Treacy, *Phys. Rev. B*, 58 (1998) 14013.
[8] N. Osakabe, K. Harada, M. I. Lutwyche, H. Kasai and A. Tonomura, *Appl. Phys. Letts.*, 70 (1997) 940.
[9] P. Poncharal, Z. L. Wang, D. Ugarte and W. A. de Heer, *Science*, 283 (1999) 1513.
[10] Z. L. Wang, P. Poncharal and W. A. De Heer, *Pure Appl. Chem.* 72 (2000) 209.
[11] Z. L. Wang, P. Poncharal, and W. A. De Heer, *J. Physics & Chemistry of Solids*, 61 (2000) 1025.
[12] L. Meirovich, *Elements of Vibration Analysis*, McGraw-Hill (New York) (1986).
[13] J. Z. Liu, Q. S. Zheng and Q. Jiang, *Phys. Rev. Letts.*, 86 (2001) 4843.
[14] W. Z. Li, S. S. Xie, L. X. Qian, B. H. Chang, B. S. Zou, W. Y. Zhou, R. A. Zhao, G. Wang, *Science*, 274 (1996) 1701.
[15] Z. F. Ren, Z. P. Huang, J. H. Xu, P. B. Wang, M. P. Siegal, P. N. Provencio, *Science*, 282 (1998) 1105.
[16] S. S. Fan, M. G. Chapline, N. R. Franklin, T. W. Tombler, A. M. Cassell, and H. J. Dai, *Science*, 283 (1999) 512.
[17] R. P. Gao, Z. L. Wang, Z. G. Bai, W. A. de Heer, L. M. Dai and M. Gao, *Phys. Rev. Letts.*, 85 (2000) 622.
[18] Z. L. Wang, Zu Rong Dai, Zhi Gang Bai, Rui Ping Gao and James Gole *Appl. Phys. Letts.*, 77 (2000) 3349.
[19] Z. W. Pan, Z. R. Dai, Z. L. Wang, *Science*, 291 (2001) 1947.
[20] Z. L. Wang, Z. W. Pan, Z. W., Z. R. Dai, *US Patent* No. 6,586,095.
[21] Z. R. Dai, Z. W. Pan, Z. L. Wang, *J. Phys. Chem. B* 106 (2002) 902.
[22] Z. R. Dai, Z. W. Pan, Z. L. Wang, *J. Am. Chem. Soc.* 124 (2002) 8673.
[23] X. Y. Kong and Z. L. Wang, *Nano Letters*, 3 (2003) 1625.
[24] X. Y. Kong and Z. L. Wang, *Appl. Phys. Letts.*, 84 (2004) 975.
[25] X. G. Wen, W. X. Zhang, S. H. Yang, *Nano Letters*, 2 (2002) 1397.
[26] Y. B. Li , Y. Bando, D. Golberg, K. Kurashima, *Applied Physics Letters*, 81 (2002) 5048.
[27] J. Zhou, N. S. Xu, S. Z. Deng, J. Chen, J. C. She, Z. L. Wang, *Adv. Mater.* 15 (2003) 1835.
[28] Y. B. Li , Y. Bando, T. Sato, *Chemical Physics Letters* 359 (2002) 141.
[29] J. Liu, J. Cai, Y. C. Sun, Q. M. Gao, S. L. Suib, M. Aindow, *J. of Physical Chemistry B*, 106 (2002) 9761.
[30] X. G. Wen, W. X. Zhang, S. H. Yang, *Langmuir*, 19 (2003) 5898.
[31] X. D. Bai, E. G. Wang, P. X. Gao, Z. L. Wang, *Appl. Phys. Letts.*, 82 (2003) 4806.
[32] G. Carlotti, G. Socino, A. Petri and E. Verona, *Appl. Phys. Letts.*, 51 (1987) 1889.
[33] S. X. Mao, M. H. Zhao, Z. L. Wang, *Applied Physics Letters*, 83 (2002) 993.
[34] H. Dai, E. Yenilmez, Q. Wang, R. J. Chen, D. Wang, *Appl. Phys. Lett.*, 80 (2002) 12.
[35] H. J. Dai, J. H. Hafner, A. G. Rinzler, D. T. Colbert, and R. E. Smalley, *Nature*, 384 (1996) 147.
[36] W. Hughes, Z. L. Wang, *Appl. Phys. Letts.*, 82 (2003) 2886.
[37] W. Z. Li, S. S. Xie, L. X. Qian, B. H. Chang, B. S. Zou, W. Y. Zhou, R. A. Zhao, G. Wang, *Science*, 274 (1996) 1701.
[38] S. S. Fan, M. G. Chapline, N. R. Franklin, T. W. Tombler, A. M. Cassell, and H. J. Dai, *Science*, 283 (1999) 512.
[39] R. P. Gao, Z. W. Pan and Z. L. Wang, *Appl. Phys. Letts.*, 78 (2001) 1757.
[40] Z. L. Wang, *Adv. Mater.*, 12 (2000) 1295.
[41] X. D. Bai, E. G. Wang, P. X. Gao, Z. L. Wang, *Nano Letters*, 3 (2003) 1147.
[42] B. G. Frost, L. F. Allard, E. Volkl, D. C. Joy, D. C. in *Electron Holography*, eds., A. Tonomura, L. F. Allard, G. Pozzi, D. C. Joy and Y. A. Ono, Elsevier Science B. V. (1995) pp. 169.
[43] J. Cumming, A. Zettl, M. R. McCartney and J. C. H. Spence, *Phys. Rev. Letts.*, 88 (2002) 56804-1.
[44] Z. L. Wang, R. P. Gao, W. A. de Heer and P. Poncharal *Appl. Phys. Letts.*, 80 (2002) 856.
[45] A. G. Rinzler, J. H. Hafner, P. Nikolaev, L. Lou, S. G. Kim, D. Tomanek, P. Nordlander, D. T. Colbert, R. E. Smalley, *Science*, 269 (1995) 1550.

[46] J. Cumings, P. G. Collins and A. Zettle, *Nature*, 406 (2000) 586.
[47] P. G. Collins and P. Avouris, *Nanoletters*, 1 (2001) 453.
[48] Y. H. Gao, Y. Bando, *Nature*, 415 (2002) 599.
[49] J. Cumings, A. Zettl, *Science*, 289 (2000) 602.
[50] For a review see C. Dekker, *Physics Today*, 22 (May, 1999).
[51] T. W. Ebbesen, H. J. Lezec, H. Hiura, J. W. Bennett, H. F. Ghaemi and T. Thio, *Nature*, 382 (1996) 54.
[52] S. J. Tans, M. H. Devoret, H. Dai, A. Thess, R. E. Smalley, L. J. Geerligs, C. Dekker, *Nature*, 386 (1997) 474.
[53] S. Frank, P. Poncharal, Z. L. Wang, W. A. de Heer, *Science*, 280 (1998) 1744.
[54] P. Poncharal, C. Berger, Yan Yi, Z. L. Wang, W. A. de Heer, J. *Phys. Cgem.* B, 106 (2002) 12104.
[55] M. S. Fuhrer, J. Nygard, L. Shih, M. Forero, Y. G. Yoon, M. S. C. Mazzoni, H. J. Choi, J. Ihm, S. G. Louie, A. Zettl, P. L. McEuen, *Science*, 288 (2000) 494.
[56] P. J. de Pablo, E. Graugnard, B. Walsh, R. P. Andres, S. Datta and R. Reifenberger, *Appl. Phys. Letts.*, 74 (1999) 323.
[57] H. J. Choi, J. Ihm, Y. G. Yoon, S. G. Louie, *Phys. Rev.* B, 60 (1999) 14009.
[58] P. Delaney, M. Di Ventra and S. T. Pantelides, *Appl. Phys. Letts.*, 75 (1999) 3787.
[59] T. W. Tombler, C. W. Zhou, L. Alexseyev, J. Kong, H. J. Dai, L. Lei, C. S. Jayanthi, M. J. Tang, S. Y. Wu, *Nature*, 405 (2000) 769.
[60] P. G. Collins, K. Bradley, M. Ishigami, A. Zettl, *Science*, 287 (2000) 1801.
[61] C. T. White and T. N. Todorov, *Nature*, 393 (1998) 240.
[62] A. Bachtold, C. Strunk, L. P. Salvetat, J. M. Bonard, L. Forro, T. Nussbaumer, C. Schonenberger, *Nature*, 397 (1999) 673.

17. ENVIRONMENTAL TRANSMISSION ELECTRON MICROSCOPY IN NANOTECHNOLOGY

RENU SHARMA AND PETER A. CROZIER

1. INTRODUCTION

Nanotechnology depends on the unique properties and behaviors of nanophase systems and the nanoparticles making up such systems often have properties that are significantly different from bulk materials. The behavior of the system may be strongly influenced by particle size, shape and the interactions between particles. In general, the configuration and evolution of the system will also be influenced by temperature, ambient atmosphere and associated gas-solid reactions. Moreover, in applications, nanoparticles are often subjected to high temperatures and pressures and as a result their structure and chemistry can dramatically change. For these reasons it is important to study nanoparticle systems under a wide range of different ambient atmospheres and temperatures. Since the invention of the transmission electron microscope (TEM), there have been continuous efforts to modify the instrument to observe biological samples in their native form (wet) and *in-situ* gas-solid reactions, e.g. corrosion, oxidation, reduction etc. These modified microscopes have been called 'controlled atmosphere transmission electron microscopes' or more recently 'environmental transmission electron microscopes' (ETEM). An ETEM can permit researchers to follow structural and chemical changes in nanophase materials, at high spatial resolution, during gas-solid or liquid-solid reactions over a wide range of different pressures. This information can be used to deduce atomic level structural mechanisms of reaction processes. With careful experimental planning, thermodynamic and kinetic data can also be obtained. An ETEM can thus be described as a nanolaboratory for the synthesis and the characterization of nanomaterials.

In order to follow the gas-solid or liquid-solid interactions at the nanometer level, we need to modify the TEM to confine gas or liquid to the area around the sample. In a transmission electron microscope (TEM), high-energy electrons (generally 100–1500 KV) are used to form an image. In order to avoid scattering from gas molecules and to increase the life of the electron source, both the column and the gun chamber are kept under high vacuum conditions (better than 10^{-6} Torr). When a field emission gun (FEG) is used as the electron source, the gun chamber should be better than 10^{-9} Torr for optimum performance and long life. However, in order to observe gas-solid reactions, or image hydrated materials (including biological samples), the environment around the sample should be typically 10^{-3} to 150 Torr. In an ETEM-our goal is to confine the reactive gas/liquid to the sample region without significantly compromising the vacuum of the rest of the microscope column. Figure 1 shows the general functioning principle of an ETEM. The ETEM allows the atmosphere around the sample to be controlled while still providing all of the high spatial resolution information (electron diffraction, bright-field images, dark-field images etc . . .) available in a regular TEM.

In this chapter we will give a brief overview of the history and development of the ETEM. This will be followed with a description of time-resolved recording techniques which are particularly important in ETEM experiments because we are interested in following the evolution of the nano-system during gas-solid reactions. Practical aspects of designing and performing controlled atmosphere experiments are discussed in section 4. In our final section on applications, we show that ETEM is useful for obtaining detailed information on nanoparticle synthesis, phase transformations pathways and nanoparticle kinetics.

2. HISTORY OF ETEM

2.1. Early Developments

The concept of controlling the sample environment during observation is almost as old as the idea of using TEM to image thin biological sections. The aim of an early ETEM design [1] was to examine biological samples in the hydrated state and to study the effect of gases on sample contamination. There was a steady development of the technique during the seventies and several review articles on the subject were published during that time [2–4]. A comprehensive review on environmental TEM and other *in-situ* techniques for TEM can be found in the book by Butler and Hale [5].

Environmental cell (E-cell) designs were based on modifying the sample area to restrict or control the gaseous flow from the sample region to the column of the microscope (Figure 1). This was achieved in two ways:

a) Window Method—gas or liquid is confined around the sample region by using thin electron transparent windows of low electron scattering power, e.g. thin amorphous carbon or SiN films.

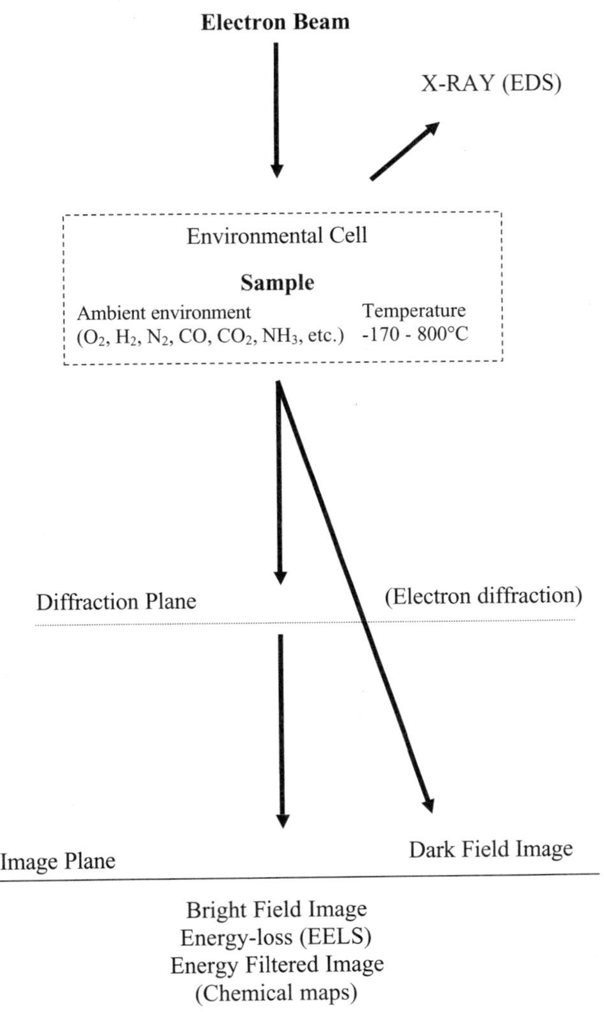

Figure 1. Schematic diagram of ETEM showing operation principle and available high resolution information. Pressures in the cell are typically 1–50 Torr.

b) Differential Pumping—a pressure difference is maintained by installing small apertures above and below the sample area and using additional pumping.

In the window method, the windows are usually placed in a TEM sample holder. The windowed design has the advantage of being able to handle high gas pressures (depending upon the strength and thickness of the window). They can also handle wet samples and are often called 'wet cell' sample holders. The main disadvantage of the window method is that high-resolution imaging is difficult due to the additional scattering from the amorphous structure of the window films. Moreover, the windows often ruptured, the increased thickness of the sample holder did not leave much room for tilting and the samples could not be heated [5].

Large objective lens pole-piece gaps were required to successfully maneuver the gas confinement system and still leave enough space for tilting and translation of the sample. Therefore, most of the early environmental cells were designed to fit into the column of a high-voltage electron microscope (HVEM; 1000–1500 KV) [2–5]. Moreover, Swan and Tighe [6] studied the loss of intensity with increasing cell pressure for different voltages. They concluded that using high voltage TEM could reduce the loss of intensity due to high gas pressures in the sample area. The use and further development of microscopes with E-cells diminished considerably in the eighties due to several problems associated with the high-voltage microscopes and controlled-atmosphere chambers. First, many materials are damaged by the high-energy electron beam and could not be studied with high-voltage microscopy. The resolution limit, after installation of the E-cell, was not suitable for atomic-level imaging and finally the high-voltage microscopes were expensive to purchase and maintain.

2.2. Later Developments and Current Status

In the early eighties, improvements in the objective lens pole-piece design led to the development of atomic-resolution medium-voltage (200–400 keV) transmission electron microscopes. This stimulated renewed interest in E-cell designs in the nineties because the pole-piece gaps (7–9 mm) were large enough to accommodate the cell while still permitting atomic resolution imaging (0.2 to 0.25 nm). The smaller pole-piece gap, 7–9 mm compared to 13–17 mm for high voltage TEM, has an added advantage of reducing the gas path through the cell and thereby reducing the amount of electron scattering from the gas or liquid. Using an intermediate voltage microscope and thin carbon windows, Parkinson was able to demonstrate atomic resolution imaging (0.31 nm) in ceria in an atmosphere of 20 Torr of N_2 [7]. Atomic resolution imaging with the differentially pumped system was demonstrated two years later [8].

In the past decade, attention has concentrated on the design of differentially pumped E-cells (Table 1, 9–18). The modern differential pumping systems are designed after the basic principles outlined by Swann and Tighe [19] and consist of two pairs of apertures with an aperture from each pair being placed above and below the sample. The first pair of apertures is placed closest to the sample and most of the gas leaking through these apertures is pumped out of the system using a turbo molecular pump. The second pair of apertures is larger than the first pair (because they see much lower gas pressure) and is used to further restrict the leakage of gases into the microscope column.

There are several factors to consider when selecting the size of the differential pumping apertures:

Table 1. Development History of ETEM Since 1991.

Year	Research Group	Microscope	Reported lattice resolution/ p\pressure/Temperature	Reference
1991	Doole, Parkinson, Hutchinson	JEOL 4000	0.31 nm/4.2 Torr H_2/670°C	9
1991	Lee, Robertson, Birnbaum	JEOL 4000	Not reported/70 Torr* H_2	10
1991	Yao, Spindler/ Gatan Inc.	Phillips CM 30	0.34 nm/20 Torr*/No reported	11
1994	Sharma *et al.*	Phillips 400T	0.42 nm/3 Torr NH_3	12
1997	Boyes & Gai	Phillips CM 30	0.23 nm/500°C/0.3 Torr N_2	13,14
1998	Sharma *et al.*	Phillips 430	0.31 nm/RT/4 Torr H_2	15
2001, 1st commercial	Hansen/Haldor Topsoe	Phillips CM 200 FEG	0.23 nm/550°C/4 Torr H_2/N_2	16,17
2003	Sharma *et al.*	Tecnai F 20 TEM/STEM	0.13/RT/4 Torr N_2	18

*Reported Pressure limit.

(1) The gas leak rate through the aperture should be comparable to the pumping rate on the high vacuum side of the aperture in order to keep the column vacuum in the 10^{-6} Torr range.
(2) The angular range in the diffraction pattern should not be severely limited by the aperture.
(3) A reasonable field of view of the sample should be preserved.

Since the most critical part requiring high vacuum is the gun area, it is desirable to have a lower leak rate from the upper aperture so this aperture may have a smaller diameter than the lower aperture. Typical aperture sizes for the first set are in the range 100–200 μm giving a good compromise between reducing the gas leak rate to the gun area while at the same time maintain high angle diffraction capabilities and large viewing areas.

Boyes and Gai [13] successfully incorporated a multilevel differential pumping system into their Philips CM 30. Recently, FEI (previously Philips Electron Optics) redesigned the vacuum system of a CM 300-FEG in order to convert it to an ETEM [16, 17]. This modification was also incorporated into the new generation Tecnai microscopes [18] and is now commercially available (Figure 2). The modifications to the objective pole-piece region of the column are shown in Figure 3. In the commercially available instrument, the first and second sets of differential pumping apertures are located at the ends of the upper and the lower objective pole-piece bores (Figure 3). The gas leaking through the first pair of apertures (Figure 3, first level pumping) is pumped out through top and bottom objective pole-pieces. The gas flow is further restricted by the second set of apertures (Figure 3, second level pumping). The region above the condenser aperture and the viewing chamber are evacuated by separate pumping systems (Figure 3, third level of pumping).

Figure 2. Tecnai F 20 field emission gun ETEM at Arizona State University operated at 200 KV and equipped with Gatan Imaging Filter.

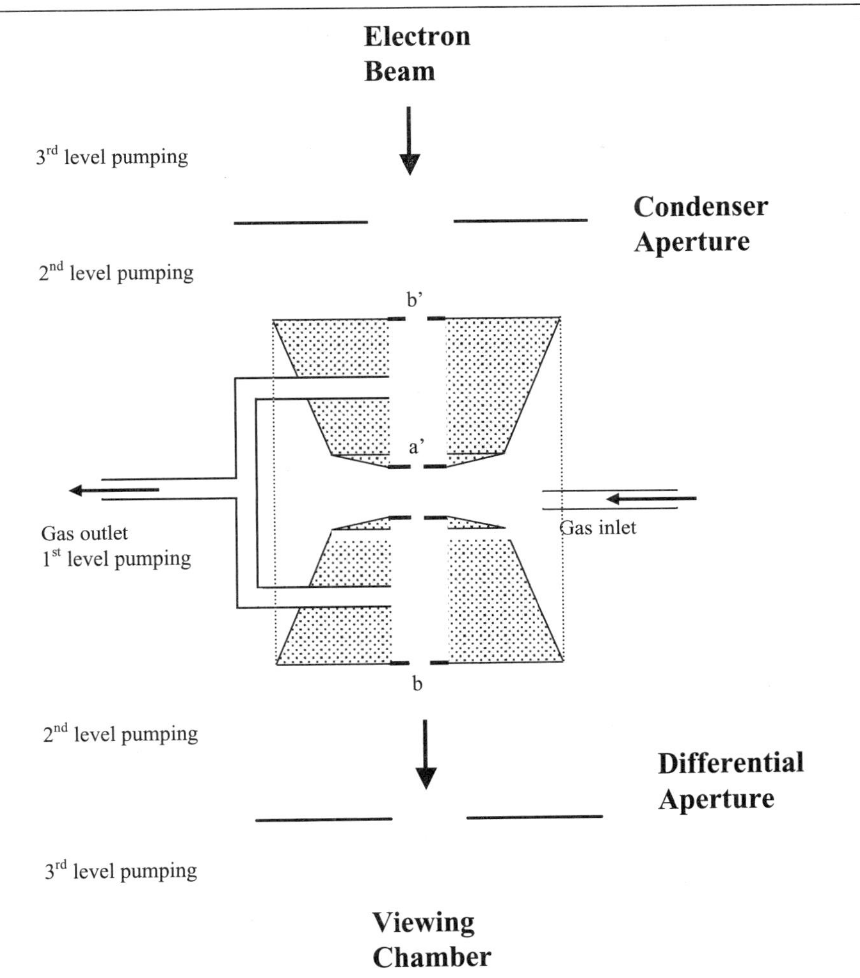

Figure 3. Block diagram showing the modifications in the objective pole-piece area to accommodate 1st level of differential pumping. The residual gases leaking out from the aperture b and b' are pumped out by 2nd level of pumping and the 3rd level of pumping is performed using separate pumps for the viewing chamber and column-section between condenser aperture and gun chamber.

The gas inlet pressure from a gas reservoir is measured outside the microscope column. A gas manifold with numerous gas inlets from various gas cylinders and one gas outlet to the sample region of the ETEM is used to handle gases. This arrangement not only makes it easy to switch between various gasses but also allows different gases to be mixed in desired ratios before leaking them into the sample area. The microscope column is isolated from the gas inlet, outlet and associated pumps using pneumatic valves. A control system can be designed to automatically open and close the valves in order to switch between high vacuum TEM and ETEM modes [18]. The ability to rapidly switch between modes is particularly important in a multi-user facility because it permits the microscope to be easily operated in the conventional high-vacuum mode. On our Tecnai F 20, we have demonstrated an information limit of 0.13 nm in 4 Torr of H_2 proving that atomic resolution capability can be easily attained [18].

To eliminate the effect of inelastic gas scattering at high pressures, a Gatan Imaging Filter (GIF) has been fitted to the ETEM. This configuration has the added advantage of permitting chemical information to be obtained through the use of electron energy-loss spectroscopy (EELS) and chemical maps by energy filtered (EFTEM) imaging. The field-emission gun permits high spatial resolution spectroscopy and scanning transmission electron microscopy (STEM) to be performed *in situ*. In the Tecnai F 20 ETEM, the electron beam can be focused down to about 0.2 nm in diameter. Annular dark-field STEM imaging can also be performed although the lower differential pumping aperture restricts the highest angle of scattering to about 50 mrad.

3. DATA COLLECTION

The data collection using an ETEM is usually performed with the same detectors used for TEM. The main difference is that the rate of data collection is directed by the rate of the reaction process of interest and often very high collection speeds are required. ETEM is usually undertaken to study dynamic processes such as phase transformations. In a typical experiment, sample temperature and pressure are varied with time in order to study gas-solid reactions at the nanometer or sub-nanometer level and extract information about reaction mechanisms and kinetics. For rapid transformation processes, it is necessary to continuously acquire and store data with good temporal resolution to ensure that the critical events are recorded. The high data collection rates result in large amounts of data being acquired during an experiment introducing practical data processing problems. In a typical experiment, many hours of data is recorded and stored although later analysis may show that only several minutes of data is scientifically interesting. Here we describe some of the considerations necessary for collecting different data types in an ETEM.

3.1. Real-Time Imaging Systems

The ideal detector for continuous image acquisition would consist of a low-noise digital camera system with a detection quantum efficiency close to unity, a large number of pixels (at least 1024^2) and the ability to perform rapid readouts (>50 frames per second). Data would be written continuously to a high-density storage media. Sophisticated image processing software would be capable of performing quantitative batch processing on extensive sequences of images and generate video output for review. Unfortunately no such system is readily available at present and most facilities use a television camera (TV) coupled to a video recording system. In the best systems, a phosphor or single crystal scintillator converts the incident electron signal to a photon signal which is then fed into an image intensifier coupled to a high-performance TV camera. The output from the TV camera is fed to a monitor and digital video recorder.

In a typical video recording set-up, several hours of data can be recorded with a time resolution of $1/30^{th}$ or even $1/60^{th}$ of a second (note the actual frame rate in the NTSC system is 29.97 Hz). For ETEM applications, the differential pumping below the sample should be reasonably good to ensure that gas products cannot deposit and react on the scintillator during electron irradiation.

The main advantage of such a video system is that the data can be recorded using reasonably priced commercially available digital cameras/recorders and the storage format can be easily transferred between labs. However this setup also suffers from a number of disadvantages which significantly compromise the data quality. The number of pixels associated with conventional television recording techniques is rather small. For example, in the NTSC system, the conventional television picture has a resolution of 480×640 pixels. In image recording, the highest spatial frequency should be sampled by at least 3 pixels so that if a resolution of d nanometers is desired in the image, the width of the field of view in the vertical direction will be $(480 \times d)/3$. For atomic resolution with 0.2 nm resolution, the width of the field of view in the vertical direction will be only 32 nm. Consequently, the field of view for real-time *in situ* observations is very much reduced making the probability of observing critical nucleation events rather small. The current development and implementation of high definition television systems (HDTV) should increase the number of pixels by about a factor of 4 and give a corresponding increase in the sampled area. However, this is still a factor of 10 less area than currently possible on conventional photographic micrographs.

Advanced cine-photography techniques could be used to record more data with improved temporal resolution (see Butler and Hale [5] for discussion of some early cine-photography setups). Improving the temporal resolution τ would be advantageous but may also be limited to low-resolution applications because of radiation damage considerations associated with atomic resolution imaging. It is common to record atomic resolution HREM images with doses of $\sim 5 \times 10^3$ e/Å^2 to obtain reasonable signal-to-noise ratios. To maintain this signal-to-noise level in each frame, the dose D that is necessary to record a sequence of length t with a temporal resolution τ is given by:

$$D(t) = 5 \times 10^3\, t/\tau$$

With 30 frames/second ($\tau = 0.0333$s) the dose rate will be 1.5×10^5 e/s/Å^2 which may result in significant damage in many materials. This simple expression shows that the dose rate has an inverse dependence on the temporal resolution; doubling the frame speed will require the electron dose to be doubled to maintain the same signal-to-noise per frame. In ETEM, it is usually desirable to run experiments with the lowest possible electron dose to minimize the impact of electron irradiation on the processes under study. It is possible to acquire high-resolution images using low-dose techniques. For example, atomic resolution images can be recorded from zeolites with 0.2 nm resolution and doses of around 100 e/Å^2 on a slow-scan CCD camera [20, 21]. However, with this dose, the information in the image is ultimately limited by counting statistics and is useful only for extracting average periodic information at the 0.2 nm level. By utilizing frame-averaging techniques, temporal resolution can be sacrificed in order to obtain improved signal-to-noise if necessary. In an ideal system, we would combine higher frame rates and low readout noise with suitable frame averaging to maximize the flexibility.

Data storage and quantitative image processing continue to be a challenge. It is necessary to convert digital video sequences into series of still frames which must be processed and re-assembled back into video format for playback. Since the data we are dealing with in ETEM has relatively high noise content, it is undesirable to utilize image compression techniques before quantitative analysis is performed. Consequently, very large volumes of digital data are generated which may consume enormous storage space. At present, some compression is often necessary to generate manageable files for presentation purposes.

For some ETEM experiments, it is not necessary to record data with high temporal resolution. For example, in metal particle sintering studies, many of the processes take place over a period of hours and data can be recorded with either a slow-scan CCD camera or using conventional photographic plates. In both cases, the image quality is better than that obtained from the TV system.

3.2. Spectroscopy and Chemical Analysis

Energy-dispersive x-ray spectroscopy (EDX) is a powerful technique for extracting elemental information in TEM. However, the EDX spectrometer is normally located in the pole-piece gap which effectively puts it in the middle of the E-cell for ETEM application. This can significantly complicate the design and implementation of the cell and spurious scattering from the windows or differential pumping apertures dramatically increases the background in the EDX spectrum. For these reasons, most of the current ETEMs rely on EELS to obtain chemical information. Detailed information about the technique can be found elsewhere [22]. In EELS, the fast electron is inelastically scattered as it passes through the thin sample resulting in significant energy transfers to the atomic electrons in the sample. The spectrum of energy losses carries detailed information about the elemental composition and electronic structure of the sample. Implementation of EELS on the ETEM is essentially identical to that on conventional microscopes because the detector is located a significant distance away from the environmental cell and, provided the differential pumping in the lower part of the column is effective, there is no negative impact on the energy-loss performance. The technique is best suited to light and medium atomic number materials in very thin samples (ideally <50 nm). The spatial resolution of the EELS analysis is limited by the probe size and instruments equipped with field emission guns can record fast spectral series from sub-nanometer areas.

It is common to install energy filters to permit energy filtering of images and electron diffraction patterns. The effect of inelastic scattering from the gases and thick samples can also be removed with zero-loss imaging. In many of the commercially available spectral processing routines, there are powerful features for batch processing of time resolved spectral data. This makes it relatively easy to study changes in composition and bonding during gas-solid reactions.

4. EXPERIMENTAL DESIGN STRATEGIES

To perform successful *in-situ* observations using an ETEM, experiments must be designed with extreme care. There are a number of parameters that must be considered mainly dependent on the following:

1. Type of data to be collected, i.e. high-resolution images, electron diffraction, electron energy-loss spectroscopy etc.
2. Type of gas and gas pressures to be used.
3. Reaction temperature.

Both the data type and the reaction rates determine the choice of recording media for data collection. While chemical processes with high reaction rates can only be observed by high-resolution imaging or electron diffraction data recorded on digital video tapes, processes with low reaction rates can be recorded using CCD cameras or photographic films. Moreover, time-resolved spectroscopic information can also be collected for high reaction rate processes provided that the very intense low-loss part of the spectrum can be utilized.

Gas contaminants may deposit in the gas delivery tubes and/or the microscope column. Therefore it is extremely important to use ultra-pure gases and keep the gas lines and samples clean for *in-situ* experiments. Contaminants can alter the reaction path and/or influence the reaction rates, for example, the presence of carbon contamination has been found to increase the reduction rate of iron oxide [23]. Similarly, if the microscope is used with different gases, cross contamination of gases could affect the reaction path. For example, we found the presence of water vapor in the system retarded the reduction rate for CeO_2 [24]. Water vapor also poisoned Ziegler-Natta catalyst used for polymerization of propylene [25]. Therefore, it is necessary to purge both the gas lines and microscope column with ultra-pure nitrogen before and after performing *in-situ* reactions. It is also advisable to purge the system with the gas that will be used for the reaction (if possible) before starting the experiment. The microscope column and gas lines should be baked (if possible) and pumped over night after purging.

Many of the experiments performed in the ETEM are conducted at elevated temperatures. Guidelines on calibration and use of heating stages are covered on the chapter on *in-situ* microscopy. However there are a number of practical situations that must be taken into account when performing heating experiments in the ETEM because of the possible reactive nature of the gases. Most of the commercially available heating holders are made from tantalum because of its high melting point and structural stability after many thermal cycles. Tantalum holders are well suited for working in vacuum and under reducing conditions; however the power required to reach a given temperature is higher under the typical gas pressures used in the ETEM compared to vacuum operation. The increased power is necessary because gases are usually admitted at room temperature and are a source considerable heat loss due to conduction. The amount of additional heating power required depends on the thermal conductivity and pressure of the gas in the cell. Table 2 gives the thermal conductivities of gases that are

Table 2. Thermal Conductivity of Various Gases.

Gas Type	Thermal Conductivity ($W\ m^{-1}\ K^{-1}$)
H_2	1684
H_2O	158
He	1415
N_2	243
O_2	151
CO	232
Ar	162
5% H_2/Ar	237
5% H_2/N_2	314

often used in ETEM experiments. This table shows that the thermal conductivity for H_2 is extremely high so that high-temperature work will require significantly higher heating currents. It is common to mix H_2 with Ar or N_2 to generate a lower thermal conductivity reducing gas. If possible it is advisable to obtain the required reaction pressures before heating the samples because introducing high gas pressures during heating will drastically cool the sample due to thermal losses to the gas.

Ta holders are not suitable for oxidizing environments because of the corrosive effects of the gas environment at elevated temperatures. Inconel alloy or Pt holders have been successfully developed (Gatan Incorporated) to provide more robust solutions when oxygen is present permitting heating up to ~1000°C. Kamino *et al.* have designed heating holders that are capable of heating powder samples to very high temperatures (1500°C) [26]. The temperatures attainable in the microscope are usually restricted by the design of the heating holder but for the ETEM it is also restricted by the material and placement of the differential pumping apertures. For example, the viton seals used to keep the apertures in place can be destroyed at high temperature due to heat transfer from the sample to the seals via the gas.

For powder samples, the choice of grid material is very important. The grid material should not react with the sample or gases used for experiments. Only Au or Pt grids should be used in oxidizing atmospheres as most other metals will be corroded in oxygen. The grid melting point should be significantly higher than the experimental temperatures. In general, atoms from the grid will become mobile when the experimental temperature exceeds the Tamman temperature (half metal melting point in degrees Kelvin). If the experimental temperatures exceed the Tamman temperature, metal atoms from the grid may diffuse over the sample and change the gas-solid reactions. For example, Cu melts at ~900°C, hence Cu grids should not be used for reaction temperatures above 300°C. In practice it is advisable to observe the behavior of grid material at the experimental temperatures and pressures using well-known samples. Carbon thin films often break due to thermal expansion and/or oxidation at elevated temperatures. Dusting bare grids with powder samples and observing the regions near grid bars give the best results. These regions also have good thermal contact with the heating furnace. The support material for cross-sectional samples should be given the same considerations as described above.

Preliminary experiments are usually performed to determine the reaction conditions, mainly temperature and pressure. The electron beam can alter the reaction rate/mechanism, therefore it is very important to monitor sample regions not exposed to the electron beam and/or confirm the results using *ex-situ* methods.

5. APPLICATIONS TO NANOMATERIALS

ETEM has been successfully used to understand many different gas-solid reactions [e.g. 27–36]. In general, one of the main advantages of performing *in-situ* measurements compared to *ex-situ* measurements is that the structural and chemical information is obtained under reaction conditions. *In situ* methods provide us with information on the reaction process, thus all the intermediate phases/steps (if any) are easily identified. Moreover, nanoparticle synthesis and characterizations can be performed and studied simultaneously. Some of the specific applications to nanotechnology are described in this section.

5.1. Transformation Mechanisms in Nanostructures Due to Gas-Solid Reactions

The majority of previous work on applying ETEM to elucidate transformation mechanisms during gas-solid reactions comes from the field of catalysis. Most heterogeneous catalysis involves gas-solid or gas-liquid reactions occurring directly on the surface of the catalyst. For this reason it is necessary to utilize nanoparticles simply to maximize the surface area. Indeed, heterogeneous catalysts may be one of the earliest technologically important areas to exploit the unique properties of nanophase materials. We now also realize that the properties of nanoparticles can be fundamentally different from their bulk counterparts and that these changes can yield unique catalytic activity. The catalytic activity may be related to structural or electronic transformations taking place as a result of the small particle size, the interaction between the particle and the support or as direct result of the interaction with the gaseous reactants. ETEM is particularly powerful for studying catalysts because it permits us to make atomic level observations of the response of the nanoparticles to gaseous environments. For catalytic research, the ETEM essentially acts as a small micro-reactor allowing us to observe the changes in nanostructure and nanochemistry under near reactor conditions.

Baker and co-workers were the first to extensively use ETEM to study heterogeneous catalysts. His first work focused on the growth of carbon filaments using Ni based catalysts [27]. Catalytic gasification of carbon and filamentous growth remained common themes for many of Baker's publications in the 80 and 90 s [28–30, 37–39]. He has also worked on a wide range of metal catalysts studying the influence of gaseous environments on particle shape and metal-support interactions [40–45]. Gai and Boyes have also developed ETEM and applied it extensively to a wide number of different heterogeneous catalysts [e.g. 31–34]. Our group and the group at Haldor Topsoe have also been studying various catalytic processes at atomic level under reaction conditions [e.g. 17, 24, 46].

5.1.1. Oxidation and Reduction Reactions

Oxidation and reduction plays a central role in many catalytic processes related to pollution control and chemical synthesis. ETEM can provide detailed information on the structural, compositional and electronic transformations associated with reduction-oxidation (redox) processes. Palladium based catalysts are of considerable importance in automotive technology for the elimination of NO_x in the exhaust gases of gasoline engines and for combustion of methane [47–48]. For both applications, oxygen is involved in the combustion and there is considerable evidence to show that the formation of palladium oxide may play a critical role in both catalytic processes. It has also been suggested that particle morphology and the Pd oxidation state are important in defining the active sites in Pd catalysts. For palladium supported on silica, it is known that under certain conditions, reduction of PdO can lead to the formation of small Pd metal particles containing central faceted voids [49–50]. Figure 4A show an example of Pd particles formed after *in-situ* reduction of PdO at 200°C in 1 Torr of H_2. These metal nanoshells form under a variety of different reducing atmospheres and at different reducing rates.

The mechanism for the void formation process was determined by following the evolution of individual nanoparticles during *in-situ* reduction [51]. A typical atomic-resolution image recorded from a particle at an intermediate state of reduction is shown in Figure 4B. Fourier analysis of this image shows that it is composed of two sets of atomic lattice planes corresponding to Pd metal and PdO confirming that the particles is indeed at an intermediate state of reduction. The coarse Moiré fringe pattern at the center of the particle arises due to overlap between the Pd and PdO fringes. By further processing this image, it is possible to reconstruct two digital dark-field images; one corresponds to the Pd metal spacing and the other to the oxide spacing. In Figure 4C these two images are overlaid using false color in order to see the distribution of Pd metal and PdO during the transformation process. The image shows that the initial reduction occurs uniformly over the surface of the particle and leads to the formation of a continuous metal shell around the oxide particle. Subsequent reduction of the particle occurs via oxygen diffusion through the metal shell. Reduction of PdO to Pd is accompanied by a 40% reduction in the overall volume of the particle. The metal shell associated with the initial reduction is rather rigid and fixes the final particle size. Consequently, part of the change in particle volume that occurs during the reduction must be accommodated by the formation of a void in the center of the metal particle. Continued heating of the sample resulted in a gradual reduction of the oxide and a steady increase in the thickness of the metal shell. Eventually the oxide disappears from the particle and a small void remains at the particle center. The voids collapse at temperatures above 500°C suggesting that voids will not form during high temperature reduction.

In the previous example, we determined the mechanism for oxide nanoparticles transforming completely to metal accompanied by a drastic change in particle morphology. In many cases, the redox processes of interest may involve transformations between different oxides states. Cerium based oxides are a technologically important

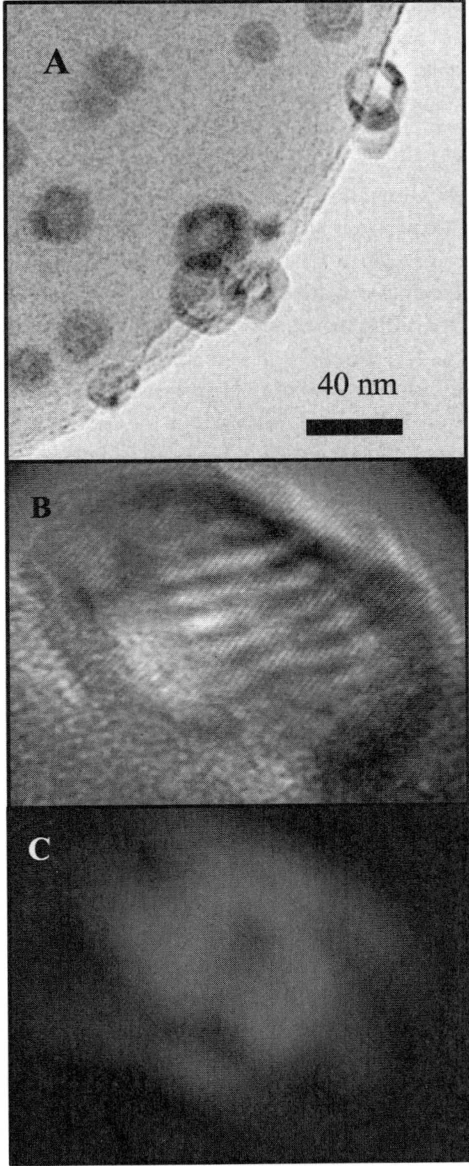

Figure 4. A) Pd metal particles formed after in situ reduction from PdO. Metal particles show evidence for void formation. B) Atomic resolution micrograph recorded from particle at an intermediate stage of transformation. C) Digital dark-field analysis of the image clearly shows that PdO (blue) gets initially reduced to Pd metal (red) on the outside of the particle. (See color plate 16.)

catalytic material for pollution control where transformations can occur between many different oxide forms depending on the reducing conditions. To follow the transformation pathways requires the combined use of imaging, spectroscopy and electron diffraction.

Three-way catalysts (TWC) are used to reduce the common pollutants CO, NO_x, and unburned hydrocarbons escaping from the exhaust system of automobiles [52]. The TWC functions by providing additional oxygen to complete combustion during fuel rich conditions. During fuel lean conditions, the catalyst can restore the oxygen reservoir by absorbing oxygen from the ambient atmosphere and reducing NO_x in the process. The effectiveness of the catalyst is directly related to the oxygen storage capacity (OSC) capacity of the material. This in turn depends on the degree and ease with which the catalysts can run through oxidation and reduction cycles. CeO_2 based catalysts are commonly used in catalytic converters because Ce can exist in both +3 and +4 oxidation states and the oxides possess high oxygen mobility at low temperature ($\approx$300°C) resulting in an easy redox cycles [53–54]. The ceria/zirconia mixed oxide systems ($Ce_{1-x}Zr_xO_2$) have been found to have higher OSC than the pure oxides and consequently are the material of choice for many TWC applications [55].

Fully oxidized CeO_2 has a fluorite structure similar to most of the rare earth oxides, e.g. PrO_2 and TbO_2. These oxides can also exist in both + 3 and + 4 oxidation states and possess high oxygen mobility. For PrO_2 and TbO_2, the oxygen vacancies introduced during reduction cause the formation of ordered superstructure phases, and a number of discrete intermediate phases with a general formula, R_nO_{2n-2m} have been identified [54]. Similar intermediate phases have been predicted to form for CeO_2 during reduction. However this system has been more difficult to study in reduced form because of its high oxygen affinity.

Recently ETEM has been used to follow the formation of superstructures under reducing environments at high temperatures [24]. Figure 5A shows the HREM image and the digital diffractogram of a single crystal oriented along $<111>_F$ (F = fluorite) zone axes at 800°C in 2 Torr of dry hydrogen. The reduction process for single crystals was observed to be very slow as expected due to their low surface area. After about 5 minutes of observation, a strained structure was observed to start forming (Figure 5B), probably due to oxygen vacancy formation. After 14 minutes, semi-ordered fringes were observed (Figure 5C) and after 40 minutes of isothermal heating at 800°C, a domain of ordered superstructure was observed to grow to ~30 nm in size (Figure 5D). Simultaneously, extra superlattice reflections could also be observed in the digital diffractogram (Figure 5D) indicating that the oxygen vacancies formed during reduction were ordering. Since the superstructure reflections observed here do not have a simple relationship with the fluorite lattice, it is not possible to determine the exact nature of the phase formed. In other words, the fluorite orientation is not a superstructure zone perpendicular to the plane in which the ordered vacancies are being formed. CeO_2 nanopartcles reduced quite easily and vacancy ordering was observed together with sintering processes. The measured reduction temperature and rates for samples with different surface areas confirmed that the fastest reductions were obtained for high surface-area nanoparticles.

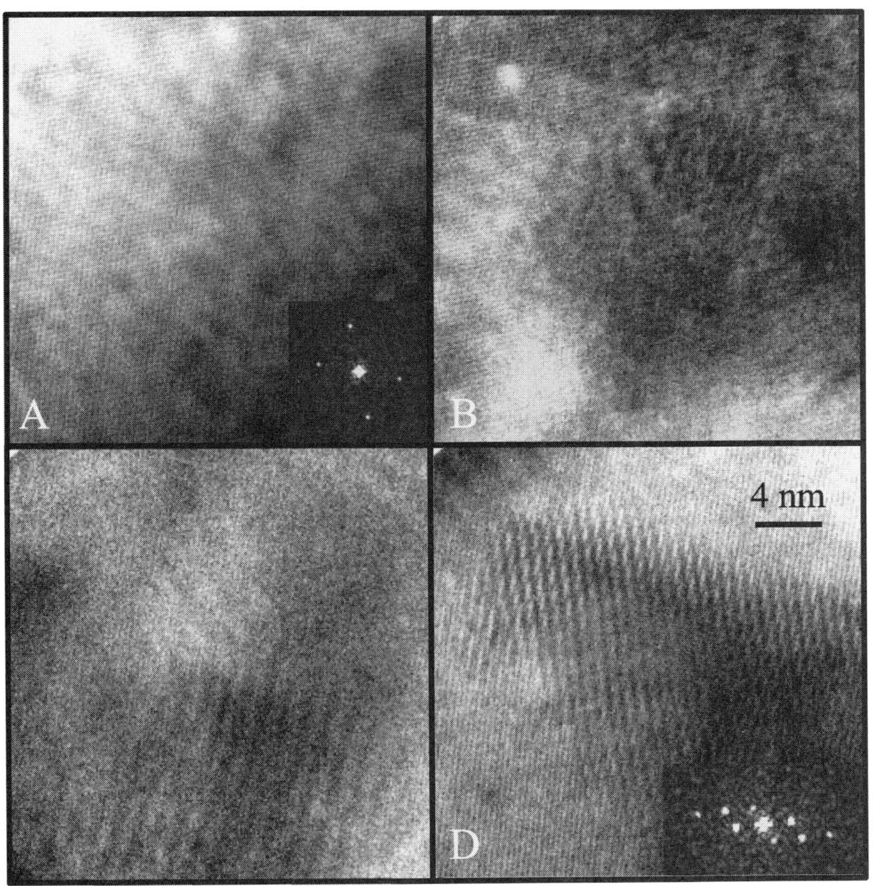

Figure 5. High resolution electron microscopy images recorded from a single crystal CeO_2 sample oriented with <110>$_F$ zone axis parallel to the electron beam, recorded at A) room temperature (RT) and B–D) at 800°C in ≈2 Torr dry H_2 showing development of defected area and (C, D) ordered superstructure after 40 minutes. A comparison of digital diffractogram inset in (A) and (D) confirms the formation of superstructure.

Electron energy-loss spectroscopy provides a convenient method to follow the average oxidation state of the Ce when exposed to different reducing conditions. Figure 6A shows a series of Ce M_{45} edges recorded from high surface-area nanoparticles of CeO_2 during *in situ* reduction in dry H_2. The large peaks at the beginning of the edge are called white lines and correspond to electron transitions from the $3d^{5/2}$ (M_5–884 eV) and the $3d^{3/2}$ (M_4–902 eV)) to unoccupied states in the 4f band. It has been shown that the relative intensity of the M_5 and M_4 peaks depends on the occupancy in the 4f band, which in turn depends on the oxidation state of Ce [56]. The occupancy of the 4f band changes from approximately 0 to 1 as the Ce transforms from an oxidation state of +4 to +3. Inspection of the Ce spectra in Figure 6A shows a reversal in the relative

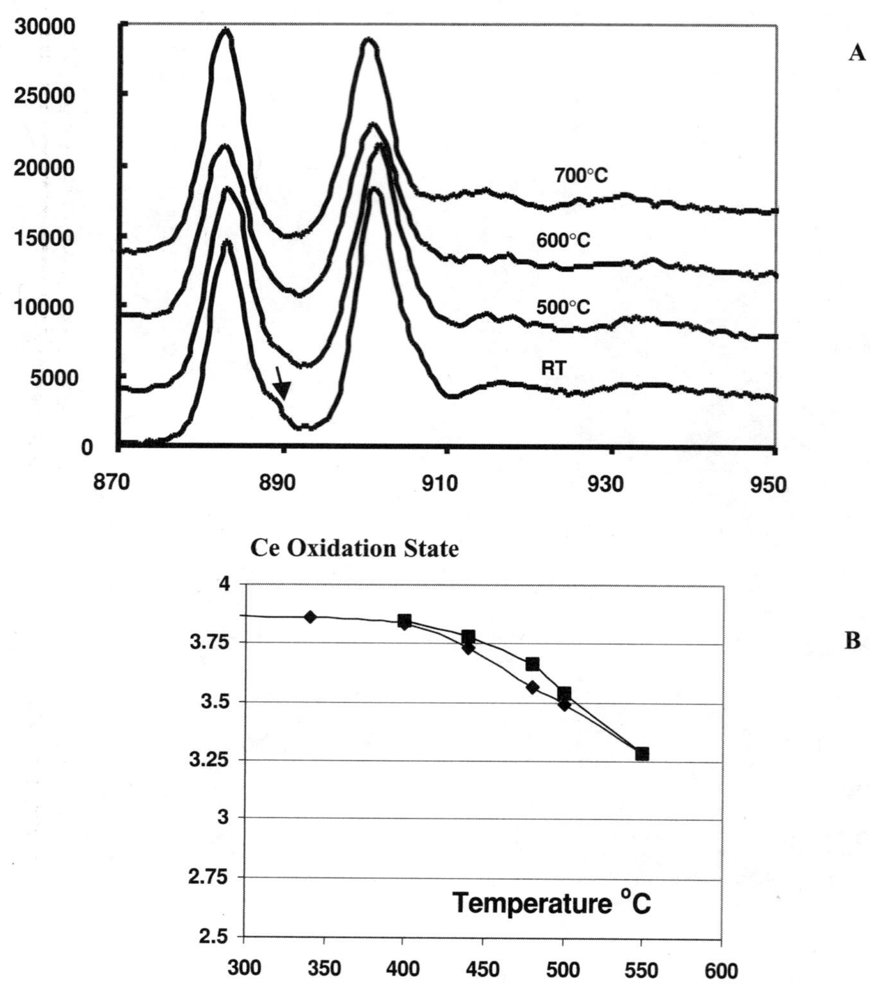

Figure 6. A) Background subtracted Ce M_{45} peaks from pure CeO_2 sample showing relative change in the white-line intensity with temperature. A small shoulder (marked by arrow) also disappeared with reduction. B) Ce oxidation state for $CeZrO_2$ samples showing reduction and oxidation of Ce with heating (diamonds) and cooling (squares) in dry H_2.

intensities of the M_5 and M_4 white lines as ceria is heated from room temperature to 700°C.

There are several different methods of extracting the variation of the white-line intensity in order to determine the occupancy of the 4f band [57]. Some of these methods require significant spectral processing and impose restrictive conditions on the data acquisition. However, in the Ce system, the change in the white-line intensity is pronounced and thus we can use a simplified procedure to quantify changes in the Ce oxidation state. Here we remove the background beneath the M_{45} edge, integrate the M_5 and M_4 intensities and determine the M_5/M_4 intensity ratio. It is not easy to calibrate the white-line ratios simply by directly measuring the O/Ce concentration because of difficulties associated with adsorbates and non-stochiometry. Instead, we obtain a calibration simply by examining the spread of white-line ratios obtained over the entire temperature range and assume that the low value obtained at room temperature corresponds to Ce^{+4} and the value obtained under severe reducing conditions (800°C in 2 Torr of H_2) corresponds to Ce^{+3}. We further assume a linear relationship between the white-line ratio and the Ce oxidation state. These assumptions seem reasonable and based on the statistical spread in data points give oxidation states that are accurate to within 5% [24].

The Ce oxidation state determined from the white-lines of a $Ce_xZr_{1-x}O_2$ mixed oxide sample is plotted as a function of temperature in Figure 6b. The mixed oxide samples consisted of grains with an average size of about 5 nm. The initial average oxidation state of this material is about 3.85 and the onset of reduction occurs between 400 and 450°C and by 550°C the average oxidation state drops to around 3.3. This reduction temperature is about 150°C lower than the equivalent reduction temperature in pure CeO_2. The cooling cycle is also shown in Figure 6B. Even though the sample was cooled in an H_2 atmosphere the ceria still re-oxidized back to the original 3.85 state by 400°C. The oxygen to re-oxidize the sample is presumable obtained from the residual background gases. HREM did reveal some very small nanometer sized superstructure domains in samples subjected to severe reduction at 800°C. However the vast majority of the material did not show evidence for ordering of oxygen vacancies. The combination of *in-situ* HREM and EELS strongly suggest that the Zr strongly retards the ordering of vacancies resulting in an increase in the mobility of oxygen vacancies leading to a significant lowering of the temperature for the onset of reduction.

Strong gas-solid interactions often rely on suitable sites on the surface which can catalyze a particular reaction. These active sites are usually associated with dangling bonds at special surface sites such as Brönstead or Lewis acid centers. In some cases, the sites may only be created when the material is "activated" by exposure to a suitable gaseous environment. The location of active sites on the surface of materials and the activation process are not well understood but in many instances the active centers may be associated with crystallographic defects. Strong interaction between such materials and their environment depends not only on the high surface area but also on extended defects with nanometer separation within the crystals. Gai *et al.* has pioneered the use of ETEM to study oxide catalyst and the relationship between crystal defects and catalytic activity. One example of this approach is the behavior of vanadyl pyrophosphate (VPO) catalyst under reducing atmospheres and its impact on selective oxidation of alkanes [58]. They conducted a series of *ex-situ* and *in-situ* experiments to elucidate the relationship between the catalytic activity and the crystal defect structure. Figure 7 shows the microstructure of VPO after reduction in butane. The low-magnification TEM image and electron diffraction pattern show that extended defects have been introduced into the material as oxygen is removed from the lattice to oxidize the butane. High densities of extended defects appear throughout the crystal running in

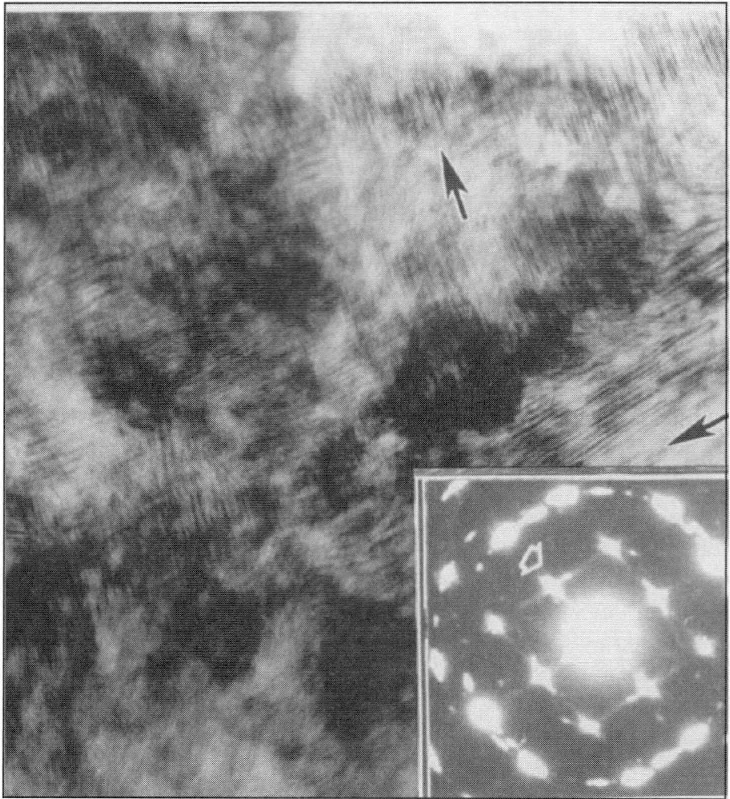

Figure 7. VPO after reducing in butane. Two sets of extended defects (arrowed) parallel to [201] and symmetry-related [20$\bar{1}$] directions in (010) orientation in diffraction contrast (~400°C several hours). The diffraction pattern reveals two sets of diffuse streaks along these directions. (Gai and Kourtakis, Science, 1995, reproduced with permission)

either the [201] direction or the symmetry related [20$\bar{1}$] direction. Under reducing conditions many oxygen vacancies are generated at the surfaces which diffuse into the crystal. In this case, the lattice minimizes strain by gliding along the [201] directions with a pure shear mechanism to reduce the misfit between the reduced surface and the adjacent matrix. The anion vacancies generated where the extended defect intersects the crystal surface are associated with the strong Lewis acid sites responsible for the high catalytic activity.

5.1.2. Evolution of Nanoparticle Systems and Sintering Mechanisms

The ambient atmosphere can strongly affect the evolution of nanoparticles giving rise to behaviors that can be much more complex than those observed under UHV conditions. Even when gross chemical transformations of the particles do not take place, gas induced surface effects can result in dramatic changes in particle shape and sintering behavior [46, 59–61]. Hansen *et al.* [46] have shown that the shape change of Cu nanoparticles on a ZnO support, formed by *in-situ* reduction of CuO, depends upon the gaseous composition of the reducing atmosphere. Model catalysts were prepared by impregnating ZnO support with copper acetate. The formation of Cu nanocrystals during reduction in an H_2 atmosphere was recorded by high-resolution imaging (Figure 8). The atomic resolution images, recorded at 280°C in 4 Torr of H_2, confirm the metallic nature of the Cu particles formed (Figure 8A). These particles are faceted and bound by (100), (110) and (111) lattice planes (Figure 8B). On the other hand round Cu particles were formed when the catalyst sample was exposed to slightly oxidizing environment generated by adding H_2O vapor to the hydrogen gas (Figure 8C). The shape of the Cu particle was observed to change in a more reducing environment, obtained by adding CO to hydrogen gas (Figure 8E). The Wulff's construction for the corresponding shapes (Figure 8B, 8D and 8F) was used to determine surface free energies. The observed dynamic restructuring of the catalyst shows that relevant active sites are generated during catalytic processes. Such information is crucial to determine the amount of H_2 adsorption on various surfaces and hence reduction rates.

Sintering and ripening processes are expected to be strongly influenced by ambient atmosphere. For example, it is well known in the catalysis literature that the presence of oxidizing agents like water can significantly alter sintering processes. Here we illustrate sintering processes for catalyst regeneration where complex interactions between the nanoparticles, the support and the ambient atmosphere result in significant noble metal sintering during low-temperatures catalyst regeneration.

Pd/alumina catalysts are widely used for hydrogenation of alkynes [62]. During the hydrogenation process, the catalysts are gradually deactivated by hydrocarbon build-up and eventually need to be regenerated. The regenerated catalysts usually exhibit less activity and different selectivity compared with the fresh catalysts. This phenomenon is often attributed to the decrease in the active metal surface area caused by sintering during the regeneration process [63]. In traditional metal particle sintering processes, significant metal atom mobility on the substrate is not achieved until the temperature reaches about half the melting point in Kelvin (the Tamman temperature). However, the temperature during catalyst regeneration is usually much lower than the Tamman temperature of Pd (~650°C) and Pd is not expected to sinter easily. The catalyst regeneration process was observed in the ETEM in order to determine the mechanism for sintering at these low temperatures.

Pd/α-alumina catalysts were run in an industrial reactor for the hydrogenation of acetylene. The catalysts were kept on-stream for 124 hours and the temperature was varied from 110 to 145°F to keep a constant acetylene conversion level. Electron diffraction and electron energy-loss spectroscopy analysis revealed that in the used catalysts, most of the Pd particles were lifted from the alumina surface and embedded in the amorphous hydrocarbon material. The regeneration process was performed *in situ* by heating the system in either steam or air to remove the hydrocarbon via either sublimation or combustion respectively. Figure 9 shows a series of images recorded from the catalyst while heating in 500 mTorr of air at 350°C. The hydrocarbon develops pores

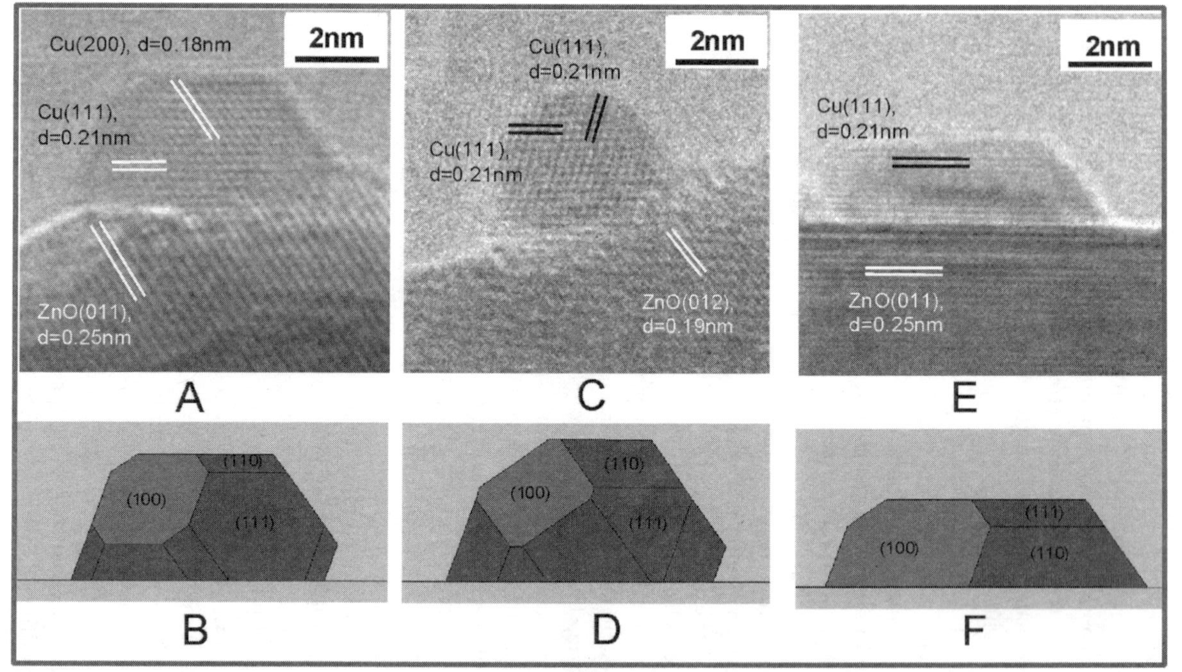

Figure 8. In situ TEM images (A, C and E) of a Cu/ZnO catalyst in various gas environments together with the corresponding Wulff construction of the Cu nanocrystals (B, D, and F). A) The image was recorded at a pressue of 1.5 mbar of H_2 at 220°C. The electron beam was parallel to [011] zone axis of Cu. (C) Obtained in a gas mixture of H_2 and H_2O, H_2:H_20 = 3;1 at a total pressure of 1.5 mbar at 220°C. (E) Obtained in a gas mixture of H_2 (95%) and CO (5%) at a total pressure of 5 mbar at 220°C. (Hansen *et al.* Science, 2002, reproduced with permission)

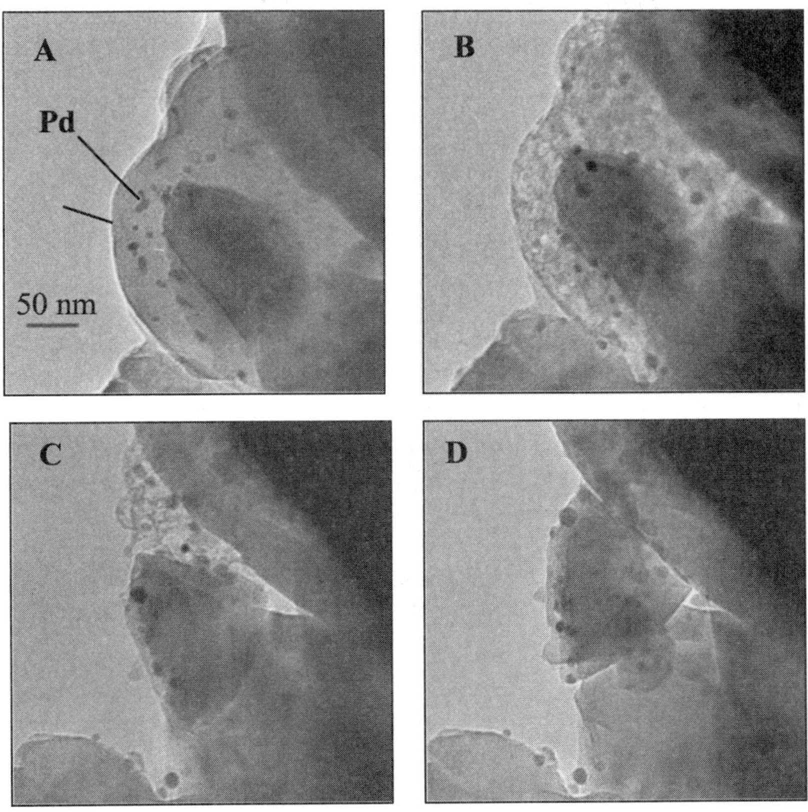

Figure 9. Time series of used Pd/Al_2O_3 catalyst heating in 500 mTorr air at 350°C for A) 0 hour; B) 1 hour C) 4 hours and D) 7 hours. Pd particles are marked and HC indicates hydrocarbon.

due in part to catalytic combustion of the hydrocarbon by the Pd particles. Calculations suggest that the local temperature rises to about 500°C. This is still significantly below the Tamman temperature (~650°C for Pd) and we would not normally expect atomic diffusion of Pd to give rise to sintering. ETEM reveals several reasons for Pd sintering in this case. First, on fresh catalyst, we have shown that major structural rearrangements can occur within the Pd particles at temperatures as low as 500°C [64]. This agrees with shape transformation measurements described earlier in this chapter in which Pd nanoshells were shown to undergo significant atomic rearrangement at 500°C. Although the atoms are mobile within the clusters, we do not see any evidence for significant diffusion of Pd atoms away from the clusters i.e. the atoms remain bonded to the metal cluster. However, during the regeneration process, particles are passively brought into physical contact via oxidation of the underlying hydrocarbon support. When physical contact between particles occurs, the high atom mobility within each Pd particle causes the two-particle cluster to rapidly reconfigure into a single crystal

Table 3. Rate of Au Particle Growth with and Without Electron Beam

Substrate Temperature	Growth Rate (atoms/cm^2/sec.)	Electron Beam Enhanced Growth rate (atoms/cm^2/sec.)	Growth Rate Enhancement
125°C	8.5×10^{10}	1.2×10^{13}	160
150°C	5×10^{11}	1.7×10^{13}	140
200°C	1.2×10^{13}	5.8×10^{13}	5

cluster. This mechanism is completely different from the traditional Ostwald ripening process and is the primary sintering mechanism during the regeneration of the Pd based catalysts.

5.2. Controlled Synthesis of Nanostructures

The E-TEM can also be used as a cold-wall chemical vapor deposition (CVD) cell. The chemical vapor deposition involves precursor adsorption and dissociation on a substrate. The materials deposited after dissociation may nucleate and grow to form nanoparticles or continuous films. The ETEM can be used for direct observation of the deposition process, and the nucleation and growth process. The dissociation of a precursor (vapor) can proceed by any of the following three main mechanisms:

a) Thermal dissociation.
b) Catalytic dissociation
c) Electron beam induced dissociation.

An advantage of performing such experiments in an electron microscope is that we can use imaging, diffraction and spectroscopy to simultaneously characterize the resulting material *in situ* allowing synthesis conditions to be varied and optimized.

Drucker *et al.* [65, 66] have used ETEM to study the CVD process of Al and Au on SiO_2 and Si surfaces respectively. They reported that the electron beam enhanced the growth rates but the effect is lower at higher temperatures (Table 3). The observed dendritic growth of Al was believed to be responsible for the frosty and non-reflective appearance of Al films reported by Beach *et al.* [67]. Surface treatment by $TiCl_4$ produced high quality conformal Al films with growth rates dependent on pressure and temperatures. The growth rates of Au films on clean Si were observed to be dependent on temperature with a higher rate at higher temperature (Table 3). These model studies have established the validity of this technique to study CVD *in situ* at near atomic level using the ETEM. Similar *in-situ* CVD studies can also be performed using specially modified ultra high vacuum (UHV) TEM [68]. For example, the *in situ* observation of the nucleation and growth mechanism of Ge islands on clean Si surfaces under low Ge_2H_6 pressures were found to follow Ostwald ripening process [69].

In an ETEM equipped with a field emission gun, nanolithography can be performed to synthesize nanostructures. The characterization techniques available in the TEM can be employed to immediately characterize the resulting structures and understand the

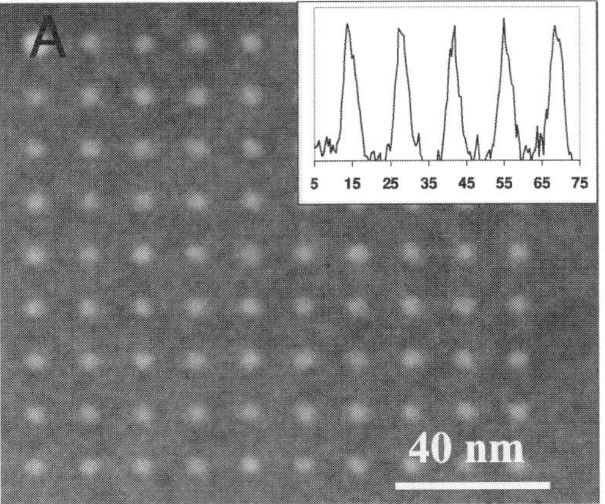

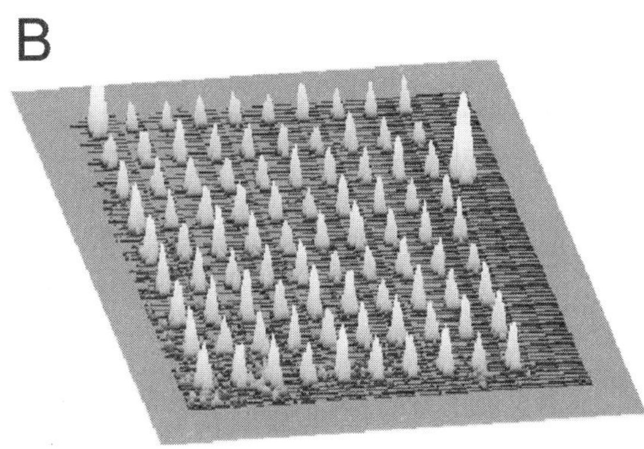

Figure 10. A) Z-contrast image of two-dimensional periodic array of GaN dots. Insert shows line profiles through the Z-contrast images showing FWHM of 4 nm. B) Surface plot of array of dots showing the height of dots derived from the ADF image intensity. Average dot height 5 nm.

deposition processes. Nanolithography proceeds by admitting a gaseous precursor into the sample area of the microscope and using the fast electron beam to locally decompose the precursor. Electron nanolithography in a modern TEM has been demonstrated for creating nanostructures from carbonaceous contamination and tungsten metal [70–72]. However, if more exotic gaseous precursors can be designed then a much wider range of materials synthesis can be performed including binary compounds.

Figure 10 shows a periodic array of uniform GaN dots generated using nanolithography in the ETEM [73]. The array was generated by exposing the SiO_x substrate to a unique inorganic and highly reactive hydride D_2GaN_3, that dissociates exothermally under electron irradiation resulting in formation of stoichiometric GaN and volatile and robust D_2 and N_2 byproducts as shown by the equation

$$D_2GaN_3 \rightarrow D_2 + N_2 + GaN.$$

This compound was previously used in GaN film growth by gas source molecular beam epitaxy to form standard heterostructures and luminescent nanostructures at extremely low temperatures, between 150°C and 450°C [74]. The dot array is generated by rastering a 0.5 nm electron probe over the substrate in well define steps. The dots are highly uniform and display an average full-width half maximum of about 4 nm, a base width of about 9 nm and a height of about 5 nm. Such arrays are small enough to manifest true quantum size effects and are likely to possess unique electronic and optical properties which may be beneficial to optoelectronic applications and information processing [75–77].

The spatial extent of the dot (5–10 nm) is considerably larger than the size of the primary electron probe (~0.5 nm). Similar results have recently been obtained on *in situ* TEM deposition of carbon nanowires and dots of tungsten metal [70, 71]. In both cases, features with dimensions in the range 2–5 nm were reported. Most processes for nanolithography rely on ionization of the valence electrons of the precursor species. The cross sections for these ionization processes are much higher for low-energy electrons and consequently the secondary electrons emitted from the surface of the substrate are more efficient at dissociation. For this reason in electron lithography, precursor decomposition is associated with the spatial distribution of emitted secondary electrons. For high-energy sub-nanometer electron probes, the spatial distribution of secondary electrons emitted from the surface of a thin film is controlled primarily by the secondary electron mean free path. Careful measurements for 100 keV electrons show that the majority of secondary electrons are emitted with energy between 2–10 eV [78, 79]. The mean free paths of secondary electrons can be estimated from universal curves to lie in the range 1–5 nm [80]. Our dot dimensions are certainly consistent with this model and provide evidence for the strong role that secondary electrons play in the electron beam assisted nanolithography.

5.3. Kinetics

The kinetics of gas-solid interactions at the nanometer level can be obtained by *in-situ* measurements of reaction rates using an ETEM. Most of the time it is possible to obtain the reaction rates from the time resolved data (mostly from video images). Sometimes it is possible to collect both time and temperature resolved data allowing us to estimate activation energies of the reaction process. Baker *et al.* [81, 82] were the first to use ETEM to make quantitative measurements of reaction kinetics. They used time resolved images to measure the growth rate of carbon nanofibers at different temperatures for different metal particles [82]. Activation energies of carbon nanofiber formation for different catalysts were obtained from Arrhenius plots (see below). This information was used to model the growth mechanisms of carbon nanofibers by comparing the activation energies obtained with carbon dissolution energies for various metal catalysts.

The reaction rate k is given by the well-known Maxwell-Boltzman equation:

$$k = Ae^{-E_a/RT}$$

where A is a constant called the frequency factor, E_a is the activation energy, R is the gas constant and T is the absolute temperature. In most reactions, the rate of the reaction increases with increasing temperature. The reaction rates can be obtained by measuring (a) the growth rates and/or (b) conversion rates. Taking logarithms and re-arranging the Maxwell-Boltzman equation gives the Arrhenius equation:

$$\ln k = \ln A - E_a/RT$$

This equation shows that plotting the reaction rate against inverse temperature (Arrhenius plot) gives a straight line and the activation energy can be determined directly from the gradient.

5.3.1. Nucleation and Growth of Cu Nanoparticles

TiN is commonly used as a barrier layer in the semiconductor industry to retard Si diffusion into the interconnect layer (Au, Al or Cu etc.). One of the ways to incorporate a TiN or CrN layer in very small integrated circuits is by depositing a thin layer of Ti or Cr on Si or SiO_2 and subjecting this layer to rapid thermal annealing in NH_3 to form the respective nitride. Au, Al or Cu is then deposited to form the interconnect layer. This two-step process could be reduced to a single step by depositing a thin layer of Cu/Ti or Cu/Cr alloy on Si or SiO_2 and performing rapid thermal annealing in NH_3 at suitable temperatures. As Ti or Cr will convert to respective nitrides, Cu will be depleted from the matrix to form nanoparticles giving the desired metal contact and nitride barrier layer. In order to understand the effect of temperature and pressure on the nitridation reaction, the process was followed by heating Cu/Ti and Cu/Cr thin films, of different compositions, in ≈3–4 Torr of high purity NH_3 up to 650°C using a modified Phillips 400-T E-TEM operated at 120 KV [35]. The nitridation temperature of Ti and Cr metals was determined by time and temperature resolved SAED patterns [35, 36]. The growth rates of Cu particles depleted during the nitridation of Cu/Ti and Cu/Cr thin films were measured using time and temperature resolved video imaging [36]. A typical sequence used for such measurements is shown in Figure 11. The images are digital still frames extracted from a video sequence recorded during growth of Cu particles at 630°C in 4 Torr of NH_3. It was also observed that the Cu grains do not grow isotropically, rather, certain facets have a preferential growth rate.

The area of two (of several measurements) Cu grains was plotted as a function of annealing time during heat treatment of the CuCr film at 630°C (Figure 12). Two growth regimes are observed for both grains. Growth is rapid initially (solid symbols, region A), whereas at longer anneal times, growth is slower and the grain area varies linearly with time (open symbols, region B). This behavior indicates that for longer annealing times, the growth is controlled by diffusion of Cu in the nitride matrix. The calculated average value of the slopes of those curves, $K = 3.0 \pm 0.4 \times 10^{-11}$ cm^2/sec, is proportional to the diffusion constant of Cu through the CrN matrix. It should be noted that for very long annealing times ($t > 10$ min), the curve becomes completely flat as the Cu grain growth process was completed. This phenomenon

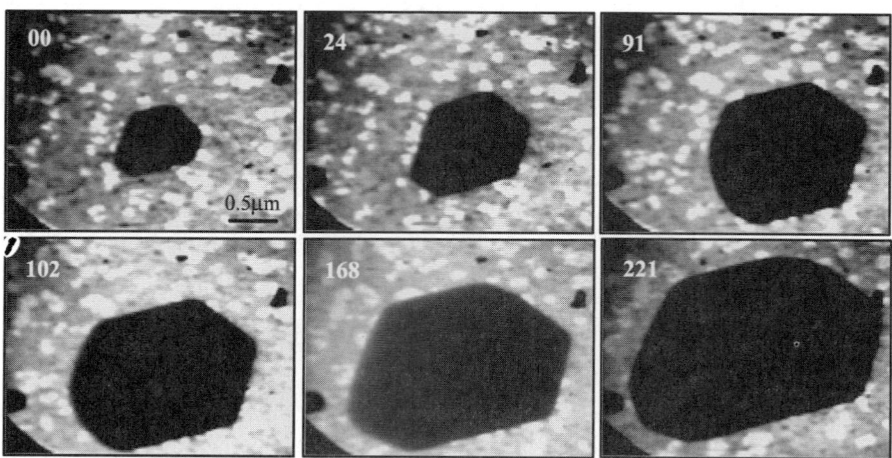

Figure 11. A time sequence of *in situ* still video frames showing the growth of a Cu particle during the nitridation process of the $Cu_{1-x}Cr_x$ ($x = 0.40$) thin film at 630°C. The corresponding video times in seconds are also shown in the top left of each picture.

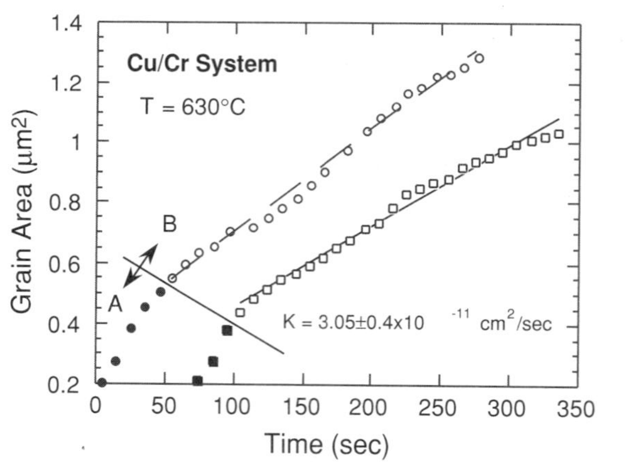

Figure 12. Cu grain area measured from the video sequence shown in Figure 12 as a function of the annealing time at 630°C when a Cu/Cr thin film was heated in ≈3 Torr of high purity NH_3 gas. The growth rate from two different particles given is obtained from the curves.

might be attributed to the fact that after long annealing times, Cu was fully depleted from the CrN matrix. Although the nitridation temperature for Ti was found to be lower (370°C) compared to Cr (580°C), the growth rate of Cu particles was an order of magnitude lower in Cr/Cu (3.05×10^{-11} cm^2/sec.) thin films than in Cu/Ti ($2.2 \times 10^{-12} - 5.0 \times 10^{-12}$ cm^2/sec.) thin films.

5.3.2. Carbon Nanotube Growth

Carbon nanotubes (CNTs) were dramatically novel nanoscale materials when they were first discovered in 1991 by Iijima using a carbon arc discharge process [83]. They have since become one of the most sought out materials for nanotechnology due to their remarkable magnetic, electronic and mechanical properties [84]. The structure of a CNT can be described in terms of a single graphite layer (graphene) rolled up to form a single cylinder or concentrically arranged cylinders. The former is referred to as a single wall nanotube (SWNT) and the latter are called multiwall nanotubes (MWNTs). Although, a number of growth mechanisms have been proposed, deduced from high-resolution electron microscopy (HREM) images and theoretical simulations [81, 85–88], there is no direct evidence to support these models.

We have been successful in recording images of the growth of CNTs at video rate. We have used the specimen area of this microscope as a chemical vapor deposition chamber [89–90]. Our preliminary observations were made using Ni/SiO_2 catalyst, and propylene and acetylene as carbon sources (precursor). Although fibrous structures were observed to grow when propylene was used as a precursor, CNTs were observed to form only when acetylene (C_2H_2) was used as the precursor. Multi-wall carbon nanotubes were often observed to form with a catalyst particle at their apex, as has been observed previously in HREM images of carbon nanotubes formed by the CVD processes.

Figures 13A–I show digitized individual frames of a typical growth process for multi-wall carbon nanotubes. A small finger shaped hollow structure (Figure 13A) moved out from the substrate, where another tube has been formed (Figure 13B), and created the tip of a multi-wall nanotube. After growing linearly for a short time, it curved and started to grow straight out again (Figure 13C). The process of changing directions continued until the apex anchored back to the substrate forming a loop (Figure 13D–F). CNT were often observed to grow in such a zigzag manner forming waves, spirals or loops. The length of the tube formed at the substrate to the end was used to measure the growth rate. Measured growth rates at 475°C and 20 m Torr of C_2H_2 pressure were 38–40 nm/second (Figure 14). It is clear from the length vs. time plot (Figure 14) that the growth of the tube is not continuous. The total growth period was observed to be in the range of 1–2.5 seconds. Nanotubes were not observed to grow after 1–2 seconds of their nucleation, on the other hand, new CNTs were observed to nucleate and grow during the first 2–5 minutes, after which significant deposition of CNTs was not observed. We observed no difference in the reaction morphology or length of CNTs formed in the area under *in-situ* observation or the area not irradiated by electron beam during conditions. The growth mechanisms for various CNT are currently being investigated.

5.3.3. Activation Energy of Nucleation and Growth of Au Nanoparticles

Drucker *et al.* [65–66] had made the first *in-situ* observations of growth and nucleation mechanism of gold CVD on Si/SiO_x from ethyl (trimethylphosphine) gold (Et $Au(PMe_3)$) at different temperatures and constant pressure with time using a

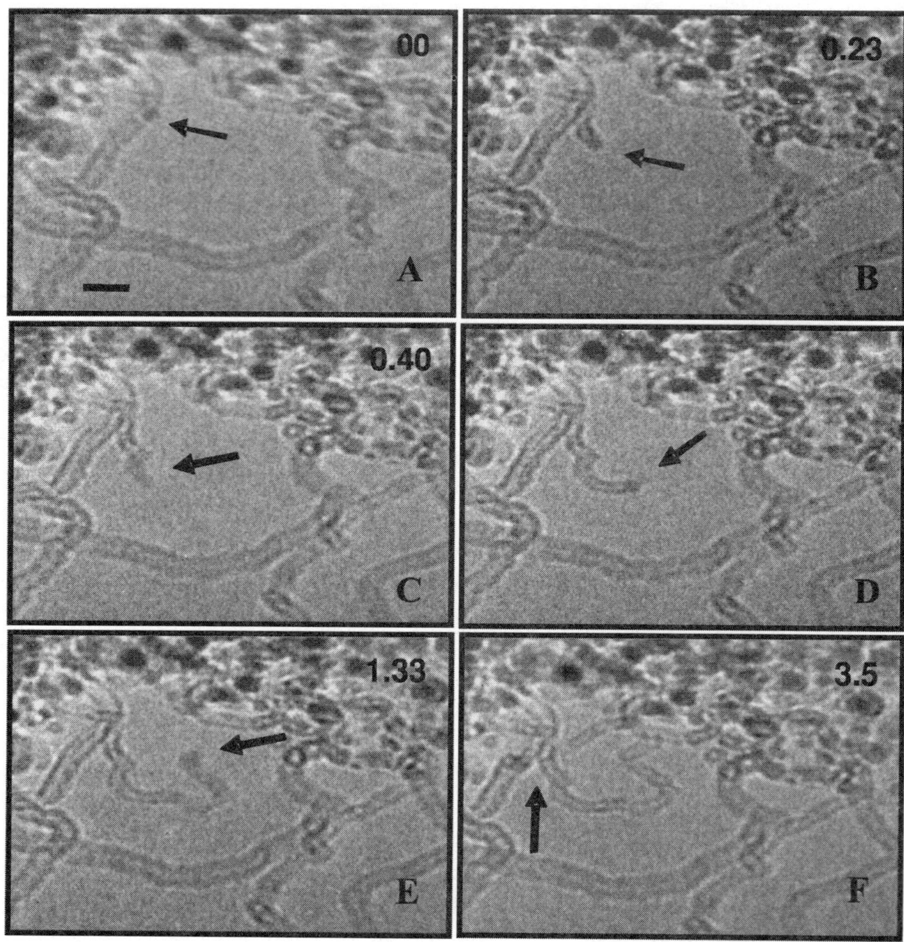

Figure 13. Individual frames digitized from a video sequence showing the nucleation and growth of a multiwall carbon nanotube. The apex is marked by arrows (A–F) showing the zigzag growth direction bending 360° (E) and finally attaching back to the substrate forming a loop (F). The bar is 10 nm and the time interval between various frames is given in the top right hand corner.

modified Philips 400T ETEM [11]. Si <111> samples were cleaned by dipping in HF and quickly transferring to the microscope in order to minimize the oxidation of the Si. The samples were heated to the deposition temperature and time-resolved images were recorded using a video recorder. Figure 15A shows that the number of nuclei formed did not increase with time but the Au nanoparticles grew in size. These particles coalesced to form continuous thin films once their growth brought them into direct contact with other particles [66]. The survey of the sample region not exposed to the electron beam indicated that the Au growth rates were lower in the areas not under direct observation (Table 3). In order to obtain growth rates without electron

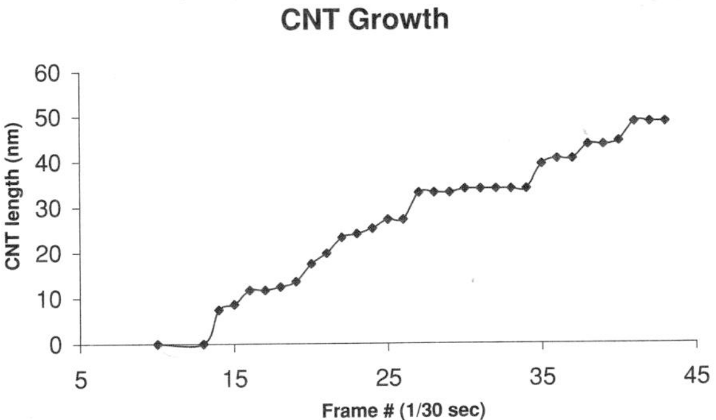

Figure 14. The discontinuous growth rates for CNT measured from individual frames (1/30 second) of the video sequence shown in Figure 14.

beam effects, the following procedure was adopted. The precursor was introduced in the sample region for 5 minutes at deposition temperature and then the ETEM column was evacuated before making TEM observations. The absence of Au particle growth during the observation confirmed that no residual precursor was present in the sample area and the deposition was not enhanced by the electron beam. The process was repeated to obtain time resolved growth rates for each temperature.

The change in particle size with time at constant temperature and pressure was used to obtain average growth rates for the Au particles. As TEM data only provides us with two-dimensional growth rates, the height of the Au particles was measured after depositions using scanning tunneling microscopy. The measure change in volume thus obtained was used to determine growth rates at three different temperatures (125°C, 150°C and 200°C) for depositions with and without electron beam effects (Table 3).

The logarithm of the growth rates (no of Au atoms/cm^2) plotted against 1/T can thus be used to obtain the activation energy (E_a) for nucleation and growth of Au nanoparticles by CVD (Figure 15B). The slope of the curve can directly be used to obtain the activation energy using Arrhenius equation (1):

$$\begin{aligned} E_a &= -(\text{slope}^*\text{R}) \\ &= 22.67\,\text{k cal/mole} \end{aligned}$$

As Au particles, once formed, were not observed to grow with time in the absence of precursor, it is safe to assume that ripening is not responsible for the growth at these low temperatures and the particles coalesced only when they were in direct contact. Moreover, the reported activation energy for Au surface diffusion on carbon is 39 kcal/mole [91] which is higher than measured here. Therefore the activation energy measured is for the nucleation and growth of Au during CVD.

Measurement of reaction kinetics thus provides us with an insight in to the reaction mechanisms and processes involved.

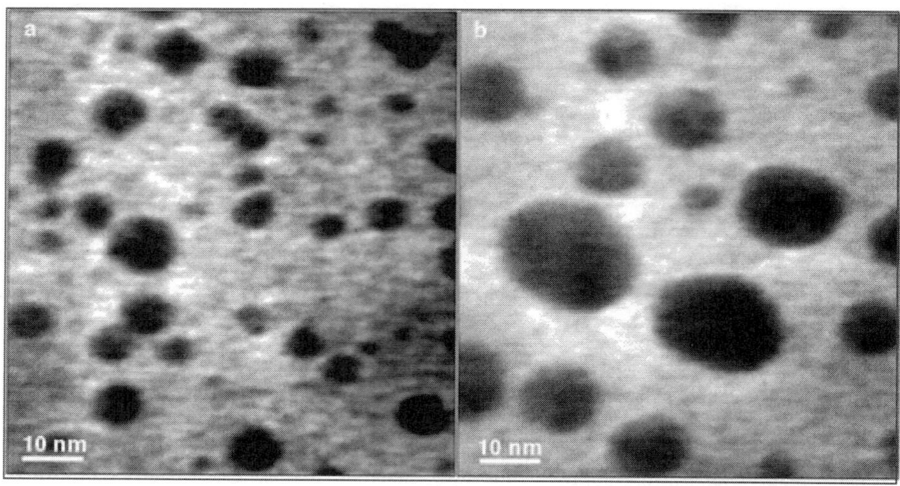

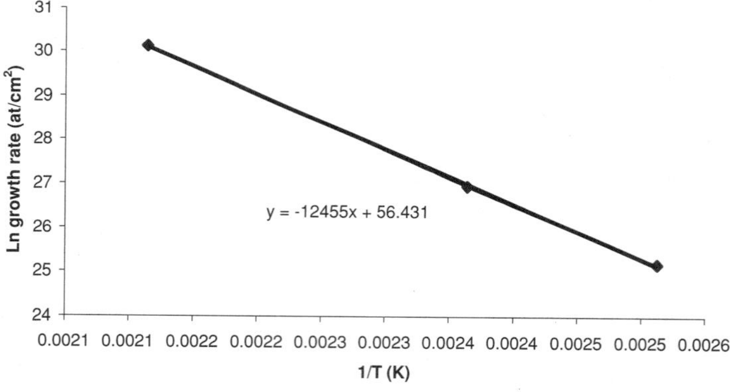

Figure 15. Bright field images showing nucleation and growth of Au particles on Si <111> surface at 125°C after exposure of A) 5 minutes and B) 15 minutes. C) Arrhenius plot showing the temperature dependence of the growth rate obtained.

6. CONCLUSIONS

We have shown that ETEM is a valuable technique for understanding the response of nanoparticle systems to a gaseous environment at near atomic-level. The modern ETEM allows the dynamic behavior of the nanoparticles to be studied in real time with atomic-resolution imaging and electron diffraction in up to 50 Torr of gas pressure. On a machine equipped with a field-emission gun, electron energy-loss spectra can be recorded using a sub-nanometer probe so that elemental and electronic structural changes occurring in individual nanoparticles can be followed *in situ*. This powerful combination of *in-situ* imaging, diffraction and spectroscopy provides detailed information about gas-solid phase transformation mechanisms in individual nanoparticles. Quantitative measurements can be used to derive reaction rates and activation energies from very small areas and should allow full reaction kinetics to be determined as a function of nanoparticles size. The ETEM can also be used to perform *in-situ* synthesis of nanophase materials. The simultaneous characterization can be performed during synthesis allowing synthesis conditions to be varied and optimized rapidly. Sub-nanometer electron probes can also permit nano-lithographic structures to be deposited and studied under a wide variety of different conditions.

REFERENCES

1. L. Marton, *Bull. Acad. R. Belg. Cl. Sci.*, 21 (1935) 553.
2. H. M. Flower, *J. Microscopy*, 97 (1973) 171.
3. D. F. Parsons, V. R. Matricardi, R. C. Moretz and J. N. Turner, in *Advances in Biological and Medical Physics*, Vol. 15, J. H. Lawrence and J. W. Gofman, Eds. (Academic Press, New York), (1974) p. 161.
4. D. L. Allinson, in *Principles and Techniques in Elelctron Microscopy, Biological Applications*, Vol. 5, M. A. Hayat, Ed. (Van Nostrand Reinhold, New York), (1975) p. 52.
5. P. Butler and K. Hale, in *Practical Methods in Electron Microscopy*, Vol. 9 (North Holland) (1981) pp. 239–308.
6. P. R. Swann and N. J. Tighe, *Proc. 5th Eur. Reg. Cong Electron Microscopy*, (1972) 436.
7. G. M. Parkinson, *Catalysis Letters*, 2 (1989) 303.
8. R. C. Doole, G. M. Parkinson, and J. M. Stead, *Inst. Phys. Conf. Ser.*, 119 (1991) 157–160.
9. R. C. Doole, G.M. Parkinson, J. L. Hutchison, M. J. Goringe and P. J. F. Harris *JEOL News* 30E, (1992) 30.
10. T. C. Lee, D. K. Dewald, J. A. Eades, I. M. Robertson, and H. K. Birnbaum, *Rev. Sci. Instrum.* 62 (1991) 1438.
11. Nan Yao, Gerard E. Spinnler, Richard A. Kemp, Don C. Guthrie, R. Dwight Cates and C. Mark Bolinger, Proc. 49th Annual; meeting of Microsc. Soc. Am. San Francisco Press (1991) 1028.
12. Renu Sharma, K. Weiss, M. McKelvy and W. Glaunsinger, *Proc. 52nd Ann. Meet. Microscopy Society of America*, (1994) 494–495P.
13. E. D. Boyes and P. L. Gai, *Ultramicroscopy*, 67 (1997) 219–232.
14. Pratibha L. Gai and Edward D. Boyes, in In *Situ Microscopy in Materials Research*, Ed. Pratibha L. Gai, Kluwer Academic Publishers, (1997) 123–146.
15. Renu Sharma and Karl Weiss, *Microscopy Research and Techniques*, 42 (1998) 270–280.
16. L. Hansen and J. B. Wagner, *Proc. 12th European Congress on Electron Microscopy*, Vol. II, (2000) 537–538.
17. Thomas W. Hansen, Jacob B. Wagner, Poul L. Hansen, Seren Dahl, Haldor Topsoe, Claus J. H. Jacobson, *Science*, 294 (2001) 1508–1510.
18. Renu Sharma, Peter A. Crozier, Ronald Marx and Karl Weiss, *Microsc. & Microanal.*, (2003) 912 CD.
19. P. R. Swann and N. J. Tighe, *Proc. 5th Eur. Reg. Cong Electron Microscopy*, (1972) 360.
20. M. Pan and P. A. Crozier, *Ultramicroscopy*, 48 (1993) 332.
21. M. Pan and P. A. Crozier, *Ultramicroscopy*, 48 (1993) 487.
22. R. F. Egerton, *Electron Energy-Loss Spectroscopy in the Electron Microscope*, Plenum Press, New York (1996) Second Edition.
23. P. R. Swann, in *Electron Microscopy and Structure of Materials*, Ed. G. Thomas, R. Fulrath and R. M. Fisher, University of California Press (1972) 878.
24. R. Sharma P. A. Crozier, Z. C. Kang, and L.Eyring, *Phil. Mag.* 84 (2004) 2731.
25. V. Oleshko, P. A. Crozier, R. Cantrell, A. Westwood, *J. of Electron Microscopy*, 151 (Supplement), (2002), S27.
26. Kamino and H. Saka, *Microsc. Microanal. Microstruct.*, 4 (1993) 127.
27. R. T. K. Baker, M. A. Barber, P. S. Harris, F. S. Feates and R. J. White, *J. Catalysis*, 26 (1972) 51.
28. R. T. K. Baker and J. J. Chludzinski, *Journal of Catalysis*, 64 (1980) 464.
29. R. T. K. Baker, J. J. Chludzinski and C. R. F. Lund, *Carbon*, 25 (1987) 295–303.

30. R. T. K. Baker, *J. Of Adhesion*, 52 (1995) 13–40 and references therein.
31. P. L. Gai, and C. C. Torardi, *Mat. Res. Soc. Symp. Proc.*, 404 (1996) 61.
32. Pratibha. L. Gai, Kostantinos Kourtakis, and Stan Zeimecki, *Microscopy and Microanalysis* (2000) 6.
33. Pratibha L. Gai, and Edward D. Boyes, *Catal. Rev. – Sci. Eng.*, 34 (1992) 1–54.
34. P. L. Gai, *Phil. Mag.*, 48 (1983) 359–371.
35. Z. Atzmon, R. Sharma, J. W. Mayer, and S. Q. Hong, *Materials Research Society Sympos.* Proceedings, 317 (1993) 245–250.
36. Z. Atzmon, R. Sharma, S. W. Russell and J. W. Mayer, *Materials Research Society Symp.* Proceedings, 337 (1994) 619–624.
37. C. A. Mims, J. J. Chludzinski, J. K. Pabst and R. T. K. Baker, *J. Catalysis*, 88 (1984) 97.
38. R. T. K. Baker, N. S. Dudash, C. R. F. Lund and J. J. Chludzinski, *Fuel*, 64 (1985) 1151.
39. R. T. K. Baker, J. J. Chludzinski and J. A. Dumesic, *J. Catalysis*, 93 (1985) 312.
40. R. T. K. Baker and J. A. Dumesic, *J. Phys. Chem.*, 90 (1986) 4730.
41. E. G. Derouane, J. J. Chludzinski and R. T. K. Baker, *J. Catalysis*, 85 (1984) 187.
42. J. A. Dumesic, S. A. Stevenson, R. D. Sherwood and R. T. K. Baker, *J. Catalysis*, 99 (1986) 79.
43. A. J. Simoens, R. T. K. Baker, D. J. Dwyer, C. R. F. Lund and R. J. Madon, *J. Catalysis*, 86 (1984) 359.
44. B. H. Upton, C. C. Chen, N. M. Rodriguez and R. T. K. Baker, *J. Catalysis*, 141 (1993) 171.
45. R. T. K. Baker and N. M. Rodriguez, *Energy and Fuels*, 8 (1994) 330.
46. P. L. Hansen, J. B. Wagner, S. Helveg, J. R. Rostrup-Nielsen, B. S. Clausen, H. Topsoe, *Science* 295 (2002) 2053.
47. R. W. McCabe and R. K. Usmen, *Studies in Surface Science and Catalysis*, 101 (1996) 355.
48. T. R. Baldwin and R. Burch. *Applied Catalysis*, 66 (1990) 359.
49. A. K. Datye, D. S. Kalakkad, E. Völkl and L. F. Allard (1995). *Electron Holography*, 199–208.
50. P. A. Crozier, R. Sharma and A. K. Datye, *Microscopy and Microanalysis*, 4 (1998) 278.
51. P. Crozier and A. K. Datye, *Studies in Surface Science and Catalysis*, 130 (2000), 3119–3124.
52. M. Shelef, and R. W. McCabe, *Catalysis Today*, 62 (2000) 35–57.
53. D. J. M. Bevan and J. Kordis, *J. Inorg. Chem.*, 26 (1964) 1509–1523.
54. Z. C. Kang, J. Zhang, and L. Eyring, *Z. anorg. Allg. Chem.*, 622 (1996) 465–472.
55. P. Fornasiero, J. Kasper, and M. Graziani, *J. of Catalysis*, 167 (1997) 576.
56. T. Manoubi, and C. Colliex, *J. of Microscopy and Related Phenomena*, 50 1990) 1–18.
57. D. H. Pearson, C. C. Ahn, and B. Fultz, Phys. Rev., B47 (1993), 8471.
58. P. L. Gai and K. Kourtakis, *Science*, 267 (1995) 661.
59. E. G. Derouane, J. J. Chludzinski and R. T. K. Baker, *J. Catalysis*, 85 (1983) 187.
60. K. Heinemann, T. Osaka, H. Poppa and M. Avalos-Borlja, *J. Catalysis* 83 (1982) 61.
61. P. L. Gai, B. C. Smith and G. Owen, *Nature* 348 (1990) 430.
62. J. Gislason W. Xia, H. Sellers *J. Phys. Chem.*, A 106 (2002):767.
63. T. B. Lin and T. C. Chou, *Ind. Eng. Chem. Res.*, 34 (1995) 128.
64. R. J. Liu *et al.*, *Microscopy and Microanalysis*, in press.
65. J. Drucker, R. Sharma, J. Kouvetakis and Karl Weiss, *J. Appl. Phys.*, 77 (1995) 2846–2848.
66. J. Drucker, R. Sharma, J. Kouvetakis and Karl Weiss, *Proc. Mat. Res. Soc.*, 404 (1996) 75–84.
67. D. B. Beach, S. E. Blum, and F. K. LeGoues, *J. Vac. Sci. Technol.*, A 7 (1989) 3117.
68. M. Hammar, F. K. LeGoues, J. Tersoff, M. C. Reuter and R. M. Tromp, *Surface Science*, 349 (1995) 129.
69. F. M. Ross, J. Tersoff and R. M. Tromp, *Microsc. Microanal.*, 4 (1998) 254.
70. N. Silvis-Cividjian, C. W. Hagen, P. Kruit, M. A. J. v. d. Stam and H. B. Groen, *Appl. Phys. Lett.*, 82 (2003) 3514.
71. N. Silvis-Cividjian, C. W. Hagen, L. H. A. Leunissen, and P. Kruit, *Microelectronic Engineering.*, 61–62 (2002) 693.
72. K. Mitsuishi, M. Shimojo, M. Han, K. Furuya, *Appl. Phys. Lett.*, 83 (2003) 2064.
73. P. A. Crozier, J. Tolle, J. Kouvetakis & Cole Ritter, *Appl. Phys. Lett.* 84 (2004) 3441.
74. L. Torrison, J. Tolle, I. S. T. Tsong, J. Kouvetakis, *Thin Solid Films*, 434 (2003) 106.
75. Y. Arakawa, T. Someya and T. Tachibana, *Phys. Stat. Sol. (b)*, 224 (2001) 1.
76. I. D'Amico, E. Biolatti, F. Rossi, S. Derinaldis, R. Rinaldis and R. Cingolani, *Superlattices and Microstructures*, 31 (2002) 117.
77. K. Kawasaki, D. Yamazaki, A., Kinoshita, H. Hirayama, K. Tsutsui and Y. Aoyagi, *Appl. Phys. Lett.*, 79 (2001) 2243.
78. J. Drucker, M. R. Scheinfein, J. Liu and J. K. Weiss, J. *Appl. Phys.*, 74(12) (1993) 7329.

79. M. R. Scheinfein, J. Drucker and J. K. Weiss, *Phys. Rev.*, *B* 47(7) (1993) 4068.
80. M. Prutton, *Introduction to Surface Physics*, Clarendon Press (1994).
81. R. T. K. Baker, R. D. Sherwood and J. A. Dumesic, *J. Catalysis*, 66 (1980) 56.
82. R. T. K. Baker, and M. A. Harris (1978), in 'Chemistry and Physics of Carbon', Eds. P. L. Walker, Jr. and P. A. Thrower, Pekker, New York. Vol. 14, 83–164.
83. S. Iijima, *Nature*, 354 (1991) 56–58.
84. M. Dresselhaus, G. Dresselhaus, and Ph. Avouris (Editors), in *Carbon Nanotubes: Synthesis, Structure, Properties and Applications*. Spriner-Verlag, Berln (2001) and references therein.
85. Jean-Christophe Charlier, Sumio Iijima in *Carbon Nanotubes: Synthesis, Structure, Properties and Applications*. Spriner-Verlag, Berln (2001) 55–79.
86. Y. Zhang, Y. Li, W. Kim, D. Wang, and H. Dai, *Appl. Phys.*, *A* 74, 325 (2002).
87. A. Maiti, C. J. Brabec, C. M. Roland, and J. Bernholc, *Phys. Rev. Lett.*, 73 (1994) 2468–2471.
88. M. Grujicic, G. Cao, and Bonnie Gersten, *Materials Sci. Eng.*, B94 (2002) 247–259.
89. Renu Sharma (2003), *Microsc. & Microanal.* 302CD.
90. Renu Sharma and Zafar Iqbal, *Appl. Phys. Lett.*, 84 (2004) 990.
91. Mona B. Mohamed, Zhong L. Wang and Mostafa A. Al-Sayed, *J. Phys. Chem. A* 103 (1999) 10255.

18. ELECTRON NANOCRYSTALLOGRAPHY

JIAN-MIN ZUO

1. INTRODUCTION

What is nanocrystallography? Just as modern crystallography has evolved from the early study of the structure of single crystals to individual molecules, liquids, quasicrystals and other complex structures, nanocrystallography is defined broadly to include the study of atomic and molecular arrangements in structural forms of the feature length from a few to hundreds nanometers. Because electrons interact strongly with matter and electrons form the very probes, electron diffraction has the potential to provide quantitative structure data for individual nanostructures in a role similar to x-ray and neutron diffraction for bulk crystals. This potential is currently being developed for reasons that electron diffraction patterns can be recorded selectively from individual nanostructure at the size as small as a nanometer using electron probe forming lenses and apertures, while electron imaging provides the selectivity.

This chapter introduces electron nanocrystallography by starting from the basic concepts of electron diffraction, the theory and then moving on to electron diffraction applications using selected examples. Emphasis is on quantitative electron diffraction (QED). The experimental techniques will be described for electron diffraction intensity recording, retrieval of structural information and inversion of diffraction patterns by solving the phase problem. We show that structure information, such as unit cell parameters, atomic positions and crystal charge distribution, can be obtained from experimental diffraction intensities by optimizing the fit between the experimental and theoretical intensities through the adjustment of structural parameters in a theoretical model. While the principle of refinement borrows from the Rietveld method in X-ray powder diffraction, its implementation in electron diffraction is much more powerful since it includes the full dynamic effect. As a part of the review, we will also describe the recently developed nano-area electron diffraction, which can record electron diffraction patterns from individual nanostructures for structural determination. Single- and double-wall carbon nanotubes will be used as application examples for this technique.

The development of quantitative electron diffraction is relatively new and happened at the convergence of several microscopy technologies. The development of the field emission gun in 70's and its adoption in conventional transmission electron microscopes (TEM) brought high source brightness, small probes and coherence to electron diffraction. The immediate impact of these technologies is our new ability to record diffraction patterns from very small (nano) structures. Electron energy-filters, such as the in-column Ω-filter, allow inelastic background from plasmon and higher electron energy losses to be removed with an energy resolution of a few eV. The development of array detectors, such as the CCD camera or imaging plates, enables a parallel recording of diffraction patterns and the quantification of diffraction intensities over a large dynamic range that simply was not available to electron microscopy before. The post-specimen lenses of the TEM give the flexibility for recording electron diffraction patterns at different magnifications. Last, but equally important, the development of efficient and accurate algorithms to simulate electron diffraction patterns and modeling the structure on a first-principle basis using fast modern computers has significantly improved our ability to interpret the experimental data.

For readers unfamiliar with electron diffraction, there are a number of books on electron diffraction for materials characterization [1,2,3,4]. The kinematical approximation for electron diffraction and diffraction geometry are required topics for materials science and engineering students. Most of these books focus on crystals since each grain of a polycrystalline material is a single crystal for electron diffraction because of the small electron probe. Full treatment of the dynamic theory of electron diffraction is given in several special topic books and reviews [5,6,7,8].

2. ELECTRON DIFFRACTION MODES AND GEOMETRY

Electron optics in a microscope can be configured for different modes of illumination from a parallel beam to convergent beams. Figure 1 illustrates three modes of electron diffraction using different illuminations, 1) selected area electron diffraction (SAED), 2) nano-area electron diffraction (NED) and 3) convergent-beam electron diffraction (CBED). Variations from these three techniques include large-angle CBED [9], convergent-beam imaging [10], electron nanodiffraction [11] and their modifications [12]. For nanostructure characterization, the electron nanodiffraction technique developed by Cowley [11, 13] and others in the late 1970s uses a scanning transmission electron microscope is particularly relevant. In this technique, a small electron probe of sizes from a few Å to a few nanometers is placed directly onto the sample.

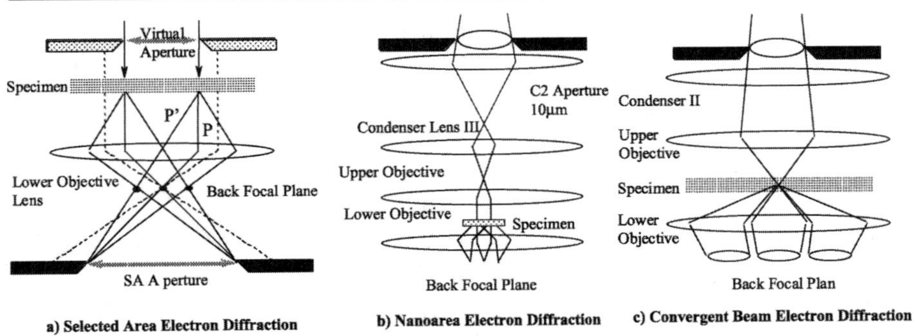

Figure 1. Three modes of electron diffraction. Both a) selected area electron diffraction (SAED) and b) nanoarea electron diffraction (NED) use parallel illumination. SAED limits the sample volume contributing to electron diffraction by using an aperture in the image plane of the image forming lens (objective). NED achieves a very small probe by imaging the condenser aperture on the sample using a third condenser lens. Convergent beam electron diffraction (CBED) uses a focused probe.

Diffraction pattern thus can be obtained from localized areas as small as a single atomic column. The diffraction pattern obtained is very sensitive to the local structure and the probe positions [14]. Readers interested in this technique can find the description and applications in Cowley's review papers.

2.1. Selected Area Electron Diffraction

Selected area electron diffraction is formed by placing an aperture in the imaging plane of the objective lens (see fig. 1a). Only rays passing through this aperture contribute to the diffraction pattern at the far field. For a perfect lens without aberrations, the diffracted rays come from an area that is defined by the back-projected image of the selected area aperture. The aperture image is typically a factor of 20 smaller because of the objective lens magnification. In a conventional electron microscope without the C_s corrector, different focuses for rays at different angles to the optic axis result in a displaced aperture image for each diffracted beams due to the objective lens spherical aberration. Take the rays marked by P and P' in Fig. 1 for an example. While the ray P parallel to the optical axis defines the back-projected aperture image, ray P' at an angle of α will move by the distance of $y = C_s\alpha^3$. For a microscope with $C_s = 1$ mm and $\alpha = 50$ mrad, $y \sim 125$ nm. The smallest area that can be selected in SAED is thus limited by the objective lens aberration.

The combination of imaging and diffraction in the SAED mode makes it particularly useful for setting diffraction conditions for imaging in a TEM. It is also one of major techniques for materials phase identifications and orientation determinations. Interpretation of SAED patterns for materials science applications is covered in Edington's book [1].

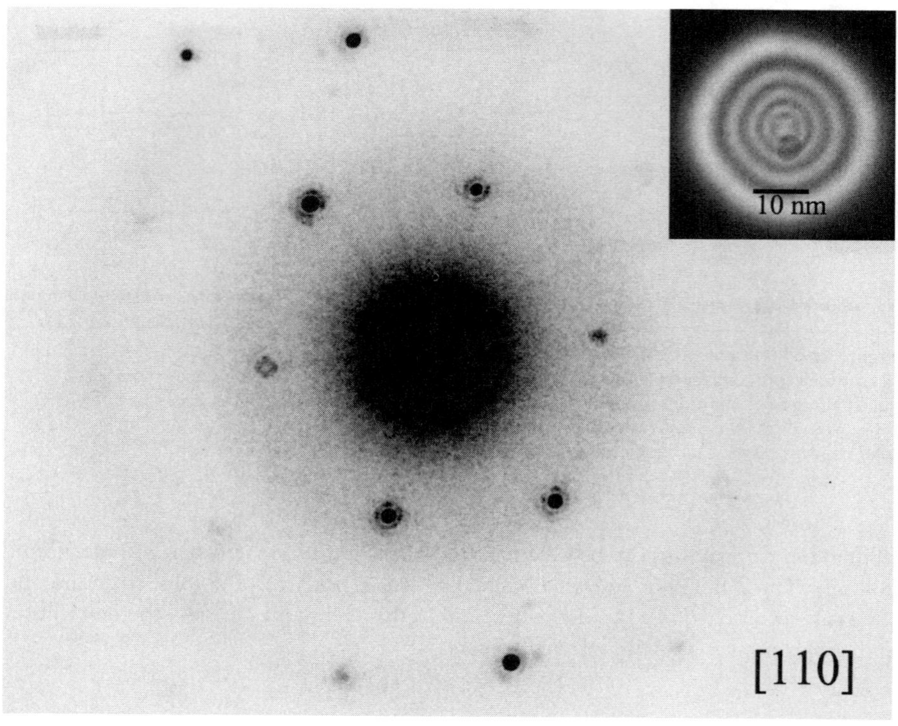

Figure 2. An example of nanoarea electron diffraction. The diffraction pattern was recorded from a single Aunanocrystal of ~4 nm near the zone axis of [110]. Around each diffraction spot, two rings of oscillation are clearly visible. The rings are not continuous because of the shape of the crystal. The electron probe and an image of the nanocrystal are shown at top-right corner.

2.2. Nano-Area Electron Diffraction

Figure 1b shows the principle of parallel-beam electron diffraction from a nanometer-sized area in a TEM. The electron beam is focused to the focal plane of the objective pre-field, which then forms a parallel beam illumination on the sample. For a condenser aperture of 10 micron in diameter, the probe diameter is ~50 nm. The beam size is much smaller than that of SAED. Diffraction patterns recorded in this mode are similar to SAED. For crystals, the diffraction pattern consists of sharp diffraction spots.

Nano-area electron diffraction in a FEG microscope also provides a higher beam intensity than SAED. The probe current intensity is $\sim 10^5$ e/s·nm^2 in the JEOL 2010 F electron microscope. The intensity is high because all electrons illuminating the sample are recorded in the diffraction pattern in NED. The small probe size allows the selection of individual nanostructures for electron diffraction.

An application of nanoarea electron diffraction for electron nanocrystallography is demonstrated in Fig. 2, which was recorded from a single Au nanocrystals close to the [110] zone axis. The 50 nm-diameter electron beam made it possible to isolate a single nanocrystal for diffraction. The parallel electron beam gives the high angular resolution for resolving the details of diffuse scattering that comes from the finite-size of the crystal and deviations from the ideal crystal structure.

The nano-area electron diffraction described here is different from electron nanodiffraction in a STEM [11]. Electron nanodiffraction, pioneered by John Cowley using a dedicated STEM, is formed using a combination of a small probe-forming aperture and convergent beams. Recorded electron diffraction is similar to CBED, but with a smaller disk size. The probe size can be as small as a few angstroms, and diffraction patterns are recorded by placing the focused probe on selected local areas. While there are several proposals to reconstruct crystal structure from a series of electron nanodiffraction patterns with overlapping disks [15], applicability to real systems so far has not been demonstrated. NED is formed by placing a focused probe at the front focal plane of pre-objective lens to form a parallel electron beam on the sample. The advantage of NED is the high angular resolution in the diffraction pattern, which when combined with a $\sim 10^1$ nm probe allows over-sampling of diffraction patterns from individual nanostructures than the so-called Nyquist frequency, which is one over the sample dimension.

The third condenser lens, or a mini-lens, provides the flexibility and demagnification for the formation of a nanometer-sized parallel beam. In electron microscopes with two condenser lenses in the illumination system, the first lens is used to demagnify the electron source and the second lens transfer the demagnified source image to the sample at focus (for probe formation) or under-focused to illuminate a large area.

2.3. Convergent Beam Electron Diffraction (CBED)

CBED is formed by focusing the electron probe at the specimen (see Fig. 1c). Compared to selected area electron diffraction, CBED has two main advantages for studying *perfect crystals* and the local structure:

1. The pattern is taken from a much smaller area with a focused probe; the smallest electron probe currently available in a high-resolution FEG-STEM is close to 1 Å. Thus, in principle and in practice, CBED can be recorded from individual atomic column. For crystallographic applications, CBED patterns are typically recorded with a probe of a few to tens of nanometers;
2. CBED patterns record diffraction intensities as a function of incident-beam directions. Such information is very useful for symmetry determination and quantitative analysis of electron diffraction patterns.

A comparison between the selected area electron diffraction and CBED is given in Fig. 3. CBED patterns consist of disks. Each disk can be divided into many pixels and each pixel approximately represents one incident beam direction. For example, let us take the beam P in Fig. 4. This particular beam gives one set of diffraction pattern shown as the full lines. The diffraction pattern by the incident beam P is the same as the selected area diffraction pattern with a single parallel incident beam. For a second

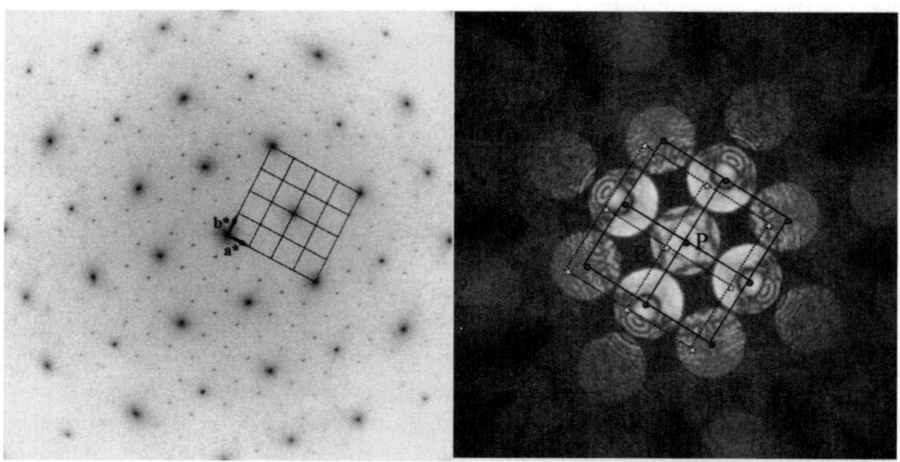

Figure 3. A comparison between CBED and SAED. Left) is a recorded diffraction pattern along [001] from magnetite cooled to liquid nitrogen temperature. There are two types of diffraction spots, strong and weak ones. The weak ones come from the low-temperature structural transformation. All diffraction spots in this pattern can be indexed based on two reciprocal lattice vectors (a^* and b^*). Right) is a recorded CBED pattern from spinel along [100] at 120 kV.

beam P', which comes at different angle compared to P, the diffraction pattern in this case is displaced from that of P by α/λ with α as the angle between the two incident beams.

Experimentally, the size of the CBED disk is determined by the condenser aperture size and the focal length of the probe-forming lenses. In a modern microscope with an additional mini-lens placed in the objective prefield, it is also possible to vary the convergence angle by changing the strength of the mini-lens. Under-focusing the electron beam also results in a smaller convergence angle. However, under-focusing leads to a bigger probe, which can be an issue for specimens with a large wedge angle.

The advantage of being able to record diffraction intensities over a range of incident beam angles makes CBED a useful technique for comparison with simulations. Also because of this, CBED is a quantitative diffraction technique. In the past 15 years, CBED has evolved from a tool primarily for crystal symmetry analysis to the most accurate technique for structure refinement, strain and structure factor measurements [16]. For crystals with defects, the large angle CBED technique is useful for characterizing individual dislocations, stacking faults and interfaces. For applications to structures without the three-dimensional periodicity, parallel-beam illumination is required for resolving details in the diffraction pattern.

3. THEORY OF ELECTRON DIFFRACTION

Electron diffraction from a nanostructure can be alternatively described by electrons interacting with an assembly of atoms (ions) or from a crystal of finite sizes and shapes. Which description is more appropriate depends on which is a better approximation for the structure. Both approaches are discussed here. We will start with the kinematical electron diffraction from a single atom, then move on to an assembly of atoms and then to crystals. Electron multiple scattering, or electron dynamic diffraction, is treated last for perfect crystals. Dynamic theory of a nanostructure, or imperfect crystals, is too complex to be treated analytically. Their treatment generally requires numerical simulations.

3.1. Kinematic Electron Diffraction and Electron Atomic Scattering

Electron interacts with an atom through the Coulomb potential of the positive nucleus and electrons surrounding the nucleus. The relationship between the potential and the atomic charge is given by the Poisson equation:

$$\nabla^2 V(\vec{r}) = -\frac{e[Z\delta(\vec{r}) - \rho(\vec{r})]}{\varepsilon_0} \tag{1}$$

If we take a small volume, $d\vec{r} = dx\,dy\,dz$, of the atomic potential at position $\vec{r}$, the exit electron wave from this small volume is approximately given by:

$$\phi_e \approx (1 + i\pi\lambda U dx\,dy\,dz)\phi_o \tag{2}$$

Here $U = 2meV(\vec{r})/h^2$, which can be treated as a constant within the small volume. Equation 2 is known as the weak-phase-object approximation. For a parallel beam of incident electrons, the incident wave is a plane wave $\exp(2\pi i\vec{k}_0 \cdot \vec{r})$. For high-energy electrons with $E \gg V$, scattering by an atom is weak and we have approximately:

$$\phi_0 \approx \exp(2\pi i\vec{k}_0 \cdot \vec{r}) \tag{3}$$

The scattered wave from the small volume is a new point source, which gives out spherical waves, the contribution from this small volume to the wave at r is:

$$d\phi_s \approx \frac{\pi U}{|\vec{r} - \vec{r}'|} e^{2\pi i k|\vec{r} - \vec{r}'|}\, d\vec{r} \tag{4}$$

The approximation here is for small scattering angles (θ) and $\cos\theta \approx 1$. Equation 4 is justified because contributions to the wave come mostly from the first Fresnel zone, which has a small angle in electron diffraction because of the short wavelength. The total scattered wave is the sum of the scattered waves over the volume of the atom:

$$\phi_s = \frac{2\pi me}{h^2} \int \frac{V(\vec{r}')}{|\vec{r} - \vec{r}'|} e^{2\pi i k|\vec{r} - \vec{r}'|} e^{2\pi i\vec{k}_0 \cdot \vec{r}'}\, d\vec{r}' \tag{5}$$

Far away from the atom, we have $|\vec{r}| \gg |\vec{r}'|$ and we replace $|\vec{r} - \vec{r}'|$ by $|\vec{r}|$ in the denominator and

$$|\vec{r} - \vec{r}'| \approx r - \frac{\vec{r}' \cdot \vec{r}}{r} \tag{6}$$

for the exponential. Thus,

$$\phi_s = \frac{2\pi me}{h^2}\int \frac{V(\vec{r}')}{|\vec{r}-\vec{r}'|}e^{2\pi ik|\vec{r}-\vec{r}'|}e^{2\pi i\vec{k}_0\cdot\vec{r}'}d\vec{r}' \approx \frac{2\pi me}{h^2}\frac{e^{2\pi i\vec{k}_0\cdot\vec{r}}}{r} \times \int V(\vec{r}')e^{2\pi i(\vec{k}-\vec{k}_0)\cdot\vec{r}'}d\vec{r}' \quad (7)$$

Here $\vec{k}$ is the scattered wave vector and the direction is taken along $\vec{r}$. The half difference between the scattered wave and incident wave, $\vec{s} = (\vec{k} - \vec{k}_0)/2$, is defined as the scattering vector. Equation 7 defines the electron atomic scattering factor

$$f(s) = \frac{2\pi me}{h^2}\int V(\vec{r}')e^{4\pi i\vec{s}\cdot\vec{r}'}d\vec{r}' \quad (8)$$

The crystal potential is related to the charge density and Fourier transform of electron charge density gives the X-ray scattering factor. The relation between electron and the X-ray scattering factors is given by the Mott formula:

$$f(s) = \frac{me^2}{8\pi\varepsilon_o h^2}\frac{(Z-f^x)}{s^2} = 0.023934\frac{(Z-f^x)}{s^2}(\text{Å}) \quad (9)$$

The X-ray scattering factor in the same unit is given by

$$f^x(s) = \left(\frac{e^2}{mc^2}\right)f^x = 2.82\times 10^{-5}f^x(\text{Å}) \quad (10)$$

For a typical value of $s \sim 0.2$ 1/Å, the ratio $f/(e^2/mc^2)f^x \sim 10^4$. Thus, electrons interacts with an atom much more strongly than X-ray.

Electron distribution of an atom in a crystal depends bonding with neighboring atoms. At sufficiently large scattering angles, we can approximate atoms in a crystal by spherical free atoms or ions. Atomic charge density and its Fourier transform can be calculated with high accuracy. Results of these calculations are published in literature and tabulated in the international table for crystallography.

3.2. Kinematical Electron Diffraction from an Assembly of Atoms

Here, we extend our treatment of kinematical electron diffraction from a single atom to an assembly of atoms. For nanostructures of a few nanometers, the treatment outlined here form the basis for electron diffraction pattern analysis and interpretation. For a large assembly of atoms in a crystal with well-defined 3-D periodicity, we will use this section to introduce the concepts of lattice and reciprocal space and lay the foundation for our treatment of crystal diffraction.

Kinematic scattering from an assembly of atoms follows the same treatment for a single atom:

$$\phi_s \approx \frac{2\pi me}{h^2}\frac{e^{2\pi i\vec{k}_0\cdot\vec{r}}}{r}\int V(\vec{r}\,')e^{2\pi i(\vec{k}-\vec{k}_0)\cdot\vec{r}\,'}d\vec{r}\,' = \frac{2\pi me}{h^2}\frac{e^{2\pi i\vec{k}_0\cdot\vec{r}}}{r}FT(V(\vec{r}\,')) \tag{11}$$

Where FT denotes Fourier Transform. The potential of an assembly of atoms can be expressed as a sum of potentials from individual atoms

$$V(\vec{r}) = \sum_i\sum_j V_i(\vec{r}-\vec{r}_j) \tag{12}$$

Here, the summation over i and j are for the type of atoms and index for atoms in each type respectively.

To see how an atomic assembly diffract differently from a single atom, we first look at a row of atoms that are separated periodically by an equal spacing of a. Each atom contributes to the potential at point $\vec{r}$. The total potential is obtained by summing up the potential of each atom. If we take the x-direction along the atomic row, then

$$V(\vec{r}) = \sum_{n=1}^{N} V_A(\vec{r}-na\hat{x}) \tag{13}$$

The sum can be considered as placing atom on a collection of points (lattice), and the potential of the atomic assembly is thus a convolution of the atomic potential and the lattice:

$$V(\vec{r}) = V_A(\vec{r}) * \sum_{n=1}^{N}\delta(\vec{r}-na\hat{x}) \tag{14}$$

The Fourier transform of this potential (see E7) is the product of FT of the atomic potential and FT of the lattice:

$$FT[V(\vec{r})] = FT[V_A(\vec{r})]\cdot FT\left[\sum_{n=1}^{N}\delta(\vec{r}-na\hat{x})\right] \tag{15}$$

The Fourier transform of atomic potential gives the atomic scattering factor. The Fourier transform of an array of delta functions gives

$$FT\left[\sum_{n=1}^{N}\delta(\vec{r}-na\hat{x})\right] = \sum_{n=1}^{N}e^{-2\pi i(\vec{k}-\vec{k}_0)\cdot\vec{x}na} = \frac{\sin[\pi(\vec{k}-\vec{k}_0)\cdot\vec{x}Na]}{\sin[\pi(\vec{k}-\vec{k}_0)\cdot\vec{x}a]} \tag{16}$$

The function (E16) has an infinite number of maximums at the condition:

$$(\vec{k}-\vec{k}_0)\cdot\hat{x} = h/a \tag{17}$$

Here h is an integer from $-\infty$ to ∞. The maxima are progressively more pronounced with an increasing N. For a sufficiently large N, (E16) reduces to a periodic array of delta functions with the spacing of $1/a$.

For a three-dimension periodic array of atoms, where the position of atoms is given by the integer displacement of the unit cell **a**, **b** and **c**. The potential is given by

$$V(\vec{r}) = V_A(\vec{r}) * \sum_{n,m,l} \delta(\vec{r} - n\vec{a} - m\vec{b} - l\vec{c}) \tag{18}$$

The Fourier transform of the three-dimensional lattice of a cube $N \times N \times N$ gives

$$FT\left[\sum_{n=1}^{N}\sum_{m=1}^{N}\sum_{l=1}^{N} \delta(\vec{r} - n\vec{a} - m\vec{b} - l\vec{c})\right] = \frac{\sin[\pi(\vec{k} - \vec{k}_0)\cdot\vec{a}N]}{\sin[\pi(\vec{k} - \vec{k}_0)\cdot\vec{a}]} \times \frac{\sin[\pi(\vec{k} - \vec{k}_0)\cdot\vec{b}N]}{\sin[\pi(\vec{k} - \vec{k}_0)\cdot\vec{b}]} \frac{\sin[\pi(\vec{k} - \vec{k}_0)\cdot\vec{c}N]}{\sin[\pi(\vec{k} - \vec{k}_0)\cdot\vec{c}]} \tag{19}$$

Similar to one-dimensional case, (E19) defines an array of peaks. The position of peaks is placed where

$$\begin{aligned} (\vec{k} - \vec{k}_0)\cdot\vec{a} &= h \\ (\vec{k} - \vec{k}_0)\cdot\vec{b} &= k \\ (\vec{k} - \vec{k}_0)\cdot\vec{c} &= l \end{aligned} \tag{20}$$

with h, k and l as integers. It can be shown that

$$\Delta\vec{k} = \vec{k} - \vec{k}_0 = h\vec{a}^* + k\vec{b}^* + l\vec{c}^* \tag{21}$$

and

$$\vec{a}^* = \frac{(\vec{b}\times\vec{c})}{\vec{a}\cdot(\vec{b}\times\vec{c})}, \quad \vec{b}^* = \frac{(\vec{c}\times\vec{a})}{\vec{a}\cdot(\vec{b}\times\vec{c})}, \quad \vec{c}^* = \frac{(\vec{a}\times\vec{b})}{\vec{a}\cdot(\vec{b}\times\vec{c})} \tag{22}$$

The vectors a^*, b^* and c^* together define the three-dimensional reciprocal lattice.

For nanostructures that has the topology of a periodic lattice, we can be describe the structure by the lattice plus a lattice-dependent displacement:

$$V(\vec{r}) = V_A(\vec{r}) * \sum_{n,m,l} \delta(\vec{r} - \vec{R} - \vec{u}(\vec{R})), \text{ with } \vec{R} = n\vec{a} - m\vec{b} - l\vec{c} \tag{23}$$

For simplicity, we will restrict the treatment to monoatomic primitive lattices. Generalization to more complex cases with a non-primitive lattice follows the same principle outlined here, but with more complex expressions. The Fourier transform of the lattice in this case gives a sum of two terms:

$$FT\left[\sum_{n,m,l} \delta(\vec{r} - \vec{R} - \vec{u}(\vec{R}))\right] = \sum_{n,m,l} \exp(2\pi i \Delta\vec{k} \cdot \vec{R}) \exp[2\pi i \Delta\vec{k} \cdot \vec{u}(\vec{R})] \tag{24}$$

For small displacement, (E24) can be expanded to the first order

$$FT\left[\sum_{n,m,l} \delta(\vec{r} - \vec{R} - \vec{u}(\vec{R}))\right] = \sum_{n,m,l} \exp(2\pi i \Delta\vec{k} \cdot \vec{R}) + \sum_{n,m,l} 2\pi i \Delta\vec{k} \cdot \vec{u}(\vec{R}) \exp(2\pi i \Delta\vec{k} \cdot \vec{R}) \tag{25}$$

The first term is the same as E19, which defines an array of diffraction peaks; the position of each peak is defined by the reciprocal lattice of the starting crystal. The second term describes the diffuse scattering around each reflections defined by the crystal reciprocal lattice. If we take the reflection as g and write

$$\Delta\vec{k} = \vec{g} + \vec{q} \text{ and } \vec{g} \cdot \vec{R} = 2n\pi \tag{26}$$

For $|g| \gg |q|$, the diffuse scattering term can be rewritten as

$$\sum_{n,m,l} 2\pi i \Delta\vec{k} \cdot \vec{u}(\vec{R}) \exp(2\pi i \Delta\vec{k} \cdot \vec{R}) \approx \sum_{n,m,l} 2\pi i \vec{g} \cdot \vec{u}(\vec{R}) \exp(2\pi i \vec{q} \cdot \vec{R}) \tag{27}$$

which is the Fourier sum of the displacements along the g direction. The intensity predicted by (E27) will increase with g with the g^2-dependence. The atomic scattering contains the Debye-Waller factor, which describes the damping of high-angle scattering because of thermal vibrations. The balance of these two terms results a maximum contribution to the diffuse scattering from deviations from the ideal crystal lattice [5].

The oscillations from the finite size of the nanocrystals are clearly visible in the diffraction pattern shown in Fig. 2. The subtle difference in the intensity oscillations for different reflections can come from several factors, including surface relaxations, the small tilt, the curvature of the Ewald sphere, and the non-negligible multiple scattering effects for heavy atoms such as Au. Surface relaxation effects can be treated using the approximation described above.

3.3. Geometry of Electron Diffraction from Perfect Crystals

We have shown that electrons diffract from a crystal under the *Laue* condition $\vec{k} - \vec{k}_0 = \vec{G}$, with $\vec{G} = h\vec{a}^* + k\vec{b}^* + l\vec{c}^*$. Thus, each diffracted beam is defined by a reciprocal lattice vector. Diffracted beams seen in an electron diffraction pattern are these close to the intersection of the Ewald sphere and the reciprocal lattice. Based on these two principles, we will develop a quantitative understanding of electron diffraction geometry.

3.3.1. The Cone of Bragg Conditions

The Bragg condition defines a cone of angle θ_{hkl} normal to the (hkl) plane. Alternatively, we use the Laue condition to specify the Bragg diffraction:

$$k_0^2 - |\vec{k} + \vec{G}|^2 = \tag{28}$$

$$2\vec{k} \cdot \vec{G} + G^2 = 0 \tag{29}$$

and

$$k_G = -G/2 \tag{30}$$

Thus irrespective to the wavelength of electrons, the wave vector along the plane normal direction is half the reciprocal lattice vector length. The Bragg angle becomes increasingly smaller as the electron energy increases, which results in a decrease in the wavelength and an increase in the wave vector in the plane.

3.3.2. Zone Axis

A typical zone axis diffraction pattern is shown in Fig. 3. The diffraction pattern can be indexed based on the two shortest G-vectors $\vec{g}$ and $\vec{h}$. The zone axis is along the direction determined by

$$\begin{aligned} \vec{g} \times \vec{h} &= (h_1\vec{a}^* + k_1\vec{b}^* + l_1\vec{c}^*) \times (h_2\vec{a}^* + k_2\vec{b}^* + l_2\vec{c}^*) \\ &= (k_1l_2 - k_2l_1)\vec{b}^* \times \vec{c}^* + (l_1h_2 - l_2h_1)\vec{c}^* \times \vec{a}^* + (h_1k_2 - h_2k_1)\vec{a}^* \times \vec{b}^* \end{aligned} \tag{31}$$

The zone axis index is taken such as

$$\vec{z} = u\vec{a} + v\vec{b} + w\vec{c} \tag{32}$$

with $[u, v, w]$ as the smallest integer satisfying the relationship of

$$u/v/w = (k_1l_2 - k_2l_1)/(l_1h_2 - l_2h_1)/(h_1k_2 - h_2k_1) \tag{33}$$

By definition, a zone axis is normal to both $\vec{g}$ and $\vec{h}$. The reciprocal lattice plane passing through the reciprocal lattice origin is called the zero-order Laue zone. A G-vector with $\vec{z} \cdot \vec{G} = n(n \neq 0)$ is said to belong to high order Laue zones.

A diffraction pattern with diffraction spots belonging to both zero-order Laue zone and higher order Laue zones can be used to determine the three-dimensional cell of the crystal.

3.3.3. The Zone Axis Coordinate and the Line Equation for Bragg Conditions

Given a zone axis, we can define an orthogonal zone axis coordinate, $(\hat{x}, \hat{y}, \hat{z})$, with the z parallel to the zone axis direction. Let us take x along the g direction and y normal to the x-direction. Expressing the condition for Bragg diffraction (E29) in this coordinate, we have

$$2\vec{k}\cdot\vec{g} + g^2 = 2(k_x g_x + k_y g_y + k_z g_z) + g^2 = 0 \qquad (34)$$

and

$$k_y = -\frac{g_x}{g_y}k_x + \frac{2g_z - g^2}{2g_y}|k_z| \qquad (35)$$

Here

$$|k_z| = \sqrt{k_0^2 - k_x^2 - k_y^2} \approx k_0 = 1/\lambda \qquad (36)$$

The approximation holds for high-energy electrons of small wavelengths and the typical acceptance angles in electron diffraction. Within this approximation, beams that satisfy the Bragg condition form a straight line.

3.3.4. Excitation Error

Deviation of the electron beam from the Bragg condition is measured by the distance from the reciprocal lattice vector to the Ewald sphere along the zone axis direction, which is approximately defined by

$$S_g = \left(k_0^2 - |\vec{k}+\vec{g}|^2\right)/2\,|k_0| \qquad (37)$$

This parameter is important for the interpretation of diffraction contrast images and electron diffraction intensities in both SAED and CBED. From (E29), let us take

$$k_G = -g/2 + \Delta \qquad (38)$$

The positive Δ corresponds to a tilt towards the zone axis center. Then we have

$$S_g = \lambda\left(\left(-\frac{g}{2}+\Delta\right)^2 - \left(\frac{g}{2}+\Delta\right)^2\right)\Big/2 = -\lambda g\,\Delta \approx -g\,\delta\theta \qquad (39)$$

Here $\delta\theta$ is the deviation angle from the Bragg condition. S_g is negative for positive Δ and positive for negative Δ.

3.4. The Geometry of a CBED Pattern

The starting point for understanding CBED is the Ewald sphere construction. Fig. 4 shows one example. By the requirement of elastic scattering, all transmitted and diffracted beam are on the Ewald sphere. Let's take the incident beam P, which

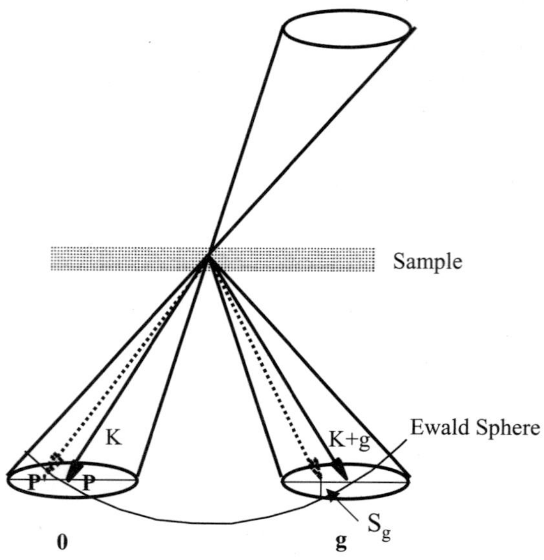

Figure 4. The geometric construction of CBED. This figure demonstrates the variation of excitation errors at different positions of the CBED disk. The beam marked by the full line (P) is at Bragg condition, while beam marked by the dashed line is associated with a positive excitation error (S_g).

satisfies the Bragg condition for g. For an incident beam P' to the left of P, the diffracted beam also moves to the left, which gives a positive excitation error. Correspondingly, a beam moving to the right of P gives a negative excitation error. The excitation error varies according to equation 39. Generally, the excitation error varies across the disk and along the g direction for each diffracted beams. The range of excitation errors within each CBED disk is proportional to the length of g and the convergence angle. Consequently, excitation error changes much faster for a HOLZ reflection than a reflection in ZOLZ close to the direct beam.

For high order reflections with a large g, the rapid increase in the excitation error away from the Bragg condition results a rapid decrease in diffraction intensities. Under the kinematical condition, the maximum diffraction intensity occurs at the Bragg condition, which appears as a straight line within a small convergence angle. These lines, often called as high order Laue zone (HOLZ) lines, are very useful for measuring lattice parameters and the local strain. The sensitivity of these lines to lattice parameters comes from the large scattering angle. The positions of these lines moves relative to each other when there is a small change in the lattice parameters [17]. For example, let us examine a cubic crystal and its Bragg condition:

$$\sin\theta = g\lambda/2 = \sqrt{h^2 + k^2 + l^2}\,\lambda/2a \tag{40}$$

A small change in a gives

$$\delta\theta \approx \sqrt{h^2 + k^2 + l^2}\,\lambda\delta a/2a^2 = 0.5g\lambda\delta a/a \tag{41}$$

The amount change in the Bragg angle is proportional to the length of g.

3.5. Electron Dynamic Theory – The Bloch Wave Method

For large nanostructures approaching several tens of nanometers or for large crystals, electron dynamic scattering must be considered for the interpretation of experimental diffraction intensities. This is because of the strong electron interactions with matter. For a perfect crystal with a relatively small unit cell, the Bloch wave method is the preferred way to calculate dynamic electron diffraction intensities and the exit-wave function because of the flexibility and accuracy of the Bloch wave method. The alternative multi-slice method is best used in the case of diffraction from strained crystals or crystals containing defects. Detailed descriptions of the multislice method can be found in a recent publication [8].

For high-energy electrons, the exchange and correlation between the beam electrons and the crystal electrons can be neglected, and the problem of electron diffraction is reduced to solve the Schrödinger equation for an independent electron in a potential field:

$$\left[K^2 - (\mathbf{k}+\mathbf{g})^2\right] C_g + \sum_h U_{\mathbf{gh}} C_g = 0 \tag{42}$$

With

$$K^2 = k_0^2 + U_0 \tag{43}$$

And

$$U_g = U_g^C + U_g'' + iU_g' \tag{44}$$

Here $U_g = 2m|e|V_g/h^2$ is the optical potential of the crystal, which consists of the crystal potential U^c, absorption U' and a correction to the crystal potential due to virtual inelastic scattering U''. The most important contribution to U' and U'' for $|g| > 0$ comes from inelastic phonon scatterings. Details on the evaluation of absorption potentials can be found in the references of Bird and King [18], Weickenmeier and Kohl [19] for atoms with isotropic Debye-Waller factors and Peng for anisotropic thermal vibrations [20]. The U'' term is significantly smaller than U', which is generally neglected in high energy electron diffraction.

For convergent electron beams with a sufficiently small probe, the diffraction geometry can be approximated by a parallel crystal slab with the surface normal $\mathbf{n}$. To satisfy the boundary condition, let us take

$$\mathbf{k} = \mathbf{K} + \gamma\mathbf{n} \tag{45}$$

Here γ is the dispersion of wavevectors inside the crystal. Inserting equation 45 into 42 and neglecting the backscattering term of γ^2, we obtain from equation 42

$$2KS_g C_g + \sum_h U_{gh} C_h = 2K_n \left(1 + \frac{g_n}{K_n}\right) \gamma C_g \tag{46}$$

Here $K_n = \mathbf{K} \cdot \mathbf{n}$ and $g_n = \mathbf{g} \cdot \mathbf{n}$. Equation 46 reduces to an eigen equation by renormalizing the eigenvector:

$$B_g = \left(1 + \frac{g_n}{K_n}\right)^{1/2} C_g \tag{47}$$

The zone axis coordinate system can be used for specifying the diffraction geometry, including both the incident beam direction and crystal orientation. In this coordinate, an incident beam of wavevector $\mathbf{K}$ is specified by its tangential component on x–y plan $\mathbf{K}_t = x\mathbf{X} + y\mathbf{Y}$, and its diffracted beam at $\mathbf{K}_t + \mathbf{g}_t$, for small angle diffraction. For each point inside the CBED disk of $\mathbf{g}$, the intensity is given by

$$I_g(x, y) = |\phi_g(x, y)|^2 = \left| \sum_i c_i(x, y) C_g^i(x, y) \exp[2\pi i \gamma^i(x, y) t] \right|^2 \tag{48}$$

Here the eigenvalue γ and eigenvector C_g are obtained from diagonalizing equation 46. And c^i is called excitation of ith Bloch wave, which is obtained from the first column of the inverse eigenvector matrix as determined by the incident beam boundary condition. The solution of equation 46 converges with the increasing number of beams included in the calculation. Solving equation 46 straightforwardly with a large number of beams is often impractical, since the matrix diagonalization is a time consuming task. The computer time needed to diagonalize a $N \times N$ matrix is proportional to N^3. A solution to this is to use the Bethe potential:

$$U_g^{eff} = U_g - \sum_h \frac{U_{g-h} U_h}{2KS_h} \tag{49}$$

and

$$2KS_g^{eff} = 2KS_g - \sum_h \frac{U_{g-h} U_{h-g}}{2KS_h} \tag{50}$$

Here the summation is over all weak beams of h. The Bethe potential accounts for the perturbational effects of weak beams on the strong beams. The criterion for a weak beam is

$$\left| \frac{KS_g}{U_g} \right| \geq \omega_{\max} \tag{51}$$

Figure 5. A simulated CBED pattern for Si[111] zone axis at 100 kV using the Bloch wave method described in the text.

The weak beams are selected based to converge theoretical calculations [21]. In practice, an initial list of beams is selected using the criteria of maximum *g* length, maximum excitation error and their perturbational strength. Additional criteria are used for selecting strong beams. These criteria should be tested for the theoretical convergence.

Figure 5 shows an example of simulated CBED patterns using the Bloch wave method described here for Si [111] zone axis and electron accelerating voltage of 100 kV. The simulation includes 160 beams in both ZOLZ and HOLZ. Standard numerical routines were used to diagonalize the complex general matrix (for a list of routines freely available for this purpose, see [22]). The whole pattern simulation on a modern PC only takes a few minutes.

4. EXPERIMENTAL ANALYSIS

4.1. Experimental Diffraction Pattern Recording

The optimum setup for quantitative electron diffraction is a combination of a flexible illumination system, an imaging filter and an array detectors with a large dynamic range. The three diffraction modes described in section 1.2 can be achieved through a three-lenses condenser system. There are two types of energy-filters that are currently employed, one is the in-column Ω-energy filter and the other is the post-column Gatan Imaging Filter (GIF). Each has its own advantages. The in-column Ω-filter takes the full advantage of the post specimen lenses of the electron microscope and can be used in combination with detectors such as films or imaging plates (IP), in addition to the CCD camera. For electron diffraction, geometric distortion, isochromaticity, and angular acceptance are the important characteristics of the energy filter [23]. Geometrical distortion complicates the comparison between experiment and theory and is best corrected by experiment. Isochromaticity defines the range of electron energies for each detector position. Ideally this should be the same across the whole detector area. Angular acceptance defines the maximum diffraction angle that can be recorded on the detector without a significant loss of isochromaticity.

Current 2-D electron detectors include the CCD cameras and imaging plates. The performance of the CCD camera and IP for electron recording has been characterized [25]. Both detectors are linear with a large dynamic range. At the low dose range, the CCD camera is limited by the readout noise and dark currents of CCD. IP has a better performance at the low dose range due to the low dark current and readout noise of the photo-multiplier used for IP readout. At medium and high dose, the IP is limited by a linear noise due to the granular variations in the phosphor and instability in the readout system. The CCD camera is limited by a linear noise in the gain image, which can be compensated by averaging over many images. The performance of CCDs varies from one to another. This makes individual characterization of CCD necessary.

Neither of the CCD camera and IP has the ideal resolution of a single pixel, and in both detectors additional noises are introduced in the detection process. The recorded image can be generally expressed as

$$I = Hf + n \tag{52}$$

Here the application of H on image f denotes the convolution between the point spread function (PSF) (encoded in H) and the image. The n is the noise in the readout image. The effects of PSF can be removed partially by deconvolution. The PSF is experimentally characterized and measured by the amplitude of its Fourier transform, which is called the modulated transfer function (MTF). However, the direct deconvolution of recorded image using the measured MTF leads to an excessive amplification of the noise. We have found two deconvolution algorithms that are particularly effective in overcoming the difficulty with noise amplifications. One uses a Wien filter and the other is the Richardson-Lucy algorithm [24].

The noise in the experimental data can be estimated using the measured detector quantum efficiency (DQE)

$$\mathrm{var}(I) = mg\, I/\mathrm{DQE}(I) \tag{53}$$

Here I is the estimated experimental intensity, var denotes the variance, m is the mixing factor defined by the point spread function and g is the gain of the detector [25]. This expression allows an estimation of variance in experimental intensity once DQE is known, which is useful in the χ^2-fitting, where the variance is used as the weight.

4.2. The Phase Problem and Inversion

In the kinematic approximation, the diffracted wave is proportional to the Fourier transform of the potential (E11). If both the amplitude and phase of the wave are known, then an image can be reconstructed, which would be proportional to the projected object potential. In a diffraction pattern, however, what is recorded is the square of the amplitude of the diffracted wave. The phase is lost. The missing phase is known as the phase problem. In case of kinematic diffraction, missing phase prevents the reconstruction of the object potential by the inverse Fourier transform. The phase is preserved in imaging up to the information limit. In electron imaging, the scattered waves recombine to form an image by the transformation of a lens and the intensity of the image is recorded, not the diffraction. A complication in imaging is the lens aberration. Spherical aberration introduces an additional phase to the scattered wave. This phase oscillates rapidly as the scattering angle increases. Additionally, the chromatic aberration and the finite energy spread of the electron source impose a damping envelop to the contrast transfer function and limits the highest resolution information (information limit) that passes through the lens. As a result, the phase of scattered waves with $\sin\theta/\lambda > 1$ Å^{-1} is typically lost in electron images and the resolution of image is $\sim$1 Å for the best microscopes currently available. Phase retrieval is a subject of great interests for both electron and X-ray diffraction. If the phase of the diffraction pattern can be retrieved, then an image can be formed without a lens.

In electron diffraction, the missing phase has not been a major obstacle to its application. The reason is that for most electron diffraction applications, electrons are multiply scattered. The missing phase is the phase of the exit wave function. Inversion of the exit wave function to the object potential is not as straightforward as the inverse Fourier transformation. Theory developed by two research groups (Spence at ASU and Allen in Australia) shows the principle of inversion using data sets of multiple thicknesses, orientations and coherent electron diffraction [26]. The inversion is based on the scattering matrix that relates the scattered wave to the incident wave. This matrix can be derived based on the Bloch wave method, which has a diagonal term of exponentials of the product of eigenvalues and thickness. Electron diffraction intensities determine the moduli of all elements of the scattering matrix. Using the properties of the scattering matrix (unitarity and symmetries), an over-determined set of non-linear equations can be obtained from these data. Solution of these equations yields the required phase information and allows the determination of a (projected) crystal potential by inversion [27].

For materials structural characterization, in many cases, the structure of materials is approximately known. What is to be determined is the accurate atomic positions and unit cell sizes. Extraction of these parameters can be done in a more efficient manner using the refinement technique, which will be introduced in the next section. Another important fact is that the phase of the object potential is actually contained in the diffraction pattern through the electron interference when electrons are multiple scattered [28].

For nanostructures such as carbon nanotubes laid horizontally, the number of atoms is few along the incident electron beam direction. Electron diffraction, to a good approximation, can be treated kinematically. Many nanostructures also have a complicated structure. Modeling, as required in the refinement technique, is difficult because the lack of knowledge about the structure. The missing phase then becomes an important issue. Fortunately, missing phase for nonperiodic objects is actually easier to retrieve than periodic crystals. The principle and technique for phase retrieval are described in 4.4.

4.3. The Refinement Technique

Crystal structure information, such as unit cell parameters, atomic positions and crystal charge distribution, can be obtained from experimental diffraction intensities using the refinement method [29, 30, 31]. The refinement method works by comparing the experimental and theoretical intensities and optimizing for the best fit. During optimization, parameters in the theoretical model are adjusted in search of the minimum difference between experiment and theory. Multiple scattering effects are taken into consideration by simulation using dynamic theory. Previously, strong electron multiple scattering effects have made it difficult to use the kinematical approximation for structure determination in the way similar to X-ray and neutron diffraction except for a few special cases [32]. In case of accurate structure factor measurements, electron interference from multiple scattering can enhance the sensitivity to small changes in crystal potential and thickness and improve the accuracy of electron diffraction measurements.

The refinement is automated by defining a goodness of the fit (GOF) parameter and using numerical optimization routines to do the search in a computer. One of the most useful GOF's for direct comparison between experimental and theoretical intensities is the χ^2

$$\chi^2 = \frac{1}{n-p-1} \sum_{i,j} \frac{1}{\sigma_{i,j}^2} \left(I_{i,j}^{exp} - c\, I_{i,j}^{Theory}(a_1, a_2, \ldots, a_p) \right)^2 \tag{54}$$

Here, I^{exp} is the experimental intensity (in unit of counts) measured from an energy-filtered CBED pattern and i and j are the pixel coordinate of the detector and n is the total number of points. I^{theory} is the theoretical intensity calculated with parameters

a_1, *to* a_p and c is the normalization coefficient. The other commonly used GOF is the R-factor

$$R = \sum_{i,j} \left| I_{i,j}^{exp} - c\, I_{i,j}^{Theory}(a_1, a_2, \ldots, a_p) \right| \Big/ \sum_{i,j} \left| I_{i,j}^{exp} \right| \quad (55)$$

The χ^2 is best used when the differences between theory and experiment are normally distributed and when the variance σ is correctly estimated. The optimum χ^2 has a value close to unity. χ^2 smaller than 1 indicates an over-estimation of the variance. On the other hand, σ is not needed for the R-factor evaluation, which measures the residual difference in percentage. The disadvantage of R-factor is that the same R factor value may not indicate the same level of fit depending on the noise in the experimental data. The other difference is that the R-factor is based on an exponential distribution of differences. This makes the R factor a more robust GOF against possible large differences between theory and experiment. The exponential distribution has a long tail compared to the normal distribution.

Large differences between experiment and theory are often the indication of systematic errors, such as, deficiency in theoretical model (as in the case of using the kinematics approximation for dynamically scattered electrons) and measurement artifacts (such as uncorrected distortions).

Another useful definition of GOF is the correlation function as defined by

$$R = \sum_{i,j} I_{i,j}^{exp} I_{i,j}^{Theory}(a_1, a_2, \ldots, a_p) \Big/ \sqrt{\sum_{i,j} I_{i,j}^{exp^2}} \sqrt{\sum_{i,j} I_{i,j}^{Theory^2}} \quad (56)$$

This measures the likeness of two patterns, which is useful for pattern matching.

To model diffraction intensities, the detector effect and the background intensity from thermal diffuse scattering must be included. A general expression for the theoretical intensity considering all of these factors is

$$I_{i,j}^{Theory} = \iint dx'dy' t(x', y') C(x_i - x', y_j - y') + B(x_i, y_j) \quad (57)$$

Here, the diffracted intensity t is convoluted with the detector response function C plus the background B. The intensity is integrated over the area of a pixel. For a pixilated detector with fixed pixel size, the electron microscope camera length determines the resolution of recorded diffraction patterns. There are two contributions to the detector response function, one is the finite size of the pixel and the other is the point spread in the detector. Point spread function can be measured and removed, or deconvoluted, numerically. Procedures for doing this have been published [25]. At sufficient large camera length, for a deconvoluted diffraction pattern, we can approximate C by a delta function. The background intensity B, in general, is slow varying, which can be subtracted or approximated by a constant.

To calculate theoretical intensities, an approximate model of the crystal potential is needed. For structure refinement, we need an estimate of cell sizes, atomic position and Debye-Waller factor. In case of bonding charge distribution measurement, crystal structure is first determined very accurately. The unknowns are the low order structure factors, absorption coefficients and experiment parameters describing the diffraction geometry and specimen thickness. The structure factor calculated from spherical atoms and ions can be used as a sufficiently good starting point. Absorption coefficients are estimated using Einstein model with known Debye-Waller factors [18]. In refinement, unknown parameters are adjusted for the best fit.

Figure 6 illustrates a typical refinement process. This process is divided into two steps. In the first step, the theoretical diffraction pattern is calculated based on a set of parameters. The pattern can be the whole, or part of the, experimental diffraction pattern. In some cases, a few line scans across the experiment diffraction contain enough data points for the refinement purpose. In the second step, the calculated pattern is placed on top of the experimental pattern and try to locate the best match by shifting, scaling and rotation. Both steps are automated by optimization. The first step optimizes structural parameters and the second step is for experimental parameters. The experimental parameters include the zone axis center (in practice the Kt for a specific pixel), length and angle of the x-axis of the zone axis coordinate used for simulation, specimen thickness, intensity normalization and backgrounds. Given a set of calculated theoretical intensities, their correspondents in experimental pattern can be found by adjusting experimental parameters without the need of dynamical calculations. Thus considerable computation time is saved and while the number of parameters in each optimization cycle is reduced. The structural parameters can be individual structure factors, lattice parameters or atomic positions. Each leads to different applications, which is discussed in [16].

For fitting using the χ^2 as the GOF parameter, the precision of measured parameters is given by

$$\sigma_{a_k}^2 = \chi^2 D_{kk}^{-1} \tag{58}$$

Here

$$D_{kl} = \sum_{i,j} \frac{1}{\sigma_{i,j}^2} \left(\frac{\partial I_{ij}^{Theory}}{\partial a_k} \right) \left(\frac{\partial I_{ij}^{Theory}}{\partial a_l} \right) \tag{59}$$

Derivative of intensity against structure parameters and thickness can be obtained using the first order perturbation method [33]. The finite difference method can also be used to evaluate the derivatives. Estimates of errors in refined parameters can also be obtained by repeating the measurement. In the case of CBED, this can also be done by using different regions of the pattern. In general, the number of experimental data points in CBED far exceeds the number required for measurement.

4.4. Electron Diffraction Oversampling and Phase Retrieval for Nanomaterials

For very small nanostructures, electron diffraction can be treated kinematically to a good approximation. Recorded electron diffraction intensities give the amplitude of

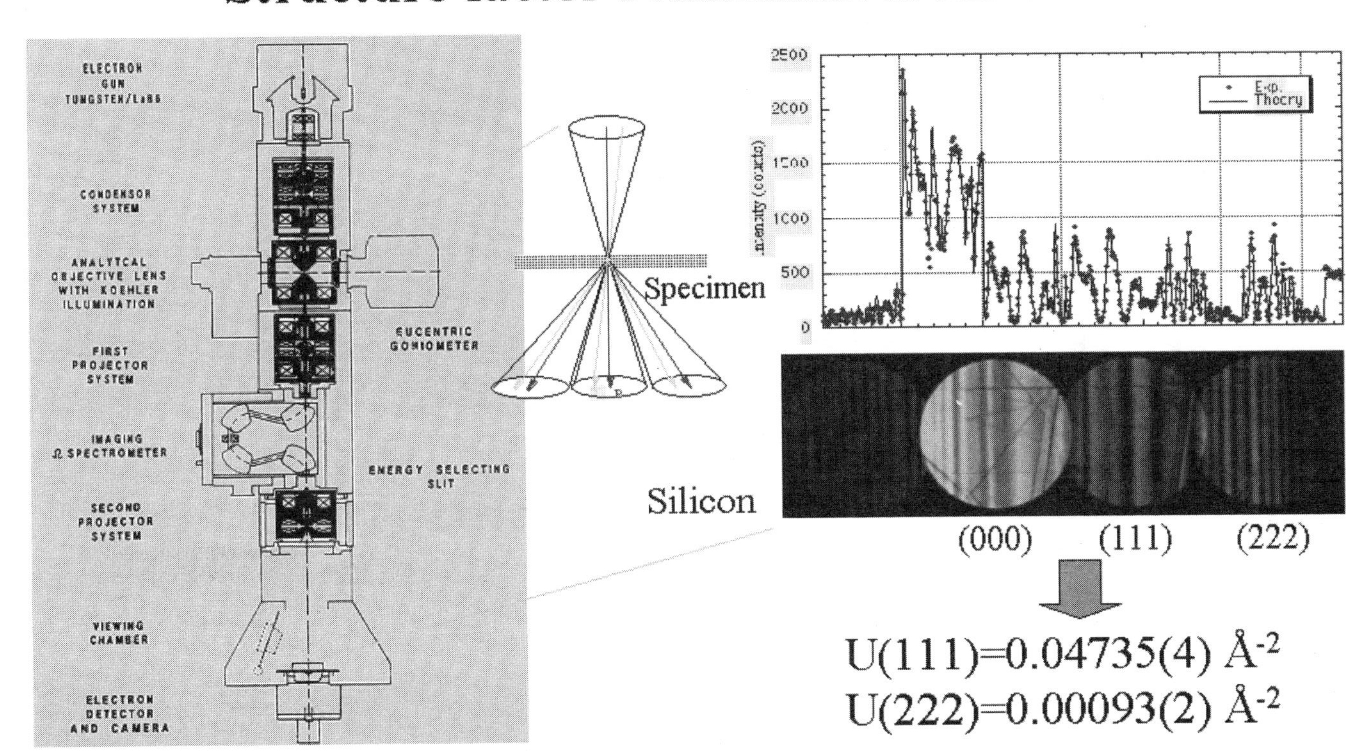

Figure 6. A schematic diagram of quantitative CBED for structure factor refinement. (From left to right) Electron diffraction pattern recording in an energy-filtering electron microscope, the schematic ray diagram of CBED, a record Si CBED pattern with (111) and (222) at Bragg condition and the intensity profile (cross) and the theoretical fit (continuous curve). The structure factors of (111) and (222) reflections are obtained from the best fit.

scattered waves but not the phase. Without the phase, inverse Fourier transformation can not be carried out to reconstruct the potential. Fortunately, most nanostructures are nonperiodic and localized. For such structure, studies have shown that in dimensions more than one, the phase problem is uniquely solvable for localized objects [34, 35, 36]. Recent breakthroughs in phasing show that the missing phase can be retrieved *ab. initio* from diffraction intensities through an iterative procedure [37, 38, 39]. The technique is capable of finding unique solutions independent of the starting phases. This opens a way to image localized objects without a lens. For example, using a soft x-rays with a wavelength of 1.7 nm, Miao *et al.* showed that an 2-D image of patterned gold dots at a resolution of ~75 nm can be reconstructed from the diffraction pattern [40] and furthermore, tomographic image at higher resolution can be reconstructed from a tilt-series of diffraction patterns by iterative phasing.

Using the combination of coherent nanoarea electron diffraction and phase retrieval, Zuo and his coworkers demonstrated for the first time that atomic resolution can be achieved from diffraction intensities without an imaging lens [41]. They applied this technique to image the atomic structure of a double-wall carbon nanotube (DWNT). Electron diffraction pattern from a single DWNT was recorded and phased. The resolution is diffraction intensity limited. The resolution obtained for the double-wall carbon nanotube was 1 Å from a microscope of nominal resolution of 2.3 Å.

The principle of phase retrieval for a localized object is based on the sampling theory. For a localized object of size S, the minimum sampling frequency (Nyquist frequency) in reciprocal space is 1/S. Sampling with a smaller frequency (over-sampling) increases the field of view (see fig. 7). Wavefunctions at these frequencies are a combination of the wavefunctions sampled at the Nyquist frequency. Because of this, phase information is preserved in over-sampled diffraction intensities. Oversampling can be achieved only for a localized object. For a periodic crystal, the smallest sampling frequency is the Nyquist frequency. The iterative phasing procedure works by imposing the amplitude of diffraction pattern in the reciprocal space and the boundary condition in real space. The procedure was first developed by Fineup [38] and improved by incorporating other constraints such as symmetry [39]. The approach of diffractive imaging, or imaging from diffraction intensities, appears to solve many technical difficulties in conventional imaging of nonperiodic objects, namely, resolution limit by lens aberration, sample drift, instrument instability and low contrast in electron images.

5. APPLICATIONS TO NANOSTRUCTURE CHARACTERIZATION

5.1. Structure Determination of Individual Single-Wall Carbon Nanotubes

Since Iijima showed the first high-resolution transmission electron microscopy (HRTEM) image and electron diffraction of multi-wall CNTs (MWNTs) [42], CNTs have attracted extraordinary attention due to their unique physical properties, from atomic structure to mechanical and electronic properties. A single wall CNT (SWNT) can be regarded as a single layer of graphite that has been rolled up into a cylindrical structure. In general, the tube is helical with the chiral vector (n, m) defined by $\vec{c} = n\vec{a} + m\vec{b}$, where $\vec{c}$ is the circumference of the tube, and $\vec{a}$ and $\vec{b}$ are the unit

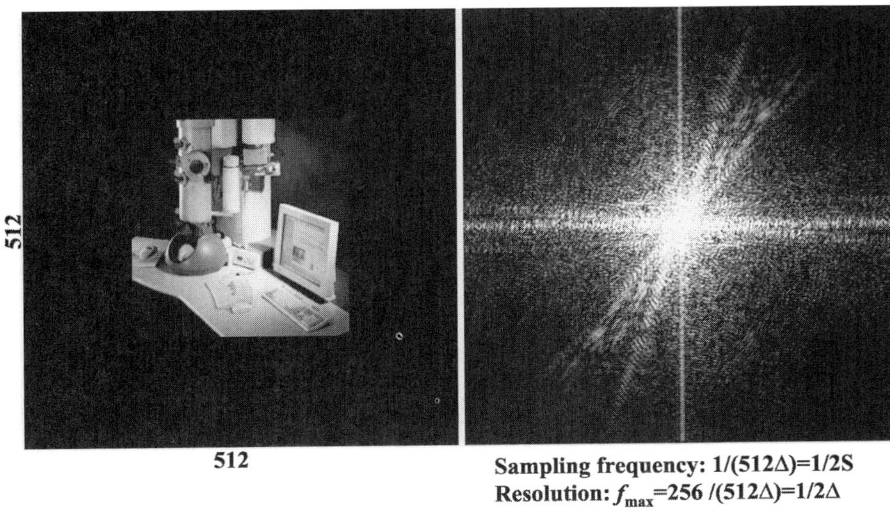

Figure 7. The principle of oversampling. Left is the image and right is the Fourier transform of the image. The image is digitized with a pixel size Δ. The Nyquist sampling frequency of the image is 1/(256 Δ). Sampling at smaller frequency (oversampling) 1/(512 Δ) is equivalent to double the field of view by padding zeros around the image. However the image itself remains the same.

vectors of the graphite sheet. A striking feature is tubes with $n - m = 3l$ (l is a integer) are metallic, while others are semiconductive [43]. This unusual property, plus the apparent stability, has made CNTs an attractive material for constructing nano-scale electronic devices. As-grown SWNTs have a dispersion of chirality and diameters. Hence, a critical issue in CNT applications is the determination of individual tube structure and its correlation to the properties of the tube. This requires a structural probe that can be applied to individual nanotubes.

Gao *et al.* has developed a quantitative structure determination technique of SWNT using NED [44]. This, coupled with improved electron diffraction pattern interpretation, allows a determination of both the diameter and chiral angle, thus the chiral vector (n, m), from an individual SWNT. The carbon nanotubes they studied were grown by chemical vapor deposition (CVD). TEM observation was carried out in a JEOL2010 F TEM with a high voltage of 200 keV.

Figure 8 shows the diffraction pattern from a SWNT. The main features of this pattern are as follows: 1: a relatively strong equatorial oscillation which is perpendicular to the tube direction; 2: some very weak diffraction lines from the graphite sheet, which are elongated in the direction normal to the tube direction [45]. The intensities of diffraction lines are very weak in this case. In their experimental setup, the strongest intensity of one pixel is about 10, which corresponds to ~12 electrons.

The diameter of the tube is determined from the equatorial oscillation, while the chiral angle is determined by measuring the distances from the diffraction lines to the equatorial line. The details are following. The diffraction of SWNT is well described by

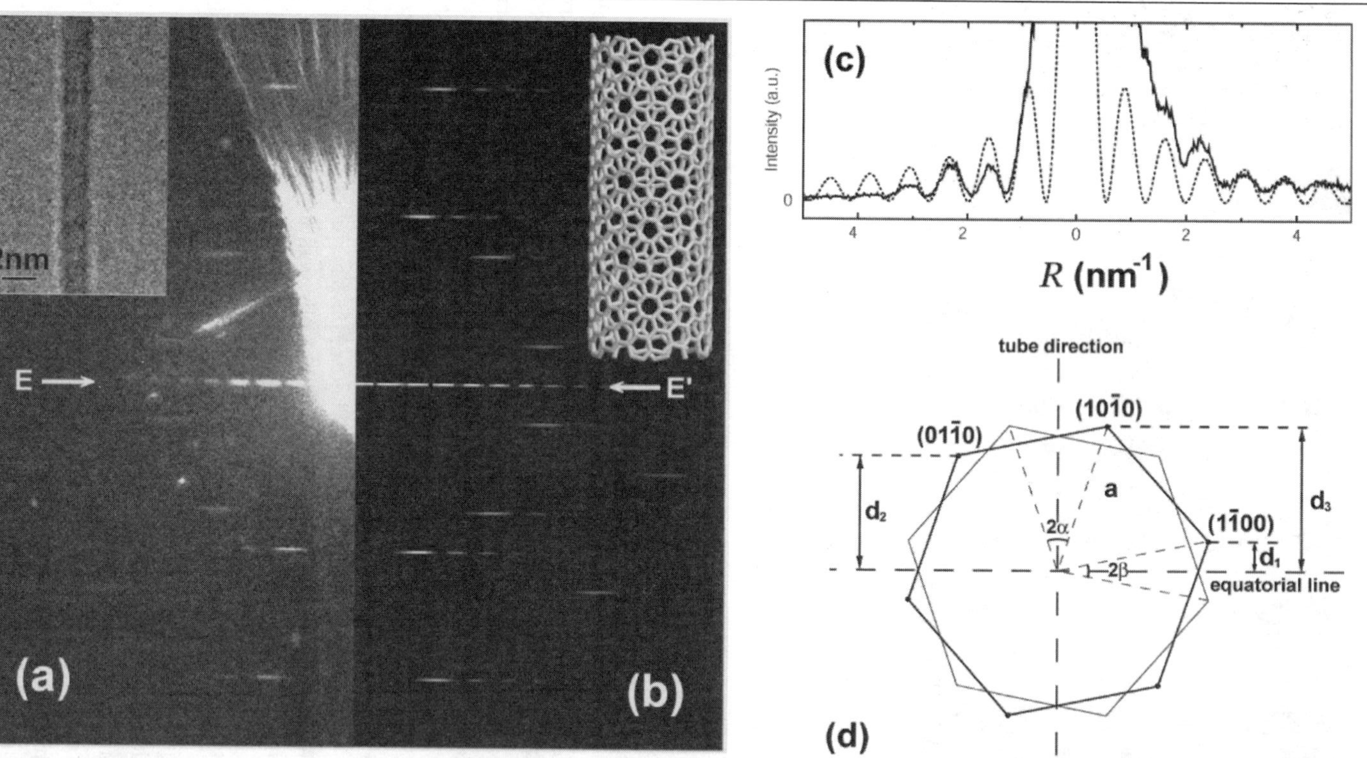

Figure 8. a) A diffraction pattern from an individual SWNT of 1.4 nm in diameter. The inset is a TEM image. The radial scattering around the saturated (000) is an artifact from aperture scattering. b) A simulated diffraction pattern of a (14,6) tube. The inset is the corresponding structure model. c) Profiles of equatorial oscillation along EE' from Fig. 2(a) and simulation for (14, 6). d) A schematic diagram of electron diffraction from an individual SWNT. The two hexagons represent the first order graphite-like {100} diffraction spots from the top and bottom of the tube.

the kinematic diffraction theory (section 3). The equatorial oscillation in the Fourier transformation of a helical structure like SWNT is a Bessel function with $n = 0$ [46] which gives:

$$I_0(X) \propto J_0^2(X) \propto \left| \int_0^{2\pi} \cos^{X\cos\Omega} d\Omega \right|^2 \tag{60}$$

Here $X = 2\pi R r_0 = \pi R D_0$, R is the reciprocal vector which can be measured from the diffraction pattern, and D_0 is the diameter of the SWNT. We use the position of $J_0^2(X)$ maxima (X_n, $n = 0, 1, 2, \ldots$) to determine the tube diameter. With the first several maxima saturated and unaccessible, X_n/X_{n-1} can be used to determine the number N for each maximum in the equatorial oscillation. Thus, by comparing the experimental equatorial oscillation with values of X_n, the tube diameter can be uniquely determined.

To measure chirality from the diffraction pattern, Fig. 8d is considered, which shows the geometry of the SWNT diffraction pattern based one the diffraction of the top-bottom graphite sheets. The distances d_1, d_2, d_3 relate to the chiral angle α by:

$$d_1 + d_2 = d_3, \alpha = a\tan\left(\frac{1}{\sqrt{3}} \cdot \frac{d_2 - d_1}{d_3}\right) = a\tan\left(\frac{1}{\sqrt{3}} \cdot \frac{2d_2 - d_3}{d_3}\right),$$
$$\text{or } \beta = a\tan\left(\sqrt{3} \cdot \frac{d_1}{d_2 + d_3}\right) = a\tan\left(\sqrt{3} \cdot \frac{d_3 - d_2}{d_2 + d_3}\right) \tag{61}$$

These relationships are not affected by the tilting angle of the tube (see below). Because d_2 and d_3 are corresponding to the diffraction lines having relatively strong intensities and are further from the equatorial line, they are used in our study instead of d_1 to reduce the error. The distances can be measured precisely done from the digitalized patterns. The errors are estimated to be $<1\%$ for the diameter determination and $<0.2°$ for the chiral angle.

Using the above methods, the SWNT giving diffraction pattern shown in Fig. 8 was determined to have a diameter of 1.40 nm (±0.02 nm) and a chiral angle of 16.9° (±0.2°). Among the possible chiral vectors, the best match is (20, 6), which has a diameter and chiral angle of 1.39 nm and 17.0° respectively. The closest alternative is (21, 6), having a diameter of 1.46 nm and chiral angle of 16.1° which is well beyond the experiment error. Figure 8b plots the simulated diffraction pattern of (20, 6) SWNT. Figure 8c compares the equatorial intensities of experiment and simulation. These results show an excellent agreement.

5.2. The Structure of Supported Small Nanoclusters and Epitaxy

Nanometer-sized structures in the forms of clusters, dots and wires have recently received considerable attention for their size-dependent transport, optical, and mechanical properties. The focus is on synthesizing nano-structures of desired shapes with narrow size distributions. For clusters or nanocrystals on crystalline substrates,

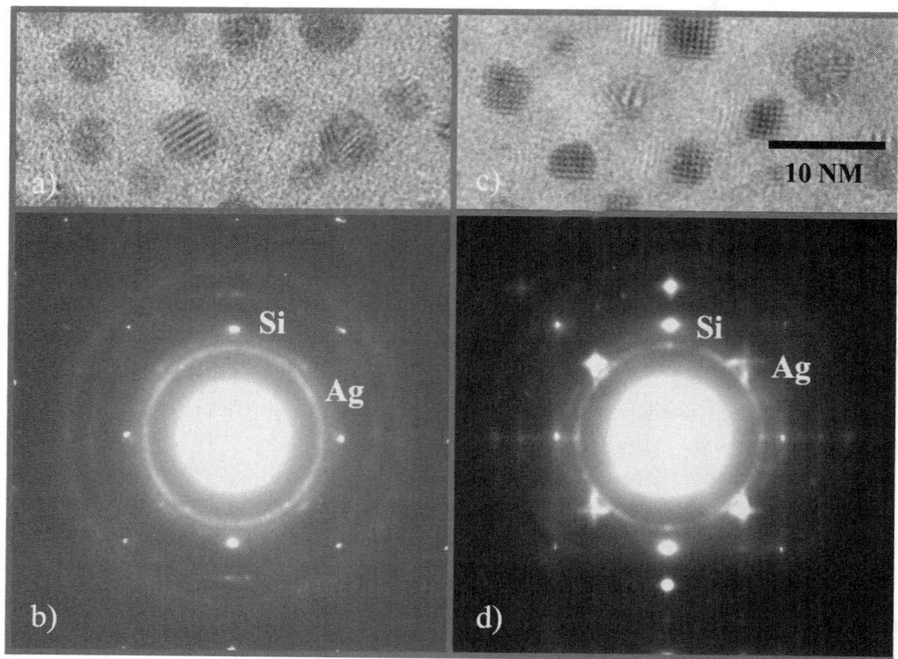

Figure 9. Combined electron diffraction and imaging characterization of supported Ag nanoclusters/nanocrystals on Si(100) substrate. a) and b) show randomly oriented nanoclusters. c) and d) show epitaxial nanocrystals and some nanoclusters.

epitaxy gives lower interface energy and can lead to enhanced stability and better control over the interfacial electronic properties. At the nanometer-scale, cluster equilibrium shape is also determined by the surface, interface, and strain energies. A challenge, thus, is how to determine the structure of individual clusters. The case highlighted below on Ag on Si(100) is taken from the experimental work by Li *et al.* at UIUC. Over past 3 years, they have carried out systematic study on nanoclusters structure and interfaces using a combination of electron diffraction and microscopy [47, 48, 49, 50].

Figure 9 shows the electron diffraction patterns for samples deposited on H-terminated Si(100) surfaces. The diffraction patterns were taken off the [100] zone axis to avoid strong multiple scattering in zone axis orientation. From Fig. 9b, the diffraction pattern of the as-deposited sample consists of strong and continuous Ag {111} ring, short Ag {200} arcs on a weak Ag {200} ring, short Ag {220} arcs on a weak Ag {220} ring and weak {311} rings. Upon annealing at 400°C, Ag (020) and (002) reflection intensities increase significant. Both have diffuse streaks along (011) and (0-11) directions. The Ag (020) and (002) are asymmetrical because of the off-zone axis orientation of diffraction pattern. Meanwhile, the diffraction intensity in the continuous ring decreases significantly, but remains visible. Fig. 9c shows high-resolution images of Ag clusters on H-Si(100) with strong Moiré fringe contrast. These images were taken at the Si [100] zone axis. At this orientation, the Si (022) and (0-22) planes are imaged. Most as-grown Ag clusters in fig. 9a show no visible Moire fringes, which is consistent with the diffraction pattern that is dominated by {111} ring. A few clusters with Moire fringes are often defective. The clusters of dark contrast with no Moire fringes contribute to the strong {111} rings in the diffraction pattern. At first sight, orientation of these clusters appears to be random. However, a close inspection of the diffraction pattern shows a much weaker {200} ring than what it would be in a powder diffraction pattern of random polycrystalline Ag. For single crystals oriented with Ag (111)//Si (100) or Ag (100)//Si (100), strong Ag {220} is expected in both cases, while a strong Ag {200} is also expected in the case of Ag (111)//Si (100). Both of these cases can be ruled out. In fig. 9c the square Ag clusters with 2D Moiré-fringes perfectly parallel to Si (220) lattice planes, in good agreement with the electron diffraction analysis. At this stage, the transformation from random orientation to epitaxial growth is not finished, because we still see weak contrast Ag clusters, supposed to be random Ag clusters. The Ag {200} reflections have the shape of a plus-sign, centered at Ag {200} position, suggestive of perfect cubic Ag nanocrystals, with their edges perfectly aligned to Si (011) and (01-1) directions.

5.3. Crystal Charge Density

A major application of quantitative electron diffraction is the accurate determination of crystal charge density. The question here is how atoms bond to form crystals. This can be approached by accurate measurement of crystal structure factors (Fourier transform of charge density) and from that to map electron distributions in crystals. The full description is beyond the scope of this review. Here the results on the study of charge density in Cu_2O are summarized with an emphasis on the significance.

Cu_2O has a cubic structure with no free internal parameters (only Ag_2O is isostructural). The copper atoms are at the points of a f.c.c. lattice with oxygen atoms in tetrahedral sites at (1/4, 1/4, 1/4) and (3/4, 3/4, 3/4) of the cubic cell. The resulting arrangement of Cu-O links is made up of two interpenetrating networks. The simplest description of Cu_2O using an ionic model with closed-shell Cu^+ and O^{2-} ions is known to be inadequate. It fails to explain the observed linear 2-coordination of Cu.

Accurate measurements of the low order structure factors were made with the quantitative CBED technique described in section 4. Using the small electron probe, a region of perfect crystal was selected for study. The measurements are made by comparing experimental intensity profiles across CBED disks (rocking curves) with calculations, as illustrated in fig. 6. The intensity was calculated using the Bloch wave method, with structure factors, absorption coefficients, the beam direction and thickness treated as refinable parameters. Structure factors for the (531) and higher-order reflections out to (14, 4, 2) were taken from X-ray measurements. Weak (ooe) (with *o* for odd and *e* for even) and very weak (eeo) reflections were also taken from X-ray work.

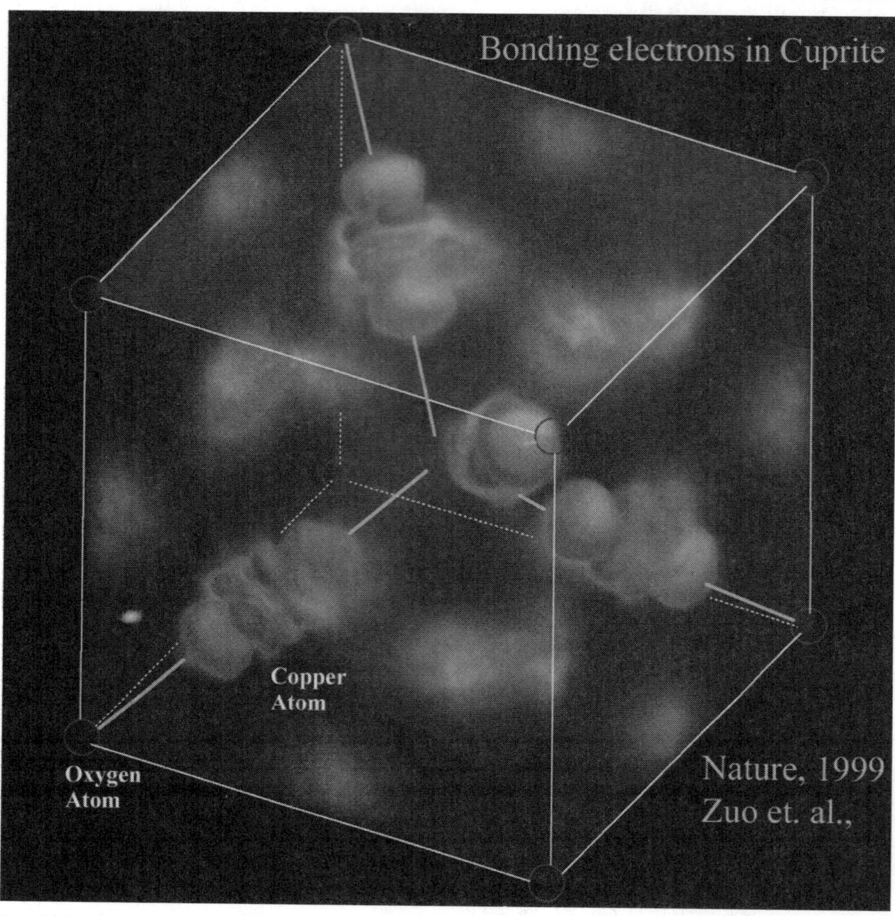

Figure 10. A 3D rendering that reveals the details of chemical bonding and d_z^2 orbital-like holes in Cu_2O. The amount of charge redistribution is very small and its detection requires a high degree of experimental accuracy. In this picture, the small charge differences between the measured crystal charge density derived from convergent-beam electron diffraction (CBED) and that derived from superimposed spherical O^{2-} and Cu^{+} ions are shown. The red and blue colors represent excess electrons and holes, respectively. (See color plate 17.)

There are two approaches to map crystal charge density from the measured structure factors; by inverse Fourier transform or by the multipole method [51]. Direct Fourier transform of experimental structure factors was not useful due to the missing reflections in the collected data set, so a multipole refinement was used to map the charge density from the measured structure factors. In the multipole method, the crystal charge density is expanded as a sum of non-spherical pseudo-atomic densities. These consist of a spherical-atom (or ion) charge density obtained from multi-configuration Dirac-Fock (MCDF) calculations [52] with variable orbital occupation factors to allow for charge transfer, and a small non-spherical part in which local symmetry–adapted spherical harmonic functions were used, which is expressed by:

$$\rho^a(\mathbf{r}) = \rho^s(\mathbf{r}) + \sum_{lm} P_{lm\pm} N_{lm\pm} R_l(r)\, y_{lm\pm}(\theta, \phi) \tag{62}$$

here $\rho^s_{Cu} = \rho^s_{Cu^+} + (1-q)\left(\rho^s_{Cu} - \rho^s_{Cu^+}\right)$ and $\rho^s_O = \rho^s_O + q\left(\rho^s_{O^{2-}} - \rho^s_O\right)$ with q as the charge transfer from Cu to O; $R_l(r) = r^{n_l}\exp(-\alpha r)\,(n_3 = 3; n_4 = 4)$ is the radial function with population coefficient $P_{lm\pm}$; $N_{lm\pm}$ is the density-normalization coefficient. In addition atomic vibrations were accounted for using the Charlier-Gram expansion [51] for the temperature factor. Refinements with and without anharmonic terms in temperature factor clearly show the importance of an anharmonic term for Cu, especially for high order reflections with $s = \sin\theta/\lambda > 1.0(\text{Å}^{-1})$. In either case the charge transfer from Cu to O refined to 1.01(5) (i.e. Cu^+ and O^{2-}).

Figure 10 shows a three-dimensional plot of the difference between the static crystal charge density obtained from the multipole fitting to experiment, and superimposed spherical O^{2-} and Cu^+ ions calculated by the MCDF method. The O^{2-} ion was calculated using a Watson sphere of 1.2 Å radius. The electron density difference shown in Fig. 10 would be zero everywhere if cuprite were purely ionic (i.e. consisted of spherical ions). The difference, here seen unambiguously for the first time, confirms earlier theoretical speculation [53] that a covalent contribution exists. The correspondence between our *experimental* map and the classical diagrams of d_z^2 orbitals sketched in textbooks is striking. All our difference maps show strong non-spherical charge distortion around the copper atoms, with the characteristic shape of d-orbitals, and excess charge in the interstitial region. There is little variation around oxygen in both the experimental and the theoretical results, which suggests that an O^{2-}anion description is valid. The most significant difference between experiment and theory is around Cu, and the charge in the interstitial region. The charge modification around Cu in the experiment is broader and larger than the theory. The experimental map also shows a large ($\sim 0.2e/\text{Å}^3$) positive peak in the unoccupied tetrahedral interstitial region of the 4 neighboring Cu atoms, which suggests a strong Cu^+–Cu^+ covalent bonding.

The non-spherical charge density around Cu^+ can be interpreted as due the hybridization of d electrons with higher-energy unoccupied s and p states, according to [53]. Among these states, hybridization is only allowed for d_z^2 and 4s by symmetry, and when this happens part of the d_z^2 state becomes unoccupied ("d hole"). These states are responsible for the spatial distribution of the deficiency in the map shown in Fig. 10. The complementary empty states are important for EELS which probe empty states. The experimental studies reveal that the unoccupied states are predominately Cu-d character for the Cu $L_{2,3}$ edge; theory shows that they originate from hybridized d_z^2 orbitals. This theoretical interpretation, based on the calculated partial DOS of the one-electron band structure, is supported by the generally good agreement with experimental spectroscopy of both occupied and unoccupied states [54]. From the charge density, we estimate the hybridization coefficient between d_z^2 and 4s, $|x| \sim 0.36$, so that about 0.22 electrons are removed from d_z^2 states.

6. CONCLUSIONS AND FUTURE PERSPECTIVES

In conclusion, this chapter describes the practice and theory of electron diffraction for structural analysis of crystals and nanomaterials. It is demonstrated that the information obtainable from electron diffraction with a small probe and the strong electron interactions complements with other characterization techniques, such as X-ray and neutron diffraction that typically samples a large volume and real space imaging by HREM with a limited resolution. The recent developments in electron energy-filtering, 2-D digital detectors and computer-based image analysis and simulations have significant improved the quantification of electron diffraction. Examples were given to demonstrate the resolution and sensitivity of electron diffraction to individual nanostructures and to highlight the remarkable achievement in the improvement of accuracy of electron diffraction and the application for crystal charge density mapping.

The future for electron nanocrystallography is very bright for two reasons. First, electron diffraction pattern can be recorded selectively from individual nanostructure at sizes as small as a nanometer using electron probe forming lenses and apertures, while electron imaging provides the selectivity. Secondly, electrons interact with matter much more strongly than X-ray and Neutron diffraction. These advantages, coupled with quantitative analysis, enable the structure determination of small, nonperiodic, structures that was not possible before.

REFERENCES

1. J. W. Edington, Practical Electron Microscopy in Materials Science, Monograph 2, Electron Diffraction in the Electron Microscope, MacMillan, Philips Technical Library (1975).
2. P. Hirsch *et al.*, Electron Microscopy of Thin Crystals, p. 19, R. E. Krieger, Florida (1977).
3. D. B. Williams and C. B. Carter, Transmission Electron Microscopy, Plenum, New York (1996).
4. B. Fultz and J. Howe, Transmission Electron Microscopy and Diffractometry of Materials, Springer Verlag, New York (2002).
5. J. M. Cowley, Diffraction Physics, North-Holland, New York (1981).
6. J. C. H. Spence and J. M. Zuo, Electron Microdiffraction, Plenum, New York (1992).
7. Z. L. Wang, Elastic and Inelastic Scattering in Electron Diffraction and Imaging, Plenum, New York (1995).
8. E. J. Kirkland, Advanced Computing in Electron Microscopy, Plenum Press, New York (1998).
9. Tanaka M, Terauchi M and Kaneyama T *Convergent-Beam Electron Diffraction, JEOL*, Tokyo (1988).
10. J. P. Morniroli, Electron Diffraction, Dedicated Software to Interpret LACBEDPatterns USTL, Lille, France (1994).
11. J. M. Cowley, Micros. Res. Tech. 46, 75 (1999).
12. L. J. Wu, Zhu YM, Tafto J, Phys. Rev. Lett. 85, 5126 (2000).
13. J. M. Cowley, Micron 35, 345 (2004).
14. J. C. H. Spence and J. M. Cowley, Optik 50, 129 (1978).
15. Rodenburg J. M, Mccallum B. C, Nellist P. D, Ultramicroscopy 48, 304 (1993).
16. J. M. Zuo, *Materials Transactions JIM 39*, 938–946 (1998).
17. J. M. Zuo, *Ultramicroscopy 41*, 211–223 (1992).
18. D. M. Bird and Q. A. King, Acta Cryst. A46, 202 (1990).
19. A. Weickenmeier and H Kohl, Acta Cryst. A47, 590 (1991).
20. L. M. Peng, Acta Cryst. A53, 663 (1997).
21. J. M. Zuo and A. L. Weickenmeier, *Ultramicroscopy 57*, 375–383 (1995).

22. http://emaps.mrl.uiuc.edu
23. H. Rose in Energy filtering Transmission Electron Microscopy, Edited by L. Reimer, Springer, Berlin (1995).
24. BH. Richardson J Opt Soc Am 62, 55–59 (1972).
25. J. M. Zuo, *Micros. Res. Tech., 49*, 245 (2000).
26. J. C. H. Spence, Acta Cryst. A54, 7 (1998).
27. L. J. Allen, T. W. Josefsson and H. Leeb, Acta Cryst. A54, 388 (1998).
28. J. M. Zuo, J. C. Spence And R. Hoier, *Phys. Rev. Lett. 62*, 547 (1989).
29. J. M. Zuo, And J. C. Spence, *Ultramicroscopy 35*, 185–196 (1991).
30. J. M. Zuo, *Ultramicroscopy 41*, 211–223 (1992).
31. J. M. Zuo, *Acta Cryst. A 49*, 429–435 (1993).
32. L. D. Marks, Phys. Rev. B 60, 2771 (1999).
33. J. M. Zuo, *Acta Cryst. A 47*, 87 (1991).
34. R. Bates, "Fourier Phase Problems Are Uniquely Solvable In More Than One Dimension.1. Underlying Theory", *Optik*, 61, 247 (1982).
35. R. P. Millane, "Phase Retrieval In Crystallography And Optics" *J. Opt. Soc. Am.* A 7, 394 (1990).
36. J. Miao, P. Charalambous, J. Kirz, D. Sayre, "Extending the methodology of X-ray crystallography to allow imaging of micrometre-sized non-crystalline specimens", *Nature* 15, 342 (1999).
37. R. W. Gerchberg ,W. O. Saxton, *Optik* 35, 237 (1972).
38. J. Fienup, *Appl. Opt.*, 21, 2758 (1982).
39. R. P. Millane, W. J. Stroud, *J. Opt. Soc. Am. A.* 14, 568 (1997).
40. J. Miao, P. Charalambous, J. Kirz, D. Sayre, Nature **15**, 342 (1999).
41. J. M. Zuo, I. Vartanyants, M. Gao, R. Zhang and L. A. Nagahara, *Science*, 300, 1419–1421 (2003).
42. S. Iijima, Nature **354**, 56 (1991).
43. J. W. Mintmire, B. I. Dunlap, C. T. White, Phys. Rev. Lett. 68, 631 (1992).
44. M. Gao, J. M. Zuo, R. D. Twesten, I. Petrov, L. A. Nagahara and R. Zhang, *Appl. Phys. Lett.* 82, 2703–06 (2003).
45. S. Amelinckx, A. Lucas, and P. Lambin, Rep. Prof. Phys. **62**, 1471 (1999).
46. D. Sherwood, *Crystal, X-rays and Proteins*. John Wiley & Sons, New York (1976).
47. B. Q. Li and J. M. Zuo, *J. Appl. Phys* 94, 743–748 (2003).
48. J. K. Bording, Y. F. Shi, B. Q. Li and J. M. Zuo, *Phys. Rev. Lett.*, accepted (2003).
49. B. Q. Li and J. M. Zuo, *Surf. Sci.* 520, 7–17 (2002).
50. J. M. Zuo and B. Q. Li, *Phys. Rev. Lett.*, 88, 255502 (2002).
51. P. Coppens, X-ray Charge Densities and Chemical Bonding, Oxford, New York (1997).
52. Rez, D., Rez, P. and Grant, I., Acta Cryst. **A50**, 481 (1994); Acta Cryst. A 53: 522 (1997).
53. Orgel, L. E., J. Chem. Soc. 4186–4190 (1958).
54. Ghijsen, J. *et al.*, Electronic structure of Cu_2O and CuO, Phys. Rev. B 38, 11322–11330 (1988).

19. TOMOGRAPHY USING THE TRANSMISSION ELECTRON MICROSCOPE

P. A. MIDGLEY

1. INTRODUCTION

The promise of nanotechnology [1] can only be fully realised if characterisation techniques are available to study structures and devices at the nanometre scale. In particular, many of the proposed nano-devices are truly three-dimensional in their design and high spatial resolution microscopy is needed to assess a device in all three dimensions. Transmission electron microscopy (TEM), be it in the form of high resolution electron microscopy (HREM) [2] or scanning transmission electron microscopy (STEM) [3] can provide images with extremely high spatial resolution (sub Å) in two dimensions. However, all TEM images are formed by the propagation of the electron beam through the specimen and as such they are two-dimensional projections of a three-dimensional object. Often such projections (particularly of cross-sectional specimens) have been sufficient to determine the structure of simple devices [4]. Where the third dimension has been recognised as important, for example in the study of dislocation networks, then 'stereo pairs' can be used in which a pair of images is recorded at $\sim 10^{\circ}$ tilt to mimic the eyes' angular separation. By viewing the two images simultaneously an illusion of three dimensions can be achieved but in reality provides very little 3D information [5].

In structural biology, there has been a need to image highly complex 3D structures at the nanoscale for many years. 3D TEM techniques have been developed to study protein structures [6], viruses [7], ribosomes [8] and larger cellular structures, such as the mitochondria [9]. Three approaches have been employed. Firstly, if the protein structure can be crystallised then standard electron crystallography techniques (involving HREM and electron diffraction) can be used to solve the crystal structure and from an electron density map retrieve the asymmetric unit that describes the unique protein structure [10]. If the crystal is sufficiently large, X-ray methods are used routinely, if not the electron microscope must be employed. Secondly, if the structure of interest is repeated many times on a specimen grid (as is the case for many biological structures, e.g. viruses), then a single TEM image will contain a large number of sub-images each of which is, in general, a projection of the structure, for example, at a different orientation. By determining the exact orientation of each sub-image, the three dimensional structure can be reconstructed [11]. The third approach is employed for unique cellular structures and involves recording a series of images (projections) of the same object at successive tilts and then reconstructing the object from the series [12]. For the purpose of this chapter and simply to aid clarity, I will describe the three methods as (i) electron crystallography, (ii) single particle analysis and (iii) electron tomography, respectively. This division is somewhat artificial and readers should bear in mind that each method is in essence an electron tomographic technique, all are based on the 'Radon transform' [13] and the 'projection requirement' [14] (described in detail later) and all use comparable reconstruction algorithms.

Similar tomographic techniques exist in the physical sciences. In materials science and engineering, X-ray tomography has been used successfully to reconstruct relatively large three-dimensional structures, such as metallic foams [15] or to probe the stress in engineering structures [16]. However, in general the wavelength of X-rays, coupled with the relatively poor quality of X-ray optics, is such that a resolution of ~2 microns is normally the best achievable. As a caveat to this, however, are recent claims that a resolution of a few 10's nm is now achievable using soft X-rays from synchrotron sources and Fresnel lenses (zone plates) [17]. Such an improvement in resolution is very exciting and promises great things. At the opposite end of the resolution scale, the atom probe field ion microscope (APFIM), designed around a time-of-flight spectrometer and a position sensitive detector, is able to reconstruct three-dimensional maps of atom positions and determine each atomic species [18]. Atom probe tomography is the only tomographic technique that allows single atom counting of a three-dimensional structure. Such remarkable sensitivity is however also a limitation in that it is very time consuming to examine large objects using this technique—for example a 100 nm cube of crystalline silicon contains 5×10^7 atoms! More problematic is the requirement for the sample to be conducting and withstand high field stresses exerted at the tip of the needle-shaped sample needed for the APFIM technique [19].

In nanotechnology, the structures and devices designed to take advantage of the mechanical, physical or chemical changes that occur at these length scales will have features at around the nanometre level but whose overall size may be tens or hundreds of nanometres. Such nanoscale design is of course already underway to a large extent in the microelectronics industry where the three-dimensionality of, for example, the metallisation or the dopant profiling becomes increasingly critical to the performance of the device [20]. In the magnetic recording industry, the magnetic 'bits' are becoming ever smaller and a need is developing to examine the composition and magnetic microstructure in three dimensions [21]. The latter requires accurate measurement of the magnetic induction and this can be achieved with a 3D form of electron holography [22]. Indeed, three-dimensional analysis will also become increasingly important not only for functional materials but also for nanoscale structural and engineering materials, such as ultra-fine cermets [23]. In the catalysis industry, heterogeneous catalysts [24] are now designed with nanometre-sized active particles distributed in three dimensions on or within a porous support structure—the tomography of such catalysts will be discussed later in the chapter.

Thus a microscope technique is needed that will allow relatively large structures and devices to be studied (say up to 500 nm in diameter) but with a 3D resolution of ~1 nm to allow the intricate detail of the internal nanostructure to be unravelled. Such requirements are remarkably similar to those demanded by structural biologists studying cellular structures and so it is natural to turn to electron tomography, used so successfully in the life sciences, as a means by which the 3D structure of nanoscale devices can be elucidated.

2. TOMOGRAPHY

2.1. A History of Tomography

Before describing the technique in detail, it is worth spending some time reviewing the birth and subsequent development of tomography, and of electron tomography in particular. The need to obtain 'structures' using data of lower dimensionality is present in many different fields of physical and life sciences. It was in the field of Astronomy that in 1956 Bracewell [25] proposed a method of reconstructing a 2D map of solar emission from a series of 1D 'fan beam' profiles measured by a radio telescope. This pioneering work covered the mathematical formulation of projection and reconstruction but despite its clear potential, this work had little impact beyond its immediate field. However, in 1963, interest in tomography was rekindled by its possible use in medical sciences [26]. This led to the development of the X-ray computerised tomography (CT) scanner [27], known more commonly as the CAT-scan (computer assisted tomography or computerised axial tomography). This remarkably successful technique is undoubtedly the most well known application of 3D tomographic reconstruction and its pioneers, Cormack and Hounsfield, were awarded jointly the Nobel Prize for Medicine in 1979. The success of the CAT-scan was mirrored by the development of similar techniques such as positron emission tomography (PET) [28], ultrasound CT [29] and zeugmatography (reconstruction from NMR imaging) [30]. Outside the medical field, tomography was applied in many other disciplines to allow, for example, 3D stress analysis [16], geophysical mapping [31] and non-destructive testing [32, 33].

Interest in three-dimensional reconstruction using electron microscopy started with the publication of three papers in 1968. The first was by de Rosier and Klug [34] in which the structure of a biological macromolecule was determined whose helical symmetry allowed a full reconstruction to be made from a single projection (micrograph). The Fourier reconstruction methods used in this paper were akin to those developed for the determination of atomic structures by X-ray crystallography [35].

While symmetry was key to these results, it was suggested in the second of these papers, by Hoppe [36], that, given a sufficient number of projections, it should be possible to reconstruct fully asymmetrical systems, i.e with no symmetry imposed. The last of the three early papers, by Hart [37], demonstrated a method of improving the signal to noise ratio in images using an 'average' re-projected image calculated from a tilt series of micrographs, a technique known as a *polytropic montage*. Used initially as a means to combat the weak contrast in biological specimens, Hart acknowledged the 3D information generated by such an approach without extending this to the possibility of full 3D reconstruction. Shortly afterwards, a number of theoretical papers were published discussing the theoretical limits of Fourier techniques [38], approaches to real space reconstruction [39] and the use of iterative reconstruction routines [40, 41].

Until recently, the advance of electron tomography was impeded by a number of technical difficulties, in particular, the poor performance of goniometers (especially at high tilt), the length of time required to acquire a series of images and the lack of computer power for image processing and reconstruction. The improvement in electron microscope design coupled with the vast improvement in computer performance has overcome all these. However, the time taken to acquire a series is still a major problem in the life sciences as specimens damage rapidly in the electron beam [42]. To increase the longevity of samples, they are often examined at liquid helium temperatures using cryo-microscopy and at high voltages, both of which reduce the effects of inelastic scattering and subsequent damage [43]. For electron tomography, as opposed to single particle analysis, many specimens are still examined in resin or plastic sections and stained to enhance contrast [44] and such specimens are relatively robust in the beam.

2.2. The Radon Transform

Although the first practical formulation of tomography was achieved in 1956 [25], it was Radon who first outlined the mathematical principles underlying the technique in 1917 [13]. In his paper a transform, known now as the *Radon transform, R*, is defined as the mapping of a function $f(x, y)$, describing a real space object D, by the projection, or line integral, through f along all possible lines L with unit length ds so that,

$$Rf = \int_L f(x, y)\, ds \tag{1}$$

The geometry of the transform is illustrated in Figure 1. A discrete sampling of the Radon transform is geometrically equivalent to the sampling of an experimental object by a projection or some form of transmitted signal. As such, the structure of an object $f(x, y)$ can be reconstructed from projections Rf by using the inverse Radon transform. All reconstruction algorithms are approximations of this inverse transform.

The Radon transform operation converts real space data into 'Radon space' (l, θ), where l is the line perpendicular to the projection direction and θ is the angle of the projection. A point in real space $(x = r\cos\phi,\ y = r\sin\phi)$ is a line in Radon space (l, θ) in which $l = r\cos(\theta - \phi)$. A single projection of the object, a discrete sampling

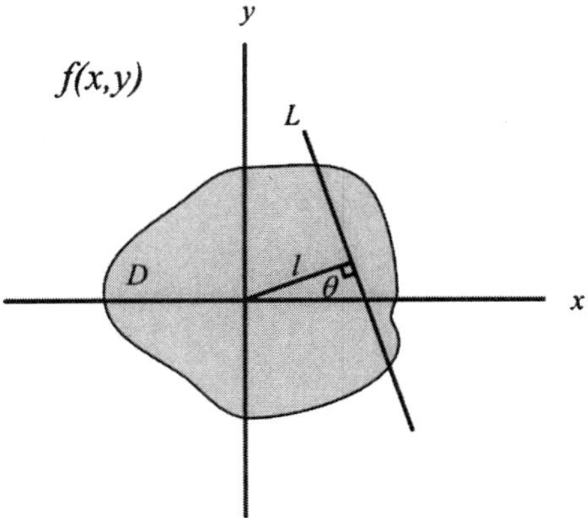

Figure 1. The Radon transform R can be visualised as the integration through a body D in real space $f(x, y)$ along all possible line integrals L, with its normal at an angle θ to the horizontal.

of the Radon transform, is a line at constant θ in Radon space. A series of projections at different angles will therefore sample Radon space and given a sufficient number of projections, an inverse Radon transform of this space should reconstruct the object. In practice the sampling of (l, θ) will be limited and any inversion will be imperfect. The goal of any reconstruction then becomes achieving the 'best' reconstruction of the object given the limited experimental data.

2.3. The Central Slice Theorem and Fourier Space Reconstruction

The relationship between real space and Radon space gives an understanding of the nature of a projection and its relationship with the original object. In addition, reconstruction from projections is aided by an understanding of the relationship between a projection in real space and Fourier space. The 'central slice theorem' or the 'projection-slice theorem' states that a projection of an object at a given angle in real space is a central section through the Fourier transform of that object. The relationship between the Fourier transform F and the Radon transform R, may be summarised in operator form as:

$$F_2 f = F_1 R f = F_1 \hat{f} \tag{2}$$

where $\hat{f}$ is the full Radon transform of object f. More detail regarding the nature of the transforms and their inter-relationship can be found in [33] and for brevity will not be explored further here.

For readers familiar with electron diffraction, the central slice theorem is of course exactly that known as the 'projection approximation' relating the intensity of Zero Order Laue Zone (ZOLZ) reflections to the crystal potential projected parallel to the zone axis.

The shape of most objects will be described only partially by the frequencies in one section but by taking multiple images (projections) at different angles many sections will be sampled in Fourier space. This will describe the Fourier transform of an object in many directions, increasing the information available in the 3D Fourier space of the object. In principle a sufficiently large number of projections taken over all angles will yield a complete description of the object.

Tomographic reconstruction is possible from an inverse Fourier transform of the superposition of a set of Fourier transformed projections: an approach known as *direct Fourier reconstruction* [39] used for the first tomographic reconstruction from electron micrographs [33]. Importantly, it provides a convenient and a logical basis to describe the effects of sampling deficiencies in the original dataset. If projections are missing from within an angular range, brought about by, for example, a limit on the maximum tilt angle, then Fourier space is under-sampled in those directions and as a consequence the back transform of the object will be degraded in the direction of this missing information.

The experimental data is always sampled at discrete angles leaving (often, regular) gaps in Fourier space. An inverse Fourier transform requires a continuous function and so radial interpolation is required to fill the gaps in Fourier space [37]; the quality of the reconstruction is greatly affected by the type of interpolation method used [45]. Although elegant, Fourier reconstruction methods have the disadvantage of being computationally intensive and difficult to implement for electron tomography. This is not the case for single particle analysis where Fourier methods are still the norm [46]. For electron tomography of unique structures, Fourier methods have been superseded by faster and easier to implement real space *backprojection* methods [47].

2.4. Real Space Reconstruction using Backprojection

The method of backprojection is based on simple reasoning: a point in space may be described uniquely by any three 'rays' passing through that point—the method of triangulation. With an increase in the object's complexity, more 'rays' are required to yield a unique description. Thus a projection of an object is an inverse of such a 'ray', and will describe some of the complexity of that object. Inverting the projection, smearing out the projection back into an object space at the angle of the original projection, generates a 'ray' that will describe uniquely an object in the projection direction: a method know as *backprojection*. Using a sufficient number of projections, from different angles, the superposition of all the backprojected 'rays' will return the original object: a reconstruction technique known as *direct backprojection*. [36, 38, 48], see Figure 2.

In principle, it is possible to reconstruct the object using backprojection in an way that is analogous to the experiment that generated the projection, i.e. by rotating the reconstruction space to the original projection angles and summing the projections along a constant reconstruction axis. However poor sampling when rotating the reconstruction will lead to artefacts. Instead the relationship between Radon space and real space, described earlier can be used to provide an algorithm that is less prone to

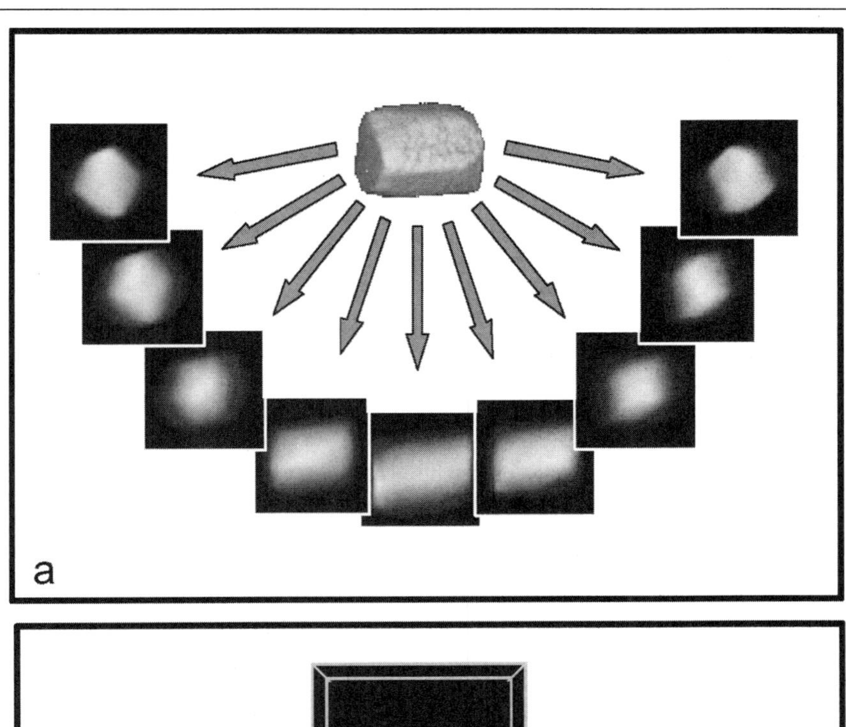

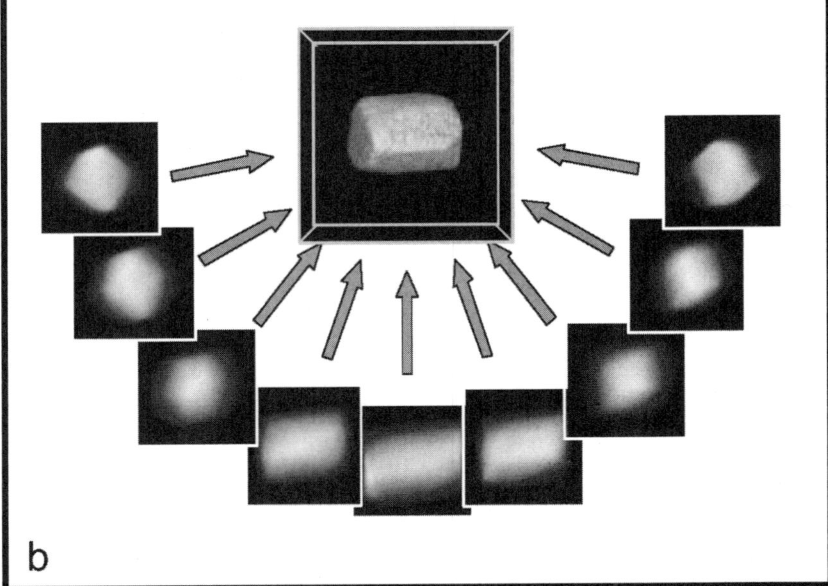

Figure 2. A schematic of tomographic reconstruction using the backprojection method. In (a) a series of images are recorded at successive tilts. These images are back-projected in (b) along their original tilt directions into a 3D object space. The overlap of all the back-projections will define the reconstructed object.

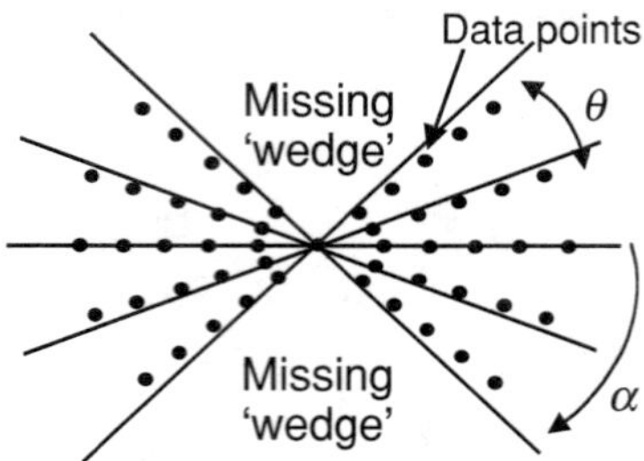

Figure 3. An illustration of the non-uniform sampling of Fourier space, brought about by the acquisition of a tilt series. The relatively small number of data points at high frequencies results in a blurred reconstruction. The angular increment between projections is θ and the maximum tilt angle, α.

error. Each projection is a sample of Radon space (l, θ) and a reconstruction should return an object in real space $f(r, \phi)$. The intensity of a real-space pixel p from a single projection, at angle θ, can be found by virtue of the relationship that such points exist within Radon space at the intersection of the line $l = r\cos(\theta - \phi)$ and θ. This can be represented by [49]:

$$[Rp](r, \phi) = \int_0^{\pi} p(r\cos(\theta - \phi), \theta)d\theta \tag{3}$$

which may be evaluated using a *Riemann sum* [49]. For all real datasets, values for every solution of $p(r\cos(\theta - \phi), \theta)$ do not exist because of the limited sampling of Radon space and an interpolation is required to determine the unknown values. The quality of the backprojection will be dependent on what form of interpolation is applied, although typically nearest neighbour or bilinear interpolation is used [50]. The detailed algorithms behind this method can be found in the books by Deans [33] and Herman [49].

Reconstructions by direct backprojection are always blurred with an enhancement of low frequencies and fine spatial detail reconstructed poorly. This is an effect of the uneven sampling of spatial frequencies in the ensemble of original projections. Described more easily in two dimensions, as illustrated in Figure 3, each of the acquired projections is a line intersecting the centre of Fourier space. Assuming a regular sampling of Fourier space in each projection this results in a proportionately greater sampling density near the centre of Fourier space compared with the periphery. This leads to an undersampling of the high spatial frequencies of the object and a 'blurred' reconstruction—see the comparison later in Figure 6(a).

Since the sampling is directly related to the position in Fourier space and the number of acquired projections, it is straightforward to correct in Fourier space using a weighting filter (a radially linear function in Fourier space, zero at the centre and a maximum at the edge). In order to avoid enhancing noise at high frequencies the filter is apodised using a Gaussian function or similar so that the Fourier transform has zero value at the Nyquist frequency [51]. This weighting filter has the effect of rebalancing the frequency distribution in Fourier space and minimising the blurring in real space; this improved reconstruction approach is known as *weighted backprojection* [52]. Weighted backprojection is now the most widely used reconstruction technique for electron tomography as it is simple to implement on large datasets and can be used for irregular sampling geometries [53].

Reconstructions using the backprojection method will always be 'imperfect' because of the limited sampling. In addition a poor reconstruction can be made worse if the number of acquired projections is small or the signal to noise ratio (SNR) is low in the original projections. However by noting that each projection is a 'perfect' reference the quality of the reconstruction can be improved. If the (imperfect) reconstruction is re- projected back along the original projection angles the re-projections, in general, will not be identical to the original projections (images). The difference between them will be characteristic of the deficiency of the reconstruction from the limited dataset. This difference can be backprojected into reconstruction space, generating a 'difference' reconstruction, which can then be used to modify the original reconstruction in order to correct the imperfections in the backprojection. This constrains the reconstruction to agree with the original projections. As the 'difference' is also being backprojected a single operation will not correct fully the reconstruction and the comparison operation must be repeated iteratively until a 'best' solution is reached [54, 55]. Such iterative methods were first developed for electron tomography in the 1970's and they have since been recognised to be part of a family of solutions by projection onto convex sets (POCS) [56], a more generalised form of the Gerchberg-Saxton algorithm [57, 58]. There have also been attempts to use maximum entropy techniques directly. These attempt to find the simplest (least complex) reconstruction taking into account the known projections, the noise in the data, the sampling artefacts in the reconstruction and the contrast limits of the original projection [58–62].

3. TOMOGRAPHY IN THE ELECTRON MICROSCOPE

3.1. Acquisition

The rest of the chapter will concentrate in particular on electron tomography, as defined previously, and in this section the problems that arise when acquiring a tomographic series of images in the TEM are discussed. For unique (non-repeating) structures, a series of images (projections) must be acquired at angular increments by tilting the specimen using the microscope goniometer. 'Single-axis tilting' is the technique normally chosen for electron tomography. The specimen is tilted about the eucentric axis of the specimen holder rod, from one extreme of the tilt range to the other. By recording images at each tilt, Fourier space is sampled in planes whose normals are perpendicular to the tilt axis.

With the capability in modern instruments of controlling the goniometer using a computer, it is now possible to fully automate the acquisition process [63]. The small movements in the position of the sample as it is tilted through the series can be minimised if the goniometer is pre-calibrated, that is the mechanical movements as a function of tilt are known and subsequently corrected. The reproducibility of the specimen position in modern goniometers is such that calibration need be done only infrequently [64]. To automate the acquisition the image must be re-focussed at each tilt, achieved through the analysis of the image as a function of defocus. Such auto-focus schemes are now well-established and relatively straightforward to implement. Further, in STEM mode it is also possible to implement a 'dynamic focus' correction, in which the probe focus is altered to account for the specimen geometry—particularly useful when the specimen is at high tilt and one part of the specimen is at a considerably different height to another [65].

3.2. Alignment

In the majority of electron tomography experiments in the life sciences, the alignment of BF images within a tilt series is made difficult by the lack of distinct contrast. Two practical methods can be used to help in this alignment: by tracking the movement of high contrast fiducial markers (typically gold particles, a few nm in diameter) [66] or by recording a STEM HAADF tilt series (which will be discussed in more detail later) [67]. If the first method is used, it means that selection of an area for tomographic reconstruction is limited only to those areas that have sufficient markers for alignment. This may be is acceptable for a specimen showing many structures dispersed on a carbon film but for specimens that have a small number of areas, perhaps only one site-specific area, suitable for analysis, such as will often be the case in nanotechnology applications, this may make alignment by fiducial markers difficult, if not impossible. An alternative is to rely on a cross-correlation alignment that, because of the change in the projection of the object through the tilt series, must be carried out image-by-image in a sequential fashion [68, 69]. It is important to correct for the tilt geometry [70], whereby each image is streched in a direction perpendicular to the tilt-axis by $1/\cos \psi$, where ψ is the angle of the projection relative to a reference zero tilt image, to improve the spatial relationship between successive projections. This action converts *orthogonal* projections, with the specimen rotated with a fixed source and detector, into *inclined* projections, which would exist if only the source was rotated.

The direction of the tilt axis for the object must also be identified accurately before any reconstruction is performed. For a single-axis tilt series, all objects through the series should follow a path that is perpendicular to that tilt axis. If an accurate spatial alignment has been achieved then a summation over all, or some, of the tilt series should highlight the movement of any objects through the series. The path of movement should be perpendicular to the tilt axis, as illustrated in Figure 4. Once the tilt axis has been determined, the entire dataset is rotated to place this axis parallel to a single image axis. The image stretch described before also has the effect of placing the tilt axis at the centre of the zero tilt image. Any misalignment of the tilt axis will 'spread' the signal from a reconstructed object and produce characteristic arcs of intensity, illustrated in Figure 5. The direction of the arc will depend on the direction of the misalignment away from the correct axis and the degree of 'spread' is dependent on the magnitude of that misalignment [71]. These distinctive distortions can provide a very sensitive method of refining the tilt axis.

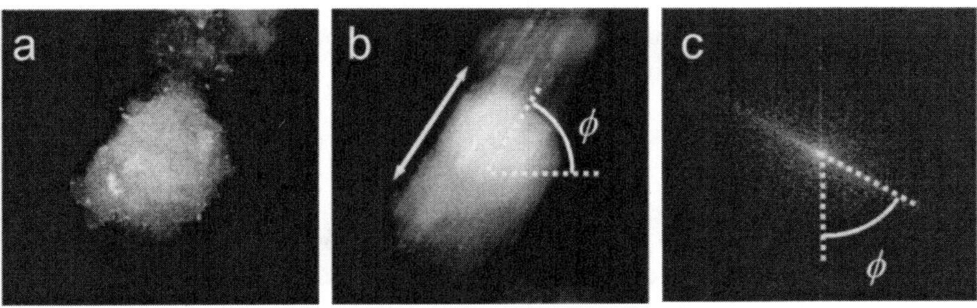

Figure 4. Tilt axis direction determination by series summation (a) A single STEM HAADF image, extracted from a tilt series, of a catalyst composed of palladium nanoparticles on a carbon matrix. (b) The summation of the entire (aligned) tilt series showing a distinct streaking in one direction at an angle ϕ to the horizontal. (c) The power spectrum allows an accurate assessment of the tilt axis.

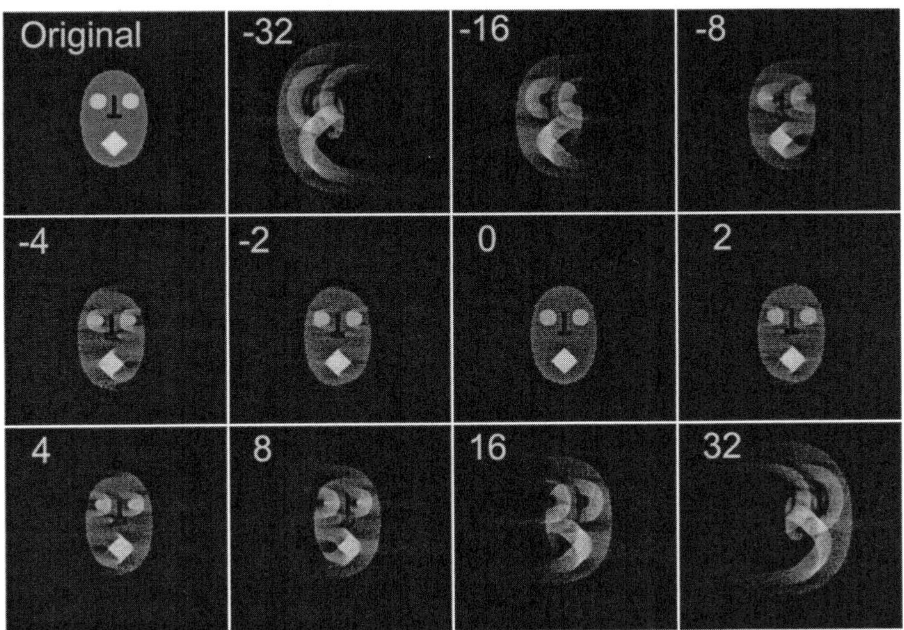

Figure 5. A demonstration of the effects of misalignment of the tilt axis on a reconstruction of a 'head' test object. The original object, from which the projections were generated, is shown top left. The number indicates the pixel misalignment, perpendicular to the tilt axis. The 'head' is 64 pixels wide.

3.3. Anisotropic Resolution

The sampling of the object controls the resolution of the tomographic reconstruction. For the single-axis tilt geometry, the resolution parallel to the tilt axis, say the x-axis, d_x, is equal to the original resolution of the projections, assuming a perfect tilt series alignment, see later. The resolution in the other perpendicular directions is controlled by the number of projections acquired, N, and the diameter, D, of the volume to be reconstructed. This is seen most easily in Fourier space [38] and is:

$$d_y = d_z = \frac{\pi D}{N} \tag{4}$$

However, this expression assumes that the N projections cover the whole angular range (i.e. ±90°). In practice the limited space between the objective lens pole pieces and the finite thickness of the specimen holder limits the tilt range, giving rise to the 'missing wedge' of information, see Figure 2. This missing information leads to the resolution in the direction parallel to the optic axis, d_z, being degraded further by an 'elongation factor' e_{yz} so that

$$d_z = d_y e_{yz} \tag{5}$$

which is related to the maximum tilt angle, α by [72]:

$$e_{yz} = \sqrt{\frac{\alpha + \sin\alpha\cos\alpha}{\alpha - \sin\alpha\cos\alpha}} \tag{6}$$

Thus in order to provide the maximum 3D information, as many projections as possible should be acquired over as wide a tilt range as possible. Figure 6 illustrates this pictorially. As an example, the polepiece gap of the FEI Super TWIN objective lens is 5.2 mm. A standard FEI single tilt holder allows a maximum tilt angle of 42°, leading to an elongation factor, e_{yz}, of 2.29 and significant blurring of the reconstruction in the z-direction (parallel to the optic axis). To improve this, slimmer, narrower holders were constructed, firstly in-house [73] and more recently by commercial manufacturers [74]. These holders can now tilt to ±70° (an elongation factor of only 1.3) or even higher without undue shadowing or problems with the polepiece gap.

Alternatively, a 'conical tilting' approach can be used, made possible by either a second perpendicular tilt axis (double-tilt electron tomography) or a tilt-rotate holder, in which the cone angle is fixed and projections are acquired throughout a full precession of the specimen [75]. With this acquisition geometry the missing volume is a cone or pyramid, rather than a wedge, and the total volume of unsampled space is lower given the same maximum tilt angle. In that regard the conical or double-tilt approach offers a higher quality reconstruction but is technically far more demanding. Nevertheless, double-axis tomography is gradually becoming more popular as better quality reconstructions are demanded.

Whilst the Crowther criterion (equation 4) is a useful guide for the expected resolution in a reconstruction it ceases to become valid for constrained reconstruction techniques [76, 77], such as the iterative POCS-based methods and maximum-entropy methods (COMET). The reconstruction resolution of such methods is dependent on

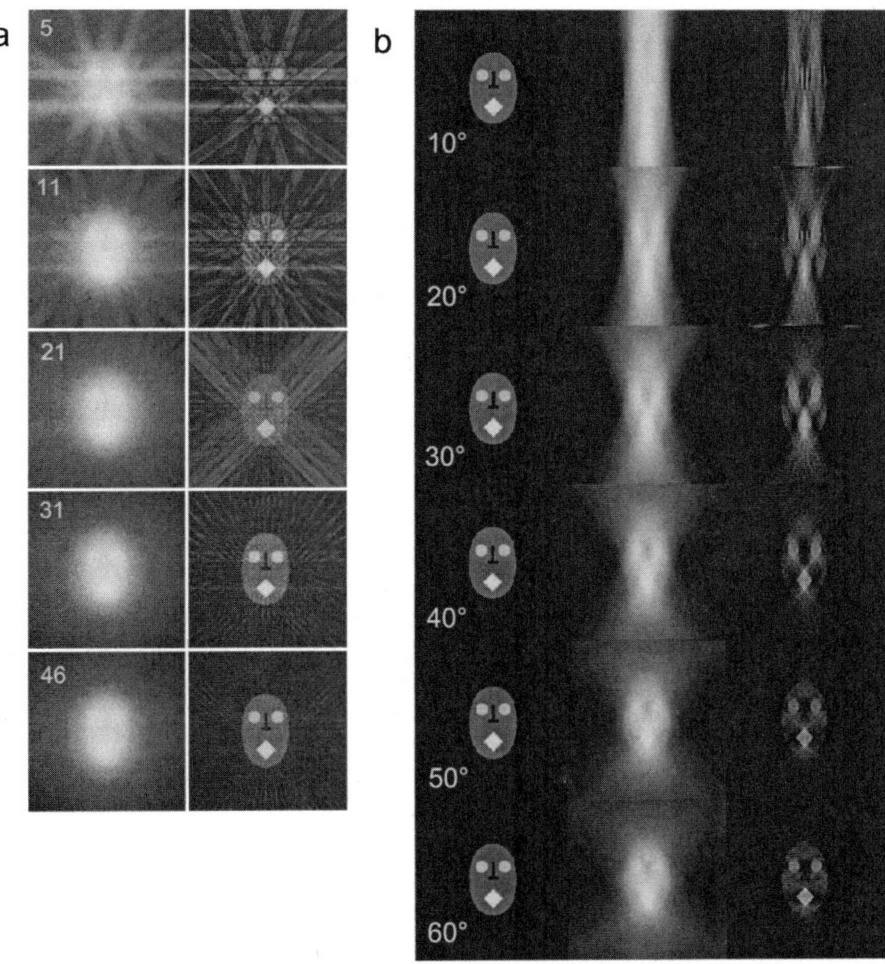

Figure 6. (a) The two columns show the result of adding successively more projections to a tilt series for reconstruction using direct backprojection in the left hand column and weighted backprojection in the right hand column. The numbers refer to the number of projections over ±90°. (b) A montage of simulations showing the original object in the left-hand column, the direct backprojection reconstruction in the middle and the weighted backprojection reconstruction in the right-hand column. The reduction in the elongation by increasing the tilt range from ±10° to ±60°, and the improvement in resolution through weighted backprojection, is quite apparent. The tilt axis is perpendicular to the page.

the noise characteristics of the original data, the shape of the object to be reconstructed and the nature of the constraints applied. The reconstruction resolution has been the subject of much debate in the literature but a recent paper [78] has elaborated on a new way to define resolution for 3D reconstructions using a spectral signal to noise ratio method.

3.4. The Projection Requirement

Any signal used for a tomographic reconstruction must meet several assumptions of which, as stated by Peter Hawkes, 'the most crucial is the belief that what is detected is some kind of projection through the structure. This 'Projection' need not be a sum or integral through the structure of some physical property of the latter; in principle a monotonically varying function would be acceptable' [79]. This is known as the *projection requirement.* Until very recently, all published electron tomography results were derived from a tilt-series of bright field (BF) TEM images, the contrast in which arises due to a combination of low angle elastic and inelastic scattering. BF tomography is based on the assumption that for sufficiently thin, weakly scattering, large unit cell crystalline or amorphous objects contrast relating to the thickness and atomic number ('mass-thickness contrast') of the specimen dominates [79]. In structural biology, BF TEM images satisfy this criterion to a very good approximation, be they unstained cryo-specimens, embedded in ice, or stained plastic or resin sections. Of course, in principle the contrast transfer function should be taken into account but for most electron tomography experiments to date the resolution achievable has not required this correction.

However, in general, for most (crystalline) specimens in the physical sciences and certainly for most specimens of nanotechnological importance, BF contrast will depend strongly upon the diffraction condition of the crystal and this will not have a monotonic relationship with the amount of material though which the beam passes. BF images of such systems cannot be used for tomography because they are not strictly projections [79]. Further, even if the specimen is amorphous or weakly diffracting, the 3D nature of the specimen coupled with the short depth of focus in the TEM ensures that Fresnel contrast will be very apparent (especially if using a FEG-based instrument) and this again cannot satisfy the projection requirement. This is true, in principle, even for specimens in the life sciences.

It is only in the last few years [80], that electron tomography has begun to be applied to nanoscale systems in the physical sciences. Although BF imaging may not be suitable in general, there are many other signals that do satisfy the projection requirement. To overcome the problem of Fresnel contrast and diffraction effects, the signals acquired must be predominantly incoherent in nature. Both STEM HAADF (Z-contrast) imaging [81] and energy-filtered TEM (EFTEM) [82] can be seen as a good basis for electron tomography in the physical sciences. Both imaging techniques are, or can be made to be, incoherent and both are chemically sensitive enabling the 3D structure and composition to be mapped simultaneously at high spatial resolution. Very recently, STEM tomography has also been recognised within the structural biology community as an ideal means of imaging 1nm gold clusters within sections of biological material [66].

With these new imaging techniques available to the electron tomographer, it is now possible to produce high spatial resolution reconstructions of nanoscale objects relevant to nanotechnology research. A variety of examples using both STEM and EFTEM tomography are shown in the next two sections and reveal how, with care, the 3D reconstruction of many structures and devices can be achieved with nanometre resolution.

4. STEM HAADF (Z-CONTRAST) TOMOGRAPHY

The low angle scattering of the electron beam is predominantly coherent in nature and as such conventional BF and DF images are prone to contrast reversals with changes in specimen thickness, orientation or defocus. On the other hand high angle scattering is predominantly incoherent, and STEM images formed using a high-angle annular dark field (HAADF) detector do not show the contrast changes associated with coherent scattering [83]. Within a classical description, such high angle scattering is associated with the interaction of the electron beam close to the nucleus of the atom and thus the cross-section for HAADF scattering approaches that for unscreened Rutherford scattering so that it is strongly dependent on the atomic number Z; in fact in the unscreened limit it is proportional to Z^2. In practice this limit is never reached and the exact dependence, particularly for crystalline specimens, is a function of many other factors, which need to be determined before any possible quantification can take place [84]. The choice of the inner angle for the HAADF detector, θ_{HAADF}, is important and must be large enough to ensure coherent effects are minimal. A guide can be obtained from $\theta_{\mathrm{HAADF}} \geq \lambda/d_{\mathrm{thermal}}$ [85] where λ is the electron wavelength and d_{thermal} is the amplitude of atomic thermal vibration. For Si at 200 kV, $\theta_{\mathrm{HAADF}} > 40$ mrad. For more information about STEM imaging and its uses in 2D nanotechnology, see the chapter by Cowley [86].

Medium-resolution (~1 nm) STEM images, formed with a HAADF detector, are very sensitive to changes in specimen composition with the intensity varying (for the most part) monotonically with composition and specimen thickness, thus satisfying the projection requirement. Although atomic resolution HAADF images depend on the excitation of Bloch states and channelling [87], in principle even 3D atomic resolution is possible given a sufficiently thin specimen and a STEM with a high resolution (perhaps aberration-corrected) probe-forming lens. Such channelling effects are also present in medium resolution STEM images and when a crystalline specimen is at or near a major zone axis there is an increase in the STEM HAADF signal that depends on the localisation of the beam onto atomic strings. The string strength [88] dictates the level of intensity enhancement. However, in general, strong channelling will occur very infrequently during a tilt series and will have little effect on the overall intensity distribution in the reconstruction.

As with all techniques used for imaging 3D objects, attention must be paid to the depth of focus. This can be maximised by using a small condenser aperture (objective aperture on a dedicated STEM) to minimise the convergence angle but ultimately the diffraction limit will dominate and a residual blurring is inevitable. This will place a lower limit on the possible resolution achievable in all STEM tomography. In practice images are re-focussed after every tilt, either manually or using computer control, to ensure optimum focus over the majority of the image. Recent results have shown how for a medium resolution STEM probe, say ~1 nm in diameter, the probe diameter,

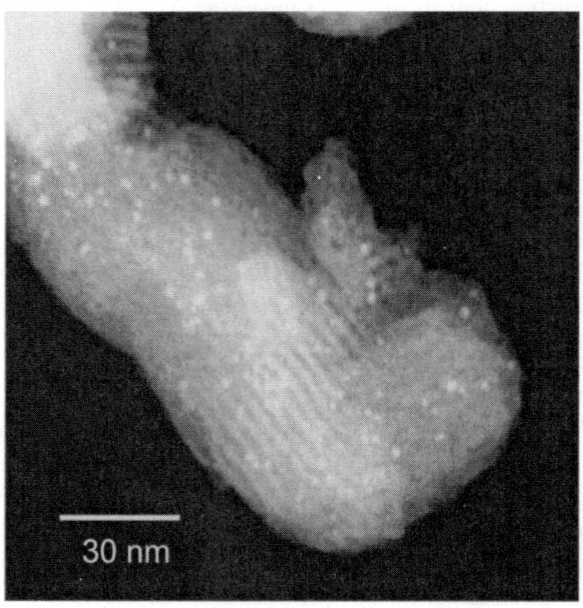

Figure 7. A typical STEM HAADF image of a heterogeneous catalyst composed of Pd_6Ru_6 nanoparticles (approximately 1 nm diameter) and an MCM-41 mesoporous silica support with mesopores of approximately 3 nm diameter.

or at least its central maximum, does not broaden significantly over a remarkably large range of defocus, particularly in an underfocus condition [89]. For STEM imaging, the tails of an underfocussed probe are not too important and in essence simply add an unwanted (and easily removed) background to the image. (This is of course not true for microanalysis where the tails of the probe can account for over half the emitted X-ray signal, for example.) The STEM image contrast can be described by a convolution of the central maximum of the probe with the object function and thus may account for the surprisingly good resolution of thick specimens in STEM mode compared to the resolution seen in an image acquired using conventional BF TEM [90].

The first example of STEM tomography is used to illustrate the resolution achievable with this technique and the ability to analyse the 3D data set in a statistical and quantitative fashion. In particular, it illustrates how the 3D distribution of nanometre-sized particles can be determined in a porous support. The specimen is a heterogeneous catalyst composed of bimetallic particles (each with a diameter of about 1 nm) within a mesoporous silica support (MCM-41) whose mesopores are hexagonal in cross-section with a diameter of about 3 nm [91]. Knowledge of the three-dimensional distribution of the metal nanoparticles, and their location at or close to the walls of the internal pores, is key to understanding the factors that govern the activity and selectivity of the nanocatalysts and their change during reaction as a possible result of sintering and coalescence [92]. Figure 7 shows a typical STEM HAADF image of one of these catalysts, recorded with a detector inner radius of 40 mrad. With this set up, the image formed will be almost totally incoherent in nature. The nanoparticles, in this case Pt_6Ru_6, stand out very well against the light SiO_2 background and some appear to lie within the mesopores. However, to ensure that this is the case it is necessary to determine a 3D reconstruction of this and similar catalysts.

As an example, consider the series of 71 STEM HAADF images of the $Pt_{10}Ru_2$-MCM 41 catalyst [93] taken at 200 kV at tilts ranging from $+70°$ to $-70°$ in $2°$ steps. This catalyst has proven to be remarkably successful in hydrogenating trans, trans-muconic acid to adipic acid, the former derived from glucose, the latter used to make nylon—a case of sugar being turned into plastic! The image series was spatially aligned using the modified cross-correlation algorithm [70]. Both a weighted back prorojection and iterative reconstruction were used with the iterative technique improving the fidelity of the reconstruction. All routines for the alignment and reconstruction were written in the *IDL* programming language [94].

Figure 8(a) shows a perspective view of the reconstruction and 8(b) and (c) show two images, displayed as multi-level voxel projections of the boxed area of the reconstructed catalyst, viewed in perpendicular directions, parallel and perpendicular to the MCM-41 pore structure. What is clear from the reconstruction is that the nanoparticles are very well resolved, in all directions, within the silica framework structure. Further the resolution is not degraded significantly in either direction. It also appears that the excellent activity of this catalyst is in part due to the relatively high filling quotient; there are a large number of particles in the pores with few if any in this view aggregated outside the pores. It is possible by sampling the 3D structure to calculate the number of particles in the volume, the internal surface area of the silica and thus the weight of active particle per unit area of support, about 20 $\mu g \cdot m^{-2}$, about 3% of the initial loading. It is also possible to measure the occurrence of particles in each pore and whether the distribution is random or not. Analysis of the reconstruction reveals the mesoporous structure of the silica has been reconstructed faithfully with little sign of beam damage despite the long acquisition time (~3 hours) needed for the series. Such silica frameworks damage rapidly when examined at low voltage and/or in fixed beam (TEM) mode [95].

A further example of how STEM HAADF tomography can be used is in the determination of the external 'shape' of a nanoscale object. To demonstrate this we focus on the magnetite (Fe_3O_4) nanocrystals found in the 'backbone' of magneto-tactic bacteria. Such organisms use this 'backbone' of magnetite crystals, which are ferromagnetically aligned, to sense the earth's geomagnetic field and thus aid navigation and feeding [96]. Recently, they have become of great interest as similar magnetite chains have been observed on the surface of martian meteorites [97–100]. To determine whether the similarity is more than superficial, enormous efforts are being made world-wide in order to characterize these crystals in particular the crystal habit and any variation in the composition within the crystal. As such, 3D analysis is vital.

Figure 9 shows a phase image reconstructed from an electron hologram of such a bacterium that illustrates quite convincingly the ferromagnetic alignment. The bottom left inset is a BF image of a typical bacterium that highlights the backbone of crystals

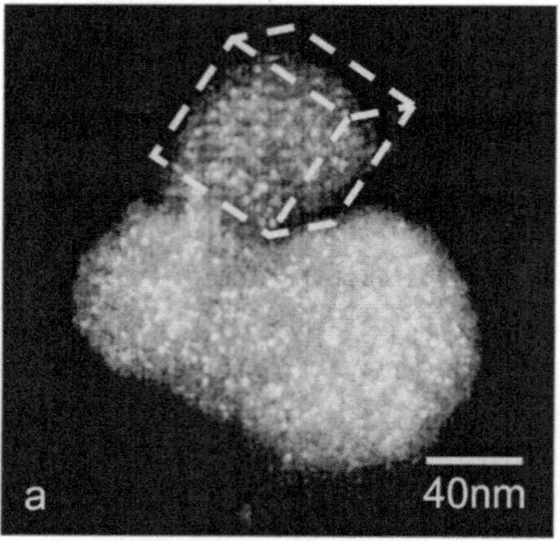

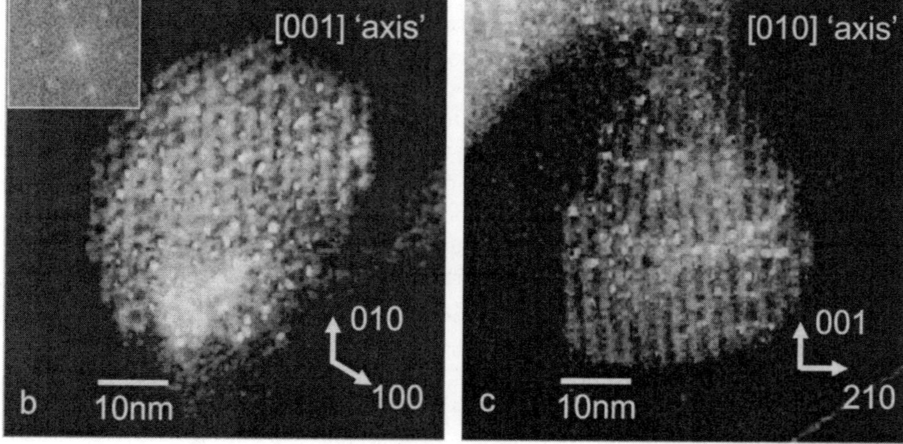

Figure 8. (a) A perspective view (voxel projection) of a reconstruction of a heterogeneous catalyst composed of $Pt_{10}Ru_2$ active nanoparticles supported within an MCM-41 framework. The reconstruction was undertaken on a series of STEM HAADF images acquired every 2° between +/−70°. (c) and (d) two perpendicular voxel projections of the reconstruction volume boxed in (b). In (c) the hexagonal order of the mesoporous silica is evident (inset shows power spectrum) and in (d) it is possible to see how the pores are filled with the nanoparticles.

surrounded by the cellular 'envelope'. Figure 10(a) shows a tomographic reconstruction (surface render) from a series of STEM HAADF images recorded between +/−76° with a 2° interval. Both the organic envelope and the backbone have been shown in the upper figure, the backbone alone in the lower figure. Note the helical arrangement of the crystals, known to exist in these systems. One of the nanocrystals has been boxed in the lower figure and two slices perpendicular to the main axis of the crystal are shown in Fig 10(b), one from near the end of the crystal, the other from near the centre.

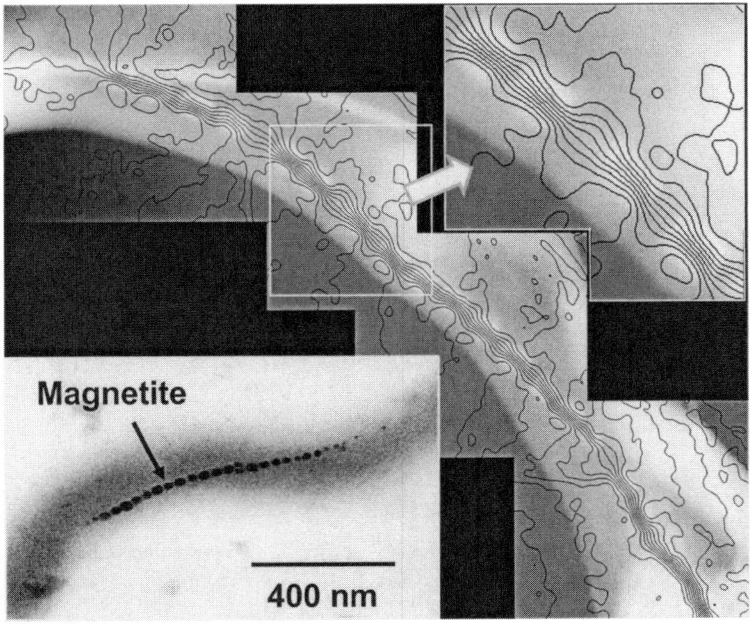

Figure 9. A reconstructed phase image from an electron hologram of a magneto-tactic bacterium showing the magnetic field lines of the ferromagnetic chain of magnetite crystals. The inset shows a BF image of the bacterium revealing the backbone of crystals within the organic membrane (figures courtesy of R. E. Dunin-Borkowski).

The cubic nature of the magnetite phase allows the facets revealed by tomography to be indexed unambiguously as shown. Further the near perfect hexagonal symmetry of the central slice is revealed with great clarity by the reconstruction, showing the 6 symmetrically equivalent $\{110\}$ facets.

Although common, this 3D morphology (or habit) is not unique to these systems and occasionally 'trigonal' prismatic crystals are seen, an example of which is shown reconstructed in Figure 11(a). In this crystal two $\{111\}$ facets are dominant. This crystal is seen, arrowed, in Figure 11(e), part of a series of images, seen in Figures 11(b)–(e) used to illustrate the effects of channeling that exist in STEM HAADF imaging of crystals. The figure is a montage of four STEM HAADF images recorded at different tilts. As the bacterium is tilted each nanocrystal will, in general, be at a different orientation to the incoming electron beam. If the crystal planes are at, or close to, a strong diffraction condition (for example close to a low order zone axis) then the strength of the scattering to high angles as recorded in the HAADF image, will increase because the incoming beam will be localised on atom strings and propagate through the crystal as Bloch states. This can be seen in the montage of figures as a sudden increase in the HAADF signal for certain crystals at certain orientations, for

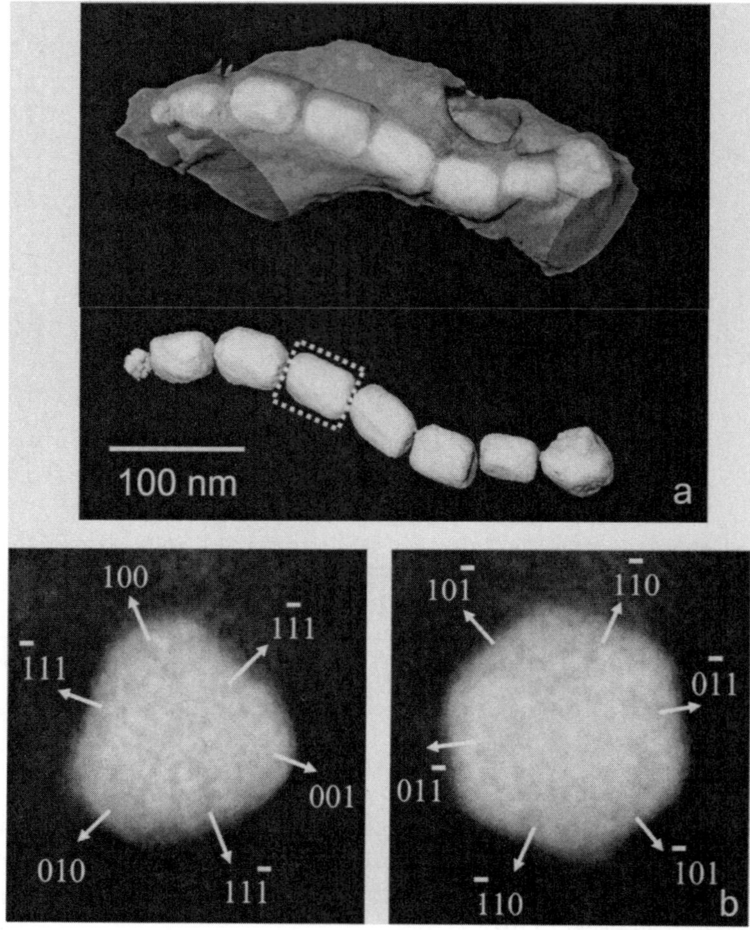

Figure 10. (a) A tomographic reconstruction of a magneto-tactic bacterium showing very clearly the backbone of magnetite crystals surrounded by the bacterium's organic 'envelope'. (b) Two slices from the crystal boxed in (a), the left hand slice is taken from the end of the crystal, the right hand crystal from the middle. Note the excellent fidelity of the reconstruction and the perfection of the crystal facetting.

example, crystal X in (b), crystal Y in (c), crystal Z in (d) and a small crystal behind that labelled Y in (e). As this dynamical enhancement occurs only once or twice in a series of perhaps 140 images it makes little difference to the final reconstruction, particularly if the exterior shape is all that is required. The prismatic crystal, shown in Figure 11(a), is seen very clearly in (e). Animations of these reconstructions can be viewed on our web site [101].

For all STEM tomography reconstructions, it is difficult to determine an overall 3-D resolution but as a rule of thumb, for large objects, $D > 100$ nm, we have found by experience that the 3-D resolution is approximately $D/100$ [73].

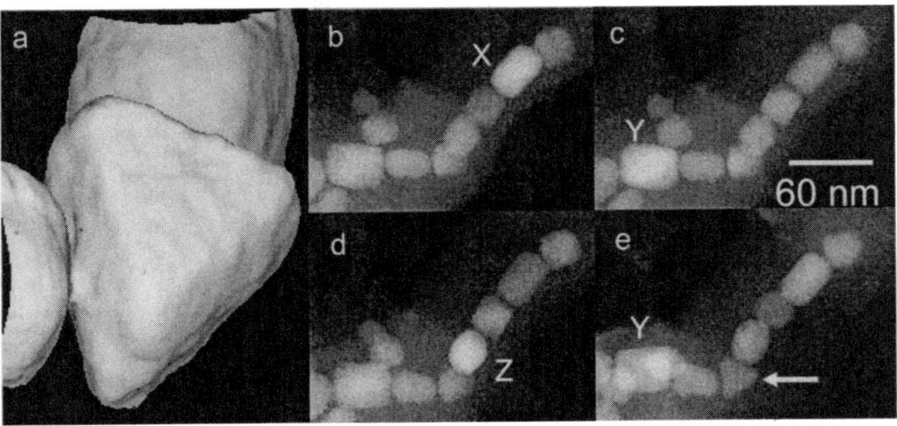

Figure 11. (a) A reconstruction of a 'prismatic' crystal. (b)–(e) HAADF images taken from a tilt series used to generate (a). Note the changes in the HAADF intensity in some crystals, labelled X, Y and Z that arise when the beam direction is close to a major zone axis. The triangular crystal arrowed in (e) is shown as a 3D reconstruction in (a).

5. EFTEM TOMOGRAPHY

With the advent of both post-column and in-column energy filters [102], energy filtered transmission electron microscopy (EFTEM) has become a routine analytical tool that allows rapid quantitative mapping of elemental species over wide fields of view with a spatial resolution of ~ 1 nm. [103–105]. If an energy slit (window) is used that allows only the zero-loss part of the spectrum to be transmitted then images can be formed using only (predominantly) elastically-scattered electrons (typically +/−5 eV). By removing electrons that have undergone inelastic scattering of greater than about 5 eV, the contrast of BF images is improved enormously, particularly for thick specimens as used often in structural biology [106], for 2D or 3D imaging. Chemical analysis by core loss imaging (using energy losses characteristic of a particular atomic species) is rarely used in biology because of the high electron doses necessarily involved [107]. However most physical sciences specimens are several orders of magnitude more beam stable than their biosciences counterparts and therefore by using a tilt series of core-loss images it should be possible to reconstruct a three-dimensional elemental distribution map [82, 108].

The intensity observed in an image formed using an energy loss window is a complex combination of inelastic scattering (through changes in composition and electronic structure) and elastic effects (via crystal thickness and orientation). The true compositional information encoded in a energy loss image may be isolated by generating either a background subtracted elemental map (from three or more images) or a jump-ratio map (from two images). Both maps will show intensity that is related to the amount of an atomic species at a given pixel. However, elemental maps often show residual diffraction contrast that will, in general, mean they do not conform to the projection requirement, in the same way as it would for a conventional BF image. Diffraction contrast can be removed partially by dividing the map by a zero-loss image, but this can also introduce artefacts associated with changes in the diffraction contrast itself as a function of energy loss: the diffraction contrast in a zero loss image is considerably sharper than that of a core-loss image. However, jump-ratio images are a convenient and simple way of removing residual diffraction contrast. They can show higher sensitivity than an elemental map but of course the intensity values of a jump-ratio map cannot be related in an absolute (quantitative) way to the composition. The jump ratio signal changes monotonically with thickness up to approximately the overall inelastic mean free path, λ [107]. Beyond this value the jump-ratio actually falls as the specimen thickness increases as the true elemental signal increases more slowly than the underlying background. Thus the jump-ratio signal can only be used for tomography so long as the specimen thickness (in projection) is less than one inelastic mean free path, typically 100 nm at 200 kV. This places some constraints on the sample preparation or the types of specimens examined in this fashion.

To illustrate the advantages of EFTEM tomography, consider a typical problem seen in the metallurgical field namely the precipitation of chromium carbides at a grain boundary in stainless steel. Although larger in scale than cases considered previously, it illustrates here the advantage of EFTEM over STEM tomography in that the atomic number contrast between the precipitates and the matrix is small. It should be possible to analyse the shape of the carbides from a tomographic reconstruction of a tilt series of chromium jump-ratio images. A series of EFTEM images were acquired using a Philips CM300, with a Gatan Imaging Filter (GIF) fitted with a $2k \times 2k$ CCD camera. The dataset was acquired at 24 tilt increments over a tilt range of $\pm 58^\circ$, an increment of just under 5°. At each tilt three energy loss images, each with a 10 eV window, were acquired over the chromium L_{23} edge (onset at 575 eV), two pre-edge at 545 eV and 565 eV and one post-edge at 580 eV. Jump-ratio and elemental maps were determined at each tilt but the latter showed considerable diffraction contrast and therefore only the jump-ratio signal was used for reconstruction.

Each group of energy loss images were corrected for any shift relative to the first pre-edge image of each group using a cross-correlation routine. Spatial and rotational alignment through the tilt series was corrected by sequential cross-correlation and series averaging. Towards the extreme ends of the tilt series there was a loss of contrast in the jump-ratio images, which was perhaps because the amount of material through which the beam passes may have increased beyond the upper thickness limit of one mean free path. The tomographic reconstruction was carried out using weighted backprojection and is shown as three perpendicular voxel projections in Figure 12. These projections clearly show that chromium carbides have complex 3D shapes and orientations, the nature of which could only be surmised from a single EFTEM elemental distribution image. For example, it becomes clear from (b) that the upper boundary between the precipitates and the matrix is (at least partially) coherent, the lower boundary predominantly incoherent in nature. An animation of this reconstruction is shown on our web site [101].

We return to the magnetotactic bacteria crystals for a second example to show the *polytropic montage* described earlier. In this case, reconstruction of the magnetite

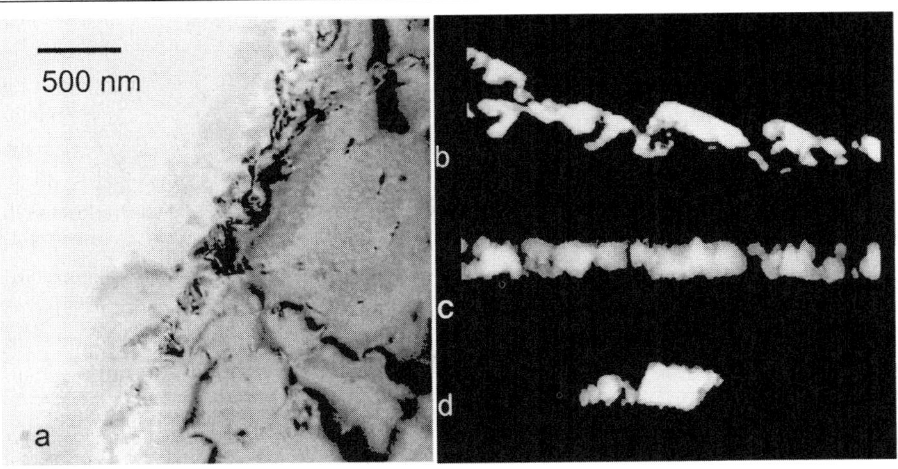

Figure 12. (a) BF zero loss image of a grain boundary in stainless steel which shows carbide precipitation at the boundary. (b)-(d) Voxel projections of a tomographic reconstruction using Cr jump-ratio images of the grain boundary carbide structure The carbides are viewed in three perpendicular directions emphasising the morphology of the precipitates. The reconstruction has been smoothed with a 2×2×2 pixel Gaussian filter prior to visualisation in order to reduce the effects of noise. In addition the voxel projections have been contrast selected to show only the chromium carbides.

'backbone' of crystals was carried out using an iterative algorithm for both the oxygen and iron datasets to help improve the reconstruction which would otherwise suffer from the low signal to noise ratio (SNR) in some of the original projections, especially in the oxygen tilt series. For more details of this reconstruction see [108]. Re-projections of both the iron and oxygen reconstructions in the zero degree direction show much higher SNR than the original projections; this increase in SNR is the *polytropic montage* effect. This is most clearly seen in the case of the oxygen data, shown in Figure 13 in which the projected reconstruction is compared with the original oxygen jump-ratio image at zero degree tilt. The improvement is remarkable.

6. CONCLUSIONS

It is evident that electron tomography offers a means to determine the three dimensional structure and composition of many different mateials at the nanometre level. In general, tomography using BF TEM for materials science applications will not yield true reconstructions because of the coherent nature of the scattering process seen in such images. BF images contain contrast that does not satisfy the projection requirement for tomography. Incoherent signals, such as those used to form STEM HAADF images or core-loss EFTEM images do satisfy the projection requirement, at least within certain limits. Further, by using these imaging techniques, it is possible to simultaneously record three-dimensional compositional information, either indirectly through the atomic number dependence of HAADF imaging or directly, by choosing a window that corresponds to a energy loss (electronic transition) within a

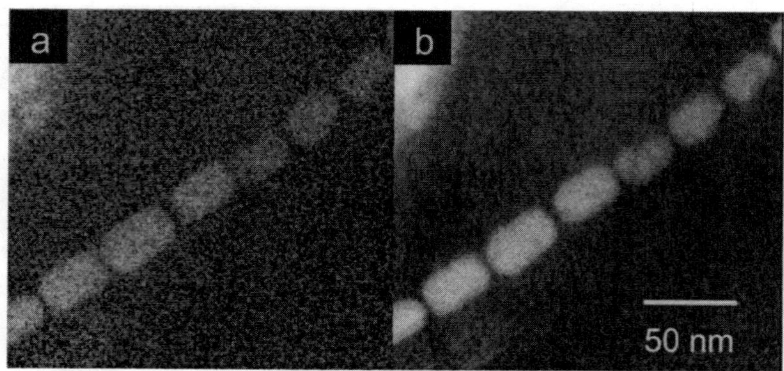

Figure 13. A comparison between (a) an original zero tilt oxygen jump-ratio image taken from a magnetite chain in a magnetotactic bacterium and (b) the zero tilt projection of the tomographic reconstruction. Note the dramatic improvement in the signal to noise ratio in the reconstruction due to the *polytropic montage* effect.

particular atomic species. This one-to-one correspondence of structure and composition in three dimensions should give the physical scientist a very powerful method to analyse nanoscale structures and devices in the future.

ACKNOWLEDGEMENTS

The author would like to acknowledge the invaluable contribution of Dr Matthew Weyland to this work. He would also like to thank Prof. Sir John Meurig Thomas, Prof Peter Buseck and Drs Chris Boothroyd, Rafal Dunin-Borkowski and Ron Broom for their help and interest. He thanks the EPSRC, FEI, Isaac Newton Trust and the Royal Commission for the Exhibition of 1851 for their financial support over a number of years.

REFERENCES

[1] K. E. Drexler, *Nanosystems: molecular machinery, manufacturing, and computation*, Wiley Interscience (1992).
[2] J. C. H. Spence, *High Resolution Electron Microscopy*, Oxford University Press (2003).
[3] A. Crewe, *Rep. Prog Phys*. 43 (1980) 621.
[4] D. B. Williams and C. B. Carter, *Transmission Electron Microscopy: a Handbook for Materials Scientists*, Plenum Press (1996).
[5] B. Hudson *J. Microscopy* 98 (1972) 396.
[6] N. Unwin and R. Henderson *J Molec Biol* 94 (1975) 425.
[7] B. Bottcher, S. A. Wynne and R. A. Crowther, *Nature* 368 (1997) 88.
[8] I. S. Gabashvili, R. K. Agrawal, C. M. T. Spahn, R. A. Grassucci, D. I. Svergun, J. Frank and P. Penczek, *Cell* 100 (2000) 537.
[9] B. J. Marsh, D. N. Mastronarde, K. F. Buttle, K. E. Howell and J. R. McIntosh, *Proc. Nat. Acad. Sci., USA*., 98 (2001) 2399.
[10] W. Kuhlbrandt *Q. Rev Biophysics* 25 (1992) 1.
[11] R. Matadeen, A. Patwardhan, B. Gowen, E.V. Orlova, T. Pape, M. Cuff, F. Mueller, R. Brimacombe and M. van Heel, *Struct Fold Design* 7 (1999) 1575.

[12] W. Baumeister, R. Grimm and J. Walz, *Trends Cell Biol* 9 (1999) 81.
[13] J. Radon, *Ber. Verh. K. Sachs. Ges. Wiss. Leipzig, Math.-Phys. Kl.*, 69 (1917) 262.
[14] H. Cramer and H. Wold, *J. London Math. Soc.* 11 (1936) 290.
[15] J. Banhart, *Progr. Mat. Sci.* 46 (2001) 559.
[16] T. Hirano, K. Usami, Y. Tanaka, C. Masuda, *J. Mater. Res.* 10 (1995) 381.
[17] Weierstall, U., Chen, Q., Spence, J. C. H., Howells, M. R., Isaacson, M. & Panepucci, R. R. *Ultramicroscopy* 90, 171–195 (2002).
[18] M. K. Miller, *Atom-Probe Tomography: Analysis at the Atomic Level*, Kluwer Academic/Plenum Press, New York (2000).
[20] J. Mardinly *Microsc Microanal* 7 (*suppl* 2) (2001) 510.
[21] R. E. Dunin-Borkowski, M. R. McCartney, B. Kardynal, S. S. P. Parkin, M. R. Scheinfein and David J. Smith *J. Microsc.* 200 (2000), 187.
[22] V. Stolojan, R. E. Dunin-Borkowski, M. Weyland, and P. A. Midgley, *Inst. Phys. Conf. Series.* 168 (2001) 243.
[23] J. Joardar, S. W. Kim and S. Kang *Materials and Manufacturing Process* 17 (2002) 567.
[24] J. M. Thomas, in: *Inorganic Chemistry: Towards the 21st century* (ed. M. H. Chisholm), A. C. S. Publication 211 (1983) 445.
[25] R. N. Bracewell, *Aust. J. Phys.* 9 (1956) 297.
[26] A. M. Cormack J. Appl. Phys. 34 (1963) 2722.
[27] G. N. Hounsfield, *A method and apparatus for examination of a body by radiation such as X or gamma radiation*, The Patent Office, London: England (1972).
[28] G. L. Brownell, C. A. Burnham, B. Hoop, D. E. Bohning, *Proceedings of the Symposium on Dynamic Studies with Radioisotopes in Medicine, Rotterdam*, 190, IAEA, Vienna (1971).
[29] K. Baba, K. Satoh, S. Sakamoto, T. Okai and S. Ishii, *J. Perinat Med.* 17 (1989) 19.
[30] D. I. Hoult *J. Magn. Reson.* 33 (1979) 183.
[31] D. Zhao and J. R. Kayal *Current Science* 79 (2000) 1208.
[32] P. Reimers, A. Kettschau and J. Goebbels. *NDT International*, 23 (1990) 255.
[33] S. R. Deans, *The Radon transform and some of its applications*, Wiley, New York, Chichester (1983).
[34] D. J. de Rosier, A. Klug, *Nature* 217 (1968) 130.
[35] W. Hoppe and R. Hegerl in: *Computer Processing of Electron Microscope Images* (Ed. P.W. Hawkes), Springer-Verlag, Berlin, Heidelberg, New York (1980).
[36] W. Hoppe, R. Langer, G. Knesch C. and Poppe, *Naturwissenschaften* 55 (1968) 333.
[37] R. G. Hart, *Science* 159 (1968) 1464.
[38] R. A., Crowther, D. J. de Rosier and A. Klug, *Proc. Roy. Soc. Lond. A.* 317 (1970) 319.
[39] G. N. Ramachandran and A.V. Lakshminarayanan, *Proc. Nat. Acad. Sci.* 68 (1971) 2236.
[40] R., Gordon, R. Bender and G. T. Herman, *J. Theor. Biol.* 29 (1970) 471.
[41] P. Gilbert, *J. Theor. Biol.* 36 (1972) 105.
[42] P. K. Luther, in: *Electron Tomography: Three-Dimensional Imaging with the Transmission Electron Microscope*, J. Frank (Ed.) pp. 39–60, Plenum Press: New York; London. (1992).
[43] C. E. Hsieh, M. Marko, J. Frank and C. Mannella, *J Structural Biol* 138 (2002) 63.
[44] G. A. Perkins, C. W. Renken, J. Y. Song, T. G. Frey, S. J. Young, S. Lamont, M. E. Martone, S. Lindsey and M. H. Ellisman *J. Structural Biol* 120 (1997) 219.
[45] P. R. Smith, T. M. Peters and R. H. T. Bates, *J. Phys. A: Math., Nucl. Gen.* 6 (1973) 361.
[46] Henderson R, Baldwin J. M., Ceska T. A., Beckman E., Zemlin F. and Downing K. *J Mol. Biol.* 213 (1990) 899.
[47] B. K. Vainshtein, *Soviet Physics–Crystallography* 15 (1970) 781.
[48] M. Radermacher, in *Electron tomography: three-dimensional imaging with the transmission electron microscope*, J. Frank (Ed.) pp. 91–116, Plenum Press, New York/London (1992).
[49] G. T. Herman, *Image Reconstruction From Projections, The Fundamentals of Computerised Tomography*, Academic Press, New York (1980).
[50] N. Grigorieff *J. Mol Biol.* 277 (1998) 1033.
[51] H. Nyquist, *Trans. AIEE*, 47 (1928), 617.
[52] P. F. C. Gilbert, *Proc. R. Soc. Lond. B.* 182 (1972) 89.
[53] M. Radermacher, T. Wagenknecht, A. Verschoor and J. Frank, *J. Microscopy* 146 (1987) 112.
[54] R. A. Crowther and A. Klug, *J. Theor. Biol.* 32 (1971) 199.
[55] S. H. Bellman, R. Bender, R. Gordon and J. E. Rowe, *J. Theor. Biol.* 32 (1971) 205.
[56] I. M. Sezan, *Ultramicroscopy* 40 (1992) 55.
[57] R. W. Gerchberg and W.O. Saxton, *Optik* 34 (1971) 275.

[58] J. Frank, *Three-Dimensional Electron Microscopy of Macromolecular Assemblies*, Academic Press, San Diego (1996).
[59] M. Barth, R. K. Bryan and R. Hegerl, *Ultramicroscopy* 31 (1989) 365.
[60] N. A. Farrow and F. P. Ottensmeyer, *Ultramicroscopy* 31 (1989) 275.
[61] M. C. Lawrence, M. A. Jaffer and B. T. Sewell, *Ultramicroscopy* 31 (1989) 285.
[62] U. Skoglund and L. Ofverstedt, *J. Struct. Biol.* 117 (1996) 173.
[63] K. Dierksen, D. Typke, R. Hegerl, A. J. Koster, and W. Baumeister, *Ultramicroscopy* 40 (1992) 71.
[64] U. Ziese, A. H. Janssen, J. L. Murk, W. J. C. Geerts, T. Krift, A. J. Verkleij and A. J. Koster, *J. Microscopy* 205 (2002) 187.
[65] M. Otten, *personal communication* (2003).
[66] U. Ziese, C. Kubel, A. Verkleij, A. Koster, *J Struct Biol.* 138 (2002) 58.
[67] M. C. Lawrence, in: *Electron tomography: three-dimensional imaging with the transmission electron microscope*, J. Frank (Ed.) pp. 197–204, Plenum Press, New York/London (1992).
[68] J. Frank and B. F. McEwen, in: *Electron tomography : three-dimensional imaging with the transmission electron microscopy* J. Frank (Ed.) pp. 205–214, Plenum Press, New York/London (1992).
[69] A. J. Koster, R. Grimm, D. Typke, R. Hegerl, A. Stoschek, J. Walz and W. Baumeister, *J. Struct. Biol.* 120 (1997) 276.
[70] R. Guckenberger, *Ultramicroscopy* 9 (1982) 167.
[71] J. C. Russ, *The Image Processing Handbook, 3rd edition*, IEEE Press, Piscataway (2000).
[72] M. Radermacher and W. Hoppe, *Proc. 7th European Congr. Electron Microscopy. Den Haag* (1980) 132.
[73] P. A. Midgley and M. Weyland *Ultramicroscopy* 96 (2003) 413.
[74] E. A. Fischione Inc. Model 2020 Advanced Tomography Holder.
[75] D. N. Mastronarde. *J. Struct. Biol.* 120 (1997) 343.
[76] P. Penczek, M. Marko, K. Buttle, J. Frank. *Ultramicroscopy*. 60 (1995) 393.
[77] J. Carazo, in: *Electron tomography : three-dimensional imaging with the transmission electron microscope*, J. Frank (Ed.) pp. 117–166, Plenum Press, New York/London (1992).
[78] P. A. Penczek *J Struct. Biol.* 138 (2002) 34.
[79] P. W. Hawkes, in: *Electron tomography : three-dimensional imaging with the transmission electron microscope*, J. Frank (Ed.) pp. 17–38, Plenum Press, New York/London (1992).
[80] A. J. Koster, U. Ziese, A. J. Verkleij, A. H. Janssen, and K. P. de Jong, *J. Phys. Chem. B*, 104 (2000) 9368.
[81] P. A. Midgley, M. Weyland J. M. Thomas and B. F. G. Johnson, *Chem. Commun.* 18 (2001) 907.
[82] G. Mobus and B. J. Inkson, *Appl. Phys. Lett.* 79 (2001) 1369.
[83] S. J. Pennycook, *Ultramicroscopy* 30 (1989) 58.
[84] D. A. Muller, B. Edwards, E. J. Kirkland, J. Silcox, *Ultramicroscopy*, 86 (2001), 371.
[85] A. Howie, *J. Microscopy*, 177 (1979) 1.
[86] J. Cowley, this volume.
[87] E. J. Kirkland, R. F. Loane, and J. Silcox, *Ultramicroscopy* 23 (1987) 77.
[88] J. W. Steeds in: *Quantitative Electron Microscopy* J. C. Chapman and A. J. Craven (Eds), SUSSP, Edinburgh (1984).
[89] T. J. V. Yates, M. Weyland, D. Zhi, R. E. Dunin-Borkowski and P. A. Midgley (2003), *Inst. Phys. Conf. Series* (MSM2003), in press.
[90] A. Beorchia, L. Heloiot, M. Menager, H. Kaplan, D. Ploton *J. Microsc.* 170 (1993) 247.
[91] D. Ozkaya, W. Z. Zhou, J. M. Thomas, P. A. Midgley, V. J. Keast and S. Hermans, *Catalysis Letters* 60 (1999) 113.
[92] J. H. Sinfelt, *Bimetallic Catalysts*, J. Wiley, New York (1983); J. M. Thomas, *Angewandte Chemie-International Edition* 38 (1999) 3589.
[93] J. M. Thomas, B. F. G. Johnson, R. Raja, G. Sankar and P. A. Midgley *Accounts of Chemical Research* 36 (2003) 20.
[94] IDL v.5.1, Research Systems, 2995 Wilderness Place, Boulder, Colorado 80301, 1988.
[95] C. F. Blanford and C. B. Carter *Microscopy and Microanalysis* 9 (2003) 245.
[96] Meldrum, F. C., Heywood, B. R., Mann, S., Frankel, R. B. & Bazylinski, D. A. *Proc. R. Soc. London Ser. B* 251 (1993) 231.
[97] K. L. Thomas-Kerpta, D. A. Bazylinski, J. L. Kirschvink, S. J. Clemett, D. S. McKay, S. J. Wentworth, H. Vali, E. K. Gibson, and C. S. Romanek, *Geochimica Et Cosmochimica Acta* 64 (2000) 4049.
[98] P. R. Buseck, R. E. Dunin-Borkowski, B. Devourard, R. B. Frankel, M. R. McCartney, P. A. Midgley, M. Posfai and M. Weyland, *Proc. Nat. Acad. Sci.* 98 (2001) 13490.

[99] D. C. Golden, D. W. Ming, C. S. Schwandt, H. V. Lauer, R. A. Socki, R. V. Morris, G. E. Lofgren, and G. A. McKay, *American Mineralogist* 86 (2001) 956.
[100] D. J. Barber and E. R. D. Scott. *Proc. Nat. Acad. Sci* 99 (2002) 6556.
[101] Department of Materials Science and Metallurgy Electron Microscopy Group: http://www-hrem.msm.cam.ac.uk/research/CETP/electron_tomography.html
[102] L. Reimer, *Energy-Filtering Transmission Electron Microscopy*, Springer-Verlag, Berlin (1995).
[103] B. Freitag and W. Mader, *J. Microsc.* 194 (1999) 42.
[104] W. Grogger, B. Schaffer, K. M. Krishnan and F. Hofer *Ultramicroscopy* 96 (2003) 481.
[105] P. J. Thomas and P. A. Midgley, *Ultramicroscopy* 88 (2001) 179, P. J. Thomas and P. A. Midgley, *Ultramicroscopy* 88 (2001) 187.
[106] R. Grimm, D. Typke and W. Baumeister, *J. Microsc.* 190 (1998) 339.
[107] P. Laquerriere, E. Kocsis, G. Zhang, T. L. Talbot and R. D. Leapman *Microsc Microanal 9 (suppl 2)* (2003) 240.
[108] M. Weyland and P. A. Midgley. *Microsc Microanal* 9 (2003) 542.

20. OFF-AXIS ELECTRON HOLOGRAPHY

MARTHA R. McCARTNEY, RAFAL E. DUNIN-BORKOWSKI AND DAVID J. SMITH

1. ELECTRON HOLOGRAPHY AND NANOTECHNOLOGY

Nanoscale electromagnetic fields are essential for the function of many nanostructured materials and devices. Important examples include elemental and compound semiconductor p–n junctions and non-volatile magnetic storage media. Theory and modeling can be used to estimate field strengths, but direct measurements are preferable for materials that have smaller dimensions, especially to understand and control the effects of local inhomogeneities on macroscopic properties. Off-axis electron holography is an ideal technique for tackling such problems.

There are many possible imaging modes in the transmission electron microscope (TEM). Most of these suffer from the drawback that the final electron micrograph is a spatial distribution of intensity so that information about the phase shift of the electron that has passed through the sample is lost. The technique of electron holography, first proposed by Gabor [1], overcomes this limitation, and allows the electron phase shift to be recovered. Since this phase shift can be related directly to the electrostatic potential and the in-plane component of the magnetic induction in the sample, an electron hologram can be interpreted to provide quantitative information about electromagnetic fields with a spatial resolution approaching the nanometer scale. The development of the field-emission gun (FEG) as an electron source for the TEM has facilitated the practical implementation of electron holography in commercially available instruments, stimulating widespread interest and activity in the technique. One particular advantage of electron holography is that unwanted contributions to the contrast that are caused by local variations in composition and sample thickness can, in principle, be removed from a recorded phase image. Electron holography has recently been applied to the characterization of a wide variety of nanostructured materials, including semiconductor devices and quantum well structures, as well as magnetic thin films and nanostructures.

This chapter provides a review of the application of off-axis electron holography to the characterization of electric and magnetic fields in nanostructured materials, which invariably behave differently from bulk structures. The chapter begins with a description of the experimental and theoretical basis of the technique. Applications of off-axis electron holography to nanostructured materials are then briefly reviewed. Examples of electrostatic fields that are described include electrostatic potentials associated with semiconductor junctions and layered structures, as well as the measurement of fringing fields around field-emitting carbon nanotubes. The characterization of nanoscale magnetic materials is then described. Results presented include nanopatterned elements and nanocrystalline chains. The interested reader is referred to several books [2–4] and review papers [5–8] for information about the development of electron holography and further applications of the technique that are beyond the scope of this chapter.

2. DESCRIPTION OF OFF-AXIS ELECTRON HOLOGRAPHY

2.1. Experimental Set-Up

More than 20 variants of electron holography have been identified [9]. By far the most fruitful of these has been off-axis electron holography which is illustrated schematically in Fig. 1. A high-brightness FEG electron source provides highly coherent electron illumination incident onto the sample. The region of interest is then positioned so that it covers approximately half of the field of view. A positive voltage, typically between 50 and 250 V, is applied to the electrostatic biprism, causing overlap between the wave that has passed through the sample with the vacuum (or reference) wave. The spacing of the resulting interference fringes is inversely proportional to the biprism voltage, and the total number of fringes is roughly proportional to the square of the biprism voltage.

A representative off-axis electron hologram from a chain of nanoscale magnetite crystals is shown in Fig. 2(a). The field of view is covered by two sets of fringes. The coarser fringes at the edges of the pattern are Fresnel fringes that originate from the edges of the biprism wire, whereas the finer fringes visible across almost the entire field of view are the holographic interference fringes. The relative changes in position and intensity of these holographic fringes can be interpreted to provide details about the phase shift and the amplitude, respectively, of the electron wave that has passed through the sample. The actual hologram reconstruction process is described in the next section.

Note that many of the results described here, such as the example shown in Fig. 2(a), have been acquired using a special minilens located below the normal objective lens, with this latter lens switched off so that the sample was located in a field-free environment. The lower magnification associated with this configuration is sufficient for

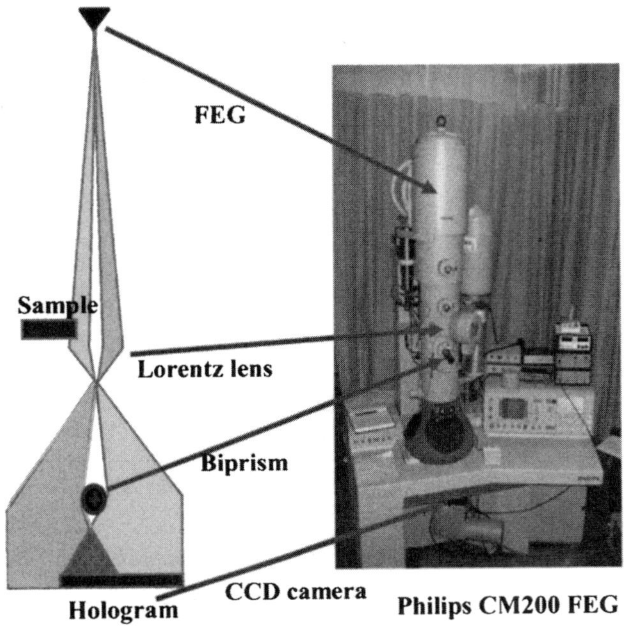

Figure 1. (Left) Schematic illustration of microscope configuration for off-axis electron holography. Field-emission-gun (FEG) electron source provides coherent electron beam incident on sample. Electrostatic biprism below sample causes overlap of vacuum (reference) wave with object wave. Modulations of hologram interference fringes recorded on CCD camera are interpretable in terms of phase/amplitude changes due to sample. (Right) Photograph of Philips CM-200 transmission electron microscope equipped with FEG electron source, a special Lorentz minilens beneath the normal objective lens enabling field-free observation of magnetic samples, an electrostatic biprism and a CCD camera for quantitative recording of holograms.

most magnetic samples, and provides an advantageous field of view for a variety of semiconductor applications.

2.2. Basic Imaging Theory and Hologram Reconstruction

Imaging in the TEM can be considered in terms of the modification of the incident electron wavefunction, first by the object, and then by the imaging objective lens. The intensity of the final image is found from the modulus squared of the resulting wavefunction [10].

At the exit-surface of the sample, the wavefunction can be expressed in the form

$$\Psi_s(\mathbf{r}) = A_s(\mathbf{r})\exp[i\phi_s(\mathbf{r})] \tag{1}$$

where A and ϕ are the amplitude and phase and $\mathbf{r}$ is a two-dimensional vector in the plane of the sample. The corresponding wavefunction at the image plane can be written

$$\Psi_i(\mathbf{r}) = A_i(\mathbf{r})\exp[i\phi_i(\mathbf{r})] \tag{2}$$

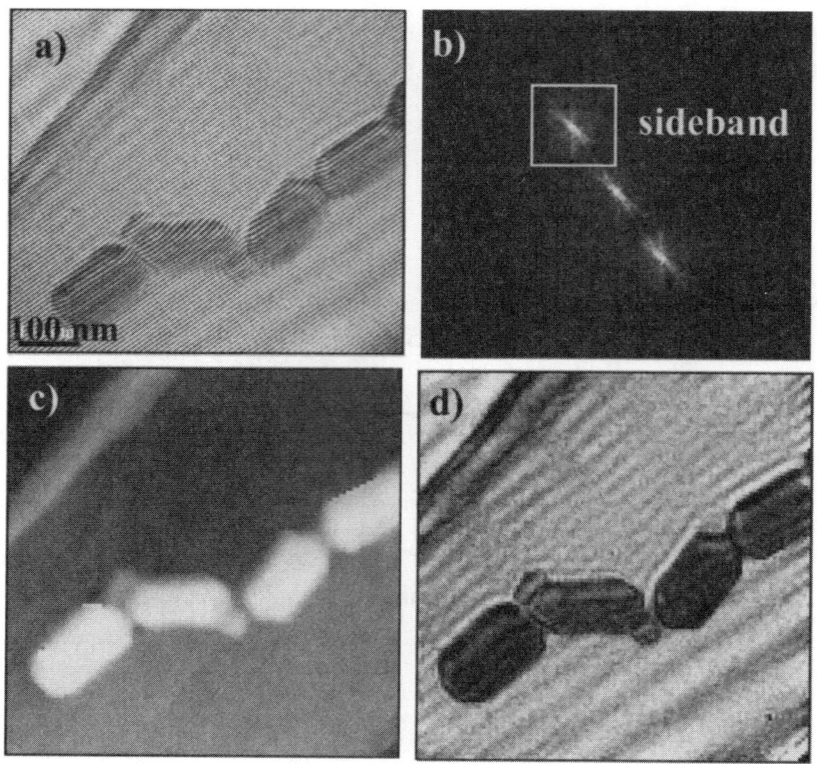

Figure 2. (a) Off-axis electron hologram showing chain of magnetite nanocrystals; (b) Fourier transform of (a), also indicating sideband used for reconstruction of complex image wave; (c) reconstructed phase image; (d) reconstructed amplitude image.

and the recorded intensity distribution is given by the expression

$$I(\mathbf{r}) = |A_i(\mathbf{r})|^2 \tag{3}$$

For a perfect thin lens, neglecting magnification and image rotation, the complex image wave $\Psi_i(\mathbf{r})$ can be regarded as equivalent to the object wave $\Psi_s(\mathbf{r})$.

In practice, unavoidable aberrations of the normal objective lens cause additional modifications to the phase and amplitude. These modifications can be represented by a phase-contrast transfer function, which takes the form

$$T(\mathbf{q}) = B(\mathbf{q})\exp[i\chi(\mathbf{q})] \tag{4}$$

where $B(\mathbf{q})$ is an aperture function in the focal plane of the objective lens, and the effects of the defocus, Δz, and spherical aberration, C_s, are included in the phase factor

$$\chi(q) = \pi\Delta z\lambda q^2 + \frac{\pi}{2}C_s\lambda^3 q^4 \tag{5}$$

The complex wavefunction in the image plane can then be written

$$\psi_i(\mathbf{r}) = \psi_s(\mathbf{r}) \otimes t(\mathbf{r}) \tag{6}$$

where $t(\mathbf{r})$ is the inverse Fourier transform of $T(\mathbf{q})$. The convolution, $\otimes$, of the object wave $\psi_s(\mathbf{r})$ with $t(\mathbf{r})$ represents an effective loss of resolution due to the lens aberrations. The intensity in the image plane can then be expressed in the form

$$I(\mathbf{r}) = A_i^2 = |\psi_s(\mathbf{r}) \otimes t(\mathbf{r})|^2 \tag{7}$$

Because of the phase oscillations introduced by $\chi(\mathbf{q})$, the image intensity is no longer related easily to the object structure. Thus, an accurate knowledge of the lens aberrations and imaging parameters is essential for high-resolution applications of electron holography.

An expression for the intensity distribution in an off-axis electron hologram can be obtained by adding a plane reference wave to the complex object wave in the image plane, with a tilt angle of $\mathbf{q} = \mathbf{q_c}$. Thus,

$$\begin{aligned} I_{\text{hol}}(\mathbf{r}) &= |\psi_s(\mathbf{r}) \otimes t(\mathbf{r}) + \exp[2\pi i q_{\mathbf{c}} \bullet \mathbf{r}]|^2 && (8) \\ &= 1 + A_i^2(\mathbf{r}) + 2A_i(\mathbf{r}) \cos[2\pi i q_c \bullet \mathbf{r} + \phi_i(\mathbf{r})] && (9) \end{aligned}$$

i.e., the hologram intensity is the sum of three terms: the intensities of the reference and image waves, and an additional set of cosinusoidal fringes having phase shifts ϕ_i and amplitudes A_i that are equivalent to the corresponding phases and amplitudes, respectively, of the image wave. These fringes correspond to the fine interference fringes that are spread across the field of view in Fig. 2(a).

The recorded hologram must be "reconstructed" in order to extract the required phase and amplitude information from the complex image wave. The hologram is first Fourier-transformed, which can be accomplished either optically or, preferably, digitally for reasons that are explained further below. This procedure is described by the equation

$$\begin{aligned} FT[I_{\text{hol}}(\mathbf{r})] &= \delta(\mathbf{q}) + FT[\mathrm{A}_i^2(\mathbf{r})] && (10a) \\ &\quad + \delta(\mathbf{q} + \mathbf{q_c}) \otimes [A_i(\mathbf{r}) \exp[i\phi_i(\mathbf{r})]] && (10b) \\ &\quad + \delta(\mathbf{q} - \mathbf{q_c}) \otimes [A_i(\mathbf{r}) \exp i\phi_i(\mathbf{r})]] && (10c) \end{aligned}$$

This expression contains four terms: the central peak at the origin corresponds to the Fourier transform of the uniform intensity of the reference image, a second peak centered at the origin represents the Fourier transform of the conventional TEM micrograph of the specimen, and the remaining two peaks centered on $\mathbf{q} = -\mathbf{q_c}$ and $\mathbf{q} = +\mathbf{q_c}$ constitute the desired image wavefunction, and its complex conjugate. In the Fourier transform of the hologram shown in Fig. 2(b), these various contributions are visible as the central "auto-correlation" peak and two "sidebands". The two sidebands contain identical information except for a change in the sign of the phase.

The final part of the hologram reconstruction procedure involves using an 'aperture', as shown in Fig. 2(b), to select one of the sidebands, which is then inverse-Fourier transformed. The sideband region must be separated from the central peak by choosing a small enough fringe spacing, and by selecting the size of the aperture, which is usually square, but can be rectangular, depending on the object being studied [11]. The reconstructed phase and amplitude images from Fig. 2(a) are shown in Figs. 2(c) and 2(d), respectively.

2.3. Phase Shifts and Mean Inner Potential

The electrons that pass through a TEM sample undergo phase shifts that depend on the electrostatic potential and the in-plane component of the magnetic induction. Neglecting dynamical diffraction effects, the phase shift can be expressed in the form [12]

$$\phi(x) = C_E \int V(x, z)\, dz - \left(\frac{e}{\hbar} \iint B_\perp(x, z) t(x)\, dx\, dz\right) \tag{11}$$

where z is the incident electron beam direction, x is a direction in the plane of the sample, V is the electrostatic potential, $B_\perp$ is the component of the magnetic induction in the sample perpendicular to both x and z, and C_E is an energy-dependent constant that has the value of 7.28×10^{-3} rad/V·nm at an electron accelerating voltage of 200 kV.

When neither V nor $B_\perp$ vary within the sample along the beam direction, then equation (11) can be simplified to

$$\phi(x) = C_E V(x) t(x) - \left(\frac{e}{\hbar}\right) \int B_\perp(x) t(x)\, dx \tag{12}$$

where t is the sample thickness. Differentiation with respect to x then yields

$$\frac{d\phi(x)}{dx} = C_E \frac{d}{dx}\{V(x)t(x)\} - \left(\frac{e}{\hbar}\right) B_\perp(x) t(x) \tag{13}$$

In a sample of uniform thickness and composition, the first term in this expression is zero, and the phase gradient is directly proportional to the in-plane component of the magnetic induction

$$\frac{d\phi(x)}{dx} = \left(\frac{et}{\hbar}\right) B_\perp(x) \tag{14}$$

By adding contours to the phase image, a direct representation of the magnetic induction in the sample is obtained. Phase differences of 2π between adjacent contours correspond to steps of $\int \mathbf{B} \cdot d\mathbf{S} = (h/e) = 4 \times 10^{-15}$ Wb, thus allowing magnetic fields to be measured and quantified on an absolute basis [8].

The mean inner potential (MIP) of the sample, V_0, which depends on the local composition and density, is usually the dominant contribution to the electrostatic potential. The MIP is defined as the volume average, or the zeroth-order Fourier coefficient, of the electrostatic potential [13]. For low-energy electrons, the MIP varies with the energy of the incident electrons [14], whereas the MIP measured using high-energy electrons [15] can be regarded as a fundamental property of the material.

In the absence of magnetic and long-range electric fields, such as those occurring at depletion regions in semiconductors, and assuming that dynamical diffraction can be neglected, then equation (11) can be re-written

$$\phi(x) = C_E \int V_0(x, z)\, dz \tag{15}$$

If the sample has uniform structure and composition in the beam direction, then this expression reduces still further to

$$\phi(x) = C_E V_0(x) t(x) \tag{16}$$

These expressions ignore multiple scattering within the sample, and therefore only apply to crystalline samples when they are tilted to weakly diffracting orientations.

Electron holography provides the most accurate technique currently available for measuring the sample MIP, by using equation (16) and making use of experimental phase profiles. The most straightforward approach is to use a cleaved wedge so that the thickness changes in a well-defined manner as a function of distance from the edge of the sample [16]. The MIP can also be determined by measuring the gradient of the phase $d\phi/dx$, and using the relation

$$V_0 = \left(\frac{1}{C_E}\right)\left(\frac{d\phi/dx}{dt/dx}\right) \tag{17}$$

The advantage of this approach is that it is independent of any contributions to the phase shift from amorphous overlayers on the sample surface, since $dt/dx = 0$ when these layers are of uniform thickness.

Dynamical corrections to the phase always need to be taken into account when determining V_0 from any crystalline sample, including a cleaved wedge. These corrections are necessary even when the electron hologram has been acquired at a weakly diffracting condition. For greatest possible accuracy, the exact orientation of the crystalline sample should be recorded using a technique such as convergent-beam electron diffraction, and the dynamical contributions should then be simulated at this angle [16, 17]. An independent measure of the sample thickness is highly recommended. Dynamical corrections to V_0 are typically between 0.1 and 0.2 V even at weakly diffracting orientations [18]. Values of the MIP measured for a wide range of materials have been compiled [17].

2.4. Quantification

Traditionally, the reconstruction of electron holograms has been carried out 'optically' using a laser bench [6]. Electron holography has recently benefitted greatly from the use of quantitative recording [19], and the charge-coupled-device (CCD) camera is nowadays the standard medium of choice for recording holograms. These devices have a linear response over several orders of magnitude, they have high detection quantum efficiency, and the recorded information is immediately accessible for digital processing [20]. Moreover, fast and inexpensive computers and sophisticated software [21] are now readily available. Digital reconstruction of holograms using computer processing has thus become widespread.

Off-axis electron holography relies on information stored in the lateral displacement of interference fringes, and thus distortions in the recording medium are liable to cause displacements that could be misinterpreted. The reconstructed phase image may have long-range modulations due to several factors, which include distortions caused by the projector lenses and the recording system, and inhomogeneities in the charge and thickness of the biprism wire. Since these distortions are usually geometrical in nature and invariant in time, their effects can be removed by the simple procedure of recording a reference hologram with the sample removed from the field of view but without any changes to the optical parameters of the microscope. Correction is then carried out by a complex division of the reconstructed (sample and reference) image waves in real space, giving the required distortion-free phase of the image wave [19, 22]. Relative phase changes within the sample can then be interpreted faithfully.

The interference-fringe spacing and the spatial resolution of the phase image depend on several geometrical factors as well as the biprism voltage and the imaging lenses [23, 24]. The phase detection limit depends on the signal-to-noise ratio in the phase image, which may be low if the recorded signal is of low intensity [25]. Since off-axis electron holograms are typically recorded at electron doses of 100–500 electrons per pixel, the recording process is dominated by Poisson noise [26]. The quality of the hologram depends on the characteristics of the recording medium (CCD camera), as well as on the fringe visibility (or contrast), which depends in turn on the illumination diameter and the biprism voltage. Loss of fringe contrast will result from the finite source size, the beam energy spread, stray electrostatic and magnetic fields, and the mechanical stability of the biprism wire.

A major problem with the digital reconstruction of phase images is that they are usually evaluated modulo 2π, meaning that 2π phase discontinuities that are unrelated to specimen features appear at positions where the phase shift exceeds this amount. Phase "unwrapping" using suitable algorithms is then required before reliable interpretation of image features becomes possible. Several phase unwrapping algorithms are available for the purpose of locating and unwrapping such discontinuities, but none of these are yet able to solve all of the problems. The most straightforward approach involves searching the image row-by-row or column-by-column for adjacent pixels with phase differences that exceed a pre-selected value, and then adding or subtracting 2π to the subsequent pixels. Alternatively, a second phase image may be calculated after multiplying the complex image wave by a phase factor $\exp(i\alpha)$ to move the phase jumps to different places in the image [27]. Advanced phase-unwrapping techniques [28] are now being introduced into the software that is used to analyze electron holograms. These and other issues relating to hologram quantification have been discussed in more detail elsewhere [29].

2.5. Practical Considerations

Electron holography requires highly coherent incident illumination, which suggests that large source demagnifications are necessary. However, the use of rotationally symmetric illumination is not advantageous since a significant improvement in coherence can be achieved by employing highly elliptical illumination. The condensor lens stigmators and focus settings are then adjusted to produce elliptical illumination at the sample level that is elongated in the direction perpendicular to the biprism wire when the condensor lens is overfocused, but relatively narrow in the parallel direction. Illumination aspect ratios of 50–100 are common for medium-resolution electron holography [30].

A severe practical limitation of off-axis electron holography is the requirement for a (vacuum) reference wave that can be overlapped onto the region of interest. Thus, off-axis electron holograms must usually be recorded from near the edge of the specimen. For high-resolution applications, which require very high sampling density, this requirement can represent a severe restriction on the size of the useful specimen regions that can be examined. For medium resolution applications, the region of interest and the reference wave can typically be separated by no more than a few microns at the very most. This restriction can be relaxed in the special case of a sample that has a thin and featureless region of electron-transparent support film, rather than vacuum, which can be overlapped onto the specimen feature of interest.

When examining magnetic materials, the conventional objective lens must be switched off, since its strong vertical field would otherwise saturate the magnetization of the sample in the electron beam direction. It is common to use a high-strength 'Lorentz' minilens located beneath the sample to provide high magnifications with the sample located in a field-free environment [31]. At the same time, a larger fringe overlap width of 1–2 μm can be obtained, with typical interference fringe spacings in the range of 1–5 nm. It should also be appreciated from equations (11) and (12) that the total phase shift for a nanostructured magnetic material will be dominated by the mean inner potential. The magnetic contribution to the phase is approximately proportional to the width of the particular magnetic element unlike the MIP contribution which is independent of its lateral dimensions. Thus, accurate quantification of the magnetic contribution to the phase becomes more challenging for elements of narrow width. Separation of the two contributions can be achieved in several ways [8]. Inversion of the sample in the microscope, which is inconvenient and not usually straightforward, would change the sign of the magnetic contribution to the phase. Subtraction of the two respective phase images would allow the MIP contribution to the phase to be removed. Alternatively, it is often possible to achieve magnetization reversal by tilting the sample *in situ* with the main objective lens slightly excited, again enabling the MIP contribution to be measured and then removed from other phase images [32]. This possibility proves to be highly useful when tracking magnetization reversal processes during entire hysteresis cycles.

3. NANOSCALE ELECTROSTATIC FIELDS

Detailed knowledge of the electrostatic potential distribution within semiconductor materials is vital to a full understanding of device performance, in particular to account for offsets that occur at heterojunctions and extended defects, such as charged dislocations. Band-bending at interfaces, in conjunction with the presence of extrinsic dopants, determines electron transport, and effectively controls device operation. The depletion field at a simple p–n junction is well-established theoretically [33], but techniques that can be used to map potential distributions across complex heterojunctions at the nanometer scale are generally unavailable. As device dimensions are further reduced into the deep-submicron regime, methods that are capable of quantitatively determining two-dimensional dopant concentration profiles with the required precision and spatial resolution become even more crucial. Off-axis electron holography is uniquely capable of accessing and quantifying nanoscale electrostatic fields. This section describes several areas where the technique has already had a substantial impact.

3.1. Dopant Profiles

Electron holography is potentially a valuable method for investigating semiconductor junctions and devices. Early low-resolution investigations of p–n junctions by electron holography revealed an electrostatic fringing field outside a biased sample [34]. We were later able to measure directly the intrinsic electrostatic potential distribution across a p- and n-doped Si/Si junction [35]. These initial results were limited in precision by signal-to-noise considerations and sample preparation methods. Much improved precision and spatial resolution have since been achieved by making use of the enhanced stability and higher electron beam coherence available with a modern FEG-TEM [36, 37]. The reconstructed hologram of a p-type field-effect transistor, which is shown in Fig. 3, illustrates this improved capability. This particular application demonstrates two-dimensional imaging of electrostatic potential distributions across devices with a sensitivity closely approaching 0.15 V and spatial resolution approaching 6 nm. Such information about doped heterojunctions is invaluable for evaluating and refining competing models for dopant diffusion.

Practical difficulties still prevent electron holography from being used routinely in device characterization. The most serious problem relates to the preparation of a uniformly thinned cross-section of the region of interest, which must also be located close to the sample edge because electron holography requires an unperturbed reference wave to facilitate phase reconstruction. Focused-ion-beam (FIB) milling is widely used in the semiconductor industry for thinning specific sample sites. However, residual Ga implanted inadvertently during FIB milling results in a surface "dead" layer [38], which decreases the potential measured by electron holography. Additional steps such

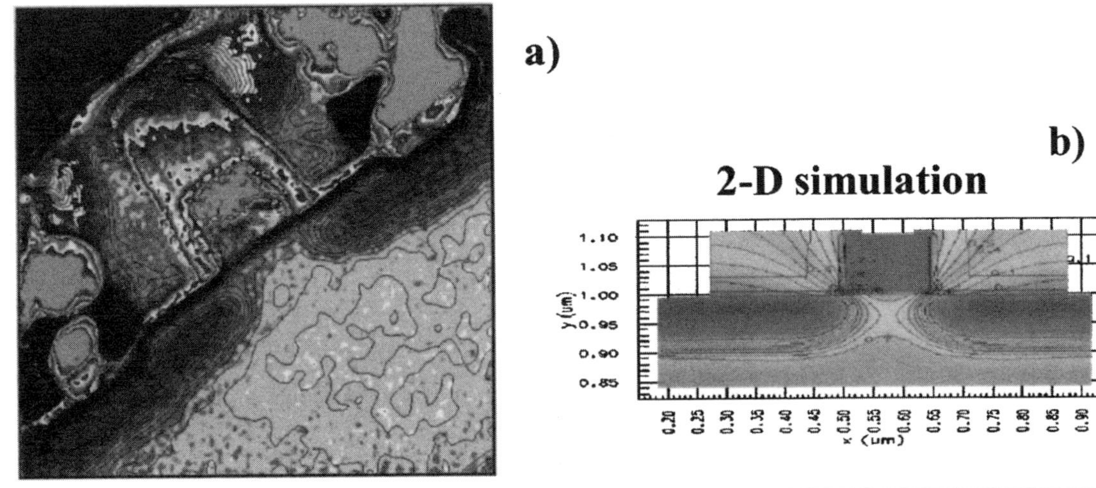

Figure 3. (a) Reconstructed off-axis electron hologram of electrostatic potential variation associated with 0.13-μm p-type field-effect transistor; and (b) corresponding simulated 2-D potential map. Phase contours of 0.1V [37]. (See color plate 18.)

as reverse biasing of the sample may then be required before reliable results can be obtained using electron holography [39].

Surface damage and depletion at the top and bottom surfaces of doped samples prepared using routine ion-milling has also been reported to create electrically "dead" layers [40], with typical thicknesses of ~25 nm. By using wedge polishing followed by low-angle, low-energy milling, we were able to mitigate these effects and obtain high measurement precision [36, 37, 41]. However, for many applications involving the characterization of 2-D dopant distributions, access to the region of interest for holography study can most conveniently be achieved only by using FIB milling. Surface depletion effects arising from sample preparation then become unavoidable and must be taken into account.

3.2. Piezoelectric Fields

Another area of recent interest and activity involving the application of off-axis electron holography to the characterization of nanoscale electrostatic fields has been the study of nitride semiconductors, which are important candidates for a range of novel optoelectronic applications [42]. For example, large internal piezoelectric fields of ~1 MV/cm have been observed in strained GaN/InGaN/GaN quantum wells [43]. Such piezoelectric fields, in addition to the spontaneous polarization field, play a major role in proposed device applications of Group III-nitride materials [42]. These fields result from the pseudomorphic growth of highly strained Ga(In)N/GaN and Ga(Al)N/GaN quantum well (QW) structures in the polar [0001] direction. Electron holography has been used in recent studies [43–46] to quantify the electrostatic potential profiles across strained QW heterostructures grown without intentional doping. Features in the measured potential profiles have been variously attributed to the accumulation of sheet charge [44, 45] and the presence of a two-dimensional gas (2DEG) near the heterointerface [46]. In our study of the electrostatic potential variation across an n-AlGaN/InGaN/p-AlGaN heterojunction diode by off-axis electron holography, we have shown that it is also possible to account for the additional effects that are caused by the overlapping built-in potential of the pn junction [47]. Figure 4(a) compares the experimental profile with the separate contributions to the electrostatic energy profile simulated for the p–n junction, V_{pn}, the spontaneous polarization, V_{sp}, and the piezoelectric field, V_{pe}. Figure 4(b) then compares the experimental profile (open circles) with a simulated profile that incorporates the piezoelectric and spontaneous polarizations, as well as the effects of charge accumulation at the n-AlGaN/InGaN and InGaN/p-AlGaN interfaces. The close agreement between the experimental and simulated profiles provides strong evidence for the presence of sheet charge at these interfaces.

3.3. Charged Defects

Optoelectronic devices based on Group III-nitride semiconductors are becoming widespread and recent activity has been directed towards improving device lifetimes and light-emitting efficiency [40]. However, fundamental questions remain unanswered or

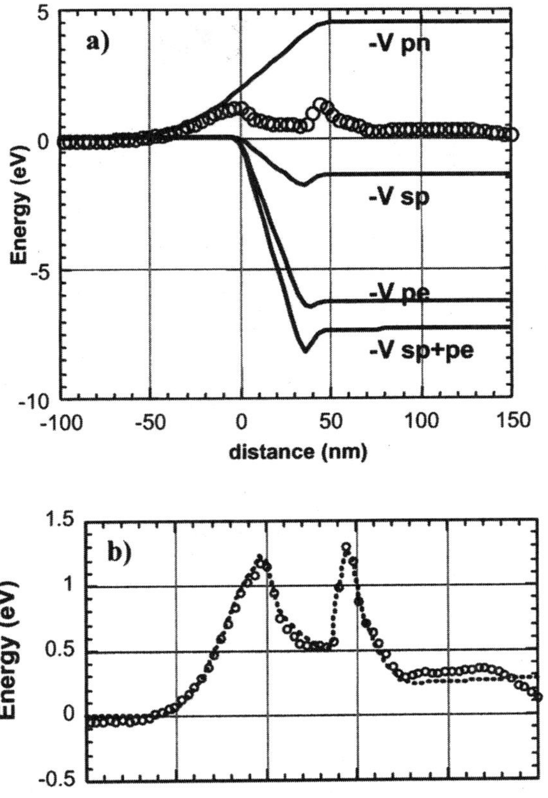

Figure 4. Analysis of electrostatic potential profile across n-AlGaN/InGaN/p-AlGaN heterojunction diode. (a) Energy profile derived from thickness-corrected phase image (open circles) for comparison with individual contributions from p–n junction, V_{pn}, spontaneous polarization, V_{sp}, and piezoelectric filed, V_{pe}. (b) Experimental energy profile (open circles) across heterojunction diode compared with simulation (dotted line) incorporating spontaneous and piezoelectric polarization as well as sheet charge at each interface [47].

results are ambiguous relating to the charge state of the ubiquitous threading dislocations found in these materials, and their effect on device performance. Electron holography has been used in attempts to resolve these issues by measuring electrostatic potential profiles across different types of dislocations [48–50]. The charge density can then be calculated by using Poisson's equation [51]. For example, based on electron holography observations, it was reported that the core regions of screw, mixed and edge dislocations in undoped GaN were negatively charged [48], whereas it was concluded from other holography studies that dislocation cores in n-GaN and p-GaN were, respectively, negatively [49] or positively [50] charged. However, these latter results have been contradicted by subsequent observations made using scanning surface potential microscopy where defects in Si-doped p-type GaN were reported to be electrically neutral [52]. Studies made using ballistic electron emission microscopy found

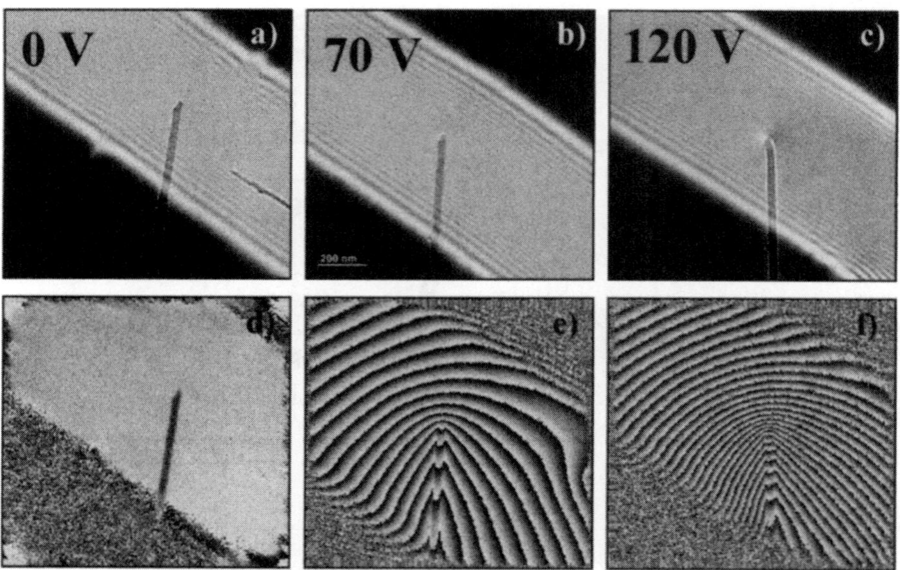

Figure 5. Observation of field-emitting carbon nanotube using off-axis electron holography: (a), (b), and (c) Electron holograms of carbon nanotube at bias voltages $V_B = 0$, 70, and 120 V, respectively. (d), (e), and (f) reconstructed phase images corresponding to (a), (b) and (c), respectively. Phase contour separation of 2π radians. Phase gradient in (f) corresponds to electric field strength at tip of 1.22 V/nm [55].

no evidence for negative dislocation charge [53]. It is conceivable that these differences result from TEM sample preparation or surface charging artefacts, or because localized strain or sample thickness variations around the dislocation core regions affects the diffraction conditions and the measured holographic phase contrast. Further systematic studies of extended defects by electron holography are required, first to eliminate these possibilities, and second to determine whether defect charging is associated with the nitride growth conditions or with the level and type of doping.

3.4. Field-Emitting Carbon Nanotubes

Single-walled and multi-walled carbon nanotubes are attracting much recent attention because of their remarkable resilience, strength and thermal conductivity [54]. The use of carbon nanotubes in arrays of field emitters is of particular interest for compact displays. Electron holography has been used to determine the electrostatic potential distribution associated with individual field-emitting carbon nanotubes [55]. A specially constructed, piezo-driven sample manipulation stage enabled nanotubes to be positioned several microns away from a small collector plate. A variable bias voltage could then be applied between the nanotubes and the plate, as off-axis electron holograms were acquired. Figure 5 shows a sequence of electron holograms from a carbon nanotube and the corresponding reconstructed phase images for bias voltages of 0, 70 and 120 V. The onset of field emission for this nanotube occurred at a threshold bias voltage close to 70 V. The closely spaced 2π phase contours in Fig. 5(c) reflect the extremely high electric field in the vicinity of the field-emitting tip during electron emission. A model based on a one-dimensional line of charge established that the electric field was concentrated at the tip of the nanotube. The field strength was calculated to be 1.22V/nm in close agreement with the experimental phase image.

3.5. Thickness and Sample Morphology

The phase image, as described by equation (16), represents a map of the projected sample thickness if the MIP is constant, and dynamical scattering is weak. Hence, it is then possible under these special circumstances to infer the surface topography and the shapes of nanoparticles using electron holography. For example, line profiles across reconstructed phase images have been used to deduce surface structures and to identify internal voids in small Pd particles [56] and cuboctahedral ZrO_2 particles [57]. Further examples are summarized elsewhere [8].

An alternative approach for specimen thickness determination makes use of the amplitude image, through the equation

$$t(x) = -2\lambda_{\text{in}} \ln\left(\frac{A_s(x)}{A_r(x)}\right) \tag{18}$$

where λ_{in} is the mean-free-path for inelastic scattering, and $A_s(x)$ and $A_r(x)$ are the measured amplitudes of the sample and reference holograms, respectively [58]. A practical problem with this approach is that amplitude images are always noisier than the corresponding phase images. Moreover, λ_{in} will invariably be different for each material in the sample, and may be unknown. Most TEM samples do not have uniform thickness profiles, and thickness determination is not straightforward when the sample has non-uniform composition. Further analysis [59] reveals yet another approach that has enabled holograms from multilayered samples with varying compositions and unknown thickness variations to be successfully interpreted [60, 61].

4. NANOSCALE MAGNETIC FIELDS

Nanoscale magnetic materials have become increasingly important in many practical applications that include non-volatile storage media, electromechanical read/write devices and mechanical sensors [62]. The future of many magnetic devices lies in their capacity for higher storage density, which necessitates further reductions in the sizes of information blocks, and the simultaneous down-scaling of the distances between them. Some of the problems that result from reducing the device dimensions are obvious. First, the magnetic domain size becomes comparable to the mean bit size. Second, individual bits may interact with neighboring blocks by way of their stray fields. In order to improve the overall performance of devices based on magnetic materials, the local microstructure and chemical composition must be tailored to optimize the magnetic properties on the nanometer scale. There are many electron microscopy techniques that are suitable for the characterization of magnetic microstructure [63], but only electron holography allows the quantification of magnetic fields in nanostructured materials at the nanometer scale. An extensive bibliography summarizing applications of the technique to magnetic materials can be found elsewhere [8]: this section provides a brief overview of some representative applications that are relevant to nanotechnology.

4.1. Patterned Nanostructures

Magnetic interactions between closely-spaced, submicron-sized magnetic elements depend sensitively on their shape, size, composition and spacing. The characterization of such interactions is of great importance if nanoscale magnetic materials are to be used in future magnetic recording and sensing applications. Such applications usually require specific values of the coercive fields and remanent magnetizations of individual magnetic nanostructures, as well as stable and reproducible magnetic domain states. To gain insight into these important issues, we have used off-axis electron holography to observe an extensive range of patterned magnetic elements, which were prepared using electron-beam lithography and lift-off processes, and supported on electron-transparent silicon nitride membranes [32, 64–67]. The element shapes included rectangles with varying aspect ratios, as well as diamonds, ellipses and rectangular bars. The fabrication processes have been described in more detail elsewhere [32]. Magnetic fields were applied to each sample by *in situ* tilting within the microscope with the conventional objective lens slightly excited in order to obtain the desired in-plane magnetic fields. The hysteresis loops of the various element shapes could be followed, despite the substantial loss of interference-fringe contrast due to the presence of the underlying silicon nitride support.

Figure 6 shows the magnetic contributions to the measured holographic phase shift for two closely-spaced polycrystalline Co rectangles over a complete magnetization reversal cycle [64]. These holograms were recorded with an average out-of-plane field of 3600 Oe for the in-plane horizontal fields indicated on the figure. The separation of the phase contours is proportional to the in-plane component of the magnetic induction integrated in the incident beam direction. The magnetic fringing fields between the elements are only minimized when the field lines are located entirely within each element, in the flux closure (or solenoidal) vortex state.

An important aspect of these studies was the use of micromagnetic simulations, based on solutions to the Landau-Lifshitz-Gilbert equations, to assist with detailed interpretation of the experimental phase images [32, 65, 66]. Although reasonable agreement with the experimental results was obtained, important differences were also noticed. For example, simulated vortex states formed at higher fields than observed experimentally, which was attributed to the influence of local defects or inhomogeneities. Moreover, slight differences in the initial, nominally saturated, state had a strong effect on subsequent domain evolution during the reversal process. The strength and direction of the out-of-plane component of the applied field also had a marked impact on the observed domain structure [65]. The simulations in Figure 7 show a comparison of two elements simulated first as a pair (left set), and then separately (but displayed together). The marked difference in the domain configurations of the smaller cell on the left

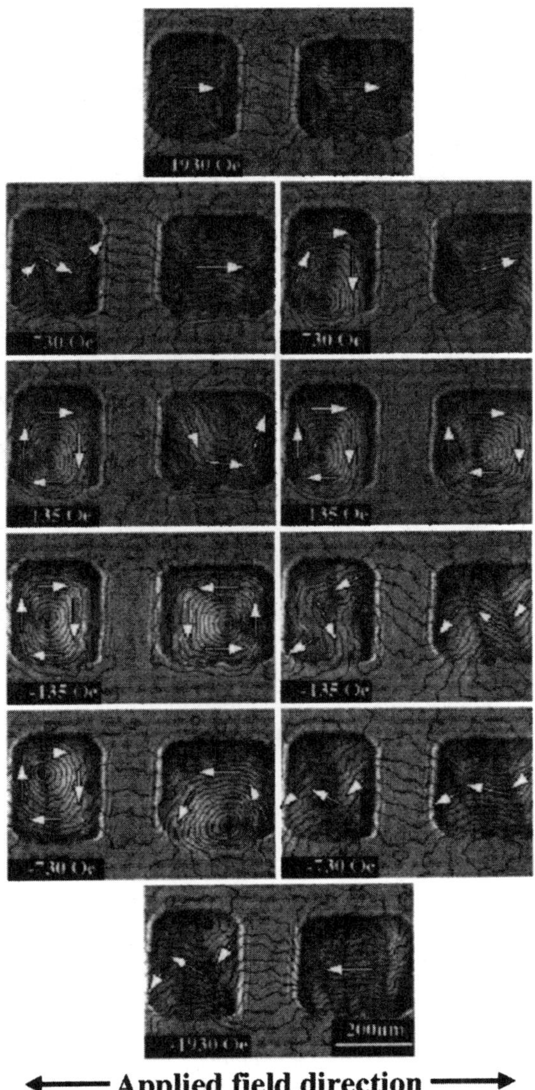

Figure 6. Magnetic contributions to phase for 30-nm-thick Co nanopatterned elements recorded during complete hysteresis cycle by tilting *in situ* within the microscope. In-plane magnetic field applied along horizontal direction as indicated. Average out-of-plane field of 3600 Oe directed into the page. Phase contours separated by 0.21π radians. Loop should be followed in counter-clockwise direction [64].

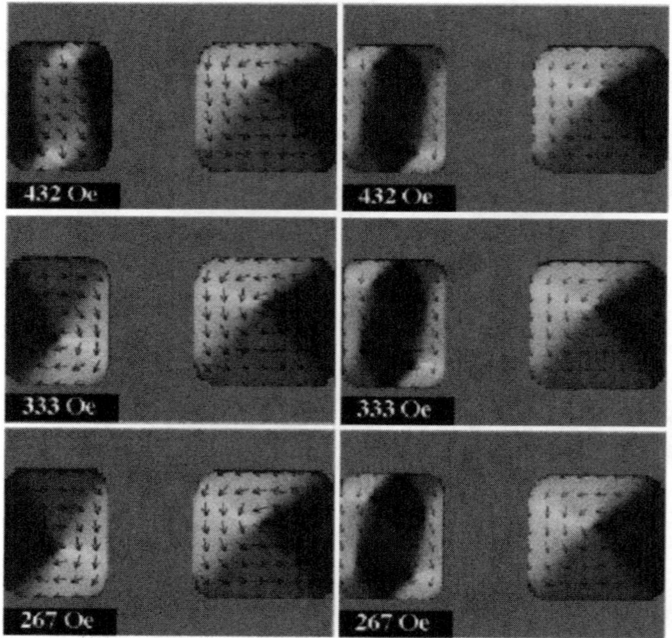

Figure 7. Micromagnetic simulations for 30-nm-thick Co elements with 3600 Oe vertical field directed into the page. Left set simulated for two cells together, whereas right set simulated for each cell separately (though shown together). Differences visible in small element at left due to close proximity of larger element [65].

clearly indicate that these adjacent Co elements are affected by their close proximity. This study emphasizes the importance of comparing experimental measurements with micromagnetic simulations. Interactions between closely-spaced magnetic nanostructures are clearly important for the design of high-density magnetic storage devices.

While these results for nanopatterned Co elements are of considerable scientific interest, the behavior of multilayered elements is of much greater relevance for many practical applications because of the large resistance changes that occur when two closely-spaced magnetic layers have either parallel or anti-parallel magnetization directions (an effect known as giant magnetoresistance [68]). Our further studies focused on submicron "spin-valve" (SV) elements consisting of Co (10 nm)/Au (5 nm)/Ni (10 nm) layers. The SV elements were shaped as rectangles, diamonds, ellipses and bars [66, 67]. Hysteresis loops for individual elements were measured experimentally, based on the density and direction of the phase contours, and compared with loops derived from micromagnetic simulations. Differences in switching fields between the data and the simulations were observed in smaller elements, possibly due to an increased practical difficulty in nucleating end domains that could initiate magnetization reversal. Another significant observation was the direct observation of two different phase spacings in different remanent states in each element. Micromagnetic simulations established that parallel or anti-parallel coupling between the Co and Ni layers was the cause of this

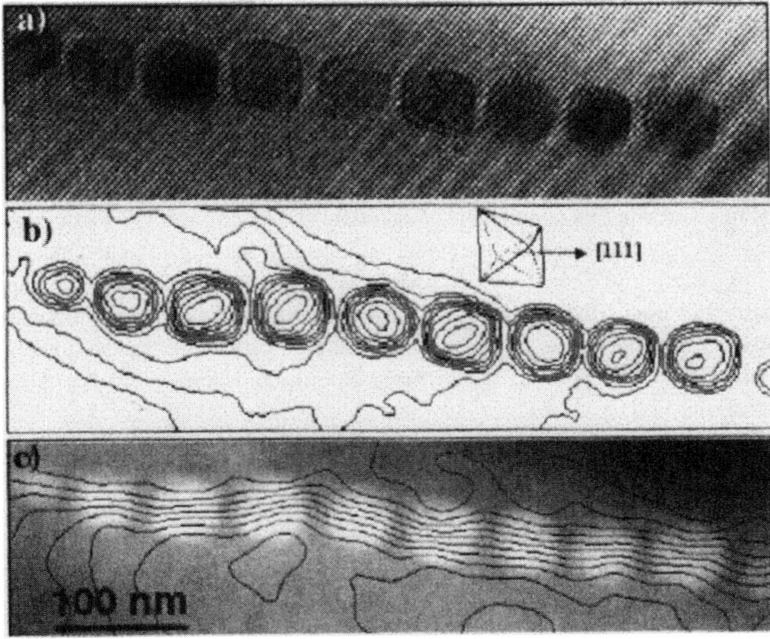

Figure 8. (a) Hologram of chain of magnetite nanocrystals from *magnetospirillum magnetotacticum*; (b) mean inner potential contribution to phase of reconstructed hologram. Thickness contours indicate that magnetite crystals have cuboctahedral shape. (c) magnetic contribution to phase.

behavior. The Ni layer in each element was found to reverse its magnetization direction well before the external field was removed due to the strong fringing field of the closely adjacent and magnetically more massive Co layer. Thus, an antiferromagnetically coupled state was always observed as the remanent state following the removal of the external field. Solenoidal states were observed experimentally for both elliptical- and diamond-shaped elements but could not be reproduced in the simulations, presumably because these states were stabilized by the presence of structural imperfections, or by the crystal grain size or crystallographic texture of the polycrystalline layers.

4.2. Nanoparticle Chains

The formation of reproducible remanent states when an applied external field is removed is desirable for nanopatterned magnetic elements intended for device applications. Because of their geometry, chains of nanocrystalline magnetic particles lend themselves to unidirectional remanent states. Moreover, magnetostatic effects in electon holographic phase images can be separated from those due to thickness and electrostatic effects simply by *in situ* tilting of the sample in the field of the conventional objective lens to achieve magnetization reversal.

Linear chains of ferrimagnetic crystals, known as magnetosomes, are contained in magnetotactic bacteria [69]. Figure 8 shows an off-axis electron hologram of a chain of magnetite crystals in a single cell of the bacterial strain *Magnetospirillum magnetotacticum*. After magnetization reversal, the MIP contribution to the phase shift can be identified, as shown in Fig. 8(b). The crystallites in this image are revealed by the thickness contours to have cuboctahedral shapes [70]. The corresponding magnetic contributions to the phase shifts, achieved by subtracting the respective pairs of holograms, are shown in Fig. 8(c). Phase contours have been overlaid onto the MIP contribution allowing the positions of the crystals and the magnetic contours to be correlated with each other. The magnetic flux within and between the crystallites can be clearly visualized. Moreover, it is straightforward to obtain an estimate for the total magnetic dipole moment of the particle chain by measuring the step in the magnetic contribution to the phase shift across the chain, as described elsewhere [8]. Further studies of several strains of magnetotactic bacteria have revealed chains of crystals that form either single magnetic domains [71], or else more complicated domain structures in larger magnetosomes [72].

In contrast, chains of FeNi nanoparticles, which have variable sizes, spherical geometry, and a mean diameter of 50 nm, show markedly different behavior [73]. Closely-spaced phase contours are observed to channel through several particles along such chains. Some very interesting domain structures arise, primarily due to the variability of the particle sizes and their relative positions, both along the chain and with respect to the chain axis. The key result, which is illustrated by the examples shown in Fig. 9 and confirmed by micromagnetic simulations, is the formation of three-dimensional magnetic vortices with complex geometries that are determined by the particle sizes and locations. These results emphasize yet again the necessity for controlling the shapes, sizes, and relative locations of closely-separated magnetic nanoparticles, and for understanding their mutual interactions.

5. FUTURE PERSPECTIVES

Off-axis electron holography has evolved over the years to the extent that it can be considered as a routine and reliable technique for characterizing nanostructured materials. As the dimensions of magnetic storage and electronic devices continue to shrink still further, electron holography will provide an increasingly valuable approach for solving industrial problems, as well as for providing advances in fundamental scientific knowledge. Unique information about nanoscale electrostatic and magnetic fields is readily obtainable, and quantification on the same scale is often possible. A major restriction of the technique is that dynamic events cannot conveniently be followed in real time because of the need for off-line processing. However, there is some realistic prospect that faster recording systems and computers may overcome this limitation. The available field of view, which depends on the geometry and the dimensions of the imaging and recording systems (e.g., the width of the fringe overlap region, the sampling of the holographic fringes, and the finite size of the CCD array) can also represent a very real limitation for some applications. Nevertheless, the results described here and elsewhere indicate that the technique will become more and more heavily used in the burgeoning field of nanotechnology.

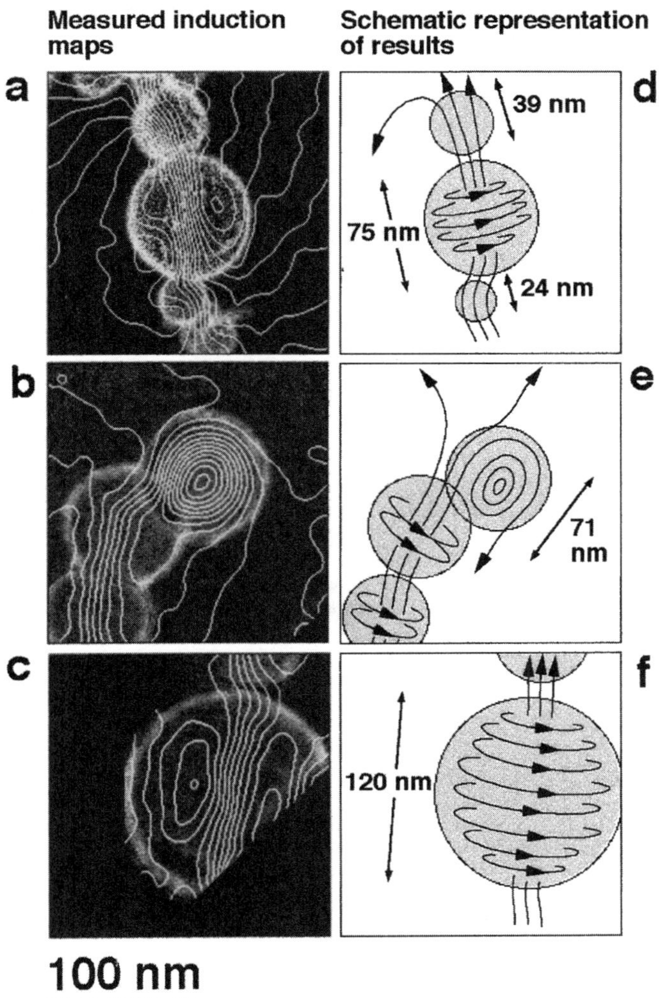

Figure 9. (a–c) Experimental phase contour maps along chains of FeNi nanoparticles after removal of mean inner potential contribution. Contour spacings of 0.083, 0.2, and 0.2 radians, respectively. (d–f) Schematic interpretation of magnetic microstructure, including magnetic vortices in each case [73].

REFERENCES

1. D. Gabor, *Proc. Roy. Soc. London*, A197 (1949) 454.
2. A. Tonomura, L. F. Allard, G. Pozzi, D. C. Joy, and Y. A. Ono (Eds.), *Electron Holography*, Elsevier, Amsterdam (1995).
3. E. Völkl, L. F. Allard, and D. C. Joy (Eds.), *Introduction to Electron Holography*, Plenum, New York (1998).
4. A. Tonomura, *Electron Holography*, Springer-Verlag, Berlin (1999).
5. H. Lichte, *Adv. Opt. Elect. Microsc.* 12 (1991) 25.
6. A. Tonomura, *Adv. Phys.* 41 (1992) 59.
7. P. A. Midgley, *Micron* 32 (2001) 167.

8. R. E. Dunin-Borkowski, M. R. McCartney, and D. J. Smith, In: *Encyclopedia of Nanoscience and Nanotechnology*, H. S. Nalwa (Ed.), American Scientific, Stevenson Ranch, CA (2004).
9. J. M. Cowley, *Ultramicroscopy* 41 (1992) 335.
10. J. M. Cowley, *Diffraction Physics*, Third Revised Edition, North Holland, Amsterdam (1995).
11. K. Ishizuka, *Ultramicroscopy* 51 (1993) 1.
12. L. Reimer, *Transmission Electron Microscopy*, Springer, Berlin (1991).
13. J. C. H. Spence, *Acta Cryst*, A49 (1993) 231.
14. M. A. van Hove, W. H. Weinberg, and C. M. Chan, *Low Energy Electron Diffraction*, Springer, Berlin (1986).
15. D. K. Saldin and J. C. H. Spence, *Ultramicroscopy* 55 (1994) 397.
16. M. Gajdardziska-Josifovska, M. R. McCartney, W. J. de Ruijter, D. J. Smith, J. K. Weiss, and J. M. Zuo, *Ultramicroscopy* 50 (1993) 285.
17. M. Gajdardziska-Josifovska and A. Carim, In: *Introduction to Electron Holography*, E. Völkl, L. F. Allard, and D. C. Joy (Eds.), Kluwer, New York (1998) pp. 267–293.
18. J. Li, M. R. McCartney, R. E. Dunin-Borkowski, and D. J. Smith, *Acta Cryst.* 55 (1999) 652.
19. W. J. de Ruijter and J. K. Weiss, *Ultramicroscopy* 50 (1993) 269.
20. W. J. de Ruijter and J. K. Weiss, *Rev. Sci. Inst.* 63 (1992) 4314.
21. W. O. Saxton, T. J. Pitt, and M. Horner, *Ultramicroscopy* 4 (1979) 343.
22. W. D. Rau, H. Lichte, E. Völkl, and U. Weierstall, *J. Comp-Assisted Microsc.* 3 (1991) 51.
23. H. Lichte, *Ultramicroscopy* 20 (1986) 293.
24. H. Lichte, *Ultramicroscopy* 51 (1993) 15.
25. A. Harscher and H. Lichte, *Ultramicroscopy* 64 (1996) 57.
26. H. Lichte, K.-H. Herrmann, and F. Lenz, *Optik* 77 (1987) 135.
27. D. J. Smith, W. J. de Ruijter, M. R. McCartney, and J. K. Weiss, In: *Introduction to Electron Holography*, E. Völkl, L. F. Allard, and D. C. Joy (Eds.), Kluwer, New York (1998) pp. 107–124.
28. D. C. Ghiglia and M. D. Pritt, *Two-Dimensional Phase Unwrapping: Theory, Algorithms and Software*, Wiley, New York (1998).
29. E. Völkl, L. F. Allard, and B. Frost, *Ultramicroscopy* 58 (1995) 97.
30. D. J. Smith and M. R. McCartney, In: *Introduction to Electron Holography*, E. Völkl, L. F. Allard, and D. C. Joy (Eds.), Kluwer, New York (1998) pp. 87–106.
31. M. R. McCartney, D. J. Smith, R. F. C. Farrow, and R. F. Marks, *J. Appl. Phys.* 82 (1997) 2461.
32. R. E. Dunin-Borkowski, M. R. McCartney, B. Kardynal, S. S. P. Parkin, M. R. Scheinfein, and D. J. Smith, *J. Microscopy* 200 (2000) 187.
33. S. M. Sze, *Physics of Semiconductor Devices*, Wiley, New York (2002).
34. S. Frabonni, G. Matteucci, G. Pozzi, and U. Valdre, *Phys. Rev. Lett.* 44 (1985) 2196.
35. M. R. McCartney, D. J. Smith, R. Hull, J. C. Bean, E. Völkl, and B. Frost, *Appl. Phys. Lett.* 65 (1994) 2603.
36. M. R. McCartney, M. A. Gribelyuk, J. Li, P. Ronsheim, J. S. McMurray, and D. J. Smith, *Appl. Phys. Lett.* 80 (2002) 3213.
37. M. A. Gribelyuk, M. R. McCartney, J. Li, C. S. Murthy, P. Ronsheim, B. Doris, J. S. McMurray, S. Hegde, and D. J. Smith, *Phys. Rev. Lett.* 89 (2002) 025502.
38. A. Twitchett, R. E. Dunin-Borkowski, and P. A. Midgley, *Phys. Rev. Lett.* 88 (2002) 238302.
39. A. Twitchett, R. E. Dunin-Borkowski, R. F. Broom, and P. A. Midgley, *J. Phys. Condens. Matt.* 16 (2004) S181.
40. W. D. Rau, P. Schwander, F. H. Baumann, W. Hoppner, and A. Ourmazd, *Phys. Rev. Lett* 82 (1999) 2614.
41. J. Li, M. R. McCartney, and D. J. Smith, *Ultramicroscopy* 94 (2003) 149.
42. H. Morkoç, *Nitride Semiconductors and Devices*, Springer, Berlin (1999).
43. J. Barnard and D. Cherns, *J. Electron Microscopy* 49 (2000) 281.
44. D. Cherns, H. Mokhtari, C. G. Jaio, R. Averback, and H. Riechert, *J. Cryst. Growth* 230 (2001) 410.
45. J. Cai and F. A. Ponce, *J. Appl. Phys.* 91 (2002) 9856.
46. J. Cai, F. A. Ponce, S. Tanaka, H. Omiya, and Y. Nakagawa, *phys. stat. sol.* (a) 188 (2001) 833.
47. M. R. McCartney, F. A. Ponce, J. Cai, and D. P. Bour, *Appl. Phys. Lett.* 76 (2000) 3055.
48. J. Cai and F. A. Ponce, *phys. stat. sol.* (a) 192 (2002) 407.
49. D. Cherns and C. G. Jiao, *Phys. Rev. Lett.* 87 (2001) 5504.
50. C. G. Jiao and D. Cherns, *J. Electron Microscopy* 51 (2002) 105.
51. I.-H. Tan, G. L. Snider, L. D. Chang, and E. L. Hu, *J. Appl. Phys.* 68 (1990) 4071.
52. A. Krtschil, A. Dadger, and A. Krosy, *Appl. Phys. Lett.* 82 (2003) 2263.

53. H.-J. Im, Y. Ding, J. P. Pelz, B. Heying, and J. S. Speck, *Phys. Rev. Lett.* 87 (2001) 106802.
54. R. Saito, G. Dresselhaus, and M. S. Dresselhaus, *Physical Properties of Carbon Nanotubes*, Imperial College Press, London (1999).
55. J. Cumings, A. Zettl, M. R. McCartney, and J. C. H. Spence, *Phys. Rev. Lett.* 88 (2002) 056804.
56. L. F. Allard, E. Völkl, D. S. Kalakkad, and A. K. Datye, J. *Mater. Sci.* 29 (1994) 5612.
57. L. F. Allard, E. Völkl, A. Carim, A. K. Datye, and R. Ruoff, *Nano. Mater.* 7 (1996) 137.
58. M. R. McCartney and M. Gajardziska-Josifovska, *Ultramicroscopy* 53 (1994) 283.
59. M. Gajdardziska-Josifovska and M. R. McCartney, *Ultramicroscopy* 53 (1994) 291.
60. J. K. Weiss, W. J. de Ruijter, M. Gajdardziska-Josifovska, M. R. McCartney, and D. J. Smith, *Ultramicroscopy* 50 (1993) 301.
61. M. Gajdardziska-Josifovska, *MSA Bulletin* 24 (1994) 507.
62. Y. Miura, *J. Magn. Magn. Mater.* 134 (1994) 209.
63. J. N. Chapman and M. R. Scheinfein, *J. Magn. Magn. Mater.* 200 (1999) 729.
64. R. E. Dunin-Borkowski, M. R. McCartney, B. Kardynal, and D. J. Smith, *J. Appl. Phys* 84 (1998) 374.
65. R. E. Dunin-Borkowski, M. R. McCartney, B. Kardynal, D. J. Smith, and M. R. Scheinfein, *Appl. Phys. Lett.* 75 (1999) 2641.
66. D. J. Smith, R. E. Dunin-Borkowski, M. R. McCartney, B. Kardynal, and M. R. Scheinfein, *J. Appl. Phys.* 87 (2000) 7400.
67. R. E. Dunin-Borkowski, M. R. McCartney, B. Kardynal, M. R. Scheinfein, D. J. Smith, and S. S. P. Parkin, *J. Appl. Phys.* 90 (2001) 2899.
68. S. S. P. Parkin, *Ann. Rev. Mater. Sci.* 25 (1995) 357.
69. D. A. Bazylinski and R. B. Frankel, In: *Biomineralization: From Biology to Biotechnology*, E. Bauerlein (Ed.), Wiley-VCH, Weinheim (2000).
70. R. E. Dunin-Borkowski, M. R. McCartney, R. B. Frankel, D. A. Bazylinski, M. Pósfai, and P. R. Buseck, *Science* 282 (1998) 1868.
71. R. E. Dunin-Borkowski, M. R. McCartney, M. Pösfai, R. B. Frankel, D. A. Bazylinski, and P. R. Buseck, *Eur. J. Mineral.* 13 (2001) 671.
72. M. R. McCartney, U. Lins, M. Farina, P. R. Buseck, and R. B. Frankel, *Eur. J. Mineral.* 13 (2001) 685.
73. M. J. Hÿtch, R. E. Dunin-Borkowski, J. Moulin, C. Duhamel, F. Mazaleyrat, and Y. Champion, *Phys. Rev. Lett.* 91 (2003) 257207.

21. SUB-NM SPATIALLY RESOLVED ELECTRON ENERGY-LOSS SPECTROSCOPY

CHRISTIAN COLLIEX AND ODILE STÉPHAN

1. INTRODUCTION: EELS AND NANOTECHNOLOGY

Several challenges face materials science at the beginning of the 21st Century. Among them particular attention should be devoted to understanding a material's behavior from the atomic/nanolevel via microstructure to macrostructure levels. Knowledge of materials at the nanoscale and control of the structural and functional properties of newly synthesized materials constitute key issues for technological progress and opening up of new markets in many major industrial domains e.g. (naming areas of economic importance) in electronics, photonics, communication, catalysis, construction, aeronautics and automobile. Most fruitful developments today involve new synthesis routes for building novel structures, with the goal of improving the presently available properties and therefore developing new fields of applications.

Within this global frame, the characterization stage constitutes a major task, as it provides the information necessary for correlating the synthesis process and the measured properties. In order to fulfill this mission, a wide range of levels must be explored, starting from the individual atom and its bonding within the condensed matter phase. Consequently, new and more powerful analytical techniques are required, combining advanced instrumentation for performing measurements at the (sub)-nanometer scale with new theoretical descriptions and increased computational support to test the validity of the models.

Transmission Electron Microscopy (TEM) with its new breathtaking developments in electron sources, monochromators and correctors, has demonstrated its unique role in providing not only knowledge about the structure at all levels but also the local properties and the ability to manipulate on the atomic scale. The importance of this performance on the local scale is increasing as we are able to access the possibilities of nanoscience and experience the need for tools for characterization, control and handling.

It combines high resolution imaging and diffraction for mapping the atomic structure of thin objects, not only as two dimensional projections but increasingly as three dimensional volumes, with probe analytical techniques for exploring the many electronic and chemical properties over a wide range of scales. It must be emphasized that the physical size of the object does not always constitute the relevant scale for the physical properties under consideration, which are more generally governed by characteristic lengths such as field penetration depth, particle mean free path, phase coherence length, etc.

Furthermore, the improvement in detectors performance and computing speed has also given access to real time measurements, so that phenomena can be monitored under the application of external constraints or fields, while keeping the high level of spatial resolution. This novel trend will therefore transform the next generation of TEMs into "nano-laboratories".

In this paper, we will focus on the role of electron energy-loss spectroscopy (EELS) for investigating the elemental composition and the electronic properties of individual nano-objects from the nm-scale down to the atomic level. In particular, the importance of correlating structural, chemical and electronic information at this level of spatial resolution will be stressed for addressing key challenges in nanosciences and nanotechnologies.

In a first part, we will provide a survey of EELS spectroscopy and underline the different types of information which it conveys, introducing therefore the developments in theoretical interpretation under course to fully exploit the power of the technique. For a more complete review of these aspects, see the formerly published reference texts such as [1–5]. In a second chapter, interest will be focused onto the instrumentation and methodology developed for practical implementation and use in the electron microscope. It will result into the introduction of the concept of spatially resolved EELS and the associated 3D data cube, the access to which can be realized either by spectrum-imaging or by energy filtered imaging. A first class of applications, namely elemental mapping at the ultimate resolution, will be described in chapter three, while the final one will be devoted to mapping electronic structures and bonding states. In both cases, examples related to individual nano-objects will be selected and the successes and limits will be stressed. Finally, we will conclude with a few arguments intended to delineate some potential routes for future applications in of these techniques: either as a type of synchrotron-type spectroscopy within an electron microscope column, or as an essential contribution to the nano-laboratory to be designed and run.

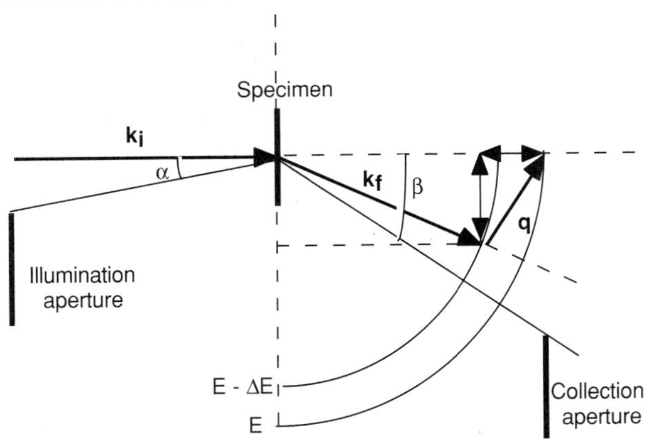

Figure 1. Schematic representation of the parameters involved in an EELS measurement and relation with the geometry of the scattering event: incident electron of wave vector $\vec{k}_i$, scattered electron wave vector $\vec{k}_f$, transferred wave vector $\vec{q}$.

2. UNDERSTANDING THE INFORMATION CONTAINED IN AN EELS SPECTRUM

2.1. Definition of an EELS Spectrum and of the Basic Information Which it Contains

In the context of a transmission electron microscope, a primary beam of high energy electrons (typically between 100 and 300 keV) and of convergence half angle α, interacts with a specimen sufficiently thin so that the number of collision events remains low. A suitable aperture defines at the level of the specimen a solid angle of collection (of acceptance half angle β of a few mrad)—see figure 1—and the velocity of the electrons contained within this aperture is analyzed with a spectrometer, of magnetic sector type in most cases. Consequently, at the exit of this prism, the distribution in energy of these electrons can be recorded and the results displayed as an electron energy loss spectrum, similar to that displayed in figure 2. As a matter of fact, this double selection in angle and energy provides access to an analysis of the occurred scattering events in terms of momentum transfer $\vec{q}$ and of energy transfer ΔE, see figure 1.

The EELS spectrum exhibits the following features: the major contribution to the detected intensity is a very intense and narrow peak (zero-loss peak) corresponding to the electrons which intersect the specimen without measurable inelastic scattering. Inelastically scattered electrons are responsible for a complex spectrum which contains two major types of excitations: the low energy-loss range or plasmon-type excitations in the 5 to 40 eV range and the core-loss range where characteristic edges are superposed on a monotonously decaying background.

The low-energy class of excitations involves the valence (and conduction) electrons which can be excited either collectively in plasmon modes, or individually as interband transitions—see figure 3—. The theoretical description of these modes is not

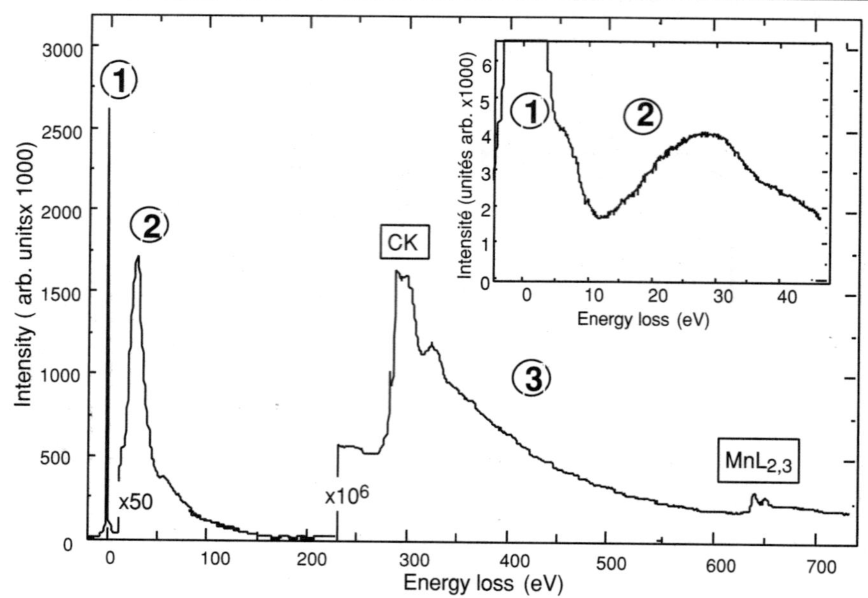

Figure 2. A typical EELS spectrum acquired on a carbon nanotube filled with manganese. The whole spectrum extending over 1000 eV covers four orders of magnitude in intensity when comparing the non-saturated zero-loss peak (1) and the details of Mn edge around 600 eV loss (the core-loss region is labeled (3). In inset: the tail of the zero-loss peak (labeled 1) and the low-loss domain, which is dominated by plasmon losses (labeled 2).

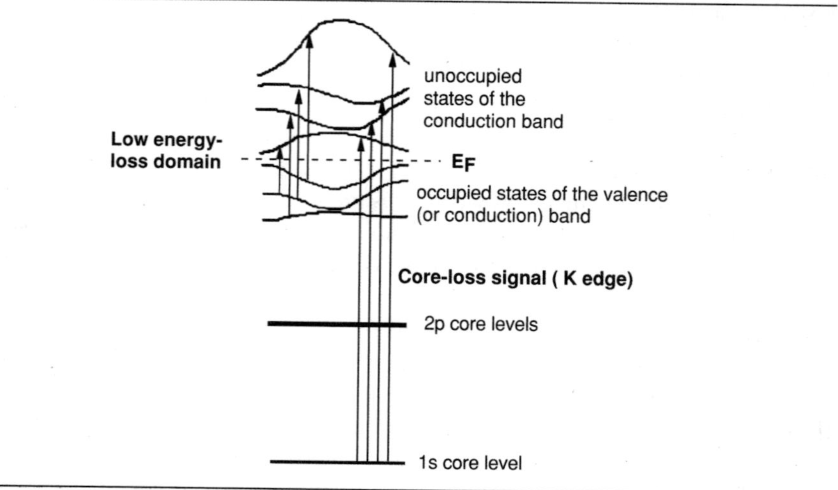

Figure 3. Schematics of the different populations of electrons involved in the different excitation processes (low energy-loss and core-loss domains) investigated in an EELS experiment.

straightforward as it encompasses a complex mixture of transitions involving a population of particles in strong Coulomb interaction. As a first stage, a phenomenological macroscopic approach, a classical electrodynamics one as we are dealing with charged particles, is commonly used. It involves free charges in a metallic case and oscillators with eigen modes in a dielectric medium. The central quantity to account for the response of any material to the effect of the impinging point charge, the incident electron, is the dielectric response $\varepsilon(\omega, \vec{q})$ where ω is related to the energy loss by $\Delta E = \hbar\omega$ and $\vec{q}$ is the wave vector transfer defined above. The inelastic cross section per incident electron for such a scattering event $(\omega, \vec{q})$ is in fact related to the dielectric response by:

$$\frac{d^2\sigma}{dE d\Omega} \propto \frac{1}{q^2} \mathrm{Im}\left(-\frac{1}{\varepsilon(\omega, \vec{q})}\right) \tag{1}$$

where ε is a complex function that can be written as the sum of a real part and an imaginary part as: $\varepsilon = \varepsilon_1 + i\varepsilon_2$. The real part ε_1 describes a dispersion process while the imaginary part ε_2 has to be related to electromagnetic absorption in the $\vec{q} = 0$ limit. Equation (1) is the basic formula for the interpretation of electron energy-loss spectra in the low-loss region in bulk materials.

Plasmon modes (polarization excitations) are defined as longitudinal solutions of the Maxwell equation $\vec{\nabla}\varepsilon\vec{E} = 0$, which is equivalent to the condition: $\varepsilon(\omega, \vec{q}) = 0$ for isotropic materials. Such condition is reached when ε_1 is 0 and ε_2 is small and is associated with the occurrence of a peak in the EELS spectrum. EELS is then a powerful technique for studying longitudinal excitations as they appear directly as intense peaks in the energy-loss spectrum. When the collection angle is small enough and centered on the optical axis, the limit $q \to 0$ is reached. A Kramers-Krönig analysis can be performed to extract $\varepsilon_1(\omega, 0)$ and $\varepsilon_2(\omega, 0)$ optical constants from the energy-loss function $\mathrm{Im}(-\frac{1}{\varepsilon(\omega,0)})$. Such a transformation relies on the causality of the dielectric response (see [6] for a review).

Generally speaking, these low-energy loss excitations reveal electron distributions governed by solid state considerations and exhibit a response of non-local character to the probing charge. This is particularly true for the bulk plasmon, the energy of which is mostly defined by an average electron density. When comparing materials of rather similar composition and structures, this plasmon energy can constitute a satisfactory criterion for monitoring variations in electron density [7]. Several groups [8–10] have demonstrated that relationships exist between the plasmon energy and physical properties such as Young's modulus, electrical and thermal conductivities. When fully understood, these correlations should open the way to a quick and reliable mapping of different types of physical property.

However, such use of plasmon loss energies remains rather limited when one deals with structures and objects of nanometer size. It has been demonstrated that its value may shift and its width rapidly increase when the size of the investigated structure becomes smaller than a few nm and completely vanish for smaller objects [11]. For these dimensions, the weight of the surface modes becomes dominant. These modes are generated when certain boundary conditions are satisfied. In particular, they are generated when an incident electron travels along a non-intersecting trajectory at a small distance from the outer surface of a nano-object, in a mode called "aloof" or "near-field" geometry [12, 13]. It has been shown that the response then depends on the nature and the shape of the surface, and in the case of empty objects (such as nanotubes) on the coupling between the outer and inner surfaces for small wall thickness [14, 15].

Another valuable information to be extracted in the vicinity of the zero-loss peak is the joint density of states at the band gap in insulating materials, and for practical reasons it has been practically tested on materials of relatively large band gap [16–18]. In these bulk materials, the nature, direct or indirect, and the energy E_g of a band gap can be deduced from the study of the first eV of the imaginary part ε_2 of the dielectric function $\varepsilon(\omega)$, which is deduced from the energy loss function $\mathrm{Im}(-\frac{1}{\varepsilon(\omega,0)})$ through a Kramers-Krönig transformation. The application of this technique to map the band gap variations in a luminescent heterostructure (preferentially in the blue) should be demonstrated within a near future.

On the contrary, absorption edges (or core-edges), occurring at higher energy losses, arise from individual electron transitions from atomic-type core levels to unoccupied states lying above the Fermi level, see figure 3. They are identified and labeled following the notations generally used in atomic physics. Their position in energy is defined by the nature of the atom and the type of level from which the excitation occurs (K for 1s electrons, L_{23} for 2p electrons, M_{45} for 3d electrons, etc). In figure 2, the insets correspond to the carbon K edge with threshold at 284 eV and to the manganese L_{23} edge with threshold at 642 eV. Such signals appearing superposed on the non-specific decaying background are useful for three reasons. When identified, they testify for the presence of the related atoms along the trajectory of the incident electrons through the specimen, this is the basis of a qualitative elemental analysis. Their total intensity after subtraction of the underlying non-characteristic background and integration over an extended energy window, can be related to the number of such atoms encountered by the impinging atoms, providing therefore access to a quantitative elemental analysis. Furthermore, as shown in the insets of figure 2, their shape varies from edge to edge and furthermore from compound to compound for the same edge. These variations of the edge fine structures is very useful for extending the analysis of the solid from elemental type to bonding type. These last two issues will be further discussed in the next paragraph.

2.2. Basic Tools Developed for Interpreting and Using Core-Loss Signals

Using basic quantum mechanics theory and in particular the Fermi golden rule, the cross section introduced in eqn. (1) can be written as:

$$\frac{\partial^2\sigma}{\partial E\partial\Omega} = \left(\frac{2\pi}{\hbar}\right)^4 m^2 \sum_f \frac{k_f}{k_i} |\langle\psi_f|\langle k_f|V|k_i\rangle|\psi_i\rangle|^2 \delta(E_i - E_f) \quad (2)$$

where $|\psi_i\rangle$ and $|\psi_f\rangle$ are respectively the initial and final states of the target solid and $|k_i\rangle$ and $|k_f\rangle$ those of the fast electron, V is the perturbing potential (Coulomb interaction potential between the incident electron and the atoms of the solid). To derive this expression, one considers the case of fast collisions (the criterion is that the incident particle velocity is "fast" relative to a mean orbital velocity of the electrons in the shells concerned with the excitation process), the influence of the incident particle upon the atoms of the solid may then be regarded as a sudden and small perturbation. This expression is very general and describes any type of inelastic scattering process (in case of a fast collision) between an incident electron and a solid.

The estimation of expression (2) is relatively simple in the case of core-loss signals. There, $|\psi_i\rangle$ is a core atomic orbital and V, the potential between the probe electron at position $\vec{r}$ and one target atom can be written as:

$$V = \frac{-Ne^2}{r} + \sum_{j=1}^{N} \frac{e^2}{|\vec{r} - \vec{r}_j|} \tag{3}$$

where the sum is performed over all N electrons centered at positions $\vec{r}_j$ of the target atom. We specify that this expression is valid as long as inner-shell excitations are considered since the target solid can be considered in that case as an arrangement of non-interacting atoms. Then, the transition matrix element $\langle k_f|V|k_i\rangle$ in eqn. (2) involves the Fourier transform of the Coulomb potential:

$$\int d^3r \frac{e^2}{|\vec{r} - \vec{r}_j|} e^{i\vec{q}\cdot\vec{r}_j} = \frac{4\pi e^2}{q^2} e^{i\vec{q}\cdot\vec{r}_j} \tag{4}$$

and expression (2) becomes:

$$\frac{\partial^2\sigma}{\partial E \partial\Omega} = \left(\frac{2me^2}{(\hbar q)^2}\right)^2 \frac{k_f}{k_i} \sum_f \sum_{j=1}^{N} |\langle\psi_f|e^{i\vec{q}\cdot\vec{r}_j}|\psi_i\rangle|^2 \delta(E_i - E_f + \hbar\omega) \tag{5}$$

where the energy loss $\Delta E = \hbar\omega = E_{k_f} - E_{k_i}$ has been introduced.

This expression of the differential cross-section actually consists of two factors of different nature. The factor $(2me^2/(\hbar q)^2)^2 \cdot k_f/k_i$ is evaluated from the observable quantities k_i, k_f, θ concerning the fast electron only. The squared matrix element concerns the target only, it represents the conditional probability that the atom of the solid makes the transition to a particular excited state upon receiving a momentum transfer $\vec{q}$. Therefore, this quantity reflects the dynamics of the involved atom. It is known as the inelastic form factor or the generalized oscillator strength. This factor together with the δ-function expressing the conservation of energy, constitutes one key factor to be extracted from an EELS experiment.

What are the major consequences of this basic theory for using EELS spectra to analyze specimens? The first one is the dipole approximation, valid for the small scattering angles generally met in the experimental conditions depicted in figure 1 with α

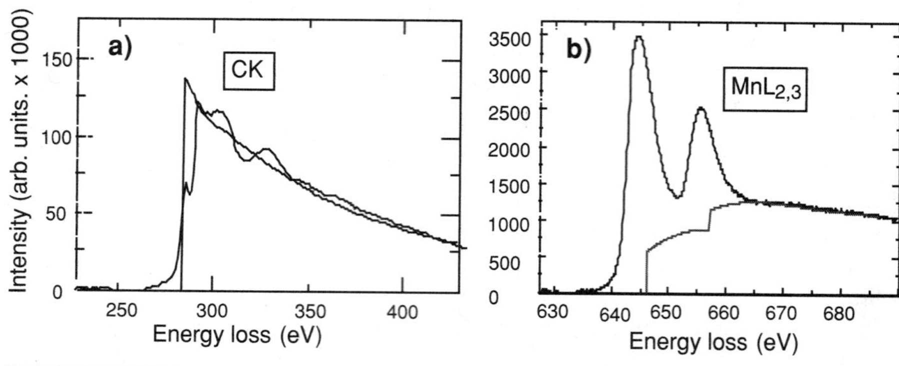

Figure 4. Detailed shapes of core-edges (C K and Mn L_{23} extracted from the whole spectrum shown in figure 2) after background subtraction fitted to calculated atomic cross-sections. Discrepancies involve solid state effects (Near Edge Structures) as well as transitions towards bound states (White Line features)

and β (illumination and collection apertures) in the range 1 to 30 mrad and primary electrons of 100 to 300 keV energy. In these conditions, $|\vec{q} \cdot \vec{r}| \ll 1$ and $|\phi_f\rangle$ and $|\phi_i\rangle$ are orthogonal, so that the matrix element is reduced to a dipolar term obtained in the zero limit value of $\vec{q}$. Consequently, the most intense atomic transitions to be searched are those corresponding mostly to $\ell \to \ell + 1$ (i.e. $s \to p$, $p \to d$ or $d \to f$) which are more intense than the other allowed transitions $\ell \to \ell - 1$. The second one is that it is relatively simple to calculate atomic cross-sections for isolated atoms, resulting in tables of $\frac{\partial^2\sigma}{\partial E \partial \Omega}$. Such cross-section computations based on the hydrogenic model or on the more sophisticated so-called Hartree-Slater method using a self-consistent atomic potential, provide profiles of the energy dependence of the cross sections and of their absolute intensities. Their major use is for quantitative analysis, as will be shown in a next chapter. If we compare such calculations with experimental profiles of edges such as shown in figure 4, one sees obvious discrepancies concerning the shapes. These are the fine structures, the interpretation of which is of highest importance to identify the coupling of the excited atom to its neighboring atoms and therefore reflects its coordination and bonding character.

For the carbon C K edge, modulations of the atomic profile account for the solid state environment of the carbon atoms within this structure, which is of graphite type. For the Mn L_{23} edge, one notes the presence of two extra lines at the edge onset, they correspond to transitions of the 2p electrons towards 3d states of bound character which are not accounted for by the simple atomic calculations. The atomic calculation generally provides a satisfactory description of the general shape of the edge, as a saw-tooth (hydrogenic) profile or as a delayed maximum induced by a centrifugal barrier within the intra-atomic potential of the excited atom.

When estimating the band structure effects in a simple one electron transition description, one shows that the probability of transition calculated by the matrix elements introduced in eqn. (5) is modulated by the respective densities of unoccupied states $D_\ell(E)$ with the right ℓ angular symmetry at the site of the excited atom:

$$\frac{d^2\sigma}{dEd\Omega} = |M_{\ell-1}(E)|^2 D_{\ell-1}(E) + |M_{\ell+1}(E)|^2 D_{\ell+1}(E) \tag{6}$$

Several methods have been developed to calculate these densities of states and it may be interesting to monitor the changes in the attained degree of refinement, as the computing power and speed have permanently been improved [19–23]. However, such a discussion obviously exceeds the scope of this review. Let us classify, in simple terms, the currently used approaches. In a first level approximation, the influence of the closest neighboring atoms can be described by building atomic clusters around the excited one, therefore involving a superposition of atomic potentials or wave functions. It has given rise to two rather equivalent sets of methods:

(i) Real space single and multiple scattering descriptions of the outgoing wave within the successive shells of these clusters, have been developed to model the EXAFS and XANES types of fine structures and the most commonly used software seems now to be the latest versions of the FEFF routines [24]. As a consequence of their shell by shell scattering path expansion, they have demonstrated a great potentiality for providing structural insight of the different peaks, to understand how they changes in position, size and shape with different nearest-neighbor environments (coordinence, bond length);
(ii) A linear combination of atomic wave functions within a molecular cluster (i.e. a molecular orbital—MO—description) constitutes another way to attribute the ELNES features to transitions towards unoccupied MOs made as combinations of atomic orbitals on the excited atom and on its nearest neighbors and possibly beyond. It has been useful in identifying structures, close to the edge threshold, to the low-energy unoccupied orbitals in molecular clusters [25].

More refined approaches to the calculation of local density of states in the band structure description require the generation of self-consistent potentials from a ground state charge density to be used for the calculation of the excited states. The one electron Schrödinger equation is solved in the density functional theory (DFT) including the local density approximation (LDA). This is basically a ground state theory and its relevance for the interpretation of excited states can only provide a reasonable framework, when one compares directly the experimental spectra to the calculated densities of states. Similar observations have been made as well when the CASTEP code with pseudo-potentials or the Wien 2k code with full linear augmented plane waves are used.

Consequently, the role of the hole created on the excited atom must be accounted for when a satisfactory description of final state is searched. The validity of using either a Z + 1 atomic description, a frozen core hole or a screened core hole, has been evaluated in many different cases (investigated atomic level, type of bonding

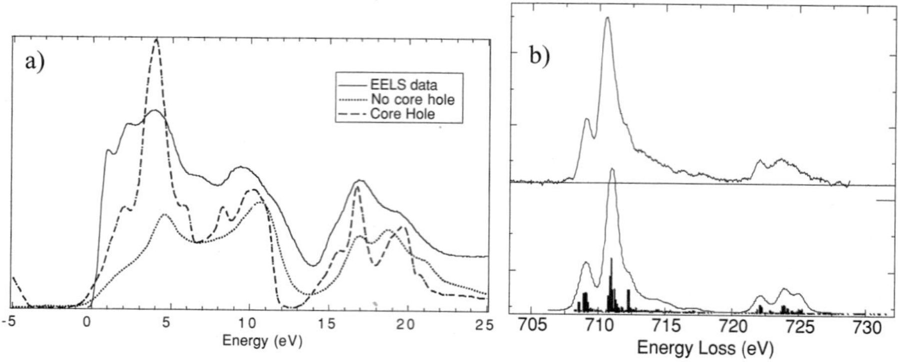

Figure 5. Comparison of high resolution EELS spectra with simulated ones: (a) C K edge for diamond carbon calculated using the CASTEP code: solid line, experiment, doted line: without core-effet, dashed line: with core-hole effect (courtesy of A. Zobelli); (b) Fe L_{23} edges in hematite α-Fe_2O_3: top, experimental spectrum recorded with a LaB_6 200 kV TEM and deconvoluted using 30 iterations of the RL procedure; bottom, atomic multiplet calculation for Fe^{3+} ions and Oh crystal field of 2 eV strength. An extra smoothing with a 0.2 eV gaussian peak has been added (courtesy of A. Gloter)

more particularly for insulating materials where screening of the hole is limited, spatial extension of the excited orbital). These effects have also been tested for the different methods used for calculating the non-perturbated density of states (real space multiple scattering cluster approach, density functional theory with a band structure supercell method, first principle Xα molecular method). The role of the size of the supercell, which is required to model the structure with one extra charge at the center, has also been considered. It is evident from all these studies, that it is necessary in many cases to introduce the core hole effect to obtain a more satisfactory agreement with the experimental data. This is demonstrated in figure 5a which compares a high resolution C K edge of diamond carbon, with CASTEP calculated spectra without and with consideration of a core hole effect on the central C atom.

When considering the detailed structures of the white lines visible on transition metal L_{23} edges such as the Mn one shown in figure 4b or the equivalent Fe one, they are mostly governed by interaction configuration effects within the potential of a given ion species (i.e. Fe^{++} or Fe^{+++}). One must then account for the strong coulomb and exchange interactions between the hole created on the 2p level and the extra electron put in the 3d level. This is another typical case where the single-electron transition description is not sufficient to interpret the data. In this case, the effect of the neighboring atoms is essentially described by an extra term dealing with the local symmetry and the associated crystal field. With the support of these two major parameters (intra-atomic correlation effects and crystal field energy), one can compute all types of white line distributions [26]. The case of the iron 2p line in hematite is shown in figure 5b.

As a conclusion, the tools required for satisfactorily interpreting the fine structures recorded on the EELS core-loss edges are quite diversified. Their capacity to provide good fits is regularly increasing with the development of new tools and the access to more powerful and fast computing facilities, even for structures of bulk materials. The next step will be to adapt these tools to the modeling of excitation spectra recorded at a single defect or within a nanometer-size cluster or nanostructure. These needs and the ways explored to progress along this direction will be outlined in the final chapter of the present review.

3. SPATIALLY RESOLVED EELS

3.1. The 3D Data Cube

The measurement of the energy loss ΔE suffered by any electron transmitted through the specimen at position (x, y) adds a new dimension of information to the current 2D spatial information provided by any type of scattering responsible for a contrast in a TEM image. One then measures an intensity $I(\Delta E, x, y)$ within an elementary volume defined in a 3D space, with two axes related to the position and one to the spectral information [27]. The knowledge of the intensity distribution within this 3D data cube enables to correlate the spectroscopic signal to the exact position of its origin, with the available spatial and energy resolution. For small objects with size in the nanometer range and probes at the angström level, it is clear that mapping the different contributions in the EELS spectrum provides an invaluable access to elemental composition and detailed electron structure at the ultimate quasi-atomic level [28–31]. Furthermore, the 3D data cube contains a collection of individual spectra corresponding to many points (typically from a few 10^2 to a few 10^4), which has the advantage of identifying the changes of fine structures between different positions on the specimen and of providing reference spectra with increased signal to noise from well defined and homogeneous areas of the specimen.

Two methods can be used to record and store the 3D data cube:

(i) A spectrometer in a scanning TEM mode with parallel acquisition of the EELS spectra provides a full $I(\Delta E)$ spectrum made of n energy-loss channels of width ∂E, for each probe position, i.e. for each pixel defined by its coordinates (x, y). All spectra are piled parallel one to another, to produce the 3D data cube, see figure 6a. The collection time Δt_1 per pixel is set by the condition that the associated dose $D = J_1 \cdot \Delta t_1$ is sufficient to produce a signal with a given signal-to-noise ratio. It depends practically on the physical parameters (thickness, composition) of the specimen and on the recorded spectral channels. This mode called "spectrum-image" has been implemented and used for about ten years [32, 33].

(ii) An energy filtering microscope (EFTEM) provides a complete 2D image made of $N \times N$ pixels, using only the electrons contained within an energy window ΔE defined by the post-filter selection slit. Similar criteria of signal-to-noise ratio apply for estimating the required recording time Δt_2. To obtain the same intensity $I(\Delta E, x, y)$ on a given area element within the same energy loss window ∂E, implies that $J_2 \cdot \Delta t_2 = J_1 \cdot \Delta t_1$, all other parameters governing the signal generation being equivalent. All filtered images are stacked parallel one to another to produce the 3D data cube., see figure 6b. This mode, called "image-spectrum", has been first implemented by Lavergne *et al.* [34].

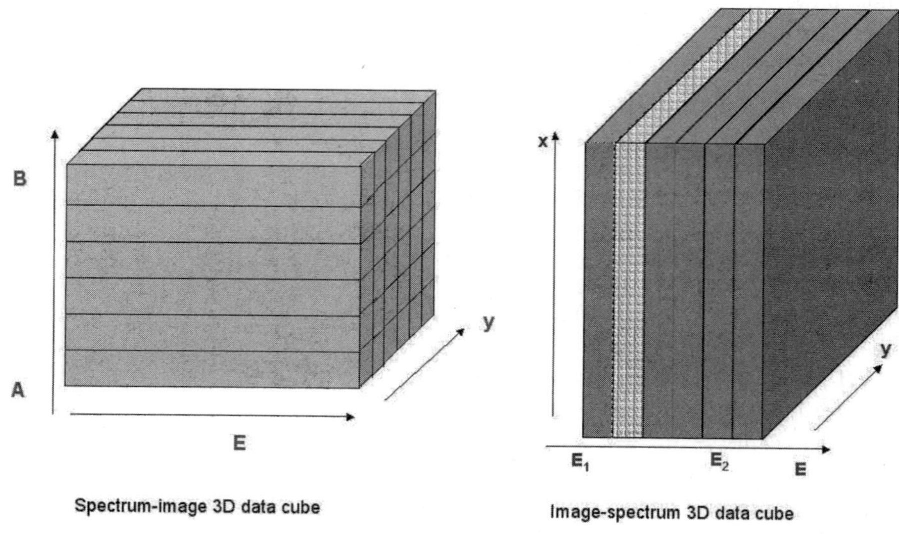

Figure 6. The mixed space-energy resolved 3D data cube recorded in a TEM equipped with an EELS spectrometer or filter: left, STEM mode and parallel acquisition EELS for the "spectrum-image" mode; right, TEM mode and energy filtered images through an energy selection slit, for the "image-spectrum" mode

The total time T required for recording the whole set of 3D data cube (i.e. $N \cdot N \cdot n$ measurements) is in the first case: $N \cdot N \cdot \Delta t_1$ and in the second one $n \cdot \Delta t_2$, but the total dose D received by the elementary area $\partial x \cdot \partial y$ is proportional to $J_1 \cdot \Delta t_1$ for the STEM and to $J_2 \cdot n \cdot \Delta t_2$ for the EFTEM. This leads to:

$$T_1/T_2 = (N^2/n) \cdot (\Delta t_1/\Delta t_2) \text{ and } D_1/D_2 = 1/n \tag{7}$$

Consequently the pros and contra of both approaches can be summarized as follows:

(i) The parallel EELS + STEM mode is always more performant in terms of dose required, because of the intrinsic superiority of the technique for recording all energy-loss channels in parallel, while the EFTEM technique requires as many images (and irradiations) as the number of required energy-loss channels. Generally speaking, the spectrum-image mode is the best approach when mapping fine structures of EELS spectra is the goal. It is also an obvious solution for recording time-dependent variations of EELS spectra, as one records sequences of spectra in fixed probe mode and as a function of time, dose and eventually of a variable external parameter;

(ii) The comparison is not as obvious when considering the total time required for the acquisition. For small numbers of pixels and high numbers of loss channels, the PEELS + STEM is better, while for large values of N and small n the EFTEM is to be preferred. Of course, for a single filtered image such as the elastic one, EFTEM is the solution.

3.2. Instrumentation Required for Recording the 3D Data Cube, Definition and Estimate of the Spatial and Energy Resolutions

The spectrum-image data cube can be recorded with any type of digitally controlled finely focused probe of electrons. For very small probes and high signal-to-noise, a field emission source is required. The size and shape of the probe depends on the high voltage, on the optical properties of the focusing lens and on the brightness of the source. With non-corrected optics, the smallest size is of the order of 1.5 to 2 angströms, the exact value depending also on the criterion used for its definition. Generally, we define it as the diameter of the circle containing 70% of the total current contained within the probe. For the VG STEM used at Orsay (100 kV and $C_s = 3.5$ mm), this diameter is only of 0.5 nm. The spectacular success in the realization of spherical aberration C_s corrected focusing lenses has recently brought this limit in spatial resolution down to below 0.1 nm [35, 36]. Although no demonstration of recorded EELS spectra at this sub-angström level has yet been demonstrated, it is clear that over the past few years the spatial resolution has decreased from the sub-nanometer to the atomic level scale. One can nowadays perform EELS measurements atomic column per atomic column and examples will be provided in a next paragraph.

These novel possibilities have fostered the elaboration of codes for simulating the propagation of narrow probes into crystalline specimens, the more as the introduction of corrected probes enables the use of increased angular convergence [37]. The spreading of the incident beam varies with the focus, the C_s value, the thickness, so that the resolution in EELS at the exit surface of the specimen, can noticeably be degraded with respect to the sub-atomic size of the incident probe. Furthermore, the exact distribution of the probe also depends on the z-dimension variation of composition, the case of depth varying composition being rather more complex to interpret. However, the recent theoretical developments seem to suggest: (i) the ability to acquire in HADF mode (High Angle Dark Field imaging mode) images which show contrast from individual impurity atoms [38]. However, the quantitative analysis of a specific column may be hindered by the elastic intensity variations in depth so that impurity atoms at given thickness might not be detected either in HADF or EELS [39]; (ii) as a consequence of the strong localization of core-loss inelastic scattering, atomic resolution core level atomic spectroscopy should be attainable in spite of the broadening due to dynamical elastic scattering [40]; (iii) with high aperture illumination, the depth of focus should be reduced to <5 nm which should allow in-depth scanning of the specimen, useful for 3D analytical investigations by STEM [41].

The experimental system in operation at Orsay is shown in figure 7. It first contains an illumination system, which delivers the probe on the specimen surface, together with its digitally controlled scanning coils. The different signals are collected simultaneously with a set of two concentric annular dark field detectors for imaging with electrons contained in two different ranges of angular scattering, and of a magnetic type spectrometer which discriminates the electrons having suffered different energy losses. The recording system for EELS has been several times upgraded, it now consists of a CCD camera (1340×100 pixels) optically coupled to the scintillator. Using the different

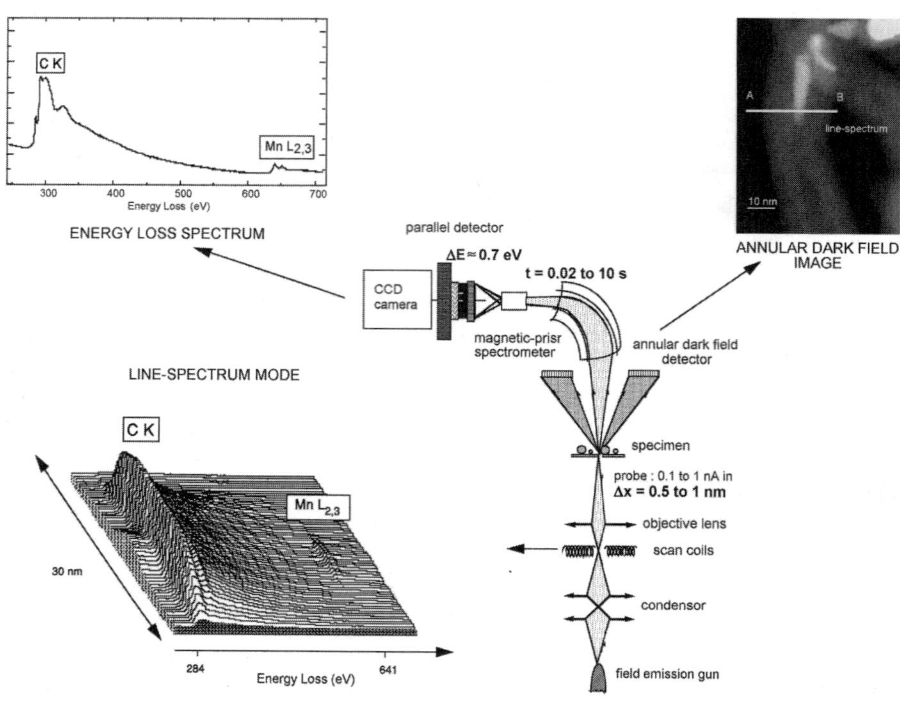

Figure 7. Experimental situation for the acquisition of spectrum-lines in the STEM + PEELS configuration in Orsay, using a VG STEM. Definition of the two useful signals acquired in parallel: the HADF which gives access to topographic maps and the PEELS spectrometer with its new parallel CCD detector for the acquisition of the EELS spectra. It also shows a typical sequence of spectra acquired when the probe is scanned from point A to point B on the HADF image, i.e. across an individual carbon nanotube filled with Mn metal.

binning capabilities, the dynamic range can be as high as 22 bits per spectrum. The rate of spectral acquisition is up to 300 per second and the detection efficiency can be varied from 6 to 24 counts/electron. These developments have been fully exploited to record individual spectra within times of a few ms to a few s, depending on the signal under investigation. Consequently, the acquisition of large quantities of spectra has become a routine procedure and spectrum-images made of 256 × 256 pixels have been processed for elemental mapping. When the interest relies on the variations of fine structures across an individual nanostructure, the recording time per spectrum is well adapted to the acquisition of sequences such as the one depicted in figure 7. It shows the evolution of the C K edge across a single nanotube. Furthermore, the presence of Mn as a filling within the inner cavity is demonstrated by the visibility of the Mn signal when the probe is located at the apex of the core of the tube only. For the acquisition of these spectra, the total energy range which is recorded, is reduced with respect to the one shown in figure 2 and contains only the part extending from about 260 to 660 eV.

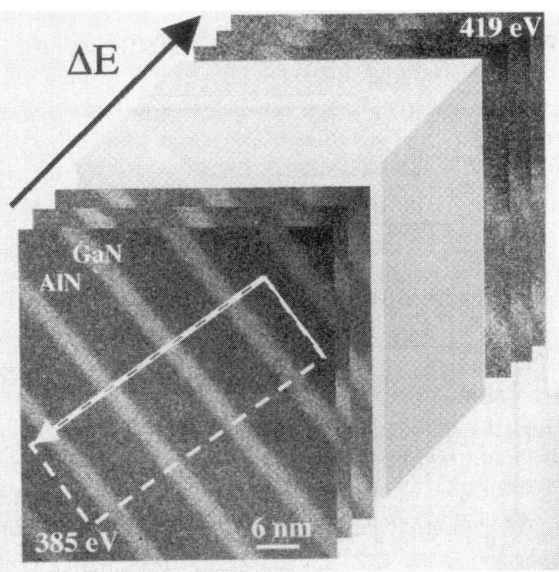

Figure 8. Example of 3D data cube recorded in the image-spectrum mode, i.e. of a series of energy filtered images recorded with a JEOL3010 with a LaB_6 filament equipped with a GIF. Acquisition time is 15 s per image. The selection slit is set at 2eV and the primary voltage is stepped with 1 V increments between each image. Consequently one can then map different types of fine structures (courtesy of Bayle-Guillemaud *et al.* [45]).

The other approach 'image-spectrum with an energy filtering microscope requires the use of a filter. In this energy-selecting mode, an imaging stage is added to the spectrometer, transforming the spectrum behind the energy-selecting slit into an image corresponding to a specific energy loss. Following the pioneering work by Henry and Castaing [42], two possibilities have been developed, the in-column filter nowadays of fully-magnetic type (generally four magnets with a symmetry plane perpendicular to the optical axis, made available on the market by LEO and JEOL) and the post-column type with a magnetic sector followed by an electron optics transfer system (Gatan Imaging Filter). With any of these instruments, full 2D energy filtered images corresponding to an energy width defined by a slit positioned in the dispersion plane at the exit of the filter, are serially acquired, aligned and stacked to reconstruct a 3D data cube. The energy resolution is mostly defined in the image plane on the CCD detector which is used for recording them. Beyond the pixel size related to a dimension on the specimen by magnification considerations, the two most important factor for limiting the spatial resolution are the signal-to-noise ratio and the blurring introduced by the chromatic aberration of the post specimen objective. Practically, it has been evaluated to be of the order of 0.4–0.6 nm depending on different instrument and specimen related parameters [43, 44].

Figure 8 shows the 3D data cube made of a series of 35 energy filtered images with a 2 eV width recorded at 1 eV step across the N K edge at 400 eV. The specimen is a cross section of A1N/GaN heterostructures and the sampling in energy is sufficiently dense to be able to distinguish the different fine structures in the two nitride phases [45].

Let us now turn to the energy resolution, which can be defined and measured in the spectroscopic working conditions, i.e. when an EELS spectrum is displayed on the final CCD detector after the spectrometer. Several factors govern it. The first one is the natural energy width of the primary beam (typically 0.35 eV for a cold FEG source, 0.6–0.7 eV from a Schottky tip and 0.8–1.0 eV for an undersaturated LaB_6 filament), the other ones are the residual aberrations of the spectrometer when focused and the point spread function (PSF) of the detector. In optimized conditions, the major factor is the first one, so that a FWHM of 0.35 eV has been routinely recorded on the zero-loss peak (with cold FEG sources) and values slightly degraded (of the order of 0.5 eV) by instabilities and signal-to-noise ratio have been measured on fine structures at core edges. Two major progresses have been recently demonstrated, which give access to increased energy resolution, even in the spatially resolved mode. The first one is the use of deconvolution techniques, several practical methods to implement them having being tested [46, 47]. The results shown by Gloter *et al.* [48], using a Richardson-Lucy algorithm applied to the 2D images of the spectra as recorded by the CCD detector, have demonstrated a clear gain in resolution down to 0.2–0.3 eV. The spectrum shown in figure 5b has benefited from such a processing. When coupled to the spectrum-image mode, this resolution has been maintained together with the subnanometer spatial resolution across individual nano-objects [49].

The second approach is the introduction of a monochromator within the electron source, which selects only a narrow part of the natural width of the emitted beam. A few years ago, Terauchi *et al.* [50] have opened the way towards sub 100-meV energy resolution with a Wien filter in a home-modified basic TEM, followed by several other realizations [51, 52]. When inserted in a commercial FEI microscope equipped with a Schottky source, this range of energy resolution around 100 meV has now been demonstrated to be accessible for a wide selection of L_{23} and O K edges [53], and in particular in 3d transition metal oxides [54] quite close to the limits imposed by the natural line widths of these edges. However, up to now, this energy resolution performance using a monochromator has not yet been demonstrated simultaneously with high spatial resolution, so that its use for the study of individual nanostructures has remained limited. Nevertheless, it seems clear that such goals of combining both high levels of spatial and energy resolutions required for the study of the electronic structure of individual nano-objects will be accessible in a near future (see figure 9).

This sketch represents, in a rather subjective manner, the general trends of development in instrumentation as compared with the typical sizes and spectral features widths, which should be accessible to recent and future instruments. In his first presentation of this graph, Batson [55] even considered the potentiality of addressing a region characteristic for mechanical, thermal and acoustic properties. In our mind, elemental mapping at the atomic level has been demonstrated. At present, the key issue is to investigate the subtle changes in optical, electronic and bonding properties in

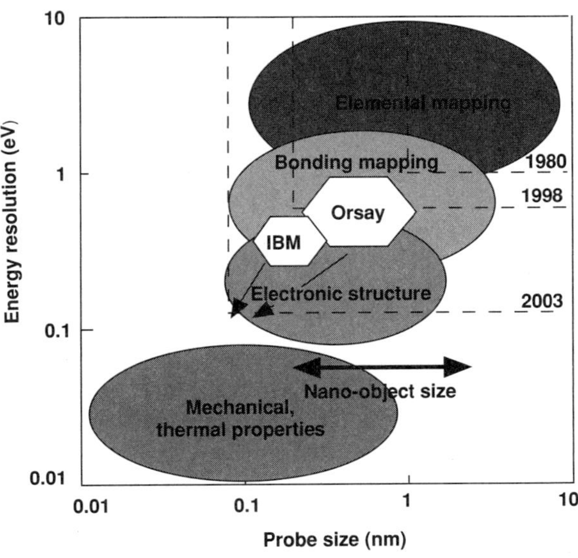

Figure 9. Representation of instrumentation typically available at different years (see dashed squares originating from the upper left corner of the figure) compared with physical properties that can be addressed (adapted from Batson). The major progress in instrumentation occurred over the past five years, have pushed the spatial resolution limit to about 0.1 nm (aberration correctors) and the energy resolution towards 0.1 eV (monochromators or /and deconvolution). The new projects under development are sketched with the arrows originating from the presently piece of equipment now running at Orsay and at IBM. The differences between the bonding mapping and the electronic structure areas are rather subjective.

individual nano-objects. As it will be demonstrated in the following chapters, important success has already been obtained regarding this challenge.

4. ELEMENTAL MAPPING OF INDIVIDUAL NANOPARTICLES USING CORE-LOSS SIGNALS

4.1. Data Processing Routines: Background Subtraction, Multiple Least Square Fitting

The measurement of the weight of a characteristic signal in an EELS spectrum provides access to quantitative elemental analysis. Using the definition of the inelastic scattering cross-section, see Egerton [1], one extracts the absolute number of atoms N of the relevant species within a cylinder with section equal to the probe area A_{probe}, from:

$$S(\beta, \Delta) = I_0(\beta, \Delta) \cdot (N/A_{probe}) \cdot \sigma(\beta, \Delta) \tag{8}$$

where $S(\beta, \Delta)$, $I_0(\beta, \Delta)$ and $\sigma(\beta, \Delta)$ are respectively the measured characteristic signal after background subtraction, the low energy-loss including the zero-loss peak and the calculated cross section for the relevant signal, all quantities being integrated over all scattering angles encompassed within the collection aperture of semi-angle of acceptance β, and over the energy window Δ above threshold.

From this formulation, it appears that such quantitative analysis relies on a good extraction of the characteristic signal S and a good estimation of the ionization cross section. The extraction of the characteristic edge implies background modeling over a fitting window before threshold and extrapolation of this model curve in the edge energy region. The most generally used model is an empirical $A \cdot E^{-R}$ power-law where A and R are parameters to be determined. The cross-section is satisfactorily estimated using a hydrogenic model for the K and L shells [56], one of the most recent version including the contribution of the white lines in the latter case using a phenomenological description. For outer shells of M, N and O types, reliable values are obtained using a self-consistent calculation based on a Hartree-Slater description [57]. We recall that such calculations are supposed to reproduce the general shape of the edge without accounting for the fine structures occurring from solid state effects. Therefore, special care has to be handled when choosing the integration energy window, so that these fine structures are properly averaged over this energy window. Another alternative is to use experimental cross-sections deduced from measurements performed in similar conditions on specimens of known composition, which may be the case in spectrum-imaging studies. In this latter situation, one is often interested by identifying local composition variations with respect to a well identified matrix or reference object.

When one is only interested in measuring relative concentrations (of different elements 1 and 2) in order to work out the stoichiometry of the sample, the above formula can be simplified as:

$$\frac{N_1}{N_2} = \frac{S_1(\beta, \Delta E)}{S_2(\beta, \Delta E)} \times \frac{\sigma_2(\beta, \Delta E)}{\sigma_1(\beta, \Delta E)} \tag{9}$$

These algorithms developed for processing individual spectra have been adapted, so that they now can process large collections of spectra such as those gathered in spectrum-image sequences [58]. In the example shown in figure 10, the sum of all acquired spectra reveals the presence of five different edges appearing when the probe is scanned over a relatively large area. Mapping the weight of each of these characteristic lines displays the localization of any of these elements, as compared with the simultaneously acquired HADF image.

For cases where the above routine for characteristic signal extraction fails (for instance deviation from the power law model, overlap of different characteristic edges, detection of small concentrations), another straightforward method has been introduced, the use of multiple least square fitting techniques. One searches for the combination of reference signals which best reproduces an experimental curve, encompassing both background and characteristic edges [59, 60]. This procedure uses reference spectra generally recorded from adjacent areas or calculated when none is directly available. It is quite versatile and may be used for different purposes. It has also been implemented for processing sequences of spectra [58]. The major limitation is due to the fact that the

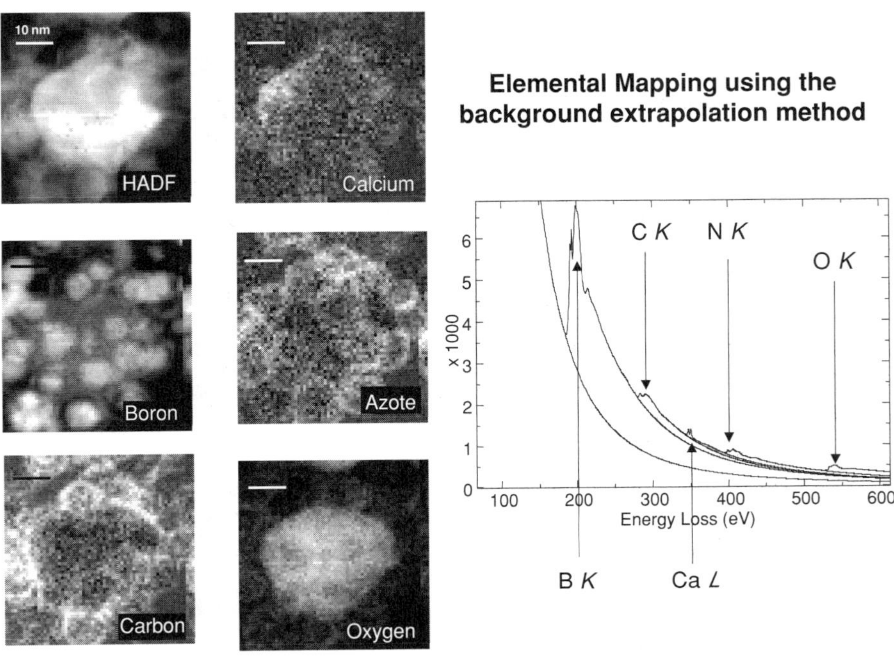

Figure 10. Elemental maps extracted from a spectrum-image set (64×64 pixels, 500 energy loss channels) using the routine background modeling and extrapolation procedure. The specimen is intended to contain BN nanotubes prepared by continuous laser ablation of BN targets.

detailed shape exhibits variable fine structures depending on the local environment. However, this effect can also be used as a new possibility for mapping fine structures depending on bonding states or on orientation [61].

4.2. A Few Examples of Elemental Mapping with EELS Core Edges

The range of nanostructures, in which the elemental distribution has been mapped, has increased quite rapidly over the past five years. Nearly all elements have been identified and used with their characteristic energy losses from helium up to uranium, although some of them display edges with smooth profiles and poorer cross sections, in the generally investigated energy loss range from 5 up to about 2000 eV range. Among the difficult cases, we can quote medium and heavy elements with a distribution of accessible unoccupied density of states made of s and p together with filled d or f states or with characteristic edges at too high energy-loss values (for instance from Hg to Bi). The different processing techniques described above, have been used to extract and measure the characteristic signals. Figure 11 is a gallery of four different types of individual nanostructures with composition analysed by EELS techniques and exhibiting quite different morphologies, with however typical nanometer dimension in at least one direction: (a) He gas in nanobubbles within a Pd-Pt alloy matrix (He 1s

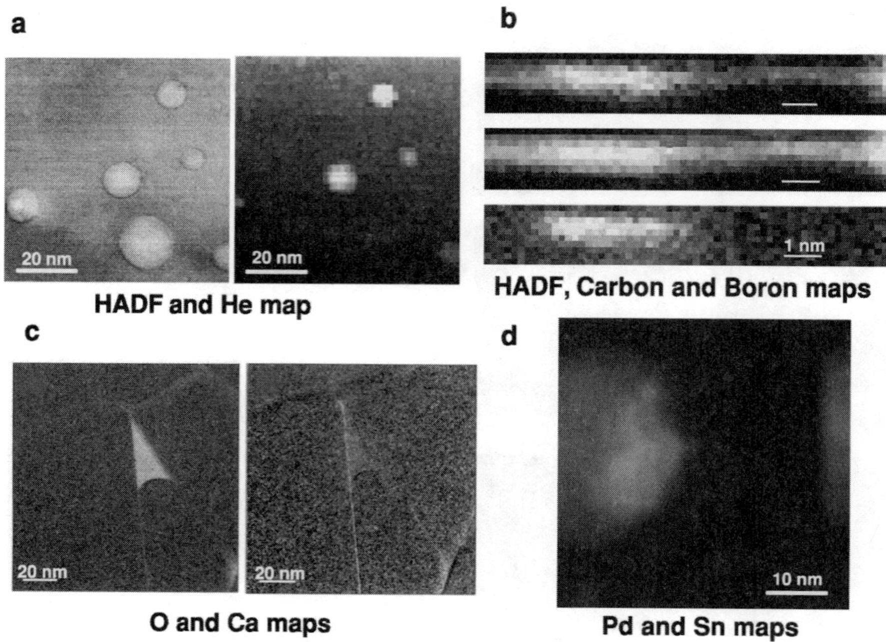

Figure 11. Gallery of elemental maps on different nano-objects made of different types of elements: (a) helium bubbles mapped with the He 1s signal at 22 eV—spectrum image technique with complex background modeling and subtraction (courtesy D. Taverna); (b) single-walled carbon nanotube filled with decaborane molecule- spectrum-image with the standard background extrapolation procedure; (c) thin intergranular amorphous films in a ceramic polycrystal exhibiting the segregation of Ca and O atoms mapped with their K edges, data acquired with the energy-filtering mode and the three-window approach (courtesy of Plitsko and Mayer); (d) phase separation in a single PdMn intermetallic nanoparticle—spectrum-image approach with the MLS fit of the Pd M_{45} at 330 eV and the Sn M_{45} at 500 eV edges (courtesy of M. Tencé)

edge at 22 eV); (b) filling with boron-rich molecules of individual single-wall carbon nanotubes (B 1s edge at 185 eV), (c) oxygen and calcium segregated in the intergranular films within a Si_3N_4 polycrystal doped with 450 ppm of Ca [62] (d) a composite Pd-Sn catalytic nanoparticle deposited on an Al_2O_3 substrate.

4.3. Sensitivity, Limits of Detection in EELS Elemental Mapping

As early as its introduction as a powerful microanalytical tool, EELS was recognized as a potential technique to tackle the identification of isolated atoms [63]. Mapping individual heavy atoms such as thorium and uranium, using their O_{45} edge at about 100 eV, has first been demonstrated in [64, 65]. These studies were performed on random distributions of these atoms on ultra-thin objects and the criterion for identification was that the signal-to-noise ratio for the characteristic signal be above a threshold value. It did not require the spatial resolution for separating different atoms. When clusters

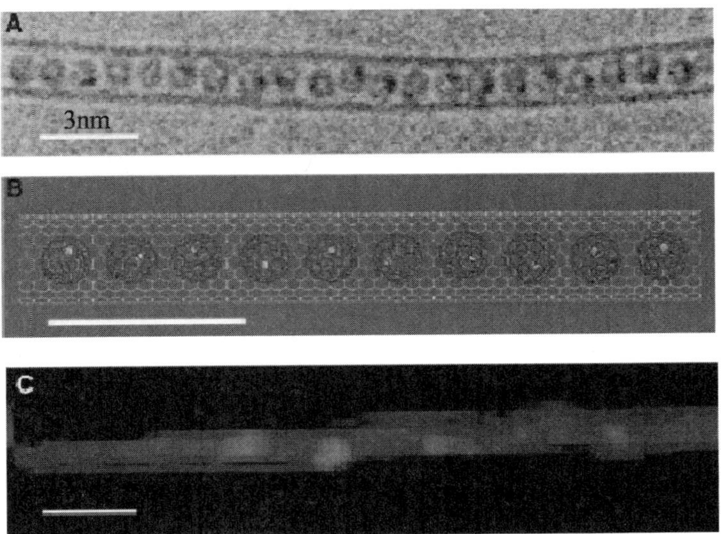

Figure 12. EELS mapping of individual Gd atoms encapsulated in C_{82} fullerene molecules inserted along the empty core of an individual SWCNT: (A) HREM image, (B) schematic representation of an analyzed nanostructure; (C) superposed C K (blue) and Gd N45 (red) maps extracted from a 32×128 pixes spectrum-image, 35 ms per pixel, bar length is 3 nm. More intense red spots have been shown to correspond to aggregated Gd clusters made of a few atoms, grown under the incident electron beam irradiation (courtesy of Suenaga *et al.* [66]) (See color plate 19.)

made of several atoms happen to lie within the incident probe area, rules have been elaborated to count them as a function of the magnitude of the signal. The first real demonstration of these ultimate capabilities for analyzing a well-defined nano-object, has been realized by Suenaga *et al.* [66] on gadolinium doped atoms encapsulated within single walled carbon nanotubes. In this case, the spectrum-image technique has been used and quantitative maps for Gd and C atoms contained in the self supported fullerene molecules and nanotube envelops have been displayed, see figure 12. The signal-to-noise for identifying one Gd atom with its M_{45} edge signal is of the order of ten and the resolution to separate two of them is reached in spite of a probe diameter of the order of 0.7 nm, because the Gd atoms are generally maintained apart from one another by more than one nanometer because they are trapped in separate adjacent C_{82} molecules.

Very recently, Leapman [67] has similarly demonstrated the detection of individual Ca atoms, using the same spectrum-image approach in a VG STEM, and shown that such capacities can also be of interest for measuring the number of atoms (iron) in a single haemoglobin molecule. As a conclusion, for small structures of nanometer typical size, the detection limits can be as low as one single atom, for optimum objects and edges. It makes this technique the only one capable to compete with the tomographic atom probes developed for identifying by field evaporation and mass spectroscopy all atoms within a given volume at the extremity of a tip. However, the EELS approach is applicable to a broader range of material and further conveys a richer information as shown in the next chapter.

5. MAPPING BONDING STATES AND ELECTRONIC STRUCTURES WITH ELNES FEATURES

The great challenge for establishing the spatially-resolved EELS approach as an unavoidable tool for testing and controlling the local properties of a given nano-object, is to map its electronic structure and bonding state at the atomic level. It involves many strategic issues to be solved, as well instrumental as theoretical, to fully benefit from the large quantities of data to be accumulated. We will first show a few examples of increased complexity before identifying a list of topics to be addressed within the next few years.

5.1. A Few Selected Examples

An interesting review of Electron energy-loss near-edge fine structures as a tool for the investigation of electronic structure on the nanometer scale, has recently been published by Keast *et al.* [68]. In order to illustrate on practical examples how this technique can be used to provide maps of different fine structures shapes and therefore to monitor their changes, let us consider first an example relying on the acquisition and processing of spectrum-images. It concerns the complex distribution of nanoparticles and nanostructures encountered in the laser-ablated deposits collected when preparing BN nanotubes. In figure 10 we have shown how elemental maps could be provided when measuring pixel per pixel over the scanned area, the weight of the different characteristic signals. One key element in this problem is boron and in particular how it is involved in bonds with the other elements: boron, nitrogen, oxygen. The idea is then to record the boron K edges near edge structures over all pixels and to compare them with those encountered in reference spectra corresponding respectively to the B-B bond (in metallic boron), to the B–N bond (a priori existing in the lamellar hexagonal structures) and B–O bond (when oxygen acts as a poisoning compound). Using a Multiple Least Square fit with these references, in which case it is referred as a "fingerprint" technique, it is then possible to map bonding states. It has thus been demonstrated that most of the BN is contained in the grown BN nanotubes. Figure 13 focuses on a single nanoparticle. The bond maps reveal a clear morphology of concentric spheres embedded within another, the core being made of metallic boron on which surface layered BN has been grown, the external surface having been poisoned with oxygen. In such a simple morphology, the distribution of bond states provides a quasi 3D view of the individual object of outer dimension typically in the 20 nm range.

With the new probes of typical diameter 0.2–0.3 nm, one can address situations where quasi-atomic resolution is essential. The first example shown in figure 14 deals with the analysis of the electronic structure across ultra-thin SiO_2 layers involved as dielectric gates in Si based field effect transistors. Batson [30] has investigated the

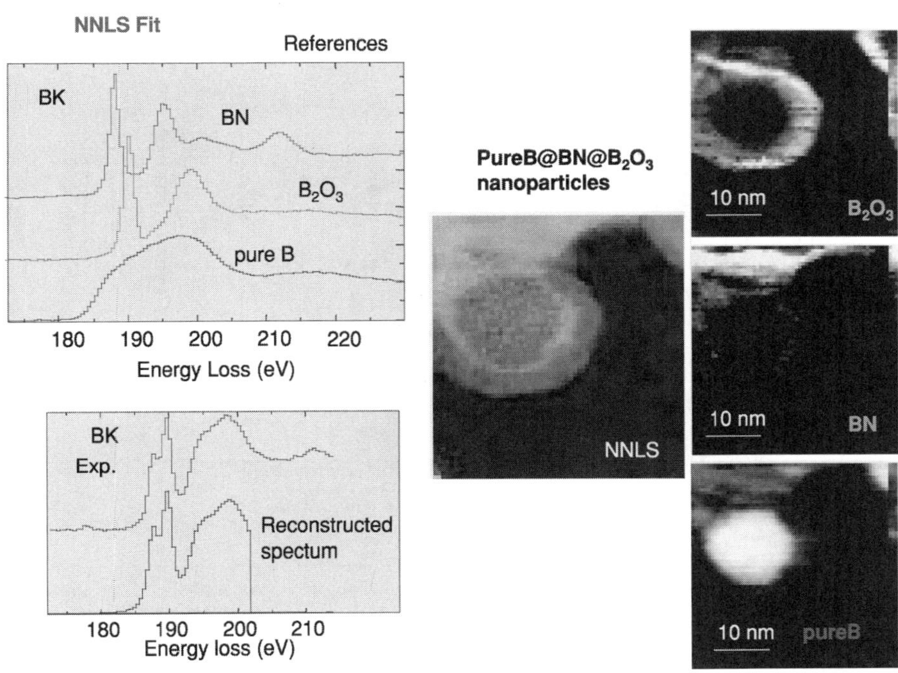

Figure 13. Mapping boron bonding states in a single composite nanoparticle using ELNES fine structures on the boron K edge: left, reference spectra for the typical BN, B_2O_3 and B amorphous phases, and comparison of an experimental spectrum with a modeled spectrum made of a linear combination of reference edges ("fingerprint approach"). Right: set of maps for the different types of bonding (B–B, B–O, B–N) and a color superposition of them which provides a clear view of the nanoparticle as made of concentric spheres, the intermediate one (B–N) corresponding to 2 or 3 BN atomic layers. (See color plate 20.)

changes of Si L_{23} edge shape when scanning the probe from an atomic column position to the next one, as seen in the ADF image. He has identified at the apex of the first crystal layer in direct contact with the amorphous oxide, a signal likely due to the existence of Si^{++} states.

In the case of ultra-thin oxide layers, Muller *et al.* [69] have monitored the variations of the oxygen K line (see figure 14). They have noticed changes for the O atoms close to the Si interfaces, revealed by a slight shift in energy and more important by a disappearance of the strong peak at the edge. They attribute it to the presence of a sub-oxide layer, corresponding to the case where the O atoms are in a Si rich environment, and therefore the strong scattering of the outgoing wave by the second neighbors does not involve other oxygen atoms (this is a simplified interpretation in terms of multiple scattering around the excited O atom). Modeling all spectra recorded across the gate, the width of the real SiO_2 layer corresponding to the well-known O-K edge shape in bulk SiO_2 is reduced by a thickness of 1 or 2 monolayers of sub-oxide silicon.

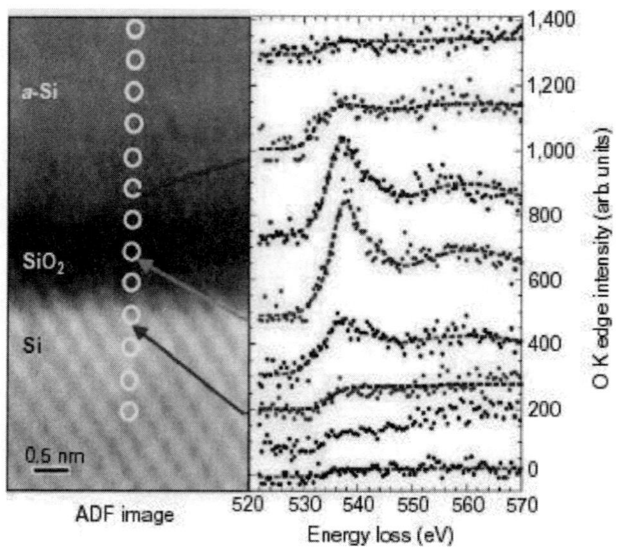

Figure 14. Sequence of oxygen K edge EELS spectra recorded point by point at the circled positions across an ultra-thin gate in a gate stack made visible in the HADF imaging mode. The background corrected O K edges are displayed on the right part of the picture and they exhibit a change of shape between those recorded close to the interfaces and those at the center of the dielectric film. The width of both SiO_2 and sub-oxide layers has been determined after fitting any spectrum in the sequence as a linear combination of the two representative profiles. It has shown that the two interfacial signals do not overlap only for gate oxides thicker than about 1.5 nm (courtesy of Muller *et al.* [69])

However, this is sufficiently important to raise doubts about the possibility of realizing good SiO_2 dielectric gates of 5 atoms thick, which would be required in the VLSI technology with gates shorter than 50 nm.

A final example deals with artificially charge modulated layers in titanate perovskites. Contrary to the above case where surface roughness increased by heat treatments can further reduce the properties of the device, it has been demonstrated that one can grow atomically flat surfaces and abrupt structures, as thin as one single atomic layer, of these oxides. Charge modulation is introduced by inserting a valence change on the titanium atoms present throughout the specimen. It is made by controlled substitution of $LaTiO_3$ layers within the sequence of $SrTiO_3$ layers, the valence on the Ti atom being then 3 + instead of 4 + and consequently an extra electron being injected. The spatially resolved EELS technique has been used by Ohtomo *et al.* [70] to monitor the change in Ti valence state across lanthanum perovskite layers of variable thickness from 1 to 6 atomic ones. Figure 15 shows one of the results obtained in this work. The Ti L_{23} white lines are monitored at atomic spacing and the weight of the two profiles, respectively corresponding to the Ti^{3+} and Ti^{4+}, is determined as a function of the probe position. The La M_5 line is simultaneously recorded, so that in a further step the authors show that the measured width of the valence change is substantially larger

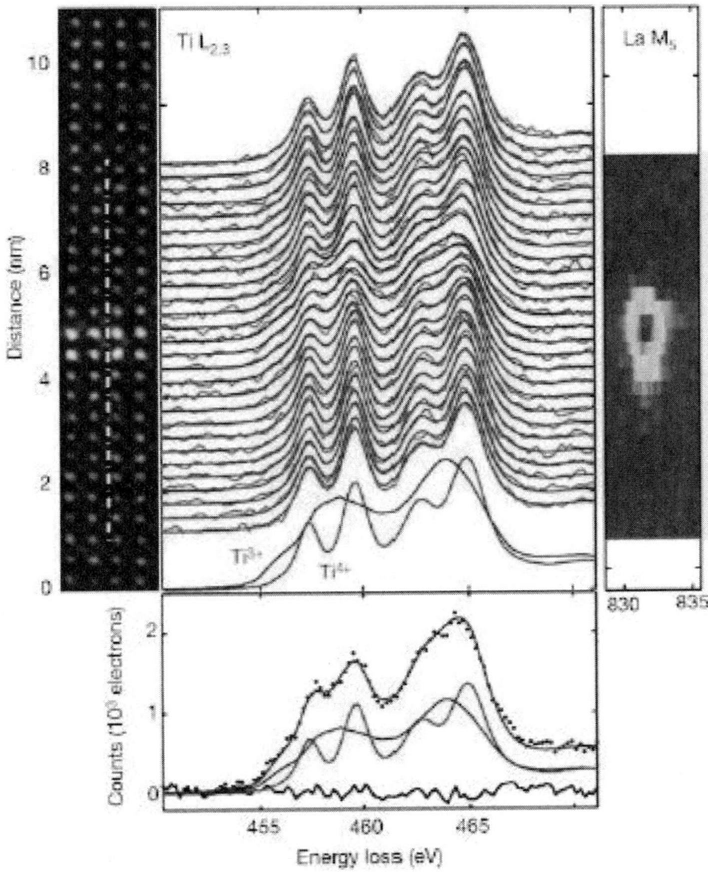

Figure 15. Simultaneously recorded Ti and La edges across a 2-unit cell La TiO_3 layer in $SrTiO_3$, as made visible in the HADF image. The Ti white lines are then decomposed as a linear combination of respective Ti^{4+} and Ti^{3+} contributions, see model curves at the bottom of the sequence of spectra. In the lower part of figure, one spectrum in the sequence is reconstructed as a linear combination of two models and the difference between experiment and modeling is roughly zero over the involved energy range. Nominal probe size for this experiment is 0.19 nm at 200 kV (courtesy Ohtomo *et al.* [70]).

than the width of the elemental La profile. They conclude that it reveals an intrinsic extended distribution of the extra electron of the order of 1.0 nm, responsible for a metallic conductivity although the superlattice structure is based on two insulators.

5.2. From Fingerprint Techniques to Interpretations Requiring Extended Theoretical Calculations

In the above examples, and in most cases published up to now, one first tries to extract and identify one fine-structure profile to be attributed to the localized feature of interest (interface, extra-atomic layer, nanostructure) and to interpret it as a combination of reference profiles acquired nearby. This is the case typically for all studies using a spatial-difference approach to discriminate an interface spectrum from its neighboring bulk ones. This is obviously a first level of processing, because it assumes that the measured environment of a given atom at the interface can be represented as a linear combination of environments encountered in well-defined phases. However in many types of solid-solid interface, non-equilibrium situations are encountered during the growth of this interface.

Consequently, new guidelines have to be proposed for improving our understanding of the local atomic structure and bonding in such extremely confined objects. First, one can check whether all spectra in a spectrum-image sequence can be built as a combination of well defined ones. Multivariate statistical analysis techniques can be of great help in this task for identifying spectrum profiles, which cannot be modeled as such a superposition [71].

The second step is to perform molecular dynamics atomistic simulations of possible structures to be encountered and to extend them with *ab initio* calculations of the electronic structures and EELS fine structures, which should be associated to them. Such DFT calculations for distorted structures are extremely time consuming and difficult to perform, see for instance [72] for the first *ab initio* time dependent DFT calculations of the optical and electron energy loss spectra of carbon nanotubes.

Using the EELS spectrum-imaging mode [61], we have identified on the π^* peak of the C-K edge the spectroscopic signature of defects trapped in the hexagonal network. Comparison with local density of states (LDOS) calculations pleads for attributing this specific feature (a doublet) to the presence of several pentagons in adjacent layers aligned vertically along the electron beam. Another explanation involving heptagons has been rejected.

As a conclusion, the increased energy resolution of the future generation of instrument appearing on the market will reach its full potentialities when supported by the appropriate progress in theoretical modeling of the electronic structures in objects of nm dimensions.

6. CONCLUSION

We have shown that the EELS-based methods developed in the environment of an electron microscope, constitute very powerful tools for understanding structural, chemical and electronic properties down to the atomic level in many new types of nanostructures. The fast development in correctors and monochromators on the new generation of microscopes to be installed within the next couple of years, will undoubtedly transform this instrumentation into lab-size synchrotron machines with the extra advantage of seeing and measuring the properties of individual nano-objects.

In order to be fully active in the field of nanotechnology, the present performance for imaging and spectroscopy must be retained while applying external constraints on the specimen and while monitoring dynamic processes. Let us mention some possibilities offered by the manufacturing of well-adapted specimen stages, such as those based on scanning tunneling microscopy (STM) where piezo-driven controls enable to manipulate objects and measure specimen properties during TEM observation. In one case, this extra movement has been used to bend *in situ* a single nanotube while recording carbon K edge EELS fine structures for different bending angles during deformation and recovering of the initial shape [73]. It was shown that the observed changes could be interpreted in terms of bond angle deformation without requiring the introduction of defects such as pentagon-heptagon rings. In a more recent study, the contact between the nano-object and the two parts of the supporting STM stage has been realized by wetting with mercury drops. Consequently, the I(V) transport characteristics have been measured on individual double-walled nanotubes, the complete atomic chiral structure of which being extracted from their electron diffraction pattern [74]. This trend towards the introduction of nano-laboratories inside a TEM column will therefore constitute another important extension of the field of use of these techniques.

As a conclusion, we fully adhere to the words from Wang [75]: "the picometer-scale science provided by HRTEM is the foundation of nanometer-scale technology."

REFERENCES

[1] R. F. Egerton, *Electron Energy-Loss Spectroscopy in the Electron Microscope*, Plenum Press, New York and London, 2nd edition (1996).

[2] M. M. Disko, C. C. Ahn and B. Fultz (Eds.), *Transmission Electron Energy Loss Spectrometry in Materials Science*, The Minerals, Metals and Materials Society, Warrendale, Pa (1992).

[3] C. Colliex, in: *Core Level Spectroscopies for Magnetic Phenomena*, P. S. Bagus, G. Pacchioni and F. Parmigiani (Eds.), pp. 213–233, NATO ASI Series B345, Plenum Press, New York and London (1995).

[4] C. Colliex, in: *Handbook of Microscopy—Applications in Materials Science, Solid-State Physics and Chemistry*, S. Amelinckx, D. van Dyck, J. van Landuyt and G. van Tendeloo (Eds.), pp. 425–445, VCH Weinheim, Germany (1997).

[5] C. Colliex, M. Kociak, O. Stéphan, K. Suenaga and S. Trasobares, in: *Nanostuctured Carbon for Advanced Applications*, G. Benedek, P. Milani and V. G. Ralchenko (Eds.), pp. 201–232, NATO ASI Series II/24, Kluwer Academic Publishers, Dordrecht, the Netherlands (2001) .

[6] J. Daniels, C. von Festenberg, H. Raether and K. Zeppenfeld, in: *Springer Tracts in Modern Physics*, Vol. 54, pp. 77–135, Springer, Berlin (1970).

[7] G. Safran, A. Kolitsch, S. Malhuitre, S. Trasobares, I. Kovacs, O. Geszti, M. Menyhard, C. Colliex and G. Radnoczi, *Diamond and Related Materials*, 11 (2000) 1552.

[8] J. J. Gilman, *Phil. Mag. B* 79 (1999) 643.

[9] L. Laffont, M. Monthioux and V. Serin, *Carbon* 40 (2002) 767.

[10] H. R. Daniels, R. Brydson, A. Brown and B. Rand, *Ultramicroscopy* 96 (2003) 547.

[11] M. Acheche, C. Colliex, H. Kohl, A. Nourtier and P. Trebbia, *Ultramicroscopy* 20 (1986) 99.

[12] A. Howie, in: *Topics in Electron Diffraction and Microscopy of Materials*, P. B. Hirsch (Ed.) pp. 79–107, Inst. of Physics Publications, Bristol UK (1999).

[13] M. Kociak, L. Henrard, O. Stéphan, K. Suenaga and C. Colliex, *Phys. Rev. B* 61 (2000) 13936.

[14] M. Kociak, O. Stéphan, L. Henrard, V. Charbois, A. Rotschild, R. Tenne and C. Colliex, *Phys. Rev. Lett.* 87 (2001) 75501.

[15] O. Stéphan, D. Taverna, M. Kociak, K. Suenaga, L. Henrard and C. Colliex, *Phys. Rev. B* 66 (2002) 155422.

[16] B. Rafferty and L. M. Brown, *Phys. Rev. B* 58 (1998) 10326.

[17] S. Lazar, G. A. Botton, M. Y. Wu, F. D. Tichelaar and H. W. Zandbergen, *Ultramicroscopy* 96 (2003) 535.

[18] S. Schamm and G. Zanchi, *Ultramicroscopy* 96 (2003) 559.

[19] P. Rez, J. Bruley, P. Brohan, M. Payne and L. A. J. Garvie, *Ultramicroscopy* 59 (1995) 159.

[20] P. Rez, J. R. Alvarez and C. Pickard, *Ultramicroscopy* 78 (1999) 175.
[21] C. Elsässer and S. Köstlmeier, *Ultramicroscopy* 86 (2001) 325.
[22] K. van Benthem, C. Elsässer and M. Rühle, *Ultramicroscopy* 96 (2003) 509.
[23] A. T. Paxton, A. J. Craven, J. M. Gregg and D. W. McComb, *J. of Microscopy* 210 (2003) 35.
[24] J. J. Rehr and R. C. Albers, *Rev. Mod. Phys.* 72 (2000) 621.
[25] H. Kurata and C. Colliex, *Phys. Rev. B* 48 (1993) 2102.
[26] F. M. F. de Groot, *J. Elec. Spec. Rel. Phen.* 67 (1994) 529.
[27] C. Jeanguillaume and C. Colliex, *Ultramicroscopy* 28 (1989) 252.
[28] N. D. Browning, M. F. Chisholm and S. J. Pennycook, *Nature* 366 (1993) 143.
[29] D. A. Muller, Y. Tsou, R. Raj and J. Silcox, *Nature* 366 (1993) 727.
[30] P. E. Batson, *Nature* 366 (1993) 727.
[31] L. M. Brown, *Nature* 366 (1993) 721.
[32] J. A. Hunt and D. B. Williams, *Ultramicroscopy* 38 (1991) 47.
[33] C. Colliex, M. Tencé, E. Lefevre, C. Mory, H. Gu, D. Bouchet and C. Jeanguillaume, *Mikrochim. Acta* 114/115 (1994) 71.
[34] J. L. Lavergne, J. M. Martin and N. Belin, *Microsc. Microanal. Microstruct.* 3 (1992) 517.
[35] O. L. Krivanek, P. D. Nellist, N. Dellby, M. F. Murfitt and Z. Szilagi, *Ultramicroscopy* 96 (2003) 229.
[36] P. E. Batson, N. Dellby and O. L. Krivanek, *Nature* 418 (2002) 617.
[37] R. F. Loane, E. J. Kirkland and J. Silcox, *Acta. Cryt. A* 44 (1988) 912.
[38] P. M. Voyles, J. L. Grazul and D. A. Muller, *Ultramicroscopy* 96 (2003) 251.
[39] C. Dwyer and J. Etheridge, *Ultramicroscopy* 96 (2003) 343.
[40] A. R. Lupini and S. J. Pennycook, *Ultramicroscopy* 96 (2003) 313.
[41] G. Möbus and S. Nufer, *Ultramicroscopy* 96 (2003) 285.
[42] R. Castaing and L. Henry, *C. R. Acad. Sci. Paris, Ser. B* 255 (1962) 76.
[43] O. L. Krivanek, M. K. Kundman and K. Kimoto, *J. Microscopy* 180 (1995) 277.
[44] G. A. Botton and M. W. Phaneuf, *Micron* 30 (1999) 109.
[45] P. Bayle-Guillemaud, G. Radtke and M. Sennour, *J. Microscopy* 210 (2003) 66.
[46] P. E. Batson, D. W. Johnson and J. C. W. Spence, *Ultramicroscopy* 41 (1992) 137.
[47] M. H. F. Overwijk and D. Reefman, *Micron* 31 (2000) 325.
[48] A. Gloter, A. Douiri, M. Tencé and C. Colliex, *Ultramicroscopy* 96 (2003) 385.
[49] A. Gloter, A. Douiri, M. Tencé, D. Imhoff, O. Stéphan and C. Colliex, *Microscopy and Microanalysis*, 9 Supt. 2 (2003) 108.
[50] M. Terauchi, R. Kuzuo, F. Satoh, M. tanaka, K. Tsuno and J. Ohyama, *Microsc. Microanal. Microstruct.* 2 (1991) 351.
[51] H. W. Mook and P. Kruit, *Ultramicroscopy* 78 (1999) 43.
[52] P. C. Tiemeijer, *Ultramicroscopy* 78 (1999) 53.
[53] S. Lazar, G. A. Botton, M. Y. Wu, F. D. Tichelaar and H. W. Zandbergen, *Ultramicroscopy* 96 (2003) 535.
[54] C. Mitterbauer, G. Kothleitner, W. Grogger, H. Zandbergen, B. Freitag, P. Tiemeijer and F. Hofer, *Ultramicroscopy* 96 (2003) 469.
[55] P. E. Batson, *Ultramicroscopy* 78 (1999) 33.
[56] R. F. Egerton, *Ultramicroscopy* 4 (1979) 169.
[57] R. D. Leapman, P. Rez and D. F. Mayers, *J. Chem. Phys.* 72 (1980) 1232.
[58] Ó. Tencé, M. Quartuccio and C. Colliex, *Ultramicroscopy* 58 (1995) 42.
[59] R. D. Leapman and C. R. Swyt, *Ultramicroscopy* 26 (1988) 393.
[60] T. Manoubi, M. Tencé, M. G. Walls and C. Colliex, *Microsc. Microanal. Microstruct.* 1 (1990) 23.
[61] O. Stéphan, A. Vlandas, R. Arenal de la Concha, A. Loiseau, S. Trasobares and C. Colliex, *Inst. Phys. Conf. Series* 179 (2004) 437.
[62] J. M. Plitzko and J. Mayer, *Ultramicroscopy* 78 (1999) 207.
[63] M. Isaacson and D. Johnson, *Ultramicroscopy* 1 (1975) 32.
[64] C. Mory and C. Colliex, *Ultramicroscopy* 28 (1989) 339.
[65] O. L. Krivanek, C. Mory, M. Tencé and C. Colliex, *Microsc. Microanal. Microstruct.* 2 (1991) 257.
[66] K. Suenaga, M. Tencé, C. Mory, C. Colliex, H. Kato, T. Okazaki, H. Shinohara, K. Hirahara, S. Bandow and S. Iijima, *Science* 290 (2000) 2280.
[67] R. D. Leapman, *J. Microscopy* 210 (2003) 5.
[68] V. J. Keast, A. J. Scott, R. Brydson, D. B. Williams and J. Bruley, *J. Microscopy* 203 (2001) 135.
[69] D. A. Muller, T. Sorsch, S. Moccio, F. H. Baumann, K. Evans-Lutterodt and G. Timp, *Nature* 399 (1999) 758.

[70] A. Ohtomo, D. A. Muller, J. L. Grazul and H. Y. Hwang, *Nature* 419 (2002) 378.
[71] N. Bonnet, N. Brun and C. Colliex, *Ultramicroscopy* 77 (1999) 97 .
[72] A. G. Marinopoulos, L. Reining, A. Robio and N. Vast, *Phys. Rev. Lett.* 91 (2003) 046402.
[73] K. Suenaga, C. Colliex and S. Iijima, *Appl. Phys. Lett.* 78 (2001) 70.
[74] M. Kociak, K. Suenaga, K. Hirahara, Y. Saito, T. Nakahoira and S. Iijima, *Phys. Rev. Lett.* 89 (2002) 155501.
[75] Z. L. Wang, *Advanced Materials* 15 (2003) 1497.

22. IMAGING MAGNETIC STRUCTURES USING TEM

TAKAYOSHI TANJI

1. INTRODUCTION

Imaging of magnetic structure is important in the investigation of magnetism. However, this area of study has been largely ignored due to poor spatial resolution and low sensitivity. Today's advanced information-oriented society requires very high-density recording media, which in turn demands materials of high fineness. Therefore, imaging at nanometer resolution has become indispensable in revealing the micro- or nano-magnetic structures and in developing noble materials and devices. Transmission electron microscopy has contributed to the investigation of crystal structures with atomic-scale resolution and can now enable imaging of magnetic structures with nano-scale resolution.

Magnetic multilayers and magnetic granular films are new materials that are attracting a great deal of attention. Giant magneto-resistance (GMR) has been discovered in both materials [1–3]. GMR drastically reduces the electric resistance in a magnetic field. These materials are expected to be applied as new magnetic sensors, and some of these materials have been applied as a magnetic head in a recording device. In addition, some of granular films have been investigated as soft magnetic materials [4]. Taking into account the magnetic integrated circuit which may be realized in the near future, fundamental research for patterning magnetic parts on thin films has begun [5, 6].

The concept of a magnetic domain structure was proposed by Weis in 1907 [7], and Bloch called the interface of domains a "domain-wall" [8]. There are several techniques by which to observe magnetic domains and/or domain walls. Electron microscopy has the advantage of high spatial resolution. In particular, Lorentz microscopy has been utilized for the observation of magnetic domain walls since the end of the 1950's [9–11]. This technique allows us to observe the domain walls, even without special devices added to the microscope, as well as the dynamic motion of magnetic domains in real time [12]. Electron holography [13, 14] developed by Gabor in 1948 is another powerful tool in the observation of magnetic structures. Although electron holography is discussed particularly in a separate chapter of this book [15], a section of this chapter is also dedicated to electron holography and is intended to be complementary and useful to the reader. The difficulty of quantitative analysis is a disadvantage of Lorentz microscopy, compared to electron holography. Some techniques of Lorentz microscopy, however, have been reported for the quantitative analysis using, for examples, a segmented detector (differential phase contrast (DPC) microscopy) [16] or a transformed-intensity equation (TIE) [17]. Several techniques using scanning transmission electron microscopy (STEM) [16, 18, 19] also have been developed. We restrict the discussion in this chapter to the application of conventional transmission microscopy (CTEM).

2. LORENTZ MICROSCOPY

2.1. Introduction

The magnetic structure first observed through microscopy was a domain wall structure with an inhomogeneous dispersion of ferromagnetic fine particles covering the specimen (Bitter method) [20]. This method has also been extended to electron microscopy [21]. However, this method does not allow iterative observation of the same specimen. Lorentz microscopy, in which the specimen need not to be decorated with magnetic particles, has been used to directly reveal domain structures in thin films as a result of the deflection of electrons caused by the magnetic field. This technique also allows the dynamic observation of domain wall motion. Here, the mechanism of contrast formation is rather different from that leading to high-resolution microscopy. Several Lorentz microscopy techniques have been developed, including Fresnel mode, Foucault mode [22], the differential phase contrast (DPC) method and Lorentz phase microscopy (transport of intensity equation (TIE) method). Although electron holography of magnetic materials may be considered to be a form of Lorentz microscopy, this technique will be explained in the latter section.

2.2. Magnetic-Shield Lens

In transmission electron microscopy (TEM), specimens are placed in a very strong magnetic field of an objective lens of more than 1 T along the vertical direction. In such a strong field, all of the specimens are magnetized in only one direction, sometimes in the vertical direction and sometimes along the surface of a tilted film. As a result, the natural characteristics of the specimen cannot be observed. The easiest way to avoid this effect is to turn off the objective lens and to focus the image using intermediate lenses. However, this mode gives only a few thousand times magnification. Therefore, special

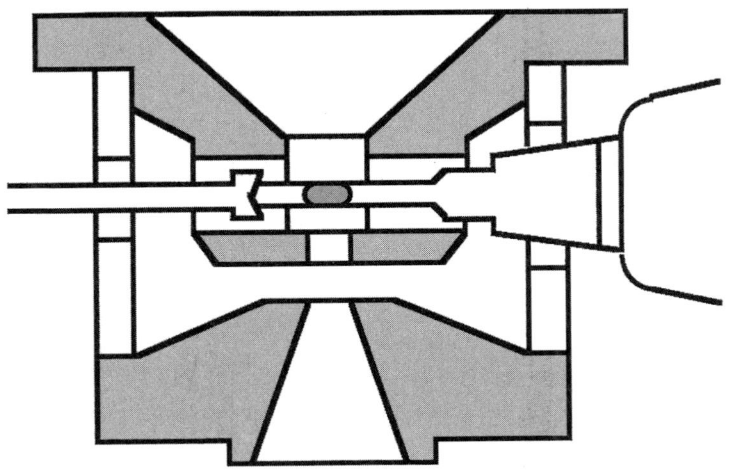

Figure 1. Cross-sectional view of the magnetic-shield lens which reduces the vertical field to 5 Oe. The specimen is set above the lens gap. The focal length is 8.6 mm [24].

objective lenses are required in order to reduce the leakage field. If a microscope has a top-entry-type specimen stage, by using a shield cylinder of permalloy and a specimen holder that is shortened by several mm, the leaking vertical field can easily be reduced below 1/100 of that at the normal position. If the microscope has a side-entry-type specimen stage, we need to introduce a magnetic-shield lens [23, 24] or an extra mini lens [25]. Figure 1 shows the magnetic-shield lens in which a specimen is not positioned in the lens gap but inside the upper piece. We can observe specimens under less than 400 A/m (~5 Oe) at the magnification of greater than 500 thousand times using this type of lens.

2.3. Deflection Angle Due to Lorentz Force

Estimating the angle at which electrons are deflected through the magnetic field is worthwhile. This deflection angle θ is represented by a Larmor radius r and a specimen thickness t as

$$\theta = \frac{t}{r}. \tag{1}$$

The radius r is determined by the motion equation of the electron in the film, expressed as

$$evB = m\frac{v^2}{r}, \tag{2}$$

where e is an elemental charge, v is the velocity of the electron, m is the mass of the electron, and B is the modulus of magnetic flux density. For example, let us consider an electron of 200 keV that passes through an iron thin film of 10 nm in thickness. Assuming that the film has almost the same value of $B = 2\text{T}$ as bulk iron, and substituting proper values of parameters to Eq. (2), we obtain the radius r as

$$r = \frac{mv}{eB} \approx 0.76 \text{ [mm]},$$

where the mass of the electron is corrected relativistically.

Therefore, we obtain the angle as

$$\theta = \frac{d}{r} = 13 \text{ [}\mu\text{rad]}.$$

The deflection angle is so small that an incident electron beam must be parallel less than 6×10^{-6} rad in order to detect the structure clearly.

2.4. Fresnel Mode

OBSERVATION OF DOMAIN WALLS Lorentz microscopy with a Fresnel mode is achieved by placing the specimen out of the magnetic field of an objective lens and observing in the plane far from the ordinary image plane (the Gaussian plane). Focusing just on the specimen plane, no contrast due to the magnetism of the specimen is observed. Properly defocusing so that the excitation of the objective lens is too strong, we have bright and dark contrast due to the converging and diverging of electrons though the adjacent domains where the spin direction alternatively changes, as shown in Fig. 2. In contrast, excitation that is too weak supplies inverse contrast at the domain walls. The amount of the defocusing should be selected in consideration of the intensity of magnetic induction, specimen thickness, thickness of domain walls and coherency of electron waves. In the case of the previous section, since an incident electron beam of a thermionic electron gun having a large source size has to be expanded widely to be parallel, the magnification available is restricted to a fairly low, 2000 times or less. Therefore, the out-of-focus value is $\delta f > 1$ mm to resolve the contrast of domain walls in an image, and is $\delta f > 10$ μm for the microscope equipped with a field emission electron gun whose source size is far smaller. The difference in the focusing values of the thermionic and field emission guns makes a great difference in the image contrast.

In pioneering work in the applications to high technology and materials science was the observation of babble domains by Watanabe *et al.* [26]. They showed wall contrast of babble domains of 1 μm in diameter by a high-voltage electron microscope.

The spin valve, a kind of magnetic multiplayer, consists of two ferromagnetic layers of different thickness separated by a non-magnetic layer, and the motion of magnetic spins in one of the ferromagnetic layers is restricted with an adjacent antiferromagnetic layer. This material shows giant magnetoresistance (GMR) effect, too, and it is appropriate for observation by Fresnel mode Lorentz microscopy. The difference in layer thickness causes a difference in the intensity and thickness of domain wall images, and as a result

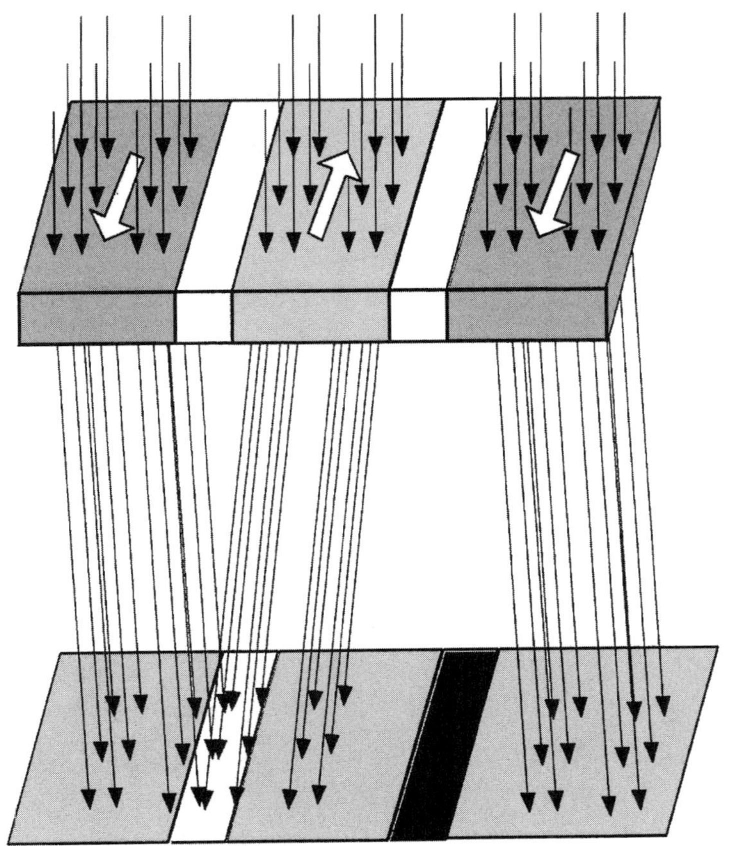

Figure 2. Fresnel mode Lorentz microscopy makes bright or dark contrast on domain walls.

the difference can be revealed individually in the motion of domain walls in each layer.

Lorentz microscopy of a spin valve film, NiFe (2 nm)/Cu/NiFe (10 nm), has been performed by a 200 kV TEM (Hitachi H-8000) with the objective-lens-off mode. Lorentz micrography shows two kinds of domain walls. Figures 3(a) and 3(c) were obtained in the underfocus and overfocus conditions, respectively. The thicker dark line in Fig. 3(a) is from the 10-nm-thick NiFe layer. Thinner lines due to the 2-nm-thick layer are also observed. When an intermediate lens was excited further (to overfocus), the contrast of the lines was reversed and the figure changed due to a faint leakage of magnetic field from the intermediate lane, as shown in Fig. 3(c). Thin extra lines appear beside the thicker bright line, and new lines have opposite contrast of thicker one. Figures 3(b) and 3(d) represent magnetization vectors (from S to N). White arrows show the induction in the thicker layer, and black arrows show that in the thinner layer. The extra domains in Fig. 3(c) are found to be in the thinner layer. That is, in Fig. 3(a)

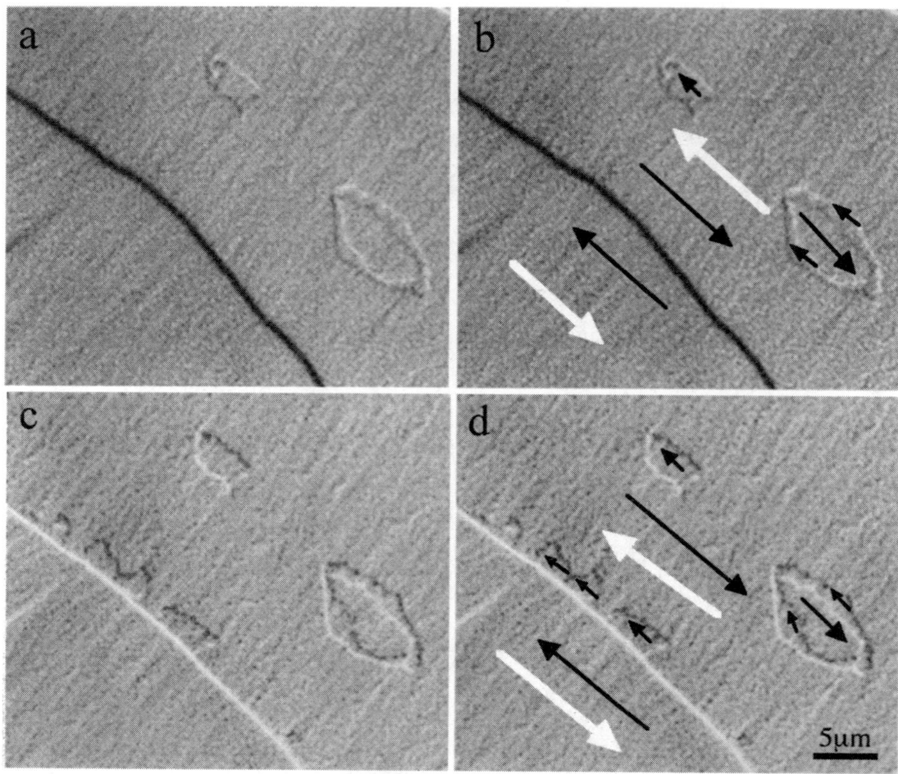

Figure 3. Lorentz microscopy of a spin valve film. Underfocused images (a, b) and overfocused images (c, d) inform the direction of induction in a thin CoFeB layer (black arrows) and that in a thick CoFeB layer which is pinned with an NiO layer (white arrows). Domain walls in the unpinned layer moved with a slight variation of the magnetic field.

both kinds of lines i.e. the thicker line and the thinner line are at the same position, and only stronger contrast (the dark line in Fig. 3(a)) can be recognized. By increasing the lens excitation, the weak leakage field affected only the domain structure of the thinner layer, in which the domain walls are easier to move, and a part of the thin line approached the side. This denotes that the domains having opposite magnetic vectors in both layers couple each other in large area.

OBSERVATION OF SINGLE MAGNETIC DOMAIN PARTICLES Magnetic domain structures were shown in the previous section. This structure is selected so that the magnetic energy, which is the sum of energies to form magnetic domains and walls, including antimagnetic field, is minimum.

The energy of the domain wall is proportional to the area of the wall (the square of size). On the other hand, the magnetostatic energy is proportional to the volume of the domain (the cube of size). Therefore, as the particle size decreases, a single-domain state

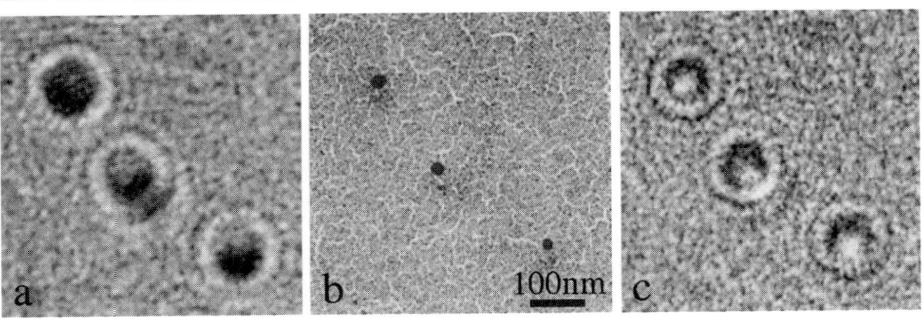

Figure 4. Through-focus images of a 30 nm thick Fe-Mo film taken δf values of (a) -2.0 mm (underfocus), (b) 0 mm (slightly underfocused), and (c) $+1.2$ mm (overfocused) [31, 32].

will become most stable below a certain critical size. This prediction was first proposed by Frenkel and Dorman [27] in 1930, and the critical size was studied theoretically by Kittel [28], Nèel [29], and Brown [30]. Single-domain particles are very important for industry as well as scientific research of magnetism, because the maximum coercive forces are expected for all permanent magnets if they consist only of single-domain particles, and the finest powders applied to the modern recording media are already single-domain particles. Kittel predicted that the critical size is about 20 nm for iron. Therefore, this single domain particle is observable by only electron microscopy. The first observation of a single magnetic domain particle was reported by Goto *et al.* [21] They applied the colloid-SEM method, which is a kind of Bitter method, to a particle of Ba-ferrite of $1 \sim 2$ μm in diameter. They showed that ferromagnetic fine powders adopted preferentially on domain walls of a multi-domain particle and uniformly on a single-domain particle. However, if a specimen, such as a metal particle, is far smaller, the resolution of the colloid-SEM method, which depends on the size of ferromagnetic powders, is no longer enough.

Fresnel mode Lorentz microscopy has been utilized to observe ferromagnetic fine particles. Figure 4 shows Lorentz microscopy of Fe single magnetic-domain particles [31, 32]. The specimens (Fe particles) were prepared by precipitation in an Fe-Mo amorphous thin film. An amorphous carbon film mounted on a Cu grid for transmission electron microscopy (TEM) was used as a substrate. The substrate was cooled to 170 K, and an amorphous Fe-Mo film (30–50 nm thick) was prepared by dc magnetron sputtering. The target was prepared so that the ratio of Fe to Mo was 80:20 in the area according to a pre-measurement of their sputter rates. $Fe_{1-x}Mo_x$ ($x < 40$) easily forms an amorphous structure below 200 K. We have fine particles of proper size precipitated by annealing the amorphous alloys in an electron microscope. Observation was performed using a 200 kV field emission TEM (HF–2000). Direct magnification of TEM was 100,000 times. Significant defocusing more than 1 mm changes the magnification of the microscope, so we have to calibrate the magnification of each image so that the distance between two particles coincides with that of the in-focus image. Figure 4 shows through-focus images of a 30 nm thick Fe-Mo film. Slightly

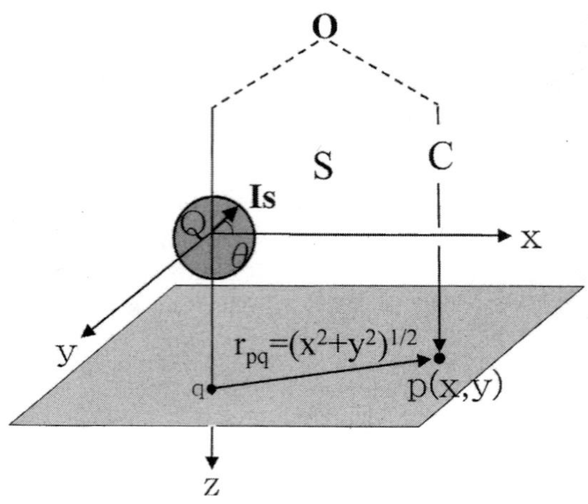

Figure 5. Simulation model of a single magnetic domain particle: Is is the magnetization in the specimen at point Q. The electron beam trajectory L is parallel to the z-axis. The observation plane is parallel to the x–z plane. Point q in the x–y plane is a projection of Q.

asymmetric contrasts of black-and-white pair of more than 100 nm in diameter are observed in the underfocused image of Fig. 4(a), which is reversed in the overfocused image of Fig. 4(c), disappears in the in-focused image Fig. 4(b). This means that the black-and-white pairs are due to the Lorentz force, that is, the magnetic contrast. The contrast, however, does not show a simple reversal of black and white between the underfocused images and the overfocused images.

Images such as magnetic domain walls in a thin film can be understood intuitively by the divergence and convergence of an electron beam due to Lorentz deflection, as shown in Fig. 2. In order to understand accurately the observed results of localized magnetic flux in Fig. 4, computer simulation using a simple model is indispensable.

Assume a single-magnetic-domain sphere having constant magnetization inside, a specimen of radius R, and an electron-beam trajectory L, which is parallel to the z-axis, as shown in Fig. 5.

The incident plane wave of time independent is given as

$$\psi_0(x, y, z) = \exp\{-ikz\}, \tag{3}$$

where k is the modulus of the wave vector $\mathbf{k}$.

The phase of the incident plane wave is modulated by the magnetic induction inside and the leaking of the particle, as well as by the inner electrostatic potential.

The phase of the electron wave shifted in the specimen is given as

$$\phi = \oint_c \left(\mathbf{k} - \frac{eV}{2E_0}\mathbf{k} - \frac{2\pi e}{h}\mathbf{A}\right) \cdot \mathrm{d}\mathbf{s}, \tag{4}$$

where E_0 is an acceleration voltage, V_0 is an electrostatic potential and $\boldsymbol{A}$ is a vector potential. Integration is performed along path C: from an electron source O to an observation point p, to the points q (a projection of the center of sphere Q) and O. The first term $\oint_c \mathbf{k} \cdot \mathbf{ds}$ shows the phase difference due to the optical path difference, which is null in this case. The second term $\phi_1 = \frac{e}{2E_0} \oint_c V\mathbf{k} \cdot \mathbf{ds}$ represents the phase shift due to the inner electrostatic potential. The final term is rewritten using Stokes' theorem as,

$$\phi_2 = \frac{2\pi e}{h} \oint\limits_c \mathbf{A} \cdot \mathbf{ds} = \frac{2\pi e}{h} \int\limits_S \mathbf{B} \cdot \mathbf{n}\, \mathrm{ds} = \frac{2\pi e}{h} \Phi, \tag{5}$$

where the second integration is on the area S surrounded by the loop C, and Φ is the total flux of $\mathbf{B}$ through the area S.

Thus, the wave shifted the phase by the inner potential and the magnetic induction is represented as

$$\psi(x, y, z) = \exp[-i\{kz - \phi_1(x, y) - \phi_2(x, y)\}]. \tag{6}$$

Assuming a specimen of single-magnetic-domain sphere has a mean inner potential $-V_0$ and a constant magnetic induction inside, we obtain $\phi_1(x, y)$ as

$$\phi_1(x, y) = -\frac{2\pi}{E_0\lambda} V_0(R^2 - x^2 - y^2)^{1/2}, \tag{7}$$

and $\phi_2(x, y)$ as

$$\begin{aligned} \phi_2(x, y) &= \frac{4\pi e R^2 I_s y}{3h(x^2 + y^2)} \left\{1 - (1 - x^2 - y^2)^{3/2}\right\} \cos\theta \qquad \text{if } x^2 + y^2 \leq R^2, \\ &= \frac{4\pi e\, R^2 I_s y}{3h(x^2 + y^2)} \cos\theta \qquad \text{elsewhere} \end{aligned} \tag{8}$$

where $\lambda = 2\pi/\mathrm{k}$ is the wavelength, I_s is the intensity of magnetization in the $x-z$ plane making an angle θ with the observation plane. Thus, the defocused image $I(x, y)$ becomes

$$I(x, y) = |\psi_i(x, y, z)|^2, \tag{9}$$

$$\psi_i(x, y, z) = \exp[-i\{kz - \phi_1(x, y) - \phi_2(x, y)\}] \otimes \exp\left\{\frac{-ik(x^2 + y^2)}{2\Delta f}\right\}, \tag{10}$$

where $\otimes$ expresses a convolution and Δf is the amount of defocus. The phase shifted in the specimen is observable by considerable defocusing in the Fresnel mode of Lorentz microscopy.

In the specimen of granular film, it is difficult to determine accurately the difference in inner potential between the precipitates and a matrix. Therefore, we treat the inner potential as a fitting parameter, and assume the intensity of the magnetization I_s to be identical to the saturation magnetization of bulk iron, its direction being parallel to the film surface. When the mean inner potential is taken into consideration, the bright area changes its position between the underfocused and overfocused images, similar to the experimental observation. In this calculation, the difference of the inner potential was assumed to be 30% of the full inner potential value of Fe.

The black-and-white contrast can be observed inside the first Fresnel zone by a considerable defocusing. Defocusing blurs the image and makes the particle appear larger than it actually is, as shown in Fig. 4. We examined the interval of the fringes and contrast in detail based on the line profiles, as shown in Fig. 6. The intensities of the image line profiles were normalized with the background in order to compare the simulation with the experimental findings. The experimental and simulated images coincide very well qualitatively.

When the intensity of magnetization is lowered to 50%, the calculated contrast decreases approximately 30% and no longer coincides with the observed result. The simulation also shows that the Fe particle having a diameter of less than 10 nm, which presents a contrast of approximately 4%, will be difficult to detect using this method. The magnetization intensity is approximately the same as the saturation magnetization, with an accuracy of 30% or better.

OBSERVATION OF SUPERCONDUCTORS Very similar contrast to the single magnetic domain particles has been observed in super conductors [33, 34]. When a magnetic field is applied to a superconductor, tiny vortices penetrate inside. A dissipation-free current is available only when these vortices are pinned anywhere against the current-induced force. Revealing the behavior of these vortices will encourage the development of new superconducting materials. Vortices inside a thin film can be observed by Lorentz microscopy. In general, magnetic flux flows inside the vortex in a direction perpendicular to the film; therefore, the specimen has to be tilted from horizontal in order to interact with electrons, as shown in Fig. 7(a). The vortex has no difference in inner potential but only magnetic flux inside. Therefore, a single domain particle changes both the amplitude and phase of an electron wave, while a vortex causes only the phase shift of the electron wave. The cross-sectional shape of magnetic flow is not rigid but penetrates gradually, so Fresnel fringes due to a large defocusing do not have strong contrast and only a black and white pair contrast of a few hundred nano meters appears, although the size of the vortices depends on the materials and film thickness. Figure 7(b) shows vortices in a high-Tc superconductor $Bi_2Sr_2CaCu_2O_{8+\delta}$ (Bi-2212) observed by a 1000 kV FE-TEM. It is found that the vortices make chains among the triangle lattice in the field of 5 mT at 50 K, and the vortex lines are perpendicular to the layer plane.

2.5. Foucault Mode

If incident electrons are parallel, an electron wave transmitted through a magnetic domain having a magnetization vector component perpendicular to the incident beam

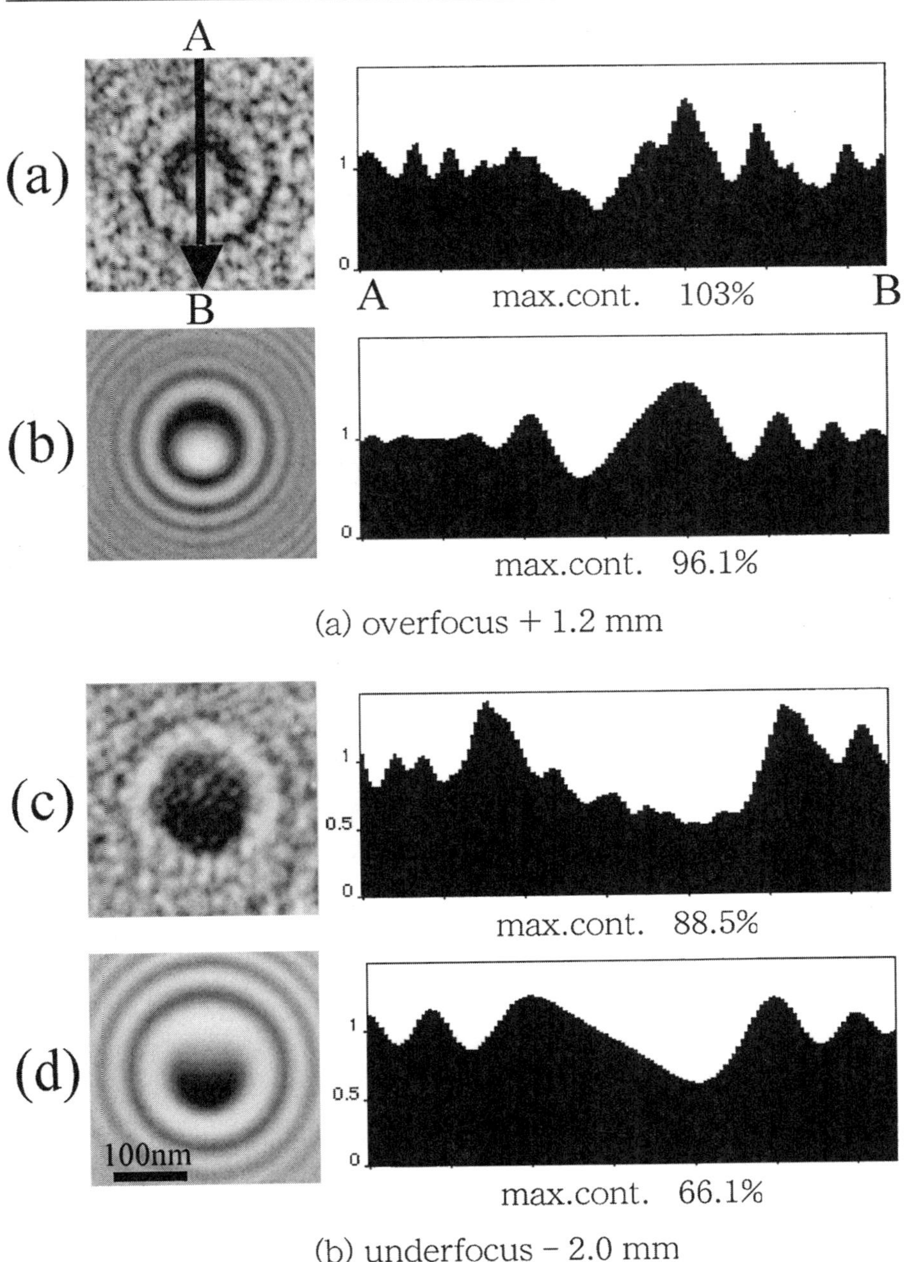

Figure 6. Line profiles [A–B] of Lorentz images. Experimental results (a, c) coincide quantitatively with simulated ones (b, d) [32].

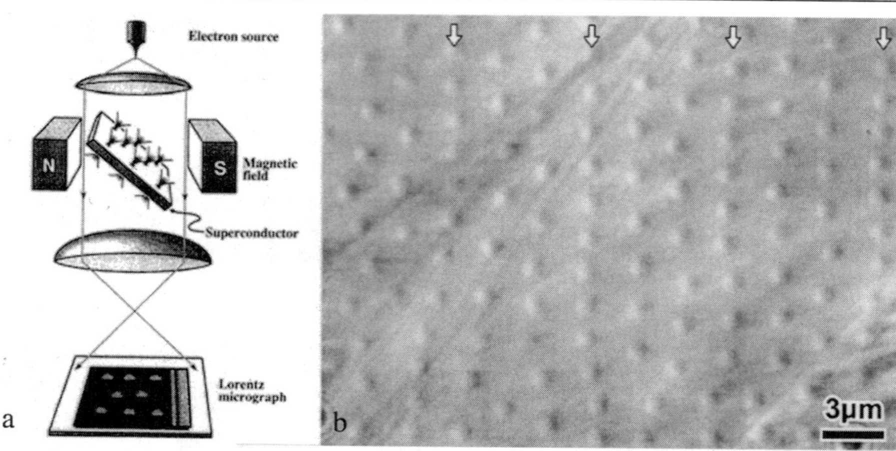

Figure 7. (a) Schematic diagram of the experimental arrangement. The specimen is tilted to both the electron beam and the externally applied magnetic field. (b) Lorentz micrographs of vortices in Bi-2212. The chain-lattice structure is formed at a magnetic field tilted by 85° to the surface normal and $T = 50$ K, $B = 5$ mT. In the chains indicated by white arrows, the vortex lines are perpendicular to the layer plane [33, 34].

deflects off the optical axis of the diffraction plane, producing a spot besides the center spot. Another wave through another domain having a different magnetization vector from the previous wave also deflects off the optical axis, but this wave produces a spot in the diffraction plane at a different point from the first spot. Selecting one of these extra spots by an objective aperture, we obtain an image with bright contrast at the domain which contributes to make the selected spot. The advantage of the Foucault mode over the Fresnel mode is that images are observable in focus, and the disadvantage is the difficulty involved in dynamic observation of the domain motion. Figure 8 shows a domain structure in an Nd-Fe-B permanent magnets of 50 nm in thickness using a 300 kV FE-TEM (JEM-3000F) with the magnetic shielded lens. In Fig. 8(b), black and dark regions correspond to magnetic domains with different magnetization vectors [35].

2.6. Lorentz Phase Microscopy

Recently a noble method of Lorenz microscopy has been developed (Transport of intensity equation (TIE) method) by Paganin and Nugent [36]. This method achieves phase retrieval from two Lorentz micrographs via computer processing.

Here, we introduce the same results in a simpler way by considering a weak amplitude and phase object. If the phase shift of an electron wave transmitted though a thin film is $\phi(x, y)$ and the absorption in the specimen is $\mu(x, y)$, then the wave function in the exit surface of the specimen is

$$\psi(x, y) = \exp\{-i\phi(x, y) - \mu(x, y)\}\cdot \tag{11}$$

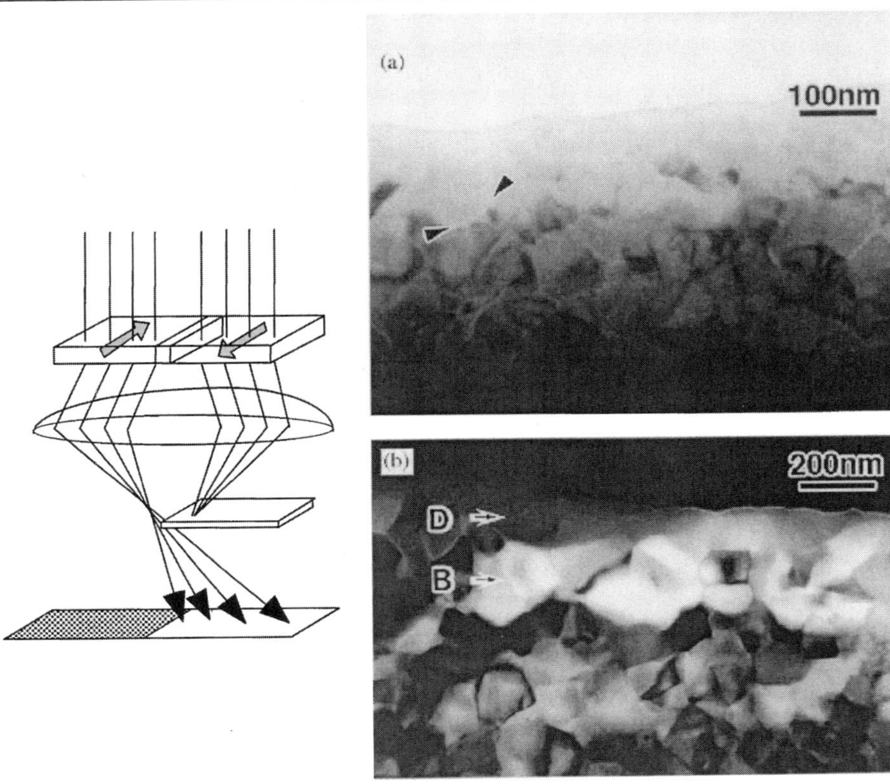

Figure 8. Foucault mode Lorentz microscopy filters a part of deflected wave (left). Only selected wave makes bright contrast on the corresponding domain. Images observed by the Fresnel (a) and the Foucault mode (b) in the Nd-Fe-B permanent magnet annealed at 843 K [35].

Let us consider only images with a spatial resolution poorer than 5 nm. Then, ignoring the effect of spherical aberration on the phase transfer function, we can represent the wave function in the image plane as

$$\begin{aligned}\psi_i(x, y, \Delta f) &= F^{-1}[F[\psi(x, y)]\exp\{-i\pi\lambda(u^2+v^2)\Delta f\}] \\ &\approx F^{-1}[F[\psi(x, y)]\{1-i\pi\lambda(u^2+v^2)\Delta f\}], \\ &= \psi(x, y) - i\pi\lambda\Delta f F^{-1}[(u^2+v^2)F[\psi(x, y)]]\end{aligned} \tag{12}$$

where u and v are the coordinates in the Fourier space, and F and F^{-1} denote the Fourier transform and inverse Fourier transform, respectively.

Using the relation

$$F\left[\frac{\mathrm{d}^n}{\mathrm{d}x^n}\varphi(x)\right] = (2\pi iu)^n F[\varphi(x)], \tag{13}$$

we can rewrite Eq. (12) as

$$\psi_i(x, y, \Delta f) = \psi(x, y) + \frac{i\lambda\Delta f}{4\pi}\left(\frac{\partial^2\psi(x, y)}{\partial x^2} + \frac{\partial^2\psi(x, y)}{\partial y^2}\right). \tag{14}$$

Using the relation

$$\begin{aligned}\frac{\partial^2\psi(x, y)}{\partial x^2} &= \frac{\partial^2}{\partial x^2}\exp\{-i\phi(x, y) - \mu(x, y)\}\\ &= \frac{\partial}{\partial x}\left[\exp\{-i\phi(x, y) - \mu(x, y)\}\left(-i\frac{\partial}{\partial x}\phi(x, y) - \frac{\partial}{\partial x}\mu(x, y)\right)\right]\\ &= \exp\{-i\phi(x, y) - \mu(x, y)\}\left\{\left(-i\frac{\partial}{\partial x}\phi(x, y) - \frac{\partial}{\partial x}\mu(x, y)\right)^2\right.\\ &\qquad \left.-\left(i\frac{\partial^2}{\partial x^2}\phi(x, y) + \frac{\partial^2}{\partial x^2}\mu(x, y)\right)\right\}\end{aligned} \tag{15}$$

and ignoring the term including λ^2, we obtain the intensity of the image as

$$\begin{aligned}I(x, y, \Delta f) &= |\psi_i|^2\\ &= |\psi|^2 + \frac{i\lambda\Delta f}{4\pi}\left\{\psi^*\left(\frac{\partial^2\psi}{\partial x^2} + \frac{\partial^2\psi}{\partial y^2}\right) - \psi\left(\frac{\partial^2\psi^*}{\partial x^2} + \frac{\partial^2\psi^*}{\partial y^2}\right)\right\},\\ &= \exp(-2\mu) + \frac{\lambda\Delta f}{2\pi}\exp(-2\mu)\left\{\left(\frac{\partial^2\phi}{\partial x^2} + \frac{\partial^2\phi}{\partial y^2}\right)\right.\\ &\qquad \left.-2\left(\frac{\partial\phi}{\partial x}\frac{\partial\mu}{\partial x} + \frac{\partial\phi}{\partial y}\frac{\partial\mu}{\partial y}\right)\right\}\end{aligned} \tag{16}$$

where * denotes a complex conjugate.

If ϕ and μ vary gradually, i.e. $\frac{\partial\phi}{\partial x}, \frac{\partial\phi}{\partial y} \ll 1$ and $\frac{\partial\mu}{\partial x}, \frac{\partial\mu}{\partial y} \ll 1$, then Eq. (16) becomes

$$I(x, y, \Delta f) = \exp(-2\mu) + \frac{\lambda\Delta f}{2\pi}\exp(-2\mu)\left(\frac{\partial^2\phi}{\partial x^2} + \frac{\partial^2\phi}{\partial y^2}\right). \tag{17}$$

As $\mu \ll 1$, considering that the product $\lambda\mu$ is minute, we have

$$I(x, y, \Delta f) \approx (1 - 2\mu) + \frac{\lambda\Delta f}{2\pi}\left(\frac{\partial^2\phi}{\partial x^2} + \frac{\partial^2\phi}{\partial y^2}\right) \tag{18}$$

and

$$\frac{\partial I}{\partial(\Delta f)} = \frac{\lambda}{2\pi}\left(\frac{\partial^2\phi}{\partial x^2} + \frac{\partial^2\phi}{\partial y^2}\right). \tag{19}$$

In order to obtain the original phase, we again use the two-dimensional relation similar to Eq. (13), that is,

$$F\left[\frac{\partial^2\phi}{\partial x^2}+\frac{\partial^2\phi}{\partial y^2}\right]=-4\pi^2(u^2+v^2)F\,[\phi]\,. \tag{20}$$

Then, we have

$$\begin{aligned}\phi&=-F^{-1}\left[\frac{F[(\partial^2\phi/\partial x^2)+(\partial^2\phi/\partial y^2)]}{4\pi^2(u^2+v^2)}\right]\\&=-F^{-1}\left[\frac{F[(2\pi/\lambda)(\partial I/\partial(\Delta f))]}{4\pi^2(u^2+v^2)}\right].\\&=-F^{-1}\left[\frac{F[\partial I/\partial(\Delta f)]}{2\pi\lambda(u^2+v^2)}\right]\end{aligned} \tag{21}$$

Note that differentiating Eq. (5) gives flux density **B** directly.

Differentiation with respect to x corresponds to selecting the loop C in Fig. 5 so that the surface normal **n** is in the y direction, and gives a y component of $\mathbf{B}'$, which is a projection of **B** onto the x–y plane

$$B_y'=\int B_y\,\mathrm{d}z=\frac{h}{2\pi e}\frac{\mathrm{d}\phi}{\mathrm{d}x}, \tag{22}$$

and similarly

$$B_x'=\int B_y\,\mathrm{d}z=\frac{h}{2\pi e}\frac{\mathrm{d}\phi}{\mathrm{d}y}. \tag{23}$$

Figures 9 shows an example of phase retrieval via TIE [17] using a conventional 200 kV TEM (JEM-200CX). Lorentz micrographs of a 1.65 μm^2 and 10 nm-thick Co dot (Figs. 9(a) and (b)) produce a differential image (d), which is converted to a phase image, shown in Fig. 9(e). Although this technique is very useful for revealing phase information, the assumptions used in Eqs. (12) and (17) produce very low-resolution results.

3. ELECTRON HOLOGRAPHY

3.1. Introduction

Electron holography, which Gabor developed for correcting aberrations of an objective lens of an electron microscope, has been used to observe vector potential. Electron holography can reveal magnetic induction quantitatively, and this has come to be one of the most important applications of this technique.

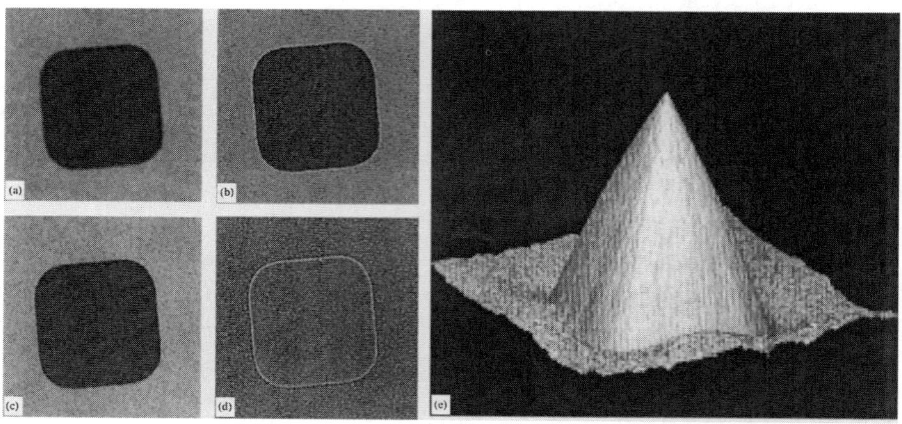

Figure 9. Lorentz phase microscopy of 1.65 × 1.65 μm cobalt squares. (a) Positive defocus image; (b) negative defocus image; (c) in-focus image; (d) difference between positive and negative defocus images after alignment; (e) a surface plot of the recovered phase [36].

In general, holography is a two-step processing for reconstructing information: the recording process by interference of an object wave and a reference wave, and the reconstruction process of the object wave on an optical bench or by digital processing.

Digital processing has the advantage of direct reconstruction of the phase over the optical reconstruction system, although the range reconstructed is only between $-\pi$ and π. A phase jump appearing at every 2π is the equiphase line, which is parallel to the line of magnetic force. In principle, the phase jumping at $2n\pi$ can be continued. In other words, phase unwrapping is possible. However, the noise in practical holograms interferes with precise unwrapping. Therefore, digital processing is often used to take the sine or cosine of the reconstructed phase, which corresponds on an interference micrograph in the optical reconstruction.

Using electron holography, we can measure quantitatively the intensity of induction. Electron holography of a permalloy thin film shows gradual bending near domain walls, which means that the magnetization in each domain is not constant as shown in Section 3.2.

In this chapter, special techniques of electron holography for observing nano-magnetic structures are primarily described. Refer to the chapter on electron holography for the fundamental principle and applications.

3.2. Observation of Single Magnetic Domain Particles

As described in section 2.2, very small ferromagnetic particles have the single-magnetic-domain structure. Figure 10 shows an example of Ba-ferrite ($BaO{\cdot}6Fe_2O_3$) particles. In this material the single-domain structure has been found to be less than 2 μm in diameter. As mentioned in Section 2.1, large single-domain particles can be observed by colloid-SEM, but particles less than 1 μm in diameter can only by TEM.

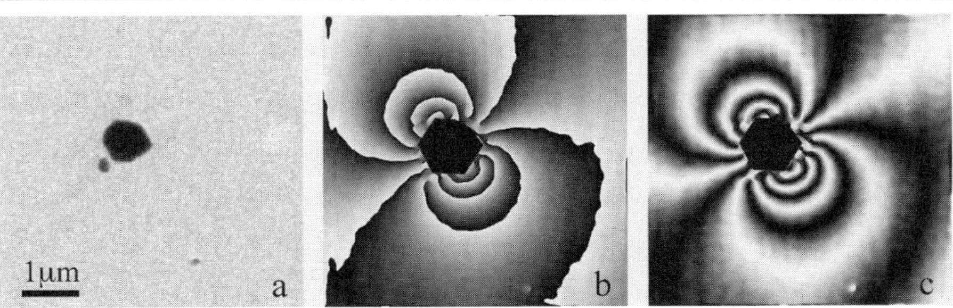

Figure 10. Electron holography of a Ba-ferrite single-magnetic-domain particle. The conventional TEM image (a) shows the particle is 800 nm in diameter. The reconstructed phase image (b) and interferogram (c) reveal the magnetic field around the particle.

The particle of 800 nm shown in Fig. 10(a) is observed by electron holography using TEM. The reconstructed phase angle on each pixel is amplified by a factor of 10 and then transferred to the brightness variation, where $[-\pi, \pi]$ corresponds to [0, 255] in Fig. 10(b), and the same phase distribution is shown as the interference micrograph in Fig. 10(c). The contour lines in Fig. 10(c) display lines of magnetic force as if they were there actually. Particles consisting of two domains have been observed even for particle diameters of less than 200 nm [37].

3.3. Real-Time Observation

REAL-TIME RECONSTRUCTION SYSTEM Holography is a type of two-step processing, as mentioned above, so real-time observation of dynamic phenomena is unsuitable. In order to overcome this weakness, some of on-line reconstruction systems have been reported.

Modern personal computers or work stations are available to reconstruct holograms of 512×512 in less than 1 second, so constructing the system using a charge coupled device (CCD) video camera and a high-performance computer is easy [38]. However, this on-line computer system does not yet have sufficient speed or resolution for continuous observation of domain motion.

Figure 11(left) shows the system combining an optical reconstruction system and an electron holographic microscope [39, 40]. A hologram recorded using a video camera is transferred to a liquid crystal (LC) panel in the optical system. The LC used is a twisted nematic liquid crystal (TNLC), which modulates the phase of transmitting light with the voltage inputted as a video signal. The linearity of the phase modulation has been confirmed in a proper region of applied voltage. An interference micrograph to display the phase distribution is formed using a Mach-Zehnder interferometer. Two plane laser waves that have been split by a half-miller are illuminated onto the LC panel receiving the signal of the hologram from the video camera. In Fig. 11(left), the incident angles of two laser waves are preset so that the true object wave of one incident wave and the complex conjugate wave of the other incident wave come to

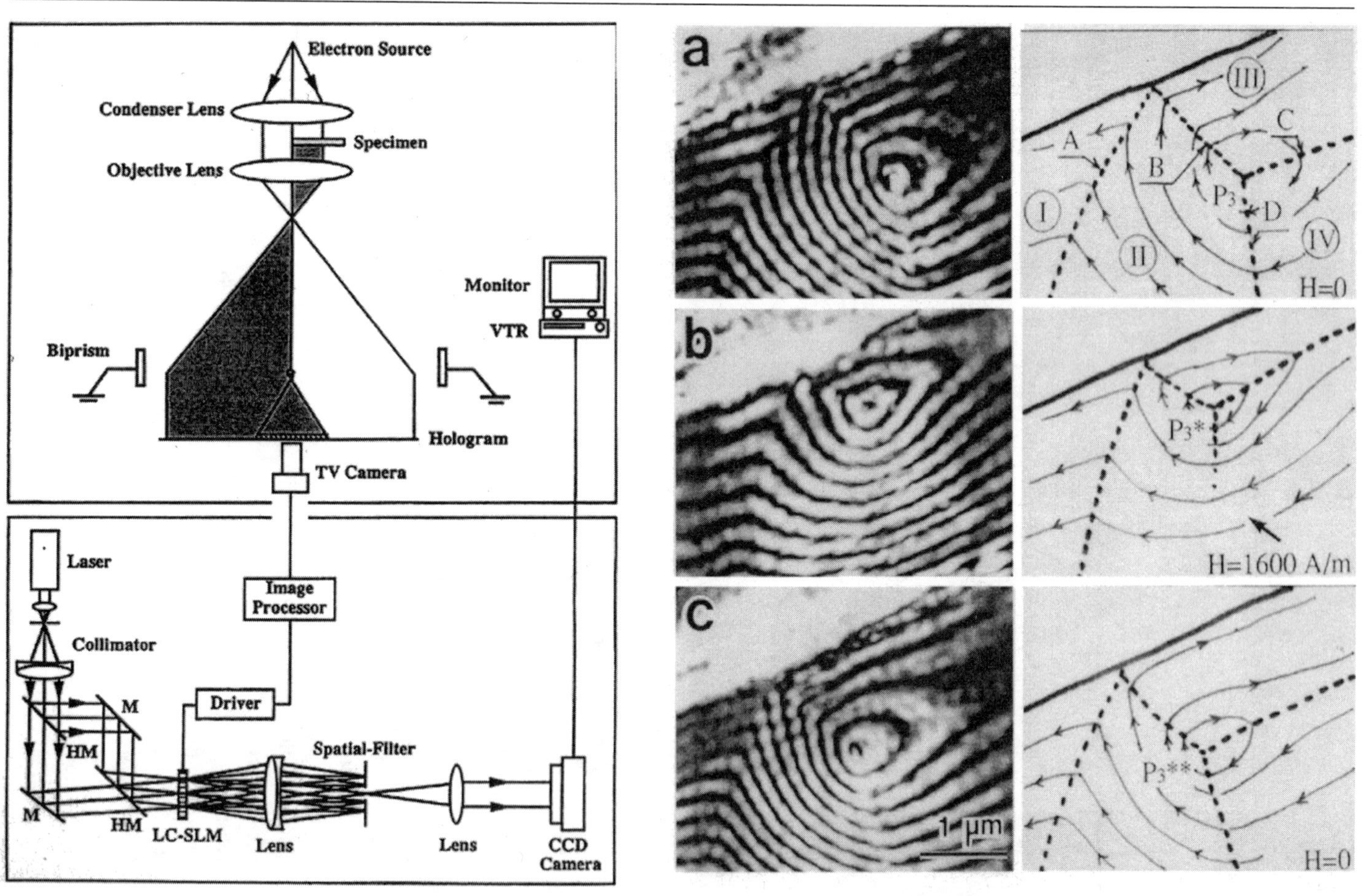
Electron Source
Condenser Lens
Specimen
Objective Lens
Monitor
VTR
Biprism
Hologram
TV Camera
Laser
Image Processor
Collimator
M
Driver
HM
Spatial-Filter
M
HM
LC-SLM
Lens
Lens
CCD Camera
a
b
c
III
A
C
B
P3
D
I
II
IV
H=0
P3*
H=1600 A/m
P3**
1 μm
H=0

be reconstructed simultaneously though the same opening in the mask in the Fourier plane and these waves interfere with one another, resulting in a double-amplified interferogram. Reconstructed interferograms are displayed on a monitor beside the microscope. The system reconstructs the phase with TV rate.

Figures 11(a)–11(c) show the motion of domains when a weak external magnetic field was applied. In order to apply the external field, a specimen was tilted by 10 degrees, and the current of the objective lens was increased from zero to 0.4 A and then decreased to zero. Interference micrographs show the demagnetization and re-magnetization processes clearly. The phase transition between ferro- and para-magnetism of a Ba-ferrite single magnetic-domain particle with heating is shown in Fig. 12. Magnetic flux leaking from the particle at room temperature decreases with specimen temperature and disappears above 700 K. Cooling the specimen, we find that the flux reappears.

MULTI-BEAM INTERFERENCE In principle, electron holography is a two-step procedure. This principle is true also in the system described above. In order to retrieve the phase information directly from interferograms, techniques using Moiré patterns are used. An interference pattern made of a free space is superimposed on a hologram which is projected on a TV monitor. If this interference pattern has a spacing coincident with that of the carrier fringes of the hologram, the pattern can act as a reference grating, allowing real-time observation.

Similar observation has been achieved by Hirayama *et al.* [41] using three-wave or four-wave interference with two biprisms. An object wave and two or three reference waves are superposed to interfere in order to produce a new type of interference pattern in which electromagnetic fields are observed directly. As this is a type of Moiré pattern with coherent waves, extra fringes due to the higher-order interference may appear. Fortunately however, such higher-order interference can be suppressed by controlling the coherence of the incident electron wave. Magnetic flux lines emerging from a Ba ferrite single-magnetic-domain particle have been observed, as shown in Fig. 13.

3.4. High-Precision Observation

3.4.1. Phase-Shifting Method

In the case that the phase difference is too small to be read out, the reconstructed phase can be amplified. In particular, digital reconstruction using a computer makes the intensification easy. The retrievable information, however, is limited to that which is included in the original hologram. In the ordinary Fourier-transform method, the sensitivity appears to be up to one-fiftieth of the electron wave length [42], because of

←

Figure 11. (left) Schematic diagram of on-line real-time electron holography system (M: mirror, HM: half mirror). The hologram formed in electron holographic microscope is transferred to the liquid-crystal spatial light modulator (LC-SLM) located at the output port of the Mach-Zehnder interferometer.
(a) Real time observation of a permalloy thin film before applying a external magnetic field H, (b) when H is increased to 1600 A/m, (c) after decreased to zero. (a'–c') show their schematic domain states.

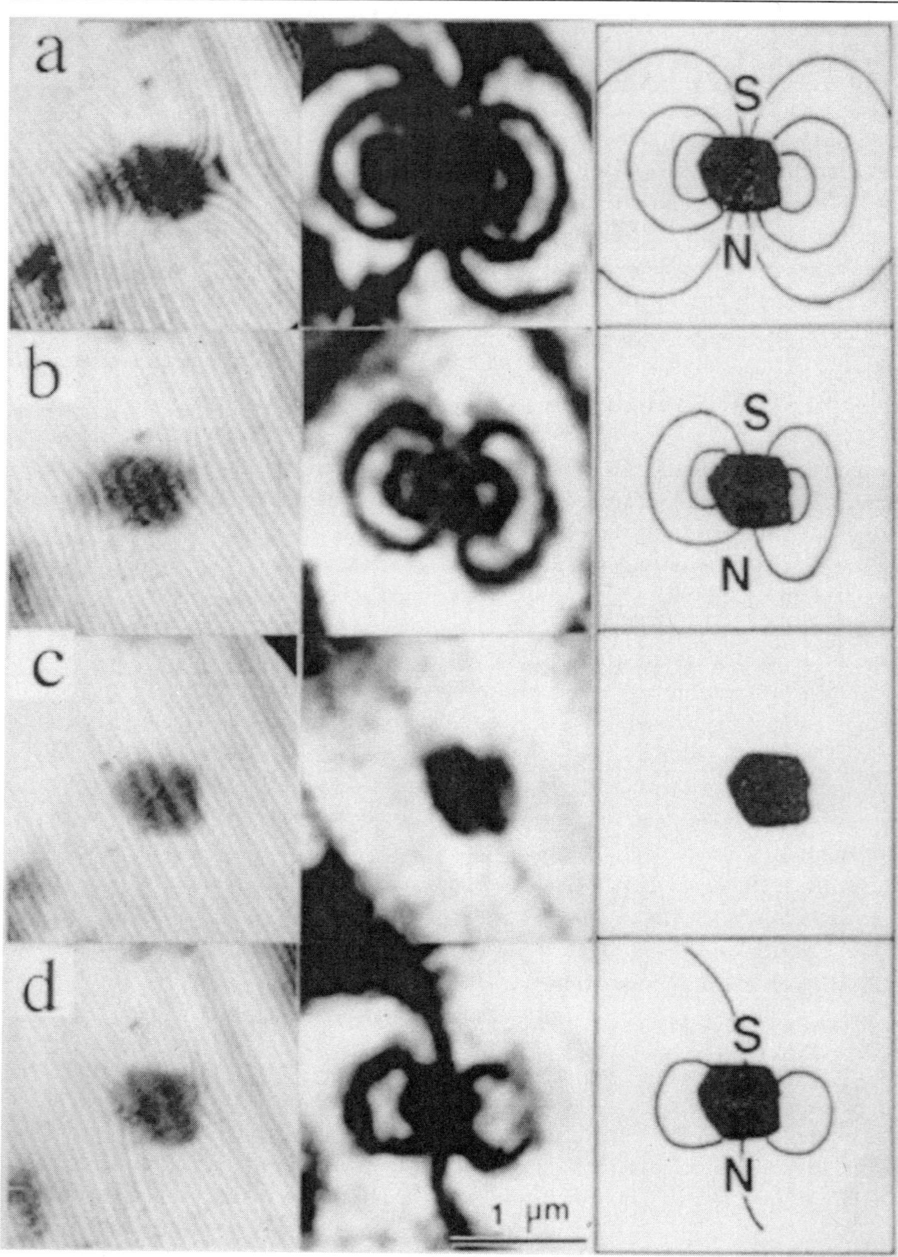

Figure 12. Real-time observation of the phase transition of a Ba-ferrite single-magnetic-domain particle from ferrimagnetism to paramagnetism at room temperature (a), 420 °C (b), 570 °C (c), and 450 °C.

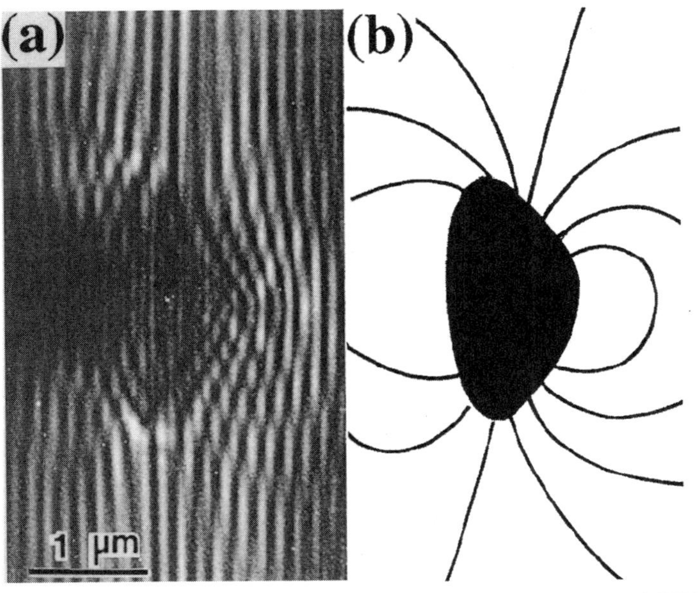

Figure 13. Single-magnetic-domain particle of Ba-ferrite: (a) interference pattern of three electron waves, (b) schematic of the particle and magnetic flux lines [41].

the numerous types of noise. A phase-shifting method has been introduced for more precise observation [43, 44]. This method uses a series of holograms. The initial-phase difference between the object and reference waves is shifted one after another; as a result, interference fringes are shifted, while the object image maintains its position as illustrated in Fig. 14. Although various methods of shifting the initial phase have been considered, thus far, two methods have been realized in electron holography. One is to shift the position of an electron biprism and the other is to tilt the incident electron wave. The reconstructed procedure of this phase-sifting electron holography was formulated by Ru *et al.* [45].

The intensity of the interference fringes is described ideally with the sum of the object wave function and the reference wave function, as

$$\begin{aligned} I(x, y, n) &= |\psi_i + \psi_r| \\ &= 1 + |A(x, y)|^2 + 2A(x, y)\cos\{2k\alpha x + \phi(x, y) - k\theta_n w\}, \end{aligned} \tag{24}$$

where $A(x, y)$ and $\phi(x, y)$ are the amplitude and phase, respectively, of the object wave modulated with a specimen, α is the angle between the electron waves deflected by the electron biprism, w is the width of the interference area, k is the wave number, and θ_n is the tilting angle of the nth incident electron wave, as shown in Fig. 14. The term $k\alpha x$ in the cosine function is due to the path difference caused by the prism and $\cos(2k\alpha x)$ expresses fundamental interference fringes produced with a normal

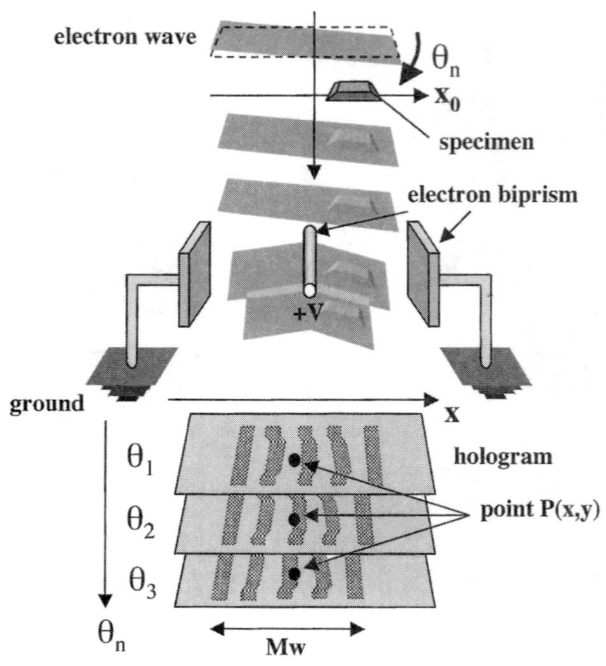

Figure 14. Schematic of phase-shifting electron holography. Interference fringes are shifted by tilting an incident electron wave [44].

incident wave. The phase term $-k\theta_n w$ denotes the initial phase difference between the object and the reference waves in the nth hologram. Hence, we can shift the fringes $2\pi/(k\theta_n w) = \lambda/(\theta_n w)$ of the spacing by tilting the incident wave at θ_n.

The intensity at a certain point (x, y) on the holograms, for example, point Z in Fig. 14, will vary sinusoidally by shifting the interference fringes. Figure 15 shows the variation of intensities plotted against the amount of the initial phase. It should be noted that the abscissa in Fig. 15 is $-kw\theta_n$ in Eq. (24). As the first phase term of Eq. (24), $2k\alpha x$, is found from the carrier frequency, and the phase change $\phi(x, y)$ showing the object at point Z is picked out from a cosine curve fitted to the intensity in Fig. 15, which leads to the reconstructed phase image. The reconstructed amplitude images are obtained from the amplitude $A(x, y)$ of the cosine curve. The curve fitting is performed using the least-square method. For convenience, Eq. (24) can be rewritten as

$$I(x, y, n) = C_1 + C_2 \exp[+i\phi_0(\theta_n)] + C_3 \exp[-i\phi_0(\theta_n)]\,, \tag{25}$$

where

$$\begin{aligned}
\phi_0(\theta_n) &= -kw\theta_n \\
C_1 &= 1 + |A(x, y)|^2 \\
C_2 &= A(x, y) \cdot \exp[+i(2k\alpha x + i\phi(x, y))] \\
C_3 &= A(x, y) \cdot \exp[-i(2k\alpha x + i\phi(x, y))]
\end{aligned}$$

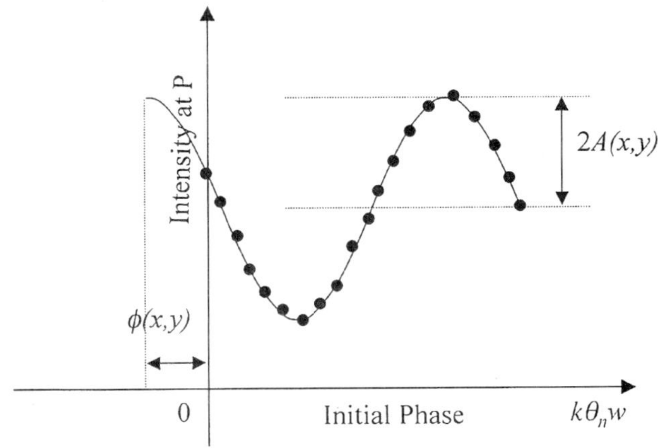

Figure 15. Procedure for reconstructing the object wave at point P(x, y). The amplitude $A(x, y)$ and phase $\phi(x, y)$ of the object wave are derived from the intensity variation against the initial phase.

We need at least three holograms that have shifted interference fringes because the cosine curve can be determined with background, amplitude, and phase. The determination of these factors is equivalent to the determination of C_1, C_2, and C_3.

Setting the measured intensity of the$n$th hologram at the position $Z(x, y)$ to$\hat{I}(x, y, n)$ and the error between the fitted and measured intensity to$Er(x, y)$, we have to minimize the error function:

$$\begin{aligned} Er(x, y) &= \sum_{n=1}^{N} (I(n) - \hat{I}(n))^2 \\ &= \sum_{n=1}^{N} (C_1 + C_2 \cdot \exp[+i\phi_0(\theta_n)] + C_3 \cdot \exp[-i\phi_0(\theta_n)] - \hat{I}(n))^2 . \end{aligned} \tag{26}$$

Partially differentiating $Er(x, y)$ by C_1,C_2, and C_3, we obtain three equations:

$$\begin{aligned} \frac{\partial Er}{\partial C_1} &= \sum_{n=1}^{N} 2(C_1 + C_2 \cdot \exp[+i\phi_0(\theta_n)] + C_3 \cdot \exp[-i\phi_0(\theta_n)] - \hat{I}(n)) = 0 \\ \frac{\partial Er}{\partial C_2} &= \sum_{n=1}^{N} 2(C_1 + C_2 \cdot \exp[+i\phi_0(\theta_n)] + C_3 \cdot \exp[-i\phi_0(\theta_n)] - \hat{I}(n)) \\ &\qquad \exp[+i\phi_0(\theta_n)] = 0. \\ \frac{\partial Er}{\partial C_3} &= \sum_{n=1}^{N} 2(C_1 + C_2 \cdot \exp[+i\phi_0(\theta_n)] + C_3 \cdot \exp[-i\phi_0(\theta_n)] - \hat{I}(n)) \\ &\qquad \exp[-i\phi_0(\theta_n)] = 0 \end{aligned} \tag{27}$$

Equation (27) can be written as a matrix equation which consists of the measured intensity and the initial phase of holograms, and C_1,C_2, and C_3 are determined by solving the matrix equation:

$$\begin{bmatrix} C_1 \\ C_2 \\ C_3 \end{bmatrix} = \begin{bmatrix} N & \sum_{n=1}^{N} \exp[+i\phi_0(\theta_n)] & \sum_{n=1}^{N} \exp[-i\phi_0(\theta_n)] \\ \sum_{n=1}^{N} \exp[-i\phi_0(\theta_n)] & N & \sum_{n=1}^{N} \exp[-2i\phi_0(\theta_n)] \\ \sum_{n=1}^{N} \exp[+i\phi_0(\theta_n)] & \sum_{n=1}^{N} \exp[+2i\phi_0(\theta_n)] & N \end{bmatrix}^{-1} \times \begin{bmatrix} \sum_{n=1}^{N} \hat{I}(n) \\ \sum_{n=1}^{N} \hat{I}(n) \exp[-i\phi_0(\theta_n)] \\ \sum_{n=1}^{N} \hat{I}(n) \exp[+i\phi_0(\theta_n)] \end{bmatrix} \tag{28}$$

A background image $1 + |A(x, y)|^2$, an amplitude image $A(x, y)$ and a phase image $\phi(x, y)$ are obtained according to Eq. (25), as

$$\begin{aligned} 1 + |A(x, y)|^2 &= C_1 \\ A(x, y) &= |C_2| = |C_3| \\ \phi(x, y) &= \tan^{-1}\left[\frac{\text{Im}[C_2]}{\text{Re}[C_2]}\right] - 2k\alpha x = \tan^{-1}\left[\frac{\text{Im}[C_3]}{\text{Re}[C_3]}\right] + 2k\alpha x, \end{aligned} \tag{29}$$

where Re and Im denote, respectively, the real and imaginary components of the complex value in brackets. The resolution of the object wave reconstructed with the phase-shifting method is independent of the interference fringe spacing, and a sharp edge of the object can be reconstructed accurately because there is no artifact due to the spatial frequency filter. These two features are different from the Fourier transform method, as mentioned in the introduction. The fitting of the intensity variation to an ideal cosine curve is another reason why object waves are reconstructed precisely with the phase-shifting method. This method requires that the distributions of $A(x, y)$ and $\phi(x, y)$ do not vary. In practice, however, holograms require long exposure times because of the electron quantum noise in recording. Therefore, when taking holograms, one care should be taken not to allow the TEM image any movement due to the tilting of the incident electron wave or the drift of the specimen. The sensitivity up to 1/300 of the electron wave length has been confirmed in this phase-sifting method [44].

3.4.2. Observation of Magnetic Multilayers

The origin of Giant magnetoresistance (GMR) is considered to be as follows: Magnetization vectors in ferromagnetic layers separated by spacer-layers of proper thickness align antiparallel to those in adjacent ferromagnetic layers in response to exchange coupling in magnetic-free layers, where both electrons with up- and down-spin

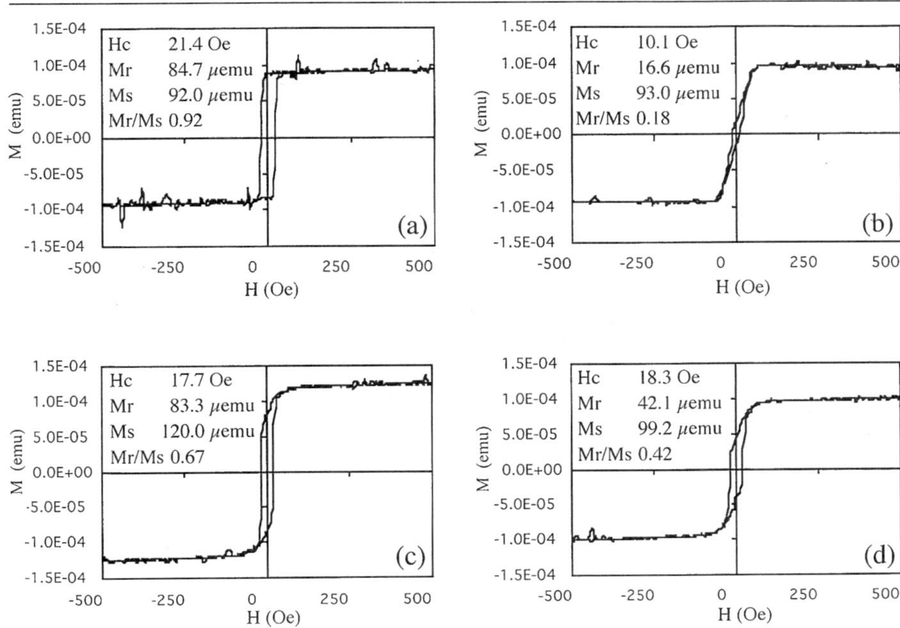

Figure 16. Magnetization characteristics of [Co(4.0 nm)/Cu(x nm)]$_4$ multilayers. In (a), (b), (c), and (d), $x = 1.5$, 2.0, 2.5, and 3.5, respectively [48].

moment are scattered at the boundaries, resulting in high electric resistance. In contrast, realignment of magnetization vectors to be parallel in a suitable external magnetic field causes scattering of only one kind of electron, with either up- or down-spin, and lowers such spin-dependent electric resistance. The intensity of the exchange coupling, which corresponds to the amplitude of GMR, depends on the thickness of spacer-layers. As the thickness of spacer-layers is increased, the ratio of magnetoresistance decreases and shows oscillative behavior [46, 47].

In the following, the magnetic structures of Co/Cu multilayered films are observed in cross-section by electron holography.

Multilayers Co(4.0 nm)/Cu(x nm), $x = 1.5$, 2.0, 2.5, and 3.5 have been prepared by RF-magnetron sputtering on Si substrates at room temperature. Figure 16 shows the magnetization characteristics of Si/[Co(4.0 nm)/Cu(x nm)]$_4$ multilayers; the sample of $x = 2.0$ nm seems to have the strongest antiferromagnetic alignment, the ratio of saturation magnetization to residual magnetization (Ms/Mr) and coercive force (Hc) of which, are the lowest among the four samples. This sample roughly corresponds to the second maximum of Co/Cu magnetoresistance ratio, $\Delta R/R_0$ (ΔR: decrease of resistance, R_0: resistance in a magnetic free space) to Cu thickness (t_{Cu}) curve, where $\Delta R/R_0$ is approximately one-half the first maximum ($t_{Cu} \sim 0.8$ nm). Both the Fourier method and the phase-shifting method have been applied to the cross-sectional observation of multilayers [48].

Figure 17 shows a TEM image of [Co (4.0 nm)/Cu (2.0 nm)]$_5$ (a) and its hologram (b). The TEM image was obtained in slight underfocus and the hologram was recorded

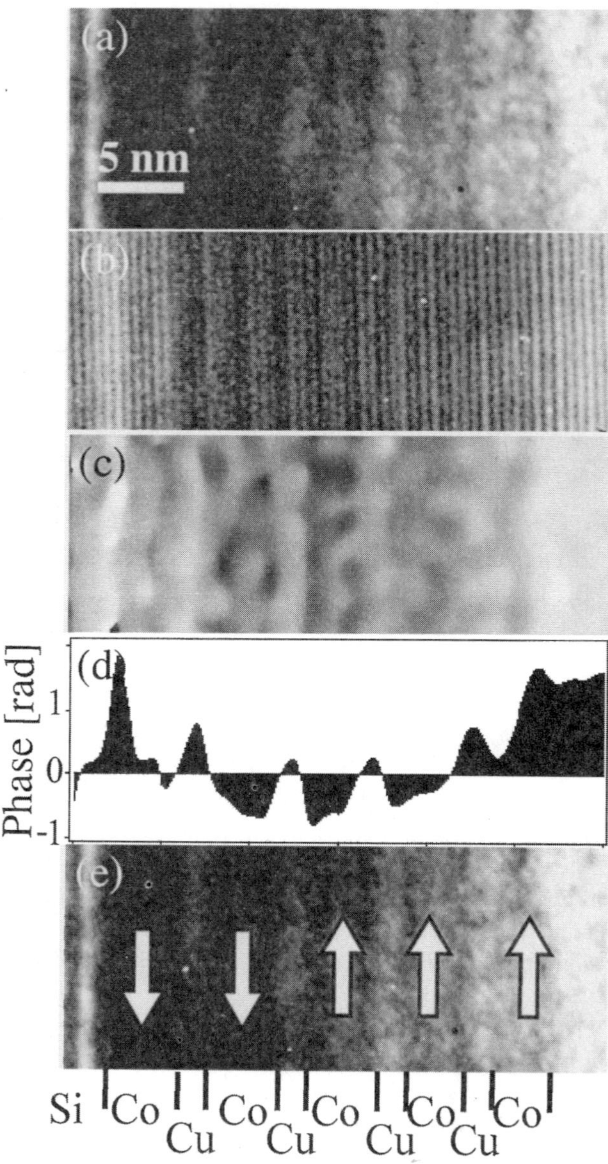

Figure 17. Electron holography of $[Co(4.0\ nm)/Cu(2.0\ nm)]_5$ obtained by the Fourier method. The TEM image (a) shows the layered structure and its hologram (b) reconstructs a phase image (c) showing magnetic induction and inner-potential difference. The line-profile of reconstructed phase angle was averaged 20 pixels along the layers (d), which shows that the Co layers are separated into two groups, in which magnetic vectors are aligned parallel (e). The interference region is about 40 nm.

after refocusing. In this hologram the spacing of interference fringes is approximately 0.8 nm with an interference width of 40 nm, and in other cases the spacing was $0.6 \sim 0.8$ nm. In order to distinguish the layers from Fresnel fringes in reconstructed images, holograms were taken so that interference fringes inclined approximately 2° toward the stacking layers. Figure 17(c) shows a phase image reconstructed by the Fourier method and line-profiles of the phase image (d). Magnetization vectors pointing N-poles are inserted in the TEM image (e). In the phase image, the influence of inner-potential difference due to the wedge shape has been subtracted, by assuming the angle of the wedge to be the same as the setting angle of the ion-milling gun, and estimating mean inner potentials as $V_{Co} = 29.6$ eV and $V_{Cu} = 22.7$ eV with forward atomic scattering factors $f_{Co}(0)$ and $f_{Cu}(0)$, respectively. Strictly speaking, the influence of the wedge shape on the magnetic phase shift remains but does not affect the determination of magnetizing direction. The influence of the wedge shape must be taken into consideration in the quantitative analysis of magnetization. Considering the direction deflected by the Lorentz force, we find that magnetization vectors align as indicated with arrowheads and they are separated into two groups; i.e. downward in the two layers on the left and upward in the three layers on the right, as shown in Fig. 17(e).

Because the spatial resolution of reconstructed images is poorer than 1.5 nm, which is limited by the carrier frequency of the hologram, boundary regions between Co and Cu layers are unreliable in these figures. The phase shifting method has a capability to resolve up to the pixel size, although a small averaging is usually applied to raw reconstructed images. Figure 18 shows the images of $[Co(14.0\ nm)/Cu(2.2\ nm)]_5$. The TEM image of Fig. 18(a) shows that the widths of Cu and Co layers coincide with expected values. Figures 18(b) and 18(c) show its electron hologram and a reconstructed phase image obtained by the phase-shifting method. The line-profile of the phase image is shown in Fig. 18(d). As before, the influence of the wedge shape has been removed. The quality of the reconstructed phase image (c) is distinctly higher than that obtained by the Fourier method. Because of the limited interference width under this electro-optical condition, only four Co layers are observed in the reconstructed image. As indicated in Fig. 18(e), in the left three layers magnetization vectors are aligned downward, and in the right-end layer, the magnetization vector is oriented upward. Approximately 25% of the investigated samples of $t_{Cu} = 2.0$ nm and 80% of samples $t_{Cu} = 2.2$ nm show the same structure in which two blocks with antiparallel magnetization vectors are separated, and the other samples show the parallel alignment of magnetization vectors. If specimens consisted of a larger number of layers, they might show more blocks with antiparallel alignment. Thus far, the alignment in which each vector is antiparallel to the adjacent ferromagnetic layers has not been obtained. This seems to be the reason why the magnetization characteristic in Fig. 16(b) does not show exact antiferromagnetism, but rather weak ferromagnetism with finite residual magnetization (Mr) and coercive force (Hc), and magnetoresistance ratio $\Delta R/R_0$ (ΔR: decrease of resistance, R_0: resistance in a magnetic free space) is approximately half the first maximum. The spatial resolution is approximately 0.3 nm in principle, which allows us to discuss the magnetic structure at the boundaries.

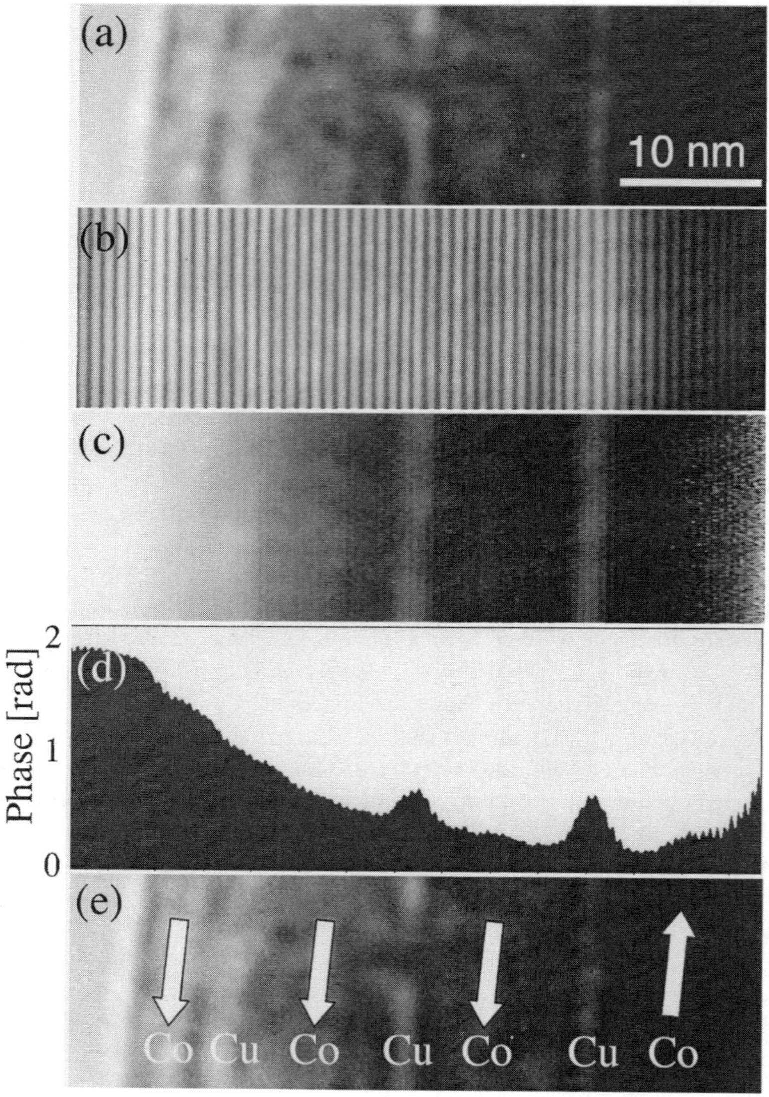

Figure 18. Phase shifting electron holography of [Co(14.0 nm)/Cu(2.2 nm)]$_5$. Only four layers of Co/Cu appear in a TEM image (a) and a hologram (b). A reconstructed phase image (c) and its line-profile (d) show two groups of the magnetic vectors as (d). The interference region is about 60 nm.

Finally, while the phase shifting method has the advantages of high spatial resolution and high sensitivity, this method also has disadvantages. In principle, the phase shifting method requires more time to acquire and processing a series of holograms, and in practice, each hologram must be aligned with the same accuracy as the desired spatial resolution, when specimen images drift for some reason.

3.4.3. *Differential Microscopy*

In off-axis holography, a well-defined reference wave is indispensable in the evaluation of interference fringes. In many cases, however, the magnetic or electric field extends beyond the lateral coherence length of electrons, so a distorted wave has to be used as the reference wave. Strictly speaking, only the phase difference between an object wave and the reference wave can be obtained using ordinary electron holography. As a result, the phase extracted from a hologram by using a distorted reference wave no longer accurately expresses the object. Moreover, the use of a distortion-free or plane reference wave restricts the observation area to the region near the edge of the specimen. This is a severe shortcoming when observing magnetic substances because the magnetic structure may differ from that inside.

In particular, in the observation of magnetic materials, the differentiation of electron phase distribution shows magnetic induction vectors projected onto an x–y image plane, as described in Section 2.5, Lorentz phase microscopy.

A simple method of differentiation is that of image processing using a single-phase map reconstructed by electron holography [49] or Lorentz phase microscopy [17]. The original phase image $\Phi(x, y)$ is shifted by a slight amount Δx in the x direction and then subtracted from itself. If Δx is sufficiently small, $\{\Phi(x, y) - \Phi(x + \Delta x, y)\}/\Delta x$ yields the y component of the induction $B_y(x, y)$. The x component, $B_x(x, y)$, is similarly obtained from $\Phi(x, y)$ and $\Phi(x, y + \Delta y)$. Combining two differentiated images, we obtain a vector map of $\mathbf{B}(x, y)$. This simple method, however, cannot correct the distortion in the reference wave.

DIFFERENTIAL MICROSCOPY USING AN ELECTRON TRAPEZOIDAL PRISM Shearing of the object wave is essential for differential microscopy. Here, we introduce a new electron prism, electron trapezoidal prism, which shears only the object wave by changing the potential applied to the prism [50]. As shown in Fig. 19, this prism has two equi-potential filament electrodes between two grounded platelet electrodes and a trapezoidal electric potential distribution. An electron wave passing between the two filament electrodes, area II in the figure, travels straight. Only the waves passing between grounded electrodes and the adjacent filament electrodes, areas I and III, are tilted, causing them to interfere with a straight-traveling wave. Here, the former wave traveling through area II was used as the reference wave and a wave traveling though area I or area III as the object wave. Only the object wave shifts when the potential of the trapezoidal prism is changed, while the reference wave maintains its position.

When the x axis is selected to be perpendicular to the prism filament, the holograms obtained at prism voltages V_0 and $V_0 + \delta V$ are written as

$$\begin{aligned} I_1(x, y) &= |a(x, y)\exp\{i\phi(x, y)\} + a(x + w_1, y)\exp[i\{\phi(x + w_1, y) - 2\pi R_1 x\}]|^2 \\ I_2(x, y) &= |a(x, y)\exp\{i\phi(x, y)\} + a(x + w_2, y)\exp[i\{\phi(x + w_2, y) - 2\pi R_2 x\}]|^2 \end{aligned}, \quad (29)$$

where ϕ is the phase of the electron wave, w_1 and w_2 are the interference distances ($\delta w = w_2 - w_1$), and R_1 and R_2 are the carrier frequencies. The first terms

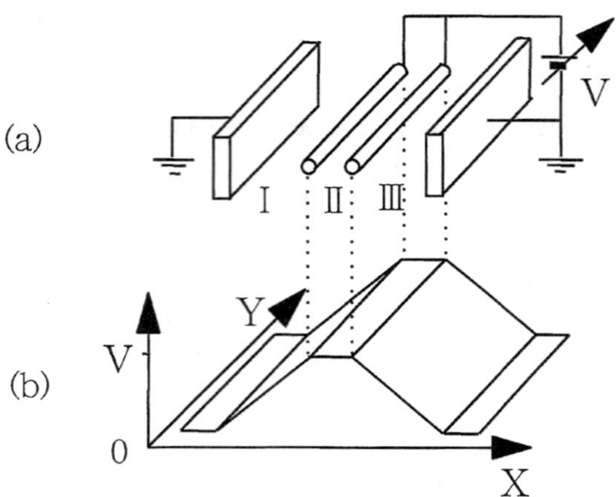

Figure 19. Schematic diagram of an electron trapezoidal prism (a) and its potential distribution (b) [50].

$a(x, y)\exp\{i\phi(x, y)\}$ in both equations correspond to the reference waves, and the second terms correspond to the object waves; the exponential factors in the second terms indicate that only the object waves are deflected. The phase term reconstructed from these holograms show only the difference between the object and reference waves.

Two holograms described by these equations are recorded, and the complex wave functions are reconstructed from both holograms as

$$\begin{aligned}\psi_1(x, y) &= a(x, y)a(x + w_1, y)\exp[i\{\phi(x + w_1, y) - \phi(x, y) - 2\pi R_1 x\}] \\ \psi_2(x, y) &= a(x, y)a(x + w_2, y)\exp[i\{\phi(x + w_2, y) - \phi(x, y) - 2\pi R_2 x\}]\end{aligned}, \tag{30}$$

We then calculate the difference between the phases of the two waves by dividing one wave by the other and obtain the phase differentiation:

$$\frac{\partial\phi(x, y)}{\partial x} = \lim_{\delta x \to 0} \frac{\delta\phi(x, y)}{\delta x}, \tag{31}$$

where

$$\begin{aligned}\delta\phi(x + w_1, y) &\equiv Arg\left[\frac{a(x + w_1, y)a(x, y)\exp[i\,\{\phi(x + w_1, y) - \phi(x, y) - 2\pi R_1 x\}]}{a(x + w_2, y)a\,(x, y)\exp[i\,\{\phi(x + w_2, y) - \phi(x, y) - 2\pi R_2 x\}]}\right] \\ &= Arg\left[\frac{a(x + w_1, y)}{a(x + w_2, y)}\exp[i\,\{\phi(x + w_1, y) - \phi(x + w_2, y) - 2\pi(R_1 - R_2)x\}]\right] \\ &= \phi(x + w_1, y) - \phi(x + w_2, y) - 2\pi(R_1 - R_2)x\end{aligned} \tag{32}$$

The effect of the distorted reference wave is thereby removed from the differentiated phase because exactly the same components in the reference waves were compensated in the two reconstructed waves.

One-dimensional differentiation gives one component, so the specimen or the prism needs to be rotated in order to obtain two-dimensional density maps [51].

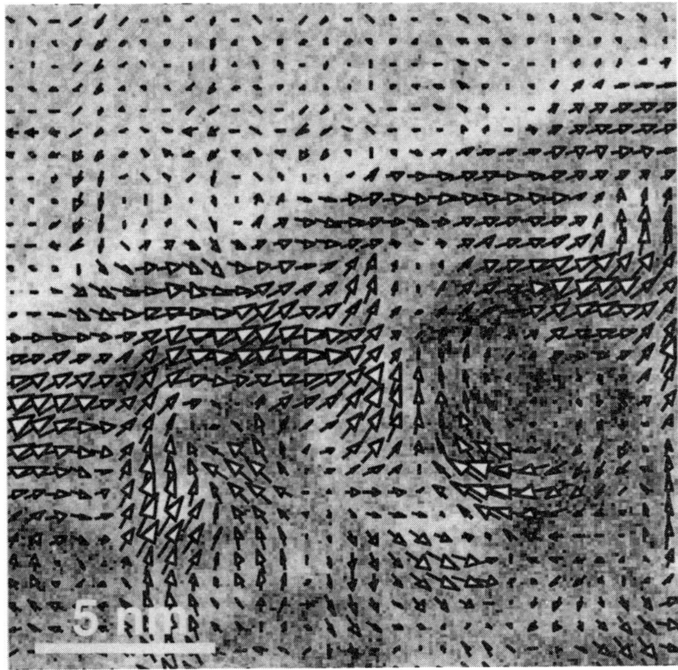

Figure 20. Magnetic induction of a Fe-MgO granular film. Arrowheads indicate projected component parallel to the film.

Figure 20 shows the projected magnetic induction of a Fe-MgO granular film [31]; Fe particles 2–6 nm in diameter were embedded in a single MgO thin crystal. The thickness of the MgO film was 20 nm. The vectors show that the magnetic flux, which has a component parallel to the film surface, flows inhomogeniously in both the inside and the outside of particles. Particles with small arrowheads seem to have flux perpendicular to the film.

In this treatment, following two points should be noted. First, the reconstruction must be remarked so that the basal phase angles of two phase maps coincide in the process of each direction, because the basal phase angle of the reconstructed phase differs according to the position of fringes in regard to the flame, which is quarried out from the original hologram for the processing. In many cases where this technique is applied we cannot find the region to be common base as a vacuum. Second, the Fresnel correction plays a very important role because of the high sensitivity of differentiation.

The electron trapezoidal prism is applicable to a real time differential interferometry by applying rectangular voltage to the prism and producing a double exposed hologram.

4. SUMMARY

We discussed the observation of micro- and nano-magnetic structures by TEM, focusing on Lorentz microscopy and electron holography. In general, Lorentz microscopy is advantageous for real time observation and disadvantageous for quantitative analysis comparing with electron holography, and vice versa. However, a number of techniques have been developed to overcome these disadvantages, which include Foucault mode, Lorentz phase microscopy, real time electron holography and multi wave interferometry. A formula for Lorentz phase microscopy, which coincides with that of TIE, was derived newly using a weak phase approximation. High-precise electron holography, phase-shifting holography and differential holography were shown with mathematical treatments and applications. The phase-shifting holography reconstructed the image of multilayers with better quality than the Fourier method. Two-dimensional differentiation of the electron phase is effective to reveal magnetic induction directly. Differential holography using an electron trapezoidal prism is also useful to reveal structures especially in the case where a well-defined reference wave is difficult to obtain. Since TEM method such as Lorentz microscopy and electron holography can observe inside the specimen, they are favorable for specimens whose inside structure might be different from the surface like magnetic structures. Improving techniques to image nano-magnetic structures should encourage the development of new magnetic materials and devices.

REFERENCES

1. M. N. Baibich, J. M. Broto, A. Fert, F. Van Dau Nguyen, F. Petroff, P. Etienne, G. Creuzet, A. Friederich, & J. Chazelas, *Phys. Rev. Lett.* 61 (1988) 2472.
2. J. Q. Xiao, J. Samuel, C. L. Chien, *Phys. Rev. Lett.* 68 (1992) 3749.
3. A. E. Berkowitz, J. R. Mitchell, M. J. Corey, A. P. Young, S. Zhang, F. T. Parker, A. Hattern and G. Thomas, *Phys. Rev. Lett.* 68 (1992) 3745.
4. B.-I-Cho. W. Win, E.-J. Yun and R. M. Walser, *IEEE Transactions on Magnetics* 31 (1995) 3859.
5. T. Ono and T. Shinjo, *Jpn. J. Phys. Soc.* 64 (1995) 363.
6. T. Ono, Y. Sugita, K. Shigeto, K. Mibu, N. Hosoito and T. Shinjo, *Phys. Rev.* B55 (1997) 14457.
7. P. Weiss, *J. Phys.* 6 (1907) 661.
8. F. Bloch, *Z. f. Phys.* 74 (1932) 285.
9. M. E. Hale, H. W. Fuller and H. Rubinstein, *J. Appl. Phys.* 30 (1959) 789.
10. H. Boersch and H. Raith, *Naturwissenschaften* 46 (1959) 574.
11. H. W. Fuller, and M. E. Hale, *J. Appl. Phys.* 31 (1960) 1699.
12. D. Watanabe, T. Sekiguchi, T. Tanaka, T. Wakiyama and M. Takahashi, *J. Appl. Phys.* 21 (1982) 179.
13. D. Gabor, *Nature* 4098 (1948) 777. *Proc. Roy. Soc.* A197-(1949) pp. 454–487.
14. N. Osakabe, K. Yoshida, Y. Horiuchi, T. Matsuda, H. Tanabe, T. Okuwaki, J. Endo, H. Fujiwara and A. Tonomura, *Appl. Phys. Lett.* 42 (1983) 746.
15. M. R. MaCartney, D. J. Smith, This Volume.
16. J. N. Chapman, R. Ploessl and D. M. Donnet, *Ultramicroscopy* 47 (1992) 331.
17. S. Bajt. A. Barty, K. A. Nugent, M. MaCartney, M. Wall. D. Paganin, *Ultramicroscopoy* 83 (2000) 67.
18. Y. Takahashi, Y. Yajima, M. Ichikawa and K. Kuroda, *Jpn. J. Apply. Phys.* 33 (1994) 1352.
19. M. Mankos, Z. J. Yang, M. R. Scheinfein and J. M. Cowley, *IEEE Transactions on Magnetics* 30 (1994) 4497.
20. F. Bitter, *Phys. Rev.* 38 (1931) 930, 41 (1932) 507.
21. K. Goto, M. Ito and T. Sakurai, *Jpn. J. Appl. Phys.* 19 (1980) 1339.
22. P. J. Grundy, *Ph. D. Dissertation* (1964).
23. K. Shirota, A. Yonezawa, K. Shibatomi and T. Yanaka, *J. Electron. Microsc.* 25 (1976) 303.
24. T. Hirayama, Q. Ru, T. Tanji and A. Tonomura, *Appl. Phys. Lett.* 63 (1993) 418.
25. J. Zweck and B. J. H. Bormans, *Philips Electron Optics Bulletin* 132 (1992) 1.
26. D. Watanabe, T. Sekiguchi and E. Aoyagi, *Proc. 7th Int. Conf. HVEM*, Berkeley, California (1983) pp. 285–286.
27. Frenkel and J. Dorfman, *Nature* 126 (1930) 274.

28. C. Kittel, *Phys. Rev.* 70 (1946) 965.
29. L. Nèel, *CR Acad. Sci.* 224 (1947) 1488.
30. W. F. Brown *Jr.* , *Phys. Rev.* 105 (1957) 1479.
31. T. Tanji, M. Maeda, N. Ishigure, N. Aoyama, K. Yamamoto and T. Hirayama, *Phys. Rev. Lett.* 83 (1999) 1038.
32. N. Aoyama, K. Yamamoto, T. Tanji and M. Hibino, *Jpn. J. Appl. Phys.* 39 (2000) 5340.
33. A. Tonomura, H. Kasai, O. Kamimura, T. Matsuda, K. Harada, T. Yoshida, T. Akashi, J. Shimoyama, K. Kishio, T. Hanaguri, K. Kitazawa, T. Masai, S. Tajima, N. Koshizuka, P. L. Gammel, D. Bishop, M. Sasase and S. Okayasu, *Phys. Rev. Lett.* 88 (2002) 237001-1.
34. K. Harada, H. Kasai, T. Matsuda, M. Yamasaki, J. E. Bonevich and A. Tonomura, *Jpn. J. Appl. Phys.* 33 (1994) 2534.
35. Y.-G. Park and D. Shindo, *J. Magnetism and Magnetic Materials.* 238 (2002) 68.
36. D. Paganin and K. A. Nugent, *Phys. Rev. Lett.* 80 (1998) 2586.
37. T. Hirayama, Q. Ru, T. Tanji and A. Tonomura, *Appl. Phys. Lett.* 63 (1993) 418.
38. M. Lehmann, E. Volkl and F. Lenz, *Ultramicroscopy* 54 (1994) 335.
39. J. Chen, T. Hirayama, G. Rai, T. Tanji, K. Ishizuka and A. Tonomura, *Opt. Lett* 18 (1993) 1887.
40. T. Hirayama, J. Chen, T. Tanji and A. Tonomura, *Ultramicroscopy* 54 (1994) 9.
41. T. Hirayama, T. Tanji and A. Tonomura, *Appl. Phys. Lett.* 67 (1995) 1185.
42. A. Tonomura, T. Matsuda, T. Kawasaki, J. Endo and N. Osakabe, *Phys. Rev. Lett.* 54 (1985) 60.
43. Q. Ru, J. Endo, T. Tanji and A. Tonomura, *Appl. Phys. Lett.* 59 (1991) 2372.
44. K. Yamamoto, I. Kawajiri, T. Tanji, M. Hibino and T. Hirayama, *J. Electron Microsc.* 49 (2000) 31.
45. Q. Ru, G. Lai, K. Aoyama, J. Endo and A. Tonomura, *Ultramicroscopy* 55 (1994) 209.
46. F. Petroff, A. Barthelemy, D. H. Mosca, D. K. Lottis, A. Fert, P. A. Schroeder, W. P. Pratt Jr., R. Loloee and S. Lequien, *Phys. Rev.* B44 (1991) 5355.
47. C. Dorner, M. Haidl and H. Hoffmann, *J. Appl. Phys.* 74 (1993) 5886.
48. T. Tanji, S. Hasebe, Y. Nakagami, K. Yamamoto and M. Ichihashi, *Microscopy and Microanalysis* 10 (2004) 146.
49. M. R. McCartney, *Proc. Microscopy and Microanalysis '97*, pp. 519–520, Springer (1997).
50. T. Tanji, S. Manabe, K. Yamamoto and T. Hirayama, *Ultramicroscopy* 75 (1999) 197.
51. T. Tanji, S. Manabe, K. Yamamoto and T. Hirayama, *Materials Characterization* 42 (1999) 183.

SUBJECT INDEX